SEP 30 1981

SAN FRANCISCO PUBLIC LIBRARY
3 1223 7740

D1707765

ELSEVIER'S DICTIONARY OF BOTANY
I. PLANT NAMES

ELSEVIER'S DICTIONARY OF BOTANY

I. PLANT NAMES

in
ENGLISH, FRENCH, GERMAN, LATIN
and RUSSIAN

compiled by

P. MACURA
Department of Foreign Languages
and Literatures
University of Nevada, Reno
Nevada, U.S.A.

ELSEVIER SCIENTIFIC PUBLISHING COMPANY
AMSTERDAM - OXFORD - NEW YORK
1979

Distribution of this book is being handled by the following publishers

ELSEVIER SCIENTIFIC PUBLISHING COMPANY
335 JAN VAN GALENSTRAAT
P.O. BOX 211, 1000 AE AMSTERDAM, THE NETHERLANDS

ELSEVIER/NORTH-HOLLAND INC.
52 VANDERBILT AVENUE
NEW YORK, NEW YORK 10017

REF
581.03
EL76
V.1
S. F. PUBLIC LIBRARY
81-46

Library of Congress Cataloging in Publication Data
Main entry under title:

Elsevier's dictionary of botany.

CONTENTS: 1. Plant names.
1. Botany--Dictionaries--Polyglot. 2. Plant names, Popular--Dictionaries--Polyglot. 3. Botany--Nomenclature. 4. Dictionaries, Polyglot. I. Macura, Paul. II. Title: Dictionary of botany.
QK9.E47 581'.03 79-15558
ISBN 0-444-41787-7 (v. 1)

ISBN 0-444-41787-7

© Elsevier Scientific Publishing Company, 1979

All rights reserved. No part of this publication may be reproduced, stored in a retrieval system, or transmitted in any form or by any means, electronic, mechanical, photocopying, recording, or otherwise, without the prior written permission of the publisher,
Elsevier Scientific Publishing Company,
P.O. Box 330, 1000 AH Amsterdam, The Netherlands

Electronic data processing:
Büro für Satztechnik W.Meyer KG
Hanau, W. Germany

Printed in The Netherlands

PREFACE

Until now there has existed no English-French-German-Latin-Russian Dictionary of Botany broad enough in concept and sufficiently detailed to serve English, French and German scholars. Most of the multilingual dictionaries are very limited in scope, ranging from less than one hundred to a maximum of over one thousand entries. The need for a more comprehensive multilingual dictionary to meet a broad range of translating requirements has been felt by nearly all who frequently use foreign language dictionaries in this field.

This Dictionary of Botany consists of two volumes. Volume I contains over 6,000 entries, constituting a listing of English, French, German, Latin and Russian names of plants, trees, shrubs, mushrooms, and lichens; Volume II will contain English, French, German and Russian terms of a general nature.

Arrangement of entries: In the basic table the English entries are given in strictly alphabetical order with their French, German and Latin equivalents. In the indexes for French, German and Latin the terms, in alphabetical order, refer to corresponding numbers in the basic table. The Russian equivalents are arranged alphabetically and numerically at the end of the dictionary.

I would be very grateful if users of the dictionary would notify me of any error that may come to their attention.

In conclusion, I would like to express my sincere gratitude to the following persons who aided me in the preparation of this dictionary. I am most of all indebted to Professor Paule-Colette Fricke, Department of Foreign Languages and Literatures, University of Nevada, for her valuable suggestions and proofreading of the French part of the manuscript, and to Professor Margarete G. Hagner, Department of Foreign Languages and Literatures, University of Nevada, for proofreading the German part. Very sincere thanks are also due to Mrs. Alice Bertalot for helping to proofread a part of the French version of the manuscript, and to Mrs. Sabine Atwell, Department of Foreign Languages and Literatures, University of Nevada, for proofreading a part of the German version. I must express further gratitude to Professor Hugh N. Mozingo, Department of Biology, University of Nevada, for his valuable suggestions. I acknowledge gratefully the financial assistance of Research Advisory Board, University of Nevada Graduate School, in the preparation of the dictionary and offer my sincere thanks to all the authors whose works I have consulted. I thank my wife Irene and daughter Renee for their help, understanding and moral support.

Paul Macura
Reno, Nevada

ABBREVIATIONS

f.	forma
ssp.	subspecies
var.	varietas
convar.	convarietas

BASIC TABLE

A

1 **Aaron's-beard**
f millepertuis à grandes fleurs
d Rose von Sharon
l *Hypericum calycinum*

* **Aaron's-beard → 1527, 3147, 5277, 5972**

* **Aaron's-rod → 2276**

2 **abaca; abaca banana**
f abaca; bananier à fibres; bananier textile; chanvre de Manille
d Manilahanf
l *Musa textilis*

3 **abelia**
f abélie
d Abelia; Abelie
l *Abelia*

4 **abroma**
f abrome
d Kakaomalve
l *Abroma*

5 **abronia; sand verbena**
f abronie; tricrate
d Abronie; Sandverbene
l *Abronia*

* **absinthium → 1564**

6 **abura mitragyna**
f tilleul d'Afrique
d Abura
l *Mitragyna stipulosa; Mitragyna macrophylla*

7 **abutilon; flowering maple**
f abutilon
d Abutilon; Samtpappel; Schmuckmalve; Schönmalve; Zimmerahorn
l *Abutilon*

8 **Abyssinian banana; ensete**
f bananier d'Abyssinie
d abessinische Banane; Ensete
l *Musa ensete*

9 **Abyssinian myrrh tree**
d Myrrahbaum
l *Commiphora abyssinica*

10 **acacia**
f acacia
d Akazie; Schotendorn
l *Acacia*

11 **acanthopanax**
f acanthopanax
d Stachelpanax; Stachelkeil; Stachelkraftwurz
l *Acanthopanax*

12 **acanthus; bear's-breech**
f acanthe
d Akanthus; Bärenklau; Löwenklau
l *Acanthus*

13 **acanthus bristle thistle**
f chardon acanthoïde
d Stacheldistel; Wegedistel
l *Carduus acanthoides*

14 **acanthus family**
f acanthacées
d Akanthazeen; Akanthusgewächse; Bärentapengewächse
l *Acanthaceae*

15 **achimenes**
f achimène
d Achimenes; Schiefteller
l *Achimenes*

16 **aconite buttercup**
f renoncule à feuilles d'aconit
d eisenhutblättriger Hahnenfuss; sturmblättriger Hahnenfuss; Sturmhut-Hahnenfuss
l *Ranunculus aconitifolius*

17 **aconite monkshood**
f aconit napel; char de Vénus; capuce de moine; casque de Jupiter
d blauer Eisenhut; echter Eisenhut
l *Aconitum napellus*

18 **acrocomia**
f acrocome
d Acrocomie; Schopfpalme; Weinpalme
l *Acrocomia*

19 **acronychia**
f acronychie
l *Acronychia*

20 **acrostichum**
f acrostiche; acrostique
d Mangrovefarn; Zeilfarn
l *Acrostichum*

21 **actinidia**
f actinidie

d Aktinidie; Strahlengriffel
l *Actinidia*

22 actinidia family
f actinidiacées
d Aktinidien; Strahlengriffelgewächse
l *Actinidiaceae*

23 actinomyces; ray fungi
f actinomyces
d Strahlenpilze
l *Actinomyces*

*** Adam's-fig → 4219**

24 Adam's-needle; Adam's-needle yucca
f yucca filamenteux
d fädige Palmlilie; fädentragende Palmlilie; Faserpalmlilie; staubfädige Palmlilie; staubfädige Yucca
l *Yucca filamentosa*

*** Adam's-needle → 6120**

25 adder's-mouth orchid
f malaxide
d Einblatt; Kleingriffel; Weichorchis; Weichwurz
l *Microstylis; Malaxis*

26 adderspit
f fougère aigle
d Adlerfarn
l *Pteridium aquilinum*

27 adder's-tongue; adder's-tongue fern
f ophioglosse
d Natter(n)farn; Natter(n)zunge
l *Ophioglossum*

28 adder's-tongue family
f ophioglossacées
d Natter(n)farne; Natter(n)farngewächse; Ophioglossazeen
l *Ophioglossaceae*

*** adder's-tongue fern → 27**

29 adenandra
f adénandre
d Kapraute
l *Adenandra*

30 adinandra
f adinandre
l *Adinandra*

31 adlai; adlay; Job's-tears
f coïx larme; herbe à chapelets; herbe à rosarie; larmes de Job; larmille
d Christustränengras; Hiobsträne; Hiobstränengras; Marientränengras; Mosestränengras
l *Coix lacrima jobi*

32 adlumia
f adlumie
d Adlumie; Doppelkappe
l *Adlumia*

33 adonis
f adonide; fleur d'Adonis
d Adonisröschen; Teufelsauge
l *Adonis*

*** Adriatic oak → 2128**

34 adsuki bean; adzuki bean
f haricot à feuilles angulaires
d Adzukibohne; Adsukibohne
l *Phaseolus angularis*

35 aechmea
f æchmée
d Lanzenrosette
l *Aechmea*

36 aethusa
f éthuse
d Gleisse; Hundspetersilie
l *Aethusa*

37 Aetna barberry
d Ätna-Berberitze
l *Berberis aetnensis*

38 afara; afara terminalia
d Limbaholz
l *Terminalia superba*

39 African afzelia
f afzélie d'Afrique
l *Afzelia africana*

40 African agapanthus
f tubéreuse bleue
d afrikanische Schmucklilie
l *Agapanthus africanus; Agapanthus umbellatus*

41 African avodire
f avodiré
l *Turraeanthus africanus*

42 African blackwood
f ébène; bois d'ébène; palissandre du Sénégal

d Ebenholz
l *Dalbergia melanoxylon*

43 African juniper
f genévrier d'Afrique; genévrier élevé
l *Juniperus procera*

*** African lily → 54**

*** African millet → 4092**

44 African myrrh tree
f bdellium d'Afrique
d afrikanischer Myrrhenstrauch
l *Commiphora africana*

45 African nitta tree
f arbre sacré
d Dourabaum; Nittabaum
l *Parkia africana*

*** African oak → 51**

46 African oil palm
f aouara d'Afrique; élais de Guinée; palmier à (l')huile
d Ölpalme; afrikanische Ölpalme; guineische Palme
l *Elaeis guineensis*

47 African padauk
f padouk
l *Pterocarpus soyauxi*

*** African peanut → 1574**

48 African piptadenia
f piptadénie d'Afrique
l *Piptadenia africana*

*** African rue → 2714**

49 African sparmannia
f sparmannie d'Afrique
d afrikanische Sparmannie; afrikanische Zimmerlinde
l *Sparmannia africana*

50 African tamarisk
f tamaris d'Afrique; tamaris africain
d afrikanische Tamariske
l *Tamarix africana*

51 African teak; African oak
f oldfieldie d'Afrique
l *Oldfieldia africana*

52 African valerian
f fédia corne d'abondance
d Schmelzkraut
l *Fedia cornucopiae*

53 afzelia
f afzélie
l *Afzelia*

54 agapanthus; African lily
f agapanthe
d Liebesblume; Schmucklilie
l *Agapanthus*

55 agar; agar-agar
f gelidium
d Gelidium
l *Gelidium*

56 agar Ceylon moss
f mousse de Ceylan; mousse de Jaffna
d Ceylon-Moos; Jaffnamoos
l *Gracilaria lichenoides*

57 agaric
f agaric
d Blätterpilz; Blätterschwamm
l *Agaricus*

58 agaricales
f agaricales
d Hutpilze
l *Agaricales*

59 agave
f agave
d Agave; Jahrhundertpflanze
l *Agave*

60 ageratum; flossflower
f agérate; agératoire
d Ageratum; Leberbalsam
l *Ageratum*

61 aglaia
f aglaie
d Glanzbaum
l *Aglaia*

62 aglaonema
f aglaonème
d Kolbenfaden
l *Aglaonema*

*** agnus castus → 3296**

63 agrimony
f aigremoine; herbe de Saint-Guillaume
d Odermennig; Leberklette
l *Agrimonia*

64 agriophyllum
f agriophylle
l *Agriophyllum*

65 ailanthus
f ailante
d Ailanthus; Götterbaum
l *Ailanthus*

66 ailanthus family
f simarubacées
d Bittereschengewächse; Bitterholzgewächse; Simarubazeen
l *Simarubaceae*

67 aipi cassawa; sweet cassawa
f manioc doux
d süsse Cassave; Süssmaniok
l *Manihot aipi*

68 airbrom; billbergia
f billbergia
d Billbergie
l *Billbergia*

69 air potato
f igname bulbifère
d Knollenyams
l *Dioscorea bulbifera*

70 aizoon
d Immergrün; Eiskraut
l *Aizoon*

71 aizoon saxifrage
f saxifrage aizoon
d traubenblütiger Steinbrech; Traubensteinbrech
l *Saxifraga aizoon*

72 ajowan caraway
f ammi
d ägyptischer Kümmel
l *Carum copticum*

73 akebia
f akébie
d Akebie
l *Akebia*

74 akee
f akée
l *Blighia*

75 akee
f akée d'Afrique
d Aki-Baum
l *Blighia sapida; Cupania sapida*

76 Alabama supplejack
f jujubier grimpant
l *Berchemia scandens*

77 Alaska bog willow
d braun werdende Weide
l *Salix fuscescens*

78 Albert rose
f rosier d'Albert
d Alberts Rose
l *Rosa alberti*

79 albizzia
f albizzie; arbre à soie
d Albizzie; Schirmakazie
l *Albizzia*

80 Alcock spruce
f épicéa bicolore
d Alcocks Fichte
l *Picea alcockiana; Picea bicolor*

81 alder
f aune; aulne
d Erle; Eller; Else
l *Alnus*

82 alder buckthorn
f nerprun à feuilles d'aune
d erlenblättriger Kreuzdorn
l *Rhamnus alnifolia*

*** alder buckthorn → 2513**

83 aldrovanda
f aldrovande
d Aldrovandie; Wasserfalle
l *Aldrovanda*

84 alectryon
f alectryon
l *Alectryon*

*** alehoof → 1429, 2647**

*** Aleppo grass → 3103**

85 Aleppo oak
f chêne à galles
d echte Galleiche
l *Quercus infectoria*

86 Aleppo pine
f pin d'Alep; pin de Jérusalem
d Aleppofichte; Aleppokiefer
l *Pinus halepensis*

87 alexanders
f maceron
d Gelbdolde; Myrrhenkraut
l *Smyrnium*

88 Alexandrian laurel
f laurier alexandrin; laurier d'Alexandrie; laurier cheval
d alexandrischer Lorbeer
l *Danae racemosa*

89 Alexandria senna
f séné d'Egypte
d Sennesblätterstrauch
l *Cassia acutifolia*

90 alfalfa
f luzerne commune; luzerne cultivée; sainfoin à fleur violette
d Luzerne; blaue Luzerne; Dauerklee; ewiger Klee; Monatsklee; Mondklee; Sinfin
l *Medicago sativa*

91 alfilaria; pin grass; stork's-bill; hemlock stork's-bill
f bec-de-grue; aiguille
d gemeiner Reiherschnabel; schierlingsblättriger Reiherschnabel; Schierlingsreiherschnabel
l *Erodium cicutarium*

92 algae
f algues
d Algen
l *Algae*

*** algal fungi → 5824**

*** algarroba → 1467**

93 Algerian ash
f frêne dimorphe
l *Fraxinus dimorpha*

94 Algerian fir
f sapin d'Algérie
d nubische Tanne
l *Abies numidica*

95 alisma family
f alismacées
d Froschlöffelgewächse
l *Alismataceae*

96 alkali grass
f misotte
d Salzschwaden
l *Atropis*

97 alkanet
f alcanna; alkanna
d Alkannawurzel
l *Alkanna*

*** alkekengi → 5275**

98 Alleg(h)any barberry
f vinettier du Canada
d kanadische Berberitze
l *Berberis canadensis*

99 Alleg(h)any chinkapin; chinquapin; dwarf chestnut
f châtaignier chincapin
d Zwergkastanie
l *Castanea pumila*

*** alligator apple → 4257**

100 alligator juniper
f genévrier gercé
l *Juniperus pachyphloea*

*** alligator pear → 183**

101 allseed
f radiole à feuilles de lin
d Haarkraut; Zwerglein
l *Radiola linoides*

*** allseed → 3479**

102 allspice; pimenta
f poivre de la Jamaïque; poivre giroflé; piment de la Jamaïque; piment giroflé; piment giroflier; piment; poivron; tout-épice(s)
d Jamaika-Pfeffer; Nelkenpfeffer; Nelkenpfefferbaum; Pimentbaum
l *Pimenta officinalis*

103 allthorn acacia
f acacia à longues épines
d Kapschotendorn; Weissdornakazie
l *Acacia horida*

104 almond
f amandier
d Bittermandelbaum; echte Mandel; echter Mandelbaum; Mandelbaum
l *Prunus amygdalus; Amygdalus communis*

105 almond cherry
f amandier à fleurs du Japon
l *Cerasus glandulosa; Prunus glandulosa*

106 almond-leaved willow
f osier amandier; osier pâle; saule à feuilles d'amandier
d dreimännige Weide; Mandelweide
l *Salix triandra; Salix amygdalina*

107 almond pear; almond-shaped pear tree
f poirier à feuilles d'amandier; poirier amandier
d mandelblättriger Birnbaum
l *Pyrus amygdaliformis*

108 alocasia
f alocasie
d Alokasie; Tropenwurz; Pfeilblatt
l *Alocasia*

109 aloe
f aloès
d Aloe; Bitterschopf
l *Aloe*

110 aloe yucca; Spanish dagger
f yucca à feuilles d'aloés
d aloeblättrige Palmlilie; Wüstenpalme
l *Yucca aloifolia*

111 alpenclock; gravel-bind
f soldanelle
d Alpenglöckchen; Troddelblume
l *Soldanella*

112 alpine anemone
f anémone des Alpes
d Alpenanemone; Teufelsbart
l *Anemone alpina*

113 alpine aster
f aster des Alpes
d Alpenaster
l *Aster alpinus*

114 alpine azalea
f azalée couchée; loiseleurie couchée; loiseleurie des Alpes
d gestreckter Felsenstrauch; liegende Azalee; Zwergazalee; Zwergfelsenröschen; Gemsheide; kriechendes Felsenröschen
l *Loiseleuria procumbens*

115 alpine bartsia
d Alpenbarts(ch)ie
l *Bartsia alpina*

116 alpine bellflower
f campanule alpine
d Alpenglockenblume
l *Campanula alpina*

117 alpine betony
f épiaire alpine
d Alpenziest
l *Stachys alpina*

118 alpine bluegrass; alpine meadow grass
f pâturin des Alpes
d Alpenrispengras
l *Poa alpina*

119 alpine-bog swertia
f swertie vivace
d ausdauernde Swertie; blaue Swertie
l *Swertia perennis*

120 alpine brook saxifrage
d Bachsteinbrech
l *Saxifraga rivularis*

121 alpine buckthorn
f nerprun des Alpes; nerprun alpin
d Alpenkreuzdorn
l *Rhamnus alpina; Rhamnus alpinus*

122 alpine cerastium
f céraiste des Alpes
d Alpenhornkraut
l *Cerastium alpinum*

123 alpine chrysanthemum
f chrysanthème des Alpes; leucanthème des Alpes
d Alpenwucherblume
l *Chrysanthemum alpinum; Leucanthemum alpinum; Pyrethrum alpinum*

124 alpine circaea
f circée des Alpes
d Alpenhexenkraut; Gebirgshexenkraut; kleines Hexenkraut
l *Circaea alpina*

125 alpine clematis
f atragène
d Alpenrebe; Alpenwaldrebe; Atragene
l *Clematis alpina*

126 alpine clover
f trèfle des Alpes
d Alpenklee
l *Trifolium alpinum*

127 alpine clubmoss
f lycopode des Alpes
d Alpenbärlapp
l *Lycopodium alpinum*

128 alpine currant
f groseillier des Alpes
d Alpenjohannisbeere; gemeiner Alpenstrauch
l *Ribes alpinum*

129 alpine dock
f patience des Alpes; rhubarbe des moines
d Alpenampfer; Mönchsrharbarber
l *Rumex alpinus*

130 alpine fleabane
f vergerette alpine
d Alpenberufkraut
l *Erigeron alpinus*

131 alpine fleeceflower
f renouée des Alpes
d Alpenknöterich; Bergknöterich
l *Polygonum alpinum*

132 alpine forget-me-not
f myosotis des Alpes
d Alpenvergissmeinnicht
l *Myosotis alpestris*

133 alpine foxtail
f alopécure des Alpes
d Alpenfuchsschwanz
l *Alopecurus alpinus*

134 alpine hair grass
d Alpenschmiele
l *Deschampsia alpina*

135 alpine hawkweed
d Alpenhabichtskraut
l *Hieracium alpinum*

* **alpine meadow grass → 118**

136 alpine meadow rue
f rue des prés
d Alpenwiesenraute
l *Thalictrum alpinum*

137 alpine milk vetch
f astragale des Alpes
d Alpentragant
l *Astragalus alpinus*

138 alpine mountain sorrel
d Säuerling
l *Oxyria digyna*

139 alpine pennycress
d Berghellerkraut
l *Thlaspi montanum*

140 alpine poppy
f pavot des Alpes
d Alpenmohn
l *Papaver alpinum*

141 alpine ptarmigan berry
f busserole alpine
d Alpenbärentraube
l *Arctostaphylos alpinus; Arctous alpinus*

142 alpine pussytoes
f pied-de-chat alpestre
d Alpenkatzenpfötchen; nordisches Katzenpfötchen
l *Antennaria alpina*

143 alpine rock cress
f arabette du Caucase; corbeille d'argent
d Alpengänsekresse; Alpenkresse
l *Arabis alpina*

144 alpine rock jasmine
d Alpenmannsschild
l *Androsace alpina*

145 alpine rose; drophip rose
f rosier des Alpes; rose alpestre
d Alpenrose; Bergrose; Gebirgsrose
l *Rosa alpina; Rosa pendulina*

146 alpine rush
f jonc des Alpes
d Alpenbinse
l *Juncus alpinus*

147 alpine saussurea; alpine sawwort
f saussurée des Alpes
d Alpenscharte; Alpenschärfling
l *Saussurea alpina*

148 alpine savory
f calament des Alpes
d Alpenquendel
l *Calamintha alpina; Satureia alpina; Acinos alpinus*

* **alpine sawwort → 147**

149 alpine silene
f silène d'Autriche
d Alpenleimkraut
l *Silene alpestris*

150 alpine speedwell
f véronique des Alpes
d Alpenehrenpreis
l *Veronica alpina*

151 **alpine spirea**
f spirée alpestre
d Alpenspierstrauch
l *Spiraea alpina*

152 **alpine sun rose**
d Alpensonnenröschen
l *Helianthemum alpestre*

153 **alpine sweet gras**
d Alpenmariengras
l *Hierochloe alpina*

154 **alpine timothy**
f fléole des Alpes
d Alpenlieschgras
l *Phleum alpinum*

155 **alpine toadflax**
f linaire alpestre
d Alpenleinkraut
l *Linaria alpina*

156 **alpine willowweed**
f épilobe des Alpes
d Gauchheilweidenröschen; gauchheilblättriges Weidenröschen
l *Epilobium alpinum*

157 **alpine woodsia; flower-cup fern**
d Alpenwimperfarn
l *Woodsia alpina*

158 **alpinia**
f alpinia
d Alpinie
l *Alpinia*

159 **alplily**
f lloïdie
d Faltenlilie
l *Lloydia*

160 **Alps anthyllis**
d Bergwundklee
l *Anthyllis montana*

161 **Alps honeysuckle**
f chèvrefeuille alpestre
d Alpengeissblatt
l *Lonicera alpigena*

162 **Alps plantain**
f plantain des Alpes
d Alpenwegerich
l *Plantago alpina*

163 **Alps willow**
d kahle Weide
l *Salix glabra*

164 **alsike clover; Swedish clover**
f trèfle bâtard; trèfle hybride
d Alsike; Alsikeklee; Bastardklee; Schwedenklee
l *Trifolium hybridum*

165 **alstonia**
f alstonie
d Pulai
l *Alstonia*

166 **alstroemeria**
f alstrémère
d Alstroemerie; Inkalilie
l *Alstroemeria*

167 **Altai hawthorn**
f aubépine altaïque
d Altaiweissdorn
l *Crataegus altaica*

168 **alternate-leaf golden saxifrage**
f dorine à feuilles alternes
d Goldmilz; Goldsteinbrech; Steinkresse; wechselblättriges Milzkraut
l *Chrysosplenium alternifolium*

169 **althaea**
f althée; guimauve
d Eibisch; Stockmalve
l *Althaea*

170 **alum bark tree**
f stryphnodendre
l *Stryphnodendron*

171 **alumroot**
f heuchère
d Purpurglöckchen
l *Heuchera*

172 **alyssum**
f alysse
d Schildkraut; Steinkraut
l *Alyssum*

173 **amanita family**
f amanitacées
d Knollenblätterpilze; Wulstlingartige
l *Amanitaceae*

174 **amaranth**
f amarante
d Amarant; Fuchsschwanz(gras); Samt-

blume
l *Amaranthus*

175 amaranth family
f amarantacées
d Amaranthazeen; Amarantgewächse; Fuchsschwanzgewächse
l *Amaranthaceae*

176 amaryllis; knight's-star
f amaryllis; amaryllide; hippéastre
d Amaryllis; Ritterstern
l *Amaryllis; Hippeastrum*

177 amaryllis family
f amaryllidacées
d Amaryllidazeen; Amaryllisgewächse; Narzissengewächse
l *Amaryllidaceae*

178 ambarella
f pomme cythère
d süsse Balsampflaume; süsse Mombinapflaume; Goldpflaume
l *Spondias cytherea; Spondias dulcis*

*** ambary hemp → 3146**

*** ambash → 179**

179 ambatch; ambash
d Ambatsch; Markbaum
l *Hermineria elaphroxylon; Aeschynomene elaphroxylon*

180 Ambatjang mango; Bachang mango
f manguier fétide
d stinkender Mangobaum
l *Mangifera foetida*

*** amber-bloom rhododendron → 4264**

181 amberboa
d Ambraflockenblume; Bisamblume
l *Amberboa*

182 amber tree
f arbrisseau ambre; anthosperme
d Amberbaum; Amberstrauch
l *Anthospermum*

*** American aloe → 1092**

*** American arborvitae → 1973**

183 American avocado; avocado; alligator pear
f avocatier
d Avocatobirne; Avocatobaum; Aquacate; Ahuaca; Alligatorbirne; Avogatenbaum
l *Persea americana*

184 American beech
f hêtre d'Amérique
d amerikanische Buche
l *Fagus grandifolia; Fagus americana*

185 American bittersweet
f bourreau des arbres; célastre grimpant
l *Celastrus scandes*

186 American black currant
f groseillier noir d'Amérique
d amerikanische Johannisbeere
l *Ribes americanum*

*** American catalpa → 5097**

187 American chestnut
f châtaignier d'Amérique
d amerikanische Kastanie; gemeine Rosskastanie
l *Castanea dentata*

188 American columbine
f ancolie du Canada
d kanadische Akelei
l *Aquilegia canadensis*

189 American elaeocarpus
f éléocarpe d'Amérique
d amerikanische Ölfrucht
l *Elaeocarpus americanus*

190 American elder
f sureau d'Amérique; sureau du Canada
d kanadischer Holunder
l *Sambucus canadensis*

191 American elm
f orme d'Amérique; orme à larges feuilles; orme blanc; orme blanc d'Amérique
d amerikanische Ulme; Weissrüster; Weissulme
l *Ulmus americana*

192 American ephedra
f éphèdre d'Amérique; éphèdre du Canada
d amerikanisches Meerträubchen
l *Ephedra americana*

193 American false hellebore
f vératre vert

d grüner Germer
l *Veratrum viride*

194 American filbert
f noisetier d'Amérique
d amerikanische Hasel
l *Corylus americana*

195 American ginseng
f ginseng à cinq feuilles
d Ginseng
l *Panax quinquefolium*

196 American holly
f houx d'Amérique
l *Ilex opaca*

197 American hop hornbeam
f ostryer de Virginie
d virginische Hopfenbuche
l *Ostrya virginiana*

198 American hornbeam
f charme d'Amérique; charme de la Caroline
d karolinische Hainbuche
l *Carpinus caroliniana; Carpinus americana*

199 American linden
f tilleul d'Amérique
d schwarze Linde; Schwarzlinde
l *Tilia americana*

200 American mangrove
f manglier; palétuvier noir
d Austerbaum; gemeiner Manglebaum; Leuchterbaum; Lichtbaum; Lichterbaum; Mangrovebaum; rote Mangrove; Wurzelbaum
l *Rhizophora mangle*

201 American millet grass
f millet diffus
d flatterige Waldhirse
l *Milium effusum*

202 American mountain ash
f sorbier des montagnes d'Amérique
d amerikanische Eberesche
l *Sorbus americana*

* **American mulberry → 4456**

203 American muskwood
f bois à balles
l *Guarea trichilioides*

204 American pitcher-plant family
f sarracéniacées
d Sarrazeniazeen
l *Sarraceniaceae*

205 American plane tree; American sycamore
f platane d'Occident
d abendländische Platane; nordamerikanische Platane; Kleiderbaum; Sykomore
l *Platanus occidentalis*

206 American plum
f prune de la Gallissonière
d amerikanische Pflaume
l *Prunus americana*

207 American potato bean
f glycine tubéreuse
d Erdbirne
l *Apios tuberosa; Apios americana*

208 American red currant; swamp red currant
f groseillier rouge des marais
l *Ribes triste; Ribes albinervium*

209 American slough grass
d Beckmannsgras
l *Beckmannia syzigachne*

210 American sweet gum
f copalme d'Amérique; liquidambar d'Amérique
d amerikanischer Amberbaum; amerikanischer Storaxbaum
l *Liquidambar styraciflua*

* **American sycamore → 205**

211 American water plantain
f alisme plantain d'eau
d gemeiner Froschlöffel; Wasserwegerich; Wegerichfroschlöffel
l *Alisma plantago-aquatica*

212 American wistaria
f glycine d'Amérique; glycine en arbre
l *Wistaria frutescens*

213 American yellowwood
f virgilier à bois jaune; cladraste à bois jaune
d amerikanisches Gelbholz
l *Cladrastis lutea; Virgilia lutea*

214 amethyst agaric; wood blewit
f pied bleu; rhodopaxille nu;

tricholome nu
d nackter Blätterschwamm; nackter Ritterling; violetter Ritterling
l *Tricholoma nudum; Lepista nuda; Agaricus nudus*

215 amethyst fescue
f fétuque améthyste; fétuque bleue
d Amethystschwingel
l *Festuca amethystina*

216 ammi
f ammi
d Knorpelmöhre
l *Ammi*

217 ammobium
f ammobium
d Papierknöpfchen; Sandimmortelle; Strohblume
l *Ammobium*

*** ammoniac plant → 731**

218 amoora
f amoora
l *Amoora*

219 amorpha honey locust
f févier d'Argentine
l *Gledits(ch)ia amorphoides*

220 amorphophallus
f amorphophallus
d Amorphophallus
l *Amorphophallus*

221 ampelopsis
f ampélopsis
d Doldenrebe; Jungfernrebe; wilder Wein
l *Ampelopsis*

222 amphibious marsh cress
d Ufersumpfkresse; Wasserkresse; Wassersumpfkresse
l *Rorippa amphibia*

223 amsonia
f amsonie
d Amsonie
l *Amsonia*

224 Amur ampelopsis
f vigne à fruits bleus
l *Ampelopsis brevipedunculata*

225 Amur barberry
f vinettier de l'Amour
d Amurberberitze
l *Berberis amurensis*

226 Amur cork tree
f arbre à liège de l'Amour; phellodendre de l'Amour
d Amurkorkbaum; Samtbaum
l *Phellodendron amurense*

227 Amur grape vine
f vigne de l'Amour
d Amurrebe
l *Vitis amurensis*

228 Amur lilac
f lilas de l'Amour
d Amurflieder
l *Syringa amurensis*

229 Amur linden
f tilleul de l'Amour
d Amurlinde
l *Tilia amurensis*

230 Amur maple
d Feuerahorn
l *Acer ginnala*

231 Amur mountain ash
f sorbier de l'Amour
d Amureberesche
l *Sorbus amurensis*

232 Amur privet
f troène de l'Amour
d Amurrainweide
l *Ligustrum amurense*

233 anabasis
f anabase
l *Anabasis*

234 anacamptis
f anacamptide
d Hundswurz; Orchis; Spitzorchis
l *Anacamptis*

235 ananas
f ananas
d Ananas; Anassa; Nanas
l *Ananas*

236 anchietea; pirageia
f anchiétée
l *Anchietea*

237 Andaman padauk
f bois rouge des Andamans
l *Pterocarpus dalbergioides*

238 Andorra creeping juniper
f genévrier de la Vallée d'Andorre
d Kriechwacholder
l *Juniperus horizontalis plumosa*

239 andrachne
f andrachné
l *Andrachne*

240 andreaea family
f andréacées
d Andreäazeen
l *Andreaeaceae*

241 andromeda
f andromède
d Gränke; Lavendelheide; Rosmarinheide
l *Andromeda*

242 anemone
f anémone
d Anemone; Windröschen; Windröslein
l *Anemone*

243 anemone; pasqueflower
f coquelourde; herbe du vent
d Kuhschelle
l *Pulsatilla*

244 anemone clematis; mountain clematis
f clématite des montagnes
d Bergwaldrebe
l *Clematis montana*

* **anemonella → 4598**

245 angelica
f angélique; archangélique
d Angelika; Brustwurz; Brustwurzel; Engelwurz
l *Angelica*

246 angelin tree; cabbage tree
f angelin; arbre à chou; bois d'angelin
d Kohlbaum
l *Andira*

* **angel's-trumpet → 247, 2304**

247 angel-tears datura; angel's-trumpet
f stramoine odorante
d Engelstrompete
l *Datura suaveolens*

248 angiosperms
f angiospermes
d Angiospermen; Bedecktsamer; Bedecktsamige; Decksamer
l *Angiospermae*

249 angle onion
d Kantenlauch; kantiger Lauch
l *Allium angulosum*

250 angle-twig poplar
f peuplier de la Caroline
d Karolina-Pappel
l *Populus angulata*

251 angostura bark tree
f galipée officinale
d Angusturabaum
l *Galipea officinalis*

* **anguria → 5869**

252 animated oat
f avoine stérile; folle avoine
l *Avena sterilis*

253 anise
f anis
d Anis
l *Pimpinella anisum*

254 anise tree
f badiane; badenier
d Sternanis
l *Illicium*

255 anisochilus
f anisochile; anisochèle
l *Anisochilus*

256 annatto tree
f rocouyer
d Annattostrauch; Orleansstrauch; Roucoustrauch
l *Bixa orellana*

257 anneslea; anneslia
f annes lée
l *Anneslea; Anneslia*

258 annual bluegrass
f pâturin annuel
d einjähriges Rispengras; jähriges Angergras; Sommerrispengras
l *Poa annua*

259 annual fleabane
f vergerette annuelle
d einjähriger Feinstrahl
l *Erigeron annuus*

260 **annual gentian**
f gentiane amarelle
d Bitterenzian
l *Gentiana amarella*

261 **annual hair grass**
d Schleiergras
l *Aira capillaris*

* **annual honesty → 1884**

262 **annual knawel**
f gnavelle
d Ackerknäuel; einjähriger Knäuel; einjähriges Knäuelkraut; grüner Knäuel; Sommerknäuel
l *Scleranthus annuus*

* **annual mercury → 2785**

* **annual pearlwort → 273**

263 **annual stock**
f giroflée des jardins
d Sommerlevkoje
l *Matthiola annua*

264 **annual stonecrop**
d einjähriger Mauerpfeffer; Sommerfetthenne
l *Sedum annuum*

265 **annual vernal grass**
d begranntes Ruchgras; Sensendüwel
l *Anthoxanthum aristatum*

266 **annual wild rice**
f riz d'eau; riz du Canada
d kanadischer Reis; nordamerikanischer Reis; Wasserreis; wilder Reis
l *Zizania aquatica*

267 **annual yellow clover**
f mélilot à petite fleur
d kleinblütiger Steinklee
l *Melilotus indica*

268 **antarctic false beech**
f hêtre antarctique
d antarktische Scheinbuche
l *Nothofagus antarctica*

269 **anthericum**
f anthéric
d Graslilie; Zaunlilie
l *Anthericum*

270 **anthurium**
f anthure
d Flamingoblume; Schwanzblume; Blütenschweif; Schwefelblume; Schweifblume
l *Anthurium*

271 **anthyllis**
f anthyllide
d Tannenklee; Wollblume; Wundblume; Wundklee
l *Anthyllis*

272 **ape's-earing**
f bois-serpent; pithécolobium
d Spiralhülse
l *Pithecolobium*

273 **apetalous pearlwort; annual pearlwort**
d kronloses Mastkraut
l *Sagina apetala*

274 **aphananthe**
f aphananthe
l *Aphananthe*

275 **aphelandra**
f aphélandre
d Glanzkölbchen
l *Aphelandra*

276 **aphyllophorales**
f aphyllophorales
d Nichtblätterpilze; Nichtlamellenpilze
l *Aphyllophorales*

277 **apio arracacia; arracacha**
f arracacia; apio; pomme de terre céleri
d essbare Arracacha
l *Arracacia xanthorrhiza; Arracacia esculenta*

278 **aponogeton family**
f aponogétonacées
d Aponogetonazeen
l *Aponogetonaceae*

279 **apophyllum**
f apophylle
l *Apophyllum*

280 **aporosella**
f aporoselle
l *Aporosella*

281 **aposeris**
f aposeris

d Stinkkohl; Stinksalat
l *Aposeris*

282 apple family
f pomacées
d Kernobstgehölze; Pomazeen; Pomeen
l *Pomaceae; Malaceae*

283 apple mint; round-leaved mint
f menthe à feuilles rondes
d rundblättrige Minze
l *Mentha rotundifolia*

284 apple of Peru
f nicandre
d Giftbeere
l *Nicandra*

285 apple of Peru
f faux coqueret; nicandre physalide
d blaue Giftbeere
l *Nicandra physaloides*

286 apple quince
d Apfelquitte
l *Cydonia oblonga* var. *maliformis*

287 apple rose
f rose pommier; rosier de pomme; rose velue
d Apfelrose; echte Hagebuttenrose; Hagerose; Rosenapfel; weichhaarige Rose
l *Rosa pomifera; Rosa villosa*

288 apple tree
f pommier
d Apfelbaum
l *Malus*

* **apricot → 1350**

289 apricot plum
f prunier abricotier
d Aprikosenpflaume
l *Prunus simoni*

290 aptandra
f aptandre
l *Aptandra*

291 apuleia
f apulée
l *Apuleia*

292 Arabian coffee
f caféier d'Arabie
d arabischer Kaffeebaum; Bergkaffee; echter Kaffeebaum
l *Coffea arabica*

293 Arabian jasmine
f jasmin d'Arabie
d arabischer Jasmin; Nachtblume
l *Jasminum sambac*

294 Arabian tea; kat
f catha
d Kathstrauch; Bügelholz
l *Catha edulis*

295 aralia
f aralie
d Aralie; Bergangelika
l *Aralia*

296 arar tree
f callitris articulé; thuya articulé
d Sandarakzypresse; Sandarakbaum
l *Tetraclinis articulata; Callitris quadrivalvis*

297 araucaria
f araucarie
d Araukarie; Andertanne; Schmucktanne; Zimmertanne
l *Araucaria*

298 araucaria family
f araucariées
d Araukariengewächse
l *Araucariaceae*

299 arborescent aloe; tree-like aloe; woody aloe
f aloès corne de bélier
d baumartige Aloe; Brandbaum
l *Aloe arborescens*

300 arborvitae
f thuya
d Biota; Lebensbaum; Thuja; Thuje; Ukkanübaum; Zypressenfichte
l *Thuya; Biota*

301 archegoniates
f archegoniates
d Archegoniaten
l *Archegoniatae*

302 archimycetes
f archimycètes
d Archimyzeten; Urpilze
l *Archimycetes*

303 archytea
f archytée
l *Archytea*

304 arctic bluegrass
d arktisches Rispengras
l *Poa arctica*

305 arctic bramble
f ronce arctique
d nordische Brombeere; nordische Himbeere
l *Rubus arcticus*

306 arctic campion
f viscaire alpine
d Alpenpechnelke
l *Viscaria alpina; Lychnis alpina*

307 arctic cassiope
f cassiopée-mousse
d Moosheide
l *Cassiope hypnoides*

308 arctic draba
d Fladnitzer Felsenblümchen
l *Draba fladnizensis*

309 arctic fleabane
d einköpfiges Berufkraut
l *Erigeron uniflorus*

310 arctic pearlwort
d Alpenmastkraut; Felsenmastkraut
l *Sagina saginoides*

311 arctic willow
f saule arctique
d arktische Weide
l *Salix arctica*

312 arctotis
f arctotide
d Bärenohr
l *Arctotis*

313 areca palm
f arec; aréquier
d Areka; Arekapalme; Betelnusspalme; Betelpalme
l *Areca*

314 arethusa
f aréthuse
d Arethuse; Runzelbart
l *Arethusa*

315 argan tree
f bois d'argan
d Arganbaum; Arganiabaum; Eisenholzbaum
l *Argania sideroxylon*

316 aricanga shadow palm
d Schotts Erdpalme
l *Geonoma schottiana*

317 Arizona cypress
f cyprès de l'Arizona
d Arizonazypresse
l *Cupressus arizonica*

318 Arizona ponderosa pine
d Arizonakiefer
l *Pinus arizonica; Pinus ponderosa arizonica*

319 Armand clematis
f clématite du Père Armand
d Armands Waldrebe
l *Clematis armandi*

320 Armand pine
f pin du Père Armand David
d Armands Kiefer
l *Pinus armandi*

321 Armenian plum
d kurdistanische Pflaume
l *Prunus curdica*

322 arnebia
f arnébie
d Arnebie
l *Arnebia*

323 arnica
f arnica; arnique
d Arnika; Engelkraut; Bergwohlverleih; Wohlverleih
l *Arnica*

324 Arnold hawthorn
f aubépine d'Arnold
d Arnolds Weissdorn
l *Crataegus arnoldiana*

* **aromatic sumac → 2358**

* **arracacha → 277**

325 arracacia
f arracacia
d Arracacha
l *Arracacia*

* **arrayan → 4141**

326 arrow arum
f peltandre
l *Peltandra*

327 arrowhead
f sagittaire; fléchière
d Pfeilkraut
l *Sagittaria*

328 arrow pod grass
f triglochin des marais
d Sumpfdreizack
l *Triglochin palustris*

329 arrowroot
f marante
d Marante; Pfeilwurz
l *Maranta*

330 arrowroot family
f maranthacées
d Pfeilwurzgewächse
l *Maranthaceae*

331 artichoke; globe artichoke
f artichaut commun
d Artischocke; Gemüseartischocke; Strobeldorn
l *Cynara scolymus*

332 artichoke betony; Chinese artichoke; Japanese artichoke; chorogi
f crosne du Japon
d japanische Artischocke; japanische Kartoffel; Japanknollen; Knollenziest
l *Stachys sieboldi*

*** artillery plant → 1259**

333 arum
f arum; gouet
d Aaronsstab; Aronstab; Aronwurzel; Zehrwurz
l *Arum*

*** arum → 5254**

334 arum family
f aracées
d Arazeen; Aroideen; Arongewächse; Aronstabgewächse
l *Araceae*

335 arundinella
f arundinelle
l *Arundinella*

336 asafetida giant fennel
f férule as(s)a fétida; férule persique; assa fétida
d Stinkasant; Steckenkraut; Teufelsdreck
l *Ferula assa foetida*

*** asarabacca → 5947**

337 ascomycetes; sac fungi
f ascomycètes
d Askomyzeten; Schlauchpilze; Schlauchschwämme
l *Ascomycetes*

338 ash
f frêne
d Esche
l *Fraxinus*

*** ash-leaved maple → 750**

339 Asia glory; silver weed
f argyreia
d Silberkraut
l *Argyreia*

340 Asian bell
f codonopside
d Glockenheide; Windenglocke
l *Codonopsis*

341 Asian Molucca balm
f mélisse des Moluques; mélisse de Constantinople
l *Molucella laevis*

342 Asian service berry
f amélanchier du Japon
d asiatische Felsenbirne
l *Amelanchier asiatica*

343 Asian toddalia
f toddalie d'Asie
l *Toddalia asiatica*

344 Asia poppy
f roemérie
d Römerie
l *Roemeria*

345 Asiatic barberry
f vinettier d'Asie
d asiatische Berberitze
l *Berberis asiatica*

346 Asiatic ginseng
f ginseng; ginsang
d Ginseng
l *Panax ginseng; Panax schinseng*

347 Asiatic moonseed
f ménisperme de Daourie
d dahurischer Mondsame
l *Menispermum dahuricum*

348 Asiatic plantain
f plantain d'Asie
d asiatischer Wegerich
l *Plantago asiatica*

349 Asiatic tree cotton
f cotonnier en arbre
l *Gossypium arboreum*

350 asparagus
f asperge
d Spargel; Spars
l *Asparagus*

* **asparagus bean → 6066**

* **aspen mushroom → 3940**

351 aspergillus
f aspergille
d Aspergillus; Giesskannenschimmel; Kolbenschimmel
l *Aspergillus*

352 asphodel
f asphodèle
d Affodill; Goldwurzel
l *Asphodelus*

353 aspidistra
f aspidistra
d Aspidistra; Schildblume; Schusterpalme
l *Aspidistra*

354 aspidium
f aspidie
d Schildfarn
l *Aspidium*

355 Assam king begonia
f bégonia rex
d Königsbegonie; Rexbegonie
l *Begonia rex*

356 Assam tea
d Assam-Tee
l *Thea sinensis* var. *assamica*

357 astelia
f astélie
l *Astelia*

358 aster
f aster
d Aster; Sternblume
l *Aster*

359 aster family
f astérées
d Korbblütler
l *Asteraceae*

360 astilbe
f astilbe
d Astilbe; Prachtspiere
l *Astilbe*

361 astrocarpus
f astrocarpe
l *Astrocarpus*

362 atalantia
f atalantie
l *Atalantia*

363 Atlantic leatherwood
f dircé des marais
d Lederholz
l *Dirca palustris*

364 Atlas cedar; Mount Atlas cedar
f cèdre d'Atlas
d Atlaszeder; Silberzeder
l *Cedrus atlantica*

365 attalea
f attalée
d Pindowapalme
l *Attalea*

366 aubrietia
f aubriétie
d Aubrietie; Blaukissen; Purpurkissen
l *Aubrietia*

367 aucuba
f aucuba
d Aukuba; Goldblatt; Schusterpalme
l *Aucuba*

368 auricula
f auricule; auricule de jardins; oreille d'ours
d Alpenaurikel; Aurikel; Bärohr
l *Primula auricula*

369 auriculariales
f auriculariales
d Ohrlappenpilzartige; Ohrlappenpilze
l *Auriculariales*

370 Australian brush-cherry eugenia
f myrte d'Australie
l *Eugenia paniculata australis; Eugenia myrtifolia*

371 Australian fan palm
f palmier éventail d'Australie
d südliche Livistone
l *Livistona australis*

372 Australian ivorywood
f siphonode d'Australie
l *Siphonodon australe*

373 Australian jelly alga; jelly plant
d Gallerttang
l *Eucheuma speciosum*

* **Australian nut → 4385**

374 Austrian brier rose
f ronce d'Autriche
d gelbe Rose; Wanzenrose
l *Rosa foetida*

375 Austrian broom
f cytise d'Autriche
d österreichischer Geissklee
l *Cytisus austriacus*

376 Austrian dragonhead
f dracocéphale d'Autriche
d österreichischer Drachenkopf
l *Dracocephalum austriacum*

377 Austrian field cress
d österreichische Kresse; österreichische Sumpfkresse
l *Rorippa austriaca*

378 Austrian flax
f lin d'Autriche
d österreichischer Lein
l *Linum austriacum*

379 Austrian leopard's-bane
f doronic d'Autriche
d österreichische Gemswurz
l *Doronicum austriacum*

380 Austrian pine
f pin noir
d österreichische Schwarzkiefer; Schwarzkiefer
l *Pinus nigra*

* **Austrian pine → 516**

381 autobasidiomycetes
f autobasidiomycètes
d Autobasidiomyzeten
l *Autobasidiomycetes*

382 autumn crocus
f colchique
d Lichtblume; Zeitlose
l *Colchicum*

383 autumn squill
f scille d'automne
d Herbstblaustern
l *Scilla autumnalis*

384 autumn water starwort
f callitriche d'automne
d Herbstwasserstern
l *Callitriche autumnalis*

385 avens
f benoîte
d Benediktenkraut; Erdrose; Nelkenwurz
l *Geum*

386 avicennia
f avicennie
d Mangrovebaum
l *Avicennia*

* **avocado → 183**

387 awlwort
f subulaire
d Pfriemenkresse
l *Subularia*

* **awnless brome → 5034**

388 axyris
f axyris
d Mengel
l *Axyris*

389 azalea
f azalée
d Alpenrose; Felsenstrauch; Azalee; Azalie
l *Azalea; Rhododendron*

390 azarole hawthorn
f azarole; azarolier; azérolier; épine d'Espagne
d Azarol-Hagedorn; italienische Mispel; welsche Mispel; pontischer Weissdorn
l *Crataegus azarolus; Crataegus pontica*

391 azolla
f azolle
d Wasserfarn
l *Azolla*

392 Azores forget-me-not
f myosotis des Açores
l *Myosotis azorica*

393 Aztec lily; Jacobean lily
f lis de St.-Jacques
d Jakobslilie; Sprekelie
l *Sprekelia*

394 Aztec marigold
f rose d'Inde
d aufrechte Studentenblume
l *Tagetes erecta*

395 Aztec tobacco
f tabac rustique
d Bauerntabak; Machorka; Veilchentabak
l *Nicotiana rustica*

B

396 **babassu**
f orbignya
l *Orbignya oleifera*

397 **babies'-breath; baby's-breath**
f gypsophile paniculée
d Rispengipskraut; Schleiergipskraut; Schleierkraut
l *Gypsophila paniculata*

398 **baboen**
d Baboen
l *Virola surinamensis*

399 **babul acacia**
f acacia d'Arabie; arbre à gomme
d arabische Akazie; Babul; Kikar; Ssant; ägyptischer Schotendorn
l *Acacia arabica*

400 **Babylon weeping willow**
f saule de Babylone; saule pleureur
d Gräberweide; Hängebaum; Hängeweide; Osterpalme; Tränenweide; Trauerweide
l *Salix babylonica*

401 **baby pondweed**
d kleines Laichkraut
l *Potamogeton pusillus*

* **baby's-breath → 397**

402 **baby's-tears**
f helxine
d Bubiköpfchen; Helxine
l *Helxine*

403 **bacaba palm**
f énocarpe
d Mostpalme
l *Oenocarpus*

404 **baccaurea**
f baccaurée
l *Baccaurea*

405 **baccharis**
f baccharide
d Kreuzstrauch
l *Baccharis*

* **Bachang mango → 180**

* **bachelor's-button → 1613**

406 **Bachofen's speedwell**
f véronique de Bachofen
d Bachofens Ehrenpreis
l *Veronica bachofeni*

407 **badderlocks; badderlocks wingkelp; henware; honeyware; murlin**
f varech alimentaire
d essbarer Flügeltang
l *Alaria esculenta*

408 **bael fruit; bel; golden apple; Bengal quince**
f bel indien; cognassier du Bengale; oranger de Malabar
d Belbaum; Bälbaum; Schleimapfelbaum
l *Aegle marmelos*

409 **Bahiagrass**
f herbe de Bahia
l *Paspalum notatum*

410 **bakupari (rheedia)**
f cirolier du Brésil
l *Rheedia brasiliensis*

411 **bakuri Guiana orange**
f grande bacury
l *Platonia insignis*

412 **balanophora family**
f balanophoracées
d Balanophorazeen; Kolbenschosser
l *Balanophoraceae*

413 **bald brome**
d traubige Trespe; Traubentrespe
l *Bromus racemosus*

414 **bald cypress**
f cyprès chauve; cyprès de la Louisiane
d amerikanische Sumpfzypresse; Eibenzypresse; Sumpfzeder; Taxodie; virginische Sumpfzypresse
l *Taxodium distichum*

415 **baldmoney; spicknel**
f méon
d Bärenwurzel; Bärwurz; Bärwurzel; Mutterwurz
l *Meum*

416 **Balfour's meadow grass**
f pâturin de Balfour
d Balfours Rispengras
l *Poa balfouri*

417 Balfour's spruce
f épicéa de Balfour
d Balfours Fichte
l *Picea balfouriana; Picea likiangensis balfouriana*

418 Balkan maple
f érable de Heldreich
d Heldreichs Ahorn
l *Acer heldreichi*

419 Balkan pine
f pin de Balkans; pin de Macédoine
d rumelische Weymoutskiefer
l *Pinus peuce*

420 ball-headed mixed-flower
f raiponce orbiculaire
d kugelige Teufelskralle; runde Teufelskralle
l *Phyteuma orbiculare*

421 ball-headed onion
d Kopflauch; rundköpfiger Lauch
l *Allium sphaerocephalum*

422 ball mustard
f neslie paniculée
d Ackernüsschen; Finkensame
l *Neslia paniculata*

423 balloonflower
f platycodone
d Ballonblume
l *Platycodon*

424 balloonflower
f campanule à grandes fleurs; platycodone à grandes fleurs
l *Platycodon grandiflorum; Campanula grandiflora*

425 ballota
f ballote
d Ballote; Gottvergess; schwarzer Andorn; Schwarznessel
l *Ballota*

426 ballroom flindersia
f flindersie d'Australie
l *Flindersia australis*

427 balm
f mélisse
d Honigkraut; Melisse
l *Melissa*

428 balm-of-gilead poplar
f peuplier baumier de Giléad; peuplier de l'Ontario
d Ontariopappel; weissliche Pappel
l *Populus candicans*

429 balm tree
f baumier du Pérou
d peruanischer Balsambaum; Perubalsambaum
l *Myroxylon*

* **balsa → 5870**

430 balsam amyris
f baumier de Jamaïque; santal du Venezuela
d Balsambaum
l *Amyris balsamifera*

431 balsam apple
f pomme de merveille; momordique balsamine; momordique baumier; pomme de Jérusalem
d Balsamapfel; Balsamgurke; Wunderbalsamapfel
l *Momordica balsamina*

432 balsam fir
f baumier de Giléad; baumier du Canada; sapin baumier
d Balsamtanne
l *Abies balsamea*

433 balsamina family
f balsaminacées
d Balsaminazeen; Balsaminengewächse; Springkrautgewächse
l *Balsaminaceae*

434 balsam pear
f margose; papareh; momordique à feuilles de vigne
d bittere Springgurke
l *Momordica charantia*

* **balsam poplar → 5100**

* **balsam shrub → 5552**

435 balsam willow
f saule baumier
l *Salix pyrolifolia*

436 Baltic rush
d baltische Binse
l *Juncus balticus*

* **Bambara groundnut → 1574**

437 bamboo
f bambou
d Bambus; Bambusgras; Bambusrohr
l *Bambusa*

438 bamboo fern
d japanische Koniogramme
l *Coniogramme japonica*

439 banana
f bananier
d Banane; Paradiesfeige; Pisang; Pisangfeige
l *Musa*

440 banana family
f musacées
d Bananengewächse; Musazeen
l *Musaceae*

441 baneberry
f actée
d Christophskraut; Hexenkraut
l *Actaea*

442 banksia
f banksia
l *Banksia*

*** Banks pine → 3001**

443 Banks rose
f rosier Banks; rosier de Banks
d Banks Rose
l *Rosa banksiae*

444 banyan fig
f figuier de Bengale; banyan
d Banyanbaum
l *Ficus bengalensis*

445 baobab; monkey-bread tree
f baobab
d Affenbrotbaum; Baobab
l *Adansonia digitata*

446 Barbados cherry
f cerise de Antilles; cerise carée; lucée; malpighier
d granatapfelblättrige Malpighie; Kirschtanne
l *Malpighia punicifolia*

447 Barbados gooseberry
f groseillier des Barbados; groseillier d'Amérique
d Barbados-Stachelbeere; westindische Stachelbeere
l *Pereskia aculeata*

448 Barbados lily
f amaryllis à bandes
l *Amaryllis vittata; Hippeastrum vittatum*

449 Barbados nut; physic nut
f médicinier; pignon d'Inde
d Purgiernuss
l *Jatropha curcas*

450 Barbary wolfberry
f lyciet de Barbarie
d Bocksdorn
l *Lycium barbarum*

451 barbatimao alum bark tree
f stryphnodendre barbatimao
l *Stryphnodendron barbatimao*

452 barberry
f épine-vinette; vinettier
d Berberitze; Sauerdorn
l *Berberis*

453 barberry family
f berbéridacées
d Berberidazeen; Berberitzengewächse; Sauerdörner; Sauerdorngewächse
l *Berberidaceae*

454 barley
f orge
d Gerste
l *Hordeum*

455 barley
f orge carrée
d mehrzeilige Gerste; vielzeilige Gerste; Saatgerste
l *Hordeum vulgare*

456 barnyard grass
f millet pied-de-coq; panis des marais
d gemeine Hühnerhirse; Stachelhirse
l *Echinochloa crusgali*

*** barnyard grass → 3057**

457 barrel bottle tree
f sterculie des rochers
l *Sterculia repestris; Brachychiton rupestris*

458 barren strawberry
f waldsteinie
d Waldsteinia
l *Waldsteinia*

459 bartonia
f bartonie
l *Bartonia*

460 bartsia
f odontite; bartschia
d Alpenhelm; Bartschie; Bartsie; Zahntrost
l *Bartsia; Odontites*

461 basidiolichens
f basidiolichens
d Basidiolichenen
l *Basidiolichenes*

462 basidium fungi
f basidiomycètes
d Basidienpilze; Basidiomyzeten; basidientragende Pilze; Sporenständerpilze; Ständerpilze
l *Basidiomycetes*

463 basil
f basilic
d Basilie; Basilienkraut; Basilikenkraut; Basilikum; Hirnkraut
l *Ocimum*

464 basil thyme
f basilic sauvage
d Quendel; Steinquendel
l *Acinos*

465 basket ivy
f cymbalaire
d Heidelberger Schlosskraut
l *Cymbalaria*

* **basket oak → 5343**

466 basket vine
f æschynanthe
d Schamblume
l *Aeschynanthus*

467 basket willow; common osier
f saule de vanniers; saule viminale; osier viminal
d Bandweide; Fahlweide; Fischweide; Flechtweide; Hanfweide; Hegeweide; Korbweide
l *Salix viminalis*

468 bassia
f bassia; arbre à beurre
d Dornmelde
l *Bassia*

469 bastard balm
f mélitte à feuilles de mélisse
d Melissenimmenblatt
l *Melittis melissophyllum*

470 bastard daisy
f paquerolle
d Bundblume
l *Bellium*

471 bastard hemp
f cannabine faux-chanvre
d gelber Hanf
l *Datisca cannabina*

* **bastard indigo → 2948**

472 bastard lucerne
f luzerne bigarrée
d Sandluzerne
l *Medicago varia*

473 bastard lupin; lupinaster
f trèfle faux-lupin
d Lupinenklee
l *Trifolium lupinaster*

474 bastard rosewood
f bois de rose
d Rosenholz
l *Synoum glandulosum*

475 bastard speedwell
f véronique bleu améthyste
d unechter Ehrenpreis
l *Veronica spuria*

476 bastard toadflax
f thésion
d Bergflachs; Leinblatt; Vermeinkraut
l *Thesium*

477 Batavian shorea
d Dammaraharzbaum
l *Shorea wisneri*

478 bath asparagus; Pyrenees star-of-Bethlehem
f aspergette; ornithogale des Pyrénées
d Pyrenäen-Milchstern
l *Ornithogalum pyrenaicum*

479 batrachospermum
f batrachosperme
d Froschlaichalge
l *Batrachospermum*

480 Bauer acronychia
f acronychie d'Australie
l *Acronychia baueri*

481 bauhinia
f bauhinier
d Affentreppen
l *Bauhinia*

482 Bavarian gentian
f gentiane de Bavière
d bayrischer Enzian
l *Gentiana bavarica*

483 bayberry; wax myrtle
f cirier; myrique
d Gagel; Gagelstrauch; Lichtmyrte; Wachsmyrte
l *Myrica*

484 bayberry family
f myricacées
d Gagelgewächse; Gagelsträucher; Myrikazeen; Myrizeen
l *Myricaceae*

485 bay boletus; Polish mushroom
f bolet bai; cèpe bai
d Kastanienpilz; Maronenpilz; Maronenröhrling
l *Boletus badius; Xerocomus badius*

* **bay willow → 3236**

486 beach cocklebur
d hakige Spitzklette; italienische Spitzklette
l *Xanthium echinatum*

487 beach grass
f ammophile
d Sandgras; Sandhalm; Sandrohr; Strandhafer
l *Ammophila*

* **beach pine → 4876**

488 beach plum
f prunier des grèves
d Strandpflaume
l *Prunus maritima*

489 beach wormwood; dusty miller
f absinthe des rivages
l *Artemisia stellariana*

490 bead lily; Clinton's lily
f clintonie
d Clintonia
l *Clintonia*

491 bead plant
f nertère humble
d Korallenbeere
l *Nertera granadensis; Nertera depressa*

492 bead-ruby
f maianthème
d Schattenblume
l *Maianthemum*

493 bead tree
f adénanthère
d Drüsenbaum; Drüsenblume
l *Adenanthera*

* **bead tree → 1160**

494 beaked chervil
f anthrisque
d Kerbel; Kerbelkraut; Klettenkerbel; Kletterkerbel
l *Anthriscus*

495 beak grain
f diarrhène
d Zweifadengras
l *Diarrhena*

496 beak rush
f rhynchospore
d Moorsimse; Schnabelried; Schnabelsimse
l *Rhynchospora*

497 bean
f haricot
d Bohne; Fisole
l *Phaseolus*

498 bean caper
f fabagelle
d Bohnenkaper; Doppelblatt; Jochblatt
l *Zygophyllum*

499 bean-caper family
f zygophyllacées
d Doppelblattgewächse; Zygophyllazeen
l *Zygophyllaceae*

* **bean tree → 3680, 5097**

500 bearberry
f busserole officinale; raisin d'ours; arbousier busserole; thé de brousse

d echte Bärentraube; gemeine Bärentraube; immergrüne Bärentraube
l *Arctostaphylos uva-ursi*

501 **bearberry cotoneaster**
f cotonéastre de Dammer
d Dammers Zwergmispel
l *Cotoneaster dammeri*

502 **bearded bellflower**
f campanule barbue
d bärtige Glockenblume
l *Campanula barbata*

503 **bearded usnea**
f usnée barbue
d Bartflechte; Bartmoos
l *Usnea barbata*

504 **beard grass**
f polypogon
d Bürstengras
l *Polypogon*

* **beardtongue → 4117**

505 **bear oak**
f chêne à feuilles de houx; chêne buisson
d Zwergeiche
l *Quercus ilicifolia*

* **bear's-breech → 12**

506 **bear's-ear sanicle**
f cortuse
d Bärsanikel; Glöckel; Heilglöckchen; Alpenglöckchen
l *Cortusa*

507 **bear's-foot**
f hellébore fétide
d stinkende Nierwurz
l *Helleborus foetidus*

* **bear's-garlic → 4415**

* **beauty-berry → 2370**

508 **beauty-leaf**
f calophylle
d Gummiapfel; Kalababaum; Schönblatt
l *Calophyllum*

* **beauty-leaf mastwood → 2943**

509 **beccabunga speedwell**
f véronique aquatique; véronique beccabunga
d Bachbunge; Bachbungen-Ehrenpreis; Bachehrenpreis
l *Veronica beccabunga*

510 **bedstraw**
f caille-lait
d Labkraut; Liebkraut; Wegekraut
l *Galium*

511 **bedstraw woodruff**
d Labmeister
l *Asperula galioides*

512 **bee balm; horsemint**
f monarde
d Bienenbalsam; Monarde; Pferdeminze
l *Monarda*

513 **beech; beech tree**
f hêtre
d Buche
l *Fagus*

514 **beech family**
f fagacées; cupulifères
d Becherfrüchtler; Buchengewächse; Fagazeen; Kupuliferen
l *Fagaceae*

* **beech tree → 513**

515 **beefsteak fungus; liver agaric**
f fistuline hépatique; glu de chêne; foie de bœuf; langue de bœuf; langue de châtaignier
d Fleischschwamm; gemeiner Leberpilz; Ochsenzunge; Zungenpilz
l *Fistulina hepatica*

516 **beefwood; Austrian pine**
f casuarina
d Känguruhbaum; Kasuarine; Keulenbaum; Streitkolbenbaum
l *Casuarina*

517 **beefwood family**
f casuarinacées
d Kasuarinazeen; Kasuarinengewächse
l *Casuarinaceae*

518 **bee larkspur**
f dauphinelle élevée
d hoher Rittersporn
l *Delphinium elatum*

519 bee orchid
d Bienenorchis; Bienenragwurz
l *Ophrys apifera*

520 beet
f betterave
d Rübe
l *Beta*

*** beggar's-lice → 2126**

521 beggarticks; bur marigold; sticktight
f bident
d Zweizahn
l *Bidens*

522 Begger rose
f rosier de Begger
d Beggers Rose
l *Rosa beggeriana*

523 begonia
f bégonia
d Begonie; Schiefblatt
l *Begonia*

524 begonia family
f bégoniacées
d Begoniazeen; Begoniengewächse
l *Begoniaceae*

*** bel → 408**

525 belladonna; deadly nightshade
f belladone; bouton noir
d Beilwurz; Belladonna; gemeine Tollkirsche; schwarze Tollkirsche; Wolfskirsche; Wolfswut
l *Atropa belladonna*

526 belladonna lily
f croix de St.-Jacques; amaryllis belladone
d Ritterstern; mexikanische Lilie
l *Amaryllis belladonna*

527 bell flambeau tree
f tulipier du gabon
l *Spathodea campanulata*

528 bellflower
f campanule
d Glockenblume; Schellenblume
l *Campanula*

529 bellflower family
f campanulacées
d Glockenblumengewächse; Kampanulazeen
l *Campanulaceae*

530 bellwort
f uvulaire
d Zäpfchenkraut
l *Uvularia*

531 belvedere summer cypres
f kochie à balais; belvédère
d Besenkraut; Besenradmelde; Besensommerzypresse
l *Kochia scoparia*

532 Bengal gambir plant
f gambier
d Klimmstrauch
l *Uncaria gambir*

533 Bengal kino; dhak tree; palas tree
f arbre à laque
d Plossobaum
l *Butea frondosa; Butea monosperma*

*** Bengal quince → 408**

*** benjamin tree → 5315**

534 bent grass
f agrostide
d Straussgras; Windhalm
l *Agrostis*

535 bergamot mint
f menthe citronnée
d Bergamottminze
l *Mentha citrata*

536 bergamot orange
f bergamotier
d Bergamottbaum; Bergamotte
l *Citrus bergamia*

537 bergenia
f bergénie
d Wickelwurz; Bergenie
l *Bergenia*

538 bergia
f bergie
l *Bergia*

539 berlinia
f berlinia
l *Berlinia*

540 Berlin poplar
f peuplier de Berlin

d Berliner Pappel
l *Populus berlinensis*

541 **Bermuda arrowroot**
f arrow-root des Antilles; marante; arrow-root de la Jamaïque
d Pfeilwurz; rohrartige Pfeilwurz; Rohrpfeilwurz; rohrartige Marante
l *Maranta arundinacea*

542 **Bermuda grass**
f gros chiendent
d Bermudagras; Fingerhundszahn; Hundshirse; Hundszahngras
l *Cynodon dactylon*

543 **Bermuda red cedar**
f cèdre des Bermudes; genévrier des Bermudes
d Bermudazeder; Florida-Zeder
l *Juniperus barbadensis; Juniperus bermudiana*

544 **berry-bearing chickweed**
d Beerentaubenkropf; beerentragender Hühnerbiss
l *Cucubalus baccifer*

* **berseem → 2000**

545 **berula**
f petite berle
d Berle; schmaler Merk; Wassermerk
l *Berula*

546 **besom heath**
f bruyère à balais
l *Erica scoparia*

547 **Bessey cherry**
f cerisier des sables
l *Cerasus besseyi*

548 **betel-nut palm; betel palm**
f noix d'arec
d Betelpalme; Betelnusspalme; Arekapalme; Arekanusspalme; Catechu-Palme
l *Areca catechu*

549 **betel pepper**
f bétel
d Betel; Betelpfeffer; Kaupfeffer
l *Piper betle*

550 **betony; woundwort**
f bétoine; épiaire
d Scheinziest; Ziest
l *Betonica; Stachys*

551 **Bhutan cypress**
f cyprès toruleux
d chinesische Zypresse; hohe Zypresse
l *Cupressus duclouxiana; Cupressus torulosa*

552 **Biberstein tulip**
f tulipe de Biberstein
d Biebersteins Tulpe
l *Tulipa bibersteiniana*

553 **Bible frankincense**
f boswellie de Carter
d Mohrmeddhu; Weihrauchbaum
l *Boswellia carteri*

554 **biennial cork oak**
f chêne occidental
d Korkeiche
l *Quercus occidentalis; Quercus suber occidentalis*

555 **bifora**
f bifora
d Hohlsame
l *Bifora*

556 **big-cone Douglas fir**
f douglas à gros cônes
l *Pseudotsuga macrocarpa*

* **big-cone pine → 1642**

557 **big-flowered broom**
f cytise capité
d kopfiger Goldregen
l *Cytisus capitatus; Cytisus supinus*

558 **big-flowered Java tea**
f thé de Java
d Javatee; Nierentee
l *Orthosiphon stamineus*

559 **big-flowered selfheal**
f brunelle à grandes fleurs
d grosse Braunelle; grossblütige Brunelle; grosse Brunelle
l *Prunella grandiflora*

560 **big-flower sage**
f sauge à grandes fleurs
d grossblättrige Salbei
l *Salvia grandiflora*

561 **big-leaf linden**
f tilleul à grandes feuilles d'Europe; tilleul de Hollande
d grossblättrige Linde; Sommerlinde
l *Tilia platyphyllos*

562 big-leaf maple; Oregon maple
f érable à grandes feuilles; érable de l'Orégon
d grossblättriger Ahorn
l *Acer macrophyllum*

563 big-leaf periwinkle; cutfinger
f grande pervenche
d grosses Immergrün
l *Vinca major*

564 big-leaf willow
f saule à grandes feuilles
d grossblättrige Weide
l *Salix grandifolia*

565 big-leaved hydrangea
f hortensia commun; hortensia des jardins; hydrangelle des jardins
d echte Hortensie
l *Hydrangea macrophylla*

566 bignay China laurel; salamander tree; Chinese laurel
f antidesme
d lorbeerblättriger Flachsbaum; Salamanderbaum
l *Antidesma bunius*

567 bignonia
f bignonier; bignone
d Trompetenblume
l *Bignonia*

568 big-pod vetch
f vesce à grosses graines
l *Vicia macrocarpa*

569 big quacking grass
f grande brize
d grosses Zittergras
l *Briza maxima*

570 bigroot geranium
f géranium à grosses racines
d grosswurzeliger Storchschnabel
l *Geranium macrorrhizum*

571 big scentless mock orange
f seringa à grandes fleurs
d grossblütiger Pfeifenstrauch
l *Philadelphus grandiflorus*

572 big-seeded false flax
f sésame d'Allemagne; caméline
d Leindotter; Saatleindotter; Butterraps; Buttersame; Flachsdotter
l *Camelina sativa*

573 big-spine honey locust
f févier à grosses épines
l *Gledits(ch)ia macrantha*

574 big-sting nettle; stinging nettle
f ortie dioïque; grande ortie
d grosse Brennessel; grosse Nessel
l *Urtica dioica*

575 big white pelargonium
f pélargonium à grandes fleurs
d grossblumige Pelargonie
l *Pelargonium grandiflorum*

* **bilberry → 3784**

576 bilimbi
f bilimbi; carambolier bilimbi
d Gurkenbaum
l *Averrhoa bilimbi*

* **billbergia → 68**

577 Billiard spirea
f spirée de Billiard
d Billiards Spierstrauch
l *Spiraea billardii*

578 birch
f bouleau
d Birke
l *Betula*

579 birch family
f bétulacées
d Betulazeen; birkenartige Gewächse; Birkengewächse
l *Betulaceae*

580 birch-leaf pear
f poirier à feuilles de bouleau
d birkenblättriger Birnbaum
l *Pyrus betulifolia*

581 birch-leaved spirea
f spirée à feuilles de bouleau
d birkenblättriger Spierstrauch
l *Spiraea betulifolia*

582 birch mushroom; rough-stalked boletus; rough-stemmed boletus
f bolet raboteux; bolet rugueux; bolet rude; cèpe rude
d Birkenpilz; Birkenröhrling; Kapuzinerpilz; Fleischling; Schafpilz
l *Boletus scaber; Leccinum scabrum*

583 birch polyporus
f polypore du bouleau

d Birkenporling
l *Polyporus betulinus*

* **bird cherry → 2078**

584 **bird-of-paradise flower**
f strélitzie
d Paradiesvogelblume
l *Strelitzia*

585 **bird rape**
f chou champêtre; rave
d Rübsen; Sommerrübsen; Rübenkohl
l *Brassica campestris*

586 **bird's-eye primrose**
f primevère farineuse
d Mehlprimel; Mehlschlüsselblume
l *Primula farinosa*

587 **bird's-foot**
f serradelle
d Vogelfuss(klee); Serradella
l *Ornithopus*

588 **bird's-foot deervetch; crowfoot**
f lotier corniculé
d gelber Honigklee; gelber Hornklee; gemeiner Hornklee; gewöhnlicher Schotenklee; Wiesenhornklee
l *Lotus corniculatus*

* **bird's-foot trefoil → 1824**

589 **bird's-nest fungi**
f nidulariées
d Nestpilze
l *Nidulariaceae*

590 **bird's-nest orchis**
f néottie
d Nestwurz; Vogelnest
l *Neottia*

591 **bird vetch; cow vetch; crow vetch**
f vesce craque
d Vogelwicke
l *Vicia cracca*

* **birthwort → 592, 1932**

592 **birthwort dutchman's-pipe; birthwort**
f aristoloche clématite
d aufrechte Osterluzei; gemeine Osterluzei
l *Aristolochia clematitis*

593 **birthwort family**
f aristolochiacées
d Artistolochiazeen; Osterluzeigewächse
l *Aristolochiaceae*

594 **Biscay heath**
f bruyère de la Méditerranée
d Mittelmeerheide
l *Erica mediterranea*

595 **bishop pine**
f pin épineux
d Stachelkiefer
l *Pinus muricata*

* **bishop's-cap → 3634**

596 **bishop's goutweed**
f herbe aux goutteux; podagraire; égopode commune
d Gänsestrenzel; Geissfuss; Gersch; Giersch; Zaungiersch
l *Aegopodium podagraria*

* **bishop's-weed → 5547**

597 **bitter almond**
f amandier amer
d Bittermandelbaum
l *Amygdalus communis* var. *amarus*

* **bitter apple → 1334**

598 **bitter boletus**
f bolet amer
d Gallenpilz; Gallenröhrling
l *Boletus felleus; Tylopilus felleus*

599 **bitterbush**
f picramnie
d Hondurasrinde
l *Picramnia*

* **bitter cassava → 1379**

600 **bitter cress**
f cardamine
d Gauchblume; Schaumkraut; Wiesenkresse; Zahnwurz
l *Cardamine*

601 **bitter dock**
f oseille à feuilles obtuses
d stumpfblättriger Ampfer
l *Rumex obtusifolius*

602 **bitter fleabane**
f érigéron âcre; vergerette âcre
d blaue Dürwurz; echtes Berufkraut;

scharfes Berufkraut
l *Erigeron acris*

603 bitter lettuce
f laitue vireuse
d Giftlattich; Giftsalat
l *Lactuca virosa*

604 bitter nightshade; bittersweet
f morelle grimpante; vigne de Judée; douce-amère
d Bittersüss; Almenkraut; Almenraute; Alpranke; Nachtschatten; Stinkteufel
l *Solanum dulcamara*

605 bitternut; bitternut hickory
f caryer amer
d Bitternuss
l *Carya cordiformis; Cary amara*

*** bitter orange → 3538**

606 bitter panus
d Eichenzwergknäueling
l *Panus stypticus; Panellus stypticus*

607 bitter pea vine
f gesse du printemps
d Frühlingsplatterbse
l *Orobus vernus; Lathyrus vernus*

608 bittersweet; staff tree
f célastre
d Baummörder; Baumwürger
l *Celastrus*

*** bittersweet → 604**

609 bitter vetch
f alliez; ers; ervilier
d Erve; Linsenwicke; knotenfrüchtige Wicke; Wicklinse
l *Vicia ervilia*

610 bitter winter cress
f barbarée; herbe de Sainte-Barbe
d Barbarakraut; echtes Barbarakraut; gemeines Barbarakraut; wildes Barbarakraut
l *Barbarea vulgaris*

611 bitterwood
f trichilie
d Mafureirabaum
l *Trichilia*

612 bixa family
f bixacées
d Bixazeen; Orleangewächse
l *Bixaceae*

*** black alder → 2072**

613 black ash
f frêne noir
d Schwarzesche
l *Fraxinus nigra*

614 black ballota; black horehound
f ballote noire; ballote fétide
d schwarzer Andorn; schwarze Ballote; schwarzer Gottvergess; Schwarznessel; Stinkandorn; stinkende Taubnessel
l *Ballota nigra; Ballota foetida*

615 black bamboo
f bambou noir
l *Bambusa nigra*

616 black baneberry; herb Christopher
f actée en épi
d schwarzfrüchtiges Christophskraut; ähriges Christophskraut; Schwarzkraut; Ährenchristophskraut
l *Actaea spicata*

617 black bent; redtop
d Fioringras; grosses Straussgras
l *Agrostis gigantea*

618 black-berried cotoneaster
f cotonéastre à fruits noirs
d schwarze Zwergmispel
l *Cotoneaster melanocarpa*

619 blackberry
f ronce
d Brombeere; Brombeerstrauch; Steinbeere
l *Rubus*

620 blackberry lily; leopard flower
f iris tigré; belamcanda de Chine
d Leopardblume; Pantherblume
l *Belamcanda chinensis*

*** black birch → 4504**

621 black bog rush
d schwarze Kopfsimse; schwarzes Kopfried
l *Schoenus nigricans*

622 black bryony
f tamier
d Schmeerwurz; Schmerwurz
l *Tamus*

623 black canary tree; black Chinese olive
f canari noir
d chinesische schwarze Olive; schwarzer Kanarienbaum
l *Canarium nigrum; Canarium pimela*

624 blackcap raspberry
f framboisier d'Amérique; ronce d'Occident
d schwarze Himbeere
l *Rubus occidentalis*

625 black centaurea
f centaurée noire
d schwarze Flockenblume
l *Centaurea nigra*

626 black cherry
f cerisier noir; cerisier tardif
d deutscher Lorbeer; spätblühende Traubenkirsche
l *Prunus serotina; Padus serotina*

* **black Chinese olive → 623**

627 black crowberry
f empètre noir; camarine noir
d schwarze Krähenbeere
l *Empetrum nigrum*

* **black cumin → 2415**

* **black currant → 2081**

628 blackening russula
f russule noircissante
d dickblättriger Schwarztäubling
l *Russula nigricans*

629 black-eyed Susan
f rudbeckie hérissée
d rauhhaariger Sonnenhut
l *Rudbeckia hirta*

630 black false hellebore
f varaire noir; vératre noir
d schwarze Nieswurz
l *Veratrum nigrum*

631 black-fruited honeysuckle
f chèvrefeuille à baie noire
d schwarze Heckenkirsche; schwarzes Geissblatt
l *Lonicera nigra*

* **black gram → 3755**

632 black haw; black-haw viburnum
f viorne à feuilles de prunier
d pflaumenblättriger Schneeball
l *Viburnum prunifolium; Viburnum pyrifolium*

* **black hellebore → 1226**

633 black henbane; sticking Roger
f jusquiame noire
d Ackerbilsenkraut; gemeines Bilsenkraut; Hühnertod; Hundskraut; Rindswurz; Schlafkraut; schwarzes Bilsenkraut; Teufelswurz; Zigeunerkorn
l *Hyoscyamus niger*

* **black horehound → 614**

634 blackjack; blackjack oak
f jaquier noir
l *Quercus marilandica*

635 black jetbead
f rhodotype faux kerria; faux-corête
d Kaimanstrauch; Scheinkerrie
l *Rhodotypos scandens; Rhodotypos kerrioides*

636 black locust; false acacia
f acacia commun; robinier faux-acacia; faux-acacia
d gemeine Robinie; Scheinakazie; unechte Akazie; unechte Robinie
l *Robinia pseudoacacia*

637 black maple
f érable noir
d Schwarzahorn
l *Acer nigrum*

638 black medic; hop clover; yellow trefoil
f lupuline; luzerne houblon; minette; minette dorée
d Gelbklee; Hirsenklee; Hopfenklee; Hopfenluzerne; Lämmerklee; Steinklee; Wolfsklee
l *Medicago lupulina*

639 black mixed-flower
d schwarze Teufelskralle
l *Phyteuma nigrum*

640 black mulberry
f mûrier noir
d schwarzer Maulbeerbaum
l *Morus nigra*

641 black mullein
f molène noire
d dunkle Königskerze; schwarze

Königskerze
l *Verbascum nigrum*

642 black mustard
f moutarde noire
d schwarzer Senf; Schwarzkohl
l *Brassica nigra*

643 black nightshade
f morelle noir; amourette
d Hühnertod; Saukraut; schwarzer Nachtschatten
l *Solanum nigrum*

644 black oak; dyer's-oak
f chêne quercitron
d amerikanische Färbereiche
l *Quercus velutina; Quercus tinctoria*

645 black pea vine
f gesse noire
d schwarze Platterbse
l *Lathyrus niger*

646 black pepper
f poivrier noir
d schwarzer Pfeffer
l *Piper nigrum*

* **black peppermint → 5969**

647 black poplar
f liardier; peuplier noir; peuplier commun
d Schwarzpappel; Feldpappel; Pappelweide; Saarbaum; Saarbuche
l *Populus nigra*

648 black rosewood; Brazilian rosewood
f bois de rose; palissandre du Brésil
d Palisanderbaum
l *Dalbergia nigra*

649 black salsify
f salsifis noir
d Gartenschwarzwurzel; Schwarzwurzel; spanische Schwarzwurzel; Winterspargel; Nattergras; Schlangengras
l *Scorzonera hispanica*

* **black sanicle → 2611**

650 black sedge
f laîche noirâtre
d schwarze Segge; Trauersegge
l *Carex atrata*

651 black spleenwort
f doradille noire
d schwarzer Streifenfarn
l *Asplenium adiantum-nigrum*

652 black-spot horn poppy
f glaucie cornue
d roter Hornmohn
l *Glaucium corniculatum*

653 black spruce
f épinette à la bière; épinette noire; sapin double; sapinette noire; épicéa noir
d Schwarzkiefer
l *Picea mariana; Picea nigra*

* **blackthorn → 5014**

654 black willow
f saule d'Amérique
d schwarze Weide; Schwarzweide
l *Salix nigra*

655 blackwood acacia
f acacia à bois noir
d Schwarzholzakazie
l *Acacia melanoxylon*

656 bladder campion
f cucubale
d Hühnerbiss; Taubenkropf; Beerennelke
l *Cucubalus*

657 bladder fern
f cystoptéris
d Blasenfarn
l *Cystopteris*

658 bladder flower
f araujie
l *Araujia*

659 bladdernut
f staphylier
d Klappernuss; Pimpernuss
l *Staphylea*

660 bladdernut family
f staphyléacées
d Klappernussgewächse; Pimpernussgewächse; Staphyleazeen
l *Staphyleaceae*

661 bladderpod
f vésicaire
d Blasenschötchen
l *Lesquerella; Vesicaria*

662 bladder senna
f baguenaudier

d Blasenschote; Blasensenne; Blasenstrauch
l *Colutea*

663 bladder silene
f silène enflé
d aufgeblasenes Leimkraut; Taubenkropf
l *Silene cucubalus; Silene inflata; Silene venosa*

664 bladderwort
f utriculaire
d Wasserschlauch
l *Utricularia*

665 bladderwort family
f lentibulariacées
d Fettkräuter; Lentibulariazeen; Wasserhalmgewächse; Wasserschlauchgewächse
l *Lentibulariaceae; Utriculariaceae*

666 bladder wrack
f varech vésiculeux
d Blasentang
l *Fucus vesiculosus*

667 blade kelp; sea tangle
f laminaire
d Blattang; Riementang
l *Laminaria*

668 bleeding heart
f dicentre; diélytre
d flammendes Herz; hängendes Herz; Herzblume; Jungfernherz
l *Dicentra; Dielytra*

669 blessed milk thistle; St. Mary thistle
f chardon argenté; chardon-Marie
d Mariendistel; Silberdistel; Frauendistel; Milchdistel
l *Silybum marianum; Carduus marianus*

670 blessed thistle
f chardon béni
d Benediktenkarde; Benediktenkraut; Heildistel
l *Cnicus*

671 blessed thistle
f chardon béni commun
d echtes Benediktenkraut; Bitterdistel; gelbes Benediktenkraut; Bernardinerkraut; Kardobenediktendistel; Spinnendistel
l *Cnicus benedictus*

672 bletilla
f blétille
d Bletilla
l *Bletilla*

673 blister buttercup
f renoncule scélérate
d Froschkraut; Froschpfeffer; Gifthahnenfuss; Giftranunkel
l *Ranunculus sceleratus*

674 blister sedge
f laîche vésiculeuse
d Blasenriedgras; Blasensegge
l *Carex vesicaria*

675 blite goosefoot; strawberry blite
f arroche-fraise; blette
d ähriger Erdbeerspinat; ähriger Gänsefuss; Beermelde; kopfiger Gänsefuss
l *Chenopodium capitatum*

* **blood currant → 6000**

676 bloodflower milkweed
f asclépiade de Curaçao
l *Asclepias curassavica*

677 bloodleaf
f irésine
d Iresine
l *Iresine*

678 blood lily
f hémanthe
d Blutblume
l *Haemanthus*

679 blood-red boletus; Satan's mushroom
f bolet satan; cèpe satan; cèpe du diable
d Blutpilz; Satanpilz
l *Boletus satanus*

680 blood-red geranium
f géranium sanguin
d Blutkraut; blutroter Storchschnabel; rote Hühnerwurz; Blutstorchschnabel
l *Geranium sanguineum*

681 bloodroot
f sanguinaire
d Blutkraut; Blutwurz
l *Sanguinaria*

682 bloodroot; puccoon
f sanguinaire du Canada

d kanadische Blutwurz
l *Sanguinaria canadensis*

683 **blood-twig dogwood; red dogwood**
f bois de pouine; bois sanguin; cornouiller femelle; cornouiller sanguin; puègne blanche; olivier de Normandie
d Hartriegel; gemeiner Hartriegel; roter Hartriegel
l *Cornus sanguinea*

684 **bloody dock**
d Blutampfer; Hainampfer
l *Rumex sanguineus*

685 **bloomy-stem rose**
d weiche Rose
l *Rosa mollis*

686 **blue and yellow russula**
f russule charbonnière; russule cyanoxanthe; russule bleue et jaune
d Frauentäubling
l *Russula cyanoxantha*

687 **blue ash**
f frêne bleu
d Blauesche
l *Fraxinus quadrangulata*

688 **blue Atlas cedar**
f cèdre bleu
d blaue Atlaszeder
l *Cedrus atlantica glauca*

689 **bluebeard**
f caryopteris
d Bartblume
l *Caryopteris*

690 **bluebell**
f campanule à feuilles rondes
d rundblättrige Glockenblume
l *Campanula rotundifolia*

691 **bluebells; smooth lungwort**
f mertensia
d Lungenwurz
l *Mertensia*

* **blue-berried honeysuckle → 5356**

692 **blueberry**
f airelle
d Heidelbeere; Moosbeere
l *Vaccinium*

693 **blueberry willow**
f saule noircissant
d schwarzwerdende Weide
l *Salix nigricans; Salix myrsinifolia*

* **bluebottle → 1613**

694 **blue broomrape**
d purpurrote Sommerwurz; violette Sommerwurz
l *Orobanche purpurea*

695 **blue Colorado spruce**
d Blaufichte; Blautanne
l *Picea pungens* var. *glauca*

696 **blue cowwheat**
f mélampyre violet
d Hainwachtelweizen
l *Melampyrum nemorosum*

697 **blue-crown passionflower**
f fleur de la passion; grenadille bleue; passiflore bleue
d gemeine Passionsblume
l *Passiflora caerulea*

698 **blue-eyed grass; satin flower**
f bermudienne
d Binsenlilie; Grasschwertel; Rüssel-schwertel; Schweinrüssel; Schwertlilie
l *Sisyrinchium*

699 **blue Gowen cypress**
f cyprès de Monterrey
l *Cupressus goveniana* var. *glauca*

700 **bluegrass; meadow gras; poa**
f pâturin
d Angergras; Rispe; Rispengras; Vieh-gras
l *Poa*

701 **blue-green algae**
f cyanophycées
d blaugrüne Algen; Blaualgen; Blaugrünalgen; Spaltalgen; Zyanophyzeen
l *Cyanophyceae*

702 **blue koeleria**
f koélérie glauque
d blaugraue Kammschmiele; grau-grünes Schillergras
l *Koeleria glauca*

703 **blue lupine**
f lupin bleu
d blaue Lupine; schmalblättrige Lupine
l *Lupinus angustifolius*

704 blue moor grass
f seslérie bleue
d blaues Kopfgras; Blaugras
l *Sesleria coerulea*

* **blue mo(u)ld → 2632**

705 blue pimpernel
f mouron bleu
d blauer Gauchheil
l *Anagallis coerulea*

* **blue spruce → 1338**

706 bluestem; broom sedge
f barbon
d Bartgras; Bartgerste; Hühnerfuss; Mannsbart
l *Andropogon*

707 bluestem wheat grass; Colorado bluestem
f chiendent de Smith
l *Agropyron smithi*

708 blue Tibetan poppy
f méconopside
d Keulenmohn; Mohnling; Scheinmohn
l *Meconopsis*

709 blue-top eryngo
f panicaut des Alpes
d Alpendistel
l *Eryngium alpinum*

710 bluets
f houstonie
d Engelsauge; Houstonie
l *Houstonia*

* **blueweed → 1555**

711 blue-white trigonella
f trèfle musque; trèfle bleu; trèfle odorant; trigonelle; baumier; sainegrain
d blauer Bockshornklee; Bisamklee; Schabziegerklee; Siebenundsiebzigerlei; Käseklee
l *Trigonella coerulea*

712 blunt-leaved pondweed
f potamot à feuilles obtuses
d stumpfblättriges Laichkraut
l *Potamogeton obtusifolius*

713 blushing amanita; reddish amanita
f amanite rougeâtre
d Perlenschwamm; Perlpilz; rötender Wulstling
l *Amanita rubescens*

714 bog angelica
f angélique palustre
d Sumpfbrustwurz; Sumpfengelwurz; Mutterwurz
l *Ostericum palustre; Angelica palustris*

715 bog asphodel
f narthèce
d Beinbrech; Knochenbrech; Ährenlilie
l *Narthecium*

716 bog bean; buck bean
f ményanthe
d Bitterklee; Dreiblatt; Fieberklee; Sumpfklee; Zottelblume; Zottenblume
l *Menyanthes*

717 bog bilberry
f airelle des marais; airelle noire
d Moorbeere; Moorheidelbeere; Rauschbeere; Trunkelbeere
l *Vaccinium uliginosum*

718 bog crane's-bill
f géranium des marais
d Sumpfstorchschnabel
l *Geranium palustre*

719 boggy spike sedge
d gemeines Sumpfried
l *Eleocharis eupalustris*

720 bog kalmia
f kalmie glauque
l *Kalmia polifolia; Kalmia glauca*

721 bog marsh cress
d gemeine Sumpfkresse; isländische Sumpfkresse
l *Rorippa palustris*

* **bog moss → 5125**

722 bog pimpernel
f mouron mignon
d zarter Gauchheil
l *Anagallis tenella*

* **bog rosemary → 723**

723 bog-rosemary andromeda; moorwort; bog rosemary
f andromède à feuilles de polium; andromède d'Europe
d Gränke; Lavendelheide; Poleigränke;

falscher Porst; Rosmarinheide
l *Andromeda polifolia*

724 bog rush
f choin
d Kopfried; Kopfsimse
l *Schoenus*

725 Bohemian serpent root
d niedrige Schwarzwurzel
l *Scorzonera humilis*

* **Bokhara clover → 5917**

726 Bokhara vine
f renouée de Boukhara; renouée de Turkestan
d Baldschuanknöterich
l *Polygonum baldschuanicum*

727 boletus
f bolet
d Dickröhrling; Röhrenschwamm; Röhrling; Steinpilz
l *Boletus*

728 boltonia
f boltonie
d Boltonie
l *Boltonia*

729 bomarea
f bomarée
d Bomarea
l *Bomarea*

730 bombax
f bombax
d Baumwollbaum
l *Bombax*

731 Bombay sumbul; ammoniac plant
f dorême ammoniac
d Ammoniakpflanze
l *Dorema ammoniacum*

732 boneset; thoroughwort
f eupatoire
d Alpkraut; Wasserdost
l *Eupatorium*

733 bony-berry
f ostéomèle
d Steinapfel
l *Osteomeles*

734 borage
f bourrache
d Borretsch
l *Borago*

735 borage family
f borraginacées
d Asperiofoliazeen; Borraginazeen; Borretschgewächse; Rauhblattgewächse; Rauhblättrige
l *Bor(r)aginaceae*

* **borecole → 3128**

736 Borneo camphor
f dryobalanops
d Flügeleiche; Kampferölbaum; ostindischer Kampferbaum
l *Dryobalanops*

737 Bornmueller fir
d Bornmüllers Tanne
l *Abies bornmuelleriana*

* **Boston ivy → 3028**

738 bo tree; sacred fig tree; pipal tree
f figuier des pagodes; figuier des banians; banian
d Asvatha; Baniane; Banyan; Götzenbaum; heiliger Feigenbaum; Pappelfeigenbaum; Pipal
l *Ficus religiosa*

739 bottle-brush buckeye
f pavier blanc
l *Aesculus parviflora*

740 bottle-brush grass
f aspérelle
d wilder Reis
l *Asperella*

* **bottle gourd → 922**

741 bottle tree
f sterculie(r)
d Stinkbaum
l *Sterculia*

742 bougainvilea
f bougainvillier
d Bougainvillea
l *Bougainvillea*

743 bouncing Bet
f saponaire officinale; saponaire savonière
d gemeines Seifenkraut; echtes Seifenkraut; Hundsnelke; rote Seifenwurzel
l *Saponaria officinalis*

744 Bourbon dombeya
f mahot rouge à petites feuilles
l *Dombeya punctata*

745 Bourbon rose
f rosier de l'Ile Bourbon
d Bourbonrose
l *Rosa borboniana*

746 Bourbon tea orchid
f angréacum
d Tropensporn
l *Angraecum*

747 bouvardia
f bouvardie
d Bouvardia
l *Bouvardia*

748 bower actinidia; tara vine
f actinidie denticulée; actinidie à feuilles dentées
d spitzblättriger Strahlengriffel
l *Actinidia arguta*

* **bowstring hemp → 4694**

749 box
f buis
d Buchs; Buchsbaum; Bux
l *Buxus*

750 box elder; ash-leaved maple; negundo
f érable à feuilles de frêne; érable négundo
d Eschenahorn; eschenblättriger Ahorn
l *Acer negundo*

751 box family
f buxacées
d Buchsbaumgewächse; Buxazeen
l *Buxaceae*

752 box-leaf azara
f azara à petites feuilles
l *Azara microphylla*

* **boxthorn → 6016**

753 boykinia
f boykinie
d Boykinie
l *Boykinia*

* **bracket fungi → 4253**

754 brake
f ptéride
d Saumfarn
l *Pteris*

755 bramble vetch
d feinblättrige Wicke; Feinblattwicke
l *Vicia tenuifolia*

756 branched wheat; branchy-eared wheat
f blé rameux
d Zweigweizen
l *Triticum vavilovianum; Triticum compositum*

757 branching asphodel
f asphodèle blanc
d weisser Affodill
l *Asphodelus albus*

* **branching larkspur → 2330**

* **branchy-eared wheat → 756**

758 brasiletto
f césalpinie
d Brasilienholz; Caesalpinie
l *Caesalpinia*

759 brass buttons
f cotule
d Fiederpolster; Laugenblume
l *Cotula*

760 Braun's holly fern
f dryoptéris de Braun
d Brauns Schildfarn
l *Polystichum brauni*

* **Brazil cherry → 2653**

761 Brazil eugenia
f eugénier du Brésil
d brasilianische Kirschmyrte
l *Eugenia brasiliensis*

762 Brazilian myrocarpus
f myrocarpe à feuilles denses
l *Myrocarpus frondosus*

763 Brazilian parrot feather
f myriophylle brésilien
d brasilianisches Tausendblatt
l *Myriophyllum brasiliense*

* **Brazilian rosewood → 648**

* **Brazilian wax palm → 1021**

764 Brazil jacaranda
f jacaranda du Brésil

d Jacarandaholz
l *Jacaranda brasiliana*

765 **Brazil nut; niggertoe**
f noyer du Brésil; noyer de Para; châtaignier du Brésil
d Amazonenmandel; Brasilnussbaum; Jurianuss; Paranussbaum
l *Bertholletia excelsa*

766 **Brazil pepper tree**
f faux-poivrier à feuilles de térébinthe
d brasilianischer Pfefferbaum
l *Schinus terebinthifolius*

767 **Brazil trithrinax palm**
d brasilianische Doppeldreizackpalme
l *Trithrinax brasiliensis*

768 **brazilwood**
d Fernambukholzbaum; Rotholzbaum
l *Caesalpinia brasiliensis*

769 **breadfruit**
f arbre à pain; artocarpe incisé
d (gemeiner) Brotbaum; (gemeiner) Brotfruchtbaum
l *Artocarpus altilis; Artocarpus incisus*

770 **breadnut tree**
f brosime
d Brotnussbaum
l *Brosimum*

771 **breeders gladiolus**
f glaïeul de Gand
d Genter Gladiole; Genter Siegwurz
l *Gladiolus gandavensis*

772 **Brewer's spruce**
f épicéa de Brewer
d Brewers Fichte
l *Picea breweriana*

773 **Briançon apricot**
f abricotier de Briançon; prunier des Alpes; marmottier
d Briançon-Aprikose
l *Prunus brigantina*

774 **brickberry cotoneaster**
f cotonéastre tomenteux; cotonéastre cotonneux
d filzige Zwergmispel
l *Cotoneaster tomentosa*

775 **bridal-wreath spiraea**
f spirée à feuilles de prunier
d pflaumenblättriger Spierstrauch
l *Spiraea prunifolia*

776 **bridelia**
f bridélie
l *Bridelia*

777 **brier grape**
f vigne épineuse
l *Vitis davidi*

778 **Brisbane box; brush box**
f tristanie à feuilles en bouquets
l *Tristania conferta*

779 **bristlecone fir**
f sapin à grandes bractées
l *Abies venusta*

780 **bristle grass; millet; pigeon grass**
f sétaire; accroche-abeille
d Borstenhirse; Fench; Fennich; Filz
l *Setaria*

781 **bristle-stem hemp nettle**
f galéolope tétrahit
d gemeiner Daun; gemeiner Hohlzahn; gewöhnlicher Hohlzahn; stechender Hohlzahn; Hanfnessel
l *Galeopsis tetrahit*

782 **bristle thistle**
f chardon
d Distel
l *Carduus*

783 **bristly oxtongue**
f picride échioïde
d natter(n)kopfähnliches Bitterkraut; Natter(n)kopfbitterkraut
l *Picris echioides*

784 **British inula**
f inule de Bretagne; inule britannique; aunée britannique
d Wiesenalant
l *Inula britannica*

785 **British timothy**
d rauhes Lieschgras
l *Phleum paniculatum*

786 **brittle bladder fern**
f cystoptéris fragile
d zerbrechlicher Blasenfarn
l *Cystopteris fragilis*

787 **brittle willow**
f saule fragile

d Bruchweide; Buschweide; Fieberweide; Knackweide
l *Salix fragilis*

* **brittleworts → 1853**

788 broad bean
f fève de marais
d Puffbohne; Ackerbohne; Feldbohne; Marschbohne; Saubohne; Viehbohne
l *Vicia faba*

789 broad caucalis
d breite Haftdolde
l *Turgenia latifolia*

790 broad-gilled collybia
f grande souchette
d breitblättriger Rübling; breitblättriger Holzritterling
l *Collybia platyphylla*

791 broadleaf saxifrage
f saxifrage à feuilles rondes
d rundblättriger Steinbrech
l *Saxifrage rotundifolia*

792 broad-leaved epipactis
d breitblättrige Sitter; breitblättrige Sumpfwurz; grüne Sitter
l *Epipactis latifolia*

793 broad-leaved euonymus
f fusain à larges feuilles
d Alpenspindelbaum; breitblättriger Spindelbaum
l *Euonymus latifolius*

794 broad-leaved grape fern
d gefiederter Rautenfarn; Kamillenrautenfarn; vielteiliger Rautenfarn
l *Botrychium multifidum*

795 broad-leaved lavender
f lavande spic
d breitblättriger Lavendel
l *Lavandula latifolia*

796 broad-leaved meadow-grass
d Bergrispengras; Waldrispengras
l *Poa chaixi*

797 broad-leaved meadowsweet
f spirée à feuilles larges
d breitblättriger Spierstrauch
l *Spiraea latifolia*

798 broad-leaved orchis
d breitblättriges Knabenkraut
l *Orchis latifolia*

799 broad-leaved Solomon's-seal
f sceau de Salomon à grandes feuilles
d breitblättrige Weisswurz
l *Polygonatum latifolium*

800 broccoli
f chou broccoli; broccoli asperge
d Broccoli; Brokkoli; Spargelkohl
l *Brassica oleracea* var. *cymosa*

801 brodiaea
f brodiée
d Brodiaea
l *Brodiaea*

802 brome; brome grass; cheat; chess
f brome
d Trespe
l *Bromus*

803 brome fescue
d Eichhörnchenfuchsschwingel; Eichhornschwanzfederschwingel
l *Vulpia bromoides; Festuca bromoides*

* **brome grass → 802**

804 bromelia
f bromélie
d Bromelie
l *Bromelia*

805 brook euonymus
f fusain d'Amérique
d amerikanischer Spindelstrauch
l *Euonymus americanus*

806 brook grass
f glycérie aquatique
d ansehnliches Mannagras; echte Mielitz; Riesensüssgras; Wasserschwaden; Wassersüssgras; zartes Quellgras
l *Glyceria aquatica; Catabrosa aquatica*

807 brookweed
f samole
d Salzbunge
l *Samolus*

808 brookweed
f samole de Valerand; pimprenelle aquatique
l *Samolus valerandi*

809 broom
f cytise
d Bohnenbaum; Bohnenstrauch; Buschgoldregen; Geissklee; Linsenbaum; Besenginster
l *Cytisus*

810 broomcorn
f sorgho à balais
d Besenhirse; Besensorghum
l *Sorghum technicum; Sorghum vulgare technicum*

* **broom corn → 5087**

* **broomcorn millet → 4315**

811 broom-leaved toadflax
f linaire à feuilles de genêt
d ginsterblättriges Leinkraut
l *Linaria genistifolia*

812 broomrape
f orobanche
d Sommerwurz; Würger
l *Orobanche*

813 broomrape family
f orobanchacées
d Orobanchazeen; Sommerwurzgewächse
l *Orobanchaceae*

* **broom sedge → 706**

814 browallia
f browallie
d Browallie
l *Browallia*

815 brown algae
f algues brunes; phéophycées
d Braunalgen; Tange
l *Phaeophyceae*

816 brown beak sedge
d braunes Schnabelried; braune Schnabelsimse
l *Rhynchospora fusca*

* **brown durra → 1929**

817 brown-scale centaurea
f centaurée jacée; jacée
d gemeine Flockenblume; Wiesenflockenblume
l *Centaurea jacea*

818 brown-top millet
f panic rameux
l *Panicum ramosum*

819 brown-twig poplar
f peuplier sombre
d dunkelblättrige Pappel
l *Populus tristis*

820 brown-yellow boletus; butter mushroom; yellow-brown boletus
f bolet jaune; cèpe jaune; cèpe annulaire; nonette voilée
d Butterpilz; Butterröhrling
l *Boletus luteus; Suillus luteus*

821 brunnera
d Kaukasusvergissmeinnicht
l *Brunnera*

* **brush box → 778**

822 brush-tipped sotol
f dasylire chevelu
d Schopflilie
l *Dasylirion acrotrichum*

823 Brussels sprouts
f chou de Bruxelles
d Brüsseler Kohl; Rosenkohl; Rosenwirsing; Sprossenkohl
l *Brassica oleracea* var. *gemmifera*

824 bryony
f bryone
d Gichtrübe; Zaunrübe; Bryonie; Stichwurz
l *Bryonia*

825 bryophytes
f bryophytes
d Bryophyten; Moose; Moospflanzen
l *Bryophyta*

826 bubinga didelotia
f kévazingo
l *Didelotia africana*

827 buchu
f barosme
d Buccostrauch; Bukkostrauch
l *Barosma*

* **buck bean → 716**

828 buckeye; horse chestnut
f marronnier; pavier
d Rosskastanie; Rosskastanienbaum;

wilder Kastanienbaum
l *Aesculus; Pavia*

829 buckhorn plantain
f plantain lancéolé
d Spitzwegerich
l *Plantago lanceolata*

* **buckhorn plantain → 1708**

830 buckthorn
f nerprun
d Färberdorn; Faulbaum; Kreuzdorn; Wegdorn
l *Rhamnus*

831 buckthorn family
f rhamnacées
d Kreuzdorngewächse; Rhamnazeen
l *Rhamnaceae*

832 buckwheat
f blé noir; sarrasin
d Blende; Buchweizen; Franzweizen; Getreidekraut; Gricken; Heidegrütze; Heidekorn; Heidenkorn
l *Fagopyrum*

833 buckwheat family
f polygonacées
d Knöterichgewächse; Polygonazeen
l *Polygonaceae*

834 buffalo berry
f shepherdia
d Büffelbeere
l *Shepherdia*

835 buffalo-bur nightshade
d Büffelklette; Schnabelnachtschatten
l *Solanum rostratum*

* **buffalo currant → 1276**

836 buffalo grass
f herbe aux bisons
d Buffalogras; Büffelgras
l *Buchloe dactyloides*

837 bugbane
f cimicaire; cimifuge
d Silberkerze; Wanzenkraut
l *Cimicifuga*

838 bugle
f bugle
d Günsel
l *Ajuga*

839 bugleweed
f lance du Christ; lycope
d Wolfstrapp
l *Lycopus*

840 bugloss
f anchuse; buglosse
d Ochsenzunge
l *Anchusa*

841 bulb buttercup
f renoncule bulbeuse
d knolliger Hahnenfuss; Knollenhahnenfuss
l *Ranunculus bulbosus*

842 bulbil lily
f lis bulbifère; lis jaune
d Feuerlilie
l *Lilium bulbiferum*

843 bulbophyllum
d Bulbophyllum
l *Bulbophyllum*

844 bulbous bluegrass
f pâturin bulbeux
d knolliges Rispengras
l *Poa bulbosa*

845 bulbous rush
d niedrige Binse; Rasenbinse; Rosenbinse
l *Juncus bulbosus*

846 bulbstem yucca
f yucca-pied d'éléphant
d guatemalische Palmlilie; riesige Palmlilie
l *Yucca elephantipes*

847 bullace plum
f prunier damas; prunier sauvage
d Haferpflaume; Haferschlehe; Krieche; Kriechenpflaume; Spelling; Spilling
l *Prunus insititia*

848 bulletwood
f mimusops
d Spitzenblume
l *Mimusops*

849 bullock's-heart; bullock's-heart custard apple
f cœur de bœuf; anone réticulée
d Netzan(n)one; netzförmiger Flaschenbaum; Ochsenherzapfel; Rahmapfel
l *Annona reticulata*

850 bull thistle
f cirse lancéolé
d gewöhnliche Kratzdistel; lanzenblättrige Kratzdistel; Lanzettkratzdistel; Wegdistel
l *Cirsium lanceolatum*

851 bulnesia
f bulnésie
l *Bulnesia*

852 bulrush
f scirpe; jonquine
d Simse; Schilf
l *Scirpus*

853 bumbum mango
d duftender Mangobaum
l *Mangifera odorata*

* **bunch-flowered narcissus → 4248**

854 bunt fungi; smut fungi
f ustilaginacées
d Brandpilze; Ustilaginazeen
l *Ustilaginaceae*

855 bunya-bunya; bunya-bunya araucaria
f araucarie d'Australie; bunya bunya; araucarie de Bidwill
d Bunya-Bunya-Baum
l *Araucaria bidwillii*

856 bur beaked chervil
f persil sauvage; anthrisque commun; cerfeuil sauvage
d Heckenkerbel; Hundskerbel; gemeiner Kerbel
l *Anthriscus vulgaris*

857 bur beggar-ticks
f bident triparti
d dreiteiliger Zweizahn
l *Bidens tripartita*

858 bur cucumber
f sicyote
d Haargurke
l *Sicyos*

* **bur cucumber → 5869**

859 burdock
f bardane; glouteron
d Klette; Klettenkraut
l *Arctium*

860 bur grass
f tragus
d Klettengras; Klettergras; Stachelgras
l *Tragus*

* **bur grass → 4682**

861 Burgundy cabbage rose
f rosier petit Saint François
d Burgunderrose
l *Rosa centifolia* var. *parvifolia*

862 burhead
f échinodore
d Igelschlauch
l *Echinodorus*

863 Burma coast padauk
f santal rouge des Indes
d echter roter Sandelholzbaum
l *Pterocarpus indicus*

864 Burma padauk
f padouk de Birmanie
l *Pterocarpus macrocarpus*

865 Burma pyingado
f arbre de jamba; bois de fer de Birmanie
d Jambea
l *Xylia xylocarpa; Xylia dolabriformis*

* **bur marigold → 521**

866 Burma toon
f cèdre de Singapour
l *Cedrela toona; Toona ciliata*

867 burnet
f petite pimprenelle
d Wiesenkopf
l *Sanguisorba*

868 burnet
f pimprenelle
d Becherblume
l *Poterium*

* **burning bush → 1866**

869 burnweed
f érechtite
d Afterkreuzkraut
l *Erechtites*

870 bur oak
f chêne à gros fruits
d Kletteneiche
l *Quercus macrocarpa*

* **bur parsley → 2753**

871 bur reed
f ruban d'eau; sparganier
d Igelkolben; Igelkopf; Igelskolben
l *Sparganium*

872 bur-reed family
f sparganiacées
d Igelkolbengewächse; Igelkopfgewächse; Sparganiazeen
l *Sparganiaceae*

873 bur sage
f gaertnère
l *Gaertneria; Franseria*

874 bursaria
f bursaire
d Beutelbaum
l *Bursaria*

875 bursera
f arbre d'encens; gommart
d Weissgummibaum
l *Bursera*

876 bursera family
f burséracées
d Burserazeen
l *Burseraceae*

877 burstwort
f herniaire
d Bruchkraut; Tausendkorn
l *Herniaria*

878 bush calceolaria
f calcéolaire à feuilles entières
d ganzblättrige Pantoffelblume
l *Calceolaria integrifolia*

879 bush cinquefoil
f potentille arbisseau
d gelber Fingerstrauch; strauchiger Fingerstrauch; strauchiges Fingerkraut
l *Potentilla fruticosa; Dasiphora fruticosa*

*** bush clover → 3279**

880 bush honeysuckle
f dierville
d Dierville; Kapselgeissblatt
l *Diervilla*

881 bush kidney bean
d Buschbohne; Kruppbohne
l *Phaseolus vulgaris humilis*

882 bush nasturtium
f petite capucine
d kleine Kapuzinerkresse
l *Tropaeolum minus*

883 bush pumpkin
f pâtisson
d Melonenkürbis
l *Cucurbita pepo* var. *melopepo*

884 bush red pepper; Spanish pepper
f piment annuel; poivre d'Espagne; poivre d'Inde
d Cayenne-Pfeffer; roter Pfeffer; spanischer Pfeffer; türkischer Pfeffer; Paprika; Schotenpfeffer
l *Capsicum annuum; Capsicum frutescens*

*** bush vetch → 2760**

885 butcher's-broom
f fragon épineux; petit houx; myrte épineux; buis piquant; épine de rat
d stachliger Mäusedorn
l *Ruscus aculeatus*

886 butea
f butée
l *Butea*

887 butter-and-eggs
f linaire commun
d echtes Leinkraut; gelbes Flachskraut; Frauenflachs; Hanfkraut; gemeines Leinkraut; gewöhnliches Leinkraut; gelbes Löwenmaul; Marienflachs; Wiesenflachs
l *Linaria vulgaris*

888 butterbur
f pétasite
d Pestilenzwurz; Pestwurz
l *Petasites*

889 buttercup; crowfoot
f renoncule
d Butterblume; Hahnenfuss; Ranunkel
l *Ranunculus*

890 buttercup family
f renonculacées
d Hahnenfussgewächse; Hahnenfusspflanzen; Ranunkulgewächse; Ranunkulazeen
l *Ranunculaceae*

*** buttercup ficaria → 2241**

891 buttercup winter hazel
f corylopsis pauciflore
d wenigblütige Scheinhasel
l *Corylopsis pauciflora*

892 butterfly bush; summer lilac
f buddléia; arbuste aux papillons; lilas d'été; budlège
d Fliederspeer; Schmetterlingsstrauch; Sommerflieder; Sommerlingsstrauch
l *Buddleia*

893 butterfly flower
f schizanthe
d Spaltblume; Schlitzblume
l *Schizanthus*

894 butterfly milkweed
f asclépiade tubéreuse
l *Asclepias tuberosa*

895 butterfly orchid
f double feuille
d Nachtzauke; kleine Waldhyazinthe; zweiblättrige Kuckucksblume; zweiblättrige Stendelwurz; zweiblättrige Waldhyazinthe
l *Platanthera bifolia*

896 butterfly palm
d Goldfruchtpalme
l *Chrysallidocarpus*

* **butterfly pea → 4178**

* **butter mushroom → 820**

897 butternut
f noyer à beurre; noyer cendré
d Graunuss; Butternussbaum; Ölnussbaum; aschgrauer Walnussbaum; grauer Walnussbaum
l *Juglans cinerea*

898 butter seed
f arbre à beurre
d Butterbaum; Schibaum
l *Butyrospermum*

899 butter tree
f arbre à beurre; lamy; tama
d Butterbaum; Talgbaum
l *Pentadesma butyracea*

900 butterwort
f grassette
d Fettkraut
l *Pinguicula*

901 buttery collybia
f collybie butyracée
d Butterrübling
l *Collybia butyracea*

902 buttonbush
f céphalante
d Kopfblume
l *Cephalanthus*

903 button clover
f luzerne orbiculaire
d Scheibenklee
l *Medicago orbicularis*

904 button flower
f hibbertie
d Hibbertia
l *Hibbertia*

905 byrsonima
f byrsonime; moureiller
l *Byrsonima*

C

906 cabbage
f chou
d Kohl
l *Brassica*

907 cabbage angelin tree
f angelin; arbre à chou
d Kohlbaum
l *Andira inermis*

908 cabbage family
f crucifères
d Kreuzblütler; Kruziferen
l *Cruciferae; Brassicaceae*

909 cabbage palmetto
f chou palmiste; palmetto
d Sabal(palme); Palmettopalme
l *Sabal palmetto; Inodes palmetto*

910 cabbage rose
f rose à cent feuilles; rosier centfeuilles; rosier de Provence
d hundertblättrige Rose; Zentifolie; Centifolie
l *Rosa centifolia*

* **cabbage tree → 246**

911 Cablin patchouli
f patchouli de Cablin
d Pogostemon
l *Pogostemon cablin*

* **cabuyao → 3538**

912 cacao
f cacaoyer
d Kakaobaum; Schokoladenbaum
l *Theobroma cacao*

913 cachrys
f cachryde
d Nussdolde
l *Cachrys*

914 cactus
f cactier
d Kaktus
l *Cactus*

915 cactus family
f cactacées
d Kaktazeen; Kakteen; Kaktusgewächse
l *Cactaceae*

916 cadia
f cadie
l *Cadia*

917 cadillo
f paka
l *Urena lobata*

918 Caesar's amanita; royal agaric
f amanite de César; oronge vraie
d Kaiserling; Kaiserschwamm
l *Amanita caesarea*

* **cainito → 919**

919 cainito star apple; cainito
d Sternapfelbaum
l *Chrysophyllum cainito*

920 cajeput tree
f cajeputier
d Niaulibaum; Silberbaum; Weissbaum; weisser Teebaum; Kajeputbaum
l *Melaleuca leucodendron*

921 Calabar bean
f physostigma
d Gottesurteilbohne
l *Physostigma*

922 calabash gourd; bottle gourd
f gourde bouteille; gourde calebasse
d Flaschenkürbis; Kalebasse
l *Lagenaria vulgaris; Lagenaria siceraria*

923 calabash nutmeg
d Kalebassenmuskatnuss
l *Monodora myristica*

924 calabash tree
f calebassier
d Kalebasbaum; Kalebassenbaum; Kürbisbaum
l *Crescentia*

* **Calabrian pine → 2052**

925 caladium
f caladion
d Buntwurz; Kaladie
l *Caladium*

926 calamint
f clinopode
d Quendel
l *Calamintha; Clinopodium*

927 calamint savory
f calament officinal
d echte Kölme; echter Quendel
l *Calamintha officinalis; Satureia calamintha*

* **calamus → 1918**

928 calathea
f calathéa
d Korbmarante
l *Calathea*

929 calendula
f calendule; souci
d Kalendel; Ringelblume; Ringelrose
l *Calendula*

* **California bayberry → 948**

930 California black walnut
d kalifornischer Walnussbaum
l *Juglans californica*

931 California buckeye
f marronnier de Californie
d kalifornische Rosskastanie
l *Aesculus californica*

932 California buckthorn
f nerprun de Californie
d kalifornischer Kreuzdorn
l *Rhamnus californica*

933 California bur clover
f luzerne hérissée
d gezähnelter Schneckenklee; steifhaariger Schneckenklee
l *Medicago hispida*

934 California ephedra
f éphèdre de Californie
d kalifornisches Meerträubchen
l *Ephedra californica*

935 California filbert
f noisetier de Californie
d kalifornischer Hasel(nuss)strauch
l *Corylus californica*

936 California incense cedar
f cèdre à encens; cèdre blanc de Californie; libocèdre décurrent; cèdre d'encens de Californie
d kalifornische Flusszeder
l *Libocedrus decurrens*

937 California juniper
f genévrier de Californie
d kalifornischer Wacholder
l *Juniperus californica*

938 California laurel; Oregon myrtle
f laurier de Californie
d kalifornischer Lorbeer
l *Umbellularia californica*

939 California live oak
d grasblättrige Eiche
l *Quercus agrifolia*

* **California nutmeg → 947**

940 California pepper tree
f arbre à résine du Pérou; faux-poivrier commun
d peruanischer Pfefferbaum
l *Schinus molle*

941 California pitcher
f darlingtonie
d Darlingtonie
l *Darlingtonia*

942 California poplar
f peuplier baumier d'Occident; peuplier cotonnier noir
d haarfrüchtige Balsampappel; haarfrüchtige Pappel; westliche Balsampappel
l *Populus trichocarpa*

943 California poppy
f eschscholtzie de Californie
d kalifornischer Mohn; Morgenröschen
l *Eschscholtzia californica*

944 California privet
f troène de Californie
l *Ligustrum ovalifolium*

* **California red fir → 4451**

945 California redwood
f séquoia à feuilles d'if; séquoia toujours vert
d Eibensequoie; Küstensequoie; immergrüne Sequoie
l *Sequoia sempervirens*

946 California saxifrage
d Rasensteinbrech; rasiger Steinbrech
l *Saxifraga caespitosa*

947 California torreya; California nutmeg
f torreya de Californie
d kalifornische Torreye
l *Torreya californica*

948 California wax myrtle; California bayberry
f cirier de Californie
d kalifornischer Gagelstrauch
l *Myrica californica*

949 calla
f calla
d Drachenwurz; Kalla; Schlangenkraut; Schlangenwurz; Schweinewurz
l *Calla*

950 calla lily; trumpet lily
f zantédesquie; richarde
d Zantedeschie
l *Zantedeschia; Richardia*

951 callianthemum
f callianthème
d Alpenschön; Jägerkraut
l *Callianthemum*

952 calligonum
f calligon
d Hackenkopf; Dshusgun; Dshusgunstrauch
l *Calligonum*

953 calotrope
f calotropis
d Kielkrone
l *Calotropis*

*** caltrap → 954**

954 caltrop; caltrap
f herse; tribule
d Bürzel; Bürzeldorn; Erdstachelnuss
l *Tribulus*

955 calumba root
d Columbo-Wurzel
l *Jatrorrhiza palmata*

956 calypso
f calypso
d Kappenstendel
l *Calypso*

957 calypso
f calypso bulbeuse
d Norne
l *Calypso bulbosa*

958 camas
f camash; camassia
d Camassie; Kamassie; Prärielilie
l *Camassia; Quamassia*

959 Cambodia whitlowwort
f badianier du Cambodge
l *Illicium cambodianum*

960 camellia
f camélia
d Kamel(l)ie
l *Camellia*

961 camel's-thorn
f manne de Perse
d Kameldorn
l *Alhagi pseudoalhagi; Alhagi camelorum*

962 cammock ononis; creeping restharrow
f bugrane; arrête-bœuf
d kriechende Hauhechel; Kriechhauhechel
l *Ononis repens*

963 camomile
f camomille; anthémis
d Afterkamille; Hundskamille; Kamille
l *Anthemis*

964 Campbell magnolia
f magnolia de Campbell
d Campbells Magnolie
l *Magnolia campbelli*

*** campeche wood → 3359**

965 camphor-fume
f camphrée
d Kampferkraut
l *Camphorosma*

966 camphor tree
f camphrier
d echter Kampferbaum; japanischer Kampferbaum; Kampferlorbeer
l *Cinnamomum camphora; Camphora officinarum*

967 campion
f coquelourde des jardins
d Lichtnelke
l *Coronaria; Lychnis*

*** campion → 1054**

968 camwood
f baphier
l *Baphia*

969 Canada blueberry
f airelle du Canada
l *Vaccinium canadense*

970 Canada bluegrass
f pâturin comprimé
d zusammengedrücktes Rispengras; flaches Rispengras; Platthalmrispe
l *Poa compressa*

971 Canada goldenrod
f verge d'or du Canada
d kanadische Goldrute; Klapperschlangenkraut
l *Solidago canadensis*

972 Canada hawthorn
f aubépine du Canada
d kanadischer Weissdorn
l *Crataegus canadensis*

973 Canada hemlock; eastern hemlock; Canadian hemlock
f sapin du Canada; tsuga du Canada; sapin ciguë; tsuga commun
d kanadische Hemlockstanne; Schierlingstanne
l *Tsuga canadensis*

974 Canada parrot feather
f myriophylle verticillé
d quirliges Tausendblatt
l *Myriophyllum verticillatum*

975 Canada thistle; creeping thistle
f cirse des champs
d Ackerdistel; Ackerkratzdistel; Haferdistel
l *Cirsium arvense*

976 Canada waterweed; Canadian waterweed
f élodée du Canada; peste d'eau
d kanadische Wasserpest
l *Elodea canadensis; Anacharis canadensis*

977 Canada wood nettle
f ortie du Canada
d kanadische Nessel
l *Laportea canadensis*

* **Canadian fleabane → 2878**

* **Canadian hemlock → 973**

978 Canadian plum
f prunier du Canada
d kanadische Pflaume
l *Prunus nigra*

979 Canadian poplar
f peuplier du Canada
d kanadische Pappel
l *Populus canadensis*

* **Canadian spruce → 5913**

* **Canadian waterweed → 976**

980 Canadian yew
f if du Canada
d kanadische Eibe
l *Taxus canadensis*

981 canaigre
f canaigre
d Canaigrewurzel
l *Rumex hymenosepalus*

982 canary clover
f dorycnie
d Backenklee
l *Dorycnium*

983 Canary date
f dattier des Canaries
d kanarische Dattelpalme; Königsdattelpalme
l *Phoenix canariensis*

984 canary grass
f alpiste; phalaris
d Glanzgras; Mariengras
l *Phalaris*

985 Canary grass
f alpiste des Canaries
d Kanariengras; Kanarienhirse
l *Phalaris canariensis*

986 Canary nasturtium
f capucine des Canaries
l *Tropaeolum peregrinum*

987 Canary oak
f chêne de Mirbeck
d Mirbecks Eiche
l *Quercus canariensis; Quercus mirbecki*

988 Canary pine
f pin des Canaries; pin de Ténériffe
d kanarische Kiefer
l *Pinus canariensis*

989 canary tree
f canari
d Kanarienbaum; Kanariennuss
l *Canarium*

990 cancerroot
f conopholide
d Sommerwurz
l *Conopholis*

* **cancerwort → 4585**

991 cancharana cabralea
f canjerana
l *Cabralea canjerana; Cabralea congerana*

992 candlenut; candlenut tree
f alévrite; noyer des Moluques
d Iguape-Nussbaum; Kerzennussbaum; Lichtnussbaum; Tungbaum
l *Aleurites moluccana*

993 candytuft
f ibéride
d Schleifenblume
l *Iberis*

994 cane
f arundinaire
d Halbrohr; Rohrgras
l *Arundinaria*

995 canella
f canelle
d Kaneelbaum; Zimtrindenbaum
l *Canella*

996 canna
f balisier; canna
d Blumenrohr
l *Canna*

997 canna family
f cannacées
d Blumenrohrgewächse; Kannazeen
l *Cannaceae*

998 cantaloupe
f cantaloup
d Kantalupe
l *Cucumis melo* var. *cantalupensis*

999 Canterbury bell
f campanule carillon
d Gartenglockenblume; Marienglocke; Marienveilchen; Mariettenveilchen
l *Campanula medium*

1000 cantharellus
f girole; chanterelle
d Kantharelle; Leistling
l *Cantharellus*

1001 cantharellus family
f cantharellacées
d Kantharellen; Leistenpilze
l *Cantharellaceae*

1002 canyon live oak
f chêne à écailles dorées
d goldschuppige Eiche
l *Quercus chrysolepis*

1003 Cape aloe
f aloès féroce
d Stachelaloe
l *Aloe ferox*

1004 Cape false olive
f éléodendre du Cap
l *Elaeodendron capense*

* **Cape gooseberry → 4156**

1005 Cape marigold
f dimorphothèque
d Kapringelblume
l *Dimorphotheca*

1006 Cape primrose
f streptocarpe
d Drehfrucht
l *Streptocarpus*

1007 caper; caperbush
f câprier
d Kap(p)ernstrauch
l *Capparis*

1008 caper euphorbia; mole plant; mole weed
f euphorbe épurge
d kreuzblättrige Wolfsmilch; Maulwurfskraut
l *Euphorbia lathyrus*

1009 caper family
f capparidacées
d Kapparidazeen; Kap(p)erngewächs; Kap(p)ernsträucher
l *Capparidaceae*

* **capillary sedge → 2678**

1010 capitate rush
d Kopfbinse
l *Juncus capitatus*

1011 capulin black cherry
f cerisier doux; cerisier du Mexique
l *Prunus serotina salicifolia*

1012 **carambola; Chinese gooseberry**
f carambolier vrai
d gestirnte Pflaume; Karambole
l *Averrhoa carambola*

1013 **caraway**
f carvi
d Kümmel; echter Kümmel; gemeiner Kümmel; Karve; Karwe; Kiem; Köhr; Wiesenkümmel
l *Carum carvi*

1014 **cardamon**
f cardamome
d Kardamompflanze
l *Elettaria cardamomum*

* **cardinal flower → 3350**

1015 **cardoon**
f cardon sauvage; chardon d'Espagne
d Kardone; Karde; Gemüseartischocke; spanische Artischocke
l *Cynara cardunculus*

1016 **carex-like kobresia**
d seggenartige Kobresie
l *Cobresia caricina*

1017 **carline thistle**
f carline
d Eberwurz; Silberdistel
l *Carlina*

1018 **carludovica**
f carludovique; carludovice
d Karl-Ludwigspalme; Panamapalme
l *Carludovica*

1019 **carnation; clove pink**
f œillet giroflée; œillet des fleuristes; œillet à bouquets
d Gartennelke; Grasnelke; Grasblume
l *Dianthus caryophyllus*

* **carnation → 4197**

1020 **carnation grass**
f laîche glauque
l *Carex glauca*

1021 **carnauba palm; Brazilian wax palm**
f copernicie à cire; palmier à cire
d Karnaubapalme; brasilianische Wachspalme
l *Copernicia cerifera*

1022 **Carniola lily**
f lis de Carniole
d Krainer Lilie
l *Lilium carniolicum*

1023 **carob; carob tree; St.-John's-bread**
f caroubier commun
d Johannisbrotbaum; Karobe; Karobenbaum; Algarova; Bockshornbaum
l *Ceratonia siliqua*

1024 **Carolina ash**
f frêne de Caroline; frêne des marais
l *Fraxinus caroliniana*

1025 **Carolina hickory**
d nordkarolinische Hickory
l *Carya carolinae-septentrionalis*

1026 **Carpathian bellflower**
f campanule des Carpathes
d Karpatenglockenblume
l *Campanula carpatica*

1027 **Carpathian currant**
f groseillier des Carpathes
d Karpatenjohannisbeere
l *Ribes carpathicum; Ribes petraeum carpathicum*

1028 **Carpathian pussytoes**
f antennaire des Carpathes
d Karpatenkatzenpfötchen
l *Antennaria carpathica*

1029 **carpesium**
f carpésium
d Kragenblume
l *Carpesium*

* **carpet bentgrass → 1665**

1030 **carpet bugle**
f bugle rampant
d kriechender Günsel; Heidegünsel; Kriechgünsel
l *Ajuga reptans*

1031 **carpet cinquefoil**
d liegendes Fingerkraut
l *Potentilla supina*

1032 **carpetweed**
f mollugine
d Weichkraut
l *Mollugo*

1033 **carpetweed family**
f aïzoacées
d Aizoazeen; Eiskrautgewächse;

Mittagsblumengewächse
l *Aizoaceae*

* **carrageen → 2961**

1034 carrion flower
f stapélie
d Aasblume; Aaspflanze; Stapelie
l *Stapelia*

1035 carrot
f carrote
d Möhre
l *Daucus*

1036 carrot family
f ombellifères
d Doldengewächse; Schirmpflanzen; Umbelliferen
l *Ammiaceae; Umbelliferae*

1037 Carthusian pink
f œillet des Chartreux
d Blutnelke; Blutströpfchen; Feldnelke; Kartäusernelke
l *Dianthus carthusianorum*

1038 cart-wrack rockweed
f varech denté
l *Fucus serratus*

1039 Cascades fir
d Purpurtanne
l *Abies amabilis*

1040 cascara buckthorn
f nerprun de Pursh
d Purshs Kreuzdorn
l *Rhamnus purshiana*

1041 cascarilla; seaside balsam
f cascarillier
d Kaskarillenstrauch
l *Croton eleuteria*

1042 cashew
f anacardier
d Espasel; Herznussbaum; Kaschubaum; Nierenbaum
l *Anacardium*

1043 cashew family; sumac family
f anacardiacées
d Balsamgewächse; Anakardiazeen; Sumachgewächse
l *Anacardiaceae*

1044 Caspian honey locust
f févier de Caspienne
l *Gledits(ch)ia caspica*

1045 Caspian lotus
d kaspische Lotosblume
l *Nelumbium caspicum*

1046 cassava; manioc
f manioc
d Cassave; Mandioka; Maniok
l *Manihot*

1047 cassia bark tree; Chinese cinnamon
f cannelier casse
d chinesischer Zimtbaum; Zimtkassie; Chinazimtbaum
l *Cinnamomum cassia*

1048 cassine
f cassine
d Safranholzbaum
l *Cassine*

1049 cassiope
f cassiopée
d Zypressenheide; Moosheide; Schuppenheide
l *Cassiope*

1050 castor bean; castor-oil plant
f ricin
d Wunderbaum
l *Ricinus*

1051 castor bean; castor-oil plant
f ricin commun
d gemeiner Wunderbaum; Christ(us)palme; Kastorbohne; Rizinus; Kreuzbaum
l *Ricinus communis*

* **castor-oil plant → 1050**

1052 Catalonian jasmine; Spanish jasmine
f jasmin à grandes fleurs
d Malabarjasmin
l *Jasminum grandiflorum*

1053 catalpa
f catalpe
d Trompetenbaum; Katalpe; Katalpenbaum
l *Catalpa*

1054 catchfly; campion
f silène
d Leimkraut; Klebnelke
l *Silene*

1055 **catchweed; catchweed bedstraw; cleavers**
f gratteron; gaillet gratteron
d Klebkraut; Kleblabkraut; Klettenlabkraut
l *Galium aparine*

1056 **catechu acacia**
f acacia à catechu
d Catechu-Akazie
l *Acacia catechu*

* **caterpillar plant → 4767**

1057 **cat grape; Missouri grape**
f vigne rouge
l *Vitis palmata*

1058 **Cathay poplar**
d chinesische Pappel
l *Populus cathayana*

1059 **catjang cowpea**
f haricot à œil noir
d Catjangbohne
l *Vigna catjang*

* **catmint → 1060**

1060 **catnip; catmint**
f cataire; herbe aux chats; népète chataire
d echtes Katzenkraut; echte Katzenminze
l *Nepeta cataria*

1061 **cat's-ear**
f porcelle
d Ferkelkraut
l *Hypochoeris*

* **cat's-eye → 2454**

* **cat's-foot → 4363**

1062 **cat's-tail; cattail**
f quenouille; massette
d Lieschkolben; Lieschrohr; Rohrkolben; Rohrpompe; Teichkolben
l *Typha*

1063 **cattail family**
f typhacées
d Rohrkolbengewächse; Typhazeen
l *Typhaceae*

* **cattail millet → 4092**

1064 **cat thyme**
f thym aux chats
d Amberkraut; Katzenkraût
l *Teucrium marum*

1065 **cattleya**
f cattleya
d Cattleya
l *Cattleya*

1066 **Caucasian alder**
f aune du Caucase
d kaukasische Erle
l *Alnus subcordata*

1067 **Caucasian centaurea**
f centaurée d'Europe orientale
d orientalische Flockenblume
l *Centaurea orientalis*

1068 **Caucasian comfrey**
f consoude du Caucase
d kaukasischer Beinwell
l *Symphytum caucasicum*

1069 **Caucasian cow parsnip**
f berce pubescente
l *Heracleum pubescens; Heracleum wilhelmi*

1070 **Caucasian daphne**
f daphné du Caucase
d kaukasischer Seidelbast
l *Daphne caucasica*

1071 **Caucasian filbert**
d pontische Hasel
l *Corylus pontica; Corylus avellana pontica*

1072 **Caucasian hackberry**
f micocoulier du Caucase
d kaukasischer Zürgelbaum
l *Celtis caucasica*

1073 **Caucasian hornbeam**
f charme du Caucase
d kaukasische Hainbuche
l *Carpinus caucasica*

1074 **Caucasian iris**
f iris du Caucase
d kaukasische Schwertlilie
l *Iris caucasica*

1075 **Caucasian leopard's-bane**
f doronic du Caucase
d kaukasische Gemswurz
l *Doronicum caucasicum*

1076 **Caucasian linden**
f tilleul dasystyla; tilleul de Crimée
d kaukasische Linde
l *Tilia caucasica; Tilia dasystyla*

* **Caucasian pyrethrum → 2306**

1077 **Caucasian wingnut**
f ptérocaryer à feuilles de sorbier; ptérocaryer à feuilles de frêne
d eschenblättrige Flügelnuss
l *Pterocarya fraxinifolia; Pterocarya sorbifolia; Pterocarya caucasica*

1078 **Caucasus mock orange**
f seringa du Caucase
d falscher Jasmin; kaukasischer Pfeifenstrauch; wilder Jasmin; wohlriechender Pfeifenstrauch; Zimtröschen
l *Philadelphus caucasicus*

1079 **cauliflower**
f chou-fleur
d Blumenkohl; Karfiol; Kardifiol
l *Brassica oleracea* var. *botrytis*

1080 **ceanothus**
f céanothe
d Säckelblume
l *Ceanothus*

1081 **Ceara rubber plant**
f caoutchoutier de Céara
d Ceara-Kautschukbaum
l *Manihot glaziovi*

1082 **cedar**
f cèdre
d Zeder; Zederbaum
l *Cedrus*

1083 **cedar of Lebanon**
f cèdre du Liban
d Libanonzeder
l *Cedrus libani*

1084 **cedrela**
f cédrèle
d Surenbaum; Zedrele; Zedrobaum
l *Cedrela*

1085 **celandine**
f chélidoine
d Schellkraut; Schellwurz; Schöllkraut
l *Chelidonium*

1086 **celery**
f céleri
d Eppich; Sellerie
l *Apium*

* **celery cabbage → 4158**

1087 **celery pine**
f phylloclade
d Blatteibe; Farneibe
l *Phyllocladus*

1088 **celsia**
f celsie
d Bartfaden; Celsie
l *Celsia*

1089 **centaurea**
f centaurée
d Flockenblume; Kornblume
l *Centaurea*

1090 **centaurium**
f grande centaurée
d Tausendgüldenkraut
l *Centaurium*

1091 **centranth**
f centranthe
d Spornblume
l *Centranthus*

1092 **century plant; century plant agave; American aloe**
f agave d'Amérique
d amerikanische Agave; hundertjährige Aloe; Saftschopf; gewöhnliche Agave; Maguah; Maguey
l *Agave americana*

1093 **cepe; edible boletus**
f bolet noble; cèpe de Bordeaux; cèpe; bolet comestible; champignon polonais
d Edelpilz; Eichpilz; Fichtenpilz; Herrenpilz; Löcherpilz; Steinpilz
l *Boletus edulis*

1094 **cephalaria**
f céphalaire
d Schuppenkopf
l *Cephalaria*

1095 **cephalocereus**
d Schopfkerzenkaktus
l *Cephalocereus*

1096 **cephalotus**
f céphalote
l *Cephalotus*

1097 ceratocarpus
f cératocarpe
d Bergmahagoni
l *Ceratocarpus*

1098 cercis; redbud
f gaînier
d Judasbaum; Judaslinde; Judenbaum; Liebesbaum
l *Cercis*

1099 cercospora
f cercospora
d Cerkospore
l *Cercospora*

1100 cereus
f cierge
d Fackeldistel; Kerzenkaktus; Säulenkaktus; Schlangenfackeldistel
l *Cereus*

1101 ceriman
d Monstera
l *Monstera deliciosa; Philodendron pertusum*

1102 ceropegia
f céropégie
d Leuchterblume; Leuchterbaum
l *Ceropegia*

1103 cetraria
f cétraire
d Schuppenflechte
l *Cetraria*

1104 Ceylon beauty-leaf
f calaba commun
l *Calophyllum calaba*

1105 Ceylon cinnamon
f cannelier de Ceylan
d ceylonischer Zimtbaum; Ceylon-Zimtbaum
l *Cinnamomum zeylanicum*

1106 Ceylon sansevieria
f sansevière de Ceylan
d Ceylon-Bogenhanf; Ceylon-Sansevieria
l *Sansevieria zeylanica*

1107 Ceylon satinwood
f bois satiné d'Inde
l *Chloroxylon swietenia*

1108 chaff-flower
f achyranthe
d Spreublume
l *Achyranthes*

1109 chaffweed
f centenille
d Kleinling
l *Centunculus*

1110 chain fern
f woodwardie
d Kettenfarn
l *Woodwardia*

1111 Chalcedonian lily
f lis de Chalcédoine
d Scharlachlilie
l *Lilium chalcedonicum*

1112 chamaedorea; pacaya
f chamédorée
d Bergpalme; Zwergrohrpalme
l *Chamaedorea*

1113 chamaedrys germander; wall germander
f germandrée petit chêne; thériaque d'Angleterre
d echter Gamander; Edelgamander; gemeiner Gamander
l *Teucrium chamaedrys*

1114 changeable mushroom; prickly cup
f pholiote changeante
d Laubholzschüppling; Stockschwämmchen; Stockschüppling; Stockschwamm
l *Pholiota mutabilis*

1115 chanterelle
f chanterelle comestible; chanterelle commune
d Eierschwamm; Dotterschwamm; Pfefferling; echter Gelbling
l *Cantharellus cibarius*

1116 chard; spinach beet; leaf beet
f poirée à couper; poirée ordinaire; bette à cardes
d Beisskohl; Beetkohl; Bete; Mangold; Schnittmangold
l *Beta cicla*

1117 Charles's sceptre
f pédiculaire à sceptre
d Karlsszepter-Läusekraut
l *Pedicularis sceptrum carolinum*

1118 charlock
f moutarde sauvage; sanve; sénevé
d Ackerkohl; Ackersenf; falscher

Hederich; wilder Senf
l *Sinapis arvensis; Brassica kaber*

1119 charophyta
f charophytes
d Armleuchteralgen; Armleuchtergewächse
l *Charophyta*

1120 chaste tree
f petit poivre; agnos; arbre au poivre; poivre sauvage
d Gewürzmüllen; Mönchspfeffer; Müllen; Rauschbaum
l *Vitex*

1121 chaulmoogra tree; sponge-berry tree
f hydnocarpe
d Schwammbeere
l *Hydnocarpus*

1122 chayote
f chayote comestible; chocho
d Chayote; Stachelgurke
l *Sechium edule; Chayota edulis*

*** cheat → 802**

1123 cheatgrass brome; downy brome; roof brome grass
f brome des toits
d Dachtrespe
l *Bromus tectorum*

1124 chebula terminalia
d Myrobalanenbaum
l *Terminalia chebula*

1125 checkerberry; checkerberry wintergreen
f gaulthérie couchée; thé de Terre-Neuve; thé de montagne
d kriechende Scheinbeere
l *Gaultheria procumbens*

1126 checkered fritillary
f fritillaire damier
d Brettspielblume; Kiebitzei; Schachblume; Schachbrettblume
l *Fritillaria meleagris*

1127 checker mallow; prairie mallow
f sidalcée
d Präriemalve; Doppelmalve; Schmuckmalve
l *Sidalcea*

*** checker tree → 1128**

1128 checker-tree mountain ash; checker tree; wild service tree
f alisier de bois; alisier tranchant; alisier torminal; sorbier de bois
d Atlasbeerbaum; Elsbeerbaum; echter Elsebeerbaum; Elsbeere
l *Sorbus torminalis*

1129 Cheddar pink
f œillet mignardise
d Felsennelke; Pfingstnelke
l *Dianthus caesius*

1130 Chee barringtonia
f ijal
l *Barringtonia acutangula*

1131 chee reed grass
f calamagrostide terrestre; calamagrostide épigeios; roseau des bois
d Landreitgras; Landrohrgras; Landschilfgras; Ostseerohr; rohrartiges Reitgras; Sandreitgras
l *Calamagrostis epigeios*

1132 cherimoya
f chérimolier; chérimolier du Pérou; anone chérimolier
d Cherimoya; Tschirimajabaum
l *Annona cherimola*

1133 cherry
f cerisier
d Kirsche; Kirschbaum
l *Cerasus*

1134 cherry-berried cotoneaster
f cotonéastre de Zabel
d Zabels Zwergmispel
l *Cotoneaster zabeli*

*** cherry birch → 5357**

1135 cherry elaeagnus
f chalef multiflore
d Beerenölweide
l *Elaeagnus multiflora*

1136 cherry emu plum
f owénie à feuilles de cerisier
l *Owenia cerasifolia*

*** cherry laurel → 4277**

1137 chervil
f cerfeuil
d Kälberkropf
l *Chaerophyllum*

* **chess → 802**

1138 chess brome; ryelike brome; rye brome
f seiglin; sélin; brome des seigles
d Gerstentrespe; Korntrespe; Roggentrespe
l *Bromus secalinus*

1139 chestnut
f châtaignier
d Kastanie; Kastanienbaum; Edelkastanie
l *Castanea*

1140 chestnut boletus
f bolet marron; cèpe marron; cèpe châtain
d Hasenröhrling
l *Boletus castaneus; Gyroporus castaneus*

1141 chestnut-leaved oak
f chêne à feuilles de châtaignier
d kastanienblättrige Eiche; Kastanieneiche
l *Quercus castaneaefolia*

1142 chestnut oak
f chêne châtaignier
d Bergeiche
l *Quercus montana*

1143 chickasaw plum
d Chicasapflaume
l *Prunus angustifolia; Prunus chicasa*

* **chicken grape → 2386**

1144 chick-pea; gram chick-pea
f pois chiche; cicerole
d echte Kicher; gemeine Kichererbse
l *Cicer arietinum*

1145 chickweed
f céraiste
d Hornkraut
l *Cerastium*

1146 chickweed
f mouron des oiseaux; moyenne stellaire
d Gänsemiere; Hühnerdarm; Hühnerkraut; Vogelkraut; Vogelmiere; Sternblume
l *Stellaria media*

1147 chicory
f chicorée
d Wegwarte; Zichorie
l *Cichorium*

* **chilauni → 1802**

1148 Chilean gunnera
f gunnère du Chili; gunnère chilienne
l *Gunnera chilensis*

1149 Chilean incense cedar
f libocèdre du Chili
d chilenische Flusszeder
l *Libocedrus chilensis*

1150 Chilean strawberry
f fraisier du Chili
d Chileerdbeere
l *Fragaria chilensis*

1151 Chilean tarweed
f madie
d einjähriges Madikraut; Madie; ölgebendes Madikraut
l *Madia sativa*

1152 Chile bells
f lapagerie
d Lapagerie
l *Lapageria*

1153 Chile dodder
f cuscute odorante
d chilenische Seide
l *Cuscuta suaveolens*

1154 Chile mayten
f maytène du Chili
l *Maytenus boaria*

1155 Chile nettle
f loasa
d Brennwinde; Loase
l *Loasa*

1156 chilghoza pine; Gerard's pine
f pin de Gérard; pin de Chilghoza; pin femelle
d Gerards Kiefer
l *Pinus gerardiana*

1157 chimney bellflower
f campanule pyramidale
l *Campanula pyramidalis*

1158 China aster
f reine-marguerite
d Gartenaster; Sommeraster
l *Callistephus*

1159 chinaberry
f mélie
d Zedrachbaum
l *Melia*

1160 chinaberry; China tree; bead tree; Persian lilac
f margousier azedarach; arbre à chapelets
d indischer Zedrachbaum; Paternosterbaum
l *Melia azedarach*

1161 China box jasmine orange
f buis de Chine
l *Murraya exotica*

1162 China cypress
f glyptostrobe
l *Glyptostrobus*

1163 China fir
f cunninghamia
d Spiesstanne
l *Cunninghamia*

1164 China laurel; Chinese laurel
f antidesme
d Schlangenbeere
l *Antidesma*

1165 China pink; Chinese pink
f œillet de Chine
d Chinesennelke; chinesische Nelke; Fischers Nelke
l *Dianthus chinensis; Dianthus fischeri*

1166 chinaroot green brier
f salsepareille squine
d Chinastechwinde
l *Smilax china*

* **China squash → 1747**

* **China tree → 1160**

* **Chinese anise → 5607**

* **Chinese apple → 4091**

1167 Chinese aralia
f angélique en arbre de Chine
d Angelikabaum; chinesische Aralie
l *Aralia chinensis*

* **Chinese arborvitae → 3948**

* **Chinese artichoke → 332**

1168 Chinese ash
f orne de Chine
d chinesische Esche
l *Fraxinus chinensis*

1169 Chinese astilbe
f astilbe de Chine
d chinesische Astilbe
l *Astilbe chinensis*

1170 Chinese azalea
d japanische Azalee
l *Rhododendron molle*

* **Chinese banana → 1936**

1171 Chinese birch
f bouleau de Chine
d chinesische Birke
l *Betula chinensis*

1172 Chinese bony-berry
d Schwerins Steinapfel
l *Osteomeles schwerinae*

1173 Chinese buckthorn
f nerprun de Chine; nerprun tinctorial
l *Rhamnus utilis*

1174 Chinese bush cherry
f cerisier du Japon
d japanische Kirsche
l *Prunus japonica*

* **Chinese cabbage → 4005, 4158**

1175 Chinese catalpa
f catalpe du Japon
d japanischer Trompetenbaum
l *Catalpa ovata*

* **Chinese cinnamon → 1047**

1176 Chinese clover
d Milchwicke
l *Astragalus sinicus*

1177 Chinese elm
f orme de Chine; thé de l'abbé Galois
l *Ulmus parvifolia*

1178 Chinese fan palm
f palmier éventail de Chine
d chinesische Livistone
l *Livistona chinensis*

1179 Chinese filbert; Chinese hazelnut
f noisetier de Chine

d chinesische Hasel
l *Corylus chinensis*

1180 Chinese flowering crab apple
f pommier à bouquets
d Prachtzierapfel
l *Malus spectabilis*

1181 Chinese flowering quince
f cognassier de Chine
d chinesischer Quittenbaum; chinesische Scheinquitte
l *Chaenomeles sinensis; Cydonia sinensis*

1182 Chinese globeflower
f trolle de Chine
d chinesische Trollblume
l *Trollius chinensis*

*** Chinese gooseberry → 1012**

1183 Chinese hackberry
f micocoulier de Chine
d chinesischer Zürgelbaum
l *Celtis sinensis*

*** Chinese hazelnut → 1179**

1184 Chinese hemlock
f tsuga de Chine
d chinesische Hemlocktanne
l *Tsuga chinensis*

1185 Chinese hibiscus
f ketmie; rose de Chine
d chinesische Rose; Roseneibisch; chinesischer Eibisch
l *Hibiscus rosa-sinensis*

1186 Chinese hickory
f caryer de Chine
d chinesische Hickorynuss
l *Carya sinensis*

1187 Chinese honey locust
f févier de Chine
d chinesischer Schotenbaum
l *Gledits(ch)ia sinensis*

1188 Chinese horse chestnut
f marronnier de Chine
d chinesische Rosskastanie
l *Aesculus chinensis*

*** Chinese lantern plant → 5275**

*** Chinese laurel → 566, 1164**

1189 Chinese lespedeza
f lespédézie soyeuse; lespédézie de Chine
d keilförmiger Buschklee
l *Lespedeza cuneata; Lespedeza sericea*

1190 Chinese lilac
f lilas de Rouen; lilas varin
d chinesischer Flieder; Rouenflieder
l *Syringa chinensis*

1191 Chinese magnolia vine
f schizandre de Chine
d chinesisches Spaltkölbchen
l *Schizandra chinensis*

1192 Chinese mahonia
f mahonia de Fortune
d Fortunes Mahonie
l *Mahonia fortunei*

1193 Chinese neillia
f neillie de Chine
d chinesische Traubenspiere
l *Neillia sinensis*

1194 Chinese parasol tree
f sterculier à feuilles de platane
d platanenblättriger Stinkbaum
l *Firmanis simplex; Stercularia platanifolia*

*** Chinese pear → 4687**

1195 Chinese pine
f pin de Chine
d chinesische Kiefer
l *Pinus sinensis; Pinus tabulaeformis*

*** Chinese pink → 1165**

1196 Chinese plum yew
f céphalotaxe de Fortune
d Fortunes Kopfeibe
l *Cephalotaxus fortunei*

1197 Chinese primrose
f primevère de Chine
d chinesische Primel; chinesische Schlüsselblume
l *Primula sinensis*

1198 Chinese privet
f troène de Chine
d chinesische Rainweide; chinesischer Liguster
l *Ligustrum sinense*

* **Chinese pyramid juniper → 4369**

1199 Chinese redbud
f gaînier de Chine
d chinesisches Judasbaumblatt
l *Cercis japonica; Cercis chinensis*

1200 Chinese rose
f rosier de l'Inde
d Edelrose; indische Rose
l *Rosa chinensis; Rosa indica* var. *vulgaris; Rosa indica*

1201 Chinese spirea
f spirée chinoise
d chinesischer Spierstrauch
l *Spiraea chinensis*

1202 Chinese stranvaesia
f stranvésie du Père David
d Davids Stranvesia
l *Stranvaesia davidiana*

1203 Chinese sumac
f sumac de Chine
d chinesischer Sumach
l *Rhus chinensis*

1204 Chinese tallow tree; Chinese vegetable tallow
f gluttier
d chinesischer Talgbaum
l *Sapium sebiferum*

* **Chinese tea → 1543**

1205 Chinese toon
f cédrèle de Chine
d chinesischer Surenbaum
l *Cedrela sinensis; Toona sinensis*

1206 Chinese torreya
f torreya de Chine; torreya élevé
d grosse Torreye
l *Torreya grandis*

1207 Chinese tulip tree
f tulipier de Chine
d chinesischer Tulpenbaum
l *Liriodendron chinense*

* **Chinese vegetable tallow → 1204**

1208 Chinese walnut
f noyer de Chine
d chinesischer Walnussbaum
l *Juglans sinensis*

1209 Chinese wax gourd
d weisser Kürbis
l *Benincasa cerifera*

1210 Chinese wing nut
f ptérocaryer de Chine
d chinesische Flügelnuss
l *Pterocarya stenoptera*

1211 Chinese wistaria
f glycine de Chine
d chinesischer Blauregen; chinesische Wistarie; chinesische Glyzinie
l *Wistaria sinensis*

1212 Chinese witch hazel
f hamamélide de Chine; hamamélis velouté
d weiche Zaubernuss
l *Hamamelis mollis*

1213 Chinese wolfberry
f lyciet de Chine
d chinesischer Bocksdorn
l *Lycium chinense*

1214 Chinese yam
f igname comestible
d essbare Yamswurzel
l *Dioscorea esculenta*

1215 Chinese yew
f if de Chine
d chinesische Eibe
l *Taxus chinensis*

1216 Chinese zelkova
d chinesische Zelkowe
l *Zelkova sinica*

1217 chingma abutilon; Indian mallow; piemarker
f fausse guimauve; abutilon ordinaire
d chinesischer Hanf; chinesische Jute; gelbe Schönmalve
l *Abutilon avicennae; Abutilon theophrasti*

* **chinquapin → 99**

1218 Chittagong chickrassy
f cèdre bâtard
l *Chukrasia tabularis*

1219 chive
f ciboulette; civette
d Breislauch; Brisslauch; Graslauch; Schnittlauch
l *Allium schoenoprasum*

1220 chloranth
f chloranthe
d Pflaumenpfeffer
l *Chloranthus*

1221 chloranth family
f chloranthacées
d Chloranthazeen
l *Chloranthaceae*

1222 chlorella
f chlorelle
l *Chlorella*

1223 chocolate family
f sterculiacées
d Sterkuliazeen; Sterkuliengewächse
l *Sterculiaceae*

1224 chokeberry
f aronie
d Apfelbeere; Apfelbeerstrauch
l *Aronia*

*** chorogi → 332**

1225 Christmas bush
f alchornée
l *Alchornea*

*** Christmas flower → 1500**

1226 Christmas rose; black hellebore
f ellébore noir; rose de Noël; herbe de feu
d Schneerose; schwarze Christblume; Christrose; schwarze Christwurz; schwarze Nieswurz; Weihnachtsrose; Winterrose
l *Helleborus niger*

1227 Christ's-thorn; Christ's-thorn paliurus
f épine du Christ; argalon; paliure épineux
d Christusdorn; gemeiner Stechdorn
l *Paliurus spina-christi*

1228 chroococca family
f chroococcacées
d Freizell-Spaltalgen
l *Chroococcaceae*

1229 chrysanthemum
f chrysanthème
d Chrysantheme; Chrysanthemum; Goldblume; Wucherblume
l *Chrysanthemum*

*** chufa → 1230**

1230 chufa flat sedge; chufa; earth almond
f amande de terre; souchet comestible
d Erdmandelzypergras; Kaffeewurzel; Chufa; Erdmandel
l *Cyperus esculentus*

1231 chusquea
f chusquéa
d Bambus
l *Chusquea*

1232 cichorium family
f chicoracées; cichoracées
d Zungenblütler; Korbblütler mit Zungenblüten
l *Cichoriaceae*

1233 cider eucalyptus
f eucalyptus de Gunn
d Mostgummibaum
l *Eucalyptus gunni*

1234 cigar-box cedrela; Spanish cedar
f acajou cédrel; cèdre de Cuba; cédrèle odorante; acajou femelle
d Cedrelabaum; Cedrelaholzbaum; Kuba-Zeder; Zedrelenbaum; Zedrelenholzbaum
l *Cedrela odorata*

1235 Cilician fir
f sapin de Cilicie; sapin d'Asie Mineure
d zilizische Tanne
l *Abies cilicica*

1236 cinchona
f quinquina
d Chinabaum; Chinarindenbaum; Fieberrindenbaum; Zinchone
l *Cinchona*

1237 cineraria
f cinéraire
d Aschenblume; Aschenkraut; Aschenpflanze; Aschkraut; Kreuzkraut; Zinerarie
l *Cineraria*

1238 cinnamon
f cannelier
d Zimtbaum; Zimtlorbeer; Zinnamom
l *Cinnamomum*

1239 cinnamonroot inula
f inule conyze; conyze
d Dürrwurz
l *Inula conyza*

1240 cinnamon rose
f rosier cannelle; rosier de mai; rosier du Saint Sacrement
d Mairose; Zimtrose
l *Rosa cinnamomea*

1241 cinnamon vine
f igname de Chine
d chinesische Yamswurzel
l *Dioscorea batatas*

1242 cinquefoil
f potentille
d Fingerkraut
l *Potentilla*

1243 citron
f cédrat; cédratier; citronnier
d Cedrat; Cedrat-Zitrone; Zitronat; Zitronat-Zitrone
l *Citrus medica*

1244 citronella grass; nard grass
d Zitronellgras
l *Cymbopogon nardus*

1245 citron melon
d Futterwassermelone
l *Citrullus vulgaris citroides*

1246 citrus
f agrume
d Agrume; Orange; Zitrone
l *Citrus*

1247 citrus plants
f agrumes
d Agrumen
l *Aurantioideae*

1248 cladanthus
f cladanthe
d Astblume
l *Cladanthus*

1249 cladonia
f cladonie
d Becherflechte; Geweihstuppe; Kladonie
l *Cladonia*

1250 clammy campion
f lychnide visqueuse
d Klebnelke; Klebraden; Pechnelke
l *Viscaria viscosa; Lychnis viscaria*

1251 clammy hop seed bush
f dodonée fébrifuge
l *Dodonaea viscosa*

1252 clammy locust
f robinier visqueux
d klebrige Robinie
l *Robinia viscosa*

1253 clarkia
f clarkie
d Armutblume
l *Clarkia; Eucharidium*

*** clary → 1254**

1254 clary sage; clary
f sclarée; orvale; sauge sclarée
d Muskatellersalbei; Muskatellerkraut; Muskatsalbei; römische Salbei; Scharlach
l *Salvia sclarea*

1255 clasping mullein
f molène fausse-phlomide
d filzige Königskerze; Filzkönigskerze
l *Verbascum phlomoides*

1256 clasping pepperweed
d durchwachsene Kresse; durchwachsenblättrige Kresse
l *Lepidium perfoliatum*

1257 clasp-leaf twisted-stalk
f streptope embrassant
d stengelumfassender Knotenfuss
l *Streptopus amplexifolius*

1258 claviceps
f ergot
d Mutterkorn
l *Claviceps*

1259 clearweed; artillery plant
f pilée
d Kanonenblume; Kanonierblume
l *Pilea*

*** cleavers → 1055**

1260 clematis
f clématite
d Clematis; Klematis; Doppelblume; Waldrebe
l *Clematis*

1261 clethra family
f cléthracées
d Klethragewächse; Klethrazeen
l *Clethraceae*

1262 cliff brake
d Pellefarn
l *Pellaea*

1263 cliff cinquefoil
f potentille des rochers
d Felsenfingerkraut; Steinfingerkraut
l *Potentilla rupestris*

1264 cliff jamesia
f jamésie américaine
d amerikanische Jamesie
l *Jamesia americana*

*** climbing cobaea → 4337**

1265 climbing entada
f liane à bœuf
d Riesenhülse
l *Entada scandens; Entada phaseoloides*

1266 climbing fern
f lygodium
d Kletterfarn
l *Lygodium*

1267 climbing fig
f figuier grimpant
d Kletterfeige
l *Ficus pumila*

1268 climbing hempweed
f mikanie grimpante
d Schnellefeu; Sommerefeu; Stubenefeu
l *Mikania scandens*

1269 climbing hydrangea
f hortensia grimpant
d Kletterhortensie
l *Hydrangea petiolaris*

*** Clinton's lily → 490**

1270 clitocybe
f clitocybe
d Trichterling
l *Clitocybe*

1271 cloak fern
d Pelzfarn
l *Notholaena*

1272 clock vine
f thunbergie
d Thunbergie
l *Thunbergia*

1273 cloudberry
f framboisier jaune; ronce faux mûrier
d Molt(e)beere; Mult(e)beere; nordische Brombeere; Schellbeere; Sumpfbrombeere; Sumpfhimbeere; Torfbeere; Zwergmaulbeere
l *Rubus chamaemorus*

1274 clouded agaric
f clitocybe nébuleux
d Graukappe; Graukopf; Herbstblatte; Nebelkappe; nebelgrauer Trichterling
l *Clitocybe nebularis; Agaricus nebularis*

1275 clove bark tree; pinkwood
f cannelier-giroflée
d Nelkenzimtbaum
l *Dicypellium caryophyllatum*

1276 clove currant; buffalo currant
f groseillier des buffles
l *Ribes odoratum; Ribes longiflorum*

1277 clove-lip toadflax
d zweiteiliges Leinkraut
l *Linaria bipartita*

*** clove pink → 1019**

1278 clover
f trèfle
d Klee; Kopfklee
l *Trifolium*

1279 clover broomrape
f orobanche mineure
d Kleeteufel; Kleetod; Kleewürger; kleine Sommerwurz
l *Orobanche minor*

1280 clover dodder
f cuscute du trèfle; cuscute mineure; perruque du diable
d Kleeseide
l *Cuscuta trifolii*

1281 clove-root plant
f colurie
l *Coluria*

1282 clove-scented broomrape
d gemeine Sommerwurz; Nelkensommerwurz
l *Orobanche caryophyllacea*

1283 clove tree
f giroflier
d Gewürznelkenbaum
l *Eugenia caryophyllata; Syzygium aromaticum*

1284 club awn grass
f corynéphore
d Silbergras
l *Corynephorus*

1285 club moss
f lycopode
d Bärlapp; Jägergrün
l *Lycopodium*

*** club moss → 4606, 4814**

1286 club-moss family
f lycopodiacées
d Bärlappgewächse; Lykopodiazeen
l *Lycopodiaceae*

*** club palm → 2470**

1287 club-stem clitocybe
d keulenfüssiger Trichterling
l *Clitocybe clavipes*

1288 club wheat
f blé compact; blé miracle
d Binkel; Kolbenweizen; Zwergweizen
l *Triticum compactum*

1289 clump speedwell
f véronique à longues feuilles
d langblättriger Ehrenpreis
l *Veronica longifolia*

1290 clusia
f clusier; palétuvier des montagnes
d Klusie; Balsamfeige
l *Clusia*

*** clusia family → 2669**

1291 Clusius gentian
f gentiane à tiges courtes
d grossblütiger Enzian
l *Gentiana clusii*

1292 cluster clover
f trèfle aggloméré
d geknäuelter Klee
l *Trifolium glomeratum*

1293 clustered dock
d Knäuelampfer; Knäuelsimse
l *Rumex conglomeratus*

1294 cluster mallow
d krause Malve
l *Malva verticillata*

1295 cluster pea
f dioclée
l *Dioclea*

1296 cluster pine
f pin de Bordeaux; pin des Landes; pin maritime
d Igelföhre; Kiefer von Bordeaux; Meerstrandkiefer; Seeföhre; Seekiefer; Sternkiefer; Strandkiefer
l *Pinus pinaster*

1297 clypeola
f clypéole
d Schildkraut
l *Clypeola*

1298 coastal-plain willow
d langfüssige Weide
l *Salix longipes*

1299 coast beefwood
f filao dressé
l *Casuarina stricta*

1300 Coatamundra wattle
f mimosa de Bailey
d Baileys Akazie
l *Acacia baileyana*

1301 cobaea
f cobée
d Kobäe; Glockenrebe
l *Cobaea*

1302 cobresia; kobresia
f kobrésie
d Kobresie; Schuppenried
l *Cobresia; Kobresia*

1303 cocaine family
f érythroxylacées
d Erythroxylazeen; Rothölzer
l *Erythroxylaceae*

1304 cocaine tree
f érythroxylum
d Kokastrauch; Rotholz
l *Erythroxylum*

1305 cochineal fig; cochineal nopal cactus
f nopal à cochenille
d mexikanische Nopalpflanze
l *Nopal coccinellifera; Nopal cochenillifer*

1306 cocklebur
f petite bardane; lampourde; xanthium

d Klopfklette; Spitzklette
l *Xanthium*

1307 **cockscomb**
f célosie
d Brandschopf; Celosie; Hahnenkamm
l *Celosia*

1308 **cockscomb rattleweed**
f rhinanthe crête de coq
l *Rhinanthus crista-galli*

1309 **cocksfoot; orchard grass**
f dactyle
d Katzengras; Katzenkraut; Knäuelgras; Knaulgras
l *Dactylis*

1310 **cockspur**
d Hühnerhirse
l *Echinochloa*

1311 **cockspur coral bean**
f érythrine crête-de-coq
d Korallenstrauch
l *Erythrina crista-gali*

1312 **cockspur flower**
f germaine en buisson
l *Plectranthus fruticosus*

1313 **cockspur hawthorn**
f épine ergot de coq
d Hahnendorn
l *Crataegus crusgalli*

1314 **coconut palm**
f cocotier
d Kokospalme; echte Kokospalme; gemeine Kokospalme
l *Cocos nucifera*

1315 **coco plum**
f chrysobalane; icaquier
d Beerenzwetsche
l *Chrysobalanus*

1316 **codium**
f codium
d Filzgrünalge
l *Codium*

1317 **coelogyne**
f cœlogyne
d Hohlnarbe
l *Coelogyne*

1318 **coffee**
f caféier
d Kaffee; Kaffeebaum; Kaffeestrauch
l *Coffea*

1319 **coffee tree**
f chicot
d Schusserbaum; Chicot; Geweihbaum
l *Gymnocladus*

* **cogon grass → 1320**

1320 **cogon satin tail; cogon grass**
d rohrartiges Silbergras
l *Imperata cylindrica*

1321 **cogwood jujube**
f jujubier de Jamaïque
l *Zizyphus chloroxylon*

1322 **cohosh**
f caulophylle
d Stammblatt
l *Caulophyllum*

1323 **cola nut**
f kolatier
d Kola; Kolabaum; Colanussbaum
l *Cola*

1324 **Colchis ivy**
f lierre de Perse; lierre du Caucase
l *Hedera colchica*

1325 **coleus; flame nettle**
f coliole; coleus
d Buntlippe; Buntblatt; Buntnessel; Ziernessel; Coleus
l *Coleus*

1326 **colewort**
f chou marin; crambe
d Meerkohl; Seekohl; Strandkohl
l *Crambe*

1327 **coliseum maple**
f érable de Cappodoce; érable de la Colchide
d freudiggrüner Ahorn
l *Acer laetum; Acer cappadocium*

1328 **collapsing puffball**
f vesse-de-loup ciselée
d Hasenbovist; Hasenstaubpilz; getäfelter Stäubling
l *Lycoperdon caelatum*

* **collard → 3128**

1329 collared marasmius
d Halsbandschwindling
l *Marasmius rotula*

1330 colletia
f collétie
d Colletie
l *Colletia*

1331 collinsia
f collinsie
d Collinsie
l *Collinsia*

1332 collomia
f collomie
d Heimsaat; Leimsaat; Knaulblume; Schleimsame
l *Collomia*

1333 collybia
f collybie
d Rübling
l *Collybia*

1334 colocynth; bitter apple
f coloquinte commune
d Koloquinthe; Koloquinte; Alhandel; Bitterzitrulle; Pomaquinte
l *Citrullus colocynthis*

1335 colonial bent grass
f agrostide commun; agrostide vulgaire
d gemeines Straussgras; Rasenstrauch; rotes Straussgras
l *Agrostis tenuis; Agrostis capillaris; Agrostis vulgaris*

*** Colorado bluestem → 707**

1336 Colorado currant
f groseillier du Colorado
d Colorado-Johannisbeere
l *Ribes coloradense*

1337 Colorado pinyon pine
f pin à amandes
l *Pinus edulis; Pinus cembroides edulis*

1338 Colorado spruce; blue spruce
f épicéa pungent; épicéa du Colorado; sapin bleu
d Blaufichte; Stechfichte; Kolorado-fichte
l *Picea pungens*

1339 coltsfoot
f pas-d'âne
d Huflattich
l *Tussilago*

1340 colubrina
f colubrina
l *Colubrina*

1341 columbine
f ancolie; columbine; aquilégie; manteau royal
d Akelei
l *Aquilegia*

1342 columbine meadow rue
f pigamon à feuilles d'ancolie
d Akeleiwiesenraute; Amselkraut; akeleiblättrige Wiesenraute
l *Thalictrum aquilegifolium*

1343 colza
f colza; navet oléifère
d Ölraps; Kohlraps
l *Brassica napus* var. *oleifera*

1344 combretum
f chigomier
d Langfaden
l *Combretum*

1345 comfrey
f consoude
d Beinwell; Beinwurz(el); Komfrey; Schwarzwurz; Wallwurz
l *Symphytum*

1346 common adder's-tongue; common adder's-tongue fern
f langue de serpent; ophioglosse commun
d Natter(n)wurz
l *Ophioglossum vulgatum*

1347 common African violet
f violette du Cap
d Usambara-Veilchen
l *Saintpaulia ionantha*

1348 common agrimony
f aigremoine eupatoire; aigremoine commune
d Odermennig; Ackermennig; Beer-kraut; Franzkraut; Leberklettenkraut; Steinwurzelkraut; gewöhnlicher Oder-mennig; Heilandstee; Steinwurz
l *Agrimonia eupatoria*

1349 **common apple; crab apple**
f pommier commun; pommier sauvage; croisier
d Apfelbaum; Gartenapfelbaum; Holzapfelbaum
l *Malus pumila; Malus sylvestris*

1350 **common apricot; apricot**
f abricotier; abricotier commun; prunier d'Arménie
d Aprikose; Marille; Aprikosenbaum
l *Armeniaca vulgaris; Prunus armeniaca*

1351 **common asparagus**
f asperge commune
d Spargel; Gemüsespargel; Gartenspargel; Stangenspargel
l *Asparagus officinalis*

1352 **common aspidistra**
f grand aspidistra
d Aspidistra; Schildblume; Schildnarbe
l *Aspidistra elatior*

1353 **common autumn crocus; meadow saffron**
f colchique d'automne; safran des prés; tue-chien; veilleuse
d Hahnenklötenwurzel; Herbstrose; Herbstzeitlose; nackte Jungfer; wilder Safran; Wiesensafran
l *Colchicum autumnale*

1354 **common avens**
f benoîte urbaine
d Igelkraut; Märzwurzel; echte Nelkenwurz; Mauernelkenwurz
l *Geum urbanum*

1355 **common balata**
f balata
d Balata; Balatabaum; gehäuftblättrige Spitzenblume; Pferdefleischholzbaum
l *Mimusops balata; Manilkara bidentata*

1356 **common balm**
f mélisse officinale
d Gartenmelisse; Mutterkraut; Zitronenkraut; Zitronenmelisse
l *Melissa officinalis*

1357 **common bamboo**
d gemeiner Bambus; gemeines Bambusrohr
l *Bambusa vulgaris*

1358 **common banana**
f bananier
d Banane; Paradiesfeige; Obst-Banane
l *Musa sapientum*

1359 **common beer yeast**
f levure de bière
d Bierhefe; Brauereihefe; Presshefe; obergärige Bierhefe
l *Saccharomyces cerevisiae*

1360 **common beet**
f bette; betterave potagère
d Dickrübe; gemeine Rübe; Speiserübe; Strandrübe
l *Beta vulgaris*

1361 **common betony**
f bétoine officinale
d echte Betonie; Heilziest; Zehrkraut
l *Betonica officinalis; Stachys officinalis*

1362 **common bladder senna**
f baguenaudier commun; faux-senné
d Blasenbaum; deutscher Blasenstrauch; Fasanenstrauch; gelber Blasenstrauch; gemeiner Blasenstrauch; Linsenbaum
l *Colutea arborescens*

1363 **common bleeding heart**
f cœur-de-Marie
d Doppelsporn; Frauenherz; tränendes Herz
l *Dicentra spectabilis*

1364 **common blue-eyed grass**
f bermudienne à feuilles étroites
d schmalblättriger Rüsselschwertel
l *Sisyrinchium angustifolium*

1365 **common bog bean**
f ményanthe trifolié; trèfle d'eau
d Biberklee; Bitterklee; dreiblättriger Fieberklee; Sumpfdreiblatt; Wasserklee; Wiesenmangold
l *Menyanthes trifoliata*

1366 **common borage**
f bourrache officinale
d Borretsch; Boretsch; Gurkenkraut; gemeiner Boretsch; Gartenboretsch
l *Borago officinalis*

1367 **common Borneo camphor**
f camphrier de Bornéo
l *Dryobalanops aromatica*

1368 **common box**
f buis commun; buis toujours vert;

buis bénit
d immergrüner Buchsbaum
l *Buxus sempervirens*

1369 common buckthorn
f nerprun purgatif
d echter Kreuzdorn; gemeiner Kreuzdorn; Amselbeere; Hirschdorn; Kreuzbeere; Purgierbeere; Rainbeere; Stechdorn
l *Rhamnus cathartica*

1370 common buckwheat
f blé noir; sarrasin
d Buchweizen; Heidenkorn
l *Fagopyrum esculentum; Fagopyrum sagittatum*

1371 common bugloss
f buglosse officinale
d echte Ochsenzunge; gebräuchliche Ochsenzunge; gemeine Ochsenzunge; gewöhnliche Ochsenzunge
l *Anchusa officinalis*

1372 common burstwort
f herbe du Turc; herniaire glabre; turquette glabre
d glattes Bruchkraut; glattes Tausendkorn; kahles Bruchkraut
l *Herniaria glabra*

1373 common calabash tree
f calebassier à feuilles longues
d Kalebassenbaum; Kürbisbaum
l *Crescentia cajete*

1374 common calla lily
f zantédesquie éthiopienne; richarde d'Afrique; calla d'Ethiopie
d äthiopische Zantedeschie
l *Zantedeschia aethiopica*

1375 common camas
f camash comestible
l *Camassia esculenta*

1376 common camellia
f camélia du Japon
d japanische Kamelie; japanische Rose
l *Camellia japonica*

1377 common capers
f câprier commun
d echter Kapernstrauch; gemeiner Kapernstrauch
l *Capparis spinosa*

1378 common cashew
f anacardier; pomme de cajou; acajou à pommes
d Acajubaum; Kaschubaum; Kajubaum; Elefantenlausbaum; westindischer Nierenbaum
l *Anacardium occidentale*

1379 common cassava; bitter cassava
f manioc amer; manioc comestible
d Cassave; Bittermaniok; bitterer Maniok; bittere Juka; bittere Cassave; Kassawe; Kassawestrauch
l *Manihot esculenta; Manihot utilissima*

1380 common cattail
f quenouille; canne de jonc; massette; roseau des étangs; typha à feuilles larges
d breitblättriger Rohrkolben
l *Typha latifolia*

1381 common chicory
f chicorée sauvage
d gemeine Wegwarte; Feldwegwarte; Kapuzinerbart; Sonnenwende; Zichorie
l *Cichorium intybus*

1382 common China aster
f reine-marguerite commune
d chinesische Gartenaster
l *Callistephus chinensis*

1383 common China fir
f cunninghamia de Chine
d chinesische Zwittertanne
l *Cunnunghamia lanceolata; Cunninghamia sinensis*

1384 common chokecherry
f cerisier de Virginie; cerisier noir
d Traubenkirsche; virginische Traubenkirsche; rote Traubenkirsche
l *Prunus virginiana; Padus virginiana*

1385 common cineraria
f cinéraire hybride
d grossblumige Cinerarie; Topfgreiskraut
l *Cineraria cruenta; Senecio cruentus*

1386 common coleus
f coleus scutellaire
d Blumes Buntlippe
l *Coleus blumei; Coleus scutellarioides*

1387 common colewort
f crambe maritime; chou marin

d Meerkohl
l *Crambe maritima*

1388 common coltsfoot
f pas-d'âne commun; tussilage pas-d'âne
d gemeiner Huflattich; Pferdefuss; Quirinkraut; Rosshuf
l *Tussilago farfara*

1389 common comfrey; woundwort
f grande consoude
d echter Beinwell; gemeiner Beinwell; gewöhnlicher Beinwell; roter Beinwell; grosse Wallwurz; Schwarzwurz
l *Symphytum officinale*

1390 common coral bean
f érythrine à corail
d Korallenbaum
l *Erythrina corallodendron*

1391 common corn cockle
f couronne des blés; nielle des blés
d Ackerkrone; Kornnelke; Kornrade; Rade; Roggenrose
l *Agrostemma githago*

1392 common cosmos
f cosmos bipenné
d Schmuckkörbchen
l *Cosmos bipinnatus*

1393 common cowpea
f haricot dolique; pois du Brésil; pois de vache
d Kuherbse; Augenbohne; Kundebohne; Langbohne
l *Dolichos sinensis; Vigna sinensis*

1394 common crape myrtle
f lagerstrémie d'Inde; lagerstrémie des Indes
l *Lagerstroemia indica*

1395 common crocus
f crocus printanier; safran printanier
d Frühlingssafran
l *Crocus vernus*

1396 common cudweed
d deutsches Filzkraut
l *Filago germanica*

1397 common daffodil
f narcisse faux-narcisse
d gelber Jakobsstab; gelbe Märzblume; gelbe Narzisse; rotrandige Narzisse; Osterglocke; Trompetennarzisse
l *Narcissus pseudonarcissus*

1398 common dandelion
f pissenlit officinal
d gemeiner Löwenzahn; Butterblume; Hundeblume; Maiblume; Pfaffenkraut; Pfaffenröhrlein; gemeine Kuhblume; echter Löwenzahn
l *Taraxacum officinale*

1399 common dill
f aneth; aneth odorant; fenouil bâtard; fenouil puant
d gemeiner Dill; Dillfenchel; Gurkenkraut; Gartendill; Kümmerlingskraut
l *Anethum graveolens*

1400 common dodder
f cuscute du thym
d Kleeseide; Quendelseide; Thymianseide; Wiesenseide
l *Cuscuta epithymum*

1401 common Douglas fir
f douglas vert; douglas d'Orégon; sapin de Douglas
d Douglasfichte; Douglasie; Douglastanne
l *Pseudotsuga taxifolia*

1402 common dracena
f cordyline à fleurs terminales
d strauchige Cordyline; strauchige Keulenlilie
l *Cordyline terminalis*

1403 common duckweed
f petite lentille d'eau
d kleine Wasserlinse
l *Lemna minor*

1404 common Dutchman's-pipe
f aristoloche siphon; pipe allemande; aristoloche de Virginie en arbre
d windende Osterluzei
l *Aristolochia durior; Aristolochia macrophylla; Aristolochia sipho*

1405 common earth-ball
f scléroderme
d gemeiner Hartbovist; dickschaliger Kartoffelbovist; falsche Trüffel
l *Scleroderma vulgare*

1406 common edelweiss
f edelweiss alpin; immortelle des Alpes; patte de lion

d Edelweiss
l *Leontopodium alpinum*

1407 common eelgrass; grass wrack
f zostère varech
d echtes Seegras
l *Zostera marina*

1408 common evening primrose
f onagre bisannuelle
d Gartenrapunzel; zweijährige Nachtkerze
l *Oenothera biennis; Onagra biennis*

1409 common falcaria
d gemeine Sichelmöhre
l *Falcaria vulgaris; Falcaria sioides*

1410 common feather cockscomb
f crête-de-coq; amarante crête-de-coq
d Hahnenkamm
l *Celosia cristata*

1411 common fennel
f aneth doux; fenouil; fenouil amer
d Fenchel; gemeiner Fenchel; echter Fenchel; Mutterwurz
l *Foeniculum vulgare*

* **common field mushroom → 1476**

1412 common fig
f figuier commun; caprifiguier
d echter Feigenbaum; gemeiner Feigenbaum
l *Ficus carica*

* **common flax → 2214**

1413 common fleabane
f herbe Saint-Roch
d grosses Flohkraut
l *Pulicaria dysenterica*

1414 common flowering quince
f cognassier à fruit en gourde
d japanische Scheinquitte
l *Chaenomeles lagenaria*

1415 common four-o'clock; marvel-of-Peru
f belle-de-nuit; faux-jalap; herbe triste; merveille du Pérou
d falsche Jalape; Vieruhrblume; Wunderblume
l *Mirabilis jalapa*

1416 common foxglove; purple foxglove
f digitale pourprée
d roter Fingerhut; purpurroter Fingerhut; purpurner Fingerhut
l *Digitalis purpurea*

1417 common garden parsley
f persil sauvage
d Gartenpetersilie; Petersilie
l *Petroselinum sativum; Petroselinum hortense; Petroselinum crispum latifolium*

1418 common giant fennel
f férule commune
d gemeines Rutenkraut; gemeines Steckenkraut
l *Ferula communis*

1419 common ginger
f gingembre; amome des Indes
d Ingwer
l *Zingiber officinale*

1420 common gladiolus
f glaïeul commun
d Gartensiegwurz; gemeine Siegwurz
l *Gladiolus communis*

1421 common globe amaranth
f amarantine globuleuse
d gemeiner Kugelamarant; rote Immortelle
l *Gomphrena globosa*

1422 common globeflower
f trolle d'Europe
d europäische Trollblume; Glotzblume
l *Trollius europaeus*

1423 common globe thistle
f boulette commune
d blaue Kugeldistel
l *Echinops sphaerocephalus*

1424 common gloxinia
f gloxinie
l *Gloxinia speciosa; Sinningia speciosa*

1425 common goat's-rue
f rue de chèvre; lavanèse
d echter Geissbart; Fleckenklee; Geissklee; Ziegenraute
l *Galega officinalis*

1426 common gorse
f grand ajonc; ajonc d'Europe
d Gaspeldorn; Stechginster
l *Ulex europaeus*

1427 common grape hyacinth
f muscari botryoïde

d kleine Bisamhyazinthe
l *Muscari botryoides; Hyacinthus botryoides*

1428 common gromwell
f grémil; herbe aux perles
d echter Steinsame
l *Lithospermum officinale*

1429 common ground ivy; alehoof; field balm
f lierre terrestre; lierre de terre
d efeublättrige Gunderrebe; Efeugundermann
l *Glec(h)oma hederacea*

1430 common groundsel
f séneçon vulgaire; herbe aux charpentiers
d Baldgreis; Goldkraut; gemeines Greiskraut; Grindkraut; gemeines Kreuzkraut
l *Senecio vulgaris*

1431 common guava
f goyave; goyavier
d Guayave; Guayabe; Guava
l *Psidium guajava*

1432 common gurjun oil tree; gurjun
f diptérocarpe à baume de gurjan
l *Dipterocarpus turbinatus*

1433 common gypsophila
f gypsophile; brouillard
l *Gypsophila elegans*

1434 common hackberry
f micocoulier occidental; micocoulier de Virginie
d amerikanischer Zürgelbaum
l *Celtis occidentalis*

1435 common hair-cup moss
d Bürstenmoos
l *Polytrichum commune*

* **common heather → 4772**

1436 common heliotrope
f héliotrope du Pérou; fleur des dames; herbe de Saint Fiacre
d peruvianische Sonnenwende; Duftheliotrop; Vanillenheliotrop; Vanillenstrauch
l *Heliotropium peruvianum*

1437 common hepatica
f hépatique des jardins
d Leberblümchen
l *Anemone hepatica*

1438 common hog's-fennel
d echter Haarstrang; Ostritzwurzel; Rosskümmel; Saufenchel
l *Peucedanum officinale*

1439 common honey locust
f arbre des escargots; carouge à miel; févier à miel; févier d'Amérique; févier à trois pointes; épine du Christ
d Christusakazie; Christusdorn; Dornbaum; dreidornige Gleditschie; dorniger Schotenbaum; Schotendorn; Zuckerschotenbaum
l *Gledits(ch)ia triacanthos*

1440 common hop
f houblon commun
d Hopfen; gemeiner Hopfen; Zaunhopfen
l *Humulus lupulus*

1441 common hop tree
f orme à trois feuilles; orme de Samarie (à trois feuilles); ptélée trifolié; trèfle de Virginie
d dreiblättrige Lederblume; gemeine Lederblume; Kleeblatt-Lederstrauch; Kleestrauch; Hopfenbaum; Hopfenstrauch
l *Ptelea trifoliata*

1442 common horehound
f marrube blanc; marrube commun
d Berghopfen; gemeiner Andorn; Mariennessel; Mauerandorn; weisser Dorant
l *Marrubium vulgare*

1443 common horse chestnut
f marronnier commun; marronnier d'Inde; marronnier blanc
d echte Rosskastanie; gemeine Rosskastanie; weisse Rosskastanie
l *Aesculus hippocastanum*

1444 common hound's-tongue
f langue de chien commune
d echte Hundszunge; gemeine Hundszunge; Venusfinger
l *Cynoglossum officinale*

1445 common hyacinth
f jacinthe commune
d Gartenhyazinthe; gemeine Hyazinthe
l *Hyacinthus orientalis*

1446 common immortelle
f xéranthème annuel

d einjährige Spreublume
l *Xeranthemum annuum*

1447 common ivory palm; tagua nut
f phytéléphas macrocarpe
d Elfenbeinpalme; grossfrüchtige Elfenbeinpalme; grossfrüchtige Steinnusspalme; Taguabaum
l *Phytelephas macrocarpa*

1448 common jack bean
f pois sabre
d Fetisch-Bohne; Jackbohne; Madagaskarbohne; Strandbohne; Schwertbohne
l *Canavalia ensiformis*

1449 common Jacob's-rod
f bâton de Jacob; bâton de St.-Jacques
d gelbe Junkerlilie
l *Asphodeline lutea*

1450 common jasmine
f jasmin commun; jasmin officinal
d echter Jasmin
l *Jasminum officinale*

1451 common Jerusalem sage
f phlomide sous-ligneuse
d strauchiges Filzkraut
l *Phlomis fruticosa*

1452 common jujube
f jujubier commun
d gemeine Jujube; kahlblättrige Jujube; Judendorn; indische Feige
l *Zizyphus jujuba*

1453 common juniper
f genévrier commun
d Dürrenstaude; Heidewacholder; gemeiner Wacholder
l *Juniperus communis*

1454 common lady's-mantle
f alchémille vulgaire; alchémille commune; pied-de-lion
d Alchemistenkraut; gemeiner Frauenmantel; Marienmantel; Sinau; Wiesenfrauenmantel
l *Alchemilla vulgaris*

1455 common lady's-tresses
f spiranthe d'automne
d Herbstdrehähre; Herbstdrehwurz
l *Spiranthes autumnalis*

1456 common laurel cherry
f laurier-cerise
d Kirschlorbeer; Lorbeerkirsche
l *Prunus laurocerasus*

1457 common lentil
f lentille; lentillon
d Ackerlinse; essbare Linse; Saatlinse
l *Lens culinaris; Ervum lens*

1458 common lespedeza; Japan clover
f trèfle du Japon; lespédézie du Japon
d japanischer Klee
l *Lespedeza striata*

1459 common licorice; Spanish liquorice
f bois sucré; réglisse; régalisse
d gemeines Süssholz; spanisches Süssholz; Lakritze
l *Glycyrrhiza glabra*

1460 common lignum vitae
f bois de vie; gaïac blanc; bois de gaïac; gaïac officinal
d echter Pockholz; Guajakbaum
l *Guaiacum officinale*

1461 common lilac
f lilas commun
d gemeiner Flieder; Jelängerjelieber; Lila; Lilak; Nagelblume; Nägelein; Nägelchenbaum; Pfeifenstrauch; spanischer Flieder; türkischer Flieder
l *Syringa vulgaris*

1462 common lungwort
f pulmonaire officinale
d buntes Lungenkraut; echtes Lungenkraut; gebräuchliches Lungenkraut; gewöhnliches Lungenkraut; Hirschkohl; Hirschmangold
l *Pulmonaria officinalis*

1463 common madder
f garance; garance tinctoriale
d Krapp; Färberröte; Färberwurzel
l *Rubia tinctorum*

1464 common mango
f manguier; arbre de mango
d indischer Mangobaum
l *Mangifera indica*

1465 common marchantia
f marchantie
l *Marchantia polymorpha*

1466 common marsh marigold
f souci d'eau; populage des marécages; populage des marais
d Sumpfdotterblume; Butterblume;

Fettblume; Kuhblume; Moosblume; Schmirgel; Schmalzblume
l *Caltha palustris*

* **common May apple → 1845**

1467 **common mesquite; honey mesquite; algarroba**
f prosope
d Mesquitebaum; Mesquitobaum; Moskitobaum
l *Prosopis juliflora; Prosopis chilensis*

1468 **common mignonette; sweet mignonette**
f mignonette; réséda odorant
d Gartenresede; wohlriechende Resede
l *Reseda odorata*

1469 **common milkweed**
f herbe à ouate; asclépiade à ouate; asclépiade de Syrie
d echte Seidenpflanze; syrische Seidenpflanze
l *Asclepias syriaca*

1470 **common milkwort**
f herbe à lait; polygala commun
d gemeine Kreuzblume; gewöhnliche Kreuzblume; Natterblümchen; Wiesenkreuzblume; Wiesenkreuzblümchen
l *Polygala vulgaris*

1471 **common monkey flower**
d gefleckte Gauklerblume
l *Mimulus guttatus*

1472 **common moonseed**
f ménisperme du Canada
d kanadischer Mondsame
l *Menispermum canadense*

1473 **common morel**
f morille comestible; morille fauve; morille à tête ronde; morille vulgaire
d Rundmorchel; Speisemorchel; gemeine Morchel; Hutmorchel
l *Morchella esculenta*

1474 **common morning-glory**
f liseron pourpre; volubilis; ipomée pourpre
d Purpurtrichterwinde; Purpurwinde
l *Ipomoea purpurea*

1475 **common motherwort**
f agripaume; cardiaque; cardiaire; léonure
d echtes Herzgespann
l *Leonurus cardiaca*

* **common mushroom → 1476**

1476 **common mushroom agaric; common mushroom; common field mushroom**
f agaric des champs; paturon blanc; pratelle; psalliote des champs
d Feldegerling; Wiesenchampignon; Angerling; Brachpilz; Edelpilz; Feldedelpilz; Herrenpilz; Trüschling; Wiesenegerling
l *Psalliota campestris; Agaricus campestris*

1477 **common myrrh tree**
f arbre à myrrhe
l *Commiphora myrrha*

1478 **common nasturtium; Indian cress**
f grande capucine; cresson d'Inde
d grosse Kapuzinerkresse
l *Tropaeolum majus*

1479 **common nipplewort**
f lampsane commune
d gemeiner Rainkohl
l *La(m)psana communis*

1480 **common nutmeg**
f muscadier commun; noix muscade
d Muskatnussbaum
l *Myristica fragrans*

1481 **common oat**
f avoine
d gemeiner Hafer; Saathafer; Rispenhafer
l *Avena sativa*

1482 **common oleander**
f laurier-rose d'Europe; nérier à feuilles de laurier
d Oleander; Lorbeerrose; Rosenlorbeer
l *Nerium oleander*

1483 **common olive**
f olivier commun
d Ölbaum; echter Ölbaum; Olive
l *Olea europaea*

1484 **common origanum; wild marjoram**
f origan; thé rouge
d echter Dost; gemeiner Dosten; brauner Dosten; wilder Dost; wilder Majoran
l *Origanum vulgare*

* **common osier → 467**

1485 common papaw
f asiminier trilobé
d dreilappiger Pappaubaum
l *Asimina triloba*

1486 common paper mulberry
f mûrier à papier; mûrier de Chine; papyrier
d Papiermaulbeerbaum
l *Broussonetia papyrifera*

1487 common pear
f poirier commun
d gemeine Birne; Holzbirnbaum; Holzbirne; wilder Birnbaum
l *Pyrus communis*

1488 common pearly everlasting
f anaphalide nacrée
d Papierblume
l *Anaphalis margaritaceae*

1489 common pennywort
f hydrocotyle commune
d Nabelkraut; Wassernabel
l *Hydrocotyle vulgaris*

1490 common peony
f pivoine officinale
d grossblumige Pfingstrose; echte Pfingstrose; Bauernrose; Gartenpfingstrose; Gichtrose
l *Paeonia officinalis*

1491 common perennial gaillardia
f gaillarde vivace
d grosse Kokardenblume
l *Gaillardia grandiflora; Gaillardia aristata*

1492 common perilla
f pérille faux-basil
d Perille
l *Perilla ocymoides; Perilla frutescens*

1493 common periwinkle
f pervenche mineure; petite pervenche
d gemeines Immergrün; kleines Immergrün
l *Vinca minor*

1494 common persimmon
f plaqueminier commun; plaqueminier de Virginie
d virginische Dattelpflaume
l *Diospyros virginiana*

1495 common petunia
f pétunia hybride; pétunie
d Bastardpetunie; Gartenpetunie
l *Petunia hybrida*

1496 common pipsissewa; prince's pine
f chimaphile en ombelle
d Doldenwinterlieb; doldiges Winterlieb
l *Chimaphila umbellata*

1497 common piptadenia
f piptadénie commune
l *Piptadenia communis*

1498 common pistache
f pistachier commun
d echte Pistazie; echte Pistacie; Pimpernuss
l *Pistacia vera*

1499 common pitcher plant
f sarracénie pourpre
d Damensattel; Jägermütze; Trompetenblatt; Wasserkrug
l *Sarracenia purpurea*

1500 common poinsettia; Christmas flower; Eastern flower
f euphorbe superbe
d Adventsstern; Weihnachtsstern
l *Euphorbia pulcherrima*

1501 common poison ivy; poison ivy; poison oak
f arbre à gale; arbre à poison; sumac vénéneux
d echter Giftsumach; Gifteiche; giftiger Efeu
l *Rhus toxicodendron; Toxicodendron radicans*

1502 common pokeberry
f raisin d'Amérique; phytolaque à 10 étamines
d Kermesbeere; Weinkermesbeere
l *Phytolacca americana; Phytolacca decandra*

1503 common polypody; wall fern
f réglisse de bois
d Engelsüss; Tüpfelfarn; wildes Süssholz
l *Polypodium vulgare*

1504 common poolmat
f alguette
d Sumpfteichfaden; Teichfaden
l *Zannichellia palustris*

1505 common portulaca
f pourpier à grandes fleurs
d Portulakröschen
l *Portulaca grandiflora*

1506 common prickly ash
f clavalier à feuilles de frêne
d Zahnwehholz
l *Zanthoxylum americanum; Zanthoxylum fraxineum*

1507 common prickly pear
f raquette commune; raquette à aiguillon unique
d kleiner Feigenkaktus; einstacheliger Feigenbaum
l *Opuntia vulgaris; Opuntia monocantha*

1508 common privet; European privet
f troène commun; troène des bois; puine blanc; frésillon
d Beinholz; gemeiner Liguster; gemeine Rainwurz; Tintenbeerstrauch
l *Ligustrum vulgare*

1509 common purslane
f pourpier potager
d Burzel(kraut); gelber Portulak; gemeiner Portulak; Kohlportulak
l *Portulaca oleracea*

1510 common pussytoes
f antennaire dioïque; pied-de-chat commun
d Engelsblümchen; Hasenpfötchen; gemeines Katzenpfötchen; Himmelfahrtsblümchen; zweihäusiges Katzenpfötchen
l *Antennaria dioica*

1511 common pygmyweed
f tillée aquatique
d Wasserdickblatt
l *Tillaea aquatica*

1512 common quince
f cognassier commun; coudonier
d echte Quitte; echter Quittenbaum; gemeiner Quittenbaum
l *Cydonia vulgaris*

1513 common ragweed
f ambroisie à feuilles d'armoise
d beifussblättriges Traubenkraut
l *Ambrosia artemisifolia*

1514 common reed
f roseau; roseau à balais; roseau commun; phragmite commun
d Ried; Rohr; gemeines Schilfrohr; gemeines Teichrohr; Teichschilf
l *Phragmites communis*

1515 common rockrose
f hysope des garigues
d Feldhysop; gemeines Sonnenröschen
l *Helianthemum nummularium*

1516 common rue; herb of grace
f rue fétide; herbe de grâce
d Feldraute; Gartenraute; Hofraute; Hofrun; Weinraute
l *Ruta graveolens*

1517 common rush
f jonc commun; jonc glauque; jonc des jardiniers
d Flatterbinse
l *Juncus effusus*

1518 common Russian thistle
f soude kali
d Kalisalzkraut; Sodakraut; russisches Salzkraut
l *Salsola kali*

1519 common sainfoin
f esparcette; sainfoin; sainfoin de prés; sainfoin cultivé
d Esper; Hasenklee; Kleberklee; Saatesparsette; Schweizer Klee; Süssklee
l *Onobrychis viciaefolia; Onobrychis sativa*

1520 common sassafras
f sassafras officinal
d Fenchelholzbaum; echter Sassafras; Sassafraslorbeer
l *Sassafras officinale; Sassafras albidum*

1521 common screw pine
f vacoua
d Schraubenpalme
l *Pandanus utilis*

1522 common scurvy weed
f cochléaria officinal; herbe au scorbut
d Arzneilöffelkraut; echtes Löffelkraut; Scharbockskraut; Skorbutkraut
l *Cochlearia officinalis*

1523 common sea buckthorn
f argousier faux-nerprun; bourdaine marine; épine marante; argasse; grisset
d Sanddorn; Rheindorn; Seedorn; Seekreuzdorn; Weidendorn; Weidensanddorn
l *Hippophae rhamnoides*

1524 common sea lettuce; green laver
f laitue de mer
d Meerlattich; Meersalat
l *Ulva lactuca*

1525 common selfheal
f brunelle commune
d kleine Braunelle; kleine Brunelle; Halskraut
l *Prunella vulgaris*

1526 common serradella
f pied-d'oiseau commun; serradelle
d grosser Krallenklee; Saarvogelfuss; grosser Vogelfuss; Serradella
l *Ornithopus sativus*

1527 common smoke tree; Aaron's-beard
f arbre à perruque; sumac fustet; fustet commun; barbe de Jupiter; bois jaune; sumac de teinturiers
d echter Perückenstrauch; Gelbholzsumach; Gelbholz; Goldholzsumach; Perückenbaum; gemeiner Perückenstrauch; Rujastrauch
l *Cotinus coggygria; Rhus cotinus*

1528 common snapdragon
f muflier des jardins
d Gartenlöwenmaul; grosses Löwenmaul
l *Antirrhinum majus*

1529 common snowberry
f arbre aux perles; symphorine à fruits blancs; symphorine commune
d Petersstrauch; traubige Schneebeere; gemeine Schneebeere
l *Symphoricarpos albus; Symphoricarpos racemosus*

1530 common snowdrop
f perce-neige; galantine; galantine d'hiver; clochette d'hiver; nivéole
d Schneeglöckchen; gemeines Schneeglöckchen; Schneetröpfchen
l *Galanthus nivalis*

1531 common sow thistle
f laiteron
d gemeine Gänsedistel; Gartengänsedistel; Kohlgänsedistel; Moosdistel; Saudistel
l *Sonchus oleraceus*

1532 common speedwell
f véronique des champs; véronique mâle
d Feldehrenpreis
l *Veronica arvensis*

1533 common spike sedge
d gemeines Sumpfried
l *Eleocharis palustris*

1534 common staghorn fern
f platycérium à corne d'élan
d Geweihfarn
l *Platycerium bifurcatum; Platycerium alcicorne*

1535 common star-of-Bethlehem
f dame-d'onze-heures
d Doldenmilchstern; doldentraubiger Milchstern; doldiger Milchstern; Judenstern; Sachsenstern
l *Ornithogalum umbellatum*

1536 common stink dragon
f gouet serpentaire; serpentaire commune
d Drachenwurz
l *Arum dracunculus; Dracunculus vulgaris*

1537 common St.-John's-wort
f herbe de Saint Jean; millepertuis commun
d durchlöchertes Hartheu; getüpfeltes Hartheu; Hexenkraut; echtes Johanniskraut; Mannsblut; Tüpfelhartheu
l *Hypericum perforatum*

1538 common stock
f matthiole blanchâtre
d Gartenlevkoje; Winterlevkoje
l *Matthiola incana*

1539 common sunflower
f tournesol; grand soleil; hélianthe annuel
d einjährige Sonnenblume; gemeine Sonnenblume
l *Helianthus annuus*

1540 common sweet shrub
f calycanthe multiflore
d Erdbeergewürzstrauch; Erdbeerstrauch; wohlriechender Gewürzstrauch; Zimmerstrauch
l *Calycanthus floridus*

1541 common sword fern
d hohe Nephrolepis
l *Nephrolepis exaltata*

1542 **common tansy**
f tanaisie vulgaire; grande marguerite
d Rainfarn; Wurmkraut
l *Tanacetum vulgare*

1543 **common tea; Chinese tea**
f thé; théier; arbre à thé
d chinesischer Tee; Teestrauch
l *Thea sinensis; Camellia sinensis*

1544 **common teak**
f tek; arbre à tek; arbre à teck
d Teakbaum; Tiekbaum; grosser Tiekbaum
l *Tectona grandis*

1545 **common thrift**
f armérie maritime
l *Armeria maritima*

1546 **common thyme**
f thym commun; thym vulgaire
d Gartenthymian; Kümmerlingskraut
l *Thymus vulgaris*

1547 **common tiger flower**
f tigridie oeil de paon
d echte Tigerblume; Pfauenlilie
l *Tigridia pavonia*

1548 **common tobacco**
f tabac commun
d echter Tabak; deutscher Tabak; gemeiner Tabak; holländischer Tabak; türkischer Tabak; Landtabak; virginischer Tabak; Virgintabak
l *Nicotiana tabacum*

1549 **common tomato**
f tomate
d Paradiesapfel; Tomate; Goldapfel; Liebesapfel; Speisetomate
l *Lycopersicon esculentum*

1550 **common trumpet creeper**
f jasmin de Virginie
d virginischer Jasmin; kletternde Trompetenblume; wurzelnde Jasmintrompete
l *Tecoma radicans; Campsis radicans*

1551 **common turmeric**
f curcuma; safran des Indes
d Gelbwurzel; Gurgemei; lange Kurkume; Turmeris; gelber Ingwer
l *Curcuma longa*

1552 **common valerian**
f valériane officinale
d gemeiner Baldrian; echter Baldrian; gebräuchlicher Baldrian; grosser Baldrian
l *Valeriana officinalis*

1553 **common velvet grass**
f houlque laineuse; blanchard velouté; avoine laineuse
d wolliges Honiggras
l *Holcus lanatus*

1554 **common vetch**
f vesce cultivée; vesce commune; vesce de printemps
d Ackerwicke; Futterwicke; Saatwicke; Feldwicke; gemeine Wicke; Kornwicke
l *Vicia sativa*

1555 **common viper's bugloss; blueweed**
f vipérine commune
d blauer Heinrich; blauer Natternkopf; gemeiner Natternkopf
l *Echium vulgare*

1556 **common wallflower**
f giroflée jaune; giroflée des murailles; giroflée commune; violier jaune
d Goldlack; Gelbveigelein; Gelbviole; gelbe Viole; Lackviole
l *Cheiranthus cheiri*

1557 **common water hyacinth**
f jacinthe d'eau
d Eichhornie; Wasserhyazinthe
l *Eichhornia crassipes*

1558 **common wax plant**
d Asklepia; Wachsblume
l *Hoya carnosa*

1559 **common white quebracho**
f quebracho blanc
d weisses Quebracho
l *Aspidosperma quebracho-blanco*

1560 **common winterberry**
f houx verticillé
l *Ilex verticillata*

1561 **common witch grass**
f panic à tiges grêles; panic capillaire
d Haarhirse
l *Panicum capillare*

1562 **common witch hazel**
f hamamélide de Virginie; noisetier de sorcière
d Zauberhasel
l *Hamamelis virginiana*

1563 common woad-waxen; dyer's greenweed
f genêt des teinturiers; herbe à jaunir
d Färberblume; Färberginster; Gilbkraut
l *Genista tinctoria*

1564 common wormwood; absinthium
f absinthe; absinthe commune; absinthe officinale; armoise amère
d Absinth; Wermut; Wermutbeifuss; Wiegekraut
l *Artemisia absinthium*

1565 common yarrow
f achillée millefeuille
d Achillenkraut; Feldschafgarbe; gemeine Schafgarbe; Gliedkraut
l *Achillea millefolium*

1566 common yellow oxalis
d steifer Sauerklee
l *Oxalis stricta*

1567 common yellowwort
f chlore perfoliée
l *Chlora perfoliata*

1568 common zinnia
f zinnia élégant
d Gartenzinnie
l *Zinnia elegans*

1569 common zoned polyporus
f polypore versicolore
d bunter Porling; Schmetterlingsporling
l *Trametes versicolor*

* **compass lettuce → 4299**

* **compass plant → 4571**

1570 composites
f composées
d Kompositen; Köpfchenblütler; Korbblütengewächse; Korbblütler; Körbchenblütler; Synanthereen; Vereinblütler; Zusammengesetztblütler
l *Compositae*

1571 condalia
f condalie
l *Condalia*

1572 condor vine; condurango
f condurango
d Kondurango
l *Marsdenia cundurango*

1573 coneflower
f rudbeckie
d Rudbeckie; Sonnenhut
l *Rudbeckia*

* **cone wheat → 4286**

1574 Congo goober; African peanut; Bambara groundnut
f voandzou
d Angolaerbse; Erderbse; Mandubi-Erbse
l *Voandzeia subterranea*

1575 Congo senna
f séné Tinnevelly
d Mekkasennes
l *Cassia angustifolia*

1576 conical silene
f silène conique
d kegeliges Leimkraut; Kegelleimkraut
l *Silene conica*

1577 conic hygrophorus
f hygrophore conique
d kegeliger Saftling; schwärzender Saftling
l *Hygrophorus conicus*

1578 conic morel
f morille conique
d Spitzmorchel
l *Morchella conica*

1579 conifers
f conifères
d Koniferen; Nadelhölzer; Zapfenbäume; Zapfenträger
l *Coniferae*

1580 conjugates
f conjuguées
d Jochalgen; Konjugaten
l *Conjugatae*

* **cooking banana → 4219**

* **coontail → 2860**

1581 coontie
f zamier
d Keulenpalme; Palmfarn; Zapfenpalme
l *Zamia*

1582 copaiba copal tree
f copaïer officinal
l *Copaifera officinalis*

* **copaifera → 1583**

1583 **copal tree; copaifera**
f copaïer; copayer; arbre à copahu
d Kopaivabaum; Kopaivabalsambaum
l *Copaifera*

1584 **copernicia**
f copernicie
d Kopernicie
l *Copernicia*

* **copper beech → 4345**

1585 **copper leaf**
f ricinelle
d Kupferblatt; Nesselblatt
l *Acalypha*

1586 **coprinus family**
f coprinacées
d Tintenpilze
l *Coprinaceae*

* **coquito → 5419**

1587 **coral bean; coral tree**
f érythrine
d Korallenbaum; Korallenbohne; Korallenstrauch
l *Erythrina*

1588 **coral fungi**
f clavariacées
d Keulenpilze
l *Clavariaceae*

1589 **coral hydnum**
f hydne coralloïde
d Bartkoralle; Eiskoralle; Igelschwamm; Korallenschwamm; Korallenstachelbart
l *Hydnum coralloides*

1590 **coral peony**
f pivoine mâle
d grossblättrige Pfingstrose
l *Paeonia corallina*

1591 **coralroot**
f coralline
d Korallenwurz
l *Corallorrhiza*

* **coral tree → 1587**

1592 **coralwort**
f dentaire bulbifère
d zwiebeltragende Zahnwurz; Zwiebelzahnwurz
l *Dentaria bulbifera*

1593 **cord grass**
f spartine
d Besengras
l *Spartina*

1594 **cordia**
f sébestier
d Brustbeerbaum; Kordie; Brustbeere
l *Cordia*

1595 **cord-rooted sedge**
d dünnwurzelige Segge
l *Carex chordorrhiza*

1596 **coreopsis; sea dahlia**
f coréopsis; leptosyne
d Scheinwanzenblume; Wanzenblume; Mädchenauge
l *Coreopsis; Leptosyne*

1597 **coriander**
f coriandre
d Koriander
l *Coriandrum*

1598 **coriander**
f coriandre
d Gartenkoriander; Gewürzkoriander; gebauter Koriander; Schwindelkorn; Wanzendill; Wanzenkraut
l *Coriandrum sativum*

1599 **coriaria**
f redoul
d Gerbermyrte; Gerberstrauch; Lederbaum
l *Coriaria*

1600 **coriaria family**
f coriariacées
d Koriariazeen; Gerberstrauchgewächse
l *Coriariaceae*

1601 **cork-barked fir; cork fir**
f sapin à liège; sapin d'Arizona
d Arizonatanne
l *Abies lasiocarpa arizonica; Abies arizonica*

* **cork-bark smooth-leaved elm → 1602**

1602 **cork elm; cork-bark smooth-leaved elm**
f orme liège; orme à liège
d Korkulme

l *Ulmus suberosa; Ulmus carpinifolia suberosa*

* **cork fir → 1601**

1603 cork oak
f chêne-liège
d Korkbaum; Korkeiche; Pantoffelbaum
l *Quercus suber*

1604 cork tree
f phellodendre-liège
d Korkbaum
l *Phellodendron*

1605 corkwood
f leitnérie
d Korkholz
l *Leitneria*

1606 corkwood duboisia
f duboisie
l *Duboisia myoporoides*

1607 corn bedstraw; rough-fruited corn bedstraw
d dreihörniges Labkraut; Hornlabkraut
l *Galium tricorne*

1608 corn buttercup
f renoncule des champs
d Ackerhahnenfuss
l *Ranunculus arvensis*

1609 corn chrysanthemum; corn marigold
f chrysanthème des moissons; chrysanthème des blés
d Feldmargerite; Saatwucherblume
l *Chrysanthemum segetum*

1610 corn cockle
f agrostemma
d Kornrade; Rade
l *Agrostemma*

1611 cornelian cherry; cornelian cherry dogwood
f cornouiller mâle
d Dürrlilitze; gelber Hartriegel; gelber Hornstrauch; Herlitze; Herlitzenbaum; Kornelkirsche
l *Cornus mas*

* **cornflag → 2499**

1612 corn flag gladiolus
f glaïeul des moissons
d Saatsiegwurz
l *Gladiolus segetum*

1613 cornflower; bachelor's-button; bluebottle
f bl(e)uet; barbeau; aubifoin
d Kornblume; Tremse; Zyane
l *Centaurea cyanus*

* **corn grass → 5981**

1614 corn gromwell
f grémil des champs; lithosperme des champs
d Ackersteinsame; Bauernschminke; Brennkraut; Steinhirse
l *Lithospermum arvense*

1615 Cornish smooth-leaved elm
f orme de Cornouailles
l *Ulmus carpinifolia* var. *cornubiensis*

* **corn marigold → 1609**

1616 corn poppy; field poppy
f coquelicot; pavot coquelicot
d Ackermohn; Feldmohn; Feuerblume; Feuermohn; Klatschmohn; Kornmohn; Schnalle; wilder Mohn
l *Papaver rhoeas*

1617 corn rocket
f buniade
d Zackenschötchen; Zackenschote
l *Bunias*

1618 corn salad
f valérianelle
d Feldsalat; Rapünzchen
l *Valerianella*

1619 corn snapdragon
f muflier oronce
d Ackerlöwenmaul; Feldlöwenmaul; kleiner Dorant
l *Antirrhinum orontium*

1620 corn spurry
f spargoule des champs
d Ackerspörgel; Ackerknöterich; Ackerspark; Feldspark; Mariengras
l *Spergula arvensis*

1621 cornutia
f cornutie
l *Cornutia*

* **corozo → 3737**

1622 corrigiole; knotgrass
f corrigiole
d Hirschsprung
l *Corrigiola*

1623 Corsican moss; worm moss
f mousse de Corse
d Korallenmoos; korsikanisches Wurmmoos
l *Gracilaria helminthochorton*

1624 Corsican pearlwort
f sagine des horticulteurs
l *Sagina subulata*

1625 Corsican pine
f pin de Calabre; pin de Corse
d kalabrische Kiefer; korsische Kiefer
l *Pinus nigra poirentiana; Pinus laricio*

1626 Corsican sandwort
f sabline des Baléares
l *Arenaria balearica*

1627 corumb rose
d Heckenrose
l *Rosa corymbifera*

1628 corydalis
f corydale
d Hohlwurz; Lerchensporn
l *Corydalis*

1629 corypha
f corypha
d Schirmpalme
l *Corypha*

1630 cosmos
f cosmos
d Schmuckkörbchen
l *Cosmos*

1631 Costa Rican guava
f goyavier à feuilles de laurier
l *Psidium friedrichsthalianum; Psidium laurifolium*

1632 costmary chrysanthemum
f balsamite
d Marienblatt
l *Chrysanthemum balsamita; Chrysanthemum majus*

1633 cotoneaster
f cotonéastre
d Steinquitte; Zwergmispel; Kutte; Quittenmispel
l *Cotoneaster*

1634 cottage rose
f rosier blanc
d weisse Rose
l *Rosa alba*

1635 cotton
f cotonnier
d Baumwolle
l *Gossypium*

1636 cotton burdock
f bardane tomenteuse
d filzige Klette; Filzklette; Wollklette
l *Arctium tomentosum*

* **cotton grass → 1637**

1637 cotton sedge; cotton grass
f jonc à coton; linaigrette; ériophoron
d Binsenseide; Daunengras; Wollgras
l *Eriophorum*

1638 cotton thistle
f chardonnette; onoporde
d Eseldistel; Krebsdistel
l *Onopordum; Onopordon*

1639 cotyledon
f cotylédon
d Nabelkraut; Nabelwurz
l *Cotyledon*

1640 couch grass; wheat grass
f agropyre; chiendent
d Hundsgras; Quecke
l *Agropyrum; Agropyron*

1641 Coulter mock orange
f seringa de Coulter
d Coulters Pfeifenstrauch
l *Philadelphus coulteri*

1642 Coulter pine; big-cone pine
f pin à gros cônes
d Coulters Kiefer
l *Pinus coulteri*

1643 Coventry bells; throatwort
f campanule à feuilles d'ortie; campanule gantée
d nesselblättrige Glockenblume; Nesselglockenblume
l *Campanula trachelium*

1644 cowage velvet bean
f mucune

d Juckbohne; Juckfasel
l *Mucuna pruriens; Stizolobium pruritum*

1645 cowberry; red bilberry; mountain cranberry
f airelle rouge; myrtille à feuilles persistantes; airelle ponctuée; canneberge ponctuée; myrtille rouge
d Hölperchen; Kränbeere; Kronsbeere; Mostjöcken; Preiselbeere; Riffelbeere; rote Heidelbeere; Steinbeere
l *Vaccinium vitis-idaea*

1646 cow breadnut tree
d Kuhbaum; Kuhmilchbaum; Milchbaum
l *Brosimum galactodendron; Brosimum utile*

1647 cow lily; spatterdock
f nufar; nuphar
d Mummel; Teichrose
l *Nuphar*

1648 cow parsley; rough chervil
f cerfeuil bâtard; cerfeuil des fous
d berauschender Kälberkropf; betäubender Kälberkropf; Eselskerbel; Taumelkälberkropf; Taumelkerbel; Wiesenkerbel
l *Chaerophyllum temulum*

1649 cow parsnip
f berce
d Bärenklau; Herkuleskraut
l *Heracleum*

1650 cowpea
f pois de vache
d Kuhbohne
l *Vigna*

1651 cowslip lungwort
f pulmonaire à feuilles étroites
d blaues Lungenkraut; schmalblättriges Lungenkraut; Schmalblattlungenkraut
l *Pulmonaria angustifolia*

1652 cowslip primrose
f primevère officinale
d duftende Schlüsselblume; gelbe Zeitlose; Frühlingsschlüsselblume; Marienschlüssel; Petersschlüssel
l *Primula officinalis; Primula veris*

1653 cow soapwort
f saponaire des vaches
d Kuhkraut; Saatkuhnelke
l *Saponaria vaccaria*

1654 cow tree tabernaemontana
f tabernémontane
l *Tabernaemontana utilis*

*** cow vetch → 591**

1655 cowwheat
f mélampyre
d Ackerweizen; Wachtelweizen
l *Melampyrum*

*** crab apple → 1349**

1656 crab cactus
f schlumbergérie
l *Schlumbergera*

*** crab grass → 2254**

1657 crab wood
f carape
d Crabbaum; Kathstrauch
l *Carapa*

1658 cranberry
f canneberge
d Moosbeere; Sumpfbeere
l *Oxycoccus*

1659 craneberry gourd
f abobra
l *Abobra*

1660 craneberry gourd
f abobra à feuilles laciniées
l *Abobra tenuifolia; Abobra viridiflora*

1661 crassula
f crassule
d Dickblatt
l *Crassula*

1662 crazyweed
f oxytropide
d Fahnenwicke; Fahnwicke
l *Oxytropis*

1663 cream scabious
d gelbe Skabiose
l *Scabiosa ochroleuca*

1664 creeper
f vigne-vierge; vigne folle
d Jungfernrebe; Jungfernwein; wilder Wein; Zaunrebe
l *Parthenocissus*

1665 creeping bentgrass; carpet bentgrass; fiorin
f agrostide des marais
d Sumpfstraussgras
l *Agrostis palustris*

1666 creeping buttercup
f renoncule rampante
d kriechender Hahnenfuss
l *Ranunculus repens*

1667 creeping cinquefoil
f potentille rampante
d kriechendes Fingerkraut
l *Potentilla reptans*

1668 creeping foxtail
d rohrartiger Fuchsschwanz; rohrartiges Fuchsschwanzgras; Rohrfuchsschwanz
l *Alopecurus ventricosus*

1669 creeping gromwell
f grémil pourpre et bleu
d Bergsteinsame; blauroter Steinsame
l *Lithospermum purpureo-caeruleum*

1670 creeping gypsophila
f gypsophile rampante
d kriechendes Schleierkraut
l *Gypsophila repens*

1671 creeping juniper
d Kriechwacholder
l *Juniperus horizontalis; Juniperus prostrata*

1672 creeping navel seed
f cynoglosse printanière
d Frühlingsgedenkemein; grosses Vergissmeinnicht; Männertreu
l *Omphalodes verna*

1673 creeping oxalis
f surelle corniculée
d gehörnter Sauerklee; Hornsauerklee
l *Oxalis corniculata*

1674 creeping pillwort
d Pillenfarn
l *Pilularia globulifera*

1675 creeping rattlesnake plantain
f goodyère rampante
d Netzblatt
l *Goodyera repens*

1676 creeping red fescue
f fétuque rouge
d Rotschwingel; Sandschwingel
l *Festuca arenaria; Festuca rubra* var. *arenaria*

* **creeping restharrow → 962**

1677 creeping St.-John's-wort; spreading St.-John's-wort; trailing St.-John's-wort
d liegendes Hartheu; niederliegendes Johanniskraut
l *Hypericum humifusum*

* **creeping thistle → 975**

1678 creeping wart cress
d gemeiner Krähenfuss; liegender Krähenfuss
l *Coronopus procumbens*

1679 creeping willow
f saule rampant
d kriechende Weide
l *Salix repens*

1680 creosote bush
f larrée
d Kreosotbusch
l *Larrea; Covillea*

1681 crested dog's-tail
f crételle commune; cynosure à crêtes; crételle des prés
d gemeiner Kammgras; Wiesenkammgras
l *Cynosurus cristatus*

1682 crested elsholtzia
d Kamminze
l *Elsholtzia cristata*

1683 crested sparassis
f sparassis crêpu; clavaire crêpue; crête-de-coq; morille des pins
d Feisterling; krauser Ziegenbart
l *Sparassis crispa*

1684 crested wheat grass
f chiendent à crête
d Kammquecke
l *Agropyron cristatum*

1685 crested wood fern
f dryoptéris à crêtes
d Kammwurmfarn; Kommafarn; hahnenkammförmiger Schildfarn
l *Dryopteris cristata*

1686 Cretan brake
f ptéris crétoise

d kretischer Saumfarn
l *Pteris cretica*

1687 **cricket-bat willow**
d Blauweide
l *Salix coerulea; Salix alba calva*

1688 **cricket rhaphis**
d Goldbartgras
l *Chrysopogon gryllus; Rhaphis gryllus*

1689 **Crimean burnet saxifrage**
d Krimbibernelle
l *Pimpinella taurica*

1690 **Crimean golden drop; long-tube gold-drop onosma**
d Krimlotwurz
l *Onosma tauricum*

1691 **Crimean ivy**
f lierre de Crimée
d Krimefeu
l *Hedera taurica*

1692 **Crimean linden**
f tilleul d'Asie
d Krimlinde
l *Tilia euchlora*

1693 **Crimean melick**
f mélique de Crimée
d Krimperlgras
l *Melica taurica*

1694 **crimson clover; scarlet clover**
f trèfle incarnat; trèfle anglais; farouche
d Blutklee; Inkarnatklee; Rosenklee; Rotklee
l *Trifolium incarnatum*

1695 **crimson flag**
f schizostyle
d Spaltgriffel
l *Schizostylis*

1696 **crinum**
f crinole; crine
d Hakenlilie
l *Crinum*

1697 **crisped helvella**
f helvelle crépue
d Herbstlorchel
l *Helvella crispa*

1698 **crispleaf mint**
d krause Minze; Krauseminze
l *Mentha crispa*

1699 **crocus**
f crocus
d Krokus
l *Crocus*

1700 **cross gentian**
f gentiane croisette
d Kreuzenzian
l *Gentiana cruciata*

1701 **cross-leaved heath**
f bruyère caminet; bruyère clarin; caminet; bruyère tétragone
d gemeine Glockenheide; Sumpfheide
l *Erica tetralix*

1702 **cross vine**
f bignone à vrilles; bignone orangée
d Kreuzrebe
l *Bignonia capreolata*

1703 **crosswort**
f crucianelle
d Kreuzblatt
l *Crucianella*

1704 **crosswort**
f croisette
d Kreuzlabkraut
l *Galium cruciatum*

1705 **croton**
f croton
d Krebsblume
l *Croton*

1706 **crowberry**
f camarine
d Rauschbeere; Krähenbeere
l *Empetrum*

1707 **crowberry family**
f empétracées
d Empetrazeen; Krähenbeerengewächse
l *Empetraceae*

* **crowfoot → 588, 889, 1708, 2447, 5163**

1708 **crowfoot plantain; crowfoot; buckhorn plantain**
d Krähenfusswegerich; schlitzblättriger Wegerich; Schlitzblattwegerich
l *Plantago coronopus*

* **crow garlic → 2223**

1709 crownbeard
f verbésine
d Gabelzahn; Verbesine
l *Verbesina*

1710 crown daisy; crown-daisy chrysanthemum
f chrysanthème des jardins
l *Chrysanthemum coronarium*

1711 crown imperial; imperial fritillary
f couronne impériale
d Kaiserkrone
l *Fritillaria imperialis*

1712 crown vetch
f coronille
d Kronwicke; Krönel
l *Coronilla*

1713 crown-vetch coronilla
f coronille bigarrée; coronille changeante
d bunte Kronenwicke; buntes Krönel; bunte Kronwicke; bunte Peltsche; Giftwicke; Schaflinse
l *Coronilla varia*

* **crow vetch → 591**

1714 crupina
f crupine
d Schlüpfsame
l *Crupina*

1715 cryptanthus
d Versteckblume
l *Cryptanthus*

1716 cryptocoryne
f cryptocoryne
d Wasserkelch
l *Cryptocoryne*

1717 cryptogamous plants
f cryptogames; plantes à spores
d Kryptogamen; Sporenpflanzen
l *Cryptogamae*

1718 cryptomeria
f cryptomérie
d Sicheltanne; Cryptomerie; Kryptomerie
l *Cryptomeria*

1719 crystal tea ledum; marsh tea
f lédon des marais
d Kienporst; Läusekraut; Mottenkraut; Porsch; Porst; Sumpfporst; Wanzenkraut; wilder Rosmarin
l *Ledum palustre*

1720 ctenanthe
d Kammarante
l *Ctenanthe*

1721 Cuban royal palm
f roi des palmiers
d Königspalme des Antillen
l *Oreodoxa regia*

1722 cubeb pepper
f cubèbe
d Kubebenpfeffer; Cubebafrucht
l *Piper cubeba*

1723 cuckoo bitter cress
f cardamine des prés; cresson des prés
d gemeine Wiesenkresse; Wiesenschaumkraut
l *Cardamine pratensis*

* **cuckoopint → 3378**

1724 cucumber
f concombre commun
d Gurke; Gartengurke
l *Cucumis sativus*

1725 cucumber-leaved sunflower
f soleil miniature
l *Helianthus cucumerifolius; Helianthus debilis*

1726 cucumber tree
f magnolia acuminé; arbre aux concombres
d Gurkenbaum; Gurkenmagnolie
l *Magnolia acuminata*

1727 cudweed
f gnaphale
d Ruhrkraut
l *Gnaphalium*

* **cudweed → 2318**

* **cultivated angelica → 2406**

1728 cumin
f cumin
d Kreuzkümmel
l *Cuminum*

1729 cumin
f cumin de Malte

d Kreuzkümmel; römischer Kümmel; Rohrkümmel; weisser Kümmel
l *Cuminum cyminum*

1730 **Cunningham araucaria**
f araucarie de Cunningham
d Cunninghams Araukarie
l *Araucaria cunnighami*

1731 **Cunningham beefwood**
f casuarina de Cunningham
d Cunninghams Keulenbaum
l *Casuarina cunninghamiana*

1732 **cupania; guara**
f châtaignier de Saint-Domingue
l *Cupania*

1733 **cupflower**
f nierembergie
d Nierembergie; Weissbecher
l *Nierembergia*

1734 **cup fungi**
f pézizacées
d Becherpilze
l *Pezizaceae*

1735 **cup grass**
f ériochloé
l *Eriochloa*

1736 **cuphea**
f saliquier; parsonsie
d Höckerkelch
l *Cuphae; Parsonsia*

1737 **Cupid's-dart**
f catananche; catanance
d Rasselblume
l *Catananche*

1738 **curculigo**
f charançonne
d Rüssellilie
l *Curculigo*

1739 **curltop lady's-thumb**
f renouée à feuilles de patience
d Ampferknöterich
l *Polygonum lapathifolium; Persicaria lapathifolia*

1740 **curly bristle thistle**
f chardon crépu
d krause Distel
l *Carduus crispus*

1741 **curly dock**
f oseille crépue
d krausblättriger Ampfer; krauser Ampfer; Krauserampfer
l *Rumex crispus*

1742 **curly-grass family**
f schizéacées
d Schizäazeen; Spaltastfarne
l *Schizaeaceae*

* **curly-leaved pondweed → 1744**

1743 **curly mallow**
f mauve crépue; mauve frisée
d Kohlmalve; Kohlpappel; krause Malve
l *Malva crispa*

1744 **curly pondweed; curly-leaved pondweed**
d Hechskraut; krauses Laichkraut
l *Potamogeton crispus*

1745 **currant**
f groseillier
d Johannisbeere; Johannisbeerstrauch
l *Ribes*

1746 **curved wood rush**
d gekrümmte Hainsimse
l *Luzula arcuata*

1747 **cushaw; China squash; musky gourd**
f courge musquée
d Bisamkürbis; Moschuskürbis
l *Cucurbita moschata*

1748 **cush-cush; cush-cush yam; yampee**
f couche-couche; igname du Brésil
d dreispaltige Yamswurzel
l *Dioscorea trifida*

1749 **cushion gypsophila**
f gypsophile des murailles
d Mauergipskraut; Mauerkraut
l *Gypsophila muralis*

1750 **cushion spurge**
f euphorbe polychrome
d bunte Wolfsmilch; vielfarbige Wolfsmilch
l *Euphorbia polychroma; Euphorbia epithymoides*

1751 **custard apple**
f corossolier; asiminier; anone
d Flaschenbaum; Ochsenherz; Anone
l *Annona*

* **custard apple → 5297**

1752 custard-apple family
f anonacées
d Anonazeen; Anonengewächse; Flaschenbaumgewächse
l *Annonaceae*

* **cutfinger → 563**

1753 cut-grass
f leersia
d Reisquecke
l *Leersia*

* **cutleaf geranium → 1757**

1754 cut-leaf stephanandra
f stéphanandre incisée
d fiederspaltige Kranzspiere
l *Stephanandra incisa; Stephanandra flexuosa*

1755 cut-leaved coneflower
f rudbeckie laciniée
d Goldball; schlitzblättriger Sonnenhut
l *Rudbeckia laciniata*

1756 cut-leaved dead nettle
d Bastardtaubnessel
l *Lamium hybridum*

1757 cut-leaved geranium; cutleaf geranium
f bec-de-grue disséqué; géranium découpé
d schlitzblättriger Storchschnabel
l *Geranium dissectum*

1758 cut-leaved germander
f germandrée femelle
d Traubengamander
l *Teucrium botrys*

1759 cut-leaved lettuce
d Schnittsalat; Rupfsalat; Stichsalat
l *Lactuca sativa* var. *secalina*

1760 cut-leaved teasel
d schlitzblättrige Karde
l *Dipsacus laciniatus*

1761 cut-scale canary grass; Harding's grass
f herbe de Harding
l *Phalaris tuberosa* var. *stenoptera*

1762 cuviera
f cuvière
l *Cuviera*

1763 cyananthus
f cyananthe
d Blaublume; Blauglöckchen
l *Cyananthus*

1764 cycad; cycas
f cycade; cycas; palmier sagou
d Farnpalme; Zykadee; Palm(en)farn; Sagopalme; Sagobaum
l *Cycas*

1765 cycad family; cycas family
f cycadacées
d Farnpalmen; Palmenfarne; Zykadazeen; Zykaden; Sagobäume
l *Cycadaceae*

* **cycas → 1764**

1766 cyclamen
f cyclamen
d Alpenveilchen; Erdscheibe; Zyklamen
l *Cyclamen*

1767 cyclantha family
f cyclanthacées
d Kolbenpalmen; Zyklanthazeen
l *Cyclanthaceae*

1768 cylcanthera
f cyclanthère
d Cyclanthera
l *Cyclanthera*

1769 cylinder colletia
f collétie épineuse
l *Colletia spinosa*

1770 cynocramba family; thelygona family
f thélygonacées
d Hundskohlgewächse
l *Cynocrambaceae; Thelygonaceae*

1771 cynomorium
f cynomoir; cynomorion
d Hundskolben; Hundsrute
l *Cynomorium*

1772 cypella
f cypelle
d Becherschwertel
l *Cypella*

1773 cypress
f cyprès
d Zypresse
l *Cupressus*

1774 **cypress euphorbia**
f euphorbe petit cyprès
d Zypressenwolfsmilch
l *Euphorbia cyparissias*

1775 **cypress family**
f cupressacées
d Zypressengewächse
l *Cupressaceae*

* **cypress grass → 2280**

1776 **cypress lavender cotton**
f santoline petit cyprès; aurone femelle; garde robe
d Gartenzypresse; Heiligenkraut; Meerwermut; zypressenartige Santolina
l *Santolina chamaecyparissus*

1777 **cypress pine**
f callitris
d Sandarakzypresse; Sandarakbaum; Schmuckzypresse
l *Callitris*

1778 **Cyprian cedar**
f cèdre de Chypre
l *Cedrus brevifolia*

1779 **Cyprian golden oak**
d erlenblättrige Eiche
l *Quercus aenifolia*

1780 **cyrtanthus**
f cyrtanthe
d Bogenlilie
l *Cyrtanthus*

1781 **cytinus**
f cytinet
d Blutschuppe
l *Cytinus*

D

1782 **dahlia**
f dahlia
d Dahlie; Georgine
l *Dahlia*

1783 **Dahurian birch**
f bouleau de Daourie
d dahurische Birke
l *Betula dahurica*

1784 **Dahurian false tamarix**
f myricaire de Daourie
d dahurische Tamariske
l *Myricaria dahurica*

1785 **Dahurian rose**
f rose de Daourie
d dahurische Rose
l *Rosa dahurica*

1786 **daimyo oak**
f chêne denté
d Kaisereiche
l *Quercus dentata*

1787 **daisy**
f pâquerette
d Gänseblümchen; Gänseblume; Maasliebe; Margaretenblume; Marguerite; Masslieb(chen); Tausendschön
l *Bellis*

1788 **daisybush**
f oléaire
d Olearie
l *Olearia*

1789 **Dallis grass**
f herbe de Dallis
d brasilianische Futterhirse
l *Paspalum dilatatum*

1790 **Dalmatian pyrethrum**
f pyrèthre de Dalmatie
d dalmatische Insektenblume; Insektenpulverpflanze
l *Pyrethrum cinerariaefolium; Chrysanthemum cinerariaefolium*

1791 **Dalmatian toadflax**
f linaire de Dalmatie
d dalmatisches Leinkraut
l *Linaria dalmatica*

1792 **Damask rose**
f rosier de Belgique; rosier de Damas; rosier des quatre saisons
d Damaszener Rose; Monatsrose
l *Rosa damascena*

1793 **damasonium**
f damasonie
d Damasonie
l *Damasonium*

1794 **dame's rocket; queen's gillyflower; dame's violet**
f julienne des dames; giroflée des dames
d Frauenkilte; Frauenviole; gemeine Nachtviole; Trauerkilte; Trauerviole; trauernde Viole
l *Hesperis matronalis*

1795 **dammar pine**
f agathis; dammara
d Dammarabaum; Kaurifichte; Kopalfichte
l *Agathis*

1796 **dandelion**
f pissenlit
d Hundeblume; Kettenblume; Kuhblume; Löwenzahn; Maiblume; Schlenke
l *Taraxacum*

1797 **Dane's-blood bellflower**
f campanule à bouquets
d Büschelglockenblume; geknäuelte Glockenblume; Knäuelglockenblume
l *Campanula glomerata*

* **Danewort → 3570**

1798 **danthonia**
f danthonie
d Traubenhafer
l *Danthonia*

1799 **Danube gentian**
d ungarischer Enzian
l *Gentiana pannonica*

1800 **daphne**
f daphné
d Kellerhals; Seidelbast
l *Daphne*

1801 **daphne willow; violet willow**
f saule noir; saule daphné; saule à feuilles de laurier
d Reifweide; Schimmelweide
l *Salix daphnoides*

1802 Darjeeling guger tree; chilauni
f schime à feuilles entières
l *Schima wallichi*

1803 dark-winged orchis
d Brandknabenkraut
l *Orchis ustulata*

1804 darnel rye grass
f ivraie vraie; ivraie énivrante
d Schwindelkorn; Taumellolch; Tollgerste
l *Lolium temulentum*

1805 Darwin barberry
f vinettier de Darwin
d Darwins Berberitze
l *Berberis darwini*

1806 dasheen
f colocase des anciens; taro; chou de Chine
d ägyptische Zehrwurzel; Taro; Wasserbrotwurzel
l *Colocasia esculenta; Colocasia antiquorum*

1807 date; date palm
f dattier; palmier dattier
d Dattelpalme
l *Phoenix*

1808 date; date palm
f dattier fruitier
d echte Dattelpalme; gemeine Dattelpalme
l *Phoenix dactylifera*

* **date palm → 1807**

1809 date-plum persimon
f plaqueminier d'Orient; plaqueminier d'Italie; plaqueminier d'Europe; plaqueminier du Levant; plaqueminier faux-lotier; prunier dattier
d Dattelpflaume; Lotuspflaume
l *Diospyros lotus*

1810 datisca
f datisque
d Steinhanf; Streichkraut; Strichkraut
l *Datisca*

1811 datura; thorn apple
f datura; stramoine
d Datura; Stachelnuss; Stechapfel
l *Datura*

1812 David keteleeria
f kételéerie du P. David
d Davids Tanne
l *Keteleeria davidiana*

1813 Davidson photinia
f photinie de Davidson
d Davidsons Glanzmispel
l *Photinia davidsoniae*

1814 dayflower
f comméline
d Kommeline; Commeline; Himmelsauge
l *Commelina*

1815 day lily
f hémérocalle
d Taglilie
l *Hemerocallis*

* **deadly amanita → 1820**

1816 deadly Calabar bean
f févier de Calabar
d Calabar-Bohne; Gottesurteilbohne
l *Physostigma venonosum*

1817 deadly carrot
f thapsie
d Böskraut; Purgierdolde
l *Thapsia*

* **deadly nightshade → 525**

1818 dead nettle
f lamier
d Bienensang; Taubnessel
l *Lamium*

1819 death camas
f zygadène
d Jochblümchen
l *Zygadenus*

1820 death cup; death-cup amanita; deadly amanita; poison amanita
f amanite phalloïde; oronge ciguë verte; amanite bulbeuse
d grüner Giftwulstling; grüner Knollenblätterpilz
l *Amanita phalloides*

* **Deccan hemp → 3146**

1821 deer fern; hardfern
f blechnum commun
d Rippenfarn
l *Blechnum spicant*

1822 deerhair; deerhair bulrush
f scirpe gazonnant
d rasige Haarsimse; rasiges Haargras; Rasensimse
l *Scirpus caespitosus; Trichophorum caespitosum*

1823 Deering velvet bean
f dolique de Floride
d Brennhülse
l *Mucuna deeringiana*

1824 deervetch; bird's-foot trefoil
f lotier; tétragonolobe
d Flügelerbse; Hornklee; Schotenklee; Spargelbohne; Spargelerbse
l *Tetragonolobus; Lotus*

1825 dehiscent common flax
d Klanglein
l *Linum usitatissimum crepitans*

1826 deinbollia
f deinbollie
l *Deinbollia*

1827 delicious lactarius; saffron milk cup
f lactaire délicieux
d echter Reizker; Wacholderschwamm; Edelreizker; Hirschling; Ritschling; Röstling
l *Lactarius deliciosus*

1828 delta maidenhair
f capillaire cunéiforme
d keilförmiger Lappenfarn
l *Adiantum cuneatum*

1829 dendrobium
f dendrobe
d Baumwucherer
l *Dendrobium*

1830 denhamia
f denhamie
l *Denhamia*

1831 dense-head mountain ash
f sorbier à feuilles d'aune
d erlenblättrige Eberesche
l *Sorbus alnifolia*

1832 dense-leaved elodea
f élodée d'Argentine
d dichtblättrige Wasserpest
l *Elodea densa*

1833 dense pondweed
d dichtes Laichkraut
l *Potamogeton densus*

1834 dentated melilot
d gezähnter Honigklee; gezähnter Steinklee
l *Melilotus dentata*

1835 dent corn
f maïs denté
d Pferdenzahnmais; Zahnmais
l *Zea mays* var. *identata; Zea mays* var. *dentiformis*

1836 deodar cedar
f cèdre de l'Himalaya
d Deodarzedar; Himalaja-Zeder
l *Cedrus deodara*

1837 Deptford pink
f œillet velu; œillet des bois
d rauhe Nelke
l *Dianthus armeria*

1838 desert candle
f érémure
d Kleopatranadel; Lilienschweif
l *Eremurus*

1839 desert rod
f érémostachys
d Einsamähre; Jerusalemkerze; Wüstenziest
l *Eremostachys*

* **desert thorn → 6016**

1840 desert willow
f saule du désert
l *Chilopsis linearis*

1841 desmids
f desmidiées
d Bandalgen; Desmidiazeen
l *Desmidiaceae*

1842 destroying angel
f amanite vireuse; oronge ciguë vireuse
d kegeliger Wulstling; spitzhütiger Knollenblätterpilz
l *Amanita virosa*

1843 deutzia
f deutzie
d Deutzie
l *Deutzia*

1844 devil pepper; rauwolfia
f rauwolfia
d Rauwolfia
l *Rauwolfia*

1845 devil's-apple; common May apple
f podophylle à feuilles peltées; podophylle américain
d Entenfuss; Fussblatt; Maiapfel; wilde Zitrone; amerikanisches Fussblatt
l *Podophyllum peltatum*

*** devil's-bit → 3561**

1846 devil's-claws
f bicorne
d Gemshorn; Proboszidea
l *Proboscidea*

1847 devil's-club
f échinopanax
d Igelkraftwurz
l *Echinopanax; Oplopanax*

1848 devil's-tongue; snake palm
f amorphophallus de Rivière
d Riviers Amorphophallus
l *Amorphophallus rivieri; Hydrosme rivieri*

*** devil's-tree → 4014**

1849 devil's-walking-stick; Hercules'-club; tree aralia
f angélique en arbre d'Amérique; aralie épineuse; bois à cassave; gourdin d'Hercule
d Angelikabaum
l *Aralia spinosa*

1850 devilwood
d Duftblume; Duftblüte
l *Osmanthus americanus*

*** dew plant → 5325**

*** dhak tree → 533**

1851 dialium
f diale
l *Dialium*

1852 diapensia
f diapensie
d Trauerblume
l *Diapensia*

1853 diatoms; brittleworts
f diatomées
d Diatomeen; Kieselalgen
l *Diatomeae*

1854 dichapetalium
f dichapétale
l *Dichapetalium*

1855 dicotyledoneae
f dicotylédones
d Blattkeimer; Dikotyledonen; Dikotylen; Zweikeimblättrige
l *Dicotyledoneae*

1856 digger pine; gray-leaf pine
f pin de J. Sabine; pin sabine
d Nusskiefer; Weisskiefer
l *Pinus sabiniana*

1857 dill
f aneth
d Dill; Dillfenchel; Dillkraut; Gurkendill; Gurkenkraut
l *Anethum*

1858 dillenia
f dillénie
d Dillenie; Rosenapfel(baum)
l *Dillenia*

1859 dingy agaric
f tricholome prétentieux; petit gris d'automne
d grauer Ritterling; schwarzfaseriger Ritterling; Russkopf; Schneepilz
l *Tricholoma portentosum; Agaricus portentosus*

*** dinkel wheat → 2029**

1860 dioecious sedge
f carex dioïque; laîche dioïque
d zweihäusiges Riedgras; zweihäusige Segge
l *Carex dioica*

1861 diosma
f diosme
d Götterduft
l *Diosma*

1862 dipterocarpus family
f diptérocarpacées
d Dipterokarpazeen
l *Diptercarpaceae*

1863 discostigma
f discostigme
l *Discostigma*

* **dishcloth gourd → 5564**

1864 distaff thistle
f atractyle
d Spindelkraut
l *Atractylis*

1865 distant sedge
f laîche distante
d entferntähriges Riedgras; entferntährige Segge
l *Carex distans*

* **dita → 4014**

* **dittander → 2118**

1866 dittany; burning bush
f dictame
d Diptam
l *Dictamnus*

1867 divi-divi
f libidibi; dividivi
d Dividivi
l *Caesalpinia coriaria*

1868 dock
f oseille
d Ampfer
l *Rumex*

* **dock cress → 3860**

1869 dodartia
f dodartie
l *Dodartia*

1870 dodder
f cuscute; barbe de moine
d Flechtgras; Seide; Teufelszwirn
l *Cuscuta*

1871 dodder family
f cuscutacées
d Seidengewächse; Teufelszwirngewächse
l *Cuscutaceae*

1872 dogbane
f apocyn
d Hundsgift; Hundswolle
l *Apocynum*

1873 dogbane family
f apocynacées
d Immergrüngewächse; Hundsgiftgewächse; Apozynazeen
l *Apocynaceae*

1874 dog nettle
f ortie brûlante
d kleine Brennessel; kleine Nessel
l *Urtica urens*

1875 dog rose
f agalancé; églantier; rose de sorcière; rosier des chiens; rose des haies
d Dornrose; Feldrose; Frauendorn; Hagebutte; Hagedorn; Heckenrose; Heideröschen; Hundsdorn; Hundsrose; wilde Rose; Zaunrose
l *Rosa canina*

1876 dog's mercury
f mercuriale des bois
d ausdauerndes Bingelkraut; Waldbingelkraut
l *Mercurialis perennis*

1877 dog's-tail grass
f crételle
d Kammgras
l *Cynosurus*

1878 dogtooth fawn lily; dogtooth violet
f dent-de-chien
d Hundszahn; Zahnlilie
l *Erythronium dens canis*

1879 dogtooth grass
f chiendent
d Hundszahngras
l *Cynodon*

* **dogtooth violet → 1878**

1880 dog violet
f violette des chiens
d Hundsveilchen
l *Viola canina*

1881 dogwood
f vornouiller
d Hartriegel; Hornstrauch
l *Cornus*

1882 dogwood family
f cornacées
d Hartriegelgewächse; Hornstrauchgewächse; Kornelkirschgewächse; Kornazeen
l *Cornaceae*

1883 dolichos
f dolique
d Fasel; Bohnenwinde; Schlingbohne
l *Dolichos*

1884 dollarplant; annual honesty
f monnaie du Pape; médaille de Judas; lunaire annuelle
d stumpfes Silberblatt
l *Lunaria annua*

1885 dombeya
f dombéya
d Dombeya
l *Dombeya*

1886 domestic langsat
f lanse de Sonde
d echter Lanzenbaum
l *Lansium domesticum*

1887 double coconut; sea coconut
f coco de mer; cocotier des Maldives
d Doppelkokos; Meerkokos; maledivische Nuss; Salomonsnuss; Seekokos; Seschellennuss; Seychellennuss; Seychellenpalme; Salomons Wundernuss
l *Lodoicea seychellarum; Lodoicea maldivica*

1888 Douglas fir
f sapin de Douglas; douglas
d Douglasie; Douglastanne
l *Pseudotsuga*

1889 Douglas spirea
f spirée de Douglas
d Douglas Spierstrauch
l *Spiraea douglasii*

1890 doum palm
f doum; doumier
d Dumpalme
l *Hyphaene*

1891 dove scabious
f scabieuse gorge de pigeon
d Taubenskabiose
l *Scabiosa columbaria*

1892 dove's-foot geranium
f géranium à feuilles molles
d weicher Storchschnabel
l *Geranium molle*

* **downy brome → 1123**

1893 downy cinquefoil
d mittleres Fingerkraut; russisches Fingerkraut
l *Potentilla intermedia*

1894 downy ground cherry
f coqueret pubescent; coqueret comestible
l *Physalis pubescens*

1895 downy hemp nettle
f galéope pubescent
d weicher Hohlzahn; weichhaariger Hohlzahn
l *Galeopsis pubescens*

* **downy oat grass → 2687**

1896 downy sedge
d filziges Riedgras; filzige Segge
l *Carex tomentosa*

* **downy serviceberry → 4838**

1897 draba; whitlowwort
f drave
d Felsenblümchen; Hungerblume
l *Draba*

1898 dracaena
f dragonnier
d Dracaene
l *Dracaena*

1899 dracena
f cordyline
d Keulenlilie; Keulenbaum; Kordyline
l *Cordyline*

1900 dragon dracaena; dragon tree
f dragonnier commun; dragonnier des Canaries; grand dragonnier
d Drachenbaum; Drachenblutbaum; Drachenlilie; Drachenpalme; gemeiner Drachenbaum; Schlangenlilie
l *Dracaena draco*

1901 dragonhead
f dracocéphale
d Drachenkopf
l *Dracocephalum*

1902 dragon's-blood padauk
f bois de corail tendre; santal rouge des Antilles
l *Pterocarpus draco*

* **dragon tree → 1900**

1903 drimys
f drimys
d Gewürzrindenbaum
l *Drimys*

* **drooping catchfly → 1905**

1904 drooping saxifrage
d nickender Steinbrech
l *Saxifraga cernua*

1905 drooping silene; drooping catchfly
f silène à fruits pendants
d hängendes Leimkraut
l *Silene pendula*

* **drophip rose → 145**

1906 dropseed
f sporobole
d Fallsame; Schleudersamengras
l *Sporobolus*

1907 dropwort
f filipendule à 6 pétales
d Filipendelwurz; Hügelmädesüss; kleines Mädesüss; Erdeichel; Haarstrang
l *Filipendula hexapetala; Spiraea filipendula*

1908 drosophyllum
f drosophyllum
d Taublatt
l *Drosophyllum*

1909 drug centaurium
f érythrée centaurée
d Biberkraut; echtes Tausendgüldenkraut; grosses Tausendgüldenkraut
l *Centaurium umbellatum; Erythraea centaurium*

1910 drug eyebright
f euphraise officinale
d gemeiner Augentrost; Wiesenaugentrost
l *Euphrasia officinalis*

1911 drug fumitory
f fumeterre officinale
d echter Erdrauch; gemeiner Erdrauch; Feldraute; Katzenkerbel; Taubenkerbel
l *Fumaria officinalis*

1912 drug fustic tree
f bois jaune; chlorophora tinctorial
l *Chlorophora tinctorial*

1913 drug hedge hyssop
f gratiole officinale; grâce-de-Dieu; herbe au pauvre homme
d echtes Gnadenkraut; Gichtkraut
l *Gratiola officinalis*

1914 drug sabadilla
d mexikanisches Läusekraut; Sabadill(a); Sebadille
l *Sabadilla officinalis; Schoenocaulon officinale*

1915 drug snowbell
f aliboufier officinal; styrax officinal
d echter Storaxbaum; gebräuchlicher Storaxbaum
l *Styrax officinalis*

1916 drug Solomon's-seal
f sceau de Salomon officinal
d Gelenkwurz; grosse Maiblume; echtes Salomonssiegel; wenigblütige Weisswurz
l *Polygonatum officinale*

1917 drug speedwell
f thé d'Europe; véronique mâle; véronique officinale
d echter Ehrenpreis; gemeiner Ehrenpreis; Grundheil; Waldehrenpreis
l *Veronica officinalis*

1918 drug sweet flag; calamus
f acore odorant; acore vrai; jonc odorant
d echter Kalmus; gemeiner Kalmus; Teichkalmus
l *Acorus calamus*

1919 Drummond phlox
f phlox de Drummond
d Drummonds Flammenblume; einjährige Flammenblume
l *Phlox drummondi*

* **drumstick tree → 2548**

1920 dryad
f dryade
d Dryade; Silberwurz
l *Dryas*

1921 dryad's saddle; scaly polyporus
f polypore écailleux
d schuppiger Porling; schuppiger Schwarzfussporling
l *Polyporus squamosus*

1922 duboisia
f duboisie
l *Duboisia*

* **duck oak → 5827**

1923 duck's-meat
f spirodèle

d Teichlinse
l *Spirodela*

1924 duckweed
f lenticule; lentille d'eau
d Entenflott; Entengrün; Entengrütze; Entenlinse; Meerlinse; Teichlinse; Teichsimse; Wasserlinse
l *Lemna*

1925 duckweed family
f lemnacées
d Lemnazeen; Wasserlinsen; Wasserlinsengewächse
l *Lemnaceae*

*** duhat → 3010**

1926 dull-seed cornbind
f renouée liseron
d Windenknöterich
l *Polygonum convolvulus; Bilderdykia convolvulus*

*** dumb cane → 5631**

1927 durian
f durio
d Durianbaum; (indischer) Zibetbaum
l *Durio*

1928 durmast oak
f chêne commun sessile; chêne a glands sessile; chêne à trouchets; chêne rouvre; chêne femelle; chêne sessile
d Steineiche; Traubeneiche; Wintereiche; Dürreiche; Eiseiche; Harzeiche; Klebeiche
l *Quercus petraea; Quercus sessiliflora*

1929 durra; brown durra
f sorgho douro
d Durra; Mohrenhirse
l *Sorghum durra; Sorghum vulgare durra*

1930 durum wheat
f blé dur
d Bartweizen; Glasweizen; Gerstenweizen; Hartweizen; Grasweizen
l *Triticum durum*

1931 dusky crane's-bill
f géranium brun
d brauner Storchschnabel
l *Geranium phaeum*

*** dusty miller → 489, 4958**

1932 Dutchman's-pipe; birthwort
f aristoloche
d Osterluzei
l *Aristolochia*

*** Dutch rush → 4777**

1933 Dutch tonka bean
f fève de tonca; févier tonka
d Tonkabaum
l *Dipteryx odorata*

1934 dwarf alpine onion
d Berglauch; trügerischer Lauch
l *Allium falax*

1935 dwarf arctic birch
f bouleau nain
d Alpenbirke; Brockenbirke; Morastbirke; Zwergbirke; nordische Zwergbirke; Polarbirke
l *Betula nana*

1936 dwarf banana; Chinese banana
f bananier de Chine
d chinesische Banane; Zwergbanane
l *Musa nana; Musa chinensis*

*** dwarf cherry → 2644, 4683**

*** dwarf chestnut → 99**

1937 dwarf cornel
f cornouiller de Suède
d schwedischer Hartriegel
l *Cornus suecica; Chamaepericlymenum suecicum*

1938 dwarf cudweed
d Zwergruhrkraut
l *Gnaphalium supinum*

1939 dwarf euonymus
f fusain nain
d Zwergspindelstrauch
l *Euonymus nanus*

1940 dwarf ginseng
f ginseng à trois feuilles
l *Panax trifolium*

1941 dwarf glorybind
f liseron tricolore; belle-de-jour
d dreifarbige Winde; Trichterwinde
l *Convolvulus tricolor*

1942 dwarf gorse
f ajonc nain
l *Ulex nanus*

1943 dwarf iris
f iris du Taurus; iris pygmée
d Krimschwertlilie; Zwergschwertlilie
l *Iris taurica; Iris pumila*

1944 dwarf lily-turf
f herbe aux turquoises; muguet du Japon
d japanischer Schlangenbart
l *Ophiopogon japonicus*

1945 dwarf mallow; running mallow
f petite mauve; mauve à feuilles rondes
d kleine Malve; nordische Käsemalve; nordische Malve; rundblättrige Malve
l *Malva pusilla; Malva rotundifolia*

1946 dwarf masterwort
f petite astrance
d Alpenstern; Stränze
l *Astrantia minor*

1947 dwarf mistletoe
f arceuthobium
d Wacholdermistel
l *Arceuthobium; Razoumowskya*

1948 dwarf palmetto
f palmier nain
d Adansons Sabal; kleine Sabal
l *Sabal minor*

1949 dwarf serviceberry
f amélanchier à feuilles rondes
d ährige Felsenbirne
l *Amelanchier spicata*

1950 dwarf spike sedge
d kleine Sumpfbinse; kleine Sumpfsimse; kleines Sumpfried
l *Eleocharis parvula*

1951 dwarf spurge
d kleine Wolfsmilch
l *Euphorbia exigua*

1952 dwarf willowweed
d dunkelgrünes Weidenröschen
l *Epilobium obscurum*

1953 dyckia
f dyckie
d Dyckie
l *Dyckia*

1954 dyer's alkanet
f orcanette; orcanète
d Alkanna; Alkannawurzel; rote Ochsenzunge; Färberochsenzunge; Schminkwurz
l *Alkanna tinctoria*

* **dyer's camomile → 2528**

* **dyer's greenweed → 1563**

* **dyer's-oak → 644**

* **dyer's rocket → 5862**

* **dyer's saffron → 4637**

1955 dyer's sawwort
f sarrette des teinturiers
d Färberdistel; Färberscharte
l *Serratula tinctoria*

* **dyer's weld → 5862**

1956 dyer's woad
f pastel cultivé; pastel des teinturiers; herbe de Saint Philippe
d Färberwaid; deutscher Indigo
l *Isatis tinctoria*

1957 dyer's woodruff
f aspérule tinctoriale
d Färbermeier; Färbermeister
l *Asperula tinctoria*

E

1958 **eaglewood**
f aquilaire
d Adlerholzbaum; Adlerbaum
l *Aquilaria*

1959 **early bugloss**
f buglosse de Barrelier
d Barreliers Ochsenzunge
l *Anchusa barrelieri*

1960 **early coralroot**
d europäische Korallenwurz
l *Corallorrhiza trifida*

1961 **early deutzia**
f deutzie à grandes fleurs
d grossblättrige Deutzie
l *Deutzia grandiflora*

1962 **early forget-me-not**
d Hügelvergissmeinnicht; rauhes Vergissmeinnicht
l *Myosotis collina*

1963 **early hair grass**
f canche précoce; foin précoce
d frühe Haferschmiele; früher Schmielenhafer
l *Aira praecox*

1964 **early pholiota; spring agaric**
f pholiote précoce
d früher Schüppling; voreilender Schüppling
l *Agrocybe praecox; Pholiota praecox*

1965 **early red clover**
d zweischäriger Klee
l *Trifolium sativum* var. *praecox*

1966 **early winter cress**
f barbarée printanière
d Frühlingsbarbenkraut
l *Barbarea verna*

1967 **ear-pod tree**
f entérolobe
l *Enterolobium*

* **earth almond → 1230**

1968 **earth moss**
f phasque
d Bartmoos
l *Phascum*

1969 **earthstar**
f géastre
d Erdstern
l *Geaster*

1970 **earthtongue**
f géoglosse
d Erdzunge; Kolbenschwamm
l *Geoglossum*

* **Easter bell → 1971**

1971 **Easter-bell starwort; Easter bell; greater stichwort**
f stellaire holostée
d Augentrostgras; grosse Sternmiere; Jungferngras
l *Stellaria holostea*

1972 **Easter lily**
f lis à longues fleurs
d langblütige Lilie
l *Lilium longiflorum*

1973 **eastern arborvitae; American arborvitae**
f arbre de vie; arbre de Paradis; thuya du Canada; thuya d'Occident; thuya américain; cèdre blanc
d abendländischer Lebensbaum
l *Thuya occidentalis*

1974 **eastern baccharis**
f bacchante de Virginie; séneçon en arbre
l *Baccharis halimifolia*

1975 **eastern black walnut**
f noyer noir d'Amérique
d Schwarznuss; schwarze Walnuss; schwarzer Walnussbaum
l *Juglans nigra*

* **eastern burning bush → 1981**

1976 **eastern fire-berry hawthorn**
d Scharlachdorn
l *Crataegus rotundifolia; Crataegus chrysocarpa phoenicea*

* **Eastern flower → 1500**

* **eastern hemlock → 973**

1977 **eastern larch; tamarack**
f mélèze d'Amérique
d amerikanische Lärche
l *Larix americana; Larix laricina*

1978 **eastern poplar**
f peuplier de la Caroline; peuplier noir du Nouveau-Monde; peuplier de Virginie
d Kanadapappel; kanadische Pappel; Rosenkranzpappel; virginische Pappel
l *Populus deltoides*

1979 **eastern redbud**
f gaînier du Canada
d kanadischer Judasbaum
l *Cercis canadensis*

1980 **eastern red cedar**
f cèdre rouge; cèdre de Virginie; genévrier de Virginie
d Bleistift-Wacholder; rote Zeder; virginischer Sadebaum; virginischer Wacholder
l *Juniperus virginiana*

1981 **eastern wahoo; eastern burning bush**
f fusain noir
l *Euonymus atropurpureus*

1982 **eastern white pine; Weymouth pine**
f pin blanc; pin de Lord Weymouth; pin Weymouth
d Strobe; Weismutskiefer; Weymouthskiefer
l *Pinus strobus*

1983 **East Indian arrowroot; tikor**
f arrow-root des Indes orientales
d schmalblättrige Kurkume
l *Curcuma angustifolia*

1984 **East Indian rosewood**
f palissandre de l'Inde
d Javapalisander; ostindischer Rosenholzbaum
l *Dalbergia latifolia*

1985 **East Indies bluestem**
d Bartgras; Bluthirse; Hühnerfussgras
l *Andropogon ischaemum*

1986 **ebony family**
f ébénacées
d Ebenholzgewächse
l *Ebenaceae*

1987 **ebony persimon**
f ébénier
d Ebenbaum; Ebenholz; echter Ebenholzbaum
l *Diospyros ebenum*

1988 **echeveria**
f échevérie
d Escheverie
l *Echeveria*

1989 **echinacea**
f échinacée
d Igelkopf
l *Echinacea*

1990 **echinocactus**
f échinocactus
d Echinokaktus; Igelkaktus
l *Echinocactus*

1991 **echinocereus**
f échinocéréus
d Igelkerzenkaktus; Igelsäulenkaktus; Säulenkaktus
l *Echinocereus*

1992 **edelweiss**
f edelweiss
d Edelweiss
l *Leontopodium*

* **edible boletus → 1093**

1993 **edible canna; Queensland arrowroot**
f balisier comestible
d essbares Blumenrohr; essbare Canna
l *Canna edulis*

* **edible granadilla → 4070**

1994 **"edible" gyromitra; false morel**
f gyromitre comestible; moricaude; morillon
d braune Faltenmorchel; Frühlorchel; Giftlorchel; Speiselorchel; Steinmorchel; Stockmorchel
l *Gyromitra esculenta; Helvella esculenta*

1995 **edible mountain ash**
d südliche Mehlbeere
l *Sorbus mougeotti*

1996 **edible snake gourd**
f trichosanthe serpentin
d Haarblume
l *Trichosanthes anguina*

1997 **eelgras**
f zostère
d Seegras
l *Zostera*

1998 **eelgrass family**
f zostéracées

d Seegrasgewächse
l *Zosteraceae*

1999 egg-shaped twayblade
f eiförmiges Zweiblatt; grosses Zweiblatt
l *Listera ovata*

*** egg squash → 6088**

*** eglantine rose → 5359**

2000 Egyptian clover; berseem
f trèfle d'Alexandrie
d Alexandriner Klee
l *Trifolium alexandrinum*

*** Egyptian cotton → 4790**

2001 Egyptian doum palm
f palmier doum
d Doompalme; Doumpalme; Dumpalme; Pfefferkuchenbaum
l *Hyphaene thebaica*

2002 Egyptian willow
f saule d'Egypte
d ägyptische Weide
l *Salix aegyptiaca*

2003 Eichler tulip
f tulipe d'Eichler
d Eichlers Tulpe
l *Tulipa eichleri*

2004 einkorn; one-grained wheat
f engrain; locular; petit épeautre; blé locular
d Einkorn
l *Triticum monococcum*

2005 ekki lophira
f azobé; lophire à grandes feuilles
l *Lophira procera; Lophira alata procera*

2006 elaeagnus
f chalef; olivier de Bohême
d Oleaster; Ölweide
l *Elaeagnus*

2007 elaeagnus willow
f pressin
d graue Weide; hechtgraue Weide; Lavendelweide; Uferweide
l *Salix incana; Salix elaeagnus*

2008 elaeocarpus
f élaeocarpus; éléocarpe
d Ölfrucht
l *Elaeocarpus*

2009 elder
f sureau; sambéquier
d Holder; Holderbaum; Holler; Hollunder; Holunder
l *Sambucus*

2010 elecampane inula
f grande aunée; aunée officinale
d grosser Heinrich; Helenakraut; Helenenkraut
l *Inula helenium*

2011 elegant boletus
f bolet élégant; bolet jaune; cèpe jaune du mélèze
d eleganter Röhrling; Goldröhrling; goldgelber Lärchenröhrling
l *Boletus grevillei; Boletus elegans; Suillus grevillei*

2012 elephant apple
f féronier; féronie géant
d Elefantenapfelbaum
l *Feronia elephantum*

*** elephant grass → 3793**

2013 elephant's-ear; taro
f colocasie; colocase
d ägyptische Zehrwurzel; Kalo; Taro; Wasserbrotwurzel
l *Colocasia*

2014 elephant's-foot
f éléphantope; pied-d'éléphant
d Elefantenfuss
l *Elephantopus*

*** elephant's-foot → 5558**

2015 eleusine
f éleusine
d Eleusine; Kreuzgras; Fingerhirse
l *Eleusine*

2016 elfin wands
f diérame
d Trichterschwertel
l *Dierama*

2017 Ellwanger hawthorn
f aubépine d'Ellwanger
d Ellwangers Weissdorn
l *Crataegus ellwangeriana*

2018 elm
f orme
d Ilme; Rüster; Ulme
l *Ulmus*

2019 elm family
f ulmacées
d Ulmazeen; Ulmengewächse
l *Ulmaceae*

2020 elm-leaved blackberry
f ronce à feuilles d'orme
d rüsterblättrige Brombeere
l *Rubus ulmifolius*

2021 elm zelkova
f orme de Sibérie
d kaukasische Zelkowe
l *Zelkova carpinifolia*

2022 elodea; waterweed
f élodée
d Wasserpest
l *Elodea*

2023 elongated sedge
f carex élongé
d verlängertes Riedgras; verlängerte Segge
l *Carex elongata*

2024 elsholtzia
f elsholtzie
d Kamminze
l *Elsholtzia*

2025 embelia
f embélie
l *Embelia*

2026 emblic leaf flower
f phyllanthe officinal
d Ambla-Baum; Ambla-Blattblume; Amla; Mirobalanenbaum
l *Phyllanthus emblica*

2027 embothrium
f embothrie
d Prachtstrauch
l *Embothrium*

2028 emetic russula
f russule émétique; colombe rouge
d kirschroter Speitäubling
l *Russula emetica*

2029 emmer; dinkel wheat
f amidonnier; blé de Jérusalem
d Ammer; Ammelkorn; Emmer; Zweikorn; Gerstendinkel
l *Triticum dicoccum*

2030 emu plum
f owénie
l *Owenia*

2031 enchanter's nightshade
f circée
d Hexenkraut
l *Circaea*

2032 endive
f chicorée endive; endive; endivie
d Bindendivie; Endivie; Escariol; Eskariol; Winterendivie
l *Cochorium endivia*

2033 Engelmann pine
f pin d'Engelmann
d Engelmannkiefer
l *Pinus engelmanni*

2034 Engelmann spruce
f sapin de l'Arizona; épicéa d'Engelmann
d Engelmannfichte
l *Picea engelmanni*

2035 Eng gurjun oil tree
f diptérocarpe élevé
l *Dipterocarpus tuberculatus*

2036 Engler beech
f hêtre d'Engler
d Englers Buche
l *Fagus engleriana*

* **English camomile → 4546**

* **English catchfly → 2372**

2037 English daisy; garden daisy
f pâquerette vivace; petite marguerite
d ausdauerndes Gänseblümchen; mehrjähriges Gänseblümchen; Gänseblümchen; Samtröschen; Tausendschön
l *Bellis perennis*

* **English elm → 5040**

2038 English hawthorn; European hawthorn
f aubépine épineuse; épine blanche; noble épine; épine fleurie
d Baselbeere; Heckdorn; gemeiner Weissdorn; zweigriffeliger Weissdorn
l *Crataegus oxyacantha*

2039 English holly
f bois franc; épine toujours verte; houx commun; houx d'Angleterre
d Hülse; Hülsdorn; Hulst; Stechdorn; Stecheiche; gemeine Stechpalme
l *Ilex aquifolium*

2040 English ivy
f lierre d'Angleterre; lierre de bois; lierre grimpant; lierre commun; lierre d'Europe
d echter Efeu; gemeiner Efeu; kleinblättriger Efeu
l *Hedera helix*

2041 English oak
f chêne rouvre; chêne à grappes; chêne commun; chêne d'Angleterre; chêne pédonculé
d Früheiche; Raseneiche; Sommereiche; starke Eiche; Stieleiche; Weideneiche
l *Quercus robur*

2042 English primrose
f primevère à grandes fleurs; primevère vulgaire
d Erdschlüsselblume; gewöhnliche Primel; schaftlose Primel; stengellose Primel; stengellose Schlüsselblume
l *Primula vulgaris*

* **English rye grass → 4135**

2043 English scurvy grass
f cochléaria d'Angleterre
d englisches Löffelkraut
l *Cochlearia anglica*

2044 English sundew
f drosère d'Angleterre
d englischer Sonnentau; langblättriger Sonnentau
l *Drosera anglica*

2045 English yew
f if commun; if d'Angleterre
d Beereneibe; gemeiner Taxbaum; Roteibe
l *Taxus baccata*

* **ensete → 8**

2046 entada
f liane à bœuf
d Riesenhülse
l *Entada*

2047 entomophthora family
f entomophthorées
d Entomophthorazeen
l *Entomophthoraceae*

2048 epacris
f épacride
d Austrahlheide
l *Epacris*

2049 epimedium
f épimède
d Bischofsmütze; Elfenblume; Säckelblume; Sockelblume; Soskonblume
l *Epimedium*

2050 epipactis
f épipactide
d Sitter; Sumpfwurz
l *Epipactis*

2051 equestrian tricholoma
f tricholome équestre
d Grünling
l *Tricholoma equestre*

2052 erect-cone Aleppo pine; Calabrian pine
f pin des Pyrénées
d italienische Kiefer
l *Pinus halepensis brutia; Pinus brutius*

2053 ergot; ergot claviceps
f ergot
d Mutterkorn
l *Claviceps purpurea*

2054 eriogonum
f ériogone
d Wollknöterich
l *Eriogonum*

2055 eriophyllum
f ériophylle
d Teppichmargerite
l *Eriophyllum*

2056 eritrichium
f éritric; mousse d'azur
d Himmelsherold; Vergissmeinnicht
l *Eritrichium*

2057 Erman's birch
f bouleau d'Erman
d Erman's Birke
l *Betula ermani*

2058 eryngo; sea holly
f panicaut

d Edeldistel; Männertreu; Mannstreu
l *Eryngium*

2059 erythea
d Erythea
l *Erythea*

2060 escallonia
f escallone
d Eskallonie
l *Escallonia*

2061 esparto grass
d Esparto(gras); Halfa; Spartgras
l *Lygeum spartum*

*** esparto grass → 2062**

2062 esparto needle grass; esparto grass
f alfa; sparte
d Espartogras; Alfagras
l *Stipa tenacissima*

2063 Ethiopian niger seed
f ramtil
d Nigersaat; Nuck; Ramtilkraut; Gingellikraut
l *Guizotia abyssinica*

2064 Ethiopian wheat
f blé d'Abyssinie
d abessinischer Weizen
l *Triticum abyssinicum*

2065 eucalyptus; gum tree
f eucalyptus; gommier
d Eukalyptus; Fieberheilbaum; Schönmütze
l *Eucalyptus*

2066 eucharis
f eucharis
d Herzenkelch
l *Eucharis*

2067 euclidium
f euclidium
d Schnabelschötchen
l *Euclidium*

2068 eugenia
f eugénier
d Eugenie; Kirschmyrte
l *Eugenia*

*** eulalia → 4956**

2069 euonymus; evonymus; spindle tree
f fusain
d Pfaffenhütchen; Pfaffenhütlein; Pfaffenpfötchen; Pfaffenröslein; Spindelbaum; Spindelstrauch
l *Euonymus*

2070 euphorbia
f euphorbe
d Wolfsmilch
l *Euphorbia*

2071 Eurasian Solomon's-seal
f sceau de Salomon multiflore
d italienische Maiblume; vielblütige Maiblume; vielblütige Weisswurz; vielblütiges Salomonssiegel
l *Polygonatum multiflorum*

2072 European alder; European black alder; black alder; sticky alder
f aune commun; aune noir; aune glutineux
d gemeine Erle; klebrige Erle; nordische Erle; rote Erle; Roterle; Schwarzerle
l *Alnus glutinosa*

2073 European ash
f frêne commun; frêne d'Europe; frêne élevé
d Esche; Edelesche; Steinesche; Geissbaum
l *Fraxinus excelsior*

2074 European aspen; trembling aspen
f peuplier tremble; peuplier tremble d'Europe
d Aspe; Espe; Zitterpappel; Zitterespe
l *Populus tremula*

2075 European barberry
f berbéride; épine-vinette commune; vinettier
d Berbesbeere; Essigdorn; gemeine Berberitze; Heckenberberitze; Sauerdorn
l *Berberis vulgaris*

2076 European beach grass
d Dünenhafer; Sandhalm; Sandrohr; Sandschilf; (gemeiner) Sandhafer; Strandsandhalm
l *Ammophila arenaria*

2077 European beech
f hêtre blanc; hêtre commun; hêtre de bois
d Blutbuche; gemeine Buche; Rotbuche
l *Fagus sylvatica*

2078 European bird cherry; bird cherry
f bois puant; cerisier à grappes; merisier à grappes
d Ahlbaum; Ahlkirsche; Almer; Faulbaum; Faulbeerbaum; Kitschbaum; Sumpfkirsche; gemeine Traubenkirsche
l *Padus racemosa; Prunus padus*

2079 European bistort
f bistorte; renouée bistorte
d Blutkraut; Blutkrautknöterich; Drachenwurz; Krebswurz; Natter(n)wurz; Otternzunge; Wiesenknöterich
l *Polygonum bistorta; Bistorta officinalis*

* **European black alder → 2072**

2080 European blackberry
f ronce des haies; mûrier des haies; mûrier de Renard; ronce commune; ronce frutescente
d Fuchsbeere; Hirschbeere
l *Rubus fruticosus*

2081 European black currant; black currant
f cassissier; groseillier noir d'Europe
d Aalbeere; Ahlbeere; Ahlbeerstrauch; Alantbeere; Albeere; Bocksbeere; Braunbeere; Gichtbeere; Pfaffenbeere; schwarze Johannisbeere; Wanzenbeere; Wendelbeere
l *Ribes nigrum*

2082 European black hawthorn
f épine à fruits noirs; épine noire
d schwarzer Weissdorn
l *Crataegus nigra*

2083 European bladdernut
f staphylier imparipenné; faux-pistachier; nez coupé; staphylier penné
d Blasennuss; Blasenstrauch; gefiederte Klappernuss; Paternosterbaum
l *Staphylea pinnata*

2084 European blue lupin
f lupin hérissé
d haarige Lupine
l *Lupinus hirsutus*

2085 European bugleweed
f lycope d'Europe; marrube aquatique; lance du Christ
d gemeiner Wolfstrapp; Uferwolfstrapp
l *Lycopus europaeus*

2086 European chestnut; Spanish chestnut
f châtaignier commun
d Edelkastanie; Esskastanie; echte Kastanie; Keste; Kestenbaum; Maronenbaum
l *Castanea sativa*

2087 European columbine
f ancolie commune; ancolie de jardins
d gemeine Akelei; Waldakelei
l *Aquilegia vulgaris*

2088 European corn salad
f doucette; mâche; blanchette
d Feldsalat
l *Valerianella olitoria*

2089 European cotoneaster
f cotonéastre à feuilles entières
d blutrote Zwergmispel; Zwergquitte; Bergmispel; rotfrüchtige Kutte; Steinmispel
l *Cotoneaster integerrima*

2090 European cow lily
f nufar jaune
d gelbe Nixblume; gelbe Seerose; gelbe Teichrose
l *Nuphar luteum*

2091 European cranberry bush
f sureau d'eau; viorne obier
d Drosselbeere; Kalinkenbeerstrauch; gemeiner Schneeball; Wasserschneeball
l *Viburnum opulus*

2092 European cyclamen
f cyclamen d'Europe
d europäisches Alpenveilchen
l *Cyclamen europaeum*

2093 European dewberry
f ronce bleue; ronce à fruits bleus; ronce bleuâtre
d Ackerbrombeere; gemeine Kratzbeere; bereifte Brombeere; Bocksbeere
l *Rubus caesius*

2094 European dune wild rye
f élyme des sables
d blauer Helm; Strandgerste; Strandroggen; Dünenhafer; Sandhaargras; Sandroggen; Strandhafer; Waldgerste; Waldroggen
l *Elymus arenarius*

2095 European elder
f sureau commun; sureau noir (d'Europe)
d Aalohrenbaum; Aashornbaum; Alhornbaum; Baumholder; Fliederbaum;

schwarzer Holunder; Schiebikenstrauch
l *Sambucus nigra*

2096 European euonymus; European spindle tree
f bois à lardoir; bois carré; bonnet carré; bonnet de prêtre; fusain d'Europe
d deutscher Buchsbaum; europäisches Pfaffenhütchen; europäischer Spindelbaum; Hundsbaum; gemeiner Spindelstrauch; kantiger Spindelstrauch
l *Euonymus europaeus*

2097 European featherfoil
f plumeau; hottonie des marais
d Sumpfprimel; Sumpfwasserfeder
l *Hottonia palustris*

*** European feather grass → 2115**

2098 European filbert
f avelinier; coudrier; noisetier; noisetier commun
d Haselnuss; gemeine Hasel(nuss); Waldhasel(nuss)strauch
l *Corylus avellana*

2099 European fly honeysuckle
f chèvrefeuille à balais; camérisier à balais; chèvrefeuille des haies; chèvrefeuille d'Europe
d Ahlkirsche; Beinholz; Frauenholz; Hundsbeere; Hundskirsche; Knochenholz; rote Heckenkirsche; Seelenholz; Zaunkirsche
l *Lonicera xylosteum*

2100 European glorybind
f liseron des champs; vrillée; petite vrillée; bédille
d Ackerwinde; Drehwurz; Drehähre; Feldwinde; Kornwinde
l *Convolvulus arvensis*

2101 European goldenrod
f verge d'or commune
d echte Goldrute; gemeine Goldrute; Sankt-Peter-Stab; heidnisches Wundkraut
l *Solidago virgaurea*

2102 European goldilocks
f aster linière
d Goldhaar(aster)
l *Linosyris vulgaris*

2103 European gooseberry
f groseillier à maquereau
d gemeiner Stachelbeerstrauch
l *Ribes reclinatum; Ribes uva-crispa; Ribes grossularia*

2104 European grape; vine grape
f vigne
d echte Weintraube
l *Vitis vinifera*

2105 European green alder
f aune vert d'Europe
d Grünerle; Alpenerle; Bergerle; Birkenerle; Drossel; Laublatsche
l *Alnus viridis*

2106 European hackberry
f micocoulier austral; micocoulier de Provence; fabrecoulier; fanabrégié
d europäischer Zürgelbaum; südlicher Zürgelbaum
l *Celtis australis*

*** European hawthorn → 2038**

2107 European hop hornbeam
f ostryer à feuilles de charme
d europäische Hopfenbuche; gemeine Hopfenbuche; Hopfenhainbuche
l *Ostrya carpinifolia*

2108 European hornbeam
f charme; charme blanc; charme commun; charme d'Europe
d Birkenhainbuche; Buchesche; gemeine Hainbuche; Hagebuche; Heister; gemeiner Hornbaum; Jochbaum; Weissbuche
l *Carpinus betulus*

2109 European lady's-slipper
f sabot de Vénus
d braungelber Frauenschuh; rotbrauner Frauenschuh; Venusschuh
l *Cypripedium calceolus*

2110 European larch
f mélèze d'Europe
d gemeine Lärche
l *Larix decidua; Larix europaea*

2111 European linden
f tilleul commun; tilleul de Hollande
d Holland-Linde; holländische Linde; Parklinde
l *Tilia vulgaris; Tilia europaea*

2112 European meadowsweet
f reine des prés
d echtes Mädesüss; Krampfkraut;

Sumpfmädesüss
l *Filipendula ulmaria*

2113 European mistletoe; white mistletoe
f gui commun; gui blanc; bois de la Sainte Croix
d weisse Mistel; Kluster; Leimmistel
l *Viscum album*

2114 European mountain ash
f sorbier des montagnes d'Europe; sorbier des oiseleurs; arbre à grives
d Eberesche; Vogelbeerbaum; Vogelbeere; Drosselbeere; Quetsche; Quicke; Quitschbeere; Quitsche; Quitschstrauch; wilde Eberesche
l *Sorbus aucuparia*

2115 European needle grass; European feather grass
f herbe aux plumets; stipe plumeuse
d federiges Pfriemengras; Steinflachs; echtes Federgras
l *Stipa pennata*

2116 European pasqueflower
f anémone pulsatille
l *Anemone pulsatilla; Pulsatilla vulgaris*

2117 European pedicularis
f pédiculaire des marais
d Sumpfläusekraut; Sumpfrodel
l *Pedicularis palustris*

2118 European pepperwort; dittander
d Breitblattkresse; Pfefferkraut
l *Lepidium latifolium*

* **European privet → 1508**

2119 European pyrola
f pyrole à feuilles rondes
d grosses Wintergrün; rundblättriges Wintergrün
l *Pyrola rotundifolia*

* **European raspberry → 4462**

2120 European red elder
f sureau à grappes; sureau rouge; sureau rouge d'Europe
d Bergholunder; Korallenholunder; Traubenholunder; Waldholunder; roter Holunder; Hirschholunder
l *Sambucus racemosa*

2121 European sanicle; wood sanicle
f sanicle d'Europe
d (gemeiner) Sanikel
l *Sanicula europaea*

2122 European scopolia
f scopolie de Carniole
d Tollrübe
l *Scopolia carniolica*

2123 European seaside plantain; sea plantain
d Meerstrandwegerich; Strandwegerich
l *Plantago maritima*

2124 European shoreweed
d Strandling
l *Littorella uniflora*

* **European spindle tree → 2096**

2125 European starflower
d europäischer Siebenstern
l *Trientalis europaea*

2126 European stickseed; beggar's-lice
d aufrechter Igelsame; gewöhnlicher Igelsame; klettenartiger Igelsame
l *Lappula echinata*

2127 European strawberry; wild strawberry
f fraisier des bois; fraisier commun
d gemeine Erdbeere; Knickbeere; wilde Erdbeere; Walderdbeere; Worbel
l *Fragaria vesca*

* **European touch-me-not → 5560**

2128 European Turkey oak; Adriatic oak
f chêne chevelu; chêne de Bourgogne; chêne velu; chêne de Turquie; chêne lombard
d burgundische Eiche; grosse Zirneiche; österreichische Eiche; Zerreiche
l *Quercus cerris*

2129 European verbena
f verveine officinale
d echtes Eisenkraut; gebräuchliches Eisenkraut
l *Verbena officinalis*

2130 European water hemlock
f cicutaire aquatique; ciguë aquatique; ciguë vireuse
d Borstenkraut; giftiger Merk; giftiger Wasserschierling; Giftwasserschierling; Parzenkraut; Scherte; Wutschierling
l *Cicuta virosa*

2131 European white birch; weeping birch
f bouleau verruqueux; bouleau

d'Europe
d Sandbirke; Weissbirke; hängende Weissbirke; hängende Sandbirke; Hängebirke; Trauerbirke; Warzenbirke
l *Betula pendula; Betula verrucosa*

* **European white elm → 4614**

2132 European white water lily; white water lily
f nénuphar blanc; lis des étangs
d Schwanenblume; Seelilie; weisse Seerose; Teichrose; Wasserlilie; Wasserrose
l *Nymphaea alba*

2133 European wild ginger
f asaret d'Europe
d braune Haselwurz; gemeine Haselwurz; Leberkraut; wilde Narde
l *Asarum europaeum*

2134 European wood anemone
f anémone des bois
d Buschwindröschen; Buschveilchen; Aprilblume; weisse Osterblume
l *Anemone nemorosa*

2135 European yellow lupin
f lupin jaune
d gelbe Lupine; Gelblupine
l *Lupinus luteus*

2136 European yellow violet
f grande pensée jaune
d gelbes Veilchen
l *Viola lutea*

2137 eurycoma
f eurycome
l *Eurycoma*

2138 euterpe palm
f euterpe
d Kohlpalme
l *Euterpe*

2139 evax
f évax
l *Evax*

2140 evening campion
f compagnon blanc; lychnide du soir
d Marienröschen; weisse Lichtnelke
l *Lychnis alba*

2141 evening primrose; sundrops
f œnothère; onagre
d Nachtkerze
l *Onagra; Oenothera*

2142 evening-primrose family
f onagracées
d Nachtkerzengewächse
l *Onagraceae*

2143 everblooming Chinese rose
f rosier de Bengale; rosier toujours fleuri
l *Rosa chinensis* var. *semperflorens*

* **evergreen buckthorn → 2974**

2144 evergreen candytuft
f corbeille d'argent; thlaspi vivace
d Schneeflocke; immergrüner Bauernsenf
l *Iberis sempervirens*

2145 evergreen chinkapin; oat chestnut
f castanopsis
d Eschenkastanie; Scheinkastanie
l *Castanopsis*

2146 evergreen euonymus
f fusain du Japon; fusain vert
d japanischer Spindelstrauch
l *Euonymus japonicus*

2147 evergreen rose
f rosier toujours vert
d immergrüne Rose
l *Rosa sempervirens*

2148 evergreen stonecrop
f orpin intermédiaire
d Bastardfetthenne
l *Sedum hybridum*

2149 everlasting
f immortelle
d Immerschön; Sonnengold; Strohblume; Trockenblume
l *Helichrysum*

* **everlasting pea → 4098**

2150 evernia
f évernie
d Bandflechte
l *Evernia*

2151 evodia
f évodie
d Stinkesche
l *Evodia*

* **evonymus → 2069**

2152 eyebright
f euphraise
d Augentrost
l *Euphrasia*

2153 eye poppy
f pavot argémone
d Sandmohn
l *Papaver argemone*

F

2154 fairy-ring marasmius; fairy-ring mushroom
f marasme d'Oréade; mousseron d'automne
d Feldschwindling; Herbstmusseron; Nelkenschwamm; Nelkenschwindling; Krösling; Nägeleinspilz; Suppenpilz
l *Marasmius oreades*

2155 falcaria
f faucillère
d Sichelmöhre
l *Falcaria*

2156 falcate hare's-ear
f buplèvre à feuilles en faux
d Sichelhasenohr
l *Bupleurum falcatum*

2157 Falconer's mock orange
f seringa de Falconer
d Falconers Pfeifenstrauch
l *Philadelphus falconeri*

2158 fall daffodil
f amaryllis jaune
d gelbe Sternbergie
l *Sternbergia lutea*

*** fall dandelion → 2159**

2159 fall hawkbit; fall dandelion
d Herbstlöwenzahn
l *Leontodon autumnalis*

*** false acacia → 636**

2160 false alyssum
f bertéroa
d Graukresse
l *Berteroa*

2161 false arborvitae
f hiba
d Hibalebensbaum
l *Thujopsis*

2162 false asphodel
f tofieldie
d Liliensimse; Simsenlilie; Tofieldie
l *Tofieldia*

2163 false beech
f nothofagus
d Scheinbuche
l *Nothofagus*

2164 false brome; false brome grass
f brachypode
d Zwenke
l *Brachypodium*

2165 false cantharellus; pseudochanterelle
f clitocybe orange
d falscher Eierschwamm
l *Clitocybe aurantiaca; Cantharellus aurantiacus*

2166 false carrot
f caucalier
d Haftdolde
l *Caucalis*

2167 false cypress
f petit cyprès; faux cyprès
d Lebensbaumzypresse; Scheinzypresse; Weisszeder
l *Chamaecyparis*

2168 false flax
f caméline
d Dotter; Leindotter; Öldotter; Rüllsaat; Schmalz
l *Camelina*

2169 false freesia
f lapeyrousie
d Lapeyrousie
l *Lapeyrousia; Lapeirousia*

2170 false hellebore
f vératre
d Germer; Nieswurz
l *Veratrum*

2171 false indigo
f faux indigo; indigo bâtard; amorphe
d Bastardindigo
l *Amorpha*

*** false indigo → 2948**

2172 false ipecac
f gillénie
d Gillenie
l *Gillenia; Porteranthus*

2173 false mallow
f malvastrum
d Scheinmalve
l *Malvastrum*

*** false morel → 1994**

2174 false nettle
f ramie

d Chinagras; Chinanessel
l *Boehmeria*

2175 false nutmeg
d unechter Muskat
l *Myristica fatua*

2176 false oat
f trisétaire
d Goldhafer(gras); Grannenhafer
l *Trisetum*

2177 false olive
f éléodendre
d Ölstrauch
l *Elaeodendron*

2178 false pennyroyal
f hédéome
d Hedeoma
l *Hedeoma*

2179 false pimpernel
f lindernia; centenille
d Büchsenkraut
l *Lindernia*

* **false saffron → 4636**

* **false Solomon's-seal → 5081**

2180 false spirea
f sorbaire
d Fiederspiere
l *Sorbaria*

2181 false tamarisk
f myricaire
d Rispenstrauch; Tamariske
l *Myricaria*

2182 fan maidenhair
f capillaire tendre
d zarter Lappenfarn
l *Adiantum tenerum*

2183 fan palm
f livistone
d Livistone
l *Livistona*

2184 fanwort
f cabombe
d Haarnixe; Fischgras; Wasserhaarnixe
l *Cabomba*

2185 farsetia
f farsétia
d Turra
l *Farsetia*

2186 fat hen
d ausgebreitete Melde; gemeine Melde; Rutenmelde; spreizende Melde
l *Atriplex patula*

2187 fatsia
f fatsie
d Fatsie
l *Fatsia*

2188 faverel
f érophile
d Hungerblümchen; Hungerblume
l *Erophila*

2189 fawn agaric; fawn-colored pluteus
f plutée couleur de cerf
d rehbrauner Dachpilz; brauner Dachpilz
l *Pluteus cervinus; Agaricus cervinus*

2190 fawn lily
f érythrone
d Zahnlilie
l *Erythronium*

2191 Faxon fir
f sapin de Faxon
d Faxons Tanne
l *Abies faxoniana*

2192 feather cockscomb
f célosie argentée
d Hahnenkamm; silberiger Brandschopf
l *Celosia argentea*

2193 featherfoil
f hottonie
d Wasserfeder; Wasserprimel
l *Hottonia*

* **feather grass → 3819**

2194 February daphne
f bois d'oreille; bois gentil; bois joli; lauréole femelle; daphné bois-joli; daphné bois-gentil
d gemeiner Kellerhals; gemeiner Seidelbast; wilder Pfefferstrauch; Zeiland
l *Daphne mezereum*

2195 feijoa
f feijoa
d Feijoa
l *Feijoa sellowiana*

2196 felicia
f félicie
d Felicie
l *Felicia*

2197 felt fern
f cyclophore
l *Cyclophorus*

2198 felwort
f pleurogyne
d Saumnarbe
l *Pleurogyna*

2199 Fendler bush
f fendlera
d Fendlere; Felsenbirne
l *Fendlera*

2200 fennel
f fenouil
d Fenchel
l *Foeniculum*

2201 fennelflower
f nigelle
d Schwarzkümmel
l *Nigella*

2202 fennel-leaved pondweed
d kammförmiges Laichkraut; Kammlaichkraut
l *Potamogeton pectinatus*

* **fenugreek → 2203**

2203 fenugreek trigonella; fenugreek
f fenugrec; sénegré
d gelber Bockshornklee; gelblicher Bockshornklee; griechisches Heu; Siebenzeit; Ziegenklee
l *Trigonella foenum-graecum*

2204 fern asparagus
f asperge plumeuse
d Federspargel
l *Asparagus plumosus*

2205 fern cycad
f stangéria
d Stangerie
l *Stangeria*

2206 ferns
f fougères
d Farne; echte Farne
l *Filicineae*

2207 ferulago
f férulage
d Birkenwurz
l *Ferulago*

2208 fescue
f fétuque
d Schwingel; Schwingelgras
l *Festuca*

2209 fetid russula
f russule fétide
d Stinktäubling
l *Russula foetens*

2210 feverfew; feverfew chrysanthemum
f chrysanthème-matricaire; grande camomille; malherbe
d Bertram(wurz); falsche Kamille; Mutterkamille; Mutterkraut; Fieberkraut
l *Chrysanthemum parthenium; Pyrethrum parthenium*

2211 fever vine
f pédérie
d Stinkknackbeere
l *Paederia*

* **feverwort → 2867**

2212 few-flowered sedge
d armblütiges Riedgras
l *Carex pauciflora*

2213 few-flowered spike sedge
d armblütige Sumpfbinse; armblütige Sumpfsimse; armblütiges Sumpfried; wenigblütiges Sumpfried
l *Eleocharis pauciflora*

2214 fiber flax; common flax; low flax
f lin à fibre longue; lin cultivée
d Faserlein; Flachs; echter Lein; gebauter Lein; Saatlein; Steppenflachs; Dresch-Lein
l *Linum usitatissimum*

2215 fiber lily
f phormier
d Flachslilie; neuseeländischer Flachs
l *Phormium*

2216 fiddle-leaved dock
d schöner Ampfer
l *Rumex pulcher*

2217 fiddle-neck
f amsinckie

d Amsinckie
l *Amsinckia*

2218 fiddlewood
f citharexylon; bois de guitare
d Geigenholz; Leierholz; Geigenholzbaum
l *Citharexylum*

* **field balm → 1429**

2219 field brome; field brome gras
f brome des champs
d Ackertrespe
l *Bromus arvensis*

2220 field camomile
f anthémis des champs
d Ackerhundskamille; falsche Kamille
l *Anthemis arvensis*

2221 field cerastium
d gemeines Hornkraut; rasiges Hornkraut
l *Cerastium caespitosum*

* **field elm → 5040**

2222 field forget-me-not
f myosotis des champs
d Ackervergissmeinnicht
l *Myosotis arvensis*

2223 field garlic; crow garlic
f ail des vignes
d Hundslauch; Weinberglauch
l *Allium vineale*

* **field garlic → 4284**

2224 field hawkweed
f épervière des prés
d Wiesenhabichtskraut
l *Hieracium pratense*

2225 field horsetail
f prêle des champs; queue-de-rat; queue-de-cheval
d Ackerschachtelhalm; Aschenkannenkraut; Kannenkraut; Pferdeschwanz; Scheuerkraut; Zinnkraut
l *Equisetum arvense*

2226 field lady's-mantle; parsley piert
f alchémille des champs
d Ackerfrauenmantel
l *Alchemilla arvensis*

* **field larkspur → 2330**

2227 field madder
f shérardie
d Ackerröte
l *Sherardia*

2228 field madder
f shérardie des champs
d Ackerröte
l *Sherardia arvensis*

2229 field mint
f menthe des champs
d Ackerminze
l *Mentha arvensis*

2230 field-nettle betony; field woundwort
f épiaire des champs
d Ackerziest
l *Stachys arvensis*

2231 field pea
f pois de pigeon
d Ackererbse; Felderbse; Futtererbse; Pel(l)uschke; wilde Erbse
l *Pisum arvense; Pisum sativum arvense*

2232 field pennycress
f monnoyère
d Ackerhellerkraut; Ackertäschelkraut; Bauernkresse; Bauernsenf; Hellerkraut; Klapper; Pfennigkraut; Täschelkraut
l *Thlaspi arvense*

2233 field pepperweed
f passerage champêtre
d Feldkresse
l *Lepidium campestre*

* **field poppy → 1616**

2234 field rose
f rosier des champs
d Ackerrose; Feldrose; grosse Hundsrose; Hundsrose; kriechende Rose
l *Rosa arvensis*

2235 field scabious
f oreille d'âne; scabieuse des champs
d Ackerknautie; Ackerskabiose; Ackerwitwenblume
l *Knautia arvensis; Scabiosa arvensis*

2236 field sow thistle
f laiteron des champs
d Ackergänsedistel; Ackersaudistel; Milchdistel
l *Sonchus arvensis*

2237 field speedwell
f véronique agreste; véronique rustique
d Ackerehrenpreis
l *Veronica agrestis*

2238 field violet
f violette des champs
d Ackerstiefmütterchen; Ackerveilchen; Feldveilchen
l *Viola arvensis*

2239 field wood rush
f luzule des champs
d Feldsimse; gemeine Hainbinse; Hasenbrot
l *Luzula campestris*

*** field wormwood → 4641**

*** field woundwort → 2230**

2240 fig
f figuier
d Feigenbaum
l *Ficus*

2241 figroot buttercup; buttercup ficaria
f éclairette; ficaire
d Eppich; Feigenwurzel; Feigwarzenkraut; Feigwurz; Scharbockskraut
l *Ficaria verna; Ranunculus ficaria*

2242 figwort
f scrofulaire
d Braunwurz; Skrofelkraut
l *Scrophularia*

2243 figwort family
f scrofulariacées
d Braunwurzgewächse; Rachenblütler; Skrofulariazeen
l *Scrophulariaceae*

2244 Fiji arrowroot; Fiji arrowroot tacca; pia
f pia; arrow-root de Tahiti
d Tahitipfeilwurz; fiederspaltige Takka
l *Tacca pinnatifida*

2245 filamentous fungi
f hyphomycètes
d Hyphomyzeteen
l *Hyphomycetes*

2246 filbert; hazel
f noisetier; avelinier; coudrier
d gemeine Hasel(nuss); Waldhasel(nuss)strauch
l *Corylus*

2247 filmy fern
f trichomane
d Haarfarn; Hautfarn
l *Trichomanes*

2248 filmy fern
f hyménophyllée
d Hautfarn; Krullfarn
l *Hymenophyllum*

2249 filmy-fern family
f hyménophyllées
d Hautfarne; Hymenophylazeen
l *Hymenophyllaceae*

2250 fine-leaved clover
d schmalblättriger Klee
l *Trifolium angustifolium*

2251 fine-leaved water dropwort
f œnanthe aquatique
d Rossfenchel; Wasserfenchel; Wasserkerbel; Wasserrebendolde
l *Oenanthe aquatica; Oenanthe phellandrium*

2252 fingered sedge
f laîche digitée
d Fingerriedgras; Fingersegge; gefingerte Segge
l *Carex digitata*

2253 fingered speedwell
f véronique digitée
d dreiteiliger Ehrenpreis; Fingerehrenpreis
l *Veronica triphyllos*

2254 finger grass; crab grass
f digitaria
d Fingerhirse
l *Digitaria*

*** finger grass → 5982**

2255 Finland pink
f œillet des sables
d Sandnelke
l *Dianthus arenarius*

2256 Finnish spruce
f épicéa de Finlande
d finnische Fichte
l *Picea fennica*

*** fiorin → 1665, 4469**

2257 fir
f sapin

d Tanne
l *Abies*

2258 fir club moss
d Tannenbärlapp
l *Lycopodium selago*

2259 fire chalice
f zauschnérie
d Kolibritrompete; Zauschnerie
l *Zauschneria*

2260 fire moss
f cassiopée tétragone
d vierkantige Heide
l *Cassiope tetragona*

2261 fire thorn
f arbre de Moïse; buisson ardent
d Feuerdorn
l *Pyracantha*

2262 fireweed; great willow herb
f antoinette; nériette; épilobe en épi; laurier de Saint-Antoine
d Antoniuskraut; Feuerkraut; Staudenkraut; Waldweidenröschen
l *Chamaenerium angustifolium; Epilobium angustifolium*

2263 fire-wheel tree
f sténocarpe
d Schmalfrucht
l *Stenocarpus*

2264 fishberry
f anamirte
d Scheinmyrte
l *Anamirta*

2265 fishtail palm
f caryote
d Brennpalme; Caryotapalme; Nusspalme
l *Caryota*

2266 fission fungi
f schizomycètes
d Schizomyzeteen; Spaltpilze
l *Schizomycetes*

2267 Fitzroya
f alerce; fitzroya
d Fitzroya
l *Fitzroya*

2268 five-leaf sumac
f sumac thézéra; sumac à 5 folioles
d fünfblättriger Sumach
l *Rhus pentaphylla*

2269 five-stamen tamarisk
f tamaris à 5 étamines
d Sommertamariske
l *Tamarix pentandra*

2270 flacourtia
f flacourtie
l *Flacourtia*

2271 flaky fir
f sapin écailleux
l *Abies squamata*

2272 flame adonis
d brennende Adonisblume; Ackerröschen; feuerrotes Adonisröschen; brennendes Teufelsauge
l *Adonis flammea*

2273 flame anemone
f anémone éclatante
l *Anemone fulgens*

2274 flame (bottle) tree
f sterculie à feuilles d'érable
l *Sterculia acerifolia; Brachychiton acerifolium*

2275 flame-leaf sumac
f sumac brillant; sumac copal
l *Rhus copallina*

* **flame nettle → 1325**

2276 flannel mullein; great mullein; Aaron's-rod
f bouillon-blanc
d echte Königskerze; kleinblättrige Königskerze
l *Verbascum thapsus*

2277 flask fungi
f pyrénomycètes
d Kernpilze; Pyrenomyzeteen
l *Pyrenomycetes*

2278 flat pea vine
f gesse des bois
d Waldplatterbse; wilde Platterbse
l *Lathyrus sylvestris*

2279 flat-pod pea vine
f gesse chiche
l *Lathyrus cicera*

2280 flat sedge; cypress grass
f souchet
d Cypergras; Zypergras
l *Cyperus*

2281 flat Siebold walnut; heartnut
f noyer cordiforme
d herzförmige Walnuss; herzförmiger Walnussbaum
l *Juglans sieboldiana cordiformis; Juglans cordiformis*

2282 flat-spine prickly ash
f clavalier de Bunge
d chinesischer Pfeffer
l *Zanthoxylum bungei; Zanthoxylum simullans*

2283 flat-stemmed pondweed
d flachstengeliges Laichkraut
l *Potamogeton zosteriformis*

2284 flattened blysmus
d flaches Quellried
l *Blysmus compressus*

2285 flaveria
f flavérie
l *Flaveria*

2286 flax
f lin
d Flachs; Lein
l *Linum*

2287 flax dodder
f cuscute du thym
d Flachsseide; Leinseide
l *Cuscuta epilinum*

2288 flax family
f linacées
d Flachsgewächse; Leingewächse
l *Linaceae*

2289 flaxseed
f radiole
d Zwergflachs
l *Radiola*

2290 flaxseed plantain; psyllium seed
f plantain psyllium
d Wegerich
l *Plantago psyllium*

2291 fleabane
f pulicaire
d Flohkraut
l *Pulicaria*

2292 fleabane
f érigéron; vergerette
d Beruf(s)kraut; Beschreikraut; Dürrkraut; Flohkraut; Feinstrahl
l *Erigeron; Stenactis*

2293 flea sedge
f laîche pulicaire
d Flohriedgras; Flohsegge
l *Carex pulicaris*

2294 flindersia
f flindersie
l *Flindersia*

2295 flint corn
f maïs corné
d Hartmais; Steinmais
l *Zea mays* convar. *vulgaris*

*** flixweed → 2296**

2296 flixweed tansy mustard; flixweed; hedge mustard
f sagesse des chirurgiens
d Besenrauke; gemeine Sophienrauke; Sophienrauke
l *Descurainia sophia*

2297 floating heart
f faux nénuphar; nympheau
d Seekanne
l *Nymphoides*

2298 floating leaf pondweed
d schwimmendes Laichkraut
l *Potamogeton natans*

*** floating moss → 4671**

2299 floating water plantain
d schwimmendes Froschkraut; flutendes Froschkraut
l *Alisma natans; Elisma natans*

2300 Florence fennel
f fenouil de Florence
d Bologneserfenchel; Pariser Anis; Zwiebelfenchel
l *Foeniculum dulce*

2301 Florentine iris
f iris de Florence
d florentinische Schwertlilie
l *Iris florentina*

2302 Florentine tulip
f tulipe sauvage

d wilde Tulpe
l *Tulipa sylvestris*

2303 Florida torreya; stinking cedar
f cèdre puant; torreya à feuilles d'if
d eibenblättrige Torreye
l *Torreya taxifolia*

2304 floripondio datura; angel's-trumpet
f stramoine en arbre
d Engelstrompete
l *Datura arborea*

2305 florists' larkspur; mongrel larkspur
f pied-d'alouette vivace hybride; dauphinelle de Tauride
d Bastardrittersporn; Gartenrittersporn
l *Delphinium cultorum; Delphinium hybridum*

2306 florists' pyrethrum; Caucasian pyrethrum
f pyrèthre rose
d kaukasische Insektenblume; persische Insektenblume
l *Chrysanthemum coccineum; Chrysanthemum roseum; Pyrethrum carneum*

*** flossflower → 60**

2307 flour corn; soft corn
f maïs tendre; maïs farineux
d Stärkemais; Weichmais
l *Zea mays* convar. *amylacea*

*** flower-cup fern → 157**

2308 flowering ash; manna ash
f frêne à fleurs; frêne à manne; orne d'Europe
d Blumenesche; Mannaesche
l *Fraxinus ornus*

2309 flowering dogwood
f bois de chien; cornouiller à grandes fleurs
d Blumenhartriegel
l *Cornus florida; Cynoxylon floridum*

*** flowering maple → 7**

2310 flowering plants
f phanérogames
d Blütenpflanzen
l *Anthophyta; Phanerogamae*

2311 flowering plum
f prunier trilobé; amandier de Chine
d dreilappige Mandel; dreilappiger Pfirsichstrauch; Mandelbäumchen
l *Amygdalus triloba; Prunus triloba*

2312 flowering quince
f chaenomeles; cognassier du Japon
d Scheinquitte
l *Chaenomeles*

2313 flowering rush
f butome
d Blumenbinse; Schwanenblume
l *Butomus*

2314 flowering rush
f jonc fleuri
d doldige Blumenbinse; doldige Schwanenblume; Schwanenkraut; Wasserliesch
l *Butomus umbellatus*

2315 flowering-rush family
f butomacées
d Blumenlieschgewächse; Schwanenblumengewächse; Wasserlieschgewächse
l *Butomaceae*

2316 flower of an hour
f fleur d'une heure
d Stundenblume; Stundeneibisch; Wetterrösel
l *Hibiscus trionum*

2317 fluellin
f kickxie
d Tännelkraut
l *Kickxia*

2318 fluffweed; cudweed
f cotonnière; filage
d Filzkraut; Schimmelkraut
l *Filago*

2319 fly agaric; fly amanita
f amanite tue-mouches; fausse oronge
d Fliegenpilz; Fliegenblätterling; Fliegenschwamm; roter Fliegenpilz
l *Amanita muscaria*

2320 fly orchid
f ophrys mouche
d Fliegenorchis; Fliegenragwurz
l *Ophrys muscifera*

2321 foamflower
f tiarelle
d Schaumblüte; Turbenkapsel
l *Tiarella*

2322 fodder beet
f betterave fourragère
d Futterrübe
l *Beta vulgaris* var. *macrorrhiza*

* **foliage beet → 3473**

2323 fontanesia
f fontanésie
d Fontanesie
l *Fontanesia*

2324 food pangium
f pang(i)e
d Pangibaum; Pitjungbaum; Samaunbaum
l *Pangium edule*

* **fool's-parsley → 2325**

2325 fool's-parsley aethusa; fool's-parsley
f petite ciguë
d Gartenschierling; Hundspetersilie
l *Aethusa cynapium*

2326 forest mushroom; wood agaric
f agaric des forêts; agaric des bois
d kleiner Blutegerling; kleiner Waldegerling; Waldchampignon; Waldedelpilz
l *Agaricus silvaticus; Psalliota silvatica*

2327 forget-me-not
f myosotis; ne-m'oublie(z)-pas
d Leuchte; Mäuseohr; Vergissmeinnicht
l *Myosotis*

* **forked catchfly → 2328**

2328 forked silene; forked catchfly
d gabelästiges Leimkraut; gabeliges Leimkraut
l *Silene dichotoma*

2329 forked spleenwort
f doradille septentrionale
d nördlicher Streifenfarn; schmaler Streifenfarn
l *Asplenium septentrionale*

2330 forking larkspur; branching larkspur; field larkspur
f éperon de chevalier; dauphinelle consoude; dauphinelle des blés; dauphinelle des champs; pied-d'alouette des champs
d Feldrittersporn; Krähenfuss; Ackerrittersporn; Hornkümmel
l *Delphinium consolida*

2331 forkmoss
f dicrane
d Gabelzahnmoos
l *Dicranum*

2332 Formosa juniper
f genévrier de Formose
l *Juniperus formosana*

2333 Forrest's birch
f bouleau de Forrest
d Forrests Birke
l *Betula forresti*

2334 Forrest's day lily; yellow day lily; tawny day lily
f lis jaune; hémérocalle jaune; hémérocalle fauve
d rotgelbe Taglilie; gelbe Taglilie
l *Hemerocallis fulva*

2335 forsteronia
f forstéronie
l *Forsteronia*

2336 Forster's wood rush
f luzule de Forster
d schlanke Hainsimse; schlanke Simse
l *Luzula forsteri*

2337 forsythia
f forsythie
d Forsythie; Goldflöckchen
l *Forsythia*

2338 Fortune keteleeria
f kételéerie de Fortune
d Fortunes Tanne
l *Keteleeria fortunei*

2339 Fortune's windmill palm
f palmier de Chine; palmier chanvre
d Hanfpalme
l *Trachycarpus fortunei; Trachycarpus excelsus*

2340 fountain humea
f humée élégante
l *Humea elegans*

2341 four-o'clock; marvel-of-Peru
f belle-de-nuit; merveille du Pérou; herbe triste
d Wunderblume; Vieruhrblume
l *Mirabilis*

2342 four-o'clock family
f nyctanginacées

d Wunderblumengewächse
l *Nyctanginaceae*

2343 four-seeded vetch; sparrow vetch
f ers tétrasperme; lentille tétrasperme
d Fadenwicke; viersamige Wicke
l *Vicia tetrasperma*

2344 four-stamen tamarisk
f tamaris à quatre étamines
d viermännige Tamariske
l *Tamarix tetrandra*

2345 fowl bluegrass; fowl meadow grass
f pâturin des marais
d spätes Rispengras; Spätrispe; Sumpfrispengras; fruchtbares Rispengras
l *Poa palustris*

2346 foxglove
f digitale
d Fingerhut
l *Digitalis*

2347 fox grape
f vigne Isabelle
d Catawbarebe; Fuchstraube; Isabellarebe
l *Vitis labrusca*

2348 fox sedge
f carex vulpin; laîche compacte
d Fuchsriedgras; Fuchssegge
l *Carex vulpina*

2349 foxtail
f vulpin
d Fuchsschwanz(gras)
l *Alopecurus*

2350 foxtail barley; squirrel-tail grass
f orge à crinière
d Mähnengerste
l *Hordeum jubatum*

2351 foxtail clover
f trèfle rougeâtre
d Fuchsklee; Purpurklee
l *Trifolium rubens*

2352 foxtail millet; Italian millet
f millet des oiseaux; sétaire d'Italie
d Kolbenhirse; italienische Hirse; Borstenhirse; Vogelhirse; welscher Fennich; Mohar(hirse); Vogelkolbenhirse
l *Setaria italica*

2353 foxtail pine
f queue-de-chien; pin queue de Renard
d Balfours Kiefer
l *Pinus balfouriana*

2354 fragile-shell almond
f amandier à coque tendre
d Krachmandelbaum; Knackmandelbaum
l *Amygdalus communis* var. *fragilis; Prunus amygdalus* var. *fragilis*

2355 fragrant agrimony
f aigremoine odorante
d Duftodermennig; grosser Odermennig
l *Agrimonia odorata*

2356 fragrant albizzia
f arbre de Kala
l *Albizzia odoratissima*

2357 fragrant dracaena
f dragonnier balsamique
d Balsamdrachenbaum; wohlriechende Dracaena
l *Dracaena fragrans*

2358 fragrant sumac; aromatic sumac
f sumac odorant
l *Rhus aromatica*

2359 fragrant thimbleberry; purple-flowering raspberry
f ronce odorante; ronce du Canada; framboisier du Canada
d wohlriechende Himbeere; Zimtbrombeere; Zimthimbeere
l *Rubus odoratus*

2360 frangipani
f frangipanier
d Frangipani; Plumeria
l *Plum(i)er(i)a*

2361 frankenia; sea heath
f franquène; frankénia; frankénie
d Seehaide
l *Frankenia*

2362 frankenia family
f frankéniacées
d Frankeniazeen
l *Frankeniaceae*

2363 frankincense; incense tree
f boswellie
d Weihrauchbaum
l *Boswellia*

2364 Fraser balsam fir
f sapin baumier du Sud des Etats-Unis; sapin de Fraser
d Frasers Tanne
l *Abies fraseri*

*** fraxinella → 2435**

2365 freesia
f fréesie
d Freesie; Kapmaiblume
l *Freesia*

*** Fremont cottonwood → 2366**

2366 Fremont poplar; Fremont cottonwood
f peuplier de Frémont
d Fremonts Pappel
l *Populus fremonti*

2367 French flax
d französischer Lein
l *Linum gallicum*

2368 French honeysuckle
f sainfoin des Alpes
d Alpensüssklee; Hahnenkamm
l *Hedysarum obscurum; Hedysarum hedysaroides*

*** French honeysuckle → 5390**

2369 French marigold; spreading marigold
f petit œillet d'Inde
d ausgebreitete Studentenblume
l *Tagetes patula*

2370 French mulberry; beauty-berry
f callicarpe
d Perlenbeerstrauch; Schönbeere
l *Callicarpa*

2371 French rose
f rosier de Provins; rose de France; rose rouge
d Apothekerrose; Essigrose; französische Rose; Provencerrose; Provinzrose; Zuckerrose
l *Rosa gallica*

2372 French silene; English catchfly
f silène de France; silène à cinq plaies
d englisches Leimkraut; französisches Leimkraut
l *Silene gallica; Silene anglica*

2373 French sorrel
f oseille ronde
d Gewürzampfer; Schildampfer
l *Rumex scutatus*

*** French sumac → 3782**

2374 French tamarisk
f tamaris commun; tamaris de France
d französische Tamariske; gallische Tamariske; gemeine Tamariske; südeuropäische Tamariske
l *Tamarix gallica*

2375 fresh-water algae
f nostocacées
d Nostokazeen
l *Nostocaceae*

2376 Fries's pondweed
f potamot de Fries
d stachelspitziges Laichkraut
l *Potamogeton friesi*

2377 fringe bell
f schizocode
d Spaltglocke
l *Schizocodon*

2378 fringed water lily
f petit-nénuphar commun
d radblättrige Seekanne
l *Nymphoides peltatum; Limnanthemum nymphaeoides*

2379 fringe tree
f arbre de neige
d Schneebeere; Schneebaum; Schneeflockenblume; Schneeflockenbaum; Schneeflockenstrauch
l *Chionanthus*

2380 fritillary
f fritillaire
d Fritillarie; Kaiserkrone; Kiebitzblume; Kronenblume
l *Fritillaria*

2381 frogbit
f hydrocharide
d Froschbiss
l *Hydrocharis*

2382 frogbit
f morène; mors de grenouille
d schwimmender Froschbiss
l *Hydrocharis morsus-ranae*

2383 frogbit family
f hydrocharidacées
d Froschbissgewächse;

Hydrocharideen; Hydrocharitazeen
l *Hydrocharitaceae*

2384 frog orchid
d grüne Hohlzunge
l *Coeloglossum viride; Habenaria viridis*

2385 frondose polyporus; hen of the woods
f polypore en touffe
l *Polyporus frondosus*

2386 frost grape; chicken grape
f vigne foxée; vigne odorante
d Duftrebe; Ufer(wein)rebe
l *Vitis vulpina; Vitis cordifolia*

2387 frostwort
f crocanthème
d Sonnenröschen
l *Crocanthemum*

2388 Fuchs groundsel
f séneçon de Fuchs
d Fuchskreuzkraut
l *Senecio fuchsi*

2389 fuchsia
f fuchsia
d Fuchsie
l *Fuchsia*

2390 fuller's teasel
f chardon à foulon; chardon à bonnetier
d Rauhkarde; Tuchkarde; Walkerdistel; Weberdistel; Weberkarde
l *Dipsacus fullonum*

2391 full-moon maple
f érable du Japon
d japanischer Ahorn
l *Acer japonicum*

2392 fumitory
f fumeterre
d Erdrauch; Erdraute
l *Fumaria*

2393 fumitory family
f fumariacées
d Erdrauchgewächse; Fumariazeen
l *Fumariaceae*

2394 funaria
f funaire
d Drehmoos; Wettermoos
l *Funaria*

* **fungus → 3758**

2395 funnel-shaped clithocybe
f clitocybe en entonnoir
d gebuckelter Trichterling; gelbbräunlicher Trichterling
l *Clitocybe infundibuliformis*

* **furze → 2566**

2396 fustic tree
f chlorophora
l *Chlorophora*

2397 fuzzy deutzia
f deutzie scabre; deutzie à feuilles crénelées
d hohe Deutzie; rauhe Deutzie
l *Deutzia scabra*

G

2398 **Gabon symphonia**
f symphonie du Gabon
l *Symphonia gabonensis*

2399 **gagea**
f gagée
d Gelbstern
l *Gagea*

2400 **gaillardia**
f gaillarde
d Gaillardie; Kokardenblume; Papageiblume
l *Gaillardia*

2401 **galax**
f galax
d Bronzenblatt; Milchblume
l *Galax*

* **galingale → 2609**

2402 **galingale flat sedge; long cyperus**
f souchet odorant
d langes Zypergras
l *Cyperus longus*

2403 **galipea**
f galipée
d Galipee
l *Galipea*

2404 **gama grass**
f tripsac
d Gamagras
l *Tripsacum*

2405 **garcinia**
f garcinie
d Garcinia
l *Garcinia*

2406 **garden angelica; cultivated angelica**
f angélique officinale; archangélique; herbe du Saint-Esprit
d Engelwurz; Erzengelwurz; echte Engelwurz; Angelika(wurzel); Brustwurzel; Gartenengelwurz; Heiliggeistwurzel; Theriakwurzel
l *Angelica archangelica; Archangelica officinalis*

2407 **garden balsam**
f balsamine des jardins
d Gartenbalsamine; Gartenspringkraut
l *Impatiens balsamina*

2408 **garden beet**
f betterave potagère; betterave rouge
d rote Rübe; Speiserübe
l *Beta vulgaris* ssp. *esculenta*

2409 **garden burnet; great burnet**
f grande pimprenelle; sanguisorbe
d Blutkraut; grosser Wiesenknopf; Sperberkraut; Wiesenbibernell
l *Sanguisorba officinalis*

2410 **garden carrot**
f carotte; racine jaune
d Gartenmöhre; gelbe Rübe; Karotte; Mohrrübe; Speisemöhre
l *Daucus carota* var. *sativus*

2411 **garden celery**
f céleri à côtes
d Bleichsellerie; Stengelsellerie
l *Apium graveolens* var. *dulce*

* **garden cress → 2412**

2412 **garden-cress peppergrass; garden cress**
f cresson alénois
d Gartenkresse
l *Lepidium sativum*

2413 **garden dahlia**
f dahlia commun
d Gartendahlie; verschiedenfarbige Dahlie
l *Dahlia variabilis*

* **garden daisy → 2037**

2414 **garden eggplant**
f aubergine; béringène; mélongène; mayenne
d Aubergine; Albergine; Eierfrucht; Eierpflanze; spanische Eier; Melanganapfel
l *Solanum melongena*

2415 **garden fennelflower; black cumin; nutmeg flower**
f nigelle de Crète; toute-épice
d Schwarzkümmel; schwarzer Kreuzkümmel; echter Schwarzkümmel; Nardensame; römischer Koriander; schwarzer Koriander
l *Nigella sativa*

2416 **gardenia**
f gardénie
d Gardenie; Jasminglanz
l *Gardenia*

2417 garden lettuce
f laitue cultivée; salade
d Gartenlattich; Gartensalat; Salat
l *Lactuca sativa*

2418 garden lovage
f ache des montagnes; livèche officinale
d Badekraut; Gartenliebstöckel; grosser Eppich; Leberstockkraut; Liebstock; gelbes Liebstöckel; Saukraut; Saustockkraut
l *Levisticum officinale*

2419 garden mushroom
f champignon de couche; agaric cultivé; champignon de fumier; champignon de Paris
d Gartenedelpilz; echter Champignon; Blätterpilz; Zuchtchampignon; zweisporiger Egerling
l *Agaricus bisporus; Psalliota bispora*

2420 garden onion
f oignon
d Bolle; Gartenzwiebel; Hauszwiebel; Knollenzwiebel; Küchenzwiebel; Sommerzwiebel; gemeine Zwiebel; Zipolle; Zwiebellauch
l *Allium cepa*

2421 garden orach
f arroche des jardins
d Gartenmelde; wilder Spinat; Zuckermelde
l *Atriplex hortensis*

2422 garden parsnip
f panais; pastenade; pastenaque
d echter Pastinak; Gartenpastinak; gebauter Pastinak; gemeiner Pastinak
l *Pastinaca sativa*

2423 garden pea
f pois cultivé
d Gartenerbse; Pflückerbse; Saaterbse; Zuckererbse
l *Pisum sativum*

2424 garden plum
f prunier commun
d Pflaume; Zwetsche(nbaum); Zwetschke; Zwetschge
l *Prunus communis; Prunus domestica*

2425 garden radish
f radis
d Eiszapfen; Monatsrettich; Gartenrettich; Radies; Rettich
l *Raphanus sativus*

2426 garden rhubarb
f rhubarbe anglaise; rhubarbe rhapontic
d Gartenrhabarber; Speiserhabarber; stumpfer Rhabarber; Rhapontik
l *Rheum rhaponticum*

2427 garden sage
f herbe sacrée; sauge officinale
d Gartensalbei; echte Salbei
l *Salvia officinalis*

2428 garden serviceberry
f amélanchier commun
d englische Mispel; gemeine Felsenbirne; rundblättrige Felsenbirne; Quantelbeerbaum
l *Amelanchier rotundifolia; Amelanchier ovalis*

2429 garden sorrel
f oseille commune; surette
d grosser Sauerampfer; französischer Spinat; Sauerampfer; Wiesensauerampfer
l *Rumex acetosa*

* **garland flower → 2491**

2430 garland rhododendron
f rhododendron hirsute
d behaarte Alpenrose; rauhblättrige Alpenrose; rauhhaarige Alpenrose
l *Rhododendron hirsutum*

2431 garland spirea
d Brautspiere; Schneespierstrauch
l *Spiraea arguta*

2432 garlic
f ail; ail blanc; ail commun
d Knoblauch
l *Allium sativum*

2433 garlic mustard; hedge garlic
f alliaire
d gemeines Lauchkraut; Knoblauchhederich; Knoblauchrauke
l *Alliaria officinalis*

2434 garlic-scented agaric
d Dürrbehndel; Dürrbein; Knoblauchpilz; Knoblauchschwamm; Knoblauchschwindling; Küchenschwindling; Lauchschwamm; Mousseron; Musseron
l *Marasmius scorodonius*

2435 gas-plant dittany; fraxinella
f dictame blanc; dictame commun; fraxinelle commune
d eschenblättriger Diptam; weisser Diptam; weisse Aschwurz
l *Dictamnus albus*

2436 gasteria
f gastérie
d Gasterie
l *Gasteria*

2437 gasteromycetes
f gastéromycètes
d Bachpilze; Gasteromyzeten
l *Gasteromycetes*

2438 gaura
f gaura
d Prachtkerze
l *Gaura*

2439 gay-feather
f liatride
d Prachtscharte; Scharte
l *Liatris*

2440 gazania
f gazanie
d Gazanie
l *Gazania*

*** gean → 3544**

2441 gelsemium
f gelsémie
d Dufttrichter; Jasminwurzel
l *Gelsemium*

2442 Geneva bugle
f bugle de Genève
d behaarter Günsel; Heidegünsel; zottiger Günsel
l *Ajuga genevensis*

2443 genipap; jagua
f génipa des Antilles
d Genipbaum; Jagua
l *Genipa americana*

2444 genip (tree)
f génipa
d Genipabaum; Genipbaum
l *Genipa*

2445 gentian
f gentiane
d Enzian
l *Gentiana*

2446 gentian family
f gentianacées
d Enziangewächse; Gentianazeen
l *Gentianaceae*

2447 geranium; crowfoot
f géranium
d Geranie; Storchschnabel
l *Geranium*

2448 geranium family
f géraniacées
d Geraniazeen; Storchschnabelgewächse
l *Geraniaceae*

2449 Gerard ephedra
f éphèdre de Gérard
d Gerards Meerträubchen
l *Ephedra gerardiana*

*** Gerard's pine → 1156**

2450 Gerard's rock cress
f arabette de Gérard
d Hainkresse
l *Arabis gerardi*

2451 gerbera
f gerbera
d Gerbera
l *Gerbera*

2452 German camomile; wild camomile
f camomille vraie; camomille ordinaire
d echte Kamille
l *Matricaria chamomilla*

2453 germander
f germandrée
d Gamander
l *Teucrium*

2454 germander speedwell; cat's-eye
f véronique germandrée
d Frauenbiss; Gamanderehrenpreis; Männertreu; wilder Gamander
l *Veronica chamaedrys*

2455 germander spirea
f spirée à feuilles de germandrée
d ulmenblättrige Spiräe; Ulmenspierstrauch
l *Spiraea chamaedryfolia*

2456 German false tamarisk
f myricaire d'Allemagne
d deutsche Tamariske; deutsche

Zypresse
l *Myricaria germanica*

2457 German gentian
f gentiane germanique
d deutscher Enzian
l *Gentiana germanica*

2458 German iris
f iris d'Allemagne; iris Armes de France; flambé; glaïeul bleu
d Gilgen; Himmelslilie; deutsche Schwertlilie
l *Iris germanica*

2459 German madwort
f bardanette; râpette
d liegendes Scharfkraut; liegendes Schlangenäuglein
l *Asperugo procumbens*

2460 German pellitory; ringflower
f anacycle; pyrèthre d'Afrique
d Ringblume; Bertramwurzel; Kreisblume; Schneckenblume
l *Anacyclus*

2461 German velvet grass
f houlque molle
d Ackerhoniggras; weiches Honiggras
l *Holcus mollis*

2462 German vetch
f vesce des buissons
d Hainwicke; Heckenwicke
l *Vicia dumetorum*

2463 German woad-waxen
f genêt allemand
d deutscher Ginster
l *Genista germanica*

2464 gesneria
f gesneria
d Gesnerie
l *Gesneria*

2465 gesneria family
f gesneriacées
d Gesneriazeen; Gesneriengewächse
l *Gesneriaceae*

* **giant arborvitae → 2484**

2466 giant arum; titan giant arum
d Titanenwurz
l *Amorphophallus titanum*

2467 giant bamboo
f dendrocalamus
d Riesenbambus
l *Dendrocalamus*

2468 giant bell
f ostrowskye
d Ostrowskie; Prachtglocke
l *Ostrowskia*

2469 giant brunsvigia
f amaryllis de Joséphine; brunsvigie de l'Impératrice Joséphine
l *Brunsvigia josephinae; Brunsvigia gigantea*

* **giant cactus → 4645**

2470 giant dracena; tikouka; club palm
f cordyline australe
d australische Cordyline; australische Keulenlilie
l *Cordyline australis; Dracaena australis*

2471 giant fennel
f férule
d Steckenkraut; Riesenfenchel
l *Ferula*

2472 giant fescue; tall fescue
f fétuque géante
d Riesenschwingel
l *Festuca gigantea*

2473 giant filbert; Lambert nut
f noisetier franc; noisetier tubuleux
d grosse Hasel(nuss); Lambertshaselnuss; Lambertsnuss
l *Corylus maxima*

2474 giant garlic
f ail rocambole
d Bergknoblauch; echte Rockenbolle; Rocambole; Graslauch; Perlzwiebel; Schlangenknoblauch; Schlangenlauch
l *Allium scorodoprasum*

2475 giant granadilla
f barbadine; grenadille vineuse
d Königsgranadille; Melonengranadille
l *Passiflora quadrangularis; Passiflora macrocarpa*

2476 giant hyssop
f lophanthe
d Büschelblume; Kammblume
l *Lophanthus*

2477 giant kelp
d Birnentang
l *Macrocystis pyrifera*

2478 giant lily
f lis gigantesque
d riesige Lilie
l *Lilium giganteum*

* **giant lily → 4209**

2479 giant puffball
f vesse-de-loup géante; tête de mort
d Flockenstreuling; Riesenbovist
l *Lycoperdon giganteum*

2480 giant ragweed
d dreilappiges Traubenkraut
l *Ambrosia trifida*

2481 giant reed
f roseau
d Pfahlrohr; Rohr; Schilf
l *Arundo*

2482 giant reed
f grand roseau; roseau à quenouille; canne de Provence
d Klarinettenrohr; italienisches Rohr; Schalmeienrohr; Riesenschilf
l *Arundo donax*

2483 giant sequoia
f séquoia géant; séquoia gigantesque; wellingtonia géant
d Mammutbaum; Riesensequoie; Wellingtonie
l *Sequoia gigantea*

2484 giant thuya; giant arborvitae
f thuya de Lobb; thuya géant de Californie; thuya géant; thuya gigantesque
d Riesenlebensbaum
l *Thuya gigantea; Thuya plicata*

* **gibbous duckweed → 5404**

2485 gibbous trametes
f tramète bossu
d Buckeltramete; gebuckelte Tramete
l *Trametes gibbosa; Trametes kalchbrenneri*

2486 gidgee
d Veilchenholz
l *Acacia homalophylla*

2487 gilia
f gilie
d Gilia
l *Gilia*

* **gill-ale → 2647**

2488 gill fungi; gill mushrooms
f agaricacées
d Blätterpilze; Blätterschwämme; Edelpilze; Agarikazeen; Champignonartige
l *Agaricaceae*

2489 ginger
f gingembre
d Ingwer; Ingber
l *Zingiber*

2490 ginger family
f zinzibéracées
d Ingwergewächse; Zingiberazeen
l *Zingiberaceae*

2491 ginger lily; garland flower
f hédychie; gandasuli
d Kranzblume; Süssschnee
l *Hedychium*

2492 ginkgo; maidenhair tree
f ginkgo
d Fächerbaum; Ginkgo(baum); Silberbaum; Mädchenhaarbaum
l *Ginkgo; Salisburia*

2493 ginkgo
f arbre aux quarante écus; arbre de Gordon; noyer du Japon; ginkgo bilobé
d echter Ginkgobaum
l *Ginkgo biloba*

2494 ginseng
f ginseng
d Gilgen; Ginseng; Kraftwurz(el)
l *Panax*

2495 ginseng family
f araliacées
d Araliazeen; Araliengewächse; Efeugewächse
l *Araliaceae*

2496 glacier alpenclock
f soldanelle alpestre
d Alpentroddelblume; echtes Alpenglöckchen
l *Soldanella alpina*

2497 glacier buttercup
f renoncule des glaciers

d Gletscherhahnenfuss
l *Ranunculus glacialis*

2498 glacier draba
f drave des glaciers
d Gletscherfelsenblümchen
l *Draba glacialis*

2499 gladiolus; cornflag
f glaïeul
d Siegwurz; Gladiole; Netzschwertel; Schwertel
l *Gladiolus*

2500 glasswort
f salicorne
d Glasschmalz; Meerkraut; Queller; Salzkraut
l *Salicornia*

2501 gleichenia
d Gabelfarn
l *Gleichenia*

2502 gleichenia family
f gleicheniacées
d Gleicheniazeen
l *Gleicheniaceae*

2503 globe amaranth
f gomphrène
d Kugelamarant; Amarantine
l *Gomphrena*

* **globe artichoke → 331**

2504 globe candytuft
f thlaspi des jardiniers; thlaspi violet; téraspic d'été
d doldiger Bauernsenf; doldige Schleifenblume
l *Iberis umbellata*

2505 globe daisies
f globulariacées
d Globulariazeen; Kugelblumengewächse
l *Globulariaceae*

2506 globe daisy
f globulaire
d Kugelblume
l *Globularia*

2507 globeflower
f trolle
d Goldranunkel; Trollblume; Trolle
l *Trollius*

2508 globe thistle
f échinope
d Kugeldistel
l *Echinops*

2509 glorybind
f calystégie; liseron
d Bärwinde; Zaunwinde
l *Convolvulus; Calystegia*

2510 glory bower; glory tree
f clérodendron
d Loosbaum; Losbaum; Priesterbaum; Schicksalbaum; Volkamerie; Volkmannie
l *Clerodendron*

2511 glory lily
f méthonique
d Prachtlilie; Ruhm(es)krone
l *Gloriosa*

2512 glory-of-the-snow
f gloire de neige
d Schneeglanz; Schneestolz
l *Chionodoxa*

* **glory tree → 2510**

2513 glossy buckthorn; alder buckthorn
f bourdaine; bourgène; nerprun bourdaine
d Faulbaum; Pulverholz; Zapfenholz
l *Rhamnus frangula*

2514 glossy privet
f troène brillant
l *Ligustrum lucidum*

2515 gloxinia
f gloxinie
d Gloxinie
l *Gloxinia; Sinningia*

2516 glutinous gomphidius
f gomphide glutineux; agaric visqueux
d grosser Schmierling; grosser Gelbfuss; Kuhmaul; Schafsnase
l *Gomphidius glutinosus; Agaricus glutinosus*

2517 goat grass
f égilope; égilops
d Walch
l *Aegilops*

* **goatroot → 5048**

2518 goatroot ononis
d Bockshauhechel; stinkende Hau-

hechel; Stinkhauhechel
l *Ononis arvensis; Ononis hircina*

2519 goatsbeard
f aronque; barbe de bouc
d Geissbart
l *Aruncus*

2520 goatsbeard
f clavaire; petite massue
d Keulenpilz; Keulenschwamm; Keule; Handpilz
l *Clavaria*

* **goatsbeard → 4662**

2521 goat's-rue
f galéga
d Geissraute
l *Galega*

2522 goat's-wheat
f atraphace
d Bocksknöterich; Bocksweizen
l *Atraphaxis*

2523 goat willow
f marsault; saule marsault; saule marsaux
d Hohlweide; Palmweide; Pfeifenholz; Sale; Salweide; Ziegenweide
l *Salix caprea*

2524 godetia
f godétie
d Atlasblume
l *Godetia*

2525 goldband lily; golden-banded lily
f lis doré du Japon
d Goldbandlilie
l *Lilium auratum*

* **golden apple → 408**

2526 golden ash
f frêne à bois jaune; frêne doré
d Goldesche
l *Fraxinus excelsior* var. *aurea*

2527 golden bamboo
f bambou doré
d Blattähre
l *Phyllostachys aurea*

* **golden-banded lily → 2525**

2528 golden camomile; dyer's camomile
f camomille des teinturiers; anthémis des teinturiers
d Färberhundskamille
l *Anthemis tinctoria*

* **golden chain → 3192**

2529 golden-chain laburnum
f bois de lièvre; cytise à grappes; cytise aubour; cytise commun; cytise faux ébénier
d gemeiner Goldregen; gemeiner Bohnenbaum; Kleestrauch
l *Laburnum anagyroides; Cytisus laburnum*

2530 golden cinquefoil
f potentille dorée
d Goldfingerkraut
l *Potentilla aurea*

2531 golden clavaria
f clavaire dorée; gallinette menotte
d goldgelbe Koralle; Hahnenkamm; gelber Ziegenbart; schwefelgelbe Koralle
l *Clavaria aurea; Ramaria aurea*

2532 golden clematis
f clématite tangoute
d Goldwaldrebe
l *Clematis tangutica*

2533 golden club
f oronce
d Goldkeule
l *Orontium*

2534 golden cup
f hunnemannie
d Hunnemannie
l *Hunnemannia*

2535 golden currant
f groseillier doré
d Goldjohannisbeere; Goldtraube; echte Goldtraube; goldgelbe Johannisbeere
l *Ribes aureum*

2536 golden dock
d Goldampfer; Meerampfer; Strandampfer
l *Rumex maritimus; Rumex persicarioides*

2537 golden elder
d Goldholunder
l *Sambucus nigra* var. *aurea*

2538 **golden flax**
f lin jaune d'or
d gelber Lein
l *Linum flavum*

2539 **golden larch**
f faux-mélèze
d Goldlärche; Scheinlärche
l *Pseudolarix*

* **golden leaf → 5211**

2540 **golden loosestrife**
f lysimaque commune
d Ackerfelberich; gemeiner Gilbweiderich; Goldfelberich
l *Lysimachia vulgaris*

2541 **golden monkey flower**
f mimule jaune
d gefleckte Gauklerblume
l *Mimulus luteus*

* **golden oat gras → 6109**

2542 **golden ray**
f ligulaire
d Bandblume; Goldkolben; Ligularie
l *Ligularia*

* **golden rhododendron → 4264**

2543 **goldenrod**
f solidage
d Goldrute; Steingünsel
l *Solidago*

2544 **goldenrod tree**
f bosée
d Goldrutenbaum
l *Bosea yervamora*

2545 **golden saxifrage**
f saxifrage dorée
d Goldmilz; Milzkraut
l *Chrysosplenium*

2546 **goldenseal**
f hydraste
d Gelbwurzel; Wasserkraut; Orangewurzel
l *Hydrastis*

2547 **goldenseal; puccoon**
f hydraste du Canada
d kanadische Orangewurz(el); kanadische Gelbwurzel
l *Hydrastis canadensis*

2548 **golden-shower senna; purging cassia; drumstick tree**
f casse fistuleuse
d Röhrenkassie
l *Cassia fistula*

2549 **golden star**
f chrysogone
l *Chrysogonum*

* **golden thistle → 3993**

2550 **goldentop**
f lamarckie
d Lamarckie; Silbergras
l *Lamarckia*

2551 **golden tuft; golden-tuft alyssum**
f alysse saxatile; corbeille d'or
d echtes Steinkraut; Felsensteinkraut; Gebirgssteinkraut; Goldkörbchen
l *Alyssum saxatile*

2552 **goldilocks**
f renoncule tête d'or
d Goldhahnenfuss
l *Ranunculus auricomus*

* **goldmoss → 2553**

2553 **goldmoss stonecrop; goldmoss**
f orpin âcre; poivre de murailles; pain d'oiseau
d gemeiner Mauerpfeffer; scharfer Mauerpfeffer; Steinpfeffer
l *Sedum acre*

2554 **gold poppy**
f eschscholtzie
d Eschscholtzie; Goldmohn
l *Eschscholtzia*

2555 **gold rain tree**
f koelreutérie
d Blasenbaum; Blasenesche
l *Koelreuteria*

2556 **goldthread**
f coptide
d Goldfaden
l *Coptis*

2557 **gomphia**
f gomphie
l *Gomphia*

2558 **gomuti sugar palm**
f areng; palmier à sucre
d molukkische Zuckerpalme; Sagwiren-

palme
l *Arenga saccharifera; Arenga pinnata*

2559 **Good-King-Henry goosefoot**
f chénopode bon-Henri; bon-Henri; épinard sauvage
d guter Heinrich
l *Chenopodium bonus-henricus*

2560 **gooseberry**
f groseillier
d Stachelbeere; Klosterbeere; Krausbeere; Krönzel; Stachelbeerstrauch
l *Grossularia*

2561 **goosefoot**
f ansérine
d Gänsefuss; Melde; Schmergel
l *Chenopodium*

2562 **goosefoot family**
f chénopodiacées
d Chenopodeen; Chenopodiazeen; Gänsefussgewächse
l *Chenopodiaceae*

2563 **goose grass**
f éleusine d'Inde
d indische Eleusine
l *Eleusine indica*

2564 **gordonia**
f gordonie
d Gordonie; Gordonstrauch
l *Gordonia*

2565 **Gordon mock orange**
f seringa de Gordon
d Gordons Pfeifenstrauch
l *Philadelphus gordonianus*

2566 **gorse; furze**
f ajonc
d Heckensame; Stachelginster; Stechginster
l *Ulex*

2567 **gourd**
f lagénaire
d Flaschenkürbis; Kalebasse; Lagenarie
l *Lagenaria*

2568 **gourd; squash**
f courge
d Kürbis
l *Cucurbita*

2569 **gourd family**
f cucurbitacées
d Gurkengewächse; Kukurbitazeen; Kürbisgewächse
l *Cucurbitaceae*

2570 **goutweed**
f égopode
d Giersch
l *Aegopodium*

* **governor's plum** → **4413**

2571 **Gowen cypress**
f cyprès de Californie; cyprès de Gowen
d Gowens Zypresse
l *Cupressus goveniana; Cupressus californica*

2572 **grama grass**
f bouteloue
d Haarschotengras
l *Bouteloua*

* **gram chick-pea** → **1144**

* **gramineous plants** → **2581**

2573 **grand fir**
f sapin de Vancouver; sapin géant
d Grosstanne
l *Abies grandis*

2574 **granulated boletus**
f bolet granulé; bolet de pin; cèpe jaune des pins; nonette jaune
d Körnchenröhrling; Schmerling
l *Boletus granulatus*

2575 **grape**
f vigne
d Traube; Rebe; Weinrebe; Weinstock
l *Vitis*

2576 **grape family**
f vitacées
d Rebengewächse; Weinrebengewächse; Vitazeen; Sarmentazeen; Ampelidazeen
l *Vitaceae*

2577 **grape fern**
f botryche
d Mondraute; Rautenfarn
l *Botrychium*

2578 **grapefruit**
f grapefruit

d Grapefruit; Pomelie
l *Citrus paradisi*

2579 grape hyacinth
f muscari; ail des champs
d Träubel
l *Muscari*

2580 grape myrtle
f lagerstrémie
d Lagerströmie; Barg-Lang
l *Lagerstroemia*

2581 grasses; gramineous plants; grass family
f graminées
d Cerealien; Gramineen; Grasartige; Gräser
l *Gramineae*

2582 grass iris
f iris à feuilles de gramen
d Grasschwertlilie; grasblättrige Schwertlilie
l *Iris graminea*

2583 grassland rose
f rosier des haies
d Ackerrose
l *Rosa agrestis*

2584 grass-leaf sandwort
d grasblättriges Sandkraut
l *Arenaria graminifolia*

2585 grass-leaved day lily
f hémérocalle graminée
d kleine Taglilie
l *Hemerocallis minor*

2586 grass-leaved goldenrod
f verge d'or à feuilles de graminée
d grasblättrige Goldrute
l *Solidago graminifolia*

* **grass-leaved pea → 2592**

* **grass of Parnassus → 4058**

2587 grass pea vine
f lentille d'Espagne; gesse cultivée; gesse domestique
d Ackerplatterbse; deutscher Kicher; essbare Platterbse; gelbe Vogelwicke; gelbe Wiesenwicke; Gemüseplatterbse; Kicherling; Saatplatterbse; spanische Erbse; spanische Linse; Wiesenplatterbse
l *Lathyrus sativus*

2588 grass pink
f œillet mignardise
d Federnelke; Pinksnelke
l *Dianthus plumarius*

2589 grass-pink orchid
f limodore
d Dingel
l *Calopogon; Limodorum*

2590 grass-pink orchid
d knolliger Dingel
l *Calopogon pulchellus; Limodorum tuberosum*

2591 grass tree
f xanthorrhée
d Harzaffodil; Gelbharzbaum; Grasbaum
l *Xanthorrhoea*

2592 grass vetchling; grass-leaved pea
d blattlose Platterbse; Grasplatterbse
l *Lathyrus nissolia*

2593 grass water plantain
f plantain d'eau
d grasblättriger Froschlöffel
l *Alisma gramineum*

* **grass wrack → 1407**

* **gravel-bind → 111**

* **gray alder → 5122**

2594 gray birch
f bouleau gris; bouleau à feuilles de peuplier
d pappelblättrige Birke
l *Betula populifolia*

2595 gray hair grass
d Keulenschmiele; Sandsilbergras; duftige Schmiele; graues Silbergras
l *Corynephorus canescens*

* **gray-leaf pine → 1856**

2596 gray-leaved willow
f saule vert-de-mer; saule glauque
d Schneeweide; seegrüne Weide
l *Salix glauca*

2597 gray poplar
f peuplier gris(ard); grisard; grisaille
d Graupappel
l *Populus canescens*

2598 gray willow
f saule gris; saule cendré
d Aschweide; aschgraue Weide; graue Weide; Grauweide
l *Salix cinerea*

2599 great bellflower
f campanule à larges feuilles
d breitblättrige Glockenblume
l *Campanula latifolia*

2600 great bulrush
f jonc des chaisiers; jonc d'eau; jonc des tonneliers; scirpe de lacs
d Flechtbinse; Pferdebinse; Seesimse; Sumpfbinse; Teichbinse
l *Scirpus lacustris*

2601 great burdock
f bardane commune; glouteron
d grosse Klette
l *Arctium lappa; Lappa major*

* **great burnet → 2409**

2602 greater alpenclock
f soldanelle montagnarde
d Bergtroddelblume; grosses Alpenglöckchen
l *Soldanella montana*

2603 greater ammi
f grande ammi
d grosse Ammei; grosse Knorpelmöhre
l *Ammi majus*

2604 greater broomrape
f orobanche majeure
d grosse Sommerwurz; hohe Sommerwurz
l *Orobanche major*

2605 greater butterfly orchid
d Bergwaldhyazinthe; grosse Waldhyazinthe; grünliche Kuckucksblume; grünliche Waldhyazinthe
l *Platanthera chlorantha*

2606 greater celandine
f grande éclaire; éclaire; herbe aux verrues; herbe de Sainte Claire; grande chélidoine; chélidoine grande éclaire
d Gilbkraut; Goldwurz; grosses Schöllkraut; Schöllwurz; Schwalbenkraut; Warzenkraut
l *Chelidonium majus*

2607 greater dodder
f cuscute d'Europe; grande cuscute
d Hopfenseide; Klebe; Nesselseide; Range; Teufelshaar; Teufelszwirn; Vogelseide
l *Cuscuta europaea*

2608 greater pimpinella
f grande pimprenelle
d grosse Bibernelle
l *Pimpinella magna*

* **greater stichwort → 1971**

2609 great galangal; Java galangal; galingale
f galanga de l'Inde; grand galanga
d grosser Galgant; Galgant-Alpinie; Galanga
l *Alpinia galanga*

2610 great horsetail
d grosser Schachtelhalm
l *Equisetum majus*

2611 great masterwort; black sanicle
f grande astrance
d grosse Sternblume; grosse Sterndolde
l *Astrantia major*

* **great mullein → 2276**

* **great plantain → 4502**

2612 great stonecrop
f grand orpin
d Geschwulstkraut; Johanniskraut; Wundkraut
l *Sedum maximum*

* **great willow herb → 2262**

2613 great wood rush
d Waldhainsimse; Waldmarbel; Waldsimse
l *Luzula sylvatica*

2614 Grecian foxglove
f digitale laineuse
d wolliger Fingerhut
l *Digitalis lanata*

2615 Grecian laurel; true bay
f laurier d'Apollon; laurier commun
d edler Lorbeer
l *Laurus nobilis*

2616 Grecian silk vine
f périploque grec
d griechische Baumschlinge
l *Periploca graeca*

2617 Greek fir
f sapin de Grèce
d griechische Tanne
l *Abies cephalonica*

2618 Greek juniper
f genévrier de Grèce
d hoher Wacholder
l *Juniperus excelsa*

2619 Greek micromeria
f micromérie de Grèce
d griechische Micromerie
l *Micromeria graeca*

* **Greek valerian → 4246**

2620 Greek-valerian polemonium; Jacob's-ladder
f polémonie bleue
d blaues Sperrkraut; griechischer Baldrian; Himmelsleiter; Jakobsleiter
l *Polemonium caeruleum*

2621 green algae
f algues vertes; chlorophycées
d Grünalgen
l *Chlorophyceae*

2622 green ash
f frêne vert
d Grünesche
l *Fraxinus pennsylvanica lanceolata; Fraxinus viridis*

2623 greenbrier
f salsepareille
d Stechwinde
l *Smilax*

2624 green bristle grass
f panic vert; sétaire vert
d grüne Borstenhirse; grüner Fennich
l *Setaria viridis*

2625 green ephedra
f éphèdre vert
d grünes Meerträubchen
l *Ephedra viridis*

2626 green felt
f vauchérie
d Blasenalge; Vaucherie
l *Vaucheria*

2627 greengage; Reine Claude
f reine-claude
d Edelpflaume; Reineclaude; Reneklode; Rundpflaume
l *Prunus italica; Prunus claudiana*

2628 green hellebore
f herbe à la bosse; herbe à setons; ellébore vert
d grüne Nieswurz; Christianswurz; grüne Christwurz
l *Helleborus viridis*

2629 greenish russula
f palomet; russule verdoyante
d gefelderter Grüntäubling; grünlicher Täubling; grünschuppiger Täubling
l *Russula virescens*

* **green laver → 1524, 4794**

2630 green-leaved hound's-tongue
d deutsche Hundszunge
l *Cynoglossum germanicum*

2631 green mo(u)ld
d grüner Kolbenschimmel
l *Aspergillus glaucus*

2632 green penicillium; blue mo(u)ld
f pénicille glauque
d grüner Pinselschimmel
l *Penicillium glaucum*

2633 green pyrola
d bleiches Wintergrün; grünblütiges Wintergrün; grünliches Wintergrün
l *Pyrola chlorantha*

2634 green spleenwort
f asplénie vert
d grüner Streifenfarn
l *Asplenium viride*

2635 green strawberry
f breslinge; fraisier vert
d Bresling; Hügelerdbeere; Knackbeere; Knackelbeere
l *Fragaria viridis*

2636 green thread
f thélesperme
d Warzensame
l *Thelesperma*

2637 green wattle; green-wattle acacia
f acacia noir
d schwarze Akazie; Grünwattle
l *Acacia decurrens*

2638 grevillea
f grévillée

d Grevillie; Seideneiche
l *Grevillea*

2639 grewia
f greuvier
d Grewie
l *Grewia*

2640 gromwell
f grémil
d Steinsame
l *Lithospermum*

2641 Gronovius dodder
f cuscute de Gronovius
d amerikanische Seide; Weidenseide
l *Cuscuta gronovi*

2642 Grossheim iris
f iris de Grossheim
d Grossheims Schwertlilie
l *Iris grossheimi*

2643 ground bamboo
d niedriger Bambus
l *Pleioblastus pumilus*

2644 ground cherry; dwarf cherry
f cerisier griottier nain
d Strauchkirsche; Zwergkirsche; Zwergkirschbaum; Zwergweichsel
l *Prunus fruticosa*

2645 ground cherry
f coqueret; physalis
d Blasenkirsche; Judenkirsche; Lampionblume
l *Physalis*

2646 ground clematis
f clématite dressée
d Brennkraut; Grensel; Grensing; aufrechte Waldrebe; steife Waldrebe
l *Clematis recta*

*** ground holly → 4202**

2647 ground ivy; gill-ale; alehoof
f gléchome
d Gunderkraut; Gundermann; Gunderrebe
l *Glec(h)oma*

2648 ground liverwort
f peltigère
d Schildflechte
l *Peltigera canina*

*** groundnut → 4088**

2649 groundnut pea vine
f gland de terre
d Ackereichel; Ackernuss; Erdeichel; Erdmandel; Erdnuss; Erdnussplatterbse; knollige Platterbse
l *Lathyrus tuberosus*

2650 ground pine
d Ackergünsel; gelber Günsel
l *Ajuga chamaepitys*

2651 groundsel; ragwort; squaw-weed
f séneçon
d Greiskraut; Kreuzkraut; Kreuzwurm
l *Senecio*

2652 grove wood rush
d Hainmarbel; schmalblättrige Hainsimse; Silbermarbel; schmalblättrige Simse
l *Luzula nemorosa*

2653 grumichama; Brazil cherry
f cerise du Brésil
l *Eugenia dombeyi*

2654 Guadalupe erythea
f palmier de la Guadeloupe
l *Erythea edulis*

*** guanabana → 5093**

*** guara → 1732**

2655 guarana paullinia
f paullinie
d Giftknippe
l *Paullinia cupana*

2656 guava
f goyavier
d Guajabenbaum; Guayave; Guavenbaum
l *Psidium*

*** guayule → 2657**

2657 guayule parthenium; guayule
f guayule
d Guayule
l *Parthenium argentatum*

2658 Guernsey flower; Guernsey lily
f lis de Guernsey; amaryllis de Guernsey
d Guernseylilie
l *Nerrine sarniensis*

2659 guger tree
f schime
l *Schima*

2660 Guiana crabwood
f carape de Guyane
d Andiroba
l *Carapa guianensis*

* **Guinea corn → 5087**

2661 Guinea grass
f herbe de Guinée
d Guinea-Gras
l *Panicum maximum*

* **Guinea grass → 3103**

2662 gum-arabic acacia
f acacia à gomme; verek; acacia du Sénégal
d Gummiarabikum-Baum; Verek; Gummiakazie; Senegal-Akazie
l *Acacia senegal*

* **gumbo → 3921**

2663 gum plant; gumweed
f grindélie
d Grindelie
l *Grindelia*

2664 gum rockrose
f ciste à gomme
l *Cistus ladaniferus*

* **gum succory → 4610**

2665 gum tree
f castilloa
d Kautschukbaum
l *Castilla; Castilloa*

* **gum tree → 2065**

2666 gumvine
f landolphie
l *Landolphia*

* **gumweed → 2663**

2667 gunnera
f gunnère
d Gunnere; Nesselschirm
l *Gunnera*

* **gurjun → 1432**

2668 gurjun oil tree
f diptérocarpe
d Zweiflügelnussbaum
l *Dipterocarpus*

2669 guttiferae; clusia family
f guttifères; clusiacées
d Gummiträger; Guttiferen; Hartheugewächse
l *Guttiferae; Clusiaceae*

2670 guzmannia
f guzmannia
d Guzmannie
l *Guzmannia*

2671 gymnosperms
f gymnospermes
d Archispermen; Freisamer; Gymnospermen; Nacktsamer
l *Gymnospermae*

2672 gypsophila
f gypsophile
d Gipskraut; Schleierkraut
l *Gypsophila*

* **gypsy → 6057**

H

2673 habenaria
f habénaria
d Habenaria; Riemenlippe; Zügelorchis
l *Habenaria*

2674 hackberry
f micocoulier
d Zürgelbaum
l *Celtis*

2675 hacquetia
f hacquétie
d Hacquetie
l *Hacquetia*

2676 haircap moss
f polytric
d Haarmoos; Widerton
l *Polytrichum*

2677 hair grass
f canche
d Haferschmiele; Schmiele; Zwergschmiele
l *Deschampsia; Aira*

2678 hairlike sedge; capillary sedge
f laîche capillaire
d haarstengelige Segge
l *Carex capillaris*

* **hair palm → 3572**

2679 hair-vein agrimony
d behaarter Odermennig
l *Agrimonia pilosa*

2680 hairy bassia
d rauhhaarige Sommerzypresse
l *Echinopsilon hirsutum; Bassia hirsuta*

* **hairy bitter cress → 4111**

2681 hairy brome
d Angertrespe; verwechselte Trespe
l *Bromus commutatus*

2682 hairy burstwort
f herniaire velue
d haariges Bruchkraut; haariges Tausendkorn
l *Herniaria hirsuta*

2683 hairy crab grass
f manne terrestre; manne sanguinale
d Blutfennich; Blutfingergras; Blutfingerhirse; Bluthirse; blutrote Fingerhirse; rotes Fingergras
l *Digitaria sanguinalis*

2684 hairy gentian
f gentiane ciliée
d Fransenenzian; gefranster Enzian
l *Gentiana ciliata*

2685 hairy greenweed
f arnigo
d behaarter Ginster; Sandginster
l *Genista pilosa*

2686 hairy lilac
f lilas de Bretschneider
l *Syringa villosa; Syringa pubescens*

2687 hairy oat; downy oat grass
f avoine pubescente
d Flaumhafer; Flaumwiesenhafer
l *Avena pubescens*

2688 hairy rock cress
f arabette hérissée
d rauhe Gänsekresse; rauhhaarige Gänsekresse
l *Arabis hirsuta*

2689 hairy rockrose
f ciste hérissé
l *Cistus hirsutus*

2690 hairy sedge
f carex hérissé; laîche velue
d behaarte Segge; behaartes Riedgras
l *Carex pilosa; Carex hirta*

2691 hairy stonecrop
f orpin velu
d behaarter Mauerpfeffer; zottige Fetthenne
l *Sedum villosum*

2692 hairy teasel
f cardère velue
d behaarte Karde; Borstenkarde
l *Dipsacus pilosus*

2693 hairy vetch; Russian vetch; winter vetch
f vesce velue; vesce de Russie
d Sandwicke; Zottelwicke; zottige Wicke; Sanderbse
l *Vicia villosa*

* **hairy vetch → 5531**

2694 hairy violet
d rauhes Veilchen; Wiesenveilchen
l *Viola hirta*

2695 hairy willowweed
f épilobe hérissé
d behaartes Weidenröschen; zottiges Weidenröschen
l *Epilobium hirsutum*

2696 hairy wood rush
d Haarhainsimse; behaarte Hainsimse
l *Luzula pilosa*

2697 halberd willow
d spiessblättrige Weide; spiessförmige Weide; Spiessweide
l *Salix hastata*

2698 Halstand rheedia
f cirolier commun
l *Rheedia laterifolia*

2699 hamelia
f hamélie
d Hamelie
l *Hamelia*

2700 hamiltonia
f hamiltonie
l *Hamiltonia*

* **hardfern → 1821**

2701 hard grass
f sclérochloa
d Hartgras
l *Sclerochloa*

2702 hard grass
d graues Hartgras
l *Sclerochloa dura*

2703 hardhack; hardhack spirea
f spirée tomenteuse
d gelbfilziger Spierstrauch
l *Spiraea tomentosa*

* **Harding's grass → 1761**

* **hard meadow grass → 5251**

* **hard pine → 4208**

2704 hard rush
d blaugrüne Binse
l *Juncus inflexus*

2705 hard-skinned puffball
f scléroderme
d Hartbovist; Fellstreuling; Kartoffelbovist
l *Scleroderma*

2706 hard-skinned puffballs
f sclérodermatacées
d Hartboviste
l *Sclerodermataceae*

2707 hardy crinum
f crinole à longues feuilles
l *Crinum longifolium*

2708 hardy rye grass
d Leinlolch
l *Lolium remotum*

2709 harebell Hungarian speedwell
d gestreckter Ehrenpreis; liegender Ehrenpreis
l *Veronica prostrata*

2710 hare's-ear
f conringia
d Ackerkohl; Conringie
l *Conringia*

* **hare's-ear → 5493**

* **hare's-foot → 4394**

2711 hare's-foot fern
f davallie
d Davallie; Krugfarn
l *Davallia*

2712 hare's-foot sedge
d Hasenriedgras; Hasensegge
l *Carex leporina*

2713 hare's-tail grass; rabbit-tail grass
f lagure
d Hasenschwanzgras; Samtgras; Sammetgras
l *Lagurus*

2714 harmel peganum; African rue
f harmale; pégane harmale; rue sauvage
d syrische Raute; Steppenraute
l *Peganum harmala*

2715 Harrison yellow rose
f rose de Harrison
d Harrisons Rose
l *Rosa harrisoni*

2716 **hart's-tongue**
f scolopendre officinale
d Hirschzunge; Zungenfarn
l *Scolopendrium vulgare; Phyllitis scolopendrium*

2717 **hart's-truffle**
f élaphomyce
d Hirschstreuling; Hirschtrüffel
l *Elaphomyces*

2718 **Hartweg pentstemon**
f pentstémon de Hartweg
d Hartwegs Bartfaden
l *Pentstemon hartwegi*

2719 **hartwort**
f tordyle
d Zirmet
l *Tordylium*

2720 **hatchet vetch**
f sécurigère
d Beilwicke
l *Securigera*

2721 **Hausa potato**
d madagaskarische Kartoffel; Numbo-Knollen
l *Coleus parviflorus*

2722 **hautbois strawberry**
f capron; fraisier caperonnier
d Moschuserdbeere; Riesenerdbeere; Zimterdbeere
l *Fragaria moschata*

2723 **hawkbit**
f dent-de-lion
d Löwenzahn
l *Leontodon*

2724 **hawk nut**
f terre-noix
d Erdkastanie
l *Bunium*

2725 **hawk's-beard**
f barkhaisie; crépide
d Pippau; Feste; Grundfeste
l *Barkhausia; Crepis*

2726 **hawkweed**
f épervière
d Habichtskraut
l *Hieracium*

2727 **hawkweed oxtongue**
d habichtskrautähnliches Bitterkraut; Habichtskrautbitterich; Habichtskraut-bitterkraut
l *Picris hieracioides*

2728 **haworthia**
f haworthie
d Haworthie
l *Haworthia*

2729 **hawthorn**
f épine
d Weissdorn
l *Crataegus*

2730 **hawthorn maple**
f érable à feuilles d'aubépine
d weissdornblättriger Ahorn
l *Acer crataegifolium*

2731 **hay plant**
f foin du Tibet
d tibetanisches Heu
l *Prangos pabularia*

* **hazel → 2246**

2732 **hazel alder**
f aune rugueux
d Haselerle; Herbsterle; Runzelerle
l *Alnus rugosa*

2733 **hazel sterculia**
f cavalam
l *Sterculia foetida*

2734 **headed cabbage**
f chou cabus; chou pommé
d Kopfkohl; Kraut; Kappes
l *Brassica oleracea* var. *capitata*

2735 **head garden lettuse; head lettuce**
f laitue pommée
d Kopfsalat
l *Lactuca sativa* var. *capitata*

2736 **heart-leaved bergenia**
f bergénie à feuilles cordiformes
d herzblättrige Wickelwurz
l *Bergenia cordifolia*

2737 **heart-leaved globe daisy**
f globulaire à feuilles en cœur
d herzblättrige Kugelblume
l *Globularia cordifolia*

2738 **heart-leaved hornbeam**
f charme à feuilles en cœur
d herzblättrige Hainbuche
l *Carpinus cordata*

2739 **heart-leaved mock orange**
f seringa à feuilles en cœur
d herzblättriger Pfeifenstrauch
l *Philadelphus cordifolius*

2740 **heart-leaved oxeye**
f télékia élégant
d grosses Rindsauge; grosse Telekie; Sonnenstern
l *Telekia speciosa; Buphthalmum speciosum*

* **heartnut → 2281**

2741 **heartseed**
f pois de cœur; cardiosperme
d Herzsame; Ballonrebe
l *Cardiospermum*

2742 **heath**
f bruyère
d Glockenheide; Heide; Heidekraut
l *Erica*

2743 **heath bedstraw**
d Felslabkraut; Harzlabkraut; Steinlabkraut
l *Galium hercynicum; Galium saxatile*

2744 **heather**
f callune; bruyère
d Besenheide; Besenkraut; Heidekraut
l *Calluna*

2745 **heath family**
f éricacées
d Erikazeen; Erizinen; Heidegewächse; Heidekräuter; Heidekrautgewächse; Moorbeerengewächse
l *Ericaceae*

2746 **heath grass**
d liegender Dreizahn
l *Sieglingia decumbens*

2747 **heath rush**
d sperrige Binse
l *Juncus squarrosus*

* **hedge apple → 3972**

* **hedge garlic → 2433**

2748 **hedge glorybind**
f liseron des haies
d deutsche Skammonie; deutsche Purgierwinde; Uferzaunwinde; Zaunwinde
l *Calystegia sepium; Convolvulus sepium*

2749 **hedgehog cactus; sea-urchin cactus**
f échinopsis
d Igelkaktus; Seeigelkaktus
l *Echinopsis*

2750 **hedgehog dog's-tail**
d stach(e)liges Kammgras
l *Cynosurus echinatus*

2751 **hedgehog holly**
f houx hérisson
l *Ilex aquifolium* var. *ferox*

2752 **hedgehog mushroom; spreading mushroom**
f hydne sinué
d gelber Stachelpilz; Semmelpilz; Semmelstachelpilz; Semmelstoppelpilz; Stoppelpilz; Süssling
l *Hydnum repandum*

2753 **hedgehog parsley; bur parsley**
d Mohrenhaftdolde
l *Caucalis daucoides*

2754 **hedge hyssop**
f gratiole
d Gnadenkraut; Gottesgnadenkraut
l *Gratiola*

2755 **hedge knotweed**
f renouée des buissons
d Heckenknöterich
l *Polygonum dumetorum*

2756 **hedge maple**
f érable champêtre
d Feldahorn; Flader; Kreuzbaum; Mas(s)holder; Masseller; Masserle
l *Acer campestre*

2757 **hedge mustard**
f herbe aux chantres
d Wegerauke
l *Sisymbrium officinale; Erysimum officinale*

* **hedge mustard → 2296**

2758 **hedge-nettle betony**
d einjähriger Ziest; Sommerziest
l *Stachys annua*

2759 **hedge parsley**
f torilis

d Klettenkerbel; Borstendolde
l *Torilis*

* **hedge-row rose → 4602**

2760 **hedge vetch; bush vetch**
f vesce des haies
d Zaunwicke
l *Vicia sepium*

2761 **Heldreich hawthorn**
f épine de Heldreich
d Heldreichs Weissdorn
l *Crataegus heldreichi*

2762 **Heldreich pine**
f pin de Heldreich
d Heldreichs Kiefer
l *Pinus heldreichi*

2763 **heliconia**
f héliconie
d Helikonie; Tafelbanane
l *Heliconia*

2764 **heliopsis**
f héliopside
d Sonnenauge
l *Heliopsis*

2765 **heliotrope**
f héliotrope
d Heliotrope; Sonnenwende
l *Heliotropium*

2766 **hellebore**
f hellébore; ellébore
d Christrose; Nieswurz; Schneerose
l *Helleborus*

2767 **helvella; turbantop; saddle fungus**
f helvelle
d Frühmorchel; Stockmorchel; Lorchel; Becherling
l *Helvella*

2768 **helvella family**
f helvellacées
d Helvellazeen; Hutmorchelpilze; Lorcheln; Morchelpilze
l *Helvellaceae*

2769 **hemlock**
f tsuga
d Hemlocktanne; Schierlingstanne
l *Tsuga*

2770 **hemlock parsley**
f coniosélin
d Schierlingssilge
l *Conioselinum*

* **hemlock stork's-bill → 91**

2771 **hemlock water dropwort**
f œnanthe safranée
l *Oenanthe crocata*

2772 **hemp**
f chanvre
d Hanf
l *Cannabis*

2773 **hemp**
f chanvre commun
d gebauter Hanf; gemeiner Hanf; Saathanf
l *Cannabis sativa*

2774 **hemp broomrape**
f orobanche rameuse
d Hanfwürger; ästige Sommerwurz; Tabakwürger
l *Orobanche ramosa*

2775 **hemp dogbane**
f apocyn chanvrin
d hanfartige Hundswolle; indischer Hanf; Hanfhundsgift
l *Apocynum cannabinum*

2776 **hemp eupatorium**
f eupatoire chanvrine; eupatoire à feuilles de chanvre
d gemeiner Wasserdost; Hirschklee; Kunigundenkraut; Wasserhanf; Wassersenf
l *Eupatorium cannabinum*

2777 **hemp family**
f cannabinacées
d Hanfgewächse
l *Cannabinaceae*

2778 **hemp nettle**
f galéope; galéopse
d Hanfnessel; Hohlzahn
l *Galeopsis*

2779 **hen-and-chickens; roof houseleek**
f joubarbe des toits
d echte Hauswurz; Jupiterbart
l *Sempervivum tectorum*

2780 **henbit dead nettle**
f lamier amplexicaule
d Ackertaubnessel; stengelumfassende

Taubnessel
l *Lamium amplexicaule*

2781 henequen agave
f agave faux-fourcroya
d Henequen-Agave; Henequen; weisser Sisal
l *Agave fourcroydes*

2782 henna
f henné
d Alhenna; echte Alkanna; Henna; Hennastrauch
l *Lawsonia inermis*

*** hen of the woods → 2385**

*** henware → 407**

2783 hepatica
f hépatique
d Leberblümchen; Lebermoos
l *Hepatica*

2784 herb canary clover
d krautiger Backenklee
l *Dorycnium herbaceum*

*** herb Christopher → 616**

2785 herb mercury; annual mercury
f mercuriale annuelle
d einjähriges Bingelkraut; Gartenbingelkraut; Hundskohl; Klistierkraut; Merkurialkraut; Mercuriuskraut; Schuttbingelkraut; Speckmelde
l *Mercurialis annua*

*** herb of grace → 1516**

2786 herb Paris
f parisette à quatre angles; parisette à quatre feuilles
d Fuchstraube; Steinbeere; vierblättrige Einbeere; Wolfsbeere
l *Paris quadrifolia*

2787 herb Robert (geranium)
f géranium herbe-à-Robert
d Gichtkraut; Robertskraut; Ruprechtskraut; Ruprechtsstorchschnabel
l *Geranium robertianum*

2788 herb tree mallow
f mauve royale
d Pappelrose; Sommerpappel; Ziermalve
l *Lavatera trimestris*

*** Hercules'-club → 1849**

*** herd's-grass → 5527, 5528**

2789 heron's-bill
f érodium
d Reiherschnabel
l *Erodium*

2790 Heuffel's saffron
f crocus d'Heufell
d Heuffels Krokus
l *Crocus heuffelianus*

2791 hevea
f hévéa
d Federharzbaum; Parakautschukbaum; Hevea
l *Hevea*

2792 Heyne patchouli; patchouli plant
f patchouli
d Patschuli(pflanze)
l *Pogostemon patchouly; Pogostemon heyneanus*

2793 hiba false arborvitae
f thuya à feuilles en herminette
d Hibalebensbaum; beilblättriger Lebensbaum; japanische Zypresse
l *Thujopsis dolabrata*

2794 hibiscus
f hibiscus; ketmie
d Eibisch
l *Hibiscus*

2795 hickory
f hickory; caryer
d Hickory(baum); Hickorynussbaum
l *Carya*

2796 high-bush blueberry
f airelle à corymbes
d amerikanische "Blueberry"
l *Vaccinium corymbosum*

2797 high mallow; wild mallow
f grande mauve; mauve sauvage
d Hanfpappel; Rossmalve; Rosspappel; Waldmalve; wilde Malve
l *Malva sylvestris*

2798 Himalaya blackberry
d hohe Brombeere
l *Rubus procerus*

2799 Himalayan camellia
d Ölkamellie
l *Camellia drupifera; Camellia oleifera*

2800 Himalayan fir
f sapin de l'Himalaya; sapin de Webb
l *Abies spectabilis; Abies webbiana*

2801 Himalayan juniper
f genévrier a rameaux arqués
l *Juniperus recurva*

2802 Himalayan larch
f mélèze de l'Himalaya; mélèze du Sikkim
l *Larix griffithi*

2803 Himalayan lilac
f lilas de l'Himalaya
d Himalaja-Flieder
l *Syringa emodi*

2804 Himalayan May apple
f podophylle indien
d indisches Fussblatt; indischer Entenfuss
l *Podophyllum emodi*

2805 Himalayan rhubarb
f rhubarbe de l'Himalaya
d Emodi-Rhabarber
l *Rheum emodi*

2806 Himalayan spruce
f épicéa de l'Himalaya
d Morindafichte
l *Picea morinda; Picea smithiana*

2807 Himalya pine; lofty pine
f pin pleureur de l'Himalaya
d Tränenkiefer; Himalaja-Weymouthskiefer
l *Pinus excelsa; Pinus griffithi*

2808 hinahina
f mélicyte touffu
l *Melicytus ramiflorus*

2809 Hinau elaeocarpus
f éléocarpe de Nouvelle-Zélande
l *Elaeocarpus dentatus*

2810 Hindu datura
f stramoine metel
d weichhaariger Stechapfel
l *Datura metel*

2811 Hindu lotus
f fève d'Egypte; lis rose des Egyptiens; nélombo d'Orient; rose du Nil
d indische Lotusblume; indischer Lotus
l *Nelumbium nuciferum; Nelumbium nelumbo; Nelumbium speciosum*

2812 Hinoki false cypress
f faux cyprès obtus
d Feuerbaum; Feuerzypresse; Hinoki der Japaner; Sonnenbaum; Sonnenzypresse
l *Chamaecyparis obtusa*

2813 hippocratea
f hippocratée
l *Hippocratea*

2814 Hispania oak
f chêne faux-liège; chêne d'Espagne
d spanische Eiche
l *Quercus hispanica*

2815 hispid althaea
d rauher Eibisch
l *Althaea hirsuta*

2816 hoarhound; horehound
f marrube
d Andorn; Dorant
l *Marrubium*

* **hoary alder → 5122**

* **hoary basil → 5354**

2817 hoary erysimum
d grauer Schötendorn; grauer Schöterich
l *Erysimum canescens*

2818 hoary false alyssum
d gemeine Graukresse
l *Berteroa incana*

2819 hoary mock orange
f seringa à grandes feuilles
d breitblättriger Pfeifenstrauch
l *Philadelphus latifolius; Philadelphus pubescens*

2820 hoary pea
f requiénie
d Surinam-Giftbaum
l *Tephrosia*

2821 hoary sun rose
d graues Sonnenröschen
l *Helianthemum canum*

2822 hoferia
f hoférie
l *Hoferia*

2823 hog bean
f jusquiame
d Bilsenkraut; Dulldill; Dullkraut; Rasenwurz
l *Hyoscyamus*

2824 hog's-fennel
f peucédan
d Haarstrang
l *Peucedanum*

2825 hogweed cow parsnip
f berce spondyle; grande berce; berce brancursine
d deutscher Bärenklau; gemeiner Bärenklau; Heilkraut; unechter Bärenklau; Wiesenbärenklau
l *Heracleum sphondylium*

2826 holly
f houx
d Stechhülse; Stechpalme
l *Ilex*

2827 holly family
f ilicacées
d Aquifoliazeen; Ilizineen; Stechhülsengewächse; stechpalmenartige Gewächse; Stechpalmengewächse
l *Ilicaceae; Aquifoliaceae*

2828 holly fern
f polystic
d Punktfarn; Schildfarn
l *Polystichum*

2829 hollyhock
f rose-trémière
d Stockmalve; Stockrose
l *Althaea rosea*

2830 hollyhock mallow
f mauve alcée; mauve musquée
d Augenpappel; Rosenmalve; Rosenpappel; Siegmarskraut; Siegmarswurz; Wetterrose
l *Malva alcea*

2831 holly-leaf mayten
f maytène à feuilles de houx
l *Maytenus ilicifolia*

2832 holly oak
f chêne vert
d Grüneiche; immergrüne Eiche; Immergrüneiche; Stecheiche; Steineiche
l *Quercus ilex*

* **holy clover → 4646**

* **holy grass → 5368**

2833 homalium
f homale
l *Homalium*

2834 honesty
f lunaire
d Mondviole; Silberblatt
l *Lunaria*

2835 honewort; stone parsley
f sison
d Herrnkümmel; Würzsilge
l *Sison*

2836 honewort
d kanadische Rispendolde
l *Cryptotaenia canadensis*

* **honey agaric → 2838**

2837 honeyberry
f mélicoque
l *Melicocca*

* **honey bush → 2839**

2838 honey-color armillaria; honey agaric
f armillaire de miel; agaric couleur de miel; tête de méduse
d Bastpilz; Honigpilz; Hallimasch; honiggelber Ringling
l *Armillaria mellea*

2839 honeyflower; honey bush
f mélianthe
d Honigstrauch
l *Melianthus*

* **honeyflower → 5308**

2840 honey locust
f févier
d Dornkronenbaum; Gleditschie; Schotenbaum
l *Gledits(ch)ia*

* **honey mesquite → 1467**

2841 honeysuckle
f chèvrefeuille
d Aalbaum; Geissblatt; Heckenkirsche;

Lonizere
l *Lonicera*

2842 honeysuckle family
f caprifoliacées
d Geissblattgewächse; Kaprifoliazeen
l *Caprifoliaceae*

* **honeyware → 407**

2843 honeywort
f mélinet
d Wachsblume
l *Cerinthe*

2844 hooked bristle grass
f sétaire verticillée
d quirlblütige Borstenhirse; Wirtelborstenhirse; Wirtelfennich
l *Setaria verticillata*

2845 Hooker's barberry
f vinettier de Hooker
d Hooker Berberitze
l *Berberis hookeri*

2846 Hooker willow
f saule de Hooker
d Hookers Weide
l *Salix hookeriana*

2847 hop
f houblon
d Hopfen
l *Humulus*

2848 hop clover
d Feldklee; liegender Klee; Goldklee; Ackergoldklee
l *Trifolium agrarium*

* **hop clover → 638**

2849 hop hornbeam
f ostryer; charme houblon
d Hopfenbaum; Hopfenbuche; Schwarzbuche
l *Ostrya*

2850 hop seed bush
f dodonée
d Dodonäe
l *Dodonaea*

2851 hop tree
f ptélée
d Kleestrauch; Lederbaum; Lederstrauch
l *Ptelea*

* **horehound → 2816**

2852 hornbeam
f charme
d Hainbuche; Hornbaum; Weissbuche
l *Carpinus*

2853 hornbeam maple
f érable à feuilles de charme
d Hainbuchenahorn
l *Acer carpinifolium*

2854 horncone
f cératozamie
d Hornpalmfarn
l *Ceratozamia*

2855 horned cucumber
f concombre à petites pyramides; concombre métulifère
d afrikanische Gurke
l *Cucumis metuliferus*

* **horned pondweed → 4266**

2856 horned violet
f violette cornue
d Hornveilchen; Pyrenäen-Stiefmütterchen
l *Viola cornuta*

2857 Hornemann's willowweed
f épilobe de Hornemann
d Hornemanns Weidenröschen; Moosweidenröschen
l *Epilobium hornemanni*

2858 hornlike craterellus
f craterelle corne d'abondance; champignon noir; trompette des morts; trompette de la mort
d Herbsttrompete; Totentrompete; Schwarzfüllhorn
l *Craterellus cornucopioides*

2859 horn poppy
f pavot cornu; glaucière; glaucion
d Hornmohn
l *Glaucium*

2860 hornwort; coontail
f cornifle; cératophylle
d Hornblatt; Hornkraut; Igellock
l *Ceratophyllum*

2861 hornwort
d gemeines Hornblatt; rauhes Hornblatt
l *Ceratophyllum demersum*

2862 hornwort family
f cératophyllacées
d Hornblattgewächse; Zeratophyllazeen
l *Ceratophyllaceae*

2863 horse balm
f collinsonia
d Kollinsonie
l *Collinsonia*

2864 horse bean
f féverole; fève de cheval
d Pferdebohne
l *Vicia faba* var. *equina*

*** horse chestnut → 828**

2865 horse-chestnut family
f hippocastanacées
d Hippokastanazeen; Rosskastaniengewächse
l *Hippocastanaceae; Aesculaceae*

2866 horse fennel
f fenouil des chevaux
d Pferdesessel; Rossfenchel
l *Hippomarathrum*

2867 horse gentian; feverwort
f triostée
d Fieberwurzel
l *Triosteum*

2868 horsehair lichen
f lichen crin-de-cheval
d Federbuschflechte
l *Alectoria jubata*

2869 horsemint
f menthe sauvage
d Rossminze
l *Mentha longifolia*

*** horsemint → 512**

2870 horse mushroom
f agaric des jachères; agaric des bruyères
d Anischampignon; Anisegerling; Ackerchampignon; Angerling; Heiderling; Schafchampignon; Schafegerling
l *Agaricus arvensis; Psalliota arvensis*

2871 horse-radish
f raifort sauvage; cranson
d Fleischkraut; Green; Kren; Marettig; Meerrettich
l *Armoracia lapathifolia; Rorippa armoracia*

2872 horse-radish tree
f moringa
d Bennussbaum
l *Moringa oleifera*

2873 horseshoe pelargonium
f géranium des horticulteurs
d Zonalpelargonie
l *Pelargonium zonale*

2874 horseshoe vetch
f fer-à-cheval; hippocrépide
d Hufeisenklee
l *Hippocrepis*

2875 horsetail
f prêle
d Schachtelhalm
l *Equisetum*

2876 horsetail beefwood
f filao à feuilles de prêle; casuarina à feuilles de prêle
d Sumpfeiche; Strandkasuarine
l *Casuarina equisetifolia*

2877 horsetail family
f équisétacées
d Equisetazeen; Schachtelhalmgewächse
l *Equisetaceae*

2878 horseweed fleabane; Canadian fleabane
f érigéron du Canada; vergerette du Canada
d Feldflohkraut; kanadisches Berufkraut
l *Erigeron canadensis*

2879 hound's-tongue
f cynoglosse; langue de chien
d Hundszunge
l *Cynoglossum*

2880 house fungus; weeping merulius
d echter Hausschwamm; Holzschwamm; Tränenschwamm; tropfender Faltenschwamm
l *Serpula lacrimans; Merulius lacrimans*

2881 house holly fern
f polystic falciforme
d Sichelschildfarn; sichelförmiger Schildfarn; Sichelpunktfarn

l *Polystichum falcatum; Cyrtonium falcatum*

2882 houseleek
f joubarbe
d Hauslaub; Hauswurz; Immergrün
l *Sempervivum*

2883 Huanuco cocaine tree
f arbre à coca
d Kokastrauch; Kokainstrauch
l *Erythroxylum coca*

2884 huckleberry
f gaylussacie
d Buckelbeere; Gaylussacie
l *Gaylussacia*

2885 hugonia
f hugonie
l *Hugonia*

2886 humea
f humée
d Humea
l *Humea*

2887 Hungarian chrysanthemum
d rundblättrige Wucherblume
l *Chrysanthemum rotundifolium*

2888 Hungarian clover
f trèfle de Dalmatie; trèfle de Pannonie
d pannonischer Klee; ungarischer Klee
l *Trifolium pannonicum*

2889 Hungarian iris
f iris panaché
d bunte Schwertlilie
l *Iris variegata*

2890 Hungarian lilac
f lilas de Hongrie
d Josika-Flieder; ungarischer Flieder
l *Syringa josikaea*

* **Hungarian oak → 2979**

2891 Hungarian sainfoin
d Sandesparsette; Schildklee; spanischer Klee; türkischer Klee
l *Onobrychis arenaria*

2892 Hungarian speedwell
f véronique germandrée
d breitblättriger Ehrenpreis; grosser Ehrenpreis
l *Veronica latifolia*

2893 Hungarian vetch
f vesce de Pannonie
d ungarische Wicke; pannonische Wicke
l *Vicia pannonica*

2894 Huon pine
f dacryde
d Schuppeneibe; Träneneibe
l *Dacrydium*

* **husk corn → 4233**

2895 hutchinsia
f hutchinsie
d Feldkresse; Gemskresse
l *Hutchinsia*

2896 hyacinth
f jacinthe
d Hyazinthe
l *Hyacinthus*

2897 hyacinth dolichos
f dolique d'Egypte
d Helmbohne; Faselbohne; Lablabbohne
l *Dolichos lablab*

2898 hybrid abutilon
f abutilon hybride
d Bastardabutilon
l *Abutilon hybridum*

2899 hydnophytum
f hydnophyte
l *Hydnophytum*

2900 hydnum
f hydne
d Stachelschwamm; Stoppelpilz
l *Hydnum*

2901 hydnum family
f hydnacées
d Stachelpilze; Stoppelpilze
l *Hydnaceae*

2902 hydrangea
f hortensia
d Hortensie
l *Hydrangea*

2903 hydrangea vine
f schizophragme
d Spalthortensie
l *Schizophragma*

2904 **hygrophorus**
f hygrophore
d Ellerling; Saftling
l *Hygrophorus*

2905 **hygrophorus family**
f hygrophoracées
d Dickblättler; Wachsblätterpilze
l *Hygrophoraceae*

2906 **hypecoum**
f hypécoon
d Lappenblume
l *Hypecoum*

2907 **hypholoma**
f hypholome
d Schwefelkopf; Saumpilz
l *Hypholoma*

2908 **hypnum**
f hypne; mousse des jardinières
d Astmoos
l *Hypnum*

2909 **hyssop**
f hysope
d Isop; Ysop
l *Hyssopus*

2910 **hyssop**
f herbe sacrée; hysope officinale
d echter Ysop; gemeiner Ysop
l *Hyssopus officinalis*

2911 **hyssop-leaved tickseed**
f corisperme à feuilles d'hysope
d ysopblättriger Wanzensame
l *Corispermum hyssopifolium*

2912 **hyssop lythrum**
f lythraire à feuilles d'hysope; salicaire à feuilles d'hysope
d violetter Weiderich; Ysopweiderich
l *Lythrum hyssopifolia*

I

2913 **Iberian iris**
f iris d'Ibérie
d iberische Schwertlilie
l *Iris iberica*

2914 **icaco coco plum**
f chrysobalane icaquier
d Guajara; Ikakopflaumenbaum; Kokospflaume
l *Chrysobalanus icaco*

2915 **icegrass**
f phippsie
l *Phippsia*

2916 **Iceland moss**
f lichen d'Islande; cétraire d'Islande
d Hungermoos; isländisches Moos; Brockenmoos; Lungenflechte; Lungenmoos; Purgiermoos; Rispal; Tartsenflechte
l *Cetraria islandica*

2917 **ice plant**
f ficoïde glaciale
d Eisblume; Eiskraut; Eispflanze
l *Mesembryanthemum crystallinum*

2918 **ifé**
f sansevière cylindrique
d Bajonettpflanze; zylindrischer Bogenhanf
l *Sansevieria cylindrica*

* **ilang-ilang → 6119**

2919 **illecebrum**
f illécèbre
d Knorpelblume; Knorpelkraut
l *Illecebrum*

2920 **illipe butter tree**
f mohwa
l *Madhuca indica; Bassia latifolia*

2921 **Illyrian buttercup**
f renoncule d'Illyrie
d illyrischer Hahnenfuss; Krainer Hahnenfuss
l *Ranunculus illyricus*

2922 **imbricated hydnum**
f hydne imbriqué
d Habichtspilz; Rehfellchen; Rehpilz; brauner Habichtsschwamm; Hirschling; Hirschzunge
l *Sarcodon imbricatus; Hydnum imbricatum*

2923 **immortelle**
f xéranthème
d Papierblume; Spreublume; Strohblume
l *Xeranthemum*

* **imperial fritillary → 1711**

2924 **incarvillea**
f incarvillée
d Inkarvillea; Staudentrompete; Trompetenblume
l *Incarvillea*

2925 **incense cedar**
f libocèdre
d Flusszeder; Schuppenzeder
l *Libocedrus*

2926 **incense juniper**
f genévrier à l'encens
d portugiesische Zeder; spanische Zeder; Weihrauchbaum
l *Juniperus thurifera*

* **incense tree → 2363**

2927 **India atalantia**
f atalantie d'Inde
l *Atalantia monophylla*

2928 **India charcoal trema**
f micocoulier d'Orient
l *Trema orientalis*

* **India date → 5302**

2929 **India fig**
f banian d'Inde
d indische Feige
l *Ficus indica*

2930 **India green star**
f arbre à mâture
l *Polyalthia longifolia*

2931 **India love grass**
d behaartes Liebesgras
l *Eragrostis pilosa*

2932 **India mock strawberry**
f fraisier des Indes
d indische Erdbeere
l *Fragaria indica; Duchesnea indica*

2933 Indian azalea; Indica azalea; Indian rhododendron
f azalée de l'Inde
d indische Alpenrose; indische Azalee
l *Rhododendron indicum*

* **Indian bean → 5097**

2934 Indian canna
f canne d'Inde
d indisches Blumenrohr
l *Canna indica*

* **Indian corn → 3434**

* **Indian cress → 1478**

* **Indian currant → 2935**

2935 Indian-currant coralberry; Indian currant
f symphorine à feuilles rondes; groseillier des Indiens
d Korallenbeere; rote Schneebeere
l *Symphoricarpos orbiculatus*

2936 Indian fig
f figuier de Barbarie; raquette figuier d'Inde
d echter Feigenkaktus; indische Feige; Feigendistel
l *Opuntia ficus-indica*

2937 Indian frankincense; salai tree
f arbre à encens
d Salai
l *Boswellia serrata*

2938 Indian lettuce
f montie
d Quellkraut
l *Montia*

* **Indian mallow → 1217**

2939 Indian mustard
f moutarde à feuilles; moutarde brune; moutarde du Loiret
d Sarepta-Senf; russischer Senf
l *Sinapis juncea; Brassica juncea*

2940 Indian pipe
f sucepin
d Fichtenspargel
l *Monotropa*

2941 Indian-plum family
f flacourtiacées
d Flacourtiengewächse; Flakourtiazeen
l *Flacourtiaceae*

* **Indian rhododendron → 2933**

2942 Indian tobacco lobelia
f tabac indien
l *Lobelia inflata*

2943 India poon; beauty-leaf mastwood
f calaba à fruits ronds
l *Calophyllum inophyllum*

* **India-rubber fig → 2944**

2944 India-rubber tree; India-rubber fig; rubber plant
f caoutchouc des jardins; figuier élastique
d Gummibaum
l *Ficus elastica*

2945 India trumpet flower
f oroxylon
l *Oroxylum*

* **Indica azalea → 2933**

2946 Indies Goa bean
f pois carré
d Flügelbohne; Goabohne
l *Psophocarpus tetragonolobus*

2947 indigo
f indigotier
d Indigostrauch; Indigopflanze
l *Indigofera*

* **indigo bush → 2948**

2948 indigo-bush amorpha; indigo bush; false indigo; bastard indigo
f amorphe arbustif; faux indigo; faux indigotier; indigo bâtard
d falscher Bastardindigo; gemeiner Bastardindigo
l *Amorpha fruticosa*

2949 inga
f inga
d Inga
l *Inga*

2950 ink-cap coprinus; ink mushroom; inky cap
f coprin noir d'encre
d Antialkoholikerpilz; grauer Faltentintling; Mistschwamm; Knotentintling;

Tintenblätterpilz; Tintenschwamm
l *Coprinus atramentarius*

2951 ink mushroom; inky cap
f coprin noir; encrier
d Mistschwamm; Tintenblätterpilz; Tintenschwamm; Tintling
l *Coprinus*

* **ink mushroom → 2950**

* **inky cap → 2950**

2952 intermediate wheatgrass
f chiendent intermédiaire
d graugrüne Quecke; blaugrüne Quecke; seegrüne Quecke
l *Agropyron intermedium*

2953 intermediate wintergreen
d mittleres Wintergrün
l *Pyrola media*

2954 interrupted club moss
d sprossender Bärlapp
l *Lycopodium annotinum*

2955 inula
f aunée; œil-de-cheval
d Alant; Heinrich
l *Inula*

2956 involute paxillus
f paxille enroulé
d empfindlicher Krempling; kahler Krempling
l *Paxillus involutus; Agaricus involutus*

2957 ipecac; ipecacuanha
f ipécacuana annelé
d Brechveilchen; echte Brechwurz; Ipekakuanha
l *Cephaelis ipecacuanha*

2958 iris
f iris
d Schwertlilie; Iris
l *Iris*

2959 iris family
f iridacées
d Iridazeen; Irideen; Schwertliliengewächse
l *Iridaceae*

2960 Irish moss
f chondre
d Knorpeltang
l *Chondrus*

2961 Irish moss; carrageen
f chondre crépu; mousse d'Irlande; mousse perlée
d Felsenmoos; irländisches Moos; Perlmoos
l *Chondrus crispus*

2962 Irkutsk anemone
d Altaiwindröschen
l *Anemone altaica*

2963 iroko fustic tree
f chlorophora élevé
l *Chlorophora excelsa*

2964 iron tree; rata tree
f arbre de rata
d Eisenholzbaum; Eisenmassbaum
l *Metrosiderosis*

2965 ironweed
f vernonie
d Vernonie
l *Vernonia*

2966 ironwood; rose chestnut
f arbre de fer; bois de fer; bois de nagas
d Eisenholzbaum; Gaugau; Nagasbaum
l *Mesua ferrea*

* **ironwort → 2967**

2967 iron woundwort; ironwort
f crapaudine; sidérite
d Gliederkraut
l *Sideritis*

2968 isidium
d Korallenflechte
l *Isidium*

2969 isoloma
f isolome
d Gleichsaum
l *Isoloma*

2970 isopyrum
f isopyre
d Muschelblümchen
l *Isopyrum*

2971 Italian alder
f aune à feuilles en cœur
d herzblättrige Erle
l *Alnus cordata*

2972 Italian arum
f gouet d'Italie
d italienischer Aronstab
l *Arum italicum*

2973 Italian aster
f aster amelle; aster œil-de-Christ
d Bergaster
l *Aster amellus*

2974 Italian buckthorn; evergreen buckthorn
f nerprun alaterne
d immergrüner Kreuzdorn
l *Rhamnus alaternus*

2975 Italian bugloss
f buglosse d'Italie
d italienische Ochsenzunge
l *Anchusa italica; Anchusa azurea*

* **Italian catchfly → 2981**

2976 Italian clematis
f clématite fausse-vigne; clématite bleue
d blaue Waldrebe; italienische Waldrebe
l *Clematis viticella*

2977 Italian corn salad
d borstiges Rapünzchen
l *Valerianella eriocarpa*

2978 Italian cypress
f cyprès commun; cyprès toujours vert
d echte Zypresse; immergrüne Zypresse
l *Cupressus sempervirens*

* **Italian millet → 2352**

2979 Italian oak; Hungarian oak
f chêne de Hongrie; chêne de Pannonie; chêne hongrois
d ungarische Eiche
l *Quercus frainetto*

2980 Italian rye grass
f ivraie multiflore; ivraie d'Italie; ivraie à fleurs nombreuses; ray-grass d'Italie
d italienisches Raygras; welsches Weidengras
l *Lolium italicum; Lolium multiflorum*

2981 Italian silene; Italian catchfly
f silène d'Italie
d italienisches Leimkraut
l *Silene italica*

2982 Italian squill
f scille d'Italie
d italienischer Blaustern; italienische Meerzwiebel
l *Scilla italica*

2983 Italian stone pine
f pin pignon; pin pinier; pin bon; pignet; pin parasol
d Pinie; italienische Steinkiefer
l *Pinus pinea*

2984 Italian toadflax
f linaire d'Italie
d italienisches Leinkraut
l *Linaria italica*

2985 Ivory Coast khaya
f acajou d'Afrique
l *Khaya ivorensis*

* **ivory-nut palm → 2986, 4644**

2986 ivory palm; ivory-nut palm
f phytéléphas
d Elfenbeinpalme; Steinnusspalme
l *Phytelephas*

2987 ivorywood
f siphonode
l *Siphonodon*

2988 ivy
f lierre
d Efeu; Immergrün
l *Hedera*

2989 ivy arum
f scindapse
d Efeutute
l *Scindapsus*

2990 ivy-leaved cyclamen; Persian cyclamen
f cyclamen de Perse
d persisches Alpenveilchen; persische Erdscheibe
l *Cyclamen persicum; Cyclamen indicum*

* **ivy-leaved duckweed → 5217**

2991 ivy-leaved pelargonium
f pélargonium à feuilles de lierre
d efeublättrige Pelargonie
l *Pelargonium hederaefolium*

2992 ivy-leaved speedwell
f véronique à feuilles de lierre
d efeublättriger Ehrenpreis; Efeuehren-

preis
l *Veronica hederaefolia*

2993 ixia
f ixia; ixie
d Abendblume
l *Ixia*

2994 ixiolirion
f ixiolire
d Ixlilie
l *Ixiolirion*

2995 ixora
f ixora
d Ixora
l *Ixora*

J

2996 jacaranda
f jacaranda
d Jacaranda; Jakaranda; Palisanderholz
l *Jacaranda*

2997 jack bean
f canavalie
d Kanavalie; Krimpbohne; Riesenbohne
l *Canavalia*

2998 jackfruit; jakfruit
f jacquier; arbre à pain; artocarpe à feuilles entières
d Brotfruchtbaum; Jackfruchtbaum; indischer Brotbaum; Jaqueira
l *Artocarpus heterophyllus; Artocarpus integer; Artocarpus integrifolius*

2999 jack-in-the-pulpit
f arisème
d Feuerkolben; Fleckenaron; Zeichenwurz
l *Arisaema*

3000 Jackman clematis
f clématite de Jackman
d Jackmans Waldrebe
l *Clematis jackmani*

3001 jack pine; Banks pine
f pin de Banks; pin de Hudson
d Banks Kiefer
l *Pinus banksiana*

3002 jacksonia
f jacksonie
d Jacksonie
l *Jacksonia*

* **Jacobean lily → 393**

* **Jacob's-ladder → 2620, 4246**

3003 Jacob's-rod
f asphodéline
d Junkerlilie; Peitschenaffodil
l *Asphodeline*

3004 jagged chickweed
d doldige Spurre
l *Holosteum umbellatum*

* **jagua → 2443**

* **jakfruit → 2998**

3005 jalap
f jalap vrai; volubilis purgatif
d Jalapawinde; Purgierwinde
l *Exogonium purga; Ipomoea jalapa*

3006 jalapa plane tree
f platane de Linden
d Lindens Platane
l *Platanus lindeniana*

3007 Jamaica cherry
f calabure
l *Muntingia calabura*

3008 Jamaica linden hibiscus
f ketmie élevée
l *Hibiscus tiliaceus elatus; Hibiscus elatus*

3009 Jamaica quassia wood
f bois de Saint-Martin; quassier de Jamaïque
l *Picrasma excelsa; Picraena excelsa*

3010 jambolan; jambolan plum; duhat; Java plum
f jambo longue; jamélonier
d Jambolanabaum; Kümmelmyrte; Wachsjambuse
l *Syzygium cumini; Eugenia cumini*

3011 jamesia
f jamésie
d Jamesie
l *Jamesia*

* **Japan cedar → 3029**

* **Japan clover → 1458**

3012 Japanese alder
f aune du Japon
d japanische Erle
l *Alnus japonica*

3013 Japanese anemone
f anémone du Japon
d japanisches Windröschen
l *Anemone japonica*

3014 Japanese anise tree
f badiane; badianier
d japanischer Sternanis; Sternanispflanze
l *Illicium anisatum*

3015 Japanese apricot
f abricotier du Japon
d japanische Aprikose; Mumebaum
l *Prunus mume; Armeniaca mume*

3016 Japanese aralia
f angélique en arbre du Japon
l *Aralia elata; Aralia chinensis glabrescens*

3017 Japanese arborvitae
f thuya du Japon
d japanischer Lebensbaum
l *Thuya japonica; Thuya standishi*

3018 Japanese ardisia
f ardisie du Japon
d japanische Ardisie
l *Ardisia japonica*

* **Japanese artichoke → 332**

3019 Japanese aucuba
f aucuba du Japon
d japanische Aukube; Goldblatt; Goldorange; Orangenbaum
l *Aucuba japonica*

3020 Japanese banana
f bananier du Japon; bananier japonais
d japanische Banane; japanische Faserbanane
l *Musa basjoo*

3021 Japanese barberry
f vinettier de Thunberg
d japanische Berberitze; Thunbergs Berberitze
l *Berberis japonica; Berberis thunbergi*

3022 Japanese beech
f hêtre du Japon
d japanische Buche
l *Fagus japonica*

3023 Japanese black pine; Thunberg pine
f pin noir du Japon; pin des temples
d japanische Schwarzkiefer
l *Pinus thunbergi*

3024 Japanese brome
f brome du Japon
d bogengrannige Trespe; Flattertrespe; japanische Trespe
l *Bromus japonicus*

3025 Japanese buckthorn
f nerprun du Japon
d japanischer Kreuzdorn
l *Rhamnus japonica*

3026 Japanese butterfly bush
f buddléia du Japon
d japanischer Sommerflieder
l *Buddleia japonica*

3027 Japanese chestnut
f châtaignier du Japon
d japanische Kastanie
l *Castanea crenata; Castanea japonica*

3028 Japanese creeper; Boston ivy
f lierre japonais; vigne-vierge du Japon; vigne-vierge à trois becs
d dreispitzige Jungfernrebe
l *Parthenocissus tricuspidata*

3029 Japanese cryptomeria; Japan cedar; sugi
f cryptomérie du Japon
d japanische Cryptomerie; japanische Kryptomerie; Japanzeder; japanische Zeder; Sugizeder
l *Cryptomeria japonica*

3030 Japanese currant
f groseillier du Japon
d japanische Johannisbeere
l *Ribes japonicum*

3031 Japanese Douglas fir
f douglas du Japon
d japanische Douglasfichte
l *Pseudotsuga japonica*

3032 Japanese evodia
f évodie du Japon
l *Evodia glauca*

3033 Japanese false brome
f brachypode penné
d gefiederte Zwenke; Fiederzwenke
l *Brachypodium pinnatum*

3034 Japanese fatsia
f aralie du Japon; fatsie
d Handaralie
l *Fatsia japonica; Aralia japonica*

3035 Japanese filbert
f noisetier du Japon
d Siebolds Hasel
l *Corylus sieboldiana*

3036 Japanese fleeceflower
f persicaire cuspidée; renouée de

Siebold
d feingespitzter Knöterich; Riesenknöterich
l *Polygonum cuspidatum*

3037 Japanese flowering crab apple
f pommier floribond
d reichblütiger Zierapfel
l *Malus floribunda*

3038 Japanese flowering quince
f cognassier du Japon; cognassier de Maule
d Feuerdorn; japanische Scheinquitte; japanischer Quittenbaum; Scharlachquitte
l *Chaenomeles japonica; Chaenomeles maulei; Cydonia maulei*

3039 Japanese garden juniper
f genévrier prostré
l *Juniperus procumbens*

3040 Japanese hemlock
f tsuga du Japon; tsuga diversifolié
d verschiedenblättrige Hemlocktanne
l *Tsuga diversifolia*

3041 Japanese honey locust
f févier du Japon
d japanische Gleditschie; japanischer Christusdorn
l *Gledits(ch)ia japonica*

3042 Japanese honeysuckle
f chèvrefeuille du Japon
d japanisches Geissblatt
l *Lonicera japonica*

3043 Japanese hop
f houblon du Japon
d Japanhopfen; japanischer Hopfen
l *Humulus japonicus*

3044 Japanese hop hornbeam
f ostryer du Japon
d japanische Hopfenbuche
l *Ostrya japonica*

3045 Japanese hornbeam
f charme du Japon
d japanische Hainbuche
l *Carpinus japonica*

3046 Japanese horse chestnut
f marronnier à fruit turbiné
l *Aesculus turbinata*

3047 Japanese iris
f iris japonais
l *Iris kaempferi*

3048 Japanese katsura tree; katsura tree
f cercidophylle du Japon
d Katsurabaum; Kuchenbaum; japanischer Judasbaum
l *Cercidiphyllum japonicum*

3049 Japanese kerria
f corête du Japon; kerrie du Japon
d Goldröschen; Nesselröschen
l *Kerria japonica*

3050 Japanese lacquer tree
f arbre à vernis; sumac vernis; sumac à laque; arbre à laque
d Firnisbaum; Firnissumach
l *Rhus vernicifera; Toxicodendron vernicifluum*

3051 Japanese larch
f mélèze du Japon
d japanische Lärche; Kämpfers Lärche
l *Larix leptolepis; Larix kaempferi; Larix japonica*

3052 Japanese lawn grass
f zoysie du Japon
l *Zoysia japonica*

3053 Japanese lily
f lis japonais; lis du Japon
d japanische Lilie
l *Lilium japonicum*

3054 Japanese little-leaf box
f buis du Japon
d japanischer Buchsbaum
l *Buxus japonica; Buxus microphylla japonica*

3055 Japanese mahonia
f mahonia du Japon
d japanische Mahonie
l *Mahonia japonica*

3056 Japanese maple
f érable palme; érable du Japon
d Fächerahorn
l *Acer palmatum*

* **Japanese medlar → 3376**

3057 Japanese millet; barnyard grass
d japanische Hirse
l *Echinochloa frumentacea; Echinochloa crusgali frumentacea*

3058 Japanese mountain ash
f sorbier du Japon
d japanische Eberesche
l *Sorbus japonica*

3059 Japanese pagoda tree
f sophora du Japon; arbre des pagodes du Japon
d japanische Sophore; japanischer Schnurbaum; Rosenkranzbaum
l *Sophora japonica*

3060 Japanese plantain lily
f funkie à feuilles lancéolées; funkie lancéolée
d lanzenblättrige Funkie
l *Funkia lancifolia; Hosta japonica; Hosta lancifolia*

3061 Japanese plum
f prunier japonais
l *Prunus salicina*

3062 Japanese poplar
f peuplier de Maximowicz
d Maximowiczs Pappel
l *Populus maximowiczi*

3063 Japanese prickly ash
f poivrier du Japon
d japanischer Pfeffer
l *Zanthoxylum piperatum*

3064 Japanese primrose
f primevère du Japon
d japanische Primel
l *Primula japonica*

3065 Japanese privet
f troène du Japon
d japanischer Liguster
l *Ligustrum japonicum*

3066 Japanese raisin tree
f hovène sucré
d Fruchtbaum
l *Hovenia dulcis*

3067 Japanese red pine
f pin rouge du Japon
d japanische Rotkiefer
l *Pinus densiflora*

3068 Japanese rose
f rosier multiflore
d Polyantherose; Büschelrose; vielblütige Rose
l *Rosa multiflora*

3069 Japanese skimmia
f skimmie du Japon
d japanische Skimmie
l *Skimmia japonica*

3070 Japanese spirea
f spirée du Japon
d japanischer Spierstrauch
l *Spiraea callosa; Spiraea japonica*

3071 Japanese Staunton vine
f stauntonie à 6 folioles
d sechsblättrige Stauntonia
l *Stauntonia hexaphylla*

3072 Japanese stone pine
d Kriechkiefer
l *Pinus pumila*

3073 Japanese sweet flag
f acore graminé
d grasartiger Kalmus
l *Acorus gramineus; Acorus japonicus*

3074 Japanese torreya
f torreya du Japon
d Nusseibe; nusstragende Stinkeibe; nusstragende Torreye
l *Torreya nucifera*

3075 Japanese tree lilac
f lilas japonais
d japanischer Flieder
l *Syringa japonica; Syringa amurensis japonica*

3076 Japanese white birch
f bouleau du Japon
d japanische Birke
l *Betula japonica*

3077 Japanese wing nut
f ptérocaryer du Japon
d japanische Flügelnuss
l *Pterocarya rhoifolia*

3078 Japanese wistaria
f glycine à longues grappes; glycine du Japon; glycine floribonde
d japanische Glyzine; Blauregen; blaue Akazie
l *Wistaria floribunda*

3079 Japanese witch hazel
f hamamélide du Japon
d japanische Zaubernuss
l *Hamamelis japonica*

3080 Japanese zelkova
f kéaki
d japanische Zelkowe
l *Zelkova acuminata; Zelkova serrata; Zelkova cuspidata; Zelkova keaki*

3081 jarrah
f jarrah
d Yarraholz
l *Eucalyptus marginata*

3082 jasione
f jasione
d Jasione; Monke; Sandknöpfchen
l *Jasione*

3083 jasmine; jessamine
f jasmin
d Jasmin
l *Jasminum*

3084 jasmine orange
f murraye
d Orangenraute
l *Murraya*

3085 Java-almond canary tree
f arbre à baume; canari commun
d Elemibaum; gemeiner Kanarienbaum
l *Canarium commune*

3086 Java bishopwood
f uriam d'Assam
l *Bischofia javanica*

3087 Java fan palm
f livistone biroo
d Saribupalme
l *Livistona rotundifolia*

*** Java galangal → 2609**

*** Java plum → 3010**

3088 Java stonewood
f tarriétie de Java
l *Tarrietia javanica*

*** jelly plant → 373**

3089 Jenny stonecrop
f orpin réfléchi
d Felsenpfeffer; gelbe Tripmadam
l *Sedum reflexum*

3090 jequirity rosary pea
f liane-réglisse
l *Abrus praecatorius*

3091 Jericho resurrection mustard
f rose de Jéricho
d Jerichorose; Weihnachtsrose
l *Anastatica hierochuntica*

3092 Jersey cudweed
f gnaphale blanc jaunâtre
d gelblichweisses Ruhrkraut
l *Gnaphalium luteo-album*

3093 Jersey orchis
d Sumpfknabenkraut
l *Orchis palustris*

3094 Jerusalem artichoke
f topinambour; artichaut d'hiver; artichaut de Jérusalem
d Erdbirne; Erdapfel; Grundbirne; Jerusalemartischocke; Topinambur
l *Helianthus tuberosus*

3095 Jerusalem cherry
f morelle faux-piment
d Judenkirsche; Korallenbaum; Korallenkirsche; Strausskirsche
l *Solanum pseudocapsicum*

*** Jerusalem oak → 3096**

3096 Jerusalem-oak goosefoot; Jerusalem oak
d klebriger Gänsefuss; Knotenkraut; Krötenkraut; Schabenkraut; Traubenkraut; Traubenschmergel
l *Chenopodium botrys*

3097 Jerusalem sage
f phlomide
d Brandkraut; Filzkraut
l *Phlomis*

3098 jessamine
f cestreau
d Hammerstrauch
l *Cestrum*

*** jessamine → 3083**

*** Jesuit's nut → 5794**

3099 jetbead
f rhodotype
d Scheinkerrie; Kaimanstrauch
l *Rhodotypos*

3100 jewelvine
f derris
l *Derris*

3101 Jew's-ear
f hirnéole oreille de Judas
d Judasohr; Judenohr; Ohrpilz
l *Auricularia auricula-judae; Hirneola auricula-judae*

* **Jew's-mallow → 4283**

3102 Jimson-weed datura; thorn apple
f stramoine commune; pomme épineuse
d Dornapfel; Fliegenkraut; gemeiner Stechapfel; Igelskolben; Krötenmelde; Teufelsapfel; Tollkraut
l *Datura stramonium; Datura tatula*

* **Job's-tears → 31**

3103 Johnson grass; Guinea grass; Aleppo grass
f houlque d'Alep; sorgho d'Alep
d Aleppobartgras; Aleppohirse
l *Sorghum halepense; Andropogon halepensis*

3104 joint fir
f raisin maritime; uvette
d Meerträubchen; Meerträubel
l *Ephedra*

3105 joint fir
f gnète
d Meerträubel
l *Gnetum*

3106 joint-fir ephedra; sea grape
f raisin de mer; uvette
d gemeines Meerträubchen
l *Ephedra distachya*

3107 joint-fir family
f gnétacées; éphédracées
d Gnetazeen; Meerträubchengewächse
l *Gnetaceae; Ephedraceae*

3108 joint-tail grass
f manisure
l *Manisuris*

3109 joint vetch
f æschynomène
d Schampflanze; Sinnpflanze
l *Aeschynomene*

3110 jonquil
f jonquille
d italienische Narzisse; Jonquille
l *Narcissus jonquilla*

3111 joyweed
f alternanthère
d Papageienblatt
l *Alternanthera*

3112 juanulloa
f juanulloa
d Juanulloa
l *Juanulloa*

3113 jubaea
f jubée
d Chilepalme; Honigpalme
l *Jubaea*

3114 Judas tree
f arbre de Judas; arbre de Judée; gaînier commun
d gemeiner Judasbaum
l *Cercis siliquastrum*

3115 jujube
f jujubier
d Judendorn
l *Zizyphus*

3116 juncagina family
f juncaginacées
d Dreizackgewächse; Dreizackpflanzen; Junkaginazeen; Scheuchzeriazeen
l *Juncaginaceae*

* **Juneberry → 4832**

3117 jungle plum
f sidéroxylon
d Eisenbaum; Eisenholzbaum
l *Sideroxylon*

3118 juniper
f genévrier
d Wacholder
l *Juniperus*

3119 Jupiter's-beard centranthus; red valerian
f centranthe rouge; barbe de Jupiter; valériane des jardins
d rote Spornblume
l *Centranthus ruber*

3120 jurinea
f jurinéa
d Filzscharte; Flockenwurz; Silberscharte
l *Jurinea*

3121 justicia
f justicie

d Justicie
l *Justicia*

3122 jute
f jute
d Jute
l *Corchorus*

K

3123 kadsura
f kadsoura
d Kugelfaden
l *Kadsura*

3124 Kaffir bean tree
f schotie
d Schotie
l *Schotia*

3125 Kaffir bread
f eucephalartos
d Brotfarnpalme; Brotpalme
l *Eucephalartos*

*** Kaffir corn → 5087**

3126 Kaffir lily
f clivie
d Klivie; Riemenblatt
l *Clivia*

3127 kaki persimmon
f plaqueminier kaki
d chinesische Dattelpflaume; Kakibaum; Kakipflaume
l *Diospyros kaki; Diospyros chinensis*

3128 kale; borecole; collard
f chou vert
d Blattkohl; Staudenkohl
l *Brassica oleracea* var. *acephala*

3129 kalmia
f kalmie
d Kalmie; Lorbeerrose; Berglorbeer
l *Kalmia*

3130 Kamchatka fritillary
f fritillaire du Kamtschatka
d Saranahlilie
l *Fritillaria camtschatcensis*

3131 kangaroo grass
f théméda; anthistirie
l *Themeda*

3132 kanniedood aloe
f aloès à gorge de perdrix; aloès panache
d bunte Aloe; vielfarbige Aloe
l *Aloe variegata*

3133 kaoliang
d japanische Mohrenhirse
l *Sorghum vulgare nervosum*

3134 kapok ceiba; silk-cotton tree; kapok tree
f arbre à bourre; arbre à kapok; kapokier; kapoquier
d Kapokbaum; Wollbaum; Baumwollbaum; Kapokwollbaum
l *Ceiba pentandra*

3135 karaka nut
f corynocarpe
d Karakabaum
l *Corynocarpus*

3136 Kashgar tamarisk
f tamaris de Kashgar
l *Tamarix hispida*

3137 Kashmir cypress
f cyprès de Cachemire
d Kaschmirzypresse
l *Cupressus cashmeriana*

*** kat → 294**

3138 Katharine blood lily
f hémanthe de Catherine Saunders
d Katharinas Blutblume
l *Haemanthus katherinae*

*** katsura tree → 3048**

3139 kauri dammar pine
f pin de cowrie
d Kaurifichte; neuseeländischer Dammarabaum
l *Agathis australis*

3140 kava pepper
f kawa; kava
d Awapfeffer; Kawapfeffer; Kawastrauch; Rauschpfeffer
l *Piper methysticum*

3141 Kawakami fir
f sapin de Kawakami
d Kawakami-Tanne
l *Abies kawakami*

3142 kayea
f kayée
l *Kayea*

3143 keeled corn salad
d gekieltes Rapünzchen; gekielter Feldsalat
l *Valerianella carinata*

3144 keeled onion
d Berglauch; gekielter Lauch
l *Allium carinatum*

3145 kelps
f laminariacées
d Laminariazeen
l *Laminariaceae*

3146 kenaf hibiscus; ambary hemp; Deccan hemp
f chanvre de Deccan; gombo chanvre
d Ambarihanf; Bombayhanf; Dekkanhanf; Gambohanf; ostindische Hanfrose
l *Hibiscus cannabinus*

3147 Kenilworth ivy; Aaron's-beard; mother-of-thousands
f cymbalaire; lierre des murailles; ruines-de-Rome
d efeublättriges Leinkraut; gemeines Zymbelkraut
l *Cymbalaria muralis; Linaria cymbalaria*

3148 kennedia
f kennedye; kennédie
d Kennedie
l *Kennedya; Kennedia*

3149 kentia
f kentie
d Kentia
l *Kentia*

3150 Kentucky bluegrass
f pâturin des prés
d Wiesenrispengras; gemeines Angergras; Wiesenrispe
l *Poa pratensis*

3151 Kentucky coffee tree
f chicot dioïque; chicot du Canada
d amerikanischer Geweihbaum; (kanadischer) Schusserbaum
l *Gymnocladus dioicus; Gymnocladus canadensis*

3152 Kerguelen cabbage
d Kerguelenkohl
l *Pringlea antiscorbutica*

3153 kermes oak
f chêne kermès
d Kermeseiche
l *Quercus coccifera*

3154 kernera
f kernéra
d Kugelschötchen
l *Kernera*

3155 Kerner bristle thistle
f chardon de Kerner
d Kerners Distel
l *Carduus kerneri*

3156 kerria
f kerrie
d Kerrie; Goldröschen; Ranunkelstrauch
l *Kerria*

3157 keteleeria
f kételéerie
d Keteleerie; Stechtanne; Zedertanne
l *Keteleeria*

3158 kidney bean
f haricot commun
d Fasel; Fasole; Fisole; Gartenbohne; Schminkbohne
l *Phaseolus vulgaris*

3159 kidney vetch anthyllis; woundwort
f vulnéraire; anthyllide vulnéraire; trèfle jaune des sables
d gelber Klee; russischer Klee; gelber Wundklee; Wollklee; Wollblume; Tannenklee; Wundklee
l *Anthyllis vulneraria*

3160 king devil
d hohes Habichtskraut
l *Hieracium praealatum*

3161 king-devil hawkweed
d Florentiner Habichtskraut
l *Hieracium florentinum*

3162 king orange
f mandarinier
d echte Mandarine; Kauchin
l *Citrus nobilis*

3163 kirengeshoma
f kirengeshoma
d Scheinglocke
l *Kirengeshoma*

*** kittool palm → 5540**

3164 Klamath plum
f prunier du Pacifique
l *Prunus subcordata*

3165 kleinhovia
f kleinhovie
l *Kleinhovia*

3166 klugia
f klugie
l *Klugia*

3167 knautia
f knautia
d Honigblume; Knautie; Witwenblume
l *Knautia*

3168 knawel
f scléranthe
d Knaul; Knäuel
l *Scleranthus*

*** knight's-star → 176**

*** knotgrass → 1622, 4317**

3169 knotted clover
d gestreifter Klee
l *Trifolium striatum*

3170 knotted hedge parsley
d knotiger Klettenkerbel
l *Torilis nodosa*

3171 knotted pearlwort
f sagine noueuse
d Knotenmastkraut; knötiges Mastkraut
l *Sagina nodosa*

3172 knotweed
f renouée
d Knöterich
l *Polygonum*

*** kobresia → 1302**

3173 koeleria
f koélérie
d Schillergras
l *Koeleria*

3174 koenigia
f koenigie
d Koenigie
l *Koenigia*

3175 kohlrabi
f chou-rave
d Oberrübe; Oberkohlrübe; Kohlrabi; Rübkohl
l *Brassica oleracea* var. *gongylodes*

3176 kolomikta actinidia
f actinidie kolomikta
d Aktinidie Kolomikta; mandschurischer Strahlengriffel
l *Actinidia kolomikta*

*** kombu → 5995**

3177 Korean arborvitae
f thuya de Corée
d koreanischer Lebensbaum
l *Thuya koraiensis*

3178 Korean barberry
f épine-vinette de Corée
d koreanische Berberitze
l *Berberis koreana*

3179 Korean chrysanthemum
f chrysanthème de Sibérie
d sibirische Wucherblume
l *Chrysanthemum sibiricum; Chrysanthemum koreanum*

3180 Korean clematis
f clématite de Corée
d koreanische Doppelblume
l *Clematis koreana*

3181 Korean fir
f sapin de Corée
d koreanische Tanne
l *Abies koreana*

3182 Korean hackberry
f micocoulier de Corée
d koreanischer Zürgelbaum
l *Celtis koraiensis*

3183 Korean lespedeza
f lespédézie stipulacée; lespédézie de Corée
d koreanischer Klee
l *Lespedeza stipulacea*

3184 Korean pine
f pin de Corée
d koreanische Kiefer
l *Pinus koraiensis*

3185 Korean poplar
f peuplier de Corée
d koreanische Pappel
l *Populus koreana*

3186 Korean spice viburnum
f viorne de Carles
d Duftschneeball
l *Viburnum carlesi*

3187 Koyama spruce
f épicéa de Corée

d koreanische Fichte
l *Picea koraiensis; Picea koyamai*

3188 kudzu bean; kudzu vine
f puéraire; kudzu
d Kopoubohne; Kudzu
l *Pueraria*

3189 kumquat
f kumquat
d Kumquat
l *Fortunella*

3190 kusso tree
f cousso; kousso
d Kosobaum
l *Hagenia abyssinica*

L

3191 **laboulbenia family**
f laboulbéniacées
d Laboulbeniazeen
l *Laboulbeniaceae*

3192 **laburnum; golden chain**
f aubour; cytise; faux ébénier
d Bohnenbaum; Goldregen; Kleebaum
l *Laburnum*

3193 **lacebark pine**
f pin à écorce en dentelle; pin Napoléon
d Bungekiefer
l *Pinus bungeana*

3194 **laceflower**
f didisque
d Blaudolde
l *Didiscus; Trachymene*

3195 **lactarius**
f lactaire
d Milchblätterschwamm; Milchling
l *Lactarius*

3196 **lacunose helvella**
f helvelle lacuneuse
d Grubenlorchel
l *Helvella lacunosa*

3197 **ladies'-tresses**
f spiranthe
d Drehwurz; Schraubenorchis; Wendelähre
l *Spiranthes*

3198 **lady bell; lady's-bell**
f adénophore
d Becherglocke; Drüsenglocke; Drüsenträger
l *Adenophora*

3199 **lady fern**
f fougère femelle commune
d falscher Wurmfarn; Waldfrauenfarn
l *Athyrium filixfemina*

3200 **lady orchis**
d purpurrotes Knabenkraut; Purpurknabenkraut
l *Orchis purpurea*

3201 **lady palm**
f rhapis
d Rutenpalme; Steckenpalme
l *Rhapis*

* **lady's-bell → 3198**

* **lady's-comb → 4861**

3202 **lady's-mantle**
f alchémille
d Frauenmantel; Sinau
l *Alchemilla*

3203 **lady's-slipper**
f sabot de Vénus; sabot de Notre-Dame; cypripède
d Frauenschuh; Venusschuh
l *Cypripedium*

3204 **lagoseris**
f lagosérìde
l *Lagoseris*

3205 **Lagos silk-rubber**
f fontumia à caoutchouc
l *Funtumia elastica*

3206 **lagotis**
f lagotis; gymnandre
d Hasenohr
l *Lagotis*

3207 **lakoocha**
f bois de dhau
l *Artocarpus lakoocha*

3208 **lallemantia**
f lallemantie
d Lallemantie
l *Lallemantia*

3209 **Lambert hawthorn**
f épine de Lambert
d Lambertsweissdorn
l *Crataegus lambertiana*

* **Lambert nut → 2473**

3210 **lambkill kalmia**
f kalmie à feuilles étroites
d schmalblättrige Lorbeerrose
l *Kalmia angustifolia*

3211 **lamb's-lettuce**
d Ackersalat; gemeiner Feldsalat; Fettmännchen; Mädchensalat; Nüsschensalat; Rabinschen; Rapünzchen; Rapunze; Salatrapünzchen
l *Valerianella dichotoma; Valerianella locusta*

3212 lamb's quarters goosefoot
f chénopode blanc; ansérine blanche
d weisser Gänsefuss; weisse Melde
l *Chenopodium album*

3213 lamb succory
f arnoséris; arnoséride
d Lämmersalat
l *Arnoseris*

3214 Lamy willowweed
f épilobe de Lamy
d graugrünes Weidenröschen
l *Epilobium lamyi*

3215 lance-leaf crotalaria
f crotalaire lancéolée
l *Crotalaria lanceolata*

3216 lancepod
f lonchocarpe
d Lonchocarpus
l *Lonchocarpus*

3217 langsat
f lanse
d Lanzenbaum
l *Lansium*

3218 lantana
f lantana
d Wandelröschen
l *Lantana*

3219 Lapland pedicularis
f pédiculaire de Laponie
d lappländisches Läusekraut
l *Pedicularis lapponica*

3220 Lapland rhododendron
f rosage de Laponie
d nordische Alpenrose
l *Rhododendron lapponicum*

3221 Lapland willow
f saule de Laponie
d lappländische Weide
l *Salix lapponum*

3222 larch
f mélèze
d Lärche(nbaum)
l *Larix*

3223 lardizabala
f lardizabale
d Lardizabala
l *Lardizabala*

* **large bird's-foot trefoil → 5872**

3224 large cranberry
f canneberge à gros fruits
d grosse Moosbeere
l *Vaccinium macrocarpum*

3225 large hop clover
f trèfle des champs
d gelber Ackerklee; liegender Klee
l *Trifolium campestre*

3226 larkspur
f dauphinelle
d Rittersporn
l *Delphinium*

3227 laserwort; woundwort
f laser
d Laserkraut
l *Laserpitium*

3228 lasia
f lasie
d Lasie
l *Lasia*

3229 latania
f latanier
d Latanie; Samtpalme
l *Latania*

3230 lathraea
f lathrée
d Schuppenwurz
l *Lathraea*

3231 laurel
f laurier
d Lorbeer(baum)
l *Laurus*

3232 laurel family
f lauracées
d Laurazeen; Laurinen; Lorbeergewächse
l *Lauraceae*

3233 laurel oak
f chêne à feuilles de laurier
d Lorbeereiche
l *Quercus laurifolia*

3234 laurel poplar
f peuplier à feuilles de laurier
d Lorbeerpappel; lorbeerblättrige Pappel
l *Populus laurifolia*

3235 laurel rockrose
f ciste à feuilles de laurier
d lorbeerblättrige Zistrose
l *Cistus laurifolius*

3236 laurel willow; bay willow
f saule laurier; saule pentandrique
d Faulweide; Lorbeerweide
l *Salix pentandra*

3237 laurestinus viburnum
f laurier-tin
d Bastardlorbeer; immergrüner Schneeball; Lorbeerschneeball; Steinlorbeer
l *Viburnum tinus*

3238 laurocerasus
f laurier-cerise
d Kirschlorbeer
l *Laurocerasus*

3239 lavender
f lavande
d Lavendel
l *Lavandula*

3240 lavender cotton
f santoline
d Heiligenkraut; Heiligenblume; Heiligenpflanze; Zypressenkraut
l *Santolina*

3241 laver
f porphyre
d Purpurtang
l *Porphyra*

3242 lawn grass
f zoysie; zoydie
l *Zoysia*

3243 Lawson false cypress
f cyprès de Lawson
d Lawsons Lebensbaumzypresse; Lawsons Scheinzypresse; Lawsons Zypresse; Ingwertanne
l *Chamaecyparis lawsoniana*

3244 lead-colored bovista
f boviste plombé
d bleigrauer Zwergbovist
l *Bovista plumbea*

3245 lead tree
d Weissfaden
l *Leucaena*

* **leadwort → 4225**

3246 leadwort family
f plumbaginacées
d Bleiwurzgewächse; Grasnelkengewächse
l *Plumbaginaceae*

* **leaf beet → 1116**

3247 leaf cactus
f épiphylle; phyllocacte
d Blattkaktus; Flügelkaktus; Phyllokaktus; Gliederkaktus; Stutzblattkaktus; Weihnachtskaktus
l *Epiphyllum; Phyllocactus*

3248 leaf cup
f polymnie
d Polymnie
l *Polymnia*

3249 leaf flower
f phyllanthe
d Blattblume
l *Phyllanthus*

3250 leafy adder's-mouth orchid
f malaxide des marais; ophrys des marais
d Weichwurz
l *Malaxis paludosa*

3251 leafy euphorbia
d Eselwolfsmilch; Rutenwolfsmilch; scharfe Wolfsmilch
l *Euphorbia virgata; Euphorbia esula*

3252 least bur reed
d kleinster Igelkolben; Zwergigelkolben
l *Sparganium minimum*

3253 least willow
f saule à feuilles rondes
d rundblättrige Weide
l *Salix rotundifolia*

3254 leather bergenia
f bergénie à feuilles charnues
d Wickelwurz; dickblättrige Wickelwurz
l *Bergenia crassifolia*

3255 leatherflower clematis
f clématite viorne
l *Clematis viorna*

3256 leather fungi
f théléphoracées

d Rindenpilze
l *Thelephoraceae*

3257 leatherleaf
f chamédaphné
d Lederblatt; Torfgränke; Zwerglorbeer
l *Chamaedaphne*

3258 leather-leaved rose
d lederblättrige Rose
l *Rosa coriifolia*

3259 leatherwood
f bois-cuir; dircé
d Lederholz
l *Dirca*

3260 Lebanon buckthorn
f nerprun du Liban
l *Rhamnus libanotica*

3261 Lebanon oak
f chêne du Liban
d Libanoneiche
l *Quercus libani*

3262 lecanora
f lécanore
d Mannaflechte; Scheibenflechte
l *Lecanora*

3263 lecoc(k)ia
f lecokie
l *Lecoc(k)ia*

3264 Ledebour lily
f lis de Ledebour
d Ledebours Lilie
l *Lilium ledebouri*

3265 ledum
f lédon; lède
d Porst
l *Ledum*

3266 leek
f poireau; porreau
d Porree; Breitlauch; spanischer Lauch; Winterlauch; Aschlauch; Fleischlauch; Preisslauch; Perlzwiebel
l *Allium porrum*

3267 legumes
f légumineuses
d Hülsenfruchtgewächse; Hülsenfrüchtige; Hülsenfrüchtler; Hülsengewächse; Leguminosen
l *Leguminosae*

3268 lemon
f citronnier; limonier; citron
d Limone(nbaum); Sauerzitrone; Zitrone(nbaum)
l *Citrus limon; Citrus limonia*

3269 lemon day lily
f lis jaune
d gelbe Taglilie
l *Hemerocallis flava*

3270 lemon grass
f verveine des Indes; herbe citron
d Lemongras; Zitronengras
l *Cymbopogon citratus*

3271 lemon-verbena lippia
f verveine-citronelle; verveine en arbre
d Punschpflanze; Zitronenkraut
l *Lippia citriodora; Aloysia citriodora; Verbena citriodora*

3272 lenten rose
f hellébore orientale; ellébore d'Orient
l *Helleborus orientalis*

3273 lentil
f lentille
d Erve; Linse; Linsenerve
l *Lens*

3274 lentinus
f lentinus
d Korkschwamm
l *Lentinus*

3275 lentisk pistache; mastic tree
f arbre à résine; arbre de mastic; pistachier lentisque
d Mastixbaum; Mastixpistazie; Mastixstrauch; Sondrio
l *Pistacia lentiscus*

* **leopard flower → 620**

3276 leopard's-bane
f doronic
d Gemswurz
l *Doronicum*

3277 leopoldia
d Bisamhyazinthe
l *Leopoldia*

3278 lepiota
f lépiote
d Parasolpilz; Schirmpilz
l *Lepiota*

3279 lespedeza; bush clover
f lespédézie; luzerne tropicale
d Buschklee; Lespedezie
l *Lespedeza*

3280 lesser honeywort
f petit mélinet
d kleine Wachsblume
l *Cerinthe minor*

3281 lettuce
f laitue
d Gartenlattich; Gartensalat; Lattich; Salat; Milchlattich
l *Lactuca; Mulgedium*

3282 leucorchis
d Weisszüngel
l *Leucorchis*

3283 Levant cotton
f cotonnier d'Asie
l *Gossypium herbaceum*

3284 Levant wormwood
d Wurmsamenkraut; Zitwersamenkraut
l *Artemisia cina*

3285 lewisia
f lewisie
d Auferstehungspflanze; Bitterwurz(el); Lewisie
l *Lewisia*

3286 libanotis
f libanotide
d Heilwurz; Hirschheil
l *Libanotis*

3287 Liberian coffee
f caféier de Libéria
d Liberiakaffee
l *Coffea liberica*

3288 libertia
f libertie
d Libertie
l *Libertia*

3289 licania
f licanie
l *Licania*

3290 lichen
f lichen
d Flechte
l *Lichen*

3291 licorice; liquorice
f réglisse
d Süssholz
l *Glycyrrhiza*

3292 licuala palm
f licuala
d Likualapalme; Strahlenpalme
l *Licuala*

3293 lignum vitae
f bois de gaïac; gaïac
d Guajakbaum
l *Guaiacum*

3294 ligusticum
f ligustique
d Mutterwurz
l *Ligusticum*

3295 lilac
f lilas
d Flieder; Syringe
l *Syringa*

3296 lilac chaste tree; agnus castus; monk's pepper tree
f arbre au poivre; gattilier; gattlier
d Keuschlamm; Mönchspfeffer; Müllen
l *Vitex agnuscastus*

3297 lilac clematis
f clématite azurée
d blaue Clematis; offenblütige Waldrebe
l *Clematis patens*

3298 lilac pink
f œillet superbe
d Prachtnelke
l *Dianthus superbus*

3299 lilac sage
f sauge verticillée
d Quirlsalbei; quirlige Salbei; vielblütige Salbei
l *Salvia verticillata*

3300 lily
f lis
d Lilie
l *Lilium*

3301 lily family
f liliacées
d Liliazeen; Liliengewächse
l *Liliaceae*

3302 **lily-leaved lady bell**
f adénophore à feuilles de lis
d Schellenblume; wohlriechende Becherglocke
l *Adenophora lilifolia*

3303 **lily leek**
f ail doré
d Goldlauch
l *Allium moly*

3304 **lily magnolia**
f magnolia à fleurs de lis
d Lilienmagnolie
l *Magnolia liliflora*

3305 **lily of the valley**
f muguet
d Maiglöckchen
l *Convallaria*

3306 **lily of the valley; May lily**
f muguet de mai
d Maiblume; Maiblümchen; Maiglöckchen; Maililie; Zauke
l *Convallaria majalis*

3307 **lily turf; snake's-beard**
f ophiopogon
d Schlangenbart
l *Ophiopogon; Mondo*

3308 **lily turf**
f liriope
d Lilienschwertel
l *Liriope*

3309 **lima bean**
f pois du Cap; haricot de Lima
d Limabohne; Madagaskarbohne; Diffinbohne; Rangoonbohne
l *Phaseolus limensis; Phaseolus lunatus* var. *macrocarpus*

3310 **limber pine**
f pin souple
d kalifornische Kiefer
l *Pinus flexilis*

3311 **lime**
f limettier; limonelle; limette acide; lime acide
d Limettenzitrone; saure Limette; Limettenbaum
l *Citrus aurantifolia; Citrus limetta*

3312 **lime prickly ash**
f bois pial
l *Zanthoxylum fagara; Zanthoxylum pterota*

3313 **lindelofia**
f lindélofie
d Lindelofie
l *Lindelofia*

3314 **linden**
f tilleul
d Linde(nbaum)
l *Tilia*

3315 **linden family**
f tiliacées
d Lindengewächse; Tiliazeen
l *Tiliaceae*

3316 **linden hibiscus**
f bois de liège des Antilles; ketmie à feuilles de tilleul
d lindenblättriger Eibisch
l *Hibiscus tiliaceus*

3317 **linden-leaf grewia**
f greuvier à feuilles de tilleul
d lindenblättrige Grewie
l *Grewia tiliaefolia*

3318 **Lindley false spirea**
f sorbaire de Lindley
d Lindleys Fiederspiere
l *Sorbaria lindleyana; Sorbaria tomentosa*

* **linseed → 3915**

3319 **lint lacebark (tree)**
f bois dentelle
d Leinwandbaum
l *Lagetta lintearia*

3320 **lion's-ear**
f léonotis
d Löwenohr; Löwenschwanz
l *Leonotis*

3321 **lion's-heart**
f physostégie
d Gelenkblume; Physostegie
l *Physostegia*

3322 **lion's-leaf**
f léontice
d Trapp
l *Leontice*

3323 **lip fern**
f chéilanthe

d Keuladerfarn; Lippenfarn
l *Cheilanthes*

3324 lippia
f lippia
d Zitronenstrauch; Lippie
l *Lippia*

* **liquorice → 3291**

* **litchi → 3397, 3398**

3325 litmus roccella
f rocelle; orseille
d Färberflechte; Lackmusflechte; Orseilleflechte
l *Roccella tinctoria*

3326 little-flower quickweed
d Choleradistel; Franzosenkraut; Goldköpfchen; kleinblütiges Knopfkraut
l *Galinsoga parviflora*

3327 little-leaf linden
f tilleul à petites feuilles
d Berglinde; kleinblättrige Linde; kleinblättrige Steinlinde; Ostlinde; Winterlinde
l *Tilia cordata; Tilia parvifolia; Tilia silvestris*

3328 little-leaved mock orange
f seringa à petites feuilles
d kleinblättriger Pfeifenstrauch
l *Philadelphus microphyllus*

3329 little-leaved pea shrub
f acaçia de Sibérie
l *Caragana microphylla*

3330 little-leaved tunic-flower
d sprossende Felsennelke
l *Tunica prolifera*

3331 little love grass
d kleines Liebesgras
l *Eragrostis minor; Eragrostis poaeoides*

3332 little medic
d kleine Luzerne; kleinster Schneckenklee; Zwergschneckenklee
l *Medicago minima*

3333 little oat
f avoine courte; avoine à fourrage
d Kurzhafer; Silberhafer
l *Avena brevis*

3334 little-pod false flax
d kleinfrüchtiger Dotter; kleinfrüchtiger Leindotter; Wilddotter
l *Camelina microcarpa*

3335 little quaking grass
f petite brize
d kleines Zittergras
l *Briza minor*

3336 little-seeded canary grass
d kleines Glanzgras
l *Phalaris minor*

3337 little serradella
d kleine Klauenschote; kleiner Vogelfuss
l *Ornithopus perpusillus*

3338 little starwort
f stellaire graminée
d Grasmiere; Grassternmiere
l *Stellaria graminea*

3339 little-tree willow
d Bäumchenweide
l *Salix arbusculoides*

3340 littonia
f littonia
d Littonie
l *Littonia*

3341 live-forever (sedum)
f orpin reprise
d breitblättrige Fetthenne; grosse Fetthenne; Schmerzwurzel
l *Sedum telephium*

3342 live oak
f chêne de Caroline
d virginische Eiche
l *Quercus virginiana; Quercus virens*

* **liver agaric → 515**

3343 liver-leaf hepatica
f hépatique trilobée
d Leberblümchen; Märzblümchen
l *Hepatica triloba; Hepatica nobilis*

3344 living rock; living-rock cactus
f anhalonie
d Wollfruchtkaktus
l *Anhalonium; Ariocarpus*

3345 lizard arum
f sauromate

d Eidechsenwurz
l *Sauromatum*

3346 lizard orchis
f himanthoglosse
d Bocksorchis; Riemenzunge
l *Himant(h)oglossum*

3347 lizard's-tail
f lézardelle
d Eidechsenschwanz; Molchschwanz
l *Saururus*

3348 lizard's-tail family
f saururacées
d Eidechsenschwanzpflanzen; Saururazeen
l *Saururaceae*

3349 lobed holly fern
f fougère femelle
d gelappter Schildfarn; Stachelfarn
l *Polystichum lobatum; Polystichum aculeatum*

3350 lobelia; cardinal flower
f lobélie
d Kardinalblume; Lobelie
l *Lobelia*

3351 lobelia family
f lobéliacées
d Lobeliengewächse
l *Lobeliaceae*

3352 loblolly pine
f pin à encens
d amerikanische Terpentinkiefer; Fackelbaum; Weihrauchkiefer
l *Pinus taeda*

3353 locoweed; milk vetch
f astragale
d Tragant; Bärenschote
l *Astragalus*

3354 locust
f faux-acacia; robinier
d Heuschreckenbaum; Robinie; Scheinakazie
l *Robinia*

3355 lodgepole pine
f pin de Murray
d Murrayakiefer
l *Pinus murrayana; Pinus contorta latifolia*

3356 Loesel twayblade
d Glanzkraut
l *Liparis loeseli*

* **lofty pine → 2807**

3357 Loganberry
f ronce-framboise
d Loganbeere
l *Rubus loganobaccus*

3358 logania family
f loganiacées
d Brechnussgewächse; Logangewächse; Loganiazeen
l *Loganiaceae*

3359 logwood; campeche wood
f bois de campêche
d Blauholz(baum); Blutbaum; Blutholzbaum; Kampeschebaum
l *Haematoxylon campechianum*

3360 Lombardy poplar
f peuplier de Lombardie; peuplier d'Italie; peuplier pyramidal
d Alleepappel; italienische Pappel; pyramidenförmige Pappel; Pyramidenpappel
l *Populus nigra* var. *italica; Populus nigra* var. *pyramidalis*

3361 London plane tree
f platane à feuilles d'érable; platane de Londres
d ahornblättrige Platane; Strauchplatane
l *Platanus acerifolia*

3362 London pride (saxifrage)
f saxifrage des ombrages
d Porzellanblümchen
l *Saxifraga umbrosa*

3363 London rocket
d Glanzrauke; langblättrige Rauke
l *Sisymbrium irio; Norta irio*

3364 long-bracted sedge
d ausgedehnte Segge
l *Carex extensa*

3365 long-cluster Japanese wistaria
f glycine à folioles nombreuses
l *Wistaria floribunda macrobotrys; Wistaria multijuga*

* **long cyperus → 2402**

3366 longleaf pine
f pin à longues feuilles; pin de Boston; pin palustre
d Sumpfkiefer
l *Pinus palustris*

3367 long-pod poppy
f pavot douteux
d Saatmohn; Schmalkopfmohn
l *Papaver dubium*

* **long pondweed → 5914**

3368 long-rooted onion
f ail de cerf; herbe à 9 chemises
d Allermannsharnisch; Siegwurz
l *Allium victorialis*

3369 long-stalked geranium
f bec-de-grue colombin; bec-de-grue pigeon; pied-de-pigeon
d Steinstorchschnabel; Taubenstorchschnabel
l *Geranium columbinum*

* **long-tube gold-drop onosma → 1690**

3370 loosestrife
f lysimaque
d Felder; Felberich; Gelbweiderich; Gilbweiderich
l *Lysimachia*

3371 loosestrife family
f lythracées
d Lythrazeen; Weiderichgewächse
l *Lythraceae*

3372 lophira
f lophire
l *Lophira*

3373 lopseed
f priva; phryma
l *Phryma*

3374 lopside oat
f avoine strigeuse; avoine nerveuse
d Rauhafer; Sandhafer
l *Avena strigosa*

3375 loquat
f bibacier
d Wollmispel; Japanmispel; Wolltraube
l *Eriobotrya*

3376 loquat; Japanese medlar
f néflier du Japon
d japanische Mispel
l *Eriobotrya japonica*

3377 loranth
f loranthe
d Eichenmispel; Riemenblume
l *Loranthus*

3378 lords-and-ladies; cuckoopint
f gouet commun; gouet maculé; gouet à feuilles maculées; pied-de-veau
d Aasblume; gefleckter Aronstab; gemeiner Aronstab; Eselsohr; deutscher gefleckter Ingwer
l *Arum maculatum*

3379 Lorentz red quebracho
f quebracho rouge
d roter Quebrachobaum
l *Quebrachia lorentzi; Schinopsis lorentzi*

3380 loropetalum
f loropétale
d Riemenblume; Riemenblüte
l *Loropetalum*

3381 lotus
f nélombo; nélumbo
d indische Seerose
l *Nelumbium; Nelumbo*

* **lotus tree → 3862**

3382 lousewort
f pédiculaire des bois
d Waldläusekraut
l *Pedicularis silvatica*

3383 lousewort; wood betony
f pédiculaire
d Läusekraut
l *Pedicularis*

3384 lovage
f livèche
d Liebstöckel; Maggikraut
l *Levisticum*

3385 love grass; teff
f éragrostide
d Liebesgras
l *Eragrostis*

3386 love-in-a-mist
f nigelle de Damas
d Braut im Haar; Gretchen im Busch; Gretel im Busch; Gretchen im Grünen; Jungfer im Busch; Jungfer im Grünen;

Jungfer in Haaren; Kapuzinerkraut; türkischer Schwarzkümmel
l *Nigella damascena*

3387 love-lies-bleeding
f amarante queue-de-renard; amarante à fleurs en queue
d echter Fuchsschwanz; Fuchsschwanz-amarant; Gartenfuchsschwanz; rispiger Amarant; roter Fuchsschwanz; Tausendschön
l *Amaranthus caudatus; Amaranthus paniculatus*

3388 lovely golden larch
f faux-mélèze de Kaempfer
d chinesische Goldlärche
l *Pseudolarix kaempferi*

3389 low bamboo
d niedriger Bambus
l *Pleioblastus humilis*

3390 low birch
f bouleau nain
l *Betula pumila*

3391 low cudweed
d Sumpfrohrkraut; Sumpfruhrkraut
l *Gnaphalium uliginosum*

*** low flax → 2214**

3392 low meadow rue
f pigamon mineur
d kleinblättrige Wiesenraute; kleine Wiesenraute
l *Thalictrum minus*

3393 lungwort
f pulmonaire
d Lungenkraut
l *Pulmonaria*

3394 lungwort sticta
d Lungenflechte; Lungenmoos
l *Lobaria pulmonaria; Sticta pulmonaria*

*** lupinaster → 473**

3395 lupine
f lupin
d Feigbohne; Lupine; Wolfsbohne
l *Lupinus*

3396 lurid boletus; underoak mushroom
f bolet blafard; cèpe blafard
d Hexenpilz; Hexenröhrling; Hexenschwamm; netzstieliger Hexenröhrling
l *Boletus luridus*

3397 lychee; litchi
f litchi
d Litschibaum
l *Litchi*

3398 lychee; litchi
f cerisier de la Chine; litchi de Chine; litchi ponceau
d Litschibaum
l *Litchi chinensis; Nephelium litchi*

3399 lycoperdon nut
d gekörnte Hirschtrüffel; warzige Hirschtrüffel
l *Elaphomyces cervinus*

*** lyme grass → 5957**

3400 lythrum
f lythraire
d Blutweiderich; Weiderich
l *Lythrum*

M

3401 **Macartney rose**
f rosier de Macartney
d Macartneys Rose
l *Rosa macartnea*

3402 **macawood**
f platymiscie
d Trebol
l *Platymiscium*

3403 **Macedonian oak**
f chêne macédonien
d makedonische Eiche
l *Quercus macedonica; Quercus trojana*

3404 **Macedonian toadflax**
f linaire de Macédonie
d makedonisches Leinkraut
l *Linaria macedonica*

3405 **MacNab cypress**
f cyprès de MacNab
d MacNabs Zypresse
l *Cupressus macnabiana*

3406 **macrozamia**
f macrozamia
d Keulenpalme; Macrozamie
l *Macrozamia*

3407 **Madagascar clove**
d madagassische Muskatnuss
l *Ravensara aromatica*

* **Madagascar plum → 4413**

3408 **Madagascar raffia palm**
d Bambuspalme; madagassische Sagopalme; Raffiabastpalme; Raffiaweinpalme
l *Raphia ruffia; Raphia pedunculata*

3409 **Madagascar traveler's-tree**
f arbre des voyageurs
d Baum der Reisenden; Quellenbaum
l *Ravenala madagascariensis*

3410 **madder**
f garance
d Färberröte; Krapp; Röte
l *Rubia*

3411 **madder family**
f rubiacées
d Labkrautgewächse; Rötegewächse; Rubiazeen
l *Rubiaceae*

3412 **Madeira bay persea**
f laurier de l'Inde
d Madeiralorbeer
l *Persea indica*

3413 **Madeira vine**
f boussingaultie
d Basellkartoffel; Boussingaultie
l *Boussingaultia*

3414 **Madeira-vine family**
f basellacées
d Basellazeen; Basellgewächse
l *Basellaceae*

3415 **madflower**
f antholise; mérianelle
d Rachenlilie; Rachenschwertel
l *Antholyza*

3416 **madia**
f madie
d Madie
l *Madia*

3417 **Madonna lily**
f lis blanc
d weisse Lilie
l *Lilium candidum*

3418 **madrone**
f arbousier
d Erdbeerbaum
l *Arbutus*

3419 **madwort**
f râpette; portefeuille
d Scharfkraut
l *Asperugo*

3420 **maerua**
f mérue
l *Maerua*

3421 **Magellan barberry**
f vinettier à feuilles de buis; épine-vinette à feuilles de buis
d Zwergkugelberberitze
l *Berberis buxifolia*

3422 **Magellan fuchsia**
f fuchsia commun
d gewöhnliche Fuchsie; Purpurfuchsie
l *Fuchsia magellanica*

3423 magnolia
f magnolier
d Magnolie
l *Magnolia*

3424 magnolia family
f magnoliacées
d Magnoliazeen; Magnoliengewächse
l *Magnoliaceae*

3425 magnolia vine
f schizandre
d Beerentraube; Spaltkölbchen
l *Schizandra*

* **maguey agave → 4328**

3426 mahaleb cherry; St. Lucy cherry
f cerisier Mahaleb; cerisier de Sainte Lucie; bois de Sainte Lucie
d Felsenkirsche; arabische Mahaleb; Mahalebkirsche; Steinweichsel; Tintenbeere; Tintenkirsche; Weichsel(kirsche)
l *Prunus mahaleb*

3427 mahogany
f swiétènie
d Mahagonibaum; Swietenie
l *Swietenia*

3428 mahogany family
f méliacées
d Meliazeen; Zedrachgewächse
l *Meliaceae*

3429 mahonia
f mahonia
d Mahonie
l *Mahonia*

3430 maidenhair
f capillaire
d Frauenhaar(farn); Venushaar
l *Adiantum*

3431 maidenhair spleenwort
f faux-capillaire
d rotes Frauenhaar; Steinfeder; brauner Streifenfarn; roter Widerton
l *Asplenium trichomanes*

* **maidenhair tree → 2492**

3432 maidenhair-tree family
f ginkgoacées
d Ginkgoazeen; Ginkgogewächse
l *Ginkgoaceae*

3433 maiden pink
f œillet à delta
d Heidenelke; Sternnelke
l *Dianthus deltoides*

3434 maize; Indian corn
f maïs; blé de Turquie
d Mais; türkischer Weizen; Welschkorn
l *Zea mays*

3435 majestic eremurus
f érémure élégant
d Steppenlilie
l *Eremurus spectabilis*

* **Malabar almond → 5600**

3436 Malabar glory lily
f lis de Malabar
l *Gloriosa superba*

3437 Malabar gourd
f courge de Siam
l *Cucurbita ficifolia; Cucurbita melanosperma*

* **Malabar nightshade → 5921**

3438 Malabar nut
f noyer des Indes
l *Adhathoda vasica*

3439 Malabar randia
f jasmin du Cap
l *Randia dumetorum*

3440 Malabar simal tree; simal
f arbre à coton; sémul
l *Bombax malabaricum; Salmalia malabarica*

3441 malanga
f xanthosome
d Goldnarbe
l *Xanthosoma*

3442 Malayan fan palm
f palmier éventail de Cochinchine
l *Livistona cochinchinensis*

* **Malay apple → 3912**

3443 Malay guttapercha nato tree
f arbre à guttapercha; arbre à gutta
d Guttaperchabaum
l *Palaquium gutta*

3444 Malcolm stock
f malcolmie

d Hundslevkoje
l *Malco(l)mia*

3445 male fern
f fougère mâle
d männliches Farnkraut; Wurmfarn; Farnkrautmännchen; Farnkrautwurzel
l *Dryopteris filixmas; Aspidium filixmas*

3446 male orchis
d männliches Knabenkraut; Mannsknabenkraut
l *Orchis mascula*

3447 mallow
f mauve
d Gänsepappel; Hasenpappel; Käsepappel; Malve
l *Malva*

3448 mallow family
f malvacées
d Käsepappelgewächse; Malvazeen; Malvengewächse
l *Malvaceae*

3449 malope
f malope
d Malope; Sommermalve
l *Malope*

3450 malpighia
f malpighier
d Malpigie; Barbados-Kirsche
l *Malpighia*

3451 Maltese cross campion
f lychnide de Chalcédoine; croix de Malte; croix de Jérusalem
d brennende Liebe; Jerusalemsblume; Jerusalemskreuz; Maltesekreuz
l *Lychnis chalcedonica*

3452 mamey; mammee
f mammée
d Mammeibaum
l *Mammea*

3453 mamey; mammee apple
f abricotier de Saint-Domingue; abricotier des Antilles
d Aprikose von St. Domingo; Mammiapfel
l *Mammea americana*

* **mammee → 3452**

3454 mammillaria
f mamillaire
d Brustwarzendistel; Kugelkaktus; Mammillarie; Warzenkaktus
l *Mammillaria*

3455 manchineel
f arbre de mort
d Manschenill(en)baum; Manzanillabaum
l *Hippomane mancinella*

3456 Manchu filbert
f noisetier de Mandchourie
d mandschurische Hasel
l *Corylus mandshurica; Corylus sieboldiana mandshurica*

3457 Manchurian apricot
f abricotier de Mandchourie
d mandschurischer Aprikosenbaum
l *Prunus mandshurica; Armeniaca mandshurica*

3458 Manchurian ash
f frêne de Mandchourie
d mandschurische Esche
l *Fraxinus mandshurica*

3459 Manchurian currant
f groseillier de Mandchourie
d mandschurische Johannisbeere
l *Ribes mandshuricum*

3460 Manchurian Dutchman's-pipe
f aristoloche de Mandchourie
d mandschurische Pfeifenblume
l *Aristolochia mandshuriensis*

3461 Manchurian elm
d japanische Bergulme; japanische Ulme
l *Ulmus laciniata*

3462 Manchurian fir
f sapin à aiguilles
l *Abies holophylla*

3463 Manchurian linden
f tilleul de Mandchourie
d mandschurische Linde
l *Tilia mandshurica*

3464 Manchurian maple
f érable de Mandchourie
d mandschurischer Ahorn
l *Acer mandshuricum*

3465 Manchurian monkshood
f aconit panaché
d bunter Eisenhut
l *Aconitum variegatum*

3466 Manchurian walnut
f noyer de Mandchourie
d mandschurischer Walnussbaum
l *Juglans mandshurica*

3467 Manchu tuber gourd
f thladianthe douteuse
d Quetschblume
l *Thladiantha dubia*

3468 mandarine orange; tangerine
f mandarinier
d Mandarine
l *Citrus nobilis deliciosa; Citrus reticulata*

3469 mandrake
f mandragore
d Alraunpflanze; Alraunwurzel
l *Mandragora*

3470 manettia
f manettie
d Manettie
l *Manettia*

3471 mangel
f betterave fourragère
d Burgunderrübe; Futterrübe
l *Beta vulgaris* ssp. *rapacea*

3472 mango
f manguier
d Mangobaum; Mangotanne
l *Mangifera*

3473 mangold; foliage beet
f poirée
d Mangold
l *Beta vulgaris* ssp. *cicla*

3474 mangosteen
f mangoustan du Malabar; mangoustanier
d Mangostane; Mangostana; Mangostanbaum
l *Garcinia mangostana*

3475 mangrove
f manglier
d Mangrovebaum; Manglebaum
l *Rhizophora*

* **manioc → 1046**

* **manna ash → 2308**

3476 manna grass; sweet grass
f glycérie; herbe à manne
d Mannagras; Schwaden; Süssgras; Viehgras
l *Glyceria*

3477 manna lecanora
f lécanore comestible; lichen de la manne
l *Lecanora esculenta*

3478 manyroot
f ruellia
d Ruellie; Rodel
l *Ruellia*

3479 manyseed; allseed
f polycarpe
d Nagelkraut
l *Polycarpon*

* **manzanillo → 6061**

3480 manzanita
f busserole
d Bärentraube
l *Arctostaphylos*

3481 maple
f érable
d Ahorn
l *Acer*

3482 maple family
f acéracées
d Ahorngewächse; Azerazeen; Azerineen
l *Aceraceae*

3483 marasca sour cherry
f griottier
d Sauerkirsche; Weichsel(kirsche)
l *Prunus cerasus marasca*

3484 marasmius
f marasme
d Lauchschwamm; Schwindling
l *Marasmius*

3485 marattia family
f marattiacées
d Marattiazeen
l *Marattiaceae*

3486 marble treebine
f liane aux voyageurs
l *Cissus gongylodes*

3487 marchantia
f marchantie
d Leberkraut
l *Marchantia*

3488 mare's-tail
f pesse
d Tannenwedel
l *Hippuris*

3489 mare's-tail
f pesse commune
d quirliger Tannenwedel
l *Hippuris vulgaris*

3490 marginate sand spurrey
d Randspärkling; flügelsamiger Spörgel
l *Spergularia marginata*

3491 margined rose
d rauhblättrige Rose
l *Rosa jundzilli; Rosa marginata*

3492 margosa
f margousier azadirachta
d indischer Flieder; Nimbaum
l *Azadirachta indica; Melia azadirachta*

3493 marigold
f tagète
d Sammetblume; Samtblume; Studentenblume; Totenblume
l *Tagetes*

3494 Mariposa; Mariposa lily; Mariposa tulip
f calochorte; tulipe de Mormons
d Mormonentulpe
l *Calochortus*

*** maritime bulrush → 4801**

3495 maritime pea vine
d Strandplatterbse
l *Lathyrus maritimus; Lathyrus japonicus*

3496 maritime wormwood
f absinthe de mer; absinthe petite; armoise maritime; sanguenite
d Meerwermut; Salzwermut
l *Artemisia maritima*

3497 marjoram
f marjolaine
d Majoran
l *Majorana*

*** marrowfat → 3498**

3498 marrow pea; marrowfat
f pois carré
d Markerbse
l *Pisum sativum* var. *medullare*

*** marrow squash → 5705**

*** marrow-type pumpkin → 5705**

3499 marsdenia
f marsdénie
d Marsdenie
l *Marsdenia*

3500 marsh bedstraw
f gaillet des marais
d Sumpflabkraut
l *Galium palustris*

3501 marsh betony
f épiaire des marais
d Sumpfziest
l *Stachys palustris*

3502 marsh cinquefoil
f comaret
d Blutauge; Sumpfblutauge
l *Potentilla palustris; Comarum palustre*

3503 marsh club moss
d Sumpfbärlapp
l *Lycopodium inundatum*

*** marsh elder → 5324**

3504 marsh fern
f thélypteris
d Sumpffarn
l *Dryopteris thelypteris*

3505 marsh hawk's-beard
d Sumpfpippau
l *Crepis paludosa*

3506 marsh horsetail
f prêle des marais; queue-de-cheval
d Sumpfschachtelhalm; Duwock
l *Equisetum palustre*

3507 marshmallow
f guimauve officinale; althée
d echter Eibisch; Apothekerstock-malve; Heilwurzel
l *Althaea officinalis*

3508 marsh marigold
f populage
d Dotterblume
l *Caltha*

3509 marsh parsley; milk parsley
d Sumpfhaarstrang
l *Peucedanum palustre*

3510 marsh pea vine
f gesse des marais
d Sumpfplatterbse
l *Lathyrus palustris*

3511 marsh-pepper smartweed; water-pepper smartweed
f poivre d'eau; renouée brûlante
d Pfefferknöterich; Wasserpfefferknöterich
l *Polygonum hydropiper*

3512 marsh purslane
d Sumpflöffelchen
l *Ludwigia palustris; Isnardia palustris*

3513 marsh ragwort
d Wassergreiskraut; Wasserkreuzkraut
l *Senecio aquaticus*

3514 marsh samphire
f salicorne
d echtes Glasschmalz; Salzhornkraut
l *Salicornia herbacea*

3515 marsh scheuchzeria
d Moorblasenbinse; Sumpfblasenbinse; Sumpfblumenbinse
l *Scheuchzeria palustris*

3516 marsh sedge
d Sumpfriedgras; Sumpfsegge
l *Carex acutiformis*

3517 marsh sow thistle
f laiteron des marais
d Sumpfgänsedistel
l *Sonchus paluster*

3518 marsh speedwell
d Schildehrenpreis
l *Veronica scutellata*

*** marsh tea → 1719**

3519 marsh thistle
d Moorkratzdistel; Sumpfkratzdistel
l *Cirsium palustre*

3520 marsh valerian
f valériane dioïque
d kleiner Baldrian; Sumpfbaldrian
l *Valeriana dioica; Valeriana palustris*

3521 marsh violet
d Sumpfveilchen
l *Viola palustris*

*** marsh willow herb → 3522**

3522 marsh willowweed; marsh willow herb
f épilobe des marais
d Sumpfweidenröschen
l *Epilobium palustre*

3523 marsilia family
f marsiliacées
d Kleefarne; Kleefarngewächse; Marsiliazeen
l *Marsiliaceae*

3524 martagon lily
f lis martagon
d Martagonlilie; Türkenbund(lilie)
l *Lilium martagon*

3525 Martens selaginella
f sélaginelle de Martens
d Martens Moosfarn
l *Selaginella martensi*

3526 martinoe
f calimérİde
d Staudenaster
l *Calimeris*

3527 martynia
f martynie
d Martynie
l *Martynia*

3528 Marumi kumquat; round kumquat
f kumquat à fruits ronds
d japanischer Kumquat
l *Fortunella japonica*

*** marvel-of-Peru → 1415, 2341**

3529 maskflower
f alonzoa
d Alonsoblume; Nesselblatt
l *Alonsoa*

3530 Masson pine
f pin de Masson
d Massons Kiefer
l *Pinus massoniana*

3531 masterwort
f astrance
d Sterndolde
l *Astrantia*

3532 masterwort hog's-fennel
f impératoire
d Meisterwurz(el); Kaiserwurzel; Magistranzwurzel
l *Peucedanum ostruthium; Imperatoria ostruthium*

* **mastic tree → 3275**

3533 matgrass
f nard raide
d steifes Borstengras
l *Nardus stricta*

3534 matico pepper
d Soldatenkraut
l *Piper angustifolium*

3535 matrimony vine
f lyciet à feuilles d'halimus; lyciet commun
d gemeiner Bocksdorn; meldenblättriger Teufelszwirn; Hexenzwirn
l *Lycium halimifolium*

3536 maurandia
f maurandie
d Maurandie
l *Maurandia*

3537 mauritia; wine palm
f mauritier
d Mauritiuspalme; Weinpalme
l *Mauritia*

3538 mauritius; bitter orange; cabuyao
f limettier hérissé
d langdorniger Orangenbaum
l *Citrus hystrix*

* **Mauritius hemp → 4209**

3539 May apple
f podophylle
d Fussblatt; Maiapfel
l *Podophyllum*

* **May lily → 3306**

3540 mayten
f maytène
l *Maytenus*

3541 mayweed
f matricaire
d Kamille
l *Matricaria*

3542 mayweed camomile
f camomille puante
d stinkende Hundskamille; Stinkhundskamille
l *Anthemis cotula*

3543 mazari palm
d Pheespalme
l *Nannorhops*

3544 mazzard cherry; gean; sweet cherry
f merisier
d Bauernkirsche; Haferkirsche; Holzkirsche; Süsskirsche(nbaum); Vogelkirsche; Waldkirsche; Weichselbaum; Zwieselbeere
l *Prunus avium; Cerasus avium*

3545 meadow beauty
f rhéxia; rhéxie
d Bruchheil
l *Rhexia*

3546 meadow brome; upright brome grass
f brome des prés
d aufrechte Trespe
l *Bromus erectus*

3547 meadow fescue
f fétuque élevée; fétuque des prés
d Wiesenschwingel
l *Festuca pratensis; Festuca elatior*

3548 meadow foam
f limnanthe
d Sumpfblume; Sumpfschnabel
l *Limnanthes*

3549 meadow foxtail
f vulpin des prés
d Wiesenfuchsschwanz(gras)
l *Alopecurus pratensis*

3550 meadow gentian
f gentiane champêtre; gentiane des champs
d Feldenzian
l *Gentiana campestris*

3551 meadow geranium
f géranium des prés
d Wiesenstorchschnabel
l *Geranium pratense*

* **meadow gras → 700**

3552 meadow horsetail
f prêle des prés
d Wiesenschachtelhalm
l *Equisetum pratense*

3553 meadow oat grass
f avoine des prés
d rauher Wiesenhafer
l *Avena pratensis; Avenastrum pratensis; Helictotrichon pratense*

3554 meadow pasqueflower
f anémone des prés; pulsatille des prés
d kleine Küchenschelle; nickende Küchenschelle; blaue Osterblume; Wiesenkuhschelle
l *Anemone pratensis; Pulsatilla pratensis*

3555 meadow pea vine
f gesse des prés
d Wiesenplatterbse
l *Lathyrus pratensis*

3556 meadow rue
f pigamon
d Wielandskraut; Wiesenraute
l *Thalictrum*

3557 meadow saffron
f bulbocode
d Lichtblume
l *Bulbocodium*

* **meadow saffron → 1353**

3558 meadow sage
f sauge des prés
d Wiesensalbei
l *Salvia pratensis*

3559 meadow salsify
f barbe de bouc
d Wiesenbocksbart
l *Tragopogon pratensis*

3560 meadow saxifrage
f saxifrage granulée
d Knöllchensteinbrech; Körnersteinbrech; körniger Steinbrech
l *Saxifraga granulata*

* **meadow saxifrage → 4836**

3561 meadow succisa; devil's-bit
f mors du diable; scabieuse succise
d Abbisskraut; gemeiner Teufelsabbiss
l *Succisa pratensis; Scabiosa succisa*

3562 meadowsweet
f filipendule
d Mädesüss; Spierstaude
l *Filipendula*

3563 Mecca myrrh tree
d Mekkabalsam
l *Commiphora opobalsamum*

3564 medeola
f médéola
d Medeola; Myrtenblatt
l *Medeola*

3565 medic
f luzerne
d Luzerne; Schneckenklee
l *Medicago*

3566 medicinal rhubarb
f rhubarbe officinale
d echter Rhabarber; chinesischer Rhabarber; türkischer Rhabarber
l *Rheum officinale*

3567 medinilla
f médinilla
d Medinilla
l *Medinilla*

3568 Mediterranean aloe
f aloès vulgaire
d echte Aloe; Barbados-Aloe
l *Aloe barbadensis; Aloe vulgaris*

3569 Mediterranean chrysanthemum
f chrysanthème à corymbe
d doldige Wucherblume; Traubenwucherblume
l *Chrysanthemum corymbosum; Pyrethrum corymbosum*

3570 Mediterranean herb elder; Danewort
f hièble; yèble
d Attich; Zwergholunder
l *Sambucus ebulus*

3571 Mediterranean mullein
d buchtige Königskerze
l *Verbascum sinuatum*

3572 Mediterranean palm; hair palm
f palmier nain
d niedrige Zwergpalme; Strauchpalme; europäische Zwergpalme; Schirmpalme;

Hanfpalme
l *Chamaerops humilis*

3573 **Mediterranean sea lavender**
f statice commun
d Strandflieder; Strandnelke; Widerstoss
l *Limonium vulgare*

3574 **Mediterranean serpent root**
d spitzblättrige Schwarzwurzel
l *Scorzonera laciniata*

3575 **Mediterranean stinkbush**
f bois puant
d Stinkbaum
l *Anagyris foetida*

3576 **Mediterranean tree mallow**
f mauve de Mauritanie
d blaue Malve
l *Malva mauritania*

3577 **medlar**
f néflier
d Mispel
l *Mespilus*

3578 **medlar**
f néflier commun
d echte Mispel; deutsche Mispel; gemeine Mispel
l *Mespilus germanica*

3579 **melaleuca**
f cajeputier
d Myrtenheide; Weissbaum; Kajeputbaum
l *Melaleuca*

3580 **melanconiales**
f mélanconiales
d Melankonialen
l *Melanconiales*

3581 **melastoma family**
f mélastomacées
d Schwarzmundgewächse; Melastomazeen
l *Melastomaceae*

3582 **melic (grass); onion grass**
f mélique
d Perlgras
l *Melica*

* **melilot → 5362**

3583 **mellitis**
f mélitte
d Bienensang; Immenblatt
l *Melittis*

3584 **melon cactus**
f mélocactus
d Melonenkaktus; Schopffackeldistel
l *Melocactus*

3585 **melon-leaved nightshade**
d melonenblättriger Nachtschatten
l *Solanum heterodoxum*

3586 **melothria**
f mélothrie
d Melothrie
l *Melothria*

3587 **mentzelia**
f mentzélie
d Mentzelie
l *Mentzelia*

3588 **merawan**
f hopée
l *Hopea*

3589 **mercury**
f mercuriale
d Bingelkraut
l *Mercurialis*

3590 **mersawa**
f anisoptère
l *Anisoptera*

3591 **mescal button peyote**
d Williams Igelkaktus
l *Lophophora williamsi*

3592 **mesembryanthemum**
f ficoïde; mésembryanthème
d Eiskraut; Mittagsblume; Nachmittagsblume; Zaserblume
l *Mesembryanthemum*

3593 **mesquite**
f prosope
d Mesquitebaum; Mesquitestrauch; Schraubenbohne
l *Prosopis*

3594 **Mexican ageratum**
f agérate du Mexique
d mexikanischer Leberbalsam
l *Ageratum mexicanum*

* **Mexican bald cypress → 3667**

3595 Mexican calceolaria
f calcéolaire du Mexique
d mexikanische Pantoffelblume
l *Calceolaria mexicana*

3596 Mexican clover
d mexikanischer Klee
l *Richardia scabra; Richardsonia scabra*

3597 Mexican cypress; Portuguese cypress
f cyprès de Lusitanie
d blaugrüne Zypresse
l *Cupressus lusitanica*

3598 Mexican drooping juniper
f genévrier flasque
l *Juniperus flaccida*

3599 Mexican juniper
f genévrier du Mexique
d mexikanischer Sadebaum
l *Juniperus mexicana*

3600 Mexican mock orange
f seringa mexicain
d mexikanischer Pfeifenstrauch
l *Philadelphus mexicanus*

3601 Mexican plane tree
f platane du Mexique
d mexikanische Platane
l *Platanus mexicana*

3602 Mexican star
f milla
d Mexikostern
l *Milla*

* **Mexican tea → 6056**

3603 Mexican teosinte
d mexikanische Teosinte
l *Euchlaena mexicana*

3604 Mexican twinbloom
f bravoa à fleurs géminées
d Zwillingsblume
l *Bravoa geminiflora*

3605 Mexican vanilla
f vanillier
d echte Vanille
l *Vanilla planifolia; Vanilla fragrans*

3606 Mexican white pine
f pin ayacahuite
l *Pinus ayacahuite*

3607 Meyer currant
f groseillier de Meyer
d Meyers Johannisbeere
l *Ribes meyeri*

3608 mezereon family
f thyméléacées; daphnoïdées
d Seidelbastgewächse; Spatzenzungengewächse; Thymeläazeen
l *Thymelaeaceae*

3609 mibora
f mibore
d Zwerggras
l *Mibora*

3610 michelia
f michélie
d Michelie
l *Michelia*

3611 micromeria
f micromérie
d Micromerie; Bartsaturei
l *Micromeria*

3612 Middendorff birch
f bouleau de Middendorff
d Middendorffs Birke
l *Betula middendorffi*

3613 Middendorff day lily
f hémérocalle de Middendorff
d Middendorffs Taglilie
l *Hemerocallis middendorffi*

3614 mignonette
f réséda
d Reseda; Wau
l *Reseda*

3615 mignonette family
f résédacées
d Resedagewächse; Resedazeen; Waugewächse
l *Resedaceae*

3616 mikania
f mikanie
d Mikanie
l *Mikania*

* **mildew → 3691**

* **military orchis → 5079**

3617 milkberry
f chiocoque

d Eisbeere; Schneebeere
l *Chiococca*

3618 milkbush
d Milchbaum
l *Synadenium*

* **milk parsley → 3509**

3619 milk vetch
f astragale glycyphylle
d Bärenschote; süsser Tragant; Süssholztragant; wildes Süssholz
l *Astragalus glycyphyllus*

* **milk vetch → 3353**

3620 milkweed; silkweed
f asclépiade
d Schwalbenschwanz; Seidenpflanze
l *Asclepias*

3621 milkweed family
f asclépiadacées
d Asklepiadazeen; Schwalbenschwanzgewächse; Seidenpflanzengewächse
l *Asclepiadaceae*

3622 milkweed gentian
f gentiane à feuilles d'asclépiade; gentiane-asclépiade
d Schwalbenwurzenzian; Würgerenzian
l *Gentiana asclepiadea*

* **milkwort → 4250**

3623 milkwort family
f polygalacées
d Kreuzblümchengewächse; Polygalazeen; Polygaleen
l *Polygalaceae*

* **millet → 780**

3624 millet grass
f millet sauvage
d Flattergras; Hirsegras; Waldhirse
l *Milium*

3625 mimosa
f mimosa
d Mimose; Sinnpflanze
l *Mimosa*

3626 mimosa family
f mimosacées
d Mimosazeen; Mimosengewächse
l *Mimosaceae*

3627 miner's lettuce
f clayton(i)e de Cuba; pourpier d'hiver
d Kuba-Spinat; Winterportulak; kubanisches Burzelkraut
l *Claytonia perfoliata; Montia perfoliata*

3628 mint
f menthe
d Minze; Münze
l *Mentha*

3629 mint family
f labiées
d Labiaten; Lippenblütler
l *Labiatae*

3630 mirabelle
f mirabelle
d Mirabelle
l *Prunus insititia syriaca*

* **Missouri grape → 1057**

3631 mistletoe
f gui
d Mistel
l *Viscum*

3632 mistletoe family
f loranthacées
d Loranthazeen; Mistelgewächse; Riemenblumengewächse
l *Loranthaceae*

3633 mistmaiden
f romanzowie
d Romanzoffie
l *Romanzoffia*

3634 miterwort; mitrewort; bishop's-cap
f mitelle
d Bischofskappe
l *Mitella*

3635 mitraria
f mitraire
d Mützenstrauch
l *Mitraria*

* **mitrewort → 3634**

3636 mixed-flower
f raiponce
d Teufelskralle
l *Phyteuma*

3637 Mlanji widdringtonia
d Whites Widdringtonia
l *Widdringtonia whitei*

3638 mnium
f mnie; mnium
d Sternmoos
l *Mnium*

3639 mock cucumber
f échinocyste
d Igelgurke; Stachelgurke
l *Echinocystis*

3640 mockernut
f caryer cotonneux
d weisse Hickory; Spottnussbaum
l *Carya alba; Carya tomentosa*

3641 mock orange
f seringa
d falscher Jasmin; Pfeifenstrauch
l *Philadelphus*

3642 mock privet
f filaria; alavert; aouret; duradeu
d Steinlinde
l *Phillyrea*

3643 mock strawberry
f duchesnée
d Steinerdbeere
l *Duchesnea*

3644 molasses grass
f mélinis à petite fleur
l *Melinis minutiflora*

3645 Moldavian dragonhead
f dracocéphale de Moldavie
d türkischer Drachenkopf
l *Dracocephalum moldavicum*

* **mole plant → 1008**

* **mole weed → 1008**

3646 molinia
f molinie
d Pfeifenbinse; Pfeifengras
l *Molinia*

3647 Molucca balm
f molucelle
d Trichterkelch; Trichtermelisse; Muschelblume
l *Moluc(c)ella*

3648 mombin
f spondias
d Mombinpflaume
l *Spondias*

3649 Momi fir
f sapin du Japon; sapin Momi
d Momitanne
l *Abies firma*

3650 monarch birch
f bouleau de Maximowicz
d Maximowiczs Birke
l *Betula maximowiczii*

3651 moneywort
f herbe aux écus
d Münzfelberich; Pfennigkraut; Wiesengeld
l *Lysimachia nummularia*

3652 Mongolian ephedra
d Schachtelhalmmeerträubchen
l *Ephedra equisetina*

3653 Mongolian oak
f chêne de Mongolie
d mongolische Eiche
l *Quercus mongolica*

3654 Mongolian poplar
f peuplier odoriférant
d wohlriechende Pappel
l *Populus suaveolens*

* **mongrel larkspur → 2305**

* **monkey-bread tree → 445**

3655 monkey flower
f mimule
d Affenblume; Gauklerblume; Larvenblume; Lochblume; Maskenblume
l *Mimulus*

3656 monkey orchis
d Affenknabenkraut
l *Orchis simia*

3657 monkeypot tree
f lécythide; quatélé
d Krukenbaum; Topffruchtbaum
l *Lecythis*

3658 monkey-puzzle araucaria
f araucarie du Chili; pin du Chili
d chilenische Araukarie; Chilefichte
l *Araucaria araucana*

3659 monkflower
f catasète
d Rasselstendel
l *Catasetum*

3660 monkshood
f aconit
d Eisenhut; Sturmhut
l *Aconitum*

*** monk's pepper tree → 3296**

3661 monocotyledons
f monocotylédonées
d Monokotyledonen; Monokotylen; einkeimblättrige Pflanzen
l *Monocotyledones*

3662 monodora
f monodora
d Kalebassenmuskat; Muskatnussbaum
l *Monodora*

3663 monstera
f monstéra
d Monstera
l *Monstera*

3664 monster flower
f rafflésia
d Riesenblume
l *Rafflesia*

*** montbretia → 5599**

3665 Monterrey cypress
f cyprès à gros fruits; cyprès de Lambert
d grossfrüchtige Zypresse; Lambertszypresse
l *Cupressus macrocarpa*

3666 Monterrey pine
f pin de Monterrey
d Monterreykiefer
l *Pinus radiata*

3667 Montezuma bald cypress; Mexican bald cypress
f cyprès de Montezuma
d mexikanische Sumpfzypresse
l *Taxodium mucronatum*

3668 Montezuma pine
f pin de Montezuma
d Montezumakiefer
l *Pinus montezumae*

3669 Montpellier maple
f érable de Montpellier
d Felsenahorn; französischer Ahorn; dreilappiger Ahorn
l *Acer monspessulanum*

3670 moonflower
f calonyction
d Mondwinde
l *Calonyction*

3671 moonseed
f ménisperme
d Mondsame
l *Menispermum*

3672 moonseed family
f ménispermacées
d Menispermazeen; Mondsamengewächse
l *Menispermaceae*

3673 moonwort
d Mondraute(nfarn); Walpurgiskraut
l *Botrychium lunaria*

3674 moor grass
f jonchée; canche bleue; molinie bleue
d Benthalm; Besenried; blaues Pfeifengras
l *Molinia caerulea*

3675 moor grass
f seslérie
d Blaugras; Kopfgras
l *Sesleria*

*** moorwort → 723**

3676 moraea
f morée
d Moräe
l *Moraea*

3677 moraine buttercup
f renoncule alpestre
d Alpenhahnenfuss
l *Ranunculus alpestris*

3678 morel
f morille
d Morchel
l *Morchella*

3679 morello
f cerisier griottier
d Sauer(weichsel)kirsche
l *Prunus cerasus austera*

3680 Moreton Bay chestnut; bean tree
d Bohnenbaum
l *Castanospermum australe*

3681 morina
f morine
d Kardendistel; Morine
l *Morina*

3682 morning-glory
f ipomée
d Prunkwinde
l *Ipomoea*

3683 morning-glory family
f convolvulacées
d Konvolvulazeen; Windengewächse
l *Convolvulaceae*

3684 moschatel family
f adoxacées
d Moschuskrautgewächse
l *Adoxaceae*

3685 moss cabbage rose
f rosier mousseux; rosier moussu; rose mousseuse
d Moosrose
l *Rosa centifolia* var. *muscosa; Rosa muscosa*

* **moss campion → 3686**

3686 moss silene; moss campion
f silène acaule
d Alpenleimkraut; stengelloses Leimkraut
l *Silene acaulis*

3687 mother chrysanthemum
f chrysanthème d'automne
d indische Wucherblume; Goldaster; Allerseelenaster; Winteraster
l *Chrysanthemum indicum*

* **mother-of-thousands → 3147**

3688 mother-of-thyme
f serpolet; thym serpolet
d Wurstkraut; Feldthymian; Hühnerpolei; schmalblättriger Quendel
l *Thymus serpyllum*

3689 motherwort
f léonure
d Herzgespann
l *Leonurus*

3690 moth mullein
f molène aux teignes; herbe aux mites
d Himmelsbrand; Schabenkönigskerze; Schabenkraut
l *Verbascum blattaria*

3691 mo(u)ld; mildew
f moisissure
d Köpfchenschimmel
l *Mucor*

3692 mound lily (yucca)
f yucca magnifique
d schöne Palmlilie; Prachtaloe
l *Yucca gloriosa*

3693 mountain alyssum
f alysse montagnarde
d Bergsteinkraut
l *Alyssum montanum*

3694 mountain arnica
f arnica des montagnes; bétoine des montagnards
d Bergwohlverleih; Gemsblume; Johannisblume; Marienkraut; Mönchswurz; Mutterwurz
l *Arnica montana*

3695 mountain ash
f sorbier
d Eberesche; Vogelbeerbaum
l *Sorbus; Micromelus*

* **mountain avens → 3726**

3696 mountain bladder fern
f cystoptéris montagnard
d Bergblasenfarn
l *Cystopteris montana*

3697 mountain bluet
f bl(e)uet vivace; centaurée de montagne
d Bergflockenblume; Bergkornblume
l *Centaurea montana*

3698 mountain bush
f kunzée; purshie
l *Kunzea*

3699 mountain buttercup
f renoncule des montagnes
d Berghahnenfuss
l *Ranunculus montanus*

3700 mountain butterwort
d Alpenfettkraut; Butterwurzel; blaues

Fettkraut; gemeines Fettkraut
l *Pinguicula alpina*

3701 mountain chick-pea
f astragale pois chiche
d Kichertragant
l *Astragalus cicer*

* **mountain clematis → 244**

3702 mountain clover
f trèfle des montagnes
d Bergklee; Spitzklee
l *Trifolium montanum*

3703 mountain common juniper
f genévrier des montagnes
d sibirischer Wacholder
l *Juniperus communis saxatilis; Juniperus montana; Juniperus sibirica*

3704 mountain coronilla
d Bergkronenwicke
l *Coronilla montana*

* **mountain cranberry → 1645**

* **mountain fern → 3884**

3705 mountain germander
f germandrée de montagne
d Berggamander
l *Teucrium montanum*

3706 mountain heath
f phyllodoce
d Blauheide; Kantenheide; Moorheide; Moosheide
l *Phyllodoce*

3707 mountain hemlock
f tsuga de Mertens
l *Tsuga mertensiana*

3708 mountain holly fern
f lonchitis
d Lanzenfarn; Lanzenschildfarn
l *Polystichum lonchitis*

3709 mountain lady's-mantle
f alchémille des Alpes
d Alpenfrauenmantel
l *Alchemilla alpina*

3710 mountain laurel
f kalmie à larges feuilles
d breitblättrige Lorbeerrose
l *Kalmia latifolia*

3711 mountain lungwort
f pulmonaire à feuilles molles
d Berglungenkraut
l *Pulmonaria montana*

3712 mountain mahogany
f cercocarpe
d Bergmahagoni
l *Cercocarpus*

3713 mountain meadow seseli
d Bergheilwurz
l *Seseli libanotis; Libanotis montana*

* **mountain melic → 3868**

3714 mountain mint
f pycnanthème
d Bergminze; Dickblume
l *Pycnanthemum*

3715 mountain parsley
f athamante; grand persil de montagne
d Berghaarstrang; Bergpetersilie; Bergsilge; Kirschwurzelkraut
l *Peucedanum oreoselinum*

3716 mountain pea vine
f gesse des montagnes
d Bergplatterbse
l *Lathyrus montanus*

3717 mountain plantain
f plantain des montagnes
d Bergwegerich
l *Plantago montana; Plantago atrata*

3718 mountain sedge
f laîche des montagnes
d Bergriedgras; Bergsegge
l *Carex montana*

3719 mountain sorrel
f oxyrie
d Säuerling
l *Oxyria*

3720 mountain speedwell
f véronique des montagnes
d Bergehrenpreis
l *Veronica montana*

3721 mountain St.-John's-wort
f millepertuis des montagnes
d Berghartheu; Bergjohanniskraut
l *Hypericum montanum*

3722 mountain valerian
f valériane de montagne
d Bergbaldrian
l *Valeriana montana*

3723 mountain willowweed
f épilobe des montagnes
d Bergweidenröschen
l *Epilobium montanum*

* **Mount Atlas cedar → 364**

3724 Mount Atlas pistache
f pistachier d'Atlas
l *Pistacia atlantica*

3725 Mount Washington bluegrass
d schlaffes Rispengras
l *Poa laxa*

3726 Mount Washington dryad; mountain avens
f dryade à 8 pétales
d achtblättrige Silberwurz
l *Dryas octopetala*

3727 mourning cypress
f cyprès des cimetières; cyprès funèbre; cyprès pleureur
d Trauerzypresse
l *Cupressus funebris*

3728 mourning iris
f iris de Suse; iris deuil
d Traueriris; Trauerschwertlilie
l *Iris susiana*

3729 mouse barley
f orge des rats
d Mäusegerste
l *Hordeum murinum*

3730 mouse-ear
f holostée
d Spurre
l *Holosteum*

3731 mouse-ear betony
f épiaire de Germanie
d deutscher Ziest; filziger Ziest; Wollziest
l *Stachys germanica*

3732 mouse-ear chickweed
d geknäueltes Hornkraut; Knäuelhornkraut
l *Cerastium glomeratum*

3733 mouse-ear cress
d Ackerschmalwand; Gänsekressling; Schmalwand; Schotenkresse
l *Arabidopsis thaliana*

3734 mouse-ear hawkweed
f piloselle
d kleines Habichtskraut; langhaariges Habichtskraut
l *Hieracium pilosella*

3735 mouse foxtail
d Ackerfuchsschwanz(gras); Mäusefuchsschwanz
l *Alopecurus myosuroides*

3736 mousetail
f ratoncule
d Mäuseschwanz; Mäuseschwänzchen
l *Myosurus*

3737 mucaja acrocomia; corozo
f mo(u)caya; palmier-canne
d Macawbaum; Macahuba; Macoya
l *Acrocomia sclerocarpa*

3738 much-good
d Hirschwurz; Schwefelwurz; starrer Haarstrang
l *Peucedanum cervaria*

3739 mucor family
f mucoracées
d Mukorazeen
l *Mucoraceae*

3740 mucuna
f mucune
d Juckbohne
l *Mucuna*

3741 mud grass
f coléanthe
d Scheidengras
l *Coleanthus*

3742 mud grass
d zartes Scheidengras
l *Coleanthus subtilis*

3743 mud plantain
f hétéranthère
d Heteranthera
l *Heteranthera*

3744 mud sedge
f laîche fangeuse
d Schlammriedgras; Schlammsegge
l *Carex limosa*

3745 mudwort
f limoselle
d Schlammkraut
l *Limosella*

3746 mugho Swiss mountain pine
f pin mugho
d Krummholzkiefer
l *Pinus montana mughus; Pinus mughus*

3747 mugwort wormwood
f armoise commune
d echter Beinfuss; Mutterkraut
l *Artemisia vulgaris*

3748 muhly
f muhlenbergie
d Muhlenbergia
l *Muehlenbergia*

3749 mulberry
f mûrier
d Maulbeerbaum
l *Morus*

3750 mulberry family
f moracées
d Maulbeergewächse; Morazeen
l *Moraceae*

3751 mulga
d Mulga
l *Acacia aneura*

3752 mullein
f molène
d Fackelblume; Fackelkraut; Königskerze; Wollkraut
l *Verbascum*

3753 multiplier onion; potato onion
f oignon patate; oignon sous terre
d Kartoffelnzwiebel
l *Allium cepa* var. *aggregatum*

3754 mung bean
f haricot-mungo; pois de Jérusalem
d Jerusalembohne; Mungbohne
l *Phaseolus aureus*

3755 mungo bean; black gram; urd
f haricot d'Angola
d Mungobohne; Urdbohne; Sansibarerbse
l *Phaseolus mungo*

* **murity palm → 5985**

* **murlin → 407, 5995**

3756 muscadine grape
f vigne musquée
d Muscadinerebe
l *Vitis rotundifolia*

3757 musenna albizzia
f albizzie anthelminthique
l *Albizzia anthelmintica*

3758 mushroom; fungus
f champignon
d Pilz
l *Fungus*

3759 musk bristle thistle
f chardon penché
d Bisamdistel; Eseldistel; nickende Distel
l *Carduus nutans*

3760 musk mallow
f mauve musquée
d Moschusmalve
l *Malva moschata*

3761 musk mallow
f ambrette
d Bisamstrauch; Sumpfeibisch
l *Hibiscus abelmoschus; Abelmoschus moschatus*

3762 muskmelon
f melon
d Melone(ngurke); Zuckermelone
l *Cucumis melo*

3763 musk orchis
f herminie
d Einorchis
l *Herminium*

3764 musk plant
f mimule musqué
d Moschusgauklerblume; Moschuskraut
l *Mimulus moschatus*

3765 muskroot
f adoxe
d Bisamkraut; Moschuskraut
l *Adoxa*

3766 muskroot
f adoxe muscatelle
d gemeines Moschuskraut; Moschusblümchen
l *Adoxa moschatellina*

3767 musk rose
f rosier musqué
d Moschusrose; Monatsrose
l *Rosa moschata*

3768 muskwood
f guaréa
l *Guarea*

*** musky gourd → 1747**

3769 musky yarrow
f achillée musquée
d Iva; Moschusschafgarbe
l *Achillea moschata*

3770 musodo manketti nut
f ricinodendre d'Afrique
l *Ricinodendron africanus*

3771 mustard
f moutarde
d Senf
l *Sinapis*

3772 mustard treacle
f érysimon
d Schöterich; Schotendotter
l *Erysimum*

3773 myagrum
f myagre
d Hohldotter
l *Myagrum*

3774 mycena
f mycène
d Helmling
l *Mycena*

3775 myoporum
f myopore
d Mäusefras
l *Myoporum*

3776 myrobalan family
f combrétacées
d Langfadengewächse
l *Combretaceae; Terminaliaceae*

3777 myrobalan plum
f prunier myrobalan
d Kirschpflaume; Myrobalane
l *Prunus cerasifera*

3778 myrocarpus
f myrocarpe
l *Myrocarpus*

3779 myrrh tree
f commiphore
d Myrrhenstrauch
l *Commiphora*

3780 myrsina family
f myrsinacées
d Myrsinengewächse
l *Myrsinaceae*

3781 myrtle
f myrte
d Myrte
l *Myrtus*

3782 myrtle coriaria; French sumac
f corroyère à feuilles de myrte
d französischer Sumach
l *Coriaria myrtifolia*

3783 myrtle family
f myrtacées
d Myrtazeen; Myrtengewächse
l *Myrtaceae*

3784 myrtle whortleberry; bilberry; wineberry
f abrétier; airelle myrtille; myrtille
d echte Heidelbeere; Besing; Bick(el)beere; Waldheidelbeere
l *Vaccinium myrtillus*

3785 myrtle willow
f saule faux-myrte
d Myrtenweide
l *Salix myrsinites*

N

3786 **naiad**
f nayade
d Nixkraut
l *Naias*

3787 **naias family**
f naïadacées
d Nixkrautgewächse; Najadazeen
l *Naiadaceae*

3788 **nailwort; whitlowwort**
f paronyque; panarine
d Mauerraute; Nagelkraut
l *Paronychia*

3789 **naked oat**
f avoine nue
d Nackthafer
l *Avena nuda*

3790 **nandina**
f nandine
d Nandine
l *Nandina*

* **nandina → 4634**

3791 **nankeen lily**
f lis couleur isabelle
d Nankinglilie
l *Lilium testaceum*

3792 **nannyberry; sheepberry**
f viorne à manchettes
d Schafbeere
l *Viburnum lentago*

3793 **Napier grass; elephant grass**
f herbe à éléphant
d Elefantengras
l *Pennisetum purpureum*

3794 **Narbonne vetch**
f vesce de Narbonne
d Ackerbohne; Narbonner Wicke; römische Wicke; schwarze Erbse
l *Vicia narbonensis*

3795 **narcissus**
f jonquille
d Narzisse
l *Narcissus*

3796 **narcissus anemone**
f anémone à feuilles de narcisse
d Berghähnlein
l *Anemone narcissiflora*

* **nard grass → 1244**

3797 **narrow beech fern**
f phégopteris
d Eichenfarn; Buchenfarn
l *Dryopteris phegopteris*

* **narrow buckler fern → 5546**

3798 **narrow-leaved ash**
f frêne oxyphylle
d spitzblättrige Esche
l *Fraxinus angustifolia*

3799 **narrow-leaved bluegrass**
f pâturin à feuilles étroites
d schmalblättriges Rispengras
l *Poa angustifolia*

3800 **narrow-leaved cattail**
f petite massette
d schmalblättriger Rohrkolben
l *Typha angustifolia*

3801 **narrow-leaved fire thorn**
f buisson ardent à feuilles étroites
d schmalblättriger Feuerdorn
l *Pyracantha angustifolia*

3802 **narrow-leaved flax**
f lin à feuilles étroites
d schmalblättriger Lein
l *Linum angustifolium*

3803 **narrow-leaved hawk's-beard**
f crépide des toits
d Dachpippau; Grundfeste; grünfester Pippau; Mauerpippau
l *Crepis tectorum*

3804 **narrow-leaved hawkweed**
f épervière ombellée
d doldiges Habichtskraut
l *Hieracium umbellatum*

3805 **narrow-leaved hemp nettle**
f galéopside à feuilles étroites; galéope à feuilles étroites
d schmalblättriger Hohlzahn
l *Galeopsis angustifolia*

3806 **narrow-leaved meadow rue**
f pigamon à feuilles étroites
d glänzende Wiesenraute; schmale Wiesenraute
l *Thalictrum angustifolium*

3807 narrow-leaved meadowsweet spirea
f spirée blanche d'Amérique
d weisser Spierstrauch
l *Spiraea alba*

3808 narrow-leaved phillyrea
f filaria à feuilles étroites
d schmalblättrige Steinlinde
l *Phillyrea angustifolia*

3809 narrow-leaved senecio
f séneçon à feuilles de roquette
d raukenblättriges Greiskraut; raukenblättriges Kreuzkraut
l *Senecio erucifolius*

3810 narrow-leaved vetch; summer vetch
f vesce à feuilles étroites
d schmalblättrige Wicke; Schmalblattwicke
l *Vicia angustifolia*

3811 narrow-leaved water parsnip
f berle à feuilles étroites
d schmalblättriger Merk
l *Sium angustifolium*

* **narrow shield fern → 5546**

3812 nasturtium
f capucine
d Kapuzinerkresse; Blumenkresse
l *Tropaeolum*

3813 nasturtium family
f tropéolacées
d Kapuzinerkressengewächse
l *Tropaeolaceae*

3814 nato tree
f palaque
d Nyatoh; Guttaperchabaum
l *Palaquium*

3815 navelseed
f omphalode
d Gedenkenmein; Nabelnuss
l *Omphalodes*

3816 navelwort
f ombilic
d Nabelkraut; Venusnabel
l *Umbilicus*

3817 naven
f navette
d Rübsen; Sommerraps
l *Brassica rapa* var. *oleifera*

* **neckweed → 4360**

3818 nectarine
f brugnonier
d Nektarine(npfirsich)
l *Prunus persica* var. *nectarina*

3819 needdle grass; feather grass
f stipe
d Federgras; Pfriemengras
l *Stipa*

3820 needle juniper
f genévrier rigide
l *Juniperus rigida*

3821 needle palm
f rhapidophylle
d Nadelpalme
l *Rhapidophyllum*

3822 needle spike sedge
d Nadelsumpfried; Nadelsumpfsimse
l *Eleocharis acicularis*

3823 needle woad-waxen
f genêt d'Angleterre
d englischer Ginster
l *Genista anglica*

* **negundo → 750**

3824 neillia
f neillie
d Neillie; Traubenspiere
l *Neillia*

3825 nemesia
f némésie
d Nemesie
l *Nemesia*

3826 nemophila
f némophile
d Hainblume
l *Nemophila*

3827 Nepal camphor tree
d drüsentragender Kampferbaum
l *Cinnamomum glanduliferum*

3828 nepenthes; pitcher plant
f népenthe
d Kannenpflanze; Kannenstrauch; Kannenstaude; Kannenträger; Krugpflanze
l *Nepenthes*

3829 nepenthes family
f népenthacées
d Nepenthazeen; Kannenstrauchgewächse
l *Nepenthaceae*

3830 nepeta
f cataire
d Katzenminze; Katzenkraut
l *Nepeta*

3831 nephroma
f néphrome
d Nierenflechte; Nierenschild
l *Nephroma*

3832 nerine
f nérine
d Nerine
l *Nerine*

3833 netleaf willow
f saule réticulé
d Netzweide; netzaderige Weide
l *Salix reticulata*

3834 netted melon
d Netzmelone
l *Cucumis melo* var. *reticulatus*

3835 nettle
f ortie
d Brennessel; Nessel
l *Urtica*

3836 nettle family
f urticacées
d Brennesselgewächse; Nesselgewächse; Urtikazeen; Urtizeen
l *Urticaceae*

3837 nettle-leaf goosefoot
f ansérine des murs
l *Chenopodium murale*

3838 nettle spurge
f médicinier
d Brechnuss; Drüsenstrauch; Purgiernuss; Maniokstrauch
l *Jatropha*

3839 New England aster
f aster de la Nouvelle-Angleterre
d Rauhblattaster; Sternblume von Neu-England
l *Aster novae-anglica*

3840 New Jersey tea; redroot
f thé de Jersey
d amerikanische Säckelblume; Rotwurzel
l *Ceanothus americanus*

3841 New Mexico locust
f robinier du Nouveau-Mexique
d neumexikanische Robinie
l *Robinia neomexicana*

3842 New York aster
f aster de la Nouvelle-Belgique
d neubelgische Aster; Glattblattaster; Herbststernblume
l *Aster novi-belgi*

3843 New Zealand fiber lily; New Zealand hemp
f lin de la Nouvelle-Zélande
d neuseeländischer Flachs; neuseeländische Flachslilie
l *Phormium tenax*

3844 New Zealand spinach
f épinard de Nouvelle-Zélande
d neuseeländischer Spinat
l *Tetragonia expansa*

3845 Niagara hawthorn
d Brainders Weissdorn
l *Crataegus brainderi asperifolia*

3846 nice hemp nettle
f galéope bigarré
d bunter Hohlzahn
l *Galeopsis speciosa; Galeopsis versicolor*

3847 nicker-nut caesalpinia
f bonduc
l *Caesalpinia crista; Caesalpinia bonducella*

3848 niepa bark tree
f samandure d'Inde
l *Samandura indica; Samandera indica*

3849 niger seed
f guizotie
d Gingellikraut
l *Guizotia*

* **niggertoe → 765**

* **night-flowering campion → 3850**

* **night-flowering petunia → 5901**

3850 night-flowering silene; night-flowering campion
f silène noctiflore; fleur de la nuit
d Ackernachtnelke; Nachtleimkraut; Nachtlichtnelke
l *Silene noctiflora*

3851 night jasmine; tree of sadness
f arbre triste
d Trauerbaum
l *Nyctanthes*

3852 night phlox
f nyctérinie
d Sternbalsam
l *Nycterinia; Zaluzianskya*

3853 nightshade
f morelle
d Eierfrucht; Nachtschatten
l *Solanum*

3854 nightshade family
f solanacées
d Nachtschattengewächse; Solanazeen
l *Solanaceae*

3855 Nikko fir
f sapin de Nikko
d Nikkotanne
l *Abies homolepis*

3856 Nikko maple
f érable Nikko
d Nikkoahorn
l *Acer nikoense*

3857 ninebark
f physocarpe
d Blasenspiere
l *Physocarpus*

3858 Ningpo clematis
f clématite laineuse
l *Clematis lanuginosa*

3859 nipa palm
f nipa
d Nipapalme
l *Nipa*

3860 nipplewort; dock cress
f lampsane
d Milchkraut; Rainkohl; Rainsalat
l *La(m)psana*

3861 nitella
f nitelle
l *Nitella*

3862 niterbush; nitrebush; lotus tree
f nitraire
d Salpeterstrauch
l *Nitraria*

3863 nit grass
f gastridie
d Nissegras
l *Gastridium*

* **nitrebush → 3862**

3864 nitta tree
f parkie
d Dourabaum
l *Parkia*

3865 noble fir
f sapin noble
d Edeltanne
l *Abies nobilis; Abies procera*

3866 nodding beggarticks
f bident penché
d nickender Zweizahn
l *Bidens cernua*

3867 nodding lilac
f lilas penché
d Bogenflieder; Hängeflieder
l *Syringa reflexa*

3868 nodding melic; mountain melic
d nickendes Perlgras
l *Melica nutans*

3869 nodding silene; Nottingham catchfly
f silène penché
d nickendes Leimkraut
l *Silene nutans*

3870 nodding star-of-Bethlehem
d nickender Milchstern
l *Ornithogalum nutans*

3871 nolana
f nolane
d Glockenwinde
l *Nolana*

* **noli-me-tangere → 5560**

3872 nolina
f noline
d Nolina
l *Nolina*

3873 nonnea
f nonnée

d Mönchskraut; Mönchskolben; Napfkraut
l *Nonnea*

3874 Nootka false cypress
f cyprès de Nutka
d Nutkalebensbaumzypresse; Nutkascheinzypresse; Nutkazypresse; Sitkazypresse
l *Chamaecyparis nootkatensis*

3875 Nootka rose
f rosier de Nutka
d Nutkarose
l *Rosa nutkana*

3876 Nordmann fir
f sapin de Crimée; sapin de Nordmann; sapin du Caucase
d Kaukasustanne; Nordmannstanne
l *Abies nordmanniana*

3877 Norfolk Island pine
f araucarie élevée; pin de l'île de Norfolk
d Norfolktanne; Zimmertanne
l *Araucaria excelsa*

3878 northern bedstraw
d nordisches Labkraut
l *Galium boreale*

3879 northern catalpa
f catalpe élégant
l *Catalpa speciosa*

3880 northern dewberry
f framboisier du Nord
d niederliegende Brombeere
l *Rubus flagellaris; Rubus humifusus*

3881 northern listera
d Herzzweiblatt; kleines Zweiblatt
l *Listera cordata*

3882 northern red currant
f castillier; groseillier à grappes; groseillier rouge
d ährige Johannisbeere; rote Johannisbeere; skandinavische Johannisbeere; Kostbeere; Ribisel; Ribitzel; nordische Johannisbeere
l *Ribes rubrum; Ribes scandicum; Ribes vulgare; Ribes spicatum; Ribes sylvestre*

3883 northern red oak; red oak
f chêne rouge d'Amérique
d Roteiche
l *Quercus borealis*

3884 northern wood fern; mountain fern
f oréoptéris
d Bergfarn
l *Dryopteris oreopteris*

3885 Norway maple
f érable plane; érable de Norvège
d Lehne; Leinbaum; Spitzahorn; spitzblättriger Ahorn
l *Acer platanoides*

* **Norway pine → 4460**

3886 Norway spruce
f épicéa commun; faux sapin
d gemeine Fichte; Pechbaum; Pechtanne
l *Picea abies; Picea excelsa*

3887 Norwegian cinquefoil
f potentille de Norvège
d norwegisches Fingerkraut
l *Potentilla norvegica*

3888 Norwegian draba
f drave de Norvège
d norwegisches Felsenblümchen
l *Draba norvegica*

3889 nostoc
f nostoc
d Nostok; Schleimalge; Schleimling
l *Nostoc*

3890 notchleaf sea lavender
f statice sinué
l *Limonium sinuatum*

3891 notchleaf willow
f saule émoussé
d gestutzte Weide; Stutzweide
l *Salix retusa*

* **Nottingham catchfly → 3869**

3892 November goldenrod
f verge d'or tardive
d spätblühende Goldrute
l *Solidago serotina*

3893 nut grass; nut-grass flat sedge
f souchet rond
d Nussgras
l *Cyperus rotundus*

3894 nutmeg
f muscadier
d Muskatnussbaum
l *Myristica*

3895 nutmeg family
f myristicacées
d Myristikazeen
l *Myristicaceae*

*** nutmeg flower → 2415**

*** nutmeg geranium → 3897**

3896 nutmeg hickory
f caryer muscade
l *Carya myristicaeformis*

3897 nutmeg pelargonium; nutmeg geranium
f pélargonium très odorant
d Zitronenpelargonie
l *Pelargonium odoratissimum*

3898 nux-vomica poison nut
f arbre à noix vomique; noix vomique
d Brechnussbaum; Krähenaugenbaum; Schlangenholzbaum
l *Strychnos nux vomica*

O

3899 **oak**
f chêne
d Eiche
l *Quercus*

3900 **oak daedalea**
d Eichenwirrling
l *Daedalea quercina; Trametes quercina*

3901 **oak fern**
f dryoptéris commun
d Eichenfarn
l *Dryopteris linnaeana*

3902 **oak-leaf mountain ash**
f alisier de Laponie; sorbier de Laponie; sorbier finlandais
d Bastardeberesche; Bastardmehlbeere
l *Sorbus hybrida*

3903 **oakleech**
d Klapperkopf
l *Aureolaria; Dasystoma*

3904 **oat**
f avoine
d Hafer
l *Avena*

* **oat chestnut → 2145**

3905 **oat grass**
f arrhénanthère
d Bandgras; Franzosengras; französisches Raigras; Glatthafer; Wiesen-hafer
l *Arrhenatherum*

3906 **obtuse rush**
d Sumpfbinse
l *Juncus obtusiflorus*

3907 **ochna**
f ochne
d Nagelbeere
l *Ochna*

3908 **ocotea**
f ocotéa
d Stinkholz
l *Ocotea*

3909 **Odessa tamarisk**
f tamaris de Russie
d russische Tamariske
l *Tamarix odessana*

3910 **odontoglossum**
f odontoglosse
d Zahnzunge
l *Odontoglossum*

3911 **oedogonium**
f œdogonie
d Sprossenschlinke
l *Oedogonium*

3912 **ohia; Malay apple**
f jambose de Malaque; poirier de Malaque; pomme de Tahiti
d grosser Rosenapfel; malabarischer Rosenapfel
l *Syzygium malaccense; Eugenia malaccensis*

3913 **Ohio buckeye**
f pavier glabre
l *Aesculus glabra*

3914 **oïdium**
f oïdium
d Faulschimmel
l *Oidium*

3915 **oil flax; linseed**
f lin à graines
d Öllein
l *Linum usitatissimum* f. *brevimulticaula*

3916 **oil palm**
f palmier à (l')huile
d Ölpalme
l *Elaeis*

3917 **oil vine**
f telfairie; kouème
d Talerkürbis
l *Telfairia*

3918 **oka oxalis**
f surelle oca; oca
d Knollensauerklee
l *Oxalis crenata; Oxalis tuberosa*

3919 **Oklahoma plum**
f prunier de l'Oklahoma
l *Prunus gracilis*

3920 **okoume**
f okoumé
l *Aucoumea*

3921 okra; gumbo
f gombo; gombaud; ketmie comestible
d Okra; Rosenpappel
l *Hibiscus esculentus; Abelmoschus esculentus*

3922 old-field toadflax
f linaire du Canada
d kanadisches Leinkraut
l *Linaria canadensis*

3923 old-man wormwood; southernwood
f aurone
d Aberraute; Abrandkraut; Eberraute; Eberreis; Grabzypresse; Stabwurz
l *Artemisia abrotanum*

3924 old-world arrowhead
f fléchière; flèche d'eau; herbe à la flèche
d spitzes Pfeilkraut; gemeines Pfeilkraut; Pfeilwurz
l *Sagittaria sagittifolia; Sagittaria sinensis; Sagittaria japonica*

3925 oleander
f laurier-rose
d Oleander; Lorbeerrose
l *Nerium*

3926 oleaster family
f éléagnacées
d Eläagnazeen; Ölweidengewächse; Silberbäume
l *Elaeagnaceae*

3927 olive
f olivier
d Ölbaum
l *Olea*

3928 olive family
f oléacées
d ölbaumartige Gewächse; Ölbaumgewächse; Oleazeen
l *Oleaceae*

3929 oncidium
f oncidier
d Oncidium
l *Oncidium*

* **one-flowered pyrola → 6037**

3930 one-flower vetch
f lentille d'Auvergne
d einblütige Wicke; einblütige Erve; Linsenwicke; Algarobas-Linse; spanische Linse
l *Vicia monanthos; Vicia articulata*

* **one-grained wheat → 2004**

* **one-sided shinleaf → 4926**

3931 onion
f oignon
d Lauch; Zwiebel
l *Allium*

* **onion grass → 3582**

3932 onosma
f onosma
d Lotwurz
l *Onosma*

3933 opium poppy
f pavot somnifère
d Gartenmohn; Schlafmohn
l *Papaver somniferum*

3934 opopanax
f opopanax
d Gummiwurz(el); Heilwurz
l *Opopanax*

3935 opposite-leaved golden saxifrage
d gegenblättriges Milzkraut; gegenständiges Milzkraut
l *Chrysosplenium oppositifolium*

* **orach(e) → 4663**

3936 orange aphelandra
f aphélandre orangée
d goldgelbe Aphelandre
l *Aphelandra aurantiaca*

3937 orange-berry pittosporum
f pittospore ondulé
l *Pittosporum undulatum*

3938 orange-brown lactarius
f lactaire grosse poire; lactaire à lait abondant; lactaire orangé doré; vachotte
d Brätling; Milchbrätling; Milchreizker; Milchschwamm; Süssling
l *Lactarius volemus*

3939 orange bulbil lily
f lis orange
d Safranlilie
l *Lilium croceum; Lilium bulbiferum croceum*

3940 orange cap boletus; aspen mushroom
f bolet orangé; gyrole rouge; cèpe orangé; bolet roux
d oranger Rauhfuss; Rotkappe; Rotkopf; Rothautröhrling; Blutpilz; Rothäutchen
l *Boletus rufus; Leccinum aurantiacum; Boletus aurantiacus; Boletus versipellis*

3941 orange-eye butterfly bush
f buddléia de David; buddléia du Père David; buddléia changeant
d Goldaugenschmetterlingsstrauch; Herbstflieder; kleinblättriger Sommerflieder; veränderliche Buddleie
l *Buddleia variabilis; Buddleia davidi*

3942 orange hawkweed
f épervière orangée
d orangerotes Habichtskraut
l *Hieracium aurantiacum*

3943 orchard grass
f dactyle pelotonné
d Hundsgras; gemeines Knäuelgras; gemeines Knaulgras; rauhes Knäuelgras
l *Dactylis glomerata*

* **orchard grass → 1309**

3944 orchid
f orchidée
d Orchidee
l *Orchis*

3945 orchid family
f orchidacées
d Knabenkrautgewächse; Kuckucksblütler; Orchidazeen; Orchideen(gewächse)
l *Orchidaceae*

3946 orchis
f orchis
d Helmblume; Knabenkraut; Kuckucksblume; Ragwurz
l *Orchis*

3947 Oregon grape
f mahonia à feuilles de houx
d gemeine Mahonie
l *Mahonia aquifolium*

* **Oregon maple → 562**

* **Oregon myrtle → 938**

3948 Oriental arborvitae; Chinese arborvitae
f thuya de Chine; thuya d'Orient
d orientalischer Lebensbaum; morgenländischer Lebensbaum
l *Thuya orientalis; Biota orientalis*

3949 Oriental beech
f hêtre d'Orient
d orientalische Buche
l *Fagus orientalis*

3950 Oriental bladder senna
f baguenaudier d'Orient
d orientalischer Blasenstrauch; roter Blasenstrauch
l *Colutea orientalis*

3951 Oriental cherry
f cerisier à feuilles dentées en scie
d gesägtblättrige Kirsche
l *Prunus serrulata*

3952 Oriental clematis
f clématite d'Orient
d orientalische Waldrebe
l *Clematis orientalis*

3953 Oriental currant
f groseillier d'Orient
d orientalische Johannisbeere
l *Ribes orientale*

3954 Oriental germander
f germandrée d'Orient
d orientalischer Gamander
l *Teucrium orientale*

3955 Oriental hackberry
f micocoulier de Tournefort
d Tourneforts Zürgelbaum
l *Celtis tourneforti*

3956 Oriental hornbeam
f charme d'Orient
d orientalische Hainbuche; orientalischer Hornbaum
l *Carpinus orientalis*

3957 Oriental maple
f érable de Crète
l *Acer creticum; Acer orientale*

3958 Oriental paper bush
f edgeworthie à papier
d Mitsumata; papierliefernde Edgeworthia
l *Edgeworthia papyrifera*

3959 Oriental plane tree
f platane d'Orient
d morgenländische Platane; orientalische Platane
l *Platanus orientalis*

3960 Oriental poppy
f pavot de Tournefort
d morgenländischer Mohn; orientalischer Mohn; Türkenmohn
l *Papaver orientale*

3961 Oriental sesame
f sésame
d Sesam; orientalischer Sesam; weisser Sesam
l *Sesamum indicum*

3962 Oriental skullcap
f toque d'Orient; scutellaire d'Orient
d orientalisches Helmkraut
l *Scutellaria orientalis*

3963 Oriental spirea
f spirée intermédiaire
d Karpatenspierstrauch
l *Spiraea media*

3964 Oriental spruce
f épicéa du Caucase; épicéa d'Orient; sapinette d'Orient
d Kaukasusfichte; morgenländische Fichte; Sapindusfichte
l *Picea orientalis*

3965 Oriental sweet gum
f copalme d'Orient
d orientalischer Amberbaum
l *Liquidambar orientalis*

3966 Oriental wormwood
f armoise à balais
d Besenbeifuss
l *Artemisia scoparia*

3967 origanum
f origan; marjolaine
d Dost
l *Origanum*

3968 Orinoco simaruba
f bois de Cayan; simarube amer
l *Simaruba amara*

3969 orpine
f télèphe
d Zierspark
l *Telephium*

3970 orpine family
f crassulacées
d Dickblattgewächse; Krassulazeen
l *Crassulaceae*

3971 Osage orange
f oranger des Osages; maclure
d Osagedorn; Osageorange
l *Maclura*

3972 Osage orange; hedge apple
f maclure épineux; maclure à feuilles d'oranger
d apfelfrüchtiger Osagedorn
l *Maclura pomifera; Maclura aurantiaca*

* **osier → 5964**

3973 osmanthus
f osmanthe
d Duftblume; Duftblüte
l *Osmanthus*

3974 osmund
f osmonde
d Königsfarn; Rispenfarn; Traubenfarn
l *Osmunda*

3975 osmund family
f osmundacées
d Osmundazeen
l *Osmundaceae*

3976 ostrich fern
f plume d'autruche
d Straussfarn
l *Pteretis nodulosa; Matteuccia struthiopteris*

3977 Otaheite gooseberry; Otaheite-gooseberry leaf flower
f cerisier de Tahiti; chérimbelier; phyllanthe sour; surette
d saure Blattblume
l *Phyllantus acidus*

3978 ottelia
f ottélie
d Ottelie
l *Ottelia*

3979 ouratea
f ouratée
l *Ouratea*

3980 oval kumquat
f kumquat à fruits oblongs
d ovaler Kumquat; Nagami Kumquat

l *Fortunella margarita; Citrus margarita*

3981 oval-leaved willow
d ovalblättrige Weide
l *Salix ovalifolia*

3982 overcup oak; swamp oak
f chêne blanc aquatique; chêne à feuilles en lyre
d Leiereiche; leierförmige Eiche
l *Quercus lyrata*

3983 ovoid spike sege
d eiförmige Sumpfbinse; eiförmiges Sumpfried
l *Eleocharis ovata*

3984 owala oil tree
f owala; bois jaune du Gabon
l *Pentaclethra macrophylla*

3985 oxalis; wood sorrel
f surelle
d Sauerklee
l *Oxalis*

3986 oxeye
f buphthalme
d Ochsenauge; Rindsauge; Telekie
l *Buphthalmum; Telekia*

3987 oxeye daisy; woundwort
f chrysanthème grande-marguerite; marguerite des champs; chrysanthème leucanthème; leucanthème vulgaire
d Gevatterblume; grosse Gänseblume; grosse Massliebe; gemeine Wucherblume; grosse Wucherblume; Johannisblume; Margaretenblume; Margerite; Marguerite; Marienblatt; Marienblume; Wiesenmargarite; Wiesenwucherblume; weisse Wucherblume
l *Chrysanthemum leucanthemum; Leucanthemum vulgare*

3988 oxlip (primrose)
f primevère élevée
d geruchlose Primel; grosse Schlüsselblume; hohe Primel; hohe Schlüsselblume
l *Primula elatior*

3989 oxtongue
f picride
d Bitterkraut
l *Picris*

3990 oxtongue gasteria
f gastérie verruqueuse
d warzige Gasterie
l *Gasteria verrucosa*

3991 oxylobus
f oxylobe
l *Oxylobus*

3992 oyster mushroom
f nouret; oreille de noyer; pleurote en coquille; pleurote ostracé; poule de bois
d Austernschwamm; Austernseitling; Austernpilz; Buchenpilz; Drehling; Muschelpilz
l *Pleurotus ostreatus*

3993 oyster plant; golden thistle
f scolyme
d Golddistel
l *Scolymus*

* **oyster plant → 5706**

P

* **pacaya → 1112**

3994 pachycentria
f pachycentrie
l *Pachycentria*

3995 pachysandra
f pachysandre
d Pachysandra
l *Pachysandra*

3996 Pacific hemlock; western hemlock
f tsuga du Nouveau-Monde
d kalifornische Hemlockstanne
l *Tsuga heterophylla*

3997 Pacific white fir
f sapin de Low
d Lows Tanne
l *Abies lowiana; Abies concolor lowiana*

3998 Pacific yew
f if de Californie
d kurzblättrige Eibe
l *Taxus brevifolia*

3999 padauk
f ptérocarpe
d Flügelfruchtbaum
l *Pterocarpus*

4000 paederota
d Mänderle
l *Paederota*

4001 pagoda tree
f sophora
d Schnurbaum; Sophore
l *Sophora*

4002 pahudia
f pahudie
l *Pahudia*

4003 painted cup
f castillèje
d Kastillea
l *Castilleja*

4004 painted Mono maple
f érable coloré
d Nipponahorn
l *Acer pictum*

4005 pakchoi; Chinese cabbage
f chou de Chine
d Chinakohl; chinesischer Kohl
l *Brassica chinensis*

4006 palafoxia
f palafoxie
d Palafoxie
l *Palafoxia*

* **palas tree → 533**

4007 pale alyssum
f alysse à calice persistant
d Kelchsteinkraut
l *Alyssum calycinum*

4008 pale meadow violet
d weisses Veilchen
l *Viola alba*

4009 pale ragwort
d Sumpfgreiskraut; Sumpfkreuzkraut
l *Senecio paludosus*

4010 pale sedge
f laîche pâle
d bleiches Riedgras; bleiche Segge
l *Carex pallescens*

4011 pale sweet shrub
f calycanthe fertile
d fruchtbarer Gewürzstrauch
l *Calycanthus fertilis*

4012 pale willowweed
d blasses Weidenröschen; rosarotes Weidenröschen; rosiges Weidenröschen
l *Epilobium roseum*

4013 pale-yellow clavaria
f clavaire jaune
d gelber Ziegenbart; goldgelbe Koralle; Hahnenkamm; schwefelgelbe Koralle
l *Clavaria flava*

4014 palimara alstonia; dita; devil's-tree
d Teufelsbaum
l *Alstonia scholaris*

4015 paliurus
f paliure
d Stechdorn
l *Paliurus*

4016 palm
f palmier
d Palme
l *Palmae*

4017 **palmetto**
f sabal
d Dachpalme; Sabal(palme)
l *Sabal*

4018 **palm family**
f palmacées; palmiers
d Palmen
l *Palmaceae*

4019 **palmyra palm**
f borasse
d Borassuspalme; Lontaropalme; Palmyrapalme; Delebpalme; Weinpalme
l *Borassus flabellifer*

4020 **pampas grass**
f herbe des Pampas; herbe à plumets
d Pampasgras
l *Cortaderia*

4021 **pampas grass**
f gynérion
d Pampasgras
l *Gynerium*

4022 **Panama gum tree**
d Holquahitl
l *Castilla elastica*

4023 **pancratium**
f pancratie
d Pankrazlilie; Trichterlilie
l *Pancratium*

* **pandanus → 4779**

4024 **pangium**
f pang(i)e
l *Pangium*

4025 **paniala**
f flacourtie de Cochinchine
l *Flacourtia jangomas; Flacourtia cataphracta*

4026 **panic grass; witch grass**
f panic
d Hirse
l *Panicum*

4027 **panicled gold rain tree**
f koelreutérie paniculée
l *Koelreuteria paniculata*

4028 **panicled hydrangea**
f hortensia en panicule
d rispige Hortensie
l *Hydrangea paniculata*

4029 **panicled monkshood**
d rispiger Eisenhut
l *Aconitum paniculatum*

4030 **panicled sedge**
f laîche paniculée
d rispiges Riedgras; rispige Segge
l *Carex paniculata*

4031 **pansy orchid**
f miltonia
d Miltonie
l *Miltonia*

* **panther amanita → 4032**

4032 **panther cap; panther amanita**
f amanite panthère
d Pantherpilz; Pantherschwamm; Königsfliegenpilz; Krötenschwamm
l *Amanita pantherina*

4033 **panus**
f panus
d Knäueling
l *Panus*

* **papaw → 4077**

4034 **papaw family**
f caricacées; papayacées
d Karikazeen; Papayazeen
l *Caricaceae; Papayaceae*

4035 **papaya; pa(w)paw**
f arbre à melon; papayer
d Mamaobaum; Melonenbaum; Papayabaum
l *Carica papaya*

4036 **paperbark maple**
f érable gris
d grauer Ahorn
l *Acer griseum*

4037 **paper birch**
f bouleau à canots; bouleau à papier
d Graubirke; Papierbirke
l *Betula papyrifera*

4038 **paper bush**
f edgeworthie
l *Edgeworthia*

4039 **paper mulberry**
f broussonétie
d Papiermaulbeerbaum
l *Broussonetia*

* **paper-white narcissus** → 4248

4040 paphinia
f paphinie
l *Paphinia*

4041 papilionaceous plants
f papilionacées
d Papilionaten; Papilionazeen; Schmetterlingsblütler
l *Papilionaceae*

4042 pappus grass
f pappophore
l *Pappophorum*

4043 papyrus
f souchet à papier
d Papierstaude; Papyrusstaude
l *Cyperus papyrus; Papyrus antiquorum*

4044 Para angelwood
f angélique
l *Dicorynia paraensis*

4045 Para cress; spot flower; Para cress spot flower
f cresson de Para
d Parakresse
l *Spilanthes oleracea*

4046 paradise apple
f pommier paradis
d Paradiesapfel; Johannisapfel
l *Malus sylvestris* var. *paradisiaca*

4047 paradise tree
d Marupa
l *Simaruba glauca*

4048 Paraguay tea; yerba maté
f arbre à maté; maté
d Matebaum; Matepflanze; Matestrauch; Paraguayteepflanze; Jesuitentee
l *Ilex paraguariensis*

4049 Parana araucaria
f araucarie du Brésil; pin de Parana
d brasilianische Araukarie; Pinheiro; Schuppentanne; Paranakiefer
l *Araucaria brasiliana; Araucaria angustifolia*

4050 Para rubber tree
f hévéa
d Parakautschukbaum
l *Hevea brasiliensis*

4051 parasol mushroom
f coulemelle; lépiote élevée
d grosser Schirmling; grosser Schirmpilz; Riesenschirmling
l *Lepiota procera*

4052 parinarium
f parinare
l *Parinarium*

4053 paris
f parisette
d Einbeere
l *Paris*

4054 Paris bedstraw
d Pariser Labkraut
l *Galium parisiense*

4055 Paris circaea
f circée parisienne
d gemeines Hexenkraut; grosses Hexenkraut
l *Circaea lutetiana*

4056 parmelia
f parmélie
d Schildflechte; Schüsselflechte
l *Parmelia*

4057 parmelia family
f parméliacées
d Parmeliazeen
l *Parmeliaceae*

4058 parnassia; grass of Parnassus
f parnassie; gazon du Parnasse
d Herzblatt
l *Parnassia*

4059 parrotbeak; parrotbill
f clianthe
d Ruhmesblume
l *Clianthus*

4060 parrot feather; water milfoil
f myriophylle
d Tausendblatt
l *Myriophyllum*

4061 parrotia
f parrotie
d Eisenholz; Parrotie
l *Parrotia*

4062 parrya
f parrye
d Parrye
l *Parrya*

4063 parsley
f persil
d Petersilie
l *Petroselinum*

* **parsley piert → 2226**

4064 parsnip
f panais
d Hammelmöhre; Pasternak; Pastinak(e)
l *Pastinaca*

4065 parthenium
f guayule
d Guayule; Mutterkraut
l *Parthenium*

4066 partridgeberry
f mitchelle
d Rebhunbeere
l *Mitchella*

4067 paspalum
f paspale
d Pfannengras
l *Paspalum*

* **pasqueflower → 243**

4068 passionflower
f passiflore
d Passionsblume; Rangapfel
l *Passiflora*

4069 passionflower family
f passifloracées
d Passiflorazeen; Passionsblumengewächse
l *Passifloraceae*

4070 passion fruit; purple granadilla; edible granadilla
f passiflore comestible; grenadille
d Purpurgranadille
l *Passiflora edulis*

4071 Patagonian fitzroya
f alerce de Patagonie
d patagonische Fitzroya
l *Fitzroya patagonica; Fitzroya cupressoides*

* **patchouli plant → 2792**

4072 patience dock
f patience des jardins; épinard oseille
d Gartenampfer; Geduldampfer; Gemüseampfer; Mönchsrhabarber; Patienzkraut
l *Rumex patientia*

4073 patrinia
f patrinie
d Vierlingskraut; Goldbaldrian
l *Patrinia*

4074 paullinia
f paullinie
d Paullinie; Giftknippe
l *Paullinia*

4075 paulownia
f paulownie
d Blauglockenbaum; Kaiserbaum; Paulownie
l *Paulownia*

4076 pavetta
f pavette
d Scheelkorn
l *Pavetta*

4077 pawpaw; papaw
f asiminier
d Papau(baum)
l *Asimina*

* **pa(w)paw → 4035**

4078 paxillus
f paxille
d Krempling
l *Paxillus*

4079 payena
f payéna
d Balam
l *Payena*

4080 pea
f pois
d Erbse
l *Pisum*

4081 peabush
d ägyptische Sesbanie
l *Sesbania aegyptiaca*

4082 peach
f pêcher
d Pfirsich(baum)
l *Prunus persica; Persica vulgaris*

4083 peach-leaved bellflower
f campanule à feuilles de pêcher
d pfirsichblättrige Glockenblume
l *Campanula persicifolia*

4084 peach-leaved willow
f saule à feuilles de pêcher
d Mandelweide
l *Salix amygdaloides*

4085 peach palm; pejibaye
f parépon
d Pfirsichpalme
l *Guilielma speciosa; Guilielma gasipaes*

4086 peak-cap mycena
d rosablättriger Helmling
l *Mycena galericulata*

4087 peanut
f arachide
d Erdnuss; Erdbohne; Erdeichel; Erdmandel; Grundeichel; Mandubi-Bohne
l *Arachis*

4088 peanut; groundnut
f arachide
d gemeine Erdnuss
l *Arachis hypogaea*

4089 pear
f poirier
d Birnbaum
l *Pyrus*

4090 pearlbush
f exochorde
d Blumenspiere; Perlbusch; Prunkspiere
l *Exochorda*

4091 pear-leaved crab apple; Chinese apple
f pommier à feuilles de prunier
d pflaumenblättriger Apfelbaum
l *Malus prunifolia; Pyrus prunifolia*

4092 pearl millet; cattail millet; African millet
f millet à chandelles; millet perle
d Bajree; Kerzenhirse; Negerhirse; Perlhirse; Pinselgras; Pinselhirse; Rohrkolbenhirse
l *Pennisetum glaucum; Pennisetum spicatum; Pennisetum typhoideum*

4093 pearlwort
f sagine
d Knebel; Mastkraut; Sternmoos
l *Sagina*

4094 pearly everlasting
f anaphalide
d Perlpfötchen; Katzenpfötchen
l *Anaphalis*

4095 pea shrub; pea tree
f caragana
d Erbsenstrauch; Erbsenbaum
l *Caragana*

*** peat moss → 5125**

4096 peat mosses
f sphagnacées
d Sphagnazeen; Torfmoose
l *Sphagnaceae*

*** pea tree → 4095**

4097 pea vetch
f vesce pisiforme; vesce à feuilles de pois
d Erbsenwicke
l *Vicia pisiformis*

4098 pea vine; everlasting pea; vetchling
f gesse; jarousse
d Kicher; Platterbse; Zierwicke
l *Lathyrus*

4099 pea vine
f orobe
d Platterbse; Walderbse
l *Orobus*

4100 pecan
f caryer pacanier
d Pekanbaum; Hickorynussbaum; Pekannussbaum
l *Carya pecan; Carya illinoensis*

4101 pedalia family
f pédaliacées
d Pedaliazeen
l *Pedaliaceae*

4102 peganum
f pégane
d Harmelkraut; Harmelraute
l *Peganum*

*** pejibaye → 4085**

*** pekea nut → 4725**

*** Peking cabbage → 4158**

4103 pelargonium; stork's-bill
f pélargonier; pélargonium
d Pelargonie
l *Pelargonium*

4104 pelican Dutchman's-pipe
f aristoloche à grandes fleurs

d grossblütige Pfeifenblume
l *Aristolochia grandiflora*

4105 pellitory
f pariétaire
d Glaskraut; Wandnessel
l *Parietaria*

4106 pencilwood
f dysoxyle
l *Dysoxylum*

4107 pendulous sedge
f laîche élevée
d grosses Riedgras; hängende Segge
l *Carex pendula*

4108 penicillaria
f pénicillaire
l *Penicillaria*

4109 penicillium
f pénicille
d Pinselschimmel
l *Penicillium*

4110 pennisetum
f pennisétum
d Borstenfedergras; Federähre; Federborstengras; Perlgras; Perlkraut
l *Pennisetum*

4111 Pennsylvania bitter cress; hairy bitter cress
f cardamine hérissée
d behaartes Schamkraut; haariges Schamkraut; rauhhaariges Schamkraut
l *Cardamine hirsuta*

4112 pennycress
f thlaspi
d Hellerkraut; Pfennigkraut; Täschelkraut
l *Thlaspi*

4113 pennyroyal
f menthe pouliot
d Polei(minze)
l *Mentha pulegium*

4114 pennywort
f hydrocotyle
d Wassernabel
l *Hydrocotyle*

4115 pentacme
f pentacme
l *Pentacme*

4116 pentapera
f pentapère
d Pentapere
l *Pentapera*

4117 pentstemon; beardtongue
f pentstémon
d Bartfaden
l *Pen(t)stemon*

4118 peony
f pivoine
d Pfingstrose; Pfingstblume; Königsrose; Pumpelrose
l *Paeonia*

4119 peperomia
f pépéromie
d Peperomie; Zwergpfeffer; Pfefferkraut
l *Peperomia*

4120 pepper
f poivrier
d Pfeffer
l *Piper*

* **pepper cress → 4126**

4121 pepper family
f pipéracées
d Pfeffergewächse; Piperazeen
l *Piperaceae*

* **peppergrass → 4126**

4122 peppermint
f menthe poivrée
d Pfefferminze; Katzenkraut
l *Mentha piperita*

4123 pepper mushroom; peppery lactarius
f lactaire poivré
d echter Pfeffermilchling; grünender Pfeffermilchling
l *Lactarius piperatus*

4124 pepper saxifrage
f persil bâtard
d Wiesensilge; Silau; Rosskümmel
l *Silaus*

4125 pepper tree
f poivrier du Pérou; faux-poivrier
d Pfefferbaum
l *Schinus*

4126 pepperweed; peppergrass; pepper cress
f passerage

d Kresse
l *Lepidium*

4127 **pepperweed whitetop**
d Pfeilkresse
l *Lepidium draba; Cardaria draba*

4128 **pepperwort**
f marsilée
d Kleefarn
l *Marsilea*

* **peppery lactarius → 4123**

4129 **perennial flax**
f lin vivace
d ausdauernder Lein
l *Linum perenne*

4130 **perennial honesty**
f lunaire vivace
d Atlasblume; ausdauerndes Silberblatt; Mondkraut; spitzes Silberblatt; Wintermondviole
l *Lunaria rediviva*

4131 **perennial knawel**
d ausdauernder Knäuel; ausdauerndes Knäuelkraut; Dauerknäuel; weisser Knäuel
l *Scleranthus perennis*

4132 **perennial lettuce**
f laitue vivace
d Blauerlattich
l *Lactuca perennis*

4133 **perennial pea vine**
f gesse à larges feuilles
d breitblättrige Platterbse; Gartenplatterbse; Staudenwicke
l *Lathyrus latifolius*

4134 **perennial quaking grass**
f brize; tremblette
d gemeines Zittergras; mittleres Zittergras
l *Briza media*

4135 **perennial rye grass; English rye grass**
f ivraie vivace; ray-grass anglais
d deutsches Weidelgras; englisches Raigras; Wiesenlolch
l *Lolium perenne*

4136 **pereskia**
f peireskia
d Laubkaktus; Peireskie
l *Peireskia*

4137 **perigord truffle**
f truffe du Périgord
d Perigord-Trüffel
l *Tuber melanospermum*

4138 **perilla**
f pérille
d Schwarznessel; Perille
l *Perilla*

4139 **peristrophe**
f péristrophe
d Peristrophe; Drehling
l *Peristrophe*

4140 **periwinkle**
f pervenche
d Immergrün; Singrün
l *Vinca*

4141 **pernettya; arrayan**
f pernettye
d Pernettya; Torfmyrte
l *Pernettya*

4142 **perovskia**
f pérowskie
d Perovskie; Perowskie
l *Perovskia; Perowskia*

4143 **perpetual begonia**
d Gottesauge; immerblühende Begonie; immerblühendes Schiefblatt
l *Begonia semperflorens*

4144 **persea**
f persée
d Fenchelholzbaum; Isabellaholz; Avocatobirne
l *Persea*

4145 **Persian berry buckthorn**
f nerprun des teinturiers
d Färberdorn
l *Rhamnus infectoria; Rhamnus infectorius*

4146 **Persian buttercup**
f renoncule d'Orient
d persischer Hahnenfuss
l *Ranunculus asiaticus*

4147 **Persian clover; shabdar; shaftal clover**
f trèfle renversé; trèfle de Perse
d persischer Klee; Wendelblumenklee
l *Trifolium resupinatum*

* **Persian cyclamen → 2990**

4148 Persian iris
f iris de Perse
d persische Schwertlilie
l *Iris persica*

4149 Persian lilac
f lilas de Perse
d persischer Flieder
l *Syringa persica*

* **Persian lilac → 1160**

4150 Persian parrotia
f parrotie de Perse
d Arganbaum; Eisenholzbaum; persische Parrotie
l *Parrotia persica*

4151 Persian walnut
f noyer commun; noyer cultivé
d edler Walnussbaum; Edelwalnuss; gemeiner Walnussbaum; welsche Nuss; Welschnussbaum
l *Juglans regia*

4152 persimon
f plaqueminier
d Dattelpflaume(nbaum); Götterpflaume; Kakipflaume
l *Diospyros*

4153 Peru balsam balm tree
f baumier du Pérou
d Perubalsambaum
l *Myroxylon pereirae; Toluifera pereirae*

4154 Peru false heath
f fabiane imbriquée
d Fabiana
l *Fabiana imbricata*

* **Peruvian almond → 4725**

4155 Peruvian elder
f sureau du Pérou
d peruanischer Holunder
l *Sambucus peruviana*

4156 Peruvian ground cherry; Cape gooseberry
f coqueret du Pérou
d Ananaskirsche; Kapstachelbeere
l *Physalis peruviana*

4157 petrea
f pétrée
d Peträa
l *Petrea*

4158 pe-tsai; celery cabbage; Peking cabbage; Chinese cabbage
f chou de Shanton
d Pekingkohl; Schantungkohl
l *Brassica pekinensis*

4159 petty euphorbia
f rhubarbe du paysan; euphorbe des vignes
d Gartenwolfsmilch
l *Euphorbia peplus*

4160 petunia
f pétunia
d Petunie
l *Petunia*

4161 phacelia
f phacélie
d Bienenbrot; Büschelblume; Büschelkraut; Büschelschön; Phazelie
l *Phacelia*

4162 phallus family
f phallacées
d Rutenpilze
l *Phallaceae*

4163 phantom orchid
f céphalanthère
d Waldvögelein; Kopforchis; Kopfständel
l *Cephalanthera*

4164 pheasant's-eye adonis
f adonide d'automne
d Herbstfeuerröschen; Blutströpfchen; Herbstadonisröschen
l *Adonis autumnalis*

4165 philodendron
f philodendron
d Baumlieb
l *Philodendron*

4166 phlox
f phlox
d Flammenblume
l *Phlox*

4167 phlox family
f polémoniacées
d Himmelsleitergewächse; Polemoniazeen; Sperrkrautgewächse
l *Polemoniaceae*

4168 Phoenicean juniper
f genévrier de Phénicie

d rotfrüchtiger Wacholder
l *Juniperus phoenicea*

4169 photinia
f photinie
d Glanzmispel; Photinie
l *Photinia*

4170 phyllostachys
f phyllostachys
d Blattähre; Phillostachys
l *Phyllostachys*

* **physic nut → 449**

4171 phytocrenum
f phytocrène
l *Phytocrenum*

4172 phytophthora
f phytophthore
d Kartoffelpilz
l *Phytophthora*

* **pia → 2244**

4173 pia family
f taccacées
d Takkazeen
l *Taccaceae*

4174 pickerelweed
f pontédérie
d Pontederie; Wasserhyazinthe
l *Pontederia*

4175 pickerelweed family
f pontédériacées
d Pontederiazeen; Pontederiengewächse
l *Pontederiaceae*

* **piemarker → 1217**

4176 pieris
f piéride
d Pieris
l *Pieris; Arcterica*

* **pigeon grass → 780**

4177 pigeon pea
f pois d'Angola
d Straucherbse; Strauchbohne; Taubenerbse
l *Cajanus cajan; Cajanus indicus*

* **pigeon vetch → 5531**

4178 pigeonwings; butterfly pea
f clitoria
d Schamblume
l *Clitoria*

4179 pignut hickory
f caryer glabre; noyer à balais; noyer des pourceaux
d Ferkelnuss
l *Carya glabra; Carya porcina*

* **pigweed → 4390**

4180 pileate fungi
f hyménomycètes
d Hautpilze; Hymenomyzeteen
l *Hymenomycetes*

4181 pilewort
f ficaire
d Feigwurz; Scharbockskraut
l *Ficaria*

4182 pill-headed sedge
d Pillenriedgras; Pillensegge
l *Carex pilulifera*

4183 pillwort
f pilulaire
d Pillenfarn; Pillenkraut
l *Pilularia*

4184 pilocarpus
f pilocarpe
d Jaborandistrauch
l *Pilocarpus*

* **pimenta → 102**

4185 pimpernel
f anagallide; mouron
d Gauchheil
l *Anagallis*

4186 pimpinella
f boucage
d Pimpinelle
l *Pimpinella*

4187 pinanga palm
f pinanga
d Pinangpalme
l *Pinanga*

4188 pin cherry
f prunier de Pennsylvanie
l *Prunus pennsylvanica*

4189 Pindrow fir
f sapin de Pindrow
d Pindrows Tanne
l *Abies pindrow*

4190 pine
f pin
d Föhre; Kiefer
l *Pinus*

4191 pineapple
f ananas; ananas cultivé
d Ananas
l *Ananas comosus; Ananas sativus*

4192 pineapple family
f broméliacées
d Ananasgewächse; Bromeliazeen; Bromelien
l *Bromeliaceae*

4193 pineapple weed
d strahllose Kamille
l *Matricaria matricaroides; Matricaria discoidea; Matricaria suaveolens*

4194 pine family
f pinacées
d Kieferngewächse
l *Pinaceae*

4195 pine strawberry
f fraisier ananas
d Ananaserdbeere
l *Fragaria ananassa*

4196 piney tree
f vatérie de l'Inde
d Talgbaum
l *Vateria indica*

*** pin grass → 91**

4197 pink; carnation
f œillet
d Nelke
l *Dianthus*

4198 pink family
f caryophyllacées
d Nelkengewächse; Nelkenpflanzen; Karyophyllazeen
l *Caryophyllaceae*

4199 pink-head knotweed
d Köpfchenknöterich
l *Polygonum capitatum*

*** pinkwood → 1275**

4200 pin oak; swamp oak
f chêne des marais
d Nadeleiche; Spiesseiche; Sumpfeiche
l *Quercus palustris*

*** pipal tree → 738**

4201 pipewort
f joncinelle
d Kugelbinse
l *Eriocaulon*

4202 pipsissewa; ground holly
f chimaphile
d Winterlieb
l *Chimaphila*

4203 piptadenia
f piptadénie
d Piptadenie
l *Piptadenia*

4204 piptanthus
f piptanthe
l *Piptanthus*

*** pirageia → 236**

4205 pistache
f pistachier
d Pistazie
l *Pistacia*

4206 pitanga; Surinam cherry
f cerisier carré; cerisier de Cayenne; eugénier à fleurs solitaires
d Pitangabaum
l *Eugenia uniflora*

4207 pitcher plant; side-saddle flower
f sarracénie
d Krugblatt; Schlauchpflanze; Trompetenblatt
l *Sarracenia*

*** pitcher plant → 3828**

4208 pitch pine; hard pine
f pin à aubier; pin noir d'Amérique; pin rigide; pin résineux; pin raide
d Pechkiefer
l *Pinus rigida*

4209 piteira furcrea; giant lily; Mauritius hemp
f fourcrée gigantesque
d Mauritiushanf
l *Furcraea gigantea*

4210 pittosporum
f pittospore
d Klebsame
l *Pittosporum*

4211 plagiochila
f plagiochile
l *Plagiochila*

4212 plains coreopsis
f coréopsis élégant
d zweifarbiges Mädchenauge
l *Coreopsis tinctoria*

4213 plains crazyweed
d gemeiner Spitzkiel; Wiesenfahnenwicke
l *Oxytropis campestris*

4214 plains poplar
f peuplier de Sargent
d Sargents Pappel
l *Populus sargenti*

4215 plane tree; sycamore
f platane
d Platane
l *Platanus*

4216 plane-tree family
f platanacées
d Platanazeen; Platanengewächse
l *Platanaceae*

4217 plane-tree maple; sycamore maple
f érable faux platane; érable sycomore
d Bergahorn; weisser Ahorn
l *Acer pseudoplatanus*

4218 plantain
f plantain
d Weg(e)breit; Wegerich; Weg(e)tritt
l *Plantago*

4219 plantain banana; Adam's-fig; cooking banana
f bananier commun; bananier du paradis; figuier d'Adam
d Banane; Mehlbanane; Paradiesfeige
l *Musa paradisiaca*

4220 plantain family
f plantaginacées
d Plantaginazeen; Wegerichgewächse
l *Plantaginaceae*

4221 plantain lily
f funkie
d Funkie; Herzlilie
l *Hosta; Funkia*

4222 platystemon
f platystémone
d Breitfaden
l *Platystemon*

4223 pluchea
f pluchée
l *Pluchea*

4224 plum
f prunier
d Pflaume(nbaum)
l *Prunus*

4225 plumbago; leadwort
f dentelaire
d Bleiwurz
l *Plumbago*

4226 plum clitopilus
f agaric à odeur de farine; clitopile petite prune
d echter Mousseron; Mehlpilz; Pflaumenpilz; rosablättriger Krempling
l *Clitopilus prunulus*

4227 plume clematis
f clématite flammèle; clématite brûlante; clématite odorante
d brennende Waldrebe; scharfer Blasenzug
l *Clematis flammula*

4228 plume grass
f érianthe
d Seidengras
l *Erianthus*

4229 plume incense cedar
f libocèdre de Nouvelle-Zélande
l *Libocedrus plumosa; Libocedrus doniana*

4230 plume poppy
f bocconie
d Federmohn
l *Macleaya; Bocconia*

4231 plum juniper
f genévrier à prunes
d grossfrüchtiger Wacholder
l *Juniperus macrocarpa*

4232 plum yew
f if à prunes; céphalotaxus; céphalotaxe

d Kopfeibe; Scheineibe
l *Cephalotaxus*

* **poa → 700**

4233 pod corn; husk corn; tunicate corn
f maïs vêtu
d Balgmais; Spelzmais
l *Zea mays* var. *tunicata*

4234 pod grass
f triglochin
d Dreizack; Salzbinse
l *Triglochin*

4235 podocarp
f podocarpe
d Steineibe
l *Podocarpus*

4236 podocarp family
f podocarpacées
d Steineibengewächse
l *Podocarpaceae*

4237 poet's narcissus
f narcisse des poètes; narcisse des jardins; œil-de-faisan
d weisse Narzisse
l *Narcissus poeticus*

4238 pogonia
f pogonie
l *Pogonia*

4239 poinsettia
f poinsettie
d Weihnachtsstern
l *Poinsettia*

* **poison amanita → 1820**

4240 poisonbush
f gastrolobe
d Bauchhülse
l *Gastrolobium*

4241 poison hemlock
f grande ciguë
d Fleckschierling; gefleckter Schierling; Tollkerbel; Wüterich
l *Conium maculatum*

* **poison ivy → 1501**

* **poison oak → 1501**

4242 poison sumac
f sumac vénéneux
d Giftsumach
l *Rhus vernix; Toxicodendron vernix*

4243 pokeberry; pokeweed
f phytolaque
d Kermesbeere; Scharlachbeere; Schminkbeere
l *Phytolacca*

* **poker plant → 5551**

4244 pokeweed family
f phytolaccacées
d Kermesbeerengewächse; Kermespflanzen; Phytolakkazeen
l *Phytolaccaceae*

4245 polar willow
d Polarweide
l *Salix polaris*

4246 polemonium; Greek valerian; Jacob's-ladder
f polémonie
d Himmelsleiter; Sperrkraut
l *Polemonium*

* **Polish mushroom → 485**

4247 Polish wheat
f blé de Pologne
d Gommer; polnischer Weizen; Riesenroggen; sibirisches Korn
l *Triticum polonicum*

4248 polyanthus narcissus; bunch-flowered narcissus; paper-white narcissus
f narcisse à bouquet; narcisse de Constantinople
d Bukettnarzisse; Tazette
l *Narcissus tazetta*

4249 polycnemum
f polycnème
d Knorpelkraut
l *Polycnemum*

4250 polygala; milkwort
f laitier
d Kreutzblümchen; Kreuzblume; Milchblume; Ramsel
l *Polygala*

4251 polypody
f polypode
d Engelsüss; Tüpfelfarn
l *Polypodium*

4252 polypody family
f polypodiacées
d Tüpfelfarngewächse
l *Polypodiaceae*

4253 polypores; bracket fungi
f polyporacées
d Löcherpilze; Löcherschwämme; Polyporazeen; Porlinge; Stielporlinge; Zähpilze
l *Polyporaceae*

4254 polyporus
f polypore destructeur
d Holzschwamm; Waldporling
l *Polyporus destructor*

4255 pomegranate
f grenadier
d Granatapfel(baum)
l *Punica granatum*

*** pompelmous → 4329**

4256 pompon cabbage rose
f rosier centfeuilles pompon; rosier d'amour
l *Rosa centifolia* var. *pomponia*

4257 pond apple; alligator apple
d Alligatorapfel
l *Annona glabra*

4258 pond dock
f grande patience
d Ampferwurzel
l *Rumex aquaticus*

4259 ponderosa pine; western yellow pine
f pin jaune du Nouveau-Monde; pin à bois lourd; pin pondéreux
d Gelbkiefer; Goldkiefer
l *Pinus ponderosa*

4260 pond scum; spirogyra
f spirogyre
d Schraubenalge
l *Spirogyra*

4261 pond water starwort
d Schlammwasserstern; Teichwasserstern
l *Callitriche stagnalis*

4262 pondweed
f épi d'eau
d Leichkraut; Samtkraut
l *Potamogeton*

4263 pondweed family
f potamogétonacées
d Laichkrautgewächse
l *Potamogetonaceae*

4264 Pontic azalea; amber-bloom rhododendron; golden rhododendron
f rhododendron pontique; azalée pontique; azalée du Pont
d gelbe Alpenrose; pontische Alpenrose; pontische Azalee
l *Azalea pontica; Rhododendron luteum; Rhododendron flavum*

4265 Pontic oak
f chêne du Pont
d pontische Eiche
l *Quercus pontica*

4266 poolmat; horned pondweed
f zannichellie
d Teichfaden; Halde; Zannichellie
l *Zannichellia*

4267 pool moss
f mayaque
l *Mayaca*

4268 popcorn
f maïs perlé
d Puffmais
l *Zea mays* var. *everta*

4269 popcorn flower
f plagiobotryde
l *Plagiobothrys*

4270 poplar
f peuplier
d Pappel
l *Populus*

4271 poplar-leaved rockrose
f ciste à feuilles de peuplier
d pappelblättrige Zistrose
l *Cistus populifolius*

4272 poppy
f pavot
d Mohn
l *Papaver*

4273 poppy anemone
f anémone couronnée; anémone des fleuristes
d Kronenanemone; Gartenanemone
l *Anemone coronaria*

4274 poppy family
f papavéracées
d Mohngewächse; Papaverazeen
l *Papaveraceae*

4275 porcupine pod tree
f centrolobe
d Zebraholzbaum
l *Centrolobium*

4276 Portuguese broom
f cytise blanc
d weisser Goldregen
l *Cytisus albus*

* **Portuguese cypress → 3597**

4277 Portuguese laurel cherry; cherry laurel
f prunier du Portugal; cerisier du Portugal; laurier du Portugal
d portugiesischer Kirschlorbeer
l *Laurocerasus lusitanica; Prunus lusitanica*

4278 Portuguese oak
f chêne du Portugal
d portugiesische Eiche
l *Quercus lusitanica*

4279 post oak
f chêne étoilé
l *Quercus stellata*

4280 Potanin birch
f bouleau de Potanin
d Potanins Birke
l *Betula potanini*

4281 potato
f pomme de terre
d Kartoffel
l *Solanum tuberosum*

4282 potato bean
f apios
d Erdbirne; Knollenerbse
l *Apios*

* **potato onion → 3753**

4283 potherb jute; Jew's-mallow
f corette potagère; mauve des Juifs
d Gemüsepappel; Judenpappel; Meluchia; langkapselige Jute
l *Corchorus olitorius*

4284 potherb onion; field garlic
f ail des champs
d Gemüselauch; Kohllauch
l *Allium oleraceum*

4285 pot marigold
f souci officinal; souci des jardins
d Gartenringelblume; Gilgenkraut; Gilke(nkraut); Goldblume; Totenblume
l *Calendula officinalis*

4286 poulard wheat; cone wheat
f blé poulard; gros blé
d englischer Weizen; welscher Weizen; Rauhweizen
l *Triticum turgidum*

4287 poverty brome
f brome stérile
d taube Trespe
l *Bromus sterilis*

4288 poverty rush; slender rush
d zarte Binse
l *Juncus tenuis*

4289 powdery mildews
f érysiphacées
d Mehltaupilze; Erysiphazeen
l *Erysiphaceae*

4290 prairie clover
f pétalostémon
d Prärieklee
l *Petalostemum; Petalostemon*

4291 prairie dogbane
f apocyn vénitien
d sibirische Hundswolle
l *Apocynum sibiricum; Apocynum venetum*

4292 prairie June grass
f koélérie à crète
l *Koeleria cristata*

* **prairie mallow → 1127**

4293 prairie pepperweed
f passerage à fleurs denses
d dichtblütige Kresse
l *Lepidium densiflorum; Lepidium apetalum*

4294 prickle grass
f crypside
d Dornengras; Sumpfgras
l *Crypsis*

4295 prickly ash
f xanthoxyle; clavalier

d Gelbholz
l *Xanthoxylum; Zanthoxylum*

4296 prickly brazilwood
f brésillet de Fernambouc; césalpinie hérissée
d echter Brasilholzbaum; echter Rotholzbaum; Fernambukholzbaum; Pernambukholzbaum; Brasilettoholz
l *Caesalpinia echinata*

4297 prickly comfrey
f consoude rugueuse
d rauhe Wallwurz; rauher Beinwell
l *Symphytum asperum*

*** prickly cup → 1114**

4298 prickly juniper; red-berried juniper
f cade; genévrier piquant; genévrier oxycèdre
d Zederwacholder; rotbeeriger Wacholder; griechische Zeder; spanische Zeder; spanischer Wacholder
l *Juniperus oxycedrus*

4299 prickly lettuce; compass lettuce
f escarole; laitue sauvage; escariole
d Leberdistel; Skariol; Stachellattich; Stachelsalad; wilder Lattich; Zaunlattich
l *Lactuca serriola; Lactuca scariola*

4300 prickly parsnip
f échinophore
d Stacheldolde
l *Echinophora*

4301 prickly pear
f figuier d'Inde
d Feigenkaktus; Feigendistel; Opuntie
l *Opuntia*

4302 prickly poppy
f argémone
d Stachelmohn
l *Argemone*

4303 prickly sedge
d dichtährige Segge; Stachelsegge
l *Carex spicata*

4304 prickly-seeded spinach
f épinard
d Binetsch; Gartenspinat; gemeiner Spinat; Gemüsespinat
l *Spinacia oleracea*

4305 prickly sow thistle
f laiteron épineux
d rauhe Gänsedistel
l *Sonchus asper*

4306 prickly thrift
f acantholimon
d Igelpolster; Stachelsandnelke; Stachelnelke; Stechnelke
l *Acantholimon*

4307 primrose
f primevère
d Himmelsschlüssel(chen); Primel; Schlüsselblume
l *Primula*

4308 primrose family
f primulacées
d Primelgewächse; Primulazeen; Schlüsselblumengewächse
l *Primulaceae*

*** prince's pine → 1496**

4309 prince's-plume lady's-thumb
f persicaire du Levant
d morgenländischer Knöterich
l *Polygonum orientale; Persicaria orientalis*

4310 prinsepia
f prinsépie
d Dornenkirsche
l *Prinsepia*

4311 pritchardia palm
f pritchardia
d Pritchardie
l *Pritchardia; Eupritchardia*

4312 privet
f troène
d Hartriegel; Liguster; Rainweide
l *Ligustrum*

4313 procumbent pearlwort
f sagine couchée
d liegendes Mastkraut; niederliegendes Mastkraut
l *Sagina procumbens*

4314 prophet flower
f macrotomia
d Prophetenblume
l *Macrotomia*

4315 proso; broomcorn millet
f millet; millet commun; panic millet
d Hirse; echte Hirse; Rispenhirse
l *Panicum miliaceum*

4316 prostrate amaranth
d weissrandiger Fuchsschwanz
l *Amaranthus blitoides*

4317 prostrate knotweed; knotgrass
f renouée des oiseaux; traînasse
d Vogelknöterich; Knicker
l *Polygonum acivulare*

4318 protea
f protée
d Proteusgewächs; Schimmerbaum
l *Protea*

4319 pseudobacterium
f pseudobactérie
d Pseudobakterie
l *Pseudobacterium*

* **pseudochanterelle → 2165**

* **psyllium seed → 2290**

4320 ptarmigan-berry
d Alpenbärentraube
l *Arctous*

4321 pteridophytes
f ptéridophytes
d Farnpflanzen; farnartige Gewächse; Pteridophyten
l *Pteridophyta*

4322 ptychosperma
f ptychosperma
d Faltensamenpalme
l *Ptychosperma*

4323 pubescent birch
f bouleau pubescent
d Haarbirke; Moorbirke
l *Betula pubescens*

4324 pubescent oak
f chêne pubescent
d Flaumeiche; weichhaarige Eiche
l *Quercus pubescens*

* **puccoon → 682, 2547**

4325 puffball
f vesse-de-loup
d Bovist(stäubling); Stäubling; Staubpilz; Staubschwamm
l *Lycoperdon*

4326 puffballs
f lycoperdacées
d Stäublinge; Staubpilze
l *Lycoperdaceae*

4327 pulasan
d Rambutan
l *Nephelium mutabile*

4328 pulque agave; maguey agave
f agave vert noirâtre
d dunkelgrüne Agave; Pulqueagave
l *Agave atrovirens*

4329 pummelo; pompelmous; shaddock
f pamplemoussier
d Lederorangenbaum; Pompelmus; Riesenorange
l *Citrus grandis; Citrus decumana; Citrus maxima*

4330 pumpkin
f citrouille; courge; pépon; giraumon(t)
d Gartenkürbis
l *Cucurbita pepo*

4331 pumpwood; snakewood tree; trumpet tree
f cécropie
d Kanonenbaum; Ameisenbaum; Trompetenbaum
l *Cecropia*

4332 puncture vine
f tribule terrestre; croix de Malte
d Erdbürzeldorn
l *Tribulus terrestris*

4333 purging agaric
d Lärchenschwamm
l *Polyporus officinalis; Fomes officinalis*

* **purging cassia → 2548**

4334 purging croton
f croton tillé; croton révulsif
d Tiglibaum; Purgierkroton
l *Croton tiglium*

4335 purging flax
f lin purgatif
d Purgierlein; Wiesenlein
l *Linum catharticum*

4336 purple apricot
f abricotier noir; abricotier pourpré; abricotier d'Alexandrie
d alexandrische Aprikose; Pflaumenaprikose; schwarze Aprikose

l *Prunus dasycarpa; Armeniaca dasycarpa*

4337 purple-bell cobaea; climbing cobaea
f cobée grimpante
d Glockenrebe; Klettertulpe
l *Cobaea scandens*

4338 purple berry barberry
f vinettier de Bretschneider
d Bretschneiders Berberitze
l *Berberis bretschneideri*

4339 purple broom
f cytise pourpré
d Purpurginster; Rosenginster
l *Cytisus purpureus*

4340 purple butterbur
f pétasite officinal
d rote Pestwurz; gemeine Pestwurz; gewöhnliche Pestwurz
l *Petasites officinalis*

4341 purple-coned spruce
d Purpurfichte
l *Picea purpurea*

4342 purple crane's-bill
f géranium pourpré
d purpurner Storchschnabel
l *Geranium purpureum*

4343 purple dead nettle
f lamier pourpré; ortie rouge
d kleine Taubnessel; rote Taubnessel
l *Lamium purpureum*

4344 purple echinacea
f rudbeckie pourprée
l *Rudbeckia purpurea; Echinacea purpurea*

4345 purple European beech; copper beech
f hêtre pourpré
d Blutbuche
l *Fagus sylvatica* var. *atropunica*

* **purple-flowering raspberry → 2359**

* **purple foxglove → 1416**

4346 purple gentian
f gentiane pourprée
d Purpurenzian; purpurner Enzian
l *Gentiana purpurea*

4347 purple-globe clover
d Alpenklee; Gebirgsklee; Waldklee
l *Trifolium alpestre*

* **purple granadilla → 4070**

4348 purpleheart
f peltogyne
l *Peltogyne*

4349 purple lythrum
f salicaire
d Blutweiderich
l *Lythrum salicaria*

4350 purple milk vetch
f astragale du Danemark
d dänischer Tragant; Trifttragant
l *Astragalus danicus; Astragalus hypoglotis*

4351 purple mullein
f molène violette; molène de Phénicie
d violette Königskerze
l *Verbascum phoeniceum*

4352 purple osier willow
f saule pourpré; saule rouge; osier rouge; verdiau
d Purpurweide; Korbweide; Bartweide
l *Salix purpurea*

4353 purple rattlesnake root
f prénanthe pourpré
d Purpurhasenlattich
l *Prenanthes purpurea*

4354 purple stonecrop
f orpin pourpré
d Purpurfetthenne; rote Fetthenne
l *Sedum purpureum*

4355 purple sweet sultan; sweet sultan
f centaurée ambrette
d Bisamflockenblume; Moschusblume
l *Amberboa turanica*

4356 purple vetch
f vesce du Bengale; vesce pourpre noire
d Purpurwicke
l *Vicia benghalensis; Vicia atropurpurea*

4357 purple viper's bugloss
f vipérine à feuilles de plantain
d violetter Natternkopf
l *Echium plantagineum; Echium violaceum*

4358 **purslane**
f pourpier
d Portulak; Burzel(kraut); Bürgelkraut
l *Portulaca*

4359 **purslane family**
f portulacacées
d Portulakazeen; Portulakgewächse
l *Portulacaceae*

4360 **purslane speedwell; neckweed**
d fremder Ehrenpreis
l *Veronica peregrina*

4361 **purslane tree**
f portulacaire
d Strauchportulak; Portulacarie
l *Portulacaria afra*

4362 **puschkinia**
f puschkinie
d Puschkinie
l *Puschkinia*

4363 **pussytoes; cat's-foot**
f antennaire
d Katzenpfötchen
l *Antennaria*

4364 **pycreus**
f souchet
d Zyperngras
l *Pycreus*

4365 **pygmyweed**
f tillée
d Dickblatt; Moosblümchen
l *Tillaea*

4366 **pygmy willow**
f saule nain
d Krautweide
l *Salix herbacea*

4367 **pyramidal Italian cypress**
f cyprès pyramidal
l *Cupressus sempervirens* var. *pyramidalis*

4368 **pyramidal mullein**
f molène pyramidale
d Pyramidenkönigskerze
l *Verbascum pyramidale*

4369 **pyramid Chinese juniper; Chinese pyramid juniper**
f genévrier de Chine
d chinesischer Sadebaum; chinesischer Wacholder
l *Juniperus chinensis*

4370 **Pyrenees aster**
f aster des Pyrénées
d pyrenäische Aster
l *Aster pyrenaeus*

4371 **Pyrenees gentian**
f gentiane des Pyrénées
d pyrenäischer Enzian
l *Gentiana pyrenaica*

4372 **Pyrenees geranium**
f géranium des Pyrénées
d pyrenäischer Storchschnabel
l *Geranium pyrenaicum*

4373 **Pyrenees monkshood**
f aconit anthora
d Giftheil; fahler Sturmhut
l *Aconitum anthora*

4374 **Pyrenees oak**
f chêne des Pyrénées
d Pyrenäen-Eiche; pyrenäische Eiche
l *Quercus pyrenaica*

4375 **Pyrenees poppy**
f pavot des Pyrénées
d pyrenäischer Mohn
l *Papaver pyrenaicum*

* **Pyrenees star-of-Bethlehem → 478**

4376 **pyrethrum**
f pyrèthre
d Edelmarguerite; Goldfliederkamille; Bertram
l *Pyrethrum*

4377 **pyrola; shinleaf; wintergreen**
f pyrole
d Birnkraut; Wintergrün
l *Pyrola; Pirola*

4378 **pyrola family; wintergreen family**
f pyrolacées
d Pirolazeen; Wintergrüngewächse
l *Pyrolaceae; Pirolaceae*

4379 **quackgrass**
f chiendent rampant
d gemeine Quecke
l *Agropyron repens*

4380 **quaking aspen**
f peuplier faux tremble; faux tremble
d amerikanische Zitterpappel
l *Populus tremuloides*

4381 **quaking grass**
f brize
d Amourettengras; Liebesgras; Zittergras
l *Briza*

4382 **quassia**
f quassier
d Bitterholz; Quassiaholzbaum
l *Quassia*

4383 **quassia wood**
f picrasme
d Bitterholz
l *Picrasma*

4384 **queen of the night**
f cierge à grandes fleurs
d Königin der Nacht
l *Cereus grandiflorus; Selenicereus grandiflorus*

* **queen's gillyflower → 1794**

* **Queensland arrowroot → 1993**

4385 **Queensland nut; Australian nut**
f noisetier du Queensland
l *Macadamia ternifolia*

4386 **quickweed**
f galinsoga
d Franzosenkraut; Knopfkraut
l *Galinsoga*

4387 **quillwort**
f isoète
d Brachsenkraut
l *Isoetes*

4388 **quillwort family**
f isoétacées
d Brachsenkrautgewächse; Brachsenkräuter; Isoetazeen; Isoeteen
l *Isoetaceae*

4389 **quince**
f cognassier
d Quitte(nbaum)
l *Cydonia*

4390 **quinoa; pigweed**
f chénopode quinoa; quinoa
d Getreidekraut; kleiner Reis von Peru; Mehlschmergel; Reismelde
l *Chenopodium quinoa*

4391 **quisqualis**
f quisqualis
d Fadenröhre; Wunderstrauch
l *Quisqualis*

4392 **quiver-tree aloe**
f aloès faux-dragonnier
d Drachenbaumaloe; Köcherbaum
l *Aloe dichotoma*

R

4393 **rabbit-ear iris**
f iris de Koempfer
d Kämpfers Iris; Kämpfers Schwertlilie
l *Iris laevigata*

4394 **rabbit-foot clover; hare's-foot**
f pied-de-lièvre; trèfle des champs
d Hasenklee; Ackerklee; Katzenklee
l *Trifolium arvense*

* **rabbit-tail grass → 2713**

4395 **radish**
f radis
d Rettich
l *Raphanus*

4396 **raffia palm**
f raphia
d Bambuspalme
l *Raphia*

4397 **rafflesia family**
f rafflésiacées
d Rafflesiazeen
l *Rafflesiaceae*

4398 **ragged robin**
f œillet des prés; lychnide des prés; fleur de coucou
d Fleischblume; Gauchraden; Kuckucksblume; Kuckucksnelke; Samtblatt
l *Lychnis flos cuculi*

* **raggee → 4399**

4399 **ragi millet; raggee; tocusso**
f coracan; dagussa; éleusine
d Korakan; Dagussa; Fingerhirse; Mandua; Marua; Radi; Telabun; Tocusso; Ragi
l *Eleusine coracana*

4400 **ragweed**
f ambroisie
d Traubenkraut; Ambrosienkraut
l *Ambrosia*

* **ragwort → 2651**

4401 **ragwort groundsel; tansy ragwort**
f jacobée; herbe de Saint Jacques; séneçon de Jacob; herbe des charpentiers
d Jakobs(greis)kraut; Jakobskreuzkraut; Wiesenkreuzkraut
l *Senecio jacobaea*

4402 **rain lily**
f cooperie
l *Cooperia*

4403 **rain orchid**
f platanthéra
d Breitkölbchen; Kuckucksblume; Stendelwurz; Waldhyazinthe
l *Platanthera*

4404 **rain tree**
f arbre à la pluie
d Genisarobaum; Regenbaum
l *Pithecolobium saman; Samanea saman*

4405 **rain tree**
f brunfelsie
d Brunfelsie
l *Brunfelsia*

4406 **raisin tree**
f hovène
d Hovenie; japanischer Rosinenbaum
l *Hovenia*

4407 **ramalina**
f ramalina
d Astflechte; Bandflechte
l *Ramalina*

4408 **rambling bellflower**
d Wiesenglockenblume
l *Campanula patula*

* **rambotang → 4409**

* **rambustan → 4409**

4409 **rambutan; rambotang; rambustan**
f ramboutan
d Rambutan
l *Nephelium lappaceum*

4410 **ramie**
f ortie blanche; ortie de Chine
d Ramie; Chinagras
l *Boehmeria nivea; Urtica utilis*

4411 **ramonda**
f ramondie
d Ramondie
l *Ramonda*

4412 ramondia
f ramondie
l *Ramondia*

4413 ramontchi; governor's plum; Madagascar plum
f flacourtie de Madagascar
d Batatopflaume; Batokpflaume; Madagaskarpflaume; Maronpflaume
l *Flacourtia ramontschi; Flacourtia indica*

4414 rampion (bellflower)
f campanule raiponce
d Rapunzelglockenblume; Ackerglockenblume
l *Campanula rapunculus*

4415 ramsons; bear's-garlic
f ail des ours
d Bär(en)lauch; Bärlapplauch
l *Allium ursinum*

4416 raoulia
f raoulia
d Raoulie
l *Raoulia*

4417 rape
f colza
d Kohlsaat; Raps
l *Brassica napus* var. *napus*

4418 rapistrum
f rapistre
d Rapsdotter
l *Rapistrum*

4419 rasamala (altingia)
d Rasamalabaum
l *Altingia excelsa; Liquidambar altingia*

4420 raspberry
f framboisier
d Himbeere; Himbeerstrauch
l *Rubus*

4421 rasp pod
f trachylobe
d Rauhhülse; Scharfhülse
l *Trachylobium*

* **rata tree → 2964**

4422 rattail cactus
d Schlangenkaktus; Peitschenkaktus
l *Aporocactus flagelliformis; Cereus flagelliformis*

4423 rattail fescue
d Mäusefuchsschwingel; Mäuseschwanzfederschwingel; Mäuseschwanzfuchsschwingel
l *Festuca myuros; Vulpia myuros*

4424 rattan palm
f rotang
d Rohrpalme; Rotangpalme; spanisches Rohr
l *Calamus*

* **rattan palm → 4574**

4425 rattlebox
f crotalaire
d Klapperhülse; Klapperschote
l *Crotalaria*

4426 rattlebox
d Klappertopf
l *Alectorolophus*

4427 rattlesnake fern
f botryche de Virginie
d virginischer Rautenfarn
l *Botrychium virginianum*

4428 rattlesnake plantain
f goodyère
d Netzblatt
l *Goodyera; Peranium*

4429 rattlesnake root
f prénanthe
d Hasenlattich; Nickwurz
l *Prenanthes*

4430 rattleweed
f cocriste; cocrête; rhinanthe
d Klappertopf
l *Rhinanthus*

* **rauwolfia → 1844**

* **ray fungi → 23**

4431 razor sedge
f sclérie
d Geisselgras; Geisselsegge
l *Scleria*

4432 red alder
f aune rouge; aune d'Orégon
d Roterle
l *Alnus rubra; Alnus oregana; Alnus oregona*

4433 red algae
f rhodophycées; algues rouges
d Rotalgen; Blütentange; Floriden; Purpuralgen; Rhodophyzeen
l *Rhodophyceae*

4434 red ash
f frêne rouge; frêne de Pennsylvanie
d Rotesche; amerikanische Grauesche
l *Fraxinus pennsylvanica*

4435 red bartsia
f odontite rouge
d roter Zahntrost
l *Odontites rubra*

4436 red bay; red-bay persea
f laurier Bourbon
l *Persea borbonia*

4437 redbead cotoneaster
f cotonéastre à grappes
d filzige Zwergmispel
l *Cotoneaster racemiflora*

4438 red beet
f betterave potagère; betterave rouge
d rote Rübe; Salatrübe
l *Beta vulgaris* ssp. *esculenta* var. *rubra; Beta vulgaris hortensis*

*** red-berried juniper → 4298**

4439 red-berry bryony
f navet du diable; couleuvrée; bryone dioïque
d rotbeerige Zaunrübe; rote Zaunrübe; rotfrüchtige Zaunrübe; zweihäusige Gichtrübe; zweihäusige Zaunrübe
l *Bryonia dioica*

4440 red-berry juniper
d Pinchots Wacholder
l *Juniperus pinchoti*

*** red bilberry → 1645**

4441 red buckeye
f pavier rouge
d rotblühende Pavie; fleischfarbige Rosskastanie; rote Rosskastanie
l *Aesculus pavia*

*** redbud → 1098**

4442 redbud maple
f érable de Trautvetter
d Trautvetters Ahorn
l *Acer trautvetteri*

4443 red bulrush
d fuchsrotes Quellried; rotes Quellried
l *Scirpus rufus; Blysmus rufus*

4444 red cabbage
f chou rouge
d Rotkohl; Blaukraut; Rotkraut
l *Brassica oleracea* var. *capitata rubra*

4445 red Chile bells
f lapagerie rose
d Rosenlapagerie
l *Lapageria rosea*

4446 red chokeberry
f aronie rouge
d rotfrüchtiger Apfelbeerstrauch
l *Aronia arbutifolia*

4447 red clover
f trèfle commun; trèfle des prés; trèfle rouge; trèfle violet
d Rotklee; Wiesenklee; roter Klee; gemeiner roter Kopfklee
l *Trifolium pratense*

4448 red-cracked boletus
f bolet à chair jaune; cèpe à pied rouge
d Rotfussröhrling; Rotfüsschen
l *Boletus chrysenteron*

*** reddish amanita → 713**

*** red dogwood → 683**

4449 red escallonia
d rote Eskallonie
l *Escallonia rubra*

4450 red fescue
f fétuque rouge
d Rotschwingel; roter Schwingel
l *Festuca rubra*

4451 red fir; California red fir
f sapin rouge de Californie
d Goldtanne
l *Abies magnifica*

*** red-flowering currant → 6000**

4452 red-haw hawthorn
f épine rouge
d Blutdorn
l *Crataegus sanguinea*

4453 red hemp nettle
f galéope commun

d Ackerhohlzahn
l *Galeopsis ladanum*

* **red-hot poker plant → 5551**

4454 red-leaf rose
f rosier à feuilles rouges
d blaugrüne Rose; Hechtrose; rotblättrige Rose
l *Rosa glauca; Rosa rubrifolia*

4455 red maple
f érable rouge
d roter Ahorn
l *Acer rubrum*

4456 red mulberry; American mulberry
f mûrier rouge
d roter Maulbeerbaum
l *Morus rubra*

* **red oak → 3883**

4457 red oat
f avoine d'Algérie
d Mittelmeerhafer
l *Avena byzantina*

4458 red osier dogwood; squawbush
f cornouiller osier; cornouiller stolonifère
d weisser Hartriegel
l *Cornus stolonifera*

4459 red pepper
f piment
d Paprika; spanischer Pfeffer
l *Capsicum*

4460 red pine; Norway pine
f pin rouge d'Amérique
d Harzföhre; Harzkiefer
l *Pinus resinosa*

4461 red quebracho
f schinopsis
d Quebrachobaum
l *Quebrachia; Schinopsis*

4462 red raspberry; European raspberry
f framboisier
d Himbeere
l *Rubus idaeus*

4463 redroot
f lachnanthe
l *Lachnanthes*

* **redroot → 3840**

4464 redroot amaranth
f amarante réfléchie; fleur de la jalousie; fleur de l'amour
d Ackerfuchsschwanz; rauhhaariger Fuchsschwanz
l *Amaranthus retroflexus*

4465 red spruce
f épicéa rouge; sapinette rouge
d amerikanische Rotfichte
l *Picea rubens; Picea rubra*

4466 red-star zinnia
f brésine
l *Zinnia multiflora*

4467 red stonecrop
f orpin rougeâtre
d rötliches Dickblatt
l *Sedum rubens*

4468 red-tipped clavaria
f clavaire en forme de chou-fleur
l *Clavaria botrytis*

4469 redtop; fiorin
f agrostide blanche; fiorin blanche
d Flechtstraussgras; kleine Quecke; weisser Windhalm; weisses Straussgras
l *Agrostis alba*

* **redtop → 617**

* **red valerian → 3119**

4470 red-veined maple
f érable à nervures roussâtres
d rotnerviger Ahorn
l *Acer rufinerve*

4471 red willow
f verdiau
l *Salix rubra*

4472 reed
f phragmite
d Rohrschilf; Schilf(rohr)
l *Phragmites*

4473 reed canary grass
f alpiste roseau
d Rohrglanzgras
l *Phalaris arundinacea*

4474 reed fescue
f fétuque roseau; fétuque faux-roseau
d Rohrschwingel; rohrartiger Schwingel

l *Festuca arundinacea; Festuca elatior arundinacea*

4475 reed grass
f gourbet; calamagrostide; calamagrostis
d Federgras; Reitgras
l *Calamagrostis*

4476 regal lily
f lis royal
d Königslilie
l *Lilium regale*

4477 Rehder wing nut
f ptérocaryer de Rehder
d Rehders Flügelnuss
l *Pterocarya rehderiana*

4478 rehmannia
f rehmannie
d Rehmannie
l *Rehmannia*

*** reindeer lichen → 4479**

4479 reindeer moss; reindeer lichen
f lichen des rennes; cladonie rangifère
d Rentierflechte; Rentiermoos
l *Cladonia rangiferina*

4480 reineckia
f reineckie
d Reineckie
l *Reineckia*

*** Reine Claude → 2627**

4481 rein orchis
f gymnadénie
d Händelwurz; Züngel
l *Gymnadenia*

4482 remote sedge
d entferntährige Segge; Winkelriedgras
l *Carex remota*

4483 rescue brome; rescue grass
f brome de Schrader
d ährengrasähnliche Trespe
l *Bromus unioloides; Bromus catharticus*

4484 restharrow
f bugrane; arrête-bœuf
d Hauhechel
l *Ononis*

4485 resurrection lily
f kæmpférie
d Gewürzlilie
l *Kaempferia*

*** resurrection plant → 4814**

4486 rhapontic; Swiss centaury
f rhapontic
d Rübendistel
l *Rhaponticum*

4487 rheedia
f cirolier; rhéédie
l *Rheedia*

4488 rhipsalis
f rhipsalide
d Binsenkaktus; Geisselkaktus; Rutenkaktus; Korallenkaktus
l *Rhipsalis*

4489 rhizoctonia
f rhizocton(i)e
d Rhizoctonia
l *Rhizoctonia*

4490 rhododendron
f azalée; rhododendron; rosage
d Alpenrausch; Alpenrose; Azalee; Rhododendron; Rosenbaum
l *Rhododendron*

4491 rhopalostylis
f rhopalostylis
d Nikanpalme
l *Rhopalostylis*

4492 rhubarb
f rhubarbe
d Rhabarber
l *Rheum*

4493 ribbon eucalyptus; ribbon gum
f eucalyptus à feuilles d'osier
d weisser Gummibaum
l *Eucalyptus viminalis*

4494 riccia
f riccie
d Wasserlasche
l *Riccia*

4495 rice
f riz
d Reis
l *Oryza*

4496 rice
f riz
d gemeiner Reis
l *Oryza sativa*

4497 rice cut-grass
d Reisquecke; wilder Reis
l *Leersia oryzoides*

4498 rice flower
f pimélée
d Glanzstrauch
l *Pimelea*

4499 rice grass
f oryzopsis
d Grannenhirse
l *Oryzopsis*

4500 rice-paper plant
f tétrapanax
d Reispapierbaum
l *Tetrapanax*

4501 Richardson willow
f saule de Richardson
d Richardsons Weide
l *Salix richardsoni*

* **ringflower → 2460**

4502 ripple-seed plantain; great plantain
f grand plantain
d Breitwegerich; grosser Wegerich
l *Plantago major*

4503 riverbank grape
f vigne de battures; vigne américaine
d Duftrebe; Uferweinrebe
l *Vitis riparia*

4504 river birch; black birch
f bouleau noir
d Schwarzbirke
l *Betula nigra*

4505 river grass
d Schwingelschilf
l *Scolochloa; Fluminea*

4506 riverweed
f podostémon
d Stielfaden
l *Podostemon*

4507 Rivini violet
d Hainveilchen
l *Viola riviniana*

4508 Robeson hawthorn
f épine de Robeson
d Robesons Weissdorn
l *Crataegus robesoniana*

4509 roccella
f rocelle
d Lackmusflechte
l *Roccella*

4510 rock beauty
f pétrocallide
d Steinschmückel
l *Petrocallis*

4511 rockbell
f wahlenbergie
d Wahlenbergie; Klingelblume; Moorglocke
l *Wahlenbergia*

4512 rock brake
f cryptogramme
d Rollfarn
l *Cryptogramma*

4513 rock buckthorn
f nerprun des rochers
l *Rhamnus saxatilis*

4514 rock candytuft
d Felsenschleifenblume
l *Iberis saxatilis*

4515 rock cotoneaster
f cotonéastre horizontal
d niedergestreckte Zwergmispel
l *Cotoneaster horizontalis*

4516 rock cress
f arabette
d Gänsekresse; Gänsegarbe; Gänsekohl; Gänsekraut
l *Arabis*

4517 rock cress
f tourette
d Turmkraut
l *Turritis*

4518 rock cress
d Sandschaumkresse
l *Cardaminopsis arenosa; Arabis arenosa*

4519 rock-cress draba
f drave alpine
d Alpenfelsenblümchen; nordisches

Felsenblümchen
l *Draba alpina*

4520 rock currant
f groseillier des rochers
d Felsenjohannisbeere
l *Ribes petraeum*

4521 rock elm; spreading elm
f orme des rochers; orme de Thomas; orme à grappes
d Felsenulme
l *Ulmus racemosa; Ulmus thomasi*

4522 rocket
f julienne
d Nachtviole
l *Hesperis*

4523 rocket candytuft
f ibéride blanche; thlaspi blanc
d bittere Schleifenblume
l *Iberis amara*

4524 rocket larkspur
f dauphinelle des jardins
d Gartenrittersporn
l *Delphinium ajacis*

4525 rocket salad
f roquette
d gemeine Rauke; Gartenrauke; Ölrauke; Raukenkohl
l *Eruca sativa*

4526 rocket weed
f fausse roquette
d Hundsrauke
l *Erucastrum*

4527 rock jasmine
f androselle; androsace
d Mannsschield(kraut)
l *Androsace*

4528 rock mat
d Rasenspiere
l *Petrophytum*

4529 rock purslane
f calandrinie
d Calandrinie
l *Calandrinia*

4530 rock rhododendron
f rosage des Alpes; rosage rouillé; rose des Alpes
d Almenrausch; Alpbalsam; rostblättrige Alpenrose; rostrote Alpenrose
l *Rhododendron ferrugineum*

4531 rockrose
f ciste
d Zistrose; Ziströschen
l *Cistus*

4532 rockrose family
f cistacées
d Zistrosengewächse; Zistazeen
l *Cistaceae*

4533 rock sedge
d Felsensegge
l *Carex rupestris*

4534 rock speedwell
f véronique sous-ligneuse
d Felsenehrenpreis; strauchiger Ehrenpreis
l *Veronica fructiculosa*

4535 rock tripe
f tripe de roche
d Kreisflechte
l *Gyrophora*

4536 rock tripe
f ombilicaire
l *Umbilicaria*

4537 rock violet
f byssus rouge
d Veilchensteinflechte; Veilchenalge; Veilchenmoos
l *Trentepohlia iolithus*

4538 rockweed; sea oak
f goémon; varech
d Fukus; Seetang; Tang
l *Fucus*

4539 rockweep
f ascophylle
d Blasenstengel
l *Ascophyllum*

4540 rock wormwood
d Felsenbeifuss
l *Artemisia rupestris*

4541 Rocky Mountain Douglas fir
f douglas bleu; douglas du Colorado
d blaue Douglasfichte
l *Pseudotsuga glauca*

* **roebuck berry → 5260**

4542 roëlla
f roëlla
d Tauglocke
l *Roella*

4543 rogeria
f rogérie
l *Rogeria*

4544 rohan soymida
f bois rouge d'Inde
l *Soymida febrifuga; Swietenia febrifuga*

4545 romaine lettuce
f laitue romaine; laitue lombarde
d Bindesalat; glatte Endivie; römischer Salat; Sommerendivie
l *Lactuca sativa* var. *romana*

4546 Roman camomile; English camomile
f anthémis noble; camomille romaine
d Edelkamille; römische Hundskamille; römische Kamille
l *Anthemis nobilis*

4547 Roman nettle
f ortie à pilules
d Pillenbrennessel
l *Urtica pilulifera*

4548 Roman wormwood
f absinthe pontique; absinthe romaine; armoise romaine
d römischer Beifuss
l *Artemisia pontica*

4549 rondeletia
f rondelétie; rondelette
d Rondeletie
l *Rondeletia*

* **roof brome grass → 1123**

* **roof houseleek → 2779**

4550 rooted collybia
d grubiger Schleimrübling; Wurzelrübling
l *Collybia radicata*

4551 rootless duckweed
d wurzellose Wasserlinse; Zwerglinse
l *Lemna arrhiza*

4552 rope grass
f restiole
d Seilgras
l *Restio*

4553 rosary pea
f arbre à chapelet
d Paternostererbse
l *Abrus*

4554 rose
f rosier
d Rose
l *Rosa*

4555 rose acacia
f acacia rose; robinier rose; robinier hispide
d rote Akazie; rosenrote Robinie; borstige Robinie
l *Robinia hispida*

4556 rose apple
f jambosier
d Rosenapfel
l *Syzygium jambos*

4557 rosebay rhododendron
f grand rosage
l *Rhododendron maximum*

4558 rose campion
f coquelourde des jardins
d Kronenlichtnelke; Kranzlichtnelke; Vexiernelke
l *Lychnis coronaria; Coronaria coriacea*

* **rose chestnut → 2966**

4559 rose daphne
f daphné camélée
d Heideröschen; flaumiges Steinrösel; Steinröschen
l *Daphne cneorum*

4560 rose family
f rosacées
d Rosengewächse; Rosazeen
l *Rosaceae*

4561 rose gentian
f sabbatie
d Bitterwurz(el)
l *Sabbatia*

4562 roselle
f roselle
d Rosellahanf
l *Hibiscus sabdariffa*

4563 rosemary
f romarin

d Rosmarin
l *Rosmarinus*

4564 **rosemary**
f romarin officinal
d Gartenrosmarin; gemeiner Rosmarin
l *Rosmarinus officinalis*

4565 **rose myrtle**
f rhodomyrte
d Rosenmyrte
l *Rhodomyrtus*

* **rose of Sharon → 4886**

4566 **rose pelargonium**
f pélargonium malodorant
d Rosengeranium
l *Pelargonium roseum; Pelargonium graveolens*

4567 **rose-root stonecrop**
f orpin rosat; orpin rose
d Rosenwurz
l *Sedum roseum; Rhodiola rosea*

4568 **rose-scented pelargonium**
f pélargonium à ombelle; géranium rosat
l *Pelargonium capitatum*

4569 **rosette valerian**
d holunderblättriger Baldrian; Holunderbaldrian
l *Valeriana excelsa*

4570 **rosewood**
f dalbergie; bois de rose
d Rosenholz
l *Dalbergia*

4571 **rosinweed; compass plant**
f plante au compas
d Kompasspflanze; Becherpflanze
l *Silphium*

4572 **Rosthorn spirea**
f spirée de Rosthorn
d Rosthorns Spierstrauch
l *Spiraea rosthornii*

4573 **rotala**
f rotale
l *Rotala*

4574 **rotang rattan palm; rattan palm**
f rotang; canne aromatique
d Rotangpalme; Schilfpalme
l *Calamus rotang*

4575 **rouge plant**
f rivinie
d Rivine; Rivinie
l *Rivina*

* **rough chervil → 1648**

4576 **rough clover**
d rauher Klee
l *Trifolium scabrum*

* **rough-fruited corn bedstraw → 1607**

4577 **rough hawk's-beard**
f crépide bisannuelle
d Wiesenfeste; Wiesenpippau; zweijähriger Pippau
l *Crepis biennis*

4578 **rough pea vine**
f gesse hérissée; pois gras
d behaarte Platterbse; behaartfrüchtige Platterbse; haarige Platterbse
l *Lathyrus hirsutus*

4579 **rough-stalk bluegrass**
f pâturin commun
d gemeines Rispengras; gewöhnliches Rispengras; rauhes Rispengras
l *Poa trivialis*

* **rough-stalked boletus → 582**

* **rough-stemmed boletus → 582**

4580 **Roulett Chinese rose**
f rosier bijou; rosier de Bengale pompon; rosier de Miss Lawrence; rosier pompon
d Liliputrose
l *Rosa chinensis* var. *minima*

4581 **round-ear willow**
f saule à oreillettes
d Öhrchenweide; Ohrweide
l *Salix aurita*

4582 **round-fruited rush**
d flache Binse; zusammengedrückte Binse
l *Juncus compressus*

* **round kumquat → 3528**

4583 **round-leaf ash**
f frêne à petites feuilles
d rundblättrige Esche
l *Fraxinus parvifolia; Fraxinus rotundifolia*

4584 round-leaved crane's-bill
f géranium à feuilles rondes
d rundblättriger Storchschnabel
l *Geranium rotundifolium*

4585 round-leaved fluellen; cancerwort
d unechtes Leinkraut; unechtes Tännelkraut
l *Kickxia spuria*

* **round-leaved hare's-ear → 4589**

4586 round-leaved henbane
f jusquiame blanche
d helles Bilsenkraut; weisses Bilsenkraut
l *Hyoscyamus albus*

* **round-leaved mint → 283**

4587 round-leaved pea vine
f gesse à feuilles rondes
d rundblättrige Platterbse
l *Lathyrus rotundifolius*

4588 round-leaved sundew
f rossolis à feuilles rondes; drosère à feuilles rondes
d rundblättriger Sonnentau
l *Drosera rotundifolia*

4589 round-leaved thoroughwax; round-leaved hare's-ear
f buplèvre à feuilles rondes
d rundblättriges Hasenohr; rundes Hasenohr
l *Bupleurum rotundifolium*

4590 round netleaf willow
d kreisförmige Weide
l *Salix orbicularis; Salix reticulata subrotunda*

4591 round-podded jute
f jute; chanvre de Calcutta
d indischer Flachs; Jutepflanze; rundkapselige Jute
l *Corchorus capsularis*

4592 roxburghia
f roxburghie
l *Roxburghia*

* **royal agaric → 918**

4593 royal fern
f osmonde royale; fougère royale
d Königsfarn
l *Osmunda regalis*

4594 royal palm
f oréodoxe; roi des palmiers
d Kohlpalme; Königspalme
l *Oreodoxa*

4595 royal paulownia
f paulownia tomenteux; paulownie impériale
d Blauglockenbaum; Kaiserpaulownie; kaiserliche Paulownie
l *Paulownia tomentosa*

4596 Royle's balsam; Royle's snapweed
f balsamine glanduleuse
d Honigspringkraut; drüsiges Springkraut
l *Impatiens roylei*

* **rubber plant → 2944**

4597 rue
f rue
d Raute
l *Ruta*

4598 rue anemone; anemonella
d Rautenanemone
l *Anemonella thalictroides; Syndesmon thalictroides*

4599 rue family
f rutacées
d Rautengewächse; Rutazeen
l *Rutaceae*

4600 rue-leaved saxifrage
d Dreifingersteinbrech; dreifingeriger Steinbrech
l *Saxifraga tridactylis*

4601 ruffle palm
f aïphane
l *Aiphanes*

4602 rugosa rose; hedge-row rose
f rosier du Japon; rosier rugueux
d Kartoffelrose; Runzelrose
l *Rosa rugosa*

4603 rugose rapistrum
d runzeliger Rapsdotter; runzeliger Windsbock
l *Rapistrum rugosum*

4604 ruizia
f ruizia
l *Ruizia*

4605 **rukam**
f flacourtie de Sonde
l *Flacourtia rukam*

* **running mallow → 1945**

4606 **running pine; club moss**
f lycopode en massue
d Drudenkraut; Gürtelkraut; Johannisgürtel; Keulenbärlapp; Kolbenbärlapp; Schlangenmoos
l *Lycopodium clavatum*

4607 **Ruprecht silene**
f silène de Ruprecht
d Ruprechts Leimkraut
l *Silene ruprechti*

4608 **rush**
f jonc
d Binse; Markbinse
l *Juncus*

4609 **rush family**
f juncasées
d Binsengewächse; Junkazeen
l *Juncaceae*

4610 **rush skeletonweed; gum succory**
d Binsenknorpellattich; grosser Knorpellattich; grosser Krümmling
l *Chondrilla juncea*

4611 **Russian almond**
f amandier nain de Géorgie; amandier nain de Russie
d grusinische Mandel; Kalmückennuss; Strauchmandel; Zwergmandel
l *Prunus tenella; Amygdalus georgica; Amygdalus nana; Prunus nana*

4612 **Russian boxthorn**
f lyciet de Russie
d russischer Bocksdorn
l *Lycium ruthenicum*

4613 **Russian dandelion**
f kok-saghyz
d Kok-Saghys; Kautschuklöwenzahn
l *Taraxacum kok-saghyz*

4614 **Russian elm; spreading elm; European white elm**
f orme blanc d'Europe; orme étalé; orme cilié
d Bastulme; Flatterrüster; Flatterulme
l *Ulmus laevis*

4615 **Russian globe thistle**
f échinope de Russie
l *Echinops exaltatus*

4616 **Russian olive**
f chalef à feuilles étroites; olivier de Bohême
d schmalblättrige Ölweide
l *Elaeagnus angustifolia*

4617 **Russian pea shrub**
f acacia de Sibérie
d kleiner Erbsenstrauch; niedriger Erbsenstrauch
l *Caragana frutex; Caragana frutescens*

4618 **Russian thistle; saltwort**
f soude
d Salzkraut
l *Salsola*

4619 **Russian truffle; summer truffle**
f truffe d'été; truffe de St. Jean
d deutsche Trüffel; Sommertrüffel
l *Tuber aestivum*

* **Russian vetch → 2693**

4620 **Russian willow**
f saule de Russie
d russische Weide
l *Salix rossica*

4621 **russula**
f russule
d Täubling; Heiderling
l *Russula*

4622 **russula family**
f russulacées
d Sprödblätterpilze
l *Russulaceae*

4623 **rust fungi**
f urédinées
d Rostpilze; Uredineen
l *Uredinaceae*

4624 **rusty bog rush**
d rostbraunes Kopfried; rostfarbiges Kopfried; rostbraune Kopfsimse
l *Schoenus ferrugineus*

4625 **rusty woodsia**
d südlicher Wimperfarn
l *Woodsia ilvensis*

4626 rutabaga; swede
f chou-navet; chou de Suède; navet jaune
d Bodenkohlrabi; Dorsche; Erdkohlrabi; Erdkohlrübe; Erdrübe; Kohlrübe; Schnittkohl; Unterkohlrabi; Wruke
l *Brassica napobrassica*

4627 Ruthenian centaurea
d russische Flockenblume
l *Centaurea ruthenica*

4628 ruyschia
f ruyschie
l *Ruyschia*

4629 Ruyschianum dragonhead
f dracocéphale de Ruysch
d nordischer Drachenkopf; schwedischer Drachenkopf
l *Dracocephalum argunense; Dracocephalum ruyschianum*

4630 rye
f seigle
d Roggen
l *Secale cereale*

* **rye brome → 1138**

4631 rye grass
f ivraie
d Lolch; Raigras
l *Lolium*

4632 rye-grass sedge
d lolchartige Segge
l *Carex loliacea*

* **ryelike brome → 1138**

S

4633 **sabadilla**
f sabadille
d Sabadill(a); Läusesamen
l *Sabadilla; Schoenocaulon*

* **saccharomyces → 6069**

* **sac fungi → 337**

4634 **sacred bamboo; nandina**
f nandine fruitière
d Gartennandine
l *Nandina domestica*

* **sacred fig tree → 738**

4635 **sacred fir**
f sapin sacré du Mexique
l *Abies religiosa*

* **saddle fungus → 2767**

4636 **safflower; false saffron**
f carthame
d Saflor
l *Carthamus*

4637 **safflower; dyer's saffron**
f safran bâtard; carthame officinal; safran d'Allemagne; carthame des teinturiers
d Bastardsafran; Bürstenkraut; Färbersaflor; falscher Safran; wilder Safran
l *Carthamus tinctorius*

4638 **saffron crocus**
f safran officinal; safran d'automne; safran du Gâtinais
d Herbstsafran; echter Safran; Safrankrokus
l *Crocus sativus*

* **saffron milk cup → 1827**

4639 **sage; salvia**
f sauge
d Salbei
l *Salvia*

4640 **sagebrush; wormwood**
f armoise
d Beifuss
l *Artemisia*

* **sage-leaved cistus → 4670**

4641 **sagewort wormwood; field wormwood**
f aurone des champs; aurone sauvage
d Feldbeifuss
l *Artemisia campestris*

4642 **sagisi palm**
d erhabene Wechselscheide
l *Heterospathe elata*

4643 **sago cycas**
f sagou du Japon; cucas enroulé du Japon
d Farnpalme; japanische Sagopalme
l *Cycas revoluta*

4644 **sago palm; ivory-nut palm**
f métroxylon
d Wassernusspalme; polynesische Steinnusspalme
l *Coelococcus; Metroxylon*

4645 **saguaro; giant cactus**
f cierge géant
d Riesenkaktus
l *Cereus giganteus; Carnegiea gigantea*

4646 **sainfoin; holy clover**
f esparcette
d Esparsette
l *Onobrychis*

4647 **St.-Andrew's-cross**
f croix de St.-André
d Andreaskraut
l *Ascyrum hypericoides; Ascyrum crux-andreae*

4648 **St.-Bernard's-lily**
f phalangère à fleur de lis; bâton de St.-Joseph
d astlose Graslilie
l *Anthericum liliago*

4649 **St.-Bruno's-lily**
f lis de St.-Bruno
d Trichterlilie; Paradieslilie
l *Paradisea*

4650 **St.-Ignatius poison nut**
f strychnos ignatier
d Ignatiusbohne
l *Strychnos ignati*

* **St.-John's-bread → 1023**

4651 **St.-John's-wort**
f millepertuis

d Hartheu; Johanniskraut
l *Hypericum*

4652 St.-John's-wort family
f hypéricacées
d Hartheugewächse; Johanniskrautgewächse
l *Hypericaceae*

*** St. Lucy cherry → 3426**

*** St. Mary thistle → 669**

4653 St.-Paul's-wort
f siégesbeckie
d Siegesbeckie
l *Siegesbeckia*

4654 St.-Thomas bauhinia
f bauhinier cotonneux
l *Bauhinia tomentosa*

4655 Sakhalin fir
f sapin de Sakhaline
d Sachalintanne
l *Abies sachalinensis*

4656 Sakhalin knotweed
f persicaire de Sakhaline
d Sachalinknöterich
l *Polygonum sachalinense*

4657 Sakhalin spruce
f épicéa de Glehn
d Glehns Fichte
l *Picea glehni*

*** salad burnet → 5018**

4658 salad chervil
f cerfeuil
d echter Kerbel; Gewürzkerbel; Gartenkerbel
l *Anthriscus cerefolium*

*** salai tree → 2937**

*** salamander tree → 566**

4659 salmon barberry
d Feuersauerdorn
l *Berberis aggregata*

4660 salmonberry
f ronce élégante
d prächtige Himbeere
l *Rubus spectabilis*

4661 sal (shorea)
f shorée robuste
d Sa(u)lbaum
l *Shorea robusta*

4662 salsify; goatsbeard
f salsifis
d Bocksbart; Bockshorn
l *Tragopogon*

4663 saltbush; orach(e)
f arroche
d Graumelde; Melde
l *Atriplex*

4664 salt-marsh mallow
f kosteletzkye
l *Kosteletzkya*

4665 salt-marsh sand spurry
d Salzspärkling
l *Spergularia salina; Spergularia marina*

4666 salt-meadow rush
d Bottenbinse; Salzbinse
l *Juncus gerardi*

4667 salt tree
f halimodendron
d Salzbaum; Salzstrauch
l *Halimodendron*

4668 saltwort
d Spatelfaden
l *Batis*

*** saltwort → 4618**

4669 salvadora
f salvadora
d Zahnbürstenbaum
l *Salvadora*

*** salvia → 4639**

4670 salvia rockrose; sage-leaved cistus
f ciste à feuilles de sauge
d salbeiblättrige Zistrose
l *Cistus salvifolius*

4671 salvinia; floating moss
f salvinie
d Schwimmblatt; Schwimmfarn
l *Salvinia*

4672 salvinia family
f salviniacées
d Schwimmblattgewächse; Schwimm-

farngewächse; Schwimmfarne; Salviniazeen
l *Salviniaceae*

4673 **samandura**
f samandure
l *Samandura; Samandera*

4674 **samphire**
f crithme
d Meerfenchel
l *Crithmum*

4675 **samphire**
f crithme marin; fenouil marin
d Meerfenchel; Seefenchel
l *Crithmum maritimum*

4676 **sandal bead tree**
f bois de corail; crête de paon
d indischer Korallenbaum; roter Sandelholzbaum
l *Adenanthera pavonina*

4677 **sandalwood**
f santal
d Sandelholzbaum
l *Santalum*

4678 **sandalwood family**
f santalacées
d Sandalgewächse; Sandelholzgewächse; Santalazeen; Santelgewächse
l *Santalaceae*

4679 **sandalwood padauk**
f bois rouge de Kodar
d roter Sandelholzbaum
l *Pterocarpus santalinus*

4680 **sand-bar willow**
f saule des bancs de sable
d langblättrige Weide
l *Salix longifolia; Salix interior*

4681 **sandbox tree**
f sablier
d gemeiner Sandbüchsenbaum
l *Hura crepitans*

4682 **sandbur; bur grass**
f râcle
d Klebgras; Stachelgras
l *Cenchrus*

4683 **sand cherry; dwarf cherry**
f cerisier des sables
d Sandkirsche
l *Prunus pumila; Cerasus pumila*

4684 **sand chickasaw plum**
f prunier de Watson
l *Prunus angustifolia* var. *watsoni; Prunus watsoni*

4685 **sandersonia**
f sandersonie
l *Sandersonia*

4686 **sand heath**
f cératiole
d Hornbusch
l *Ceratiola*

4687 **sand pear; Chinese pear**
f poirier des sables
d chinesische Birne; Sandbirne
l *Pyrus pyrifolia*

4688 **sand sedge**
f laîche des sables
d Sandriedgras; Sandsegge
l *Carex arenaria*

4689 **sand spurry**
f spergulaire
d Spörgel; Spärkling
l *Spergularia*

4690 **sand timothy**
f fléole des sables
d Sandlieschgras
l *Phleum arenarium*

* **sand verbena → 5**

4691 **sand violet**
f violette des sables
d Sandveilchen
l *Viola arenaria*

4692 **sandwort**
f arénaire; sabline
d Sandkraut
l *Arenaria*

4693 **sanicle**
f sanicle
d Sanikel; Heil aller Schäden; Heilkraut
l *Sanicula*

4694 **sansevieria; bowstring hemp**
f sansevière
d Bogenhanf; afrikanischer Hanf
l *Sansevieria*

4695 **santol**
f sandoric

d Sandoribaum
l *Sandoricum*

4696 santol
f sandoric d'Inde
d indischer Sandoribaum
l *Sandoricum indicum; Sandoricum koetjape*

4697 sanvitalia
f sanvitalie
d Sanvitalie
l *Sanvitalia*

4698 sapele
f entandrophragme
d Bastardmahagoni
l *Entandrophragma*

4699 sapium
f gluttier
d Klebbaum
l *Sapium*

4700 sapodilla; sapote
f sapotier; sapotil(l)ier
d Sapotillbaum; Breiapfelbaum
l *Achras zapota*

4701 sapodilla family
f sapotacées
d Sapotengewächse; Sapotazeen
l *Sapotaceae*

*** sapote → 4700**

4702 sappan caesalpinia
f bois de sappan; bois de Sibucoa; brésillet des Indes
d Sappanholzbaum; ostindischer Rotholzbaum
l *Caesalpinia sappan*

4703 saprolegnia family
f saprolégniées
d Saprolegniazeen
l *Saprolegniaceae*

4704 saraca
f saraca; jonésie
d Zauberblume
l *Saraca*

4705 sargasso
f sargasse
d Beerentang
l *Sargassum*

4706 Sargent barberry
f vinettier de Sargent
d Sargents Berberitze
l *Berberis sargentiana*

4707 Sargent's juniper
f genévrier de Sargent
d Sargents Wacholder
l *Juniperus sargentii*

4708 Sargent spirea
f spirée de Sargent
d Sargents Spierstrauch
l *Spiraea sargentiana*

4709 Sargent spruce
f épicéa de Sargent
d Sargents Fichte
l *Picea sargentiana; Picea brachytyla*

4710 sasa
d Zwergbambus
l *Sasa*

4711 sasanqua camellia
f camélia sasanqua
d kleine Kamelie
l *Camellia sasanqua*

4712 sassafras
f sassafras; laurier des Iroquois
d Fieberbaum; Sassafras
l *Sassafras*

*** Satan's mushroom → 679**

*** satin flower → 698**

4713 satin tail
f impérate
d Silbergras; Dschungelgras
l *Imperata*

4714 Satsuma mock orange
f seringa du Japon
l *Philadelphus satsumanus; Philadelphus satsumi*

4715 saucer magnolia
f magnolia de Soulange
d Soulanges Magnolie
l *Magnolia soulangeana*

4716 sausage tree
f kigélie; saucissonnier
d Kigelie; Leberwurstbaum; Götzenholz
l *Kigelia*

4717 saussurea
f saussurée
d Alpenscharte; Schärfling
l *Saussurea*

4718 savanna flower
f échitès
d Klammerstrauch
l *Echites*

4719 savin(e) juniper
f genévrier sabine; sabine; sabinier
d Alpensadebaum; Sabinerbaum; gemeiner Sadebaum; Sadewacholder; sabinischer Wacholder; Stinkwacholder
l *Juniperus sabina*

4720 savory
f calament
d Bergminze; Steinquendel; Steinminze; Schönminze
l *Calamintha*

4721 savory
f sarriette
d Bergminze; Bohnenkraut; Kölle; Pfefferkraut; Saturei
l *Satureia; Satureja*

4722 Savoy cabbage
f chou de Milan; chou pommé frisé; chou frisé pommé; chou de Savoie
d Savoyerkohl; Welschkohl; Wirsingkohl
l *Brassica oleracea* var. *sabauda*

4723 Savoy hawkweed
f épervière de Savoie
d Savoyer Habichtskraut
l *Hieracium sabaudum*

4724 sawara false cypress
f cyprès à fruits de pois; cyprès du Sawara; faux cyprès pisifère
d Sawarazypresse; erbsenfrüchtige Lebensbaumzypresse; Sawarascheinzypresse
l *Chamaecyparis pisifera*

4725 sawari nut; pekea nut; souvari nut; Peruvian almond
f caryocar nucifère
d Butternussbaum; Suarinussbaum
l *Caryocar nucifera*

4726 saw fern
f blechnum
d Rippenfarn
l *Blechnum*

4727 saw grass; twig rush
f cladion
d Schneide(gras)
l *Cladium; Mariscus*

4728 sawwort
f sarrette
d Scharte; Färberscharte
l *Serratula*

4729 saxaul
f haloxylon; saxaoul
d Salzsteppenstrauch; Saxaul
l *Haloxylon*

4730 saxifrage
f saxifrage
d Steinbrech
l *Saxifraga*

4731 saxifrage
f peltiphylle
d Schildblatt
l *Peltiphyllum*

4732 saxifrage family
f saxifragacées
d Steinbrechgewächse; Saxifragazeen
l *Saxifragaceae*

4733 saxifrage pimpinella
f pied-de-chèvre
d kleine Bibernelle; Bockspetersilie; Steinbibernell; Steinbrechbibernelle
l *Pimpinella saxifraga*

4734 scabiosa centaurea
f centaurée scabieuse
d Skabiosenflockenbaum; Grundflockenbaum
l *Centaurea scabiosa*

4735 scabious
f scabieuse
d Grundkraut; Skabiose
l *Scabiosa*

4736 scabious jasione
f jasione vivace
d ausdauernde Jasione
l *Jasione perenis*

4737 scale fern
f cétérach
d Schriftfarn
l *Ceterach*

4738 scalloped salpiglossis
f salpiglossis des jardins

d Trompetenzunge
l *Salpiglossis sinuata*

4739 scaly lentinus
f lentin joli; lentin squameux
d Anissägeblättling; schuppiger Sägeblättling
l *Lentinus lepideus; Lentinus squamosus*

4740 scaly pholiota
d sparriger Schüppling
l *Pholiota squarrosa*

*** scaly polyporus → 1921**

4741 scammony glorybind
f liseron scammonée; scammonée du Levant
d Purgierwinde
l *Convolvulus scammonia*

4742 Scarborough lily
f amaryllis pourpre
l *Vallota speciosa; Vallota purpurea*

*** scarlet clover → 1694**

4743 scarlet fire thorn
f buisson ardent écarlate; petit corail
d gemeiner Feuerdorn
l *Pyracantha coccinea*

*** scarlet hawthorn → 5485**

4744 scarlet kadsura
f kadsoura du Japon
d japanischer Kugelfaden
l *Kadsura japonica*

4745 scarlet Kaffir lily
f clivie vermillon
d Klivie
l *Clivia miniata*

4746 scarlet oak
f chêne écarlate; chêne cocciné
d Scharlacheiche
l *Quercus coccinea*

4747 scarlet pimpernel
f anagallide des champs; mouron des champs
d Ackergauchheil; roter Gauchheil
l *Anagallis arvensis*

4748 scarlet runner bean
f haricot écarlat; haricot des Indes; haricot-fleur; haricot d'Espagne
d Feuerbohne; Prunkbohne; türkische Bohne
l *Phaseolus coccineus; Phaseolus multiflorus*

4749 scarlet sage
f sauge éclatante (du Brésil)
d scharlachrote Salbei; rote Salbei; Prachtsalbei
l *Salvia splendens*

4750 scented grass
f chrysopogon
d Goldbartgras
l *Chrysopogon*

4751 scentless mayweed
f matricaire inodore
d falsche Kamille; geruchlose Kamille
l *Matricaria inodora; Chrysanthemum inodorum*

4752 scentless mock orange
f seringa inodore
l *Philadelphus inodorus*

4753 scentless rose
d Duftrose; elliptische Rose; keilblättrige Rose
l *Rosa inodora; Rosa elliptica; Rosa caryophyllacea*

4754 scheuchzeria
f scheuchzeria
d Scheuchzerie; Blasenbinse; Blumenbinse; Blumensimse
l *Scheuchzeria*

4755 scheuchzeria family
f scheuchzeriacées
d Blumenbinsengewächse; Scheuchzeriazeen
l *Scheuchzeriaceae*

4756 Scheuchzer's bellflower
f campanule de Scheuchzer
d einblütige Glockenblume; Scheuchzers Glockenblume
l *Campanula scheuchzeri*

4757 schmidelia
f schmidélie
l *Schmidelia*

4758 Schmidt's birch
f bouleau de Schmidt
d Eisenbirke
l *Betula schmidti*

4759 **Schneider deutzia**
f deutzie de Schneider
d Schneiders Deutzie
l *Deutzia schneideriana*

4760 **Schrenk mock orange**
f seringa de Schrenk
d Schrencks Pfeifenstrauch
l *Philadelphus schrenkii*

4761 **Schrenk spruce**
f épicéa de Schrenk
d Tienschan-Fichte
l *Picea schrenkiana*

4762 **Schubert onion**
f ail de Schubert
d Schuberts Lauch
l *Allium schuberti*

4763 **Schwarz widdringtonia**
f widdringtonia de Schwarz
d Schwarzs Widdringtonia
l *Widdringtonia schwarzi*

4764 **sclerotinia**
f sclérotinia
d Eichelbecherling
l *Sclerotinia*

4765 **scopolia**
f scopolie
d Tollkraut; Skopolie
l *Scopolia*

4766 **scorpion senna**
f coronille des jardins; coronille arbisseau
l *Coronilla emerus*

4767 **scorpion's-tail; caterpillar plant**
f chenille; chenillette; scopiure
d Raupenkraut
l *Scorpiurus*

4768 **Scotch asphodel; Scottish asphodel**
d Sumpfliliensimse; Sumpfsimsenlilie
l *Tofieldia palustris*

4769 **Scotch broom**
f genêt à balais
d Besenginster
l *Cytisus scorparius; Genista scoparia; Spartium scoparium*

4770 **Scotch cotton thistle**
f onoporde à feuilles d'acanthe; onoporde acanthe; acanthe sauvage; chardon aux ânes; chardon onoporde
d gewöhnliche Eseldistel; Krampfdistel; Wolldistel
l *Onopordum acanthium*

4771 **Scotch elm; wych elm**
f orme blanc; orme de montagne
d Bergrüster; Bergulme
l *Ulmus glabra; Ulmus scabra; Ulmus montana*

4772 **Scotch heather; common heather**
f bruyère commune
d Besenheide; gemeines Heidekraut; Immerschönkraut; Sandheide
l *Calluna vulgaris*

4773 **Scotch laburnum**
f cytise d'Ecosse; cytise des Alpes; cytise faux ébénier
d Alpengoldregen; Alpenbohnenbaum
l *Laburnum alpinum; Cytisus alpinus*

4774 **Scotch ligusticum**
d schottische Mutterwurz
l *Ligusticum scoticum*

4775 **Scotch pine**
f pin sylvestre; pin de Genève; pin de Russie; pin rouge du Nord; pin sauvage
d gemeine Kiefer; Fackelbaum; Waldkiefer
l *Pinus sylvestris*

4776 **Scotch rose**
f rosier pimprenelle; rosier d'Ecosse
d Bibernellrose; stachelige Rose; Stachelrose
l *Rosa spinosissima; Rosa illinoensis*

* **Scottish asphodel → 4768**

4777 **scouring rush; Dutch rush**
f prêle d'hiver; prêle des tourneurs
d Winterschachtelhalm
l *Equisetum hyemale*

4778 **scratchbush**
f uréra
l *Urera*

4779 **screw pine; pandanus**
f baquois; vacoua
d Pandane; Pandang; Schraubenbaum; Schraubenpalme
l *Pandanus*

4780 **screw-pine family**
f pandanacées; pandanées
d Pandanazeen; Schraubenbaum-

gewächse; Schraubenpalmengewächse
l *Pandanaceae*

4781 screw tree
f hélictère
d Isorabaum; Schraubenbaum
l *Helicteres*

* **scrub pine → 5750**

4782 scurf(y) pea
f psoralée
d Harzklee; Drüsenklee
l *Psoralea*

4783 scurvy weed; spoonwort
f cochléaria
d Löffelkraut
l *Cochlearia*

4784 Scythian lamb
f agneau de Scythie
d skytisches Lamm
l *Cibotium barometz*

* **sea aster → 5597**

4785 sea beet
f betterave sauvage
d Seemangold; Wildbete
l *Beta maritima*

4786 seaberry
f halorage
d Seebeere
l *Haloragis*

* **sea bindweed → 4804**

* **sea blite → 4802**

4787 sea buckthorn
f argousier
d Sanddorn
l *Hippophae*

* **sea bulrush → 4801**

* **sea coconut → 1887**

* **sea dahlia → 1596**

4788 sea grape
f raisinier
d Meertraube; Seetraube; Lappenbeere; Traubenbaum
l *Coccoloba*

* **sea grape → 3106**

* **sea heath → 2361**

* **sea holly → 2058**

4789 sea-holly eryngo
f panicaut maritime
d Männertreu; Mannstreu; Meerbrachdistel; Meerstrandstreu; Meerwurzeldistel; Seemannstreu; Strandmännertreu; Strandmannstreu
l *Eryngium maritimum*

4790 sea-island cotton; Egyptian cotton
f cotonnier d'Egypte
l *Gossypium barbadense*

4791 sea kale
f rhodyménie
l *Rhodymenia*

4792 sea lace
d Meersaite
l *Chorda filum*

4793 sea lavender
f lavande de mer; immortelle bleue
d Strandflieder; Widerstoss; Meerlavendel
l *Limonium; Goniolimon*

4794 sea lettuce; green laver
f ulve
d Meersalat
l *Ulva*

4795 sealing-wax palm
f cyrtostachys
d Krummähre
l *Cyrtostachys*

4796 sea milkwort
f glauce
d Milchkraut; Mutterkraut
l *Glaux*

4797 sea oak
f halidryde
d Meereiche
l *Halidrys*

* **sea oak → 4538**

4798 sea onion
f scille maritime
d Meerzwiebel
l *Urginea*

* **sea plantain → 2123**

4799 **sea rocket**
f roquette de mer
d Meersenf
l *Cakile*

4800 **sea rush**
f jonc maritime
d Meerstrandbinse; Strandbinse
l *Juncus maritimus*

4801 **sea scirpus; sea bulrush; maritime bulrush**
f scirpe maritime
d Meersimse; Strandsimse
l *Scirpus maritimus; Bulboschoenus maritimus*

4802 **sea seepweed; sea blite**
d Schmalzmelde; Strandsode
l *Suaeda maritima*

4803 **seashore alkali grass**
d Strandsalzschwaden
l *Puccinellia maritima*

4804 **seashore glorybind; sea bindweed**
f liseron soldanelle
d Meerkohlwinde; Meerstrandwinde; Strandwinde
l *Calystegia soldanella; Convolvulus soldanella*

* **seaside balsam → 1041**

4805 **sea spurge**
d dickblättrige Wolfsmilch
l *Euphorbia paralias*

4806 **sea staff; sea wand**
f laminaire palmé
d Fingertang; handförmiger Riementang
l *Laminaria digitata*

4807 **sea tangle**
f laminaire du Japon
d japanischer Palmentang
l *Laminaria japonica*

* **sea tangle → 667**

* **sea-urchin cactus → 2749**

* **sea wand → 4806**

4808 **seccomaria**
f drymaire
d Eichenwald
l *Drymaria*

4809 **securinega**
f sécurinéga
d Hartholz
l *Securinega*

4810 **sedge**
f laîche; carex
d Segge
l *Carex*

4811 **sedge family**
f cypéracées
d Halbgräser; Riedgräser; Riedgrasgewächse; Sauergräser; Scheingräser; Zyperazeen
l *Cyperaceae*

4812 **seedbox**
f ludwigia
d Heusenkraut; Ludwigie
l *Ludwigia*

4813 **seepweed**
f suaéda
d Salzmelde; Sode; Strandsode
l *Suaeda*

4814 **selaginella; resurrection plant; club moss**
f sélaginelle
d Moosfarn; Mooskraut
l *Selaginella*

4815 **selaginella family**
f sélaginellacées
d Moosfarngewächse; Mooskrautgewächse; Selaginellazeen
l *Selaginellaceae*

4816 **selfheal**
f brunelle
d Braunelle; Braunheil; Prunelle
l *Prunella*

4817 **selinum**
d Kümmelsilge
l *Selinum*

4818 **selloa pampas grass**
f gynérion argenté
d Pampasgras
l *Cortaderia selloana; Gynerium argenteum*

4819 **semaphore grass**
f pleuropogon
l *Pleuropogon*

4820 Senegal date
f dattier de Zanzibar
l *Phoenix reclinata*

4821 Senegal khaya
f cailcedrat du Sénégal; acajou du Sénégal
l *Khaya senegalensis*

4822 Senegal soapberry
f savonnier du Sénégal
l *Sapindus senegalensis*

4823 senna
f casse; séné
d Kassie; Sennesstrauch; Gewürzrinde
l *Cassia*

4824 senna family
f césalpiniacées
d Johannisbrotgewächse
l *Caesalpiniaceae*

4825 sensitive brier
f neptunia
d Neptunie
l *Neptunia*

4826 sensitive fern
f onocléa délicate
d Perlfarn; Straussfarn
l *Onoclea sensibilis*

4827 sensitive plant
f sensitive pudique
d keusche Sinnpflanze; schamhafte Sinnpflanze
l *Mimosa pudica*

4828 sentry palm
f howée
d Lord Howe Palme
l *Howea*

4829 sequoia
f séquoia
d Mammutbaum; Küstensequoie; Redwood
l *Sequoia*

4830 Serbian spruce
f épicéa de Serbie
d Omorikafichte; serbische Fichte
l *Picea omorika*

4831 serpent root
f scorsonère
d Schwarzwurzel; Skorzonere
l *Scorzonera*

4832 serviceberry; Juneberry; shadebush
f amélanchier
d Felsenbirne
l *Amelanchier*

4833 service tree; service-tree mountain ash
f sorbier domestique; cormier
d Speierling; Sperbe; Hauseberesche; zahme Eberesche
l *Sorbus domestica*

4834 sesame
f sésame
d Sesam
l *Sesamum*

4835 sesbania
f sesbanie
d Sesbanie
l *Sesbania*

4836 seseli; meadow saxifrage
f séséli
d Sesel; Bergfenchel; Heilwurz
l *Seseli*

4837 seven-leaved cinquefoil
d rötliches Fingerkraut; trübes Fingerkraut; Thüringer Fingerkraut
l *Potentilla heptaphylla*

* **Seville orange → 5092**

* **shabdar → 4147**

4838 shad-blow serviceberry; downy serviceberry; shadbush serviceberry
f amélanchier du Canada
d kanadische Felsenbirne
l *Amelanchier canadensis*

* **shaddock → 4329**

* **shadebush → 4832**

4839 shade fescue
f fétuque hétérophylle
d Borstenschwingel; verschiedenblättriger Schwingel
l *Festuca heterophylla; Festuca rubra heterophylla*

4840 shadow palm
f géonoma
d Erdpalme
l *Geonoma*

* **shaftal clover → 4147**

4841 shagbark hickory
f caryer blanc
d Schuppenrindenhickory
l *Carya ovata; Carya blanc*

4842 shaggy hawkweed
f épervière à toison
d zottiges Habichtskraut
l *Hieracium villosum*

4843 shaggy-mane
f coprin chevelu
d grosser Tintenpilz; Schopftintenpilz; Schopftintling
l *Coprinus comatus*

4844 shallot
f échalotte
d Schalotte; askalonische Zwiebel
l *Allium ascalonicum*

4845 shallow-cup Mongolian oak
d krause Eiche
l *Quercus crispula; Quercus mongolica grosseserrata*

4846 shallow-pored boletus
f bolet des bouviers; cèpe des bouviers
d Kuhpilz; Kuhröhrling; Ochsenpilz
l *Suillus bovinus; Boletus bovinus; Ixocomus bovinus*

4847 shamrock pea
f parochète
d Blauklee
l *Parochetus*

4848 sharp-leaved asparagus
f asperge à feuilles aiguës
d strauchartiger Spargel
l *Asparagus acutifolius*

4849 sharp-leaved jacaranda
f jacaranda à feuilles de mimosa
d mimosenblättrige Jacaranda
l *Jacaranda acutifolia; Jacaranda mimosaefolia; Jacaranda mimosifolia; Jacaranda ovalifolia*

4850 sharp-leaved willow
f saule à feuilles aiguës
d spitzblättrige Weide; kaspische Weide
l *Salix acutifolia*

4851 sharp-pointed fluellin
d echtes Tännelkraut; Spiessleinkraut
l *Kickxia elatine; Linaria elatine*

* **shea butter seed → 4852**

4852 shea butter tree; shea butter seed
d Butterbaum; Schinussbaum
l *Butyrospermum parkii*

4853 sheathed amanitopsis
f amanite engaînée; amanite à étui
d ringloser Wulstling; brauner Wulstblätterschwamm; Scheidenpilz; Scheidenschwamm; Scheidenstreifling
l *Amanita vaginata; Amanitopsis vaginata*

* **sheepberry → 3792**

4854 sheep bur
f acéna
d Stachelnüsschen
l *Acaena*

4855 sheepbush
f pentzie
l *Pentzia*

4856 sheep's-beard
f urosperme
d Schwanzsame
l *Urospermum*

4857 sheep's-bit jasione; shepherd's-scabious
f jasione des montagnes
d Bergheilkraut; Bergjasione; Bergsandknöpfchen; Sandglöckchen; Schafskabiose
l *Jasione montana*

4858 sheep's fescue
f fétuque ovine; fétuque des moutons
d Schafschwingel
l *Festuca ovina*

4859 sheep sorrel
f oseille des brebis; petite oseille
d Feldampfer; kleiner Ampfer; kleiner Sauerampfer
l *Rumex acetosella*

4860 shellseed
f cochlosperme
d Schneckensame
l *Cochlospermum*

4861 shepherd's-needle; lady's-comb; Venus's-comb
f aiguille de berger; peigne de Vénus
d Nadelkerbel; Venuskamm
l *Scandix pecten-veneris*

4862 shepherd's-purse
f bourse à pasteur; bourse de berger
d Hirtentäschel(kraut)
l *Capsella bursa pastoris*

* **shepherd's-scabious → 4857**

4863 shield lepiota
d wolliggestiefelter Schirmling
l *Lepiota clypeolaria*

4864 shieldwort
f peltaire
d Scheibenkraut; Scheibenschötchen
l *Peltaria*

4865 shingle oak
f chêne à lattes
d Schindeleiche
l *Quercus imbricaria*

4866 shining pondweed
d spiegelndes Laichkraut
l *Potamogeton lucens*

4867 shining white clitocybe
d wachsstieliger Trichterling
l *Clitocybe candicans*

* **shinleaf → 4377**

4868 shinleaf yellow-horn; shiny-leaved yellow-horn
f xanthocère à feuilles de sorbier
d eschenblättriger Gelbhorn
l *Xanthoceras sorbifolium*

4869 shiny camwood
f bois de Cam; bois de corail dur
l *Baphia nitida*

* **shiny-leaved yellow-horn → 4868**

4870 shoalweed
f halodule
l *Halodule*

4871 shooting star
f gyroselle
d Götterblume
l *Dodecatheon*

4872 shorea
f shorée
d Salbaum
l *Shorea*

4873 shore gumweed
f grindélie robuste
l *Grindelia robusta*

4874 shore juniper
f genévrier des rivages
l *Juniperus conferta; Juniperus litoralis*

4875 shore orach
d Strandmelde
l *Atriplex litoralis*

4876 shore pine; beach pine
f pin vrillé
d Küstenkiefer
l *Pinus contorta*

4877 shore pod grass
f triglochin maritime
d Meerstranddreizack; Sechszack; Stranddreizack
l *Triglochin maritima*

4878 shore sea onion
f scille maritime; scille officinale
d echte Meerzwiebel
l *Urginea maritima*

4879 shoreweed
f littorelle
d Strandling
l *Littorella*

4880 shorthusk
f brachyélytre
l *Brachyelytrum*

4881 shortia
f shortie
d Shortie; Winterblatt
l *Shortia*

4882 short-leaf pine
f pin jaune
l *Pinus echinata; Pinus mitis*

4883 showy clover
f trèfle élégant
l *Trifolium elegans*

4884 showy crotalaria
f crotalaire remarquable
l *Crotalaria spectabilis*

4885 showy vetch
d grossblütige Wicke
l *Vicia grandiflora*

4886 shrub(by) althea; rose of Sharon
f ketmie de Syrie
d syrischer Eibisch
l *Hibiscus syriacus*

4887 shrub harebell
f isotome
d Gleichzipfel
l *Isotoma*

4888 shrub thoroughwax
f buplèvre ligneux
l *Bupleurum fruticosum*

4889 sibbaldia
f sibbaldia
d Sibbaldie; Gelbling
l *Sibbaldia*

4890 Sibbald's potentilla
d Alpengelbling; Gelbling
l *Sibbaldia procumbens*

4891 Siberian apricot
f abricotier de Sibérie
d sibirischer Aprikosenbaum
l *Prunus sibirica; Armeniaca sibirica*

4892 Siberian aster
f aster de Sibérie
d sibirische Aster
l *Aster sibiricus*

4893 Siberian barberry
f vinettier de Sibérie
d sibirische Berberitze
l *Berberis sibirica*

4894 Siberian bellflower
f campanule de Sibérie
d sibirische Glockenblume
l *Campanula sibirica*

4895 Siberian campion
f silène de Sibérie
d sibirisches Leimkraut
l *Silene sibirica*

4896 Siberian clematis
f clématite de Sibérie
d sibirische Doppelblume
l *Clematis alpina* var. *sibirica*

4897 Siberian corydalis
f corydale noble
d sibirischer Lerchensporn
l *Corydalis nobilis; Corydalis sibirica*

4898 Siberian cow parsnip
f berce de Sibérie
d sibirisches Futterkraut; sibirischer Bärenklau
l *Heracleum sibiricum*

4899 Siberian crab apple
f pommier à baies; pommier de Sibérie
d Beerenapfelbaum; Beerenzierapfel; sibirischer Apfelbaum
l *Malus baccata*

4900 Siberian currant
d Felsenjohannisbeere
l *Ribes saxatilis; Ribes diacanthum*

4901 Siberian dogwood
f cornouiller de Sibérie
d sibirischer Hartriegel; tatarischer Hartriegel
l *Cornus sibirica; Cornus alba sibirica*

4902 Siberian draba
f drave de Sibérie
d sibirisches Felsenblümchen
l *Draba sibirica*

4903 Siberian elm
f orme de Sibérie; orme nain
d Zwergulme
l *Ulmus pumila; Ulmus sibirica*

4904 Siberian fir
f sapin de Sibérie
d sibirische Tanne
l *Abies sibirica*

4905 Siberian geranium
f géranium de Sibérie
d sibirischer Storchschnabel
l *Geranium sibiricum*

4906 Siberian globeflower
f trolle d'Asie
d asiatische Trollblume
l *Trollius asiaticus*

4907 Siberian iris
f iris de Sibérie
d sibirische Schwertlilie; Wiesen-schwertlilie
l *Iris sibirica*

4908 Siberian larch
f mélèze de Sibérie
d sibirische Lärche
l *Larix sibirica*

4909 Siberian milkwort
f polygala de Sibérie
d sibirische Kreuzblume
l *Polygala sibirica*

4910 Siberian pea tree
f caragana arborescent
d baumartiger Erbsenstrauch; gemeiner grosser Erbsenstrauch; Erbsenstrauch
l *Caragana arborescens*

4911 Siberian phlox
f phlox de Sibérie
d sibirische Flammenblume
l *Phlox sibirica*

4912 Siberian St.-John's-wort
d zierliches Hartheu
l *Hypericum elegans*

4913 Siberian salt tree
f caragana argenté
d Silbersalzstrauch
l *Halimodendron argenteum; Halimodendron halodendron*

4914 Siberian sea lavender
f statice de Gmelin
d Gmelins Strandflieder
l *Statice gmelini*

4915 Siberian spruce
f épicéa de Sibérie; sapinette de Sibérie
d Altaifichte; sibirische Fichte
l *Picea obovata*

4916 Siberian squill
f scille de Sibérie
d nickender Blaustern; sibirische Meerzwiebel; sibirischer Blaustern
l *Scilla sibirica*

4917 Siberian stone pine
f pin de Sibérie; pin cembro de Sibérie
d sibirische Kiefer
l *Pinus sibirica; Pinus cembra sibirica*

4918 sibirea
f sibérienne
d Blauspiere
l *Sibiraea*

4919 Sibthorp sage
f sauge de Sibthorp
d Sibthorps Salbei
l *Salvia sibthorpi*

4920 Sicilian sumac
f sumac des corroyeurs
d echter Essigbaum; Gerberstrauch; Gerbersumach; sizilianischer Sumach
l *Rhus coriaria*

4921 sickle alfalfa; sickle medic
f luzerne en faux
d deutsche Luzerne; gelbe Luzerne; schwedische Luzerne; Sichelluzerne; Sichelklee
l *Medicago sativa falcata*

4922 sickle euphorbia
d Sichelwolfsmilch
l *Euphorbia falcata*

4923 sickle-grass
d Dünnschwanz; Schuppenschwanz
l *Pholiurus*

4924 sickle-grass
d gekrümmter Dünnschwanz
l *Pholiurus incurvus*

* **sickle medic → 4921**

4925 sida
f sida
d Samtpappel
l *Sida*

4926 side-bells pyrola; one-sided shinleaf
f pyrole unilatérale
d Birnbäumchen; einseitswendiges Wintergrün; nickendes Wintergrün
l *Pyrola secunda; Ramischia secunda*

4927 side oat
f avoine d'Orient
l *Avena orientalis; Avena sativa orientalis*

4928 siderodendron
f sidérodendron
l *Siderodendron*

* **side-saddle flower → 4207**

4929 Siebold aspen
f peuplier tremble du Japon; tremble du Japon
d Siebolds Pappel
l *Populus sieboldi*

4930 Siebold barberry
f vinettier de Siebold
d Siebolds Berberitze
l *Berberis sieboldi*

4931 Siebold beech
f hêtre de Siebold
d Siebolds Buche
l *Fagus sieboldi*

4932 Siebold deutzia
f deutzie de Siebold
d Siebolds Deutzie
l *Deutzia sieboldiana*

4933 Siebold hemlock
f tsuga de Siebold; tsuga du Japon
d Siebolds Hemlocktanne
l *Tsuga sieboldi*

4934 Siebold maple
f érable de Siebold
d Siebolds Ahorn
l *Acer sieboldianum*

4935 Siebold viburnum
f viorne de Siebold
d Siebolds Schneeball
l *Viburnum sieboldi*

4936 Siebold walnut
f noyer de Siebold
d Siebolds Walnussbaum
l *Juglans sieboldiana*

4937 sieglingia
f siéglingie
d Dreizahn
l *Sieglingia*

4938 Sierra juniper
f genévrier d'Occident
l *Juniperus occidentalis*

4939 Sierra Leone copal tree
d Kobobaum
l *Copaifera copallifera*

4940 Sieva bean
f haricot de Lima; pois du Cap; haricot de Madagascar
d Mondbohne; Duffinbohne; Madagaskarbohne; Rangoonbohne; Sievabohne
l *Phaseolus lunatus*

4941 sieversia
f siéversie
d Petersbart
l *Sieversia*

4942 Sikkim spruce
f sapin du Sikkin; épicéa spinuleux
l *Picea spinulosa*

4943 Silesian willow
f saule de Silésie
d schlesische Weide
l *Salix silesiaca*

4944 silk-cotton family
f bombacacées; bombacées
d Bombakazeen; Bombazeen; Wollbäume
l *Bombacaceae*

* **silk-cotton tree → 3134**

4945 silk oak; silk oak grevillea
f fougère en arbre
d australische Seideneiche
l *Grevillea robusta*

4946 silk-rubber
f funtumia
d Seidenkautschuk
l *Funtumia*

4947 silk-tassel tree
f garrye
d Becherkätzchen
l *Garrya*

4948 silk tree; silk-tree albizzia
f acacia de Constantinople
d Seidenakazie; persische Akazie
l *Albizzia julibrissin*

4949 silk vine
f périploque
d Baumschlinge
l *Periploca*

* **silkweed → 3620**

4950 silver bell; snowdrop tree
f halésie; arbre aux cloches d'argent
d Schneeglöckchenbaum; Maiglöckchenbaum; Maiblumenbaum
l *Halesia*

4951 silverberry
f chalef argenté
d amerikanischer Silberbaum; breitblättrige Ölweide; Silberweide
l *Elaeagnus argentea; Elaeagnus commutata*

4952 silver buffalo berry
d Büffelbeere
l *Shepherdia argentea*

4953 silverbush
f argythamne
l *Argythamnia*

4954 silver cinquefoil
f potentille argentée
d Silberfingerkraut
l *Potentilla argentea*

4955 silver fir
f sapin argenté; sapin pectiné; sapin de Lorraine; sapin de Normandie
d europäische Edeltanne; Weisstanne
l *Abies alba*

4956 silver grass; sword grass; eulalia
f miscanthe
d Stielblütengras
l *Miscanthus*

4957 silvergreen-wattle acacia
f acacia blanchâtre; mimosa blanchâtre; mimosa blanchissant
d australische Silberakazie; australischer Gummibaum; echte Akazie
l *Acacia dealbata*

4958 silver groundsel; dusty miller
f séneçon cinéraire; cinéraire maritime
d Strandkreuzkraut
l *Cineraria maritima; Senecio cineraria*

4959 silver hair grass
f canche caryophyllée
d gemeine Haferschmiele; Nelken-hafer(schmiele); Nelkenschmielenhafer
l *Aira caryophyllea*

4960 silver hawthorn
f épine d'Orient
d orientalischer Weissdorn
l *Crataegus orientalis*

4961 silverleaf
f leucophylle
l *Leucophyllum*

4962 silverleaf
f argyrolobe
d Silberhülse; Silberkappe
l *Argyrolobium*

4963 silver-leaf sunflower
f soleil à feuilles argentées
d silberblättrige Sonnenblume
l *Helianthus argophyllus*

4964 silver leucadendron; silver tree
f arbre d'argent
d Silberbaum
l *Leucadendron argenteum*

4965 silver linden
f tilleul argenté
d Silberlinde; ungarische Silberlinde
l *Tilia tomentosa; Tilia argentea*

4966 silver maple
f érable blanc
d Silberahorn
l *Acer saccharinum; Acer dasycarpum*

4967 silver pendent linden
f tilleul argenté pleureur; tilleul d'Amérique pleureur
d Hängesilberlinde
l *Tilia petiolaris*

* **silver tree → 4964**

4968 silvervine actinidia
f actinidie polygame
d Silberwein
l *Actinidia polygama*

* **silver weed → 339**

4969 silverweed cinquefoil
f potentille ansérine
d Gänsefingerkraut; Gänsekohl; Gänserich
l *Potentilla anserina*

4970 silvery sedge; whitish sedge
f laîche blanchâtre
d graues Riedgras; graue Segge; weiss-graues Riedgras; weissgraue Segge
l *Carex canescens*

* **simal → 3440**

4971 simaruba
f simarube
d Bitteresche; Ruhrrindenbaum
l *Simaruba*

4972 single-leaf ash
f frêne à feuilles simples
l *Fraxinus anomala*

4973 single-seed hawthorn
f aubépine à un style; aubépine monogyne
d eingriffeliger Weissdorn
l *Crataegus monogyna*

4974 single-seed juniper
f genévrier écailleux
d Blauzederwacholder
l *Juniperus squamata*

4975 siphonia
f siphonia
l *Siphonia*

4976 sisal agave
f agave sisal
d Sisalagave; Sisalhanf
l *Agave sisalana*

4977 sissoo
f arbre de Shisham
d Sissoo
l *Dalbergia sissoo*

4978 sisymbrium
f sisymbre
d Rauke; Raukensenf
l *Sisymbrium*

4979 Sitka spruce
f épicéa de Sitka
d Sitkafichte
l *Picea sitchensis*

4980 Sitka willow
f saule de Sitka
d Sitkaweide
l *Salix sitchensis*

4981 six-rowed barley
f escourgeon; orge à six rangs
d sechszeilige Gerste; Stockgerste
l *Hordeum hexastichon*

4982 six-stamened waterwort
d sechsmänniger Tännel; Stieltännel
l *Elatine hexandra*

4983 skeletonweed
f chondrille
d Knorpellattich; Krümmling
l *Chondrilla*

4984 skimmia
f skimmie
d Skimmie
l *Skimmia*

4985 Skinner columbine
f ancolie de Skinner
d Skinners Akelei
l *Aquilegia skinneri*

4986 skirret water parsnip
f chervi(s); perle des potagères
d Zuckerwurzel; Süsswurzel
l *Sium sisarum*

4987 skullcap
f scutellaire; toque
d Helmkraut; Schildträger
l *Scutellaria*

4988 skunk bugbane
d stinkendes Wanzenkraut
l *Cimicifuga foetida*

4989 skunkbush
f menziésie
d Menziesie
l *Menziesia*

4990 skunk cabbage
f symplocarpc
d Stinkkohl
l *Symplocarpus*

4991 skunk currant
f groseillier fétide
l *Ribes glandulosum; Ribes prostratum*

4992 sky-flower
f durante
l *Duranta*

4993 slash pine
f pin de Cuba
d karaibische Kiefer
l *Pinus caribaea; Pinus ellioti*

4994 slender cudweed
d kleines Filzkraut; Zwergfilzkraut
l *Filago minima*

4995 slender deutzia
f deutzie grêle
d zierliche Deutzie
l *Deutzia gracilis*

4996 slender false brome
f brachypode des bois
d Waldzwenke
l *Brachypodium sylvaticum*

4997 slender fritillary
d Bergkaiserkrone
l *Fritillaria tenella; Fritillaria montana*

4998 slender hare's-ear
f buplèvre menu
d feinblättriges Hasenohr; feines

Hasenohr; zartes Hasenohr
l *Bupleurum tenuissimum*

4999 slender knotgrass
d kleiner Knöterich
l *Polygonum minus*

5000 slender-leaf crotalaria
f crotalaire intermédiaire
l *Crotalaria intermedia*

5001 slender-leaved flax; slim-leaved flax
d zarter Lein
l *Linum tenuifolium*

5002 slender naiad
d biegsames Nixkraut
l *Naias flexilis*

5003 slender oat
d Barthafer
l *Avena barbata*

*** slender rush → 4288**

5004 slender vetch
f vesce grêle
d schlanke Wicke; zierliche Wicke
l *Vicia gracilis*

5005 slim amaranth
f amarante paniculée
d Bastardfuchsschwanz
l *Amaranthus hybridus*

5006 slime fungi; slime mo(u)lds
f myxomycètes
d Myxomyzeten; Schleimpilze
l *Myxomycetes*

*** slim-leaved flax → 5001**

5007 slim-leaved wall rocket
f diplotaxe à feuilles ténues
d schmalblättriger Doppelsame; schmaler Doppelsame
l *Diplotaxis tenuifolia*

5008 slim-stem reed grass
d Moorreitgras; übersehenes Reitgras
l *Calamagrostis neglecta*

5009 slim-top meadow rue
d einfache Wiesenraute
l *Thalictrum simplex*

5010 slipperplant
f pédilanthe
d Pantoffelstrauch
l *Pedilanthus*

5011 slipperwort
f calcéolaire; pantoufle
d Pantoffelblume; Kalzeolarie
l *Calceolaria*

5012 slippery elm
f orme roux
d Rotulme
l *Ulmus fulva*

5013 sloanea
f quapalier
l *Sloanea*

5014 sloe; blackthorn
f épine noire; prunellier; prunier épineux
d Dorn(schlehe); Schleh(busch); Schlehdorn; Schlehe(ndorn); Schwarzdorn; Stechdorn
l *Prunus spinosa*

*** slogwood → 5016**

5015 slough grass
f beckmannie
d Beckmannsgras; Raupenähre
l *Beckmannia*

5016 slugwood; slogwood
f beilschmiedie
l *Beilschmiedia*

5017 small bedstraw
d dreispaltiges Labkraut
l *Galium trifidum*

5018 small burnet; salad burnet
f petite pimprenelle; pimprenelle sanguisorbe
d Becherblume; Blutkraut; Gartenpimpinelle; kleines Nagelkraut; kleiner Wiesenknopf; Sperberkraut
l *Poterium sanguisorba; Sanguisorba minor*

5019 small cranberry
f canneberge commune
d kleine Moosbeere; Torfbeere
l *Oxycoccus palustris; Vaccinium oxycoccos*

5020 smaller burdock
d kleine Klette
l *Arctium minus*

* **small-flowered balsam → 5023**

5021 small-flowered bitter cress
f cardamine à petites fleurs
d kleinblütiges Schaumkraut; Teichschaumkraut
l *Cardamine parviflora*

5022 small-flowered evening primrose
d stachelige Nachtkerze
l *Oenothera muricata; Oenothera parviflora*

5023 small-flowered snapweed; small-flowered balsam
f balsamine à petites fleurs
d kleinblütiges Springkraut; kleines Springkraut
l *Impatiens parviflora*

* **small-flowered willow herb → 5024**

5024 small-flowered willowweed; small-flowered willow herb
f épilobe à petites fleurs
d Bachweidenröschen; kleinblütiges Weidenröschen
l *Epilobium parviflorum*

5025 small-flower rose
f rose à petites fleurs
d kleinblütige Rose
l *Rosa micrantha*

5026 small geranium
f bec-de-grue mauvin; géranium nain
d kleinblütiger Storchschnabel; kleiner Storchschnabel
l *Geranium pusillum*

5027 small globe thistle
f échinope boule azurée
l *Echinops ritro*

5028 small horse bean
d Eselsbohne; gemeine Feldbohne
l *Vicia faba minor*

5029 small lamb succory
d kleiner Lämmersalat
l *Arnoseris minima*

5030 small radish
f petit radis
d Radies(chen)
l *Raphanus sativus* var. *sativus*

5031 smilax family
f smilacées
d Stechwindengewächse
l *Smilacaceae*

5032 smilo grass
f faux-millet; petit millet
d Grannenhirse
l *Oryzopsis miliacea*

5033 smoke tree
f arbre à perruque; fustet
d Perückenstrauch
l *Cotinus*

5034 smooth brome; awnless brome
f brome des prés; brome inerme
d Queckentrespe; grannenlose Trespe; unbegrannte Trespe; unbewehrte Trespe; wehrlose Trespe
l *Bromus inermis*

5035 smooth carlina
f carline acaule
d grosse Eberwurz; Karlsdistel; Sonnendistel; Wetterdistel
l *Carlina acaulis*

5036 smooth cat's-ear
d kahles Ferkelkraut; Sandferkelkraut
l *Hypochoeris glabra*

5037 smooth crab grass
d Fadenfingerhirse; Fadenhirse
l *Digitaria ischaemum; Digitaria linearis*

5038 smooth hawk's-beard
d grüner Pippau; dünnästiger Pippau; kleinköpfiger Pippau
l *Crepis capillaris; Crepis virens*

5039 smooth hydrangea; wild hydrangea
f hortensia en arbre; hortensia de Virginie; hydrangelle arborescente
d nordamerikanische Hortensie; baumartige Hortensie
l *Hydrangea arborescens*

5040 smooth-leaved elm; field elm; English elm
f orme champêtre; orme à feuilles de charme; orme rouge
d Feldulme; Feldrüster; glattblättrige Rüster; Rotulme; Rotrüster
l *Ulmus campestris; Ulmus foliacea; Ulmus carpinifolia; Ulmus procera*

* **smooth lungwort → 691**

5041 smooth sago palm
f sagoutier
d Sagobaum; Sagopalme
l *Metroxylon sagu*

5042 smooth sumac
f sumac à bois glabre
d glattblättriger Sumach; kahler Sumach
l *Rhus glabra*

5043 smooth woodsia
d kahler Wimperfarn
l *Woodsia glabella*

* **smut fungi → 854**

5044 snail-seed
f cocculus
d Kokkelstrauch
l *Cocculus*

5045 snake gourd
f trichosanthe
d Haarblume; Schlangenhaargurke
l *Trichosanthes*

5046 snake melon
f melon serpent
d Schlangenmelone
l *Cucumis melo flexuosus*

* **snake palm → 1848**

5047 snakeroot eryngo
f chardon-Roland; panicaut champêtre; panicaut des champs
d Brachdistel; Elend; Feldmännertreu; Feldmannstreu; Krausdistel; Radendistel
l *Eryngium campestre*

5048 snakeroot ononis; yellow-flowered restharrow; goatroot
d gelbe Hauhechel
l *Ononis natrix*

* **snake's-beard → 3307**

5049 snakewood
f bois de chat
l *Piratinera guianensis; Brosimum aubleti*

* **snakewood tree → 4331**

5050 snapdragon
f muflier
d Dorant; Löwenmaul
l *Antirrhinum*

5051 snapweed
f balsamine
d Balsamine; Springkraut
l *Impatiens*

5052 sneezeweed
f hélénie
d Sonnenbraut; Staudensonnenblume
l *Helenium*

5053 sneezewood
f ptéroxyle
d Niesholz
l *Ptaeroxylon*

5054 sneezewort yarrow
f achillée sternutatoire; herbe à éternuer
d Bertramsgarbe; Sumpfgarbe; weisser Dorant
l *Achillea ptarmica*

* **snowball → 5733**

5055 snowball saxifrage
d Schneesteinbrech
l *Saxifraga nivalis*

5056 snowbell; styrax; storax
f aliboufier
d Storaxbaum
l *Styrax*

5057 snowberry
f symphorine
d Schneebeere
l *Symphoricarpus; Symphoricarpos*

5058 snowberry mountain ash
f sorbier discolore
l *Sorbus discolor*

5059 snow buttercup
f renoncule de neige
d Schneehahnenfuss
l *Ranunculus nivalis*

5060 snowdrop
f perce-neige
d Schneeglöckchen; Märzblume; nackte Jungfrau
l *Galanthus*

5061 snowdrop anemone
f anémone sauvage
d Steppenwindröschen; Waldanemone; Waldwindröschen; grosses Windröschen
l *Anemone sylvestris*

* **snowdrop tree → 4950**

5062 snowflake
f nivéole
d Knotenblume
l *Leucojum*

5063 snow gentian
f gentiane de neige
d Schnee-Enzian
l *Gentiana nivalis*

5064 snow-in-summer
f argentine; céraiste cotonneux; oreille-de-souris; céraiste tomenteux
d filziges Hornkraut
l *Cerastium tomentosum*

5065 snowline pyrole
d kleines Wintergrün
l *Pyrola minor*

5066 snow pear
f poirier de neige
d Schneebirne
l *Pyrus nivalis*

5067 snow spirea
f spirée crénelée
d kerbblättriger Spierstrauch
l *Spiraea crenata*

5068 snow wreath
f néviusie
d Schneeglocke
l *Neviusia*

5069 snowy cinquefoil
f potentille de neige
d Schneefingerkraut
l *Potentilla nivea; Potentilla nivalis*

5070 soapbark tree
f quillaya; quillai
d Seifenbaum; Quillajabaum
l *Quillaja*

5071 soapbark tree
f quillai savonneux
d Seifenbaum; Quillajabaum
l *Quillaja saponaria*

5072 soapberry
f savonnier
d Seifenbaum; Seifenblume; Seifennussbaum
l *Sapindus*

5073 soapberry family
f sapindacées
d Sapindazeen; Sapindengewächse; Seifenbaumgewächse
l *Sapindaceae*

5074 soapwort
f saponaire
d Seifenkraut
l *Saponaria*

5075 Socotra begonia
d Bastardbegonie
l *Begonia socotrana*

5076 Socotrine aloe
f aloès soccotrin
d Sokotraaloe
l *Aloe succotrina; Aloe soccotrina*

5077 soft acanthus
f acanthe molle
d weicher Akanthus
l *Acanthus mollis*

5078 soft brome
f brome mou
d Samttrespe; weiche Trespe
l *Bromus mollis*

* **soft corn → 2307**

5079 soldier orchis; military orchis
f orchis militaire
d Helmknabenkraut
l *Orchis militaris*

5080 solitary clematis
f clématite à feuilles simples
d blaue Waldrebe
l *Clematis integrifolia*

5081 Solomon's-plume; false Solomon's-seal
f smilacine
d Schattenblume
l *Smilacina; Vagnera*

5082 Solomon's-seal
f sceau de Salomon
d Salomonssiegel; Weisswurz
l *Polygonatum*

5083 solorina
f solorina
l *Solorina*

5084 sorbaronia
d Ebereschenspiere; Fiederspiere
l *Sorbaronia*

5085 sorgho; sweet sorghum
f sorgho sucré
d chinesisches Zuckerrohr; Zuckerhirse; Zuckersorgho
l *Sorghum vulgare saccharatum*

5086 sorghum
f sorgho
d Sorgho; Sorghum; Mohrenhirse; Negerhirse; Durra
l *Sorghum*

5087 sorghum; broom corn; Kaffir corn; Guinea corn
f sorgho d'Afrique; grand millet
d Dari; Dourg; Durra(hirse); Durragras; Durrha; Kaffernhirse; Kaffernkorn; Mohrenhirse
l *Sorghum vulgare; Andropogon sorghum*

5088 sorrel family
f oxalidacées
d Oxalidazeen; Sauerkleegewächse
l *Oxalidaceae*

5089 sorrel rhubarb
f rhubarbe palmée
d handlappiger Rhabarber
l *Rheum palmatum*

* **sorrel tree → 5095**

5090 sotol
f dasylire
d Rauhlilie; Rauhschopf
l *Dasylirion*

5091 sour cherry
f griottier
d Baumweichsel; Glaskirsche; Sauerkirschbaum; Sauerkirsche
l *Prunus cerasus*

5092 sour orange; Seville orange
f bigardier; orange amer
d Bigardie; Bitterorange; Pomeranze(nbaum)
l *Citrus aurantium; Citrus amara; Citrus bigardia; Citrus vulgaris*

5093 soursop; guanabana
f anone muriquée; anone hérissée
d Sauersack; Sauerapfel; sauere Sobbe; Stachelan(n)one; stachliger Flaschenbaum
l *Annona muricata*

5094 sourwood
f oxydendre
d Sauerbaum
l *Oxydendron*

5095 sourwood; sorrel tree
f andromède en arbre; oxydendre en arbre
l *Oxydendrum arboreum*

5096 South American lupin
f lupin muant; lupin changeant
d veränderliche Lupine
l *Lupinus mutabilis*

5097 southern catalpa; American catalpa; bean tree; Indian bean
f catalpe commun
d amerikanischer Trompetenbaum; gemeiner Trompetenbaum; Zigarrenbaum
l *Catalpa bignonioides*

5098 southern magnolia
f magnolia à grandes fleurs
d grossblütige Magnolie; Riesenlorbeer
l *Magnolia grandiflora*

5099 southern maidenhair; Venus's-hair
f capillaire de Montpellier; cheveu de Vénus
d Frauenhaar; Jungfernhaar; Venushaar
l *Adiantum capillus-veneris*

5100 southern poplar; balsam poplar
f peuplier du Missouri; peuplier baumier
d Balsampappel
l *Populus balsamifera; Populus deltoides missouriensis*

5101 southern red oak; Spanish oak
f chêne rouge; chêne d'Espagne
d Roteiche
l *Quercus rubra*

5102 southern soapberry
f arbre à savonnette; savonnier des Antilles
d Seifenbaum
l *Sapindus saponaria*

5103 southern succisa
d südlicher Abbiss
l *Succisa australis; Scabiosa australis*

5104 southern wax myrtle
f cirier de Louisiane; arbre à cire
d Kerzenbeerstrauch; Wachsbaum;

Wachsgagel
l *Myrica cerifera*

* **southernwood → 3923**

* **souvari nut → 4725**

5105 **sow thistle**
f laiteron
d Gänsedistel; Saudistel
l *Sonchus*

5106 **soybean**
f soja; soya
d Ölbohne; Sojabohne
l *Glycine soja; Glycine hispida; Glycine max; Soja max*

5107 **spadiciflorae**
f spadiciflores
d Spadizifloren
l *Spadiciflorae*

5108 **Spanish box**
f buis de Mahon
l *Buxus balearica*

5109 **Spanish brome**
f brome madrilène; brome de Madrid
d Mittelmeertrespe
l *Bromus madritensis*

* **Spanish broom → 5857**

* **Spanish cedar → 1234**

* **Spanish chestnut → 2086**

* **Spanish dagger → 110**

5110 **Spanish fir**
f sapin d'Espagne
d spanische Tanne
l *Abies pinsapo*

5111 **Spanish iris**
f iris d'Espagne
d spanische Iris; spanische Schwertlilie
l *Iris hispanica; Iris xiphium*

* **Spanish jasmine → 1052**

* **Spanish liquorice → 1459**

* **Spanish moss → 5576**

* **Spanish oak → 5101**

5112 **Spanish oyster plant**
f cardon d'Espagne; épine jaune; scolyme d'Espagne
d spanische Golddistel
l *Scolymus hispanicus*

* **Spanish pepper → 884**

5113 **Spanish plane tree**
f platane d'Espagne
d spanische Platane
l *Platanus hispanica; Platanus acerifolia hispanica*

5114 **Spanish stonecrop**
f orpin d'Espagne
d spanische Fetthenne
l *Sedum hispanicum*

5115 **sparmannia**
f sparmannie
d Sparmannie; Zimmerlinde
l *Sparmannia*

* **sparrow vetch → 2343**

5116 **sparrowwort**
f thymélée
d Purgierstrauch; Spatzenzunge
l *Thymelaea*

* **spatterdock → 1647**

5117 **spearflower**
f ardisie
d Ardisie; Spitzenblume
l *Ardisia*

5118 **spear-leaved fat hen saltbush**
d Spiessmelde
l *Atriplex hastata; Atriplex patula hastata*

5119 **spear lily**
f doryanthe
d Speerblume
l *Doryanthes*

5120 **spearmint**
f menthe verte
d grüne Minze
l *Mentha spicata; Mentha viridis*

5121 **spearwort buttercup**
f petite douve; renoncule flammule; renoncule flammette
d Brennhahnenfuss; brennender Hahnenfuss
l *Ranunculus flammula*

5122 speckled alder; gray alder; hoary alder
f aune blanc; aune blanchâtre
d Grauerle; Weisserle
l *Alnus incana*

5123 speedwell
f véronique
d Ehrenpreis
l *Veronica*

5124 spelt
f épeautre
d Dinkel(weizen); Himmelsgerste; Schwabenkorn; Spelt(weizen); Spelz
l *Triticum spelta*

5125 sphagnum; bog moss; peat moss
f sphaigne
d Bleichmoos; Sphagnum; Sumpfmoos; Torfmoos; Weissmoos
l *Sphagnum*

5126 sphere fungi
f sphaeriales
d Kugelkernpilze; Kugelpilzartige
l *Sphaeriales*

5127 spice bush
f benjoin
d Benzoinbaum; Fieberstrauch
l *Benzoin; Lindera*

* **spicknel → 415**

5128 spiderflower
f cléome
d Pillenbaum; Spinnenpflanze
l *Cleome*

5129 spider herb
f gynandropsis
d Gynanderpflanze
l *Gynandropsis*

5130 spider lily
f hyménocallide
d Schönhäutchen
l *Hymenocallis*

5131 spiderling
f boerhaavie
d Boerhaavie
l *Boerhaavia*

5132 spider orchid
f brassie
l *Brassia*

5133 spider-web houseleek
f joubarbe à toile d'araignée
l *Sempervivum arachnoideum*

5134 spiderwort
f tradescantie
d Dreimasterblume; Tradeskantie
l *Tradescantia*

5135 spiderwort family
f commélinacées
d Kommelinazeen; Himmelsaugengewächse
l *Commelinaceae*

5136 spigelia
f spigélie
d Nelkenwurz
l *Spigelia*

5137 spike broom
f cytise en épis
d schwarzwerdender Geissklee; schwarzer Geissklee; schwarzer Goldregen
l *Cytisus nigricans*

5138 spiked milfoil
f myriophylle à épis
d Ährentausendblatt; ährenblütiges Tausendblatt; ähriges Tausendblatt
l *Myriophyllum spicatum*

5139 spiked mixed-flower
f raiponce en épi
d ährige Teufelskralle; weisse Teufelskralle
l *Phyteuma spicatum*

5140 spiked speedwell
f véronique en épi
d ähriger Ehrenpreis
l *Veronica spicata*

5141 spiked wood rush
f luzule en épi
d Ährensimse; ährige Hainsimse
l *Luzula spicata*

5142 spike heath
f bruckenthalie
d Ährenheide
l *Bruckenthalia*

5143 spike pilocarpus
d fiederblättriger Jaborandistrauch
l *Pilocarpus pennatifolius*

5144 spike primrose
f boisduvalie
l *Boisduvalia*

5145 spike sedge
f éléocharis; héléocharis
d Sumpfried; Sumpfbinse
l *Eleocharis*

5146 spike trisetum
d ähriger Goldhafer; ähriger Grannenhafer
l *Trisetum spicatum*

5147 spike winter hazel
f corylopsis en épis
d ährige Scheinhasel
l *Corylopsis spicata*

5148 spinach
f épinard
d Spinat
l *Spinacia*

*** spinach beet → 1116**

5149 spinach joint fir
d Gnemonbaum
l *Gnetum gnemon*

*** spindle tree → 2069**

5150 spine palm
d Stachelphoenix
l *Acanthophoenix*

5151 spiny acanthus
f acanthe épineuse
d dorniger Akanthus
l *Acanthus spinosus*

5152 spiny alyssum
f alysse épineuse
l *Alyssum spinosum*

5153 spiny amaranth
f épinard malabre; épinard piquant
d Malabarspinat
l *Amaranthus spinosus*

5154 spiny club palm
f bactris; parépon
d Pechpalme; Pfirsichpalme
l *Bactris*

5155 spiny cocklebur
d Choleradistel; dornige Spitzklette; stachelige Spitzklette
l *Xanthium spinosum*

5156 spiny naiad
d Meernixenkraut
l *Naias marina*

*** spiny restharrow → 5492**

5157 spiny sago palm
d echte Sagopalme
l *Metroxylon rumphii*

5158 spiral flag
f coste
d Kostwurz
l *Costus*

5159 spiral wild celery
f vallisnérie spiralée
d Schraubenstengel; Schraubenvallisnerie
l *Vallisneria spiralis*

5160 spirea
f spirée; reine des prés
d Spierstrauch
l *Spiraea*

*** spirogyra → 4260**

5161 spleenwort
f doradille; asplénie
d Streifenfarn; Milzfarn; Mauerraute
l *Asplenium*

*** sponge-berry tree → 1121**

*** spoonwort → 4783**

5162 spot flower
f spilanthe(s)
d Fleckblume
l *Spilanthes*

*** spot flower → 4045**

5163 spotleaf orchis; spotted orchis; crowfoot
d geflecktes Knabenkraut
l *Orchis maculata*

5164 spotted cat's-ear
f porcelle enracinée
d gewöhnliches Ferkelkraut
l *Hypochoeris radicata*

5165 spotted centaurea
d gefleckte Flockenblume; rispige Flockenblume
l *Centaurea maculosa*

5166 spotted dead nettle
f lamier maculé
d gefleckte Taubnessel
l *Lamium maculatum*

5167 spotted euphorbia
d gefleckte Wolfsmilch
l *Euphorbia maculata*

5168 spotted lady's-thumb
f persicaire douce; renouée persicaire
d Flohknöterich; Flohkraut
l *Polygonum persicaria*

5169 spotted loosestrife
f lysimaque ponctuée
d punktierter Gilbweiderich; Punktfelberich
l *Lysimachia punctata*

5170 spotted medic
f luzerne tachetée; luzerne maculée
d arabischer Schneckenklee; arabische Luzerne
l *Medicago arabica; Medicago maculata*

*** spotted orchis → 5163**

5171 sprangle-top
f leptochloé
d Dünngras; Feingras
l *Leptochloa*

*** spreading anemone → 5177**

5172 spreading creosote bush
f jarilla
l *Larrea divaricata*

5173 spreading dogbane
f apocyn gobe-mouches
l *Apocynum androsaemifolium*

5174 spreading elm
f orme à longues pédoncules; orme diffus
d Flatterrüster; Flatterulme
l *Ulmus effusa*

*** spreading elm → 4521, 4614**

5175 spreading erysimum
d sperriger Schotendotter; sperriger Schöterich; Spreizschöterich
l *Erysimum repandum*

5176 spreading hedge parsley
d Ackerborstendolde; Ackerklettenkerbel
l *Torilis arvensis*

*** spreading marigold → 2369**

*** spreading mushroom → 2752**

5177 spreading pasqueflower; spreading anemone
f anémone pulsatille
d ausgebreitete Kuhschelle; Heidekuhschelle
l *Anemone patens; Pulsatilla patens*

*** spreading St.-John's-wort → 1677**

5178 Sprenger asparagus
f asperge de Sprenger
d Sprengers Spargel
l *Asparagus sprengeri*

5179 spring adonis
f adonide printanière
d Frühlingsadonisröschen; Frühlingsteufelsauge
l *Adonis vernalis*

*** spring agaric → 1964**

5180 spring amanita
f amanite printanière; oronge ciguë blanche
d Frühlingswulstling; weisser Knollenblätterpilz
l *Amanita verna*

5181 spring beauty
f clayton(i)e
d Claytonie
l *Claytonia*

5182 spring cinquefoil
f potentille printanière
d Frühlingsfingerkraut
l *Potentilla verna*

5183 spring draba
f drave printanière
d Frühlingshungerblümchen
l *Draba verna*

5184 spring figwort
f scrofulaire printanière
d Frühlingsbraunwurz
l *Scrophularia vernalis*

5185 spring gentian
f gentiane printanière

d Frühlingsenzian
l *Gentiana verna*

5186 spring heath
f bruyère des neiges
d Alpenheiderich; Frühlingsheide
l *Erica carnea*

5187 spring savory
f sarriette des champs
d Steinquendel
l *Satureia acinos; Calamintha acinos*

5188 spring snowflake
f nivéole printanière
d Frühlingsknotenblume; Märzbecher; Märzblume; Schneeglöckchen; Schneelilie
l *Leucojum vernum*

5189 spruce
f épicéa
d Fichte
l *Picea*

5190 spurflower
f plectranthe
d Hahnensporn; Harfenstrauch; Mottenpflanze; Mottenkönig
l *Plectranthus*

5191 spurge family
f euphorbiacées
d Euphorbiazeen; Wolfsmilchgewächse
l *Euphorbiaceae*

5192 spurge laurel (daphne)
f daphné lauréole; lauréole mâle; lauréole des bois
d gelber Seidelbast; Lorbeerkraut; Lorbeerseidelbast; Zeiland; Zindelbast
l *Daphne laureola*

5193 spur gentian
f halénie
l *Halenia*

5194 spurge olive
f camélée
d Zeiland; Zwergölbaum
l *Cneorum*

5195 spurred coral root
d Widerbart
l *Epipogon aphyllum*

5196 spurry
f spargoule; spergule
d Spark; Spergel; Spörgel
l *Spergula*

5197 square-pod deervetch; winged pea
f lotier rouge
d englische Erbse; rote Spargelbohne; Spargelklee
l *Tetragonolobus purpureus; Lotus tetragonolobus*

5198 square-stalked St.-John's-wort
d Kantenhartheu
l *Hypericum quadrangulum*

5199 square-stemmed St.-John's-wort
d Flügelhartheu
l *Hypericum acutum*

* **squash → 2568**

* **squawbush → 4458, 6016**

* **squaw-weed → 2651**

5200 squill
f scille
d Blaustern
l *Scilla*

5201 squinancy
f herbe à l'esquinancie; aspérule des sables
d Hügelmeister; Hügelmeier
l *Asperula cynanchica*

* **squirrel-tail grass → 2350**

5202 squirting cucumber
f concombre d'âne; concombre d'attrape; giclet
d Eselsgurke; Spritzgurke; Vexiergurke
l *Ecballium elaterium*

* **staff tree → 608**

5203 staff-tree family
f célastracées
d Spindelbaumgewächse; Spindelsträucher
l *Celastraceae*

5204 staghorn evernia
f évernie des pruniers; lichen de prunellier
d Pflaumenbandflechte
l *Evernia prunastri*

5205 staghorn fern
f platycérium
d Geweihfarn
l *Platycerium*

5206 staghorn sumac
f sumac amarante
d Hirschkolbensumach
l *Rhus hirta; Rhus typhina*

5207 stalked bur grass
d traubenblütiges Klettengras
l *Tragus racemosus*

5208 stalked orach
d Salzmelde
l *Atriplex pedunculata*

5209 stalky berula
d Berle
l *Berula erecta*

5210 standard crested wheatgrass
f chiendent désertique
l *Agropyron desertorum*

5211 star apple; golden leaf
f jacquinie
d Goldblatt; Sternapfel
l *Chrysophyllum*

5212 starbush
f turraée
d Turräe
l *Turraea*

5213 star cactus
f astrophyte
d Sternkaktus
l *Astrophytum*

5214 starch grape hyacinth
f muscari à grappe
d Strauchhyazinthe
l *Muscari racemosum*

5215 starchy-sweet corn
d zuckersüsser Stärkemais
l *Zea mays* convar. *amyleosaccharata*

5216 star clusters
f pentas
d Fünfling
l *Pentas*

5217 star duckweed; ivy-leaved duckweed
d dreifurchige Wasserlinse; untergetauchte Wasserlinse
l *Lemna trisulca*

5218 starflower
f trientale
d Siebenstern
l *Trientalis*

5219 starfruit; thrumwort
f damasonie commune
l *Damasonium stellatum*

5220 star glory
f quamoclit
d Sternwinde; Prunkwinde
l *Quamoclit*

5221 star grass
f alétris
d Fieberwurzel
l *Aletris*

5222 star-headed sedge
d Sternriedgras
l *Carex stellulata*

5223 star hyacinth (squill)
f scille aimable
d schöne Meerzwiebel; schöner Blaustern
l *Scilla amoena*

5224 star jasmine
f trachélosperme
d Sternjasmin
l *Trachelospermum*

5225 star jelly
f gélatine de terre; nostoc commune
d gemeine Gallertalge
l *Nostoc commune*

5226 star lily
f leucocrine; weldénie
d Koloradosandlilie; Colorado-Sandlilie
l *Leucocrinum*

5227 star magnolia
f magnolia étoilé
d Sternmagnolie
l *Magnolia stellata*

5228 star-of-Bethlehem
f ornithogale
d Milchstern; Vogelmilch
l *Ornithogalum*

5229 star pink
f chironie
d Chironie
l *Chironia*

5230 starry cerastium
f céraiste des champs
d Ackerhornkraut
l *Cerastium arvense*

5231 star saxifrage
f saxifrage étoilée
d Sternsteinbrech
l *Saxifraga stellaris*

5232 star thistle
f centaurée chausse-trappe; chardon étoilé
d Sternflockenblume
l *Centaurea calcitrapa*

5233 starthorn
f hygrophile
d Wasserfreund
l *Hygrophila*

5234 star tree
f astrone
l *Astronium*

5235 starwort
f stellaire
d Sternmiere; Sternkraut
l *Stellaria; Alsine*

5236 starwort chickweed
d dreigriffeliges Hornkraut
l *Cerastium cerastioides*

5237 Staunton vine
f stauntonie
d Stauntonia
l *Stauntonia*

5238 stemless thistle
f cirse acaule
d Erdkratzdistel; stengellose Distel; stengellose Kratzdistel
l *Cirsium acaule*

5239 stenanthium
f sténanthe
l *Stenanthium*

5240 stenogynum
f sténogyne
l *Stenogynum*

5241 stephanandra
f stéphanandre
d Kranzspiere
l *Stephanandra*

5242 sternbergia
f sternbergie
d Gewitterblume; Sternbergie
l *Sternbergia*

5243 Steven gypsophila
f gypsophile de Steven
d Stevens Gipskraut
l *Gypsophila stevenii*

5244 stevia
f stévie
d Palafoxie; Stevie
l *Stevia*

5245 stewartia
f stuartie
d Scheinkamelie; Stewartie
l *Stewartia; Stuartia*

* **sticking Roger → 633**

5246 stickseed
f bardanette
d Igelsame
l *Lappula*

* **sticktight → 521**

* **sticky alder → 2072**

5247 sticky groundsel
f séneçon visqueux
d klebriges Greiskraut; klebriges Kreuzkraut
l *Senecio viscosus*

5248 sticky gypsophila
f gypsophile visqueuse
d klebriges Schleierkraut
l *Gypsophila viscosa*

5249 sticky sage
f sauge glutineuse
d Klebkraut; klebrige Salbei
l *Salvia glutinosa*

5250 sticta
f sticta
d Lungenflechte
l *Sticta*

5251 stiff grass; hard meadow grass
f fétuque raide
d Steifgras; Starrgras
l *Scleropoa*

5252 stiff-hair wheat grass
f chiendent chevelu
d Flaumquecke
l *Agropyron trichophorum*

5253 stilt palm
f iriartée

d Stelzenpalme
l *Iriartea*

* **stinging nettle → 574**

5254 stink dragon; arum
f draconcule
d Drachenwurz
l *Dracunculus*

5255 stink grass
d grosses Liebesgras
l *Eragrostis cilianensis; Eragrostis megastachya; Eragrostis major*

5256 stinkhorn
f satyre puant; satyre fétide; morille fétide; phalle puant
d Gichtmorchel; Gichtschwamm; Rutenmorchel; Stinkmorchel
l *Phallus impudicus; Ithyphallus impudicus*

* **stinking cedar → 2303**

5257 stinking wall rocket
f diplotaxe des murailles
d Mauerdoppelsame; Mauerrampe
l *Diplotaxis muralis*

5258 stock
f giroflée; matthiole
d Levkoje
l *Matthiola*

5259 stokesia
f stokésie
d Stokesie
l *Stokesia*

5260 stone bramble; roebuck berry
f ronce des rochers
d Steinbeere; echte Steinbeere
l *Rubus saxatilis*

5261 stone cress
f éthionème
d Steintäschel
l *Aethionema*

5262 stonecrop
f rosulaire
d Dickröschen
l *Rosularia*

5263 stonecrop
f orostachys
d Sternwurz
l *Orostachys*

5264 stonecrop
f orpin
d Fetthenne; Mauerpfeffer
l *Sedum*

5265 stone mint
f cunile
l *Cunila*

* **stone parsley → 2835**

5266 stonewood
f tarriétie
l *Tarrietia*

5267 stonewort
f chara; charagne
d Armleuchtergewächs
l *Chara*

5268 stool iris
f iris à tige sans feuilles
d nacktstengelige Schwertlilie
l *Iris aphylla*

* **storax → 5056**

5269 storax family
f styracacées
d Styrakazeen; Styraxgewächse
l *Styracaceae*

* **stork's-bill → 91, 4103**

5270 stranvaesia
f stranvésie
d Stranvesie
l *Stranvaesia*

5271 strapwort
f corrigiole littorale
d Strandhirschsprung; Uferhirschsprung
l *Corrigiola littoralis*

5272 strawberry
f fraisier
d Erdbeere
l *Fragaria*

5273 strawberry blite
f épinard fraise
d echter Erdbeerspinat
l *Chenopodium foliosum*

* **strawberry blite → 675**

5274 strawberry clover
f trèfle porte-fraise

d Erdbeerklee
l *Trifolium fragiferum*

5275 strawberry ground cherry; Chinese lantern plant; alkekengi
f coqueret alkékenge
d Alkekengi; gemeine Blasenkirsche; wilde Blasenkirsche; Boberelle; Judenkirsche; Korallenkirsche; Teufelspuppe
l *Physalis alkekengi; Physalis francheti*

5276 strawberry madrone; strawberry tree
f arbousier commun; arbousier des Pyrénées; arbre aux fraises
d Erdbeerbaum
l *Arbutus unedo*

5277 strawberry saxifrage; Aaron's-beard
f saxifrage sarmenteuse
d Judenbart
l *Saxifraga sarmentosa*

* **strawberry shrub → 5386**

5278 strawberry-shrub family
f calycanthacées
d Gewürzstrauchgewächse; Kalykanthazeen
l *Calycanthaceae*

* **strawberry tree → 5276**

5279 strawflower
f immortelle à bractées
d Gartenstrohblume
l *Helichrysum bracteatum*

5280 straw foxglove
f digitale jaune
d gelber Fingerhut
l *Digitalis lutea*

5281 streaky-leaved garlic
d steifer Lauch
l *Allium strictum*

5282 stream-bank sedge
f laîche des rivages
d Uferriedgras; Ufersegge
l *Carex riparia*

5283 striped crotalaria
f crotalaire striée
l *Crotalaria mucronata; Crotalaria striata*

5284 striped maple
f érable jaspé; érable de Pennsylvanie
d pennsylvanischer Ahorn
l *Acer pennsylvanicum*

5285 strombosia
f strombosie
l *Strombosia*

5286 strongbark
f bourrérie
l *Bourreria*

5287 strophanthus
f strophanthus
d Strophant
l *Strophanthus*

5288 stropharia
f stropharia
d Träuschling
l *Stropharia*

5289 stylewort
f stylidier
d Griffelblume; Säulenblume
l *Stylidium*

* **styrax → 5056**

5290 Suakwa vegetable sponge
f louffa cylindrique
d Luffa-Gurke; Schwammgurke
l *Luffa cylindrica*

* **submentose mushroom → 6081**

5291 subterranean clover
f trèfle souterrain
d unterirdischer Klee; bodenfrüchtiger Klee
l *Trifolium subterraneum*

5292 suckling clover
f trèfle douteux
d Fadenklee; kleiner Klee
l *Trifolium dubium*

5293 Sudan cola nut
f arbre à kola
d Kolabaum
l *Cola acuminata*

5294 Sudan grass
f sorgho menu; sorgho du Soudan
d Sudangras
l *Sorghum vulgare sudanense; Holcus sorghum sudanensis; Holcus sudanensis*

5295 Sudetic pedicularis
f pédiculaire des Sudètes

d Sudetenläusekraut
l *Pedicularis sudetica*

5296 Sudetic wood rush
f luzule des Sudètes
d schwärzliche Marbel; Sudetenhainsimse
l *Luzula sudetica*

5297 sugar apple; sweetsop; custard apple
f corossolier à fruit hérissé; grand corossolier
d Rahmapfel; Süssack; Zuckerapfel; Zimtapfel
l *Annona squamosa*

5298 sugar beet
f betterave sucrière; betterave à sucre
d Zuckerrübe
l *Beta saccharifera*

5299 sugar blade kelp
f laminaire à sucre
d Zuckertang; Zuckerriementang
l *Laminaria saccharina*

* **sugar bush → 5308**

5300 sugar cane
f canne à sucre
d Zuckerrohr; echtes Zuckerrohr
l *Saccharum officinarum*

5301 sugar corn; sweet corn
f maïs sucré
d Zuckermais
l *Zea mays* var. *saccharata*

5302 sugar date; India date
f dattier sauvage
d indische Waldpalme; Walddattelpalme
l *Phoenix sylvestris*

5303 sugar grass
f pollinia
d Zuckergras
l *Pollinia*

5304 sugar maple
f érable à sucre
d Zuckerahorn
l *Acer saccharum*

5305 sugar palm
f arenga
d Zuckerpalme
l *Arenga; Saguerus*

5306 sugar pine
f pin à sucre; pin de Lambert
d Lambertskiefer; Zuckerkiefer
l *Pinus lambertiana*

5307 sugarplum tree
f lagunaire
l *Lagunaria*

5308 sugar protea; honeyflower; sugar bush
f protée mellifère
d Zuckerbusch
l *Protea mellifera*

5309 sugar sorghum; sweet sorghum
f sorgho sucré
d Zuckersorghum
l *Sorghum saccharatum*

* **sugi → 3029**

* **sulla → 5310**

5310 sulla sweet-vetch; sulla
f sainfoin d'Espagne; sainfoin d'Italie
d italienischer Hahnenkopf
l *Hedysarum coronaria*

5311 sulphur clover
d gelblichweisser Klee
l *Trifolium ochroleucum*

5312 sulphurous mushroom
f polypore soufré
d Schwefelporling
l *Polyporus sulphureus*

5313 sumac
f sumac
d Essigbaum
l *Rhus*

* **sumac family → 1043**

5314 Sumatra-dragonsblood
f rotang sang-dragon
d Drachenblutpalme
l *Daemonorops draco*

5315 Sumatra snowbell; benjamin tree
f aliboufier du Sumatra; aliboufier benzoïn
d Benzoë-Storaxbaum
l *Styrax benzoin*

5316 sumbul
f dorême
d Ammoniakpflanze
l *Dorema*

5317 **summer adonis**
f adonide d'été
d Sommeradonisröschen; Sommerblutströpfchen; Feuerröschen
l *Adonis aestivalis*

5318 **summer cypress**
f kochie
d Besenkraut; Radmelde; Sommerzypresse
l *Kochia*

5319 **summer hyacinth**
f galtonia
d Riesenhyazinthe; Sommerhyazinthe
l *Galtonia*

* **summer lilac → 892**

5320 **summer phlox**
f phlox paniculé
d Staudenphlox; Staudenflammenblume; Herbstflieder
l *Phlox paniculata*

5321 **summer savory**
f sarriette de jardins
d Bohnenkraut; Gartenpfefferkraut; Gartenquendel; Kölle; Pfefferkraut; wilder Ysop
l *Satureia hortensis*

5322 **summer snowflake**
f nivéole d'été
d Sommerknotenblume
l *Leucojum aestivum*

5323 **summer sweet; white alder**
f cléthra
d Laubheide; Scheineller; Scheinerle
l *Clethra*

* **summer truffle → 4619**

* **summer vetch → 3810**

5324 **sump weed; marsh elder**
f faux quinquina
d Sumpfholunder
l *Iva*

5325 **sundew; dew plant**
f drosère; rossolis
d Sonnenkraut; Sonnentau
l *Drosera*

5326 **sundew family**
f droséracées
d Droserazeen; Sonnentaugewächse
l *Droseraceae*

5327 **sundial lupine**
f lupin vivace
d perennierende Lupine
l *Lupinus perennis*

* **sundrops → 2141**

5328 **sun euphorbia; sun spurge**
f euphorbe des vignes; euphorbe réveille-matin; réveille-matin
d Sonnenwolfsmilch
l *Euphorbia helioscopia*

5329 **sunflower**
f tournesol
d Sonnenblume; Sonnenrose
l *Helianthus*

5330 **Sungari clematis**
f clématite de Soungari
l *Clematis songarica*

5331 **sunn crotalaria; sunn hemp**
f crotalaire; chanvre du Bengale
d bengalischer Hanf; Bombayhanf; Madrashanf; ostindischer Hanf; Sunn-Hanf
l *Crotalaria juncea*

5332 **sunpitcher**
d Sumpfkrug
l *Heliamphora*

5333 **sunray**
f héliptère
d Sonnenflügel
l *Helipterum*

5334 **sun rose**
f hélianthème
d Sonnenröschen
l *Helianthemum*

* **sun spurge → 5328**

5335 **supplejack**
f serjania
d Serjania
l *Serjania*

5336 **supplejack**
f berchémie
d Berchemie
l *Berchemia*

5337 surf grass
f phyllospadix
l *Phyllospadix*

* **Surinam cherry → 4206**

5338 Surinam quassia
f quassier amer; bois de Surinam
d brasilianischer Quassiabaum
l *Quassia amara*

5339 Swainson pea
f swainsonie
d Swainsonie
l *Swainsona*

5340 swallowwort
f cynanque
d Hundswürger; Schwalbenwurz
l *Cynanchum*

5341 swamp bedstraw
d Moorlabkraut
l *Galium uliginosum*

5342 swamp blackberry
f framboisier des marais
l *Rubus hispidus*

5343 swamp chestnut oak; basket oak
f chêne prin; chêne châtaignier des marais
d Korbeiche
l *Quercus michauxi; Quercus prinus*

5344 swamp horsetail
f prêle des bourbiers
d Schlammschachtelhalm; Teichschachtelhalm
l *Equisetum limosum*

5345 swamp milkweed
f asclépiade incarnate
d fleischrote Seidenpflanze
l *Asclepias incarnata*

* **swamp oak → 3982, 4200**

5346 swamp pink
f hélonie
d Helonie
l *Helonias*

* **swamp red currant → 208**

5347 swamp white oak
f chêne bicolore
d zweifarbige Eiche
l *Quercus bicolor*

5348 swan orchid; swan-plant
f cycnoque; cycnoche
d Schwanblume
l *Cycnoches*

* **swede → 4626**

* **Swedish clover → 164**

5349 Swedish coffee
d Kaffeewicke; Kaffeetragant; schwedischer Kaffee
l *Astragalus boeticus*

5350 Swedish mountain ash
f sorbier intermédiaire
d schwedische Eberesche
l *Sorbus intermedia; Sorbus scandica*

5351 sweet acacia
f acacia de Farnèse
d Antillenakazie; Antillenkassie
l *Acacia farnesiana*

5352 sweet alyssum
f alysse maritime; alysse odorante
d Duftsteinrich
l *Lobularia maritima; Alyssum maritimum*

5353 sweet autumn clematis
f clématite à panicules
d rispenblütige Waldrebe; rispige Waldrebe
l *Clematis paniculata*

5354 sweet basil; hoary basil
f basilic commun; herbe royale; oranger des savetiers
d echtes Basilienkraut; Gartenhirnkraut; Basilikum
l *Ocimum basilicum*

5355 sweet bay magnolia
f magnolia de Virginie
d Biberbaum; Sumpfsassafras; Weisslorbeer
l *Magnolia virginiana; Magnolia glauca*

5356 sweet-berry honeysuckle; blue-berried honeysuckle
f chèvrefeuille à fruits bleus
d blaue Heckenkirsche; blaues Geissblatt
l *Lonicera coerulea*

5357 sweet birch; cherry birch
f bouleau bleu; bouleau merisier

d Hainbirke; Zuckerbirke
l *Betula lenta*

5358 **sweet brahea palm**
f brahée
l *Brahea dulcis*

5359 **sweetbrier (rose); eglantine rose**
f églantier odorant; églantier rouge; rosier églantine; rosier rouillé
d schottische Zaunrose; Weinrose
l *Rosa eglanteria*

5360 **sweet cassawa**
f manioc doux
d süsser Maniok
l *Manihot dulcis*

* **sweet cassawa → 67**

* **sweet cherry → 3544**

5361 **sweet cicely**
f myrrhis odorant
d Süssdolde; Myrrhenkerbel
l *Myrrhis odorata*

5362 **sweet clover; melilot**
f mélilot
d Honigklee; Steinklee; Melilotenklee
l *Melilotus*

* **sweet corn → 5301**

5363 **sweet devil's-claws**
f bicorne odorante
d wohlriechendes Gemshorn
l *Proboscidea fragrans; Martynia fragrans*

5364 **sweet fern**
f comptonie
d Farnmyrte; Farnstrauch
l *Comptonia*

5365 **sweet flag**
f acore
d Kalmus
l *Acorus*

5366 **sweet gale**
f galé
d echter Gagel; Gagelstrauch; Wachsstrauch
l *Myrica gale*

5367 **sweet grass**
f hiérochloé
d Darrgras; Mariengras
l *Hierochloe*

5368 **sweet grass; sweet-scented grass; holy grass**
f houlque odorante
d wohlriechendes Mariengras
l *Hierochloe odorata*

* **sweet grass → 3476**

5369 **sweet gum (tree)**
f liquidambar
d Amberbaum
l *Liquidambar*

5370 **sweet honeysuckle**
f chèvrefeuille des jardins
d italienisches Geissblatt; wohlriechendes Geissblatt; Jelängerjelieber; echtes Geissblatt; südliches Geissblatt
l *Lonicera caprifolium*

5371 **sweet iris**
f iris pâle
d blasse Schwertlilie
l *Iris pallida*

5372 **sweetish lactarius**
f lactaire douceâtre
d süsslicher Milchling
l *Lactarius subdulcis*

5373 **sweetleaf**
f symplocos
d Rechenblume; Saphirbeere
l *Symplocos*

5374 **sweetleaf family**
f symplocacées
d Symplokazeen
l *Symplocaceae*

5375 **sweet marjoram**
f marjolaine
d Mairan; Marjoran; Wurstkraut; Sommermajoran
l *Majorana hortensis; Origanum majorana*

* **sweet mignonette → 1468**

5376 **sweet mock orange**
f seringa à bouquets; seringa commun; jasmin bâtard
d wohlriechender Pfeifenstrauch; falscher Pfeifenstrauch; Zimtröschen
l *Philadelphus coronarius*

5377 sweet orange
f oranger doux
d Apfelsine; Orange
l *Citrus sinensis*

5378 sweet pea
f gesse odorante; pois de senteur
d spanische Wicke; wohlriechende Platterbse; Edelwicke; Gartenwicke
l *Lathyrus odoratus*

5379 sweet plantain
f plantain moyen
d mittlerer Wegerich; Weidenwegerich
l *Plantago media*

5380 sweet potato
f batate; patate douce
d Batate; Camote; indische Kartoffel; Knollenwinde; süsse Kartoffel; Süsskartoffel
l *Ipomoea batatas*

5381 sweetroot
f osmorrhize
l *Osmorrhiza*

5382 sweet sansevieria
d Guinea-Bogenhanf
l *Sansevieria guineensis; Sansevieria thyrsiflora*

5383 sweet scabious
f scabieuse des jardins
d Purpurskabiose
l *Scabiosa atropurpurea*

5384 sweet-scented bedstraw
d dreiblütiges Labkraut
l. *Galium triflorum*

*** sweet-scented bush → 5386**

5385 sweet-scented clitocybe
f clitocybe odorant
d Anistrichterling; grüner Anistrichterling
l *Clitocybe odora*

*** sweet-scented grass → 5368**

5386 sweet shrub; sweet-scented bush; strawberry shrub
f calycanthe
d Gewürzstrauch
l *Calycanthus*

*** sweetsop → 5297**

*** sweet sorghum → 5085, 5309**

5387 sweet spire
f itéa
d Rosmarienweide
l *Itea*

5388 sweet sultan
f centaurée ambrette
d Bisamflockenblume; Moschusblume
l *Centaurea moschata*

*** sweet sultan → 4355**

5389 sweet vernal grass
f flouve odorante
d gemeines Ruchgras; Wiesenruchgras; wohlriechendes Ruchgras
l *Anthoxanthum odoratum*

5390 sweet vetch; French honeysuckle
f sainfoin
d Süssklee; Hahnenkopf
l *Hedysarum*

5391 sweet viburnum
f viorne odorante
l *Viburnum odoratissimum*

5392 sweet violet
f violette odorante
d wohlriechendes Veilchen; Märzveilchen
l *Viola odorata*

5393 sweet William
f bouquet parfait; jalousie; œillet de poète
d Bartnelke; Büschelnelke; Studentenblume
l *Dianthus barbatus*

*** sweet-william campion → 5394**

5394 sweet-william silene; sweet-william campion
f silène à bouquets
d Gartenleimkraut; Morgenröschen
l *Silene armeria*

5395 sweet woodruff
f aspérule odorante; petit muguet
d echter Waldmeister
l *Asperula odorata*

5396 sweet wormwood
f armoise annuelle
d einjähriger Beifuss; gemeiner Bei-

fuss
l *Artemisia annua*

5397 swertia
f swertie
d Tarant; Sumpfstern
l *Swertia*

* **swine's-cress → 5783**

* **Swiss centaury → 4486**

5398 Swiss chard; white beet
f poirée à carde
d Rippenmangold; römischer Kohl; römische Bete; Schweizer Mangold; Stengelmangold; Stielmangold
l *Beta vulgaris* var. *cicla*

5399 Swiss forget-me-not
f myosotis à grandes fleurs
l *Myosotis dissitiflora*

5400 Swiss mountain pine
f pin de montagne
d Bergkiefer; Krummholzkiefer
l *Pinus mugo; Pinus montana*

5401 Swiss rock jasmine
f androsace de Suisse
d Schweizer Mannsschield
l *Androsace helvetica*

5402 Swiss stone pine
f pin cembro; pin des rochers de Suisse
d Zierbelkiefer; Zierbelnusskiefer; Zierbelnusstanne
l *Pinus cembra*

5403 switch grass
f panic érigé
d rutenförmige Hirse
l *Panicum virgatum*

5404 swollen duckweed; gibbous duckweed
d buckelige Wasserlinse
l *Lemna gibba*

* **sword bean → 5406**

5405 sword fern
f néphrolépis
d Nephrolepis
l *Nephrolepis*

* **sword grass → 4956**

5406 sword jack bean; sword bean
d Schwertbohne
l *Canavalia gladiata*

5407 sword-leaved inula
f aunée ensiforme
d schwertblättriger Alant
l *Inula ensifolia*

* **sword plant → 6084**

* **sycamore → 4215**

* **sycamore maple → 4217**

5408 sycomore (fig)
f figuier sycomore
d Eselfeige; Maulbeerfeige(nbaum); ägyptischer Feigenbaum; Sykomore
l *Ficus sycomorus*

5409 Sydney acacia
f acacia à longues feuilles
d langblättrige Akazie
l *Acacia longifolia*

5410 sylvan goatsbeard
f barbe de bouc commune
d Johanniswedel; Waldgeissbart; Ziegenbart
l *Aruncus sylvestris; Spiraea aruncus*

5411 sylvan horsetail
d Waldschachtelbaum
l *Equisetum sylvaticum*

5412 sylvan violet
d Waldveilchen
l *Viola sylvestris*

5413 sympetalae
f sympétales
d Sympetalen; Verwachsenkronblättrige
l *Sympetalae*

5414 symphonia
f symphonie
d Schweinsgummibaum
l *Symphonia*

5415 Syria ixiolirion
f ixiolire montagnard
d Bergixlilie
l *Ixiolirion montanum*

5416 Syrian ash
f frêne de Syrie

d syrische Esche
l *Fraxinus syriaca*

5417 Syrian cephalaria
f céphalaire de Syrie
d syrischer Schuppenkopf
l *Cephalaria syriaca*

5418 Syrian juniper
f genévrier de Syrie
d pflaumfrüchtiger Wacholder
l *Juniperus drupacea*

5419 syrup jubaea; coquito
f cocotier du Chili
d Coquito
l *Jubaea spectabilis*

T

5420 **Tabernaemontanus bulrush**
d rauhe Simse; Salzsimse
l *Scirpus tabernaemontani*

5421 **tacca**
f tacca
d Tacca
l *Tacca*

* **tagua nut → 1447**

5422 **tail grape**
f artabotrys
d Birnentraube; Brottraube; Klimm-Ylang-Ylang
l *Artabotrys*

5423 **talauma**
f talauma
d Champak
l *Talauma*

5424 **talipot palm**
f talipot
d Tallipotbaum; Tallipotpalme; Schattenpalme; Fächerpalme
l *Corypha umbraculifera*

5425 **tall buttercup**
f bouton d'or; renoncule âcre
d Brennkraut; scharfer Hahnenfuss; Goldknöpfchen
l *Ranunculus acris; Ranunculus acer*

* **tall fescue → 2472**

5426 **tall melilot**
f grand mélilot
d hoher Honigklee; hoher Steinklee
l *Melilotus altissimus*

5427 **tall oat grass**
f avoine élevée; fromental
d französisches Raigras; hoher Glatthafer; hoher Wiesenhafer
l *Arrhenatherum elatius*

5428 **tallowwood**
f oranger des falaises; ximénie d'Amérique
l *Ximenia americana*

5429 **tall skullcap**
f toque élevée
d hohes Helmkraut
l *Scutellaria altissima*

5430 **tall stewartia**
f stuartie monodelphe
l *Stewartia monadelpha*

5431 **tall violet**
f violette élevée
d hochwüchsiges Veilchen
l *Viola elatior*

5432 **tall wheatgrass**
f chiendent allongé
d langährige Quecke
l *Agropyron elongatum*

* **tamarack → 1977**

5433 **tamarind**
f tamarinier
d Tamarinde
l *Tamarindus*

5434 **tamarind**
f tamarinier
d Tamarinde
l *Tamarindus indica*

5435 **tamarisk**
f tamaris
d Tamariske
l *Tamarix*

5436 **tamarisk family**
f tamaricacées
d Tamarikazeen; Tamariskgewächse
l *Tamaricaceae*

5437 **Tanaka stephanandra**
f stéphanandre de Tanaka
d gelappte Kranzspiere
l *Stephanandra tanakae*

* **tangerine → 3468**

5438 **Tangier pea vine**
f gesse de Tanger
d tangerische Platterbse; afrikanische Wicke
l *Lathyrus tingitanus*

5439 **tan oak**
f pasanie
d Südeiche
l *Pasania; Lithocarpus*

5440 **tansy**
f tanaisie
d Rainfarn
l *Tanacetum*

5441 tansy chrysanthemum
f chrysanthème à grandes feuilles
d grossblättrige Wucherblume
l *Pyrethrum macrophyllum; Chrysanthemum macrophyllum*

5442 tansy mustard
f sagesse des chirurgiens
d Besenrauke; Raukensenf; Sophienkraut
l *Descurainia*

5443 tansy phacelia
f phacélie à feuilles de tanaisie
d rainfarnblättrige Büschelblume
l *Phacelia tanacetifolia*

*** tansy ragwort → 4401**

*** tara vine → 748**

5444 tarbush
f flourensie
l *Flourensia*

*** tare vetch → 5531**

*** taro → 2013**

5445 tarragon
f dragone; estragon; armoise petit-dragon
d Dragon; Estragon
l *Artemisia dracunculus*

*** Tartar dogwood → 5454**

5446 Tartary buckwheat
f sarrasin de Tartarie
d tatarischer Buchweizen; tatarisches Heidekorn
l *Fagopyrum tataricum*

5447 Tasmanian blue eucalyptus
f gommier bleu; eucalyptus commun
d Eisenveilchenbaum; blauer Gummibaum; Fieberbaum; Fieberheilbaum
l *Eucalyptus globulosus*

5448 Tasmanian Fitzroya
f alerce de Tasmanie
d Archers Fitzroya
l *Fitzroya archeri*

5449 Tasmanian podocarpus
f podocarpe montagnard
d Alpensteineibe
l *Podocarpus alpinus*

5450 tassel(ed) grape hyacinth
f ail à toupet; ail des chiens; muscari à toupet; muscari chevelu
d schopfige Bisamhyazinthe; Schopf-träubel
l *Muscari comosum*

5451 tassel flower
f émilie
d Emilie
l *Emilia*

5452 Tatarian bread
f chou marin de Tartarie
d tatarischer Meerkohl
l *Crambe tatarica*

*** Tatarian catchfly → 5459**

5453 Tatarian cephalaria
f céphalaire de Tartarie
d tatarischer Schuppenkopf
l *Cephalaria tatarica*

5454 Tatarian dogwood; Tartar dogwood
f cornouiller blanc; cornouiller à fruits blancs
d Weissbeerenhartriegel
l *Cornus alba*

5455 Tatarian honeysuckle
f chèvrefeuille de Tartarie
d tatarische Heckenkirsche; Tatarengeissblatt
l *Lonicera tatarica*

5456 Tatarian maple
f érable de Tartarie
d tatarischer Ahorn
l *Acer tataricum*

5457 Tatarian saltbush
f arroche de Tartarie
d tatarische Melde; Tatarenmelde
l *Atriplex tatarica*

5458 Tatarian sea lavender
f statice de Tartarie
d tatarischer Strandflieder
l *Goniolimon tataricum*

5459 Tatarian silene; Tatarian catchfly
f silène de Tartarie
d Tatarenleimkraut; tatarisches Leimkraut
l *Silene tatarica*

5460 Taurus candytuft
d Krimschleifenblume
l *Iberis taurica*

5461 Taurus cerastium
f céraiste de Bieberstein
d Biebersteins Hornkraut
l *Cerastium biebersteini*

5462 Taurus cotton thistle
f onoporde de Crimée
d Krimeseldistel
l *Onopordum tauricum*

*** tawny day lily → 2334**

5463 taxodium family
f taxodiacées
d Taxodiazeen; Sumpfzypressen-gewächse
l *Taxodiaceae*

5464 tea family
f théacées
d Teegewächse; Teestrauchgewächse
l *Theaceae*

5465 teak
f teck
d Teakbaum; Tiekbaum
l *Tectona*

5466 tea-leaved willow
d zweifarbige Weide
l *Salix phylicifolia*

5467 tea (plant)
f théier
d Teebaum; Teestrauch
l *Thea*

5468 tea rose
f rosier à odeur de thé; rosier thé
d Teerose
l *Rosa odorata*

5469 teasel
f cardère
d Kardendistel; Karde
l *Dipsacus*

5470 teasel clover
d kleinblütiger Klee
l *Trifolium parviflorum*

5471 teasel family
f dipsacacées
d Dipsazeen; Kardengewächse
l *Dipsacaceae*

5472 tea tree
f leptosperme
d Südseemyrte
l *Leptospermum*

5473 teff
f teff
d abessinisches Liebesgras; Taf(gras)
l *Eragrostis abyssinica*

*** teff → 3385**

*** telegraph plant → 5474**

5474 telegraph tick clover; telegraph plant
f trèfle oscillant; sainfoin oscillant
d Telegraphenpflanze; Wandelklee
l *Desmodium gyrans*

5475 teosinte
f téosinte
d Teosinte
l *Euchlaena*

5476 terebinth pistache
f térébinthe; pistachier térébinthe
d Terpentinpistazie
l *Pistacia terebinthus*

5477 tetracera
f tétracère
d Raspelstrauch
l *Tetracera*

5478 Texas bean
d Teparybohne
l *Phaseolus acutifolius*

5479 Texas black walnut
f noyer des rochers
d Felsenwalnuss
l *Juglans rupestris*

5480 thatch palm
f thrinace; thrinax
d Schilfpalme; Thrinaxpalme
l *Thrinax*

*** thelygona family → 1770**

5481 thelygonum
f thélygone
d Hundskohl
l *Thelygonum*

5482 thelymitra
f thélymitre
l *Thelymitra*

5483 thermopsis
f thermopside
d Fuchsbohne
l *Thermopsis*

5484 thicket creeper
d gemeine Jungfernrebe; wilder Wein
l *Parthenocissus inserta*

5485 thicket hawthorn; scarlet hawthorn
f épine écarlate
d Scharlachdorn
l *Crataegus coccinea*

5486 thingan merawan
f hopée odorante
l *Hopea odorata*

5487 thistle
f cirse; chardon
d Elfenbeindistel; Kratzdistel
l *Cirsium; Chamaepeuce*

5488 thistle broomrape
d Distelsommerwurz
l *Orobanche reticulata*

5489 thitka
f pentace
l *Pentace*

* **thorn apple → 1811, 3102**

5490 thornless blackberry
f ronce du Canada
d kanadische Brombeere
l *Rubus canadensis*

5491 thorny bamboo
f bambou commun
l *Bambusa arundinacea*

5492 thorny ononis; spiny restharrow
f ononis épineux; arrête-bœuf; mâche noire
d Hauhechel
l *Ononis spinosa*

5493 thoroughwax; thorowax; hare's-ear
f buplèvre
d Hasenohr
l *Bupleurum*

* **thoroughwort → 732**

5494 thoroughwort pennycress
d durchwachsenes Hellerkraut; Öhrenhellerkraut; durchwachsenblättriges Täschelkraut
l *Thlaspi perfoliatum*

* **thorowax → 5493**

5495 thorowort pondweed
d durchwachsenes Laichkraut
l *Potamogeton perfoliatus*

5496 thouinia
f thouinier
l *Thouinia*

5497 thread clover
f trèfle filiforme
l *Trifolium filiforme*

5498 thread rush
d Fadenbinse; fadenförmige Binse
l *Juncus filiformis*

5499 three-awn; three-awned grass
f aristide
d Borstengras
l *Aristida*

5500 three-flowered gladiolus
f glaïeul des marais
d Sumpfsiegwurz
l *Gladiolus paluster*

5501 three-stamened waterwort
f élatine à trois étamines
d dreimänniger Tännel; Kreuztännel
l *Elatine triandra*

5502 thrift
f armérie
d Grasblume; Grasnelke
l *Armeria*

5503 throatwort
f trachélie
d Halskraut
l *Trachelium*

* **throatwort → 1643**

* **thrumwort → 5219**

5504 Thunberg day lily
f hémérocalle de Thunberg
d Thunbergs Taglilie
l *Hemerocallis thunbergi*

5505 Thunberg kudzu bean; Thunberg kudzu vine
f puéraire hirsute; kudzu du Japon
d Kopoubohne; Thunbergs Pueraria

l *Pueraria thunbergiana; Pueraria hirsuta; Dolichos japonicus*

* **Thunberg pine → 3023**

5506 **Thunberg spirea**
f spirée de Thunberg
d Thunbergs Spierstrauch
l *Spiraea thunbergi*

5507 **Thuringian (tree) mallow**
f lavatère de Thuringe
d thüringische Strauchmalve; Thüringer Strauchpappel
l *Lavatera thuringiaca*

5508 **Thurlow weeping willow**
f saule pleureur du Japon
d Trauerweide
l *Salix elegantissima*

5509 **thyme**
f thym
d Quendel; Thymian
l *Thymus*

5510 **thyme-flowered dragonhead**
f dracocéphale à fleurs de thym
d thymianblättriger Drachenkopf
l *Dracocephalum thymiflorum*

5511 **thyme-leaved sandwort**
f sabline à feuilles de serpolet
d quendelblättriges Sandkraut; Quendelsandkraut
l *Arenaria serpyllifolia*

5512 **thyme-leaved speedwell**
f véronique à feuilles de serpolet
d quendelblättriger Ehrenpreis; thymianblättriger Ehrenpreis; Quendelehrenpreis
l *Veronica serpyllifolia*

* **thyrse-flowered loosestrife → 5629**

5513 **Tibetan filbert**
f noisetier du Tibet
d tibetische Hasel
l *Corylus tibetica*

5514 **Tibet neillia**
f neillie du Tibet
d tibetische Traubenspiere
l *Neillia thibetica*

5515 **tibourbou**
f tibourbou
l *Apeiba tiburbu*

5516 **tick clover; tick trefoil**
f desmodie
d Büschelkraut; Telegraphenpflanze; Wandelklee
l *Desmodium*

5517 **tickseed**
f corisperme
d Wanzensame
l *Corispermum*

* **tick trefoil → 5516**

5518 **tide-marsh water hemp; Virginian hemp**
f acnide; chanvre de Virginie
d virginischer Hanf
l *Acnida cannabina*

5519 **tidytips**
f layie
d Laya
l *Layia*

5520 **tiger flower**
f tigridie
d Tigerblume; Tigerlilie
l *Tigridia*

5521 **tiger grass**
d Tigergras
l *Thysanolaena*

5522 **tiger lily**
f lis tigré
d Tigerlilie
l *Lilium tigrinum*

5523 **tiger-tail spruce**
f épicéa du Japon; sapin (à) queue de tigre
d Tigerschwanzfichte
l *Picea polita*

5524 **tikoki alectryon**
f tikoki
l *Alectryon excelsum*

* **tikor → 1983**

* **tikouka → 2470**

5525 **tillandsia**
f tillandsia
d Tillandsie; Haarananas
l *Tillandsia*

5526 **tilletia family**
f tillétiacées

d Tilletiazeen
l *Tilletiaceae*

5527 timothy; herd's-grass
f fléole
d Lieschgras; Hirtengras
l *Phleum*

5528 timothy; herd's-grass
f fléole des prés
d Timotheegras; Wiesenlieschgras
l *Phleum pratense*

5529 tinder fungus
f amadouvier
d Buchenpilz; Buchenschwamm; echter Feuerschwamm; echter Zunderschwamm
l *Fomes fomentarius*

5530 tiny mousetail
d Ackermäuseschwänzchen; Zwergmäuseschwanz
l *Myosurus minimus*

5531 tiny vetch; pigeon vetch; tare vetch; hairy vetch
f ers velu; petit vesceron
d behaarte Wicke; rauhhaarige Wicke
l *Vicia hirsuta*

5532 tirite
f ischnosiphon
l *Ischnosiphon*

*** titan giant arum → 2466**

5533 toadflax
f linaire
d Leinkraut
l *Linaria*

5534 toad lily
f tricyrtis
d Dreihöckerblume; Höckerblume; Krötenlilie
l *Tricyrtis*

5535 toad rush
f jonc de crapaud
d Krötenbinse
l *Juncus bufonius*

5536 tobacco (plant)
f tabac
d Tabak
l *Nicotiana*

5537 tobira family
f pittosporacées
d Klebsamengewächse; Pittosporazeen
l *Pittosporaceae*

5538 tobira pittosporum
f pittospore tobira; pittospore de Chine
d chinesischer Klebsame; Pechsame
l *Pittosporum tobira*

*** tocusso → 4399**

5539 toddalia
f toddalie
l *Toddalia*

5540 toddy fishtail palm; kittool palm
f caryote brûlante
d Brennpalme; Kit(t)upalme
l *Caryota urens*

5541 Tolu balsam balm tree
f baumier du Tolu
l *Myroxylon toluifera; Myroxylon balsamum*

5542 tomato
f tomate
d Tomate
l *Lycopersicon*

5543 tomentose rose
f rosier tomenteux
d Filzrose
l *Rosa tomentosa*

5544 tongue buttercup
f renoncule-langue
d grosser Hahnenfuss; Zungenhahnenfuss
l *Ranunculus lingua*

5545 tonka bean
f fève de tonca
d Tonkabaum
l *Dipteryx*

5546 toothed wood fern; narrow buckler fern; narrow shield fern
f dryoptéris spinuleux
d Dornfarn
l *Dryopteris spinulosa*

5547 toothpick ammi; bishop's-weed
f herbe aux cure-dents
d Zahnstocherkraut
l *Ammi visnaga*

5548 toothwort
f dentaire

d Zahnwurz
l *Dentaria*

5549 toothwort
f clandestine écailleuse; lathrée écailleuse
d rötliche Schuppenwurz
l *Lathraea squamaria*

5550 top primrose
f primevère à coupes
d verkehrtkegelförmige Schlüsselblume; Becherprimel
l *Primula obconica*

5551 torch lily; poker plant; red-hot poker plant
f kniphofie; tritome
d Fackellilie; Raketenblume; Tritome
l *Kniphofia; Tritoma*

5552 torchwood; balsam shrub
d Balsambaum; Balsamstrauch
l *Amyris*

5553 torenia
f torénie
d Torenie
l *Torenia*

5554 Toringo crab apple
f pommier de Siebold
d japanischer Zierapfel
l *Malus sieboldi*

5555 tormentilla cinquefoil
f tormentille; potentille dressée
d Blutwurz; aufrechtes Fingerkraut; hohes Fingerkraut; Ruhrwurz
l *Potentilla erecta; Potentilla tormentilla; Potentilla silvestris*

5556 torreya
f torreya
d Nusseibe
l *Torreya*

5557 Torrey pine
f pin de Torrey
d Torreys Kiefer
l *Pinus torreyana*

5558 tortoise plant; elephant's-foot
f testudinaire; pied-d'éléphant
d Schildkrötenpflanze; Elefantenfuss
l *Testudinaria*

5559 torus herb
f dorsténie
d Dorstenie
l *Dorstenia*

5560 touch-me-not; European touch-me-not; noli-me-tangere
f balsamine des bois; ne-me-touchez-pas
d Nolimetangere; Rührmichnichtan; echtes Springkraut
l *Impatiens noli tangere*

5561 toulicia
f toulicie
l *Toulicia*

5562 tournefortia
f pittone
d Tournefortie; Strauchsonnenwende
l *Tournefortia*

5563 Tournefort speedwell
f véronique de Tournefort
d persischer Ehrenpreis; Tourneforts Ehrenpreis
l *Veronica persica; Veronica tourneforti*

5564 towel gourd; dishcloth gourd; vegetable sponge
f éponge végétable; louffa
d Schwammkürbis; Schwammgurke
l *Luffa*

5565 tower cress
d Turm(gänse)kresse
l *Arabis turrita*

5566 tower-mustard rock cress
f tourette glabre
d Turmkraut
l *Arabis glabra; Turritis glabra*

5567 tozzia
f tozzie
d Alpenrachen
l *Tozzia*

5568 tragacanth milkvetch
f astragale à gomme-adragante
d Gummitraganth
l *Astragalus gummifer*

5569 trailing arbatus
f épigée
d Bodenlorbeer; Primelstrauch
l *Epigaea*

* **trailing St.-John's-wort → 1677**

5570 **Transylvanian melic**
f mélique de Transylvanie
d siebenbürgisches Perlgras
l *Melica transilvanica*

5571 **traveler's-joy**
f clématite des bois; clématite des haies; herbe aux gueux
d gemeine Waldrebe; gewöhnliche Waldrebe; echte Waldrebe; deutsche Waldrebe
l *Clematis vitalba*

5572 **traveler's-tree**
f ravenala
d Baum der Reisenden
l *Ravenala*

5573 **treacle erysimum**
f fausse-giroflée
d Ackerschotendotter; Ackerschöterich; lackartiger Schotendotter
l *Erysimum cheiranthoides*

5574 **treacle hare's-ear**
f roquette d'Orient
d morgenländischer Ackerkohl; weisser Ackerkohl; weisser Schöterich
l *Conringia orientalis*

5575 **treculia**
f tréculie
d Okwabaum
l *Treculia*

* **tree alfalfa → 5583**

* **tree aralia → 1849**

5576 **treebeard tillandsia; Spanish moss**
f cheveux du roi
d Bartflechten-Tillandsie
l *Tillandsia usneoides*

5577 **treebine**
f cisse
d Klimme
l *Cissus*

5578 **tree buckthorn**
f alphitonie
l *Alphitonia*

5579 **tree buckthorn**
f alphitonie élevée
l *Alphitonia excelsa*

5580 **tree fern**
f cyathée
d Becherfarn
l *Cyathea*

5581 **tree heath**
f bruyère arborescente; bruyère en arbre
d Baumerika; Baumheide
l *Erica arborea*

* **tree-like aloe → 299**

5582 **tree mallow**
f lavatère
d Baummalve; Strauchmalve; Strauchpappel
l *Lavatera*

5583 **tree medic; tree alfalfa**
f luzerne arborescente; luzerne en arbre
d baumartiger Schneckenklee
l *Medicago arborea*

5584 **tree of heaven ailanthus**
f ailante glanduleux; vernis du Japon
d drüsiger Götterbaum; hoher Götterbaum; Himmelsbaum; chinesischer Götterbaum
l *Ailanthus altissima; Ailanthus glandulosa*

* **tree of sadness → 3851**

5585 **tree peony**
f pivoine en arbre
d baumartige Pfingstrose; strauchartige Päonie; strauchige Pfingstrose; Strauchpäonie; Strauchpfingstrose
l *Paeonia arborea; Paeonia suffruticosa*

5586 **tree phillyrea**
f filaria à larges feuilles; gros daraden; putine
d breitblättrige Steinlinde
l *Phillyrea latifolia*

5587 **tree rhododendron**
f rhododendron en arbre; rosage en arbre
l *Rhododendron arboreum*

5588 **tree scarlet elder**
f sureau arborescent
l *Sambucus arborescens; Sambucus pubens arborescens*

5589 **tree tomato**
f cyphomandre bétacé

d Baumtomate
l *Cyphomandra betacea*

* **trembling aspen → 2074**

5590 triangular scirpus
d dreikantige Simse
l *Scirpus triqueter*

5591 tricolor Cape cowslip
f lachenalie tricolore
l *Lachenalia tricolor*

5592 trifoliate orange
f citronnier trifolié; citronnier à trois folioles
d dreiblättrige Zitrone; Dreiblatt-zitrone
l *Citrus trifoliata; Poncirus trifoliata*

5593 trigonella
f trigonelle
d Bockshornklee
l *Trigonella*

5594 trillium; wake-robin
f trillie
d Dreiblatt; Waldlilie
l *Trillium*

5595 Trinia
f trinie
d Faserschirm; Trinie
l *Trinia*

5596 triodia
f triodie
d Dreizahn
l *Triodia*

5597 Tripoli aster; sea aster
f aster maritime
d Strandaster
l *Aster tripolium*

5598 trithrinax palm
d Doppeldreizackpalme
l *Trithrinax*

5599 tritonia; montbretia
f tritonie
d Gartenmontbretie
l *Tritonia; Montbretia*

5600 tropical almond; Malabar almond
f badamier
d Katappenbaum
l *Terminalia catappa*

5601 true aloe
f aloès vrai; aloès officinal
d echte Aloe
l *Aloe vera*

* **true bay → 2615**

5602 true forget-me-not
f myosotis des marais; myosotis palustre
d Sumpfvergissmeinnicht
l *Myosotis scorpioides; Myosotis palustris*

5603 true indigo
f indigo; indigotier
d Indigostrauch
l *Indigofera tinctoria*

5604 true lavender
f lavande commune; lavande femelle; lavande vraie
d echter Lavendel; Gartenlavendel
l *Lavandula officinalis; Lavandula spica; Lavandula vera*

5605 true mosses
f mousses
d Birnmoose; Laubmoose
l *Bryales*

5606 true myrtle
f myrte commun
d echte Myrte; gemeine Myrte
l *Myrtus communis*

5607 true-star anise tree; Chinese anise
f anis étoilé vrai
d Sternanis
l *Illicium verum*

5608 truffle
f truffe
d Trüffel; Speisetrüffel; echte Trüffel
l *Tuber*

5609 truffles
f tubérales
d Trüffelartige; Trüffeln; Trüffelpilze; Tuberaeen
l *Tuberales*

5610 trumpetbush
f técoma
d Trompetenwinde; Jasmintrompete
l *Tecoma*

5611 trumpet creeper
f campsis

d Trompetenblume; Jasmintrompete; Klettertrompete
l *Campsis*

5612 trumpet-creeper family
f bignoniacées
d Bignoniazeen; Bignoniengewächse; Trompetenbaumgewächse; Trompetenblütler
l *Bignoniaceae*

5613 trumpet honeysuckle
f chèvrefeuille toujours vert
d immergrünes Geissblatt
l *Lonicera sempervirens*

*** trumpet lily → 950**

5614 trumpet tree
f tabébuia
l *Tabebuia*

*** trumpet tree → 4331**

5615 tuber flower
f siphonanthe
l *Siphonanthus*

5616 tuber gourd
f thladianthe
d Quetschblume; Quetschgurke
l *Thladiantha*

5617 tuber nasturtium
f capucine tubéreuse
d Knollenkapuzinerkresse
l *Tropaeolum tuberosum*

5618 tuberose
f tubéreuse
d Tuberose
l *Polianthes*

5619 tuberous comfrey
f consoude tubéreuse
d knolliger Beinwell; knotiger Beinwell; knollige Wallwurz
l *Symphytum tuberosum*

5620 tuberous Jerusalem sage
f phlomide tubéreuse
d Knollenbrandkraut
l *Phlomis tuberosa*

5621 tuberous-rooted begonia
f bégonia tuberculeux hybrides
d Knollenbegonie
l *Begonia tuberhybrida*

5622 tuber ullucus; ulluco
f ulluque; melloco
d Knollenbaselle
l *Ullucus tuberosus; Basella tuberosa*

5623 tuber vine
f rajania
d Rajanie
l *Rajania*

5624 tube-tongue
f salpiglossis
d Trompetenzunge
l *Salpiglossis*

5625 tubular water dropwort
f œnanthe fistuleuse
d röhrige Nebendolde; hohle Pferdesaat; röhrige Rebendolde; Röhrenrosskümmel
l *Oenanthe fistulosa*

5626 tucuma
f astrocaryum
d Sternnusspalme
l *Astrocaryum*

5627 tufted forget-me-not
f myosotis gazonnant
d rasiges Vergissmeinnicht; Wiesenvergissmeinnicht
l *Myosotis caespitosa*

5628 tufted hair grass
f canche cespiteuse
d gemeine Schmiele; Rasenschmiele
l *Deschampsia caespitosa*

5629 tufted loosestrife; water loosestrife; thyrse-flowered loosestrife
f corneille en bouquets
d Straussfelberich
l *Naumburgia thyrsiflora; Lysimachia thyrsiflora*

5630 tufted sedge
f laîche en gazon
d Rasenriedgras; rasige Segge
l *Carex caespitosa*

5631 tuft root; dumb cane
f dieffenbachie
d Dieffenbachie
l *Dieffenbachia*

5632 tulbaghia
f tulbaghie
d Tulbaghie; Kranzlauch
l *Tulbaghia*

5633 tulip
f tulipe
d Tulpe
l *Tulipa*

5634 tulip orchid
f cattleya jaune citron
l *Cattleya citrina*

5635 tulip tree
f tulipier
d Tulpenbaum
l *Liriodendron*

5636 tulip tree; yellow poplar
f tulipier de Virginie
d gemeiner Tulpenbaum
l *Liriodendron tulipifera*

5637 tumbleweed amaranth
d weisser Amarant; weisser Fuchsschwanz
l *Amaranthus graecizans; Amaranthus albus*

5638 tumbling mustard
d ungarische Rauke
l *Sisymbrium altissimum*

5639 tumbling orach
d Rosenmelde
l *Atriplex rosea*

5640 tung oil tree
f alévrite; bois de Chine
d Tung(öl)baum; Holzölbaum
l *Aleurites fordii*

* **tunicate corn → 4233**

5641 tunic flower
f tunique
d Felsennelke
l *Tunica*

5642 tupelo
f toupélo
d Tupelobaum
l *Nyssa*

5643 tupelo gums
f nyssacées
d Nyssazeen
l *Nyssaceae*

* **turbantop → 2767**

5644 Turkestan ash
d turkestanische Eberesche
l *Fraxinus potamophila*

5645 Turkestan pearlbush
f exochorde de Korolkow
d Alberts Blumenspiere
l *Exochorda alberti; Exochorda korolkowi*

5646 Turkish filbert
f noisetier de Byzance; noisetier de Constantinople; noisetier de Turquie
d Baumhasel; türkische Hasel; türkische Baumhasel
l *Corylus colurna*

5647 Turk terebinth pistache
d Terpentinpistazie
l *Pistacia mutica*

5648 turmeric
f curcuma
d Gelbwurz; Safranwurz
l *Curcuma*

5649 turnera
f turnère
d Turnera
l *Turnera*

5650 turnip
f rave
d Herbstrübe; Saatrübe; Stoppelrübe; Turnip; Wasserrübe; weisse Rübe
l *Brassica rapa*

* **turnip garden parsley → 5652**

5651 turnip-rooted chervil
f carotte de Philadelphie; cerfeuil bulbeux
d Kerbelrübe; Knollenkerbel; Rübenkerbel
l *Chaerophyllum bulbosum*

5652 turnip-rooted parsley; turnip garden parsley
d Wurzelpetersilie
l *Petroselinum crispum* var. *radicosum*

5653 turnsole
f crozophore
d Krebskraut; Lackmuskraut
l *Chrozophora*

5654 turtle grass
f thalassie
l *Thalassia*

5655 turtlehead
f galane
d Schildblume; Schildkraut; Schlangenkraut
l *Chelone*

5656 tussock sedge
f laîche élevée
d steifes Riedgras; steife Segge
l *Carex elata*

5657 twayblade
f listère
d Zweiblatt
l *Listera*

5658 twayblade
f liparis
d Glanzkraut
l *Liparis*

* **twig rush → 4727**

5659 twinebark
f plagianthe
l *Plagianthus*

5660 twinflower
f linnée
d Moosglöckchen
l *Linnaea*

5661 twin-flowered violet
f violette biflore
d zweiblütiges Veilchen
l *Viola biflora*

5662 twinleaf
f jeffersonie
d Jeffersonie; Zwillingsblatt
l *Jeffersonia*

5663 twinleaf saxifrage
f saxifrage à feuilles opposées
d bunter Steinbrech; gegenblättriger Steinbrech; roter Steinbrech
l *Saxifraga oppositifolia*

5664 twin-leaved squill
f scille à deux feuilles
d zweiblättriger Blaustern; zweiblättrige Meerzwiebel
l *Scilla bifolia*

5665 twinspur
f diascie
l *Diascia*

5666 twisted arum
f hélicodicéros
d Hornwurz
l *Helicodiceros; Megotigea*

5667 twisted draba
d weissgraues Felsenblümchen
l *Draba incana*

5668 twisted heath
f bruyère cendrée
d graue Heide
l *Erica cinerea*

5669 twisted-stalk
f streptope
d Knotenfuss
l *Streptopus*

5670 two-ball nitta tree
f parkie du Soudan
l *Parkia biglobosa*

5671 two-leaved bead-ruby
f maianthème à deux feuilles
d zweiblättrige Schattenblume
l *Maianthemum bifolium*

5672 two-rowed barley
f orge à deux rangs; paumelle
d zweizeilige Gerste; Sommergerste; Staudengerste
l *Hordeum distichon*

5673 two-row stonecrop
f orpin bâtard
d kaukasische Fetthenne; Kaukasusfetthenne
l *Sedum spurium*

U

5674 **uapaca**
f uapaca
l *Uapaca*

5675 **udo**
f aralie du Japon
d japanische Bergangelika
l *Aralia cordata; Aralia edulis*

5676 **Ukrainian catnip**
f cataire d'Ukraine
d ukrainische Katzenminze
l *Nepeta ucrainica*

* **ulluco → 5622**

5677 **umbrella flat sedge**
f souchet à feuilles alternes
d wechselblättriges Zypergras
l *Cyperus alternifolius*

5678 **umbrella leaf**
f diphyllée
d Schirmblatt
l *Diphylleia*

5679 **umbrella magnolia**
f magnolia parasol
d Schirmmagnolie
l *Magnolia tripetala*

5680 **umbrella palm**
f hédyscèpe
d Säulenschirmpalme
l *Hedyscepe*

5681 **umbrella pine**
f sciadopitys
d Schirmtanne
l *Sciadopitys*

5682 **umbrella pine**
f sapin parasol du Japon
d japanische Schirmtanne; quirlblättrige Schirmtanne; Schirmfichte
l *Sciadopitys verticillata*

* **underoak mushroom → 3396**

5683 **unicorn-plant family**
f martyniacées
d Martyniengewächse
l *Martyniaceae*

5684 **upas tree**
f upas antiar
d Upasbaum; Giftbaum
l *Antiaris toxicaria*

5685 **upland cotton**
f cotonnier américain
l *Gossypium hirsutum*

5686 **upland geranium**
f géranium des collines
d Hügelstorchschnabel
l *Geranium collinum*

* **upright brome grass → 3546**

5687 **upright hedge nettle**
d Bergziest
l *Stachys recta*

5688 **upright hedge parsley**
d gemeiner Klettenkerbel; Waldklettenkerbel
l *Torilis japonica*

5689 **upright spurge**
d steife Wolfsmilch
l *Euphorbia stricta*

5690 **Ural crazyweed**
f oxytropide de l'Oural
d Uralfahnenwicke
l *Oxytropis uralensis*

5691 **Ural false spirea**
f sorbaire à feuilles de sorbier
d ebereschenblättrige Fiederspiere; Ebereschenfiederspiere
l *Spiraea sorbifolia; Sorbaria sorbifolia*

* **urd → 3755**

5692 **ursinia**
f sphénogyne
d Ursinie
l *Ursinia; Sphenogyne*

5693 **usnea**
f usnée
d Bartflechte
l *Usnea*

5694 **Ussurian pear**
f poirier de Chine
l *Pyrus ussuriensis*

5695 **valerian**
f valériane
d Baldrian
l *Valeriana*

5696 **valerian family**
f valérianacées
d Baldriangewächse; Valerianazeen
l *Valerianaceae*

5697 **Valonia oak**
f chêne vélani
d Walloneneiche
l *Quercus macrolepis; Quercus aegilops; Quercus graeca*

5698 **vanda**
f vanda
d Vanda
l *Vanda*

5699 **vanilla**
f vanillier
d Vanille
l *Vanilla*

5700 **variegated boletus**
f bolet tacheté; bolet moucheté
d Sandpilz; Sandröhrling; Hirsepilz; Semmelpilz; Zitronenpilz
l *Boletus variegatus; Suillus variegatus*

5701 **variegated horsetail**
d bunter Schachtelhalm
l *Equisetum variegatum*

5702 **various-leaved pondweed**
d grasartiges Laichkraut
l *Potamogeton heterophyllus*

5703 **Vasey grass**
f herbe de Vasey; paspale d'Urville
l *Paspalum urvillei*

5704 **vateria**
f vatérie
d Kopalbaum
l *Vateria*

5705 **vegetable marrow; marrow squash; marrow-type pumpkin**
f giraumon(t)
d Melonenkürbis
l *Cucurbita pepo* var. *giraumontia*

5706 **vegetable-oyster salsify; oyster plant**
f salsifis à feuilles de poireau
d Haferwurzel
l *Tragopogon porrifolius*

* **vegetable sponge → 5564**

5707 **Veitch fir**
f sapin de Veitch
d Veitchs Tanne
l *Abies veitchi*

5708 **velvet bent grass**
f agrostide des chiens
d Hundsstraussgras
l *Agrostis canina*

5709 **velvet-footed collybia; velvet-stemmed collybia**
f collybie à pied velouté
d Samtfussrübling; Samtfusswinterpilz
l *Collybia velutipes*

5710 **velvet grass**
f houque; houlque
d Honiggras
l *Holcus*

5711 **velvetleaf**
f limnochare
d Sumpflieb
l *Limnocharis*

5712 **velvet maple**
d Höherahorn
l *Acer velutinum*

5713 **velvet plant**
f gynure
d Fettkörbchen
l *Gynura; Crassocephalum*

5714 **velvetseed**
f guettarde; guettardie
d Guettarde
l *Guettarda*

* **velvet-stemmed collybia → 5709**

5715 **velvet tamarind**
f diale de Guinée
l *Dialium guineense*

5716 **velvet tree mallow**
f lavatère en arbre
l *Lavatera arborea*

5717 **Venus-cup teasel**
f cardère sauvage; chardon à foulon

d wilde Karde
l *Dipsacus sylvestris*

* **Venus's-comb → 4861**

5718 Venus's-flytrap
f dionée
d Venusfliegenfalle
l *Dionaea*

* **Venus's-hair → 5099**

5719 Venus's-looking-glass
f miroir de Vénus
d Frauenspiegel; Venusspiegel
l *Legouzia; Specularia*

5720 verbena sage
f sauge fausse-verveine
l *Salvia verbenacea*

5721 vernal anemone
f anémone printanière; pulsatille printanière
d Frühlingsküchenschelle; Frühlingskuschelle
l *Anemone vernalis; Pulsatilla vernalis*

5722 vernal barberry
d Frühlingsberberitze
l *Berberis vernae*

5723 vernal clematis
f clématite à vrilles
l *Clematis cirrhosa*

5724 vernal grass
f flouve
d Ruchgras
l *Anthoxanthum*

5725 vernal sedge
f laîche précoce
d Frühsegge; Frühlingsriedgras; frühes Riedgras
l *Carex praecox*

5726 vernal speedwell
f véronique printanière
d Frühlingsehrenpreis
l *Veronica verna*

5727 vernal witch hasel
f hamamélis printanier
d Frühlingszaubernuss
l *Hamamelis vernalis*

5728 vervain
f verveine
d Eisenkraut
l *Verbena*

5729 vervain family
f verbénacées
d Eisenkrautgewächse
l *Verbenaceae*

5730 vessel fern
f angioptère
d Bootfarn
l *Angiopteris*

5731 vetch
f vesce
d Wicke
l *Vicia*

* **vetchling → 4098**

5732 vetiver
f vétiver
d Vetiverwurzel
l *Vetiveria*

5733 viburnum; snowball
f viorne
d Schneeball(strauch)
l *Viburnum*

5734 villaresia
f villarésie
l *Villaresia*

* **vine grape → 2104**

5735 vine maple
f érable à feuilles rondes
d rundblättriger Ahorn
l *Acer circinatum*

5736 vine spinach
f baselle
d Baselle
l *Basella*

5737 violet
f violette
d Veilchen
l *Viola*

5738 violet bush
f iochrome
d Veilchenstrauch
l *Iochroma*

5739 violet cortinarius
f cortinaire violet
d violetter Dickfuss; dunkelvioletter

Dickfuss
l *Cortinarius violaceus*

5740 violet family
f violacées
d Veilchengewächse
l *Violaceae*

5741 violet horned poppy
d Bastardrömerie
l *Roemeria hybrida*

5742 violet petunia
f pétunia violet
d violette Petunie
l *Petunia violaceae*

5743 violet sage
f sauge des bois
d Hainsalbei
l *Salvia nemorosa*

* **violet willow → 1801**

5744 viper's bugloss
f herbe aux vipères; vipérine
d Natter(n)kopf
l *Echium*

5745 viraru
f ruprechtie
l *Ruprechtia*

5746 virescent Anzac wood
f pomaderre d'Australie
l *Pomaderris apetala*

5747 virgilia
f virgilier
d Virgilie
l *Virgilia*

5748 Virginia creeper
f vigne-vierge grimpante; vigne-vierge à 5 folioles
d fünfblättrige Jungfernrebe; kanadische Rebe; selbstkletternde Jungfernrebe; wilder Wein
l *Parthenocissus quinquefolia; Ampelopsis quinquefolia*

* **Virginian hemp → 5518**

5749 Virginia pepperweed
f passerage de Virginie
d virginische Kresse
l *Lepidium virginicum*

5750 Virginia pine; scrub pine
f pin de Virginie; pin de Jersey
d virginische Kiefer
l *Pinus virginiana*

5751 Virginia rose
f rosier de Virginie
d virginische Rose
l *Rosa virginiana*

5752 Virginia spiderwort
f éphémère de Virginie
d virginische Dreimasterblume
l *Tradescantia virginiana; Tradescantia montana*

5753 Virginia spirea
f spirée de Virginie
d virginischer Spierstrauch
l *Spiraea virginiana*

5754 Virginia strawberry
f fraisier de Virginie
d virginische Erdbeere; Himbeererdbeere
l *Fragaria virginiana*

5755 Virginia sweet spire; Virginia willow
f itéa de Virginie
d virginischer Moorstrauch
l *Itea virginica*

5756 Virgin's-bower
f clématite de Virginie
d virginische Waldrebe
l *Clematis virginiana*

5757 viscous campion; viscous catchfly
d klebrige Nachtnelke; klebriges Leimkraut
l *Silene viscosa*

5758 vismia
f vismie
d Vismie
l *Vismia*

5759 viviparous bistort
f renouée vivipare
d Knöllchenknöterich; Knollenknöterich
l *Polygonum viviparum; Bistorta vivipara*

5760 Voss calceolaria
f calcéolaire hybride herbacée; calcéolaire pantoufle
d Bastardpantoffelblume
l *Calceolaria herbeohybrida*

W

5761 wachendorfia
f wachendorfie
d Wachendorfie
l *Wachendorfia*

5762 Wahlenberg wood rush
f luzule de Wahlenberg
d Wahlenbergs Simse
l *Luzula wahlenbergi*

* **wake-robin → 5594**

5763 walking fern; walking leaf
f camptosore
d wanderndes Blatt
l *Camptosorus*

5764 wallaba tree
f épéru
d Wallabarholzbaum
l *Eperua*

5765 wall bur cucumber
f sicyote anguleux
d eckige Haargurke
l *Sicyos angulatus*

5766 wall draba; wall whitlow grass
d Mauerfelsenblümchen
l *Draba muralis*

5767 wallenia
f wallénie
l *Wallenia*

* **wall fern → 1503**

5768 wallflower
f giroflée
d Goldlack
l *Cheiranthus*

* **wall germander → 1113**

5769 wall hawkweed
f épervière des murs
d Mauerhabichtskraut; Waldhabichtskraut
l *Hieracium murorum*

5770 Wallich palm
f wallichie
l *Wallichia*

5771 Wallich's juniper
f genévrier de Wallich
d Wallichs Wacholder
l *Juniperus wallichiana*

5772 wall lettuce
d Mauerlattich
l *Mycelis muralis; Lactuca muralis*

5773 wall pellitory
f épinard de muraille; pariétaire officinale; perce-muraille
d aufrechtes Glaskraut
l *Parietaria officinalis*

5774 wall rock cress
f arabette du Caucase
d Gartengänsekresse; kaukasische Gänsekresse
l *Arabis caucasica*

5775 wall rocket
f diplotaxe
d Doppelsame
l *Diplotaxis*

5776 wall rue
f rue de muraille
d Mauerrauke; Mauerstreifenfarn
l *Asplenium ruta-muraria*

* **wall whitlow grass → 5766**

5777 walnut
f noyer
d Walnussbaum
l *Juglans*

5778 walnut family
f juglandacées
d Walnussbaumgewächse; Juglandazeen
l *Juglandaceae*

5779 wampee
f clausène
d Wampibaum
l *Clausena*

5780 wandering Jew
f éphémère de Rio
d Tradeskantie aus Rio de Janeiro; Dreimasterblume aus Rio de Janeiro
l *Tradescantia fluminensis*

5781 wandflower
f sparaxide
d Fransenschwertel
l *Sparaxis*

5782 **wand lythrum**
f salicaire effilée
d Rutenweiderich
l *Lythrum virgatum*

5783 **wart cress; swine's-cress**
f coronope
d Krähenfuss
l *Coronopus*

5784 **warted puffball**
f vesse-de-loup perlée
d Flaschenbovist; Flaschenstäubling; Warzenstäubling
l *Lycoperdon perlatum*

5785 **wartwort**
f verrucaire
d Warzenflechte
l *Verrucaria*

5786 **warty barberry**
f vinettier subverruqueux
d Warzenberberitze
l *Berberis verruculosa*

5787 **warty-barked euonymus**
f fusain verruqueux
d Warzenspindelbaum; warziger Spindelbaum
l *Euonymus verrucosa*

5788 **warty mock orange**
f seringa verruqueux
d Warzenpfeifenstrauch
l *Philadelphus verrucosus*

5789 **Washington lupine**
f lupin polyphylle
d Dauerlupine; vielblättrige Lupine; vielzählige Lupine
l *Lupinus polyphyllus*

5790 **Washington palm**
f washingtonia
d Washingtonie
l *Washingtonia*

* **water arum → 5941**

5791 **water avens**
f benoîte aquatique; benoîte de rivage; benoîte d'eau
d Bachnelkenwurz; Ufererdröschen
l *Geum rivale*

5792 **water awlwort**
f subulaire aquatique
d Wasserpfriemenkresse
l *Subularia aquatica*

5793 **water-bug-trap**
f aldrovande vésiculeuse
d blasige Wasserfalle
l *Aldrovanda vesiculosa*

5794 **water chestnut; Jesuit's nut**
f châtaigne d'eau; marron d'eau; mâcre; cornuelle
d Jesuitennuss; schwimmende Wassernuss; Wasserkastanie
l *Trapa natans*

5795 **water chickweed**
f céraiste aquatique
d Wasserdarm; Wasserweichkraut
l *Malachium aquaticum*

5796 **water cress**
f cresson de fontaine; cresson officinal; cresson d'eau
d Brunnenkresse; echte Brunnenkresse; gemeine Brunnenkresse; Bornkresse
l *Nasturtium officinale; Rorippa nasturtium-aquaticum*

5797 **water crowfoot**
f renoncule aquatique
d Wasserhahnenfuss
l *Ranunculus aquatilis*

5798 **water dock**
f oseille aquatique
d Flussampfer; Teichampfer
l *Rumex hydrolapathum*

5799 **water dropwort**
f œnanthe
d Rebendolde; Wasserfenchel
l *Oenanthe*

5800 **water elm**
f planère
d Wasserulme
l *Planera; Zelkova*

5801 **water elm**
f orme aquatique
l *Planera aquatica*

5802 **water fennel**
f callitriche printanier
d Frühlingswasserstern; Sumpfwasserstern
l *Callitriche verna*

5803 water fern
f cératoptéris
d Hornfarn
l *Ceratopteris*

5804 water figwort
f scrofulaire aquatique
d orientalische Braunwurz
l *Scrophularia aquatica*

5805 water foxtail
f vulpin genouillé
d geknieter Fuchsschwanz; geknietes Fuchsschwanzgras
l *Alopecurus geniculatus*

5806 water germander
d Knoblauchgamander; Lauchgamander; Skordienkraut
l *Teucrium scordium*

5807 water hawthorn
f aponogéton
d Wasserähre
l *Aponogeton*

5808 water hemlock
f cicutaire
d Wasserschierling; Wüterich
l *Cicuta*

5809 water hickory
f carya aquatique
d Wasserhickory
l *Carya aquatica*

5810 water hyacinth
f eichhornie
d Eichhornie; Wasserhyazinthe
l *Eichhornia*

5811 water hyssop
f bacope
d Bakope; Fettblatt
l *Bacopa*

5812 water lady's-thumb
d Wasserknöterich
l *Polygonum amphibium*

5813 waterleaf
f hydrophylle
d Wasserblatt
l *Hydrophyllum*

5814 waterleaf family
f hydrophyllacées
d Wasserblattgewächse; Hydrophyllazeen
l *Hydrophyllaceae*

5815 water lemon; yellow granadilla
f pomme-liane
d lorbeerblättrige Granadille; lorbeerblättrige Grenadille; lorbeerblättrige Passionsblume; Wasserlimone
l *Passiflora laurifolia*

5816 water lettuce
f pistie
d Muschelblume; Pistie
l *Pistia*

5817 water lily
f nénuphar; nénufar
d Seerose; Wasserrose
l *Nymphaea*

5818 water-lily family
f nymphéacées
d Seerosengewächse; Nymphäazeen; Teichrosengewächse; Wasserrosengewächse
l *Nymphaeaceae*

5819 water lobelia
f lobélie de Dortmann
d Wasserlobelie
l *Lobelia dortmanna*

* **water loosestrife → 5629**

5820 water manna grass
f glycérie flottante; herbe à la manne; manne de Pologne
d flutender Schwaden; Grützschwaden; Himmelstau; Mannaschwaden; Entengras
l *Glyceria fluitans*

5821 watermelon
f melon d'eau; pastèque
d Arbuse; Wassermelone
l *Citrullus vulgaris*

* **water milfoil → 4060**

5822 water mint
f menthe aquatique
d Wasserminze
l *Mentha aquatica*

5823 water moss
f fontinale
d Quellmoos
l *Fontinalis*

5824 water mo(u)lds; algal fungi
f phycomycètes
d Algenpilze; Pilzalgen; Phykomyzeteen
l *Phycomycetes*

5825 water mudwort
d kleines Schlammkraut
l *Limosella aquatica*

5826 water net
f hydrodicte
d Wassernetzalge
l *Hydrodictyon*

5827 water oak; duck oak
f chêne aquatique; chêne noir (d'Amérique)
d Schwarzeiche; Wassereiche
l *Quercus nigra*

5828 water parsnip
f berle
d Merk; Wassermerk
l *Sium*

*** water-pepper smartweed → 3511**

5829 water plantain
f flûteau
d Froschlöffel
l *Alisma*

5830 water platter
f victoria (regia)
d Viktoria
l *Victoria*

5831 water poppy
f hydrocléis
d Wassermohn; Wasserschlüssel
l *Hydrocleis*

5832 water primrose
f jussiée
d Jussiea
l *Jussiaea*

5833 water purslane
f péplide
d Sumpfquendel; Zipfelkraut
l *Peplis*

5834 water sedge
f laîche aquatique
d Wassersegge
l *Carex aquatilis*

5835 water shield
f brasénie
d Schleimkraut; Wasserschild
l *Brasenia*

5836 water soldier
f stratiote
d Krebsschere; Meeraloe; Wasseraloe
l *Stratiotes*

5837 water soldier
f stratiote faux-aloès
d Aloekrebsschere; gemeine Wasseraloe
l *Stratiotes aloides*

5838 water speedwell
d Wasserehrenpreis
l *Veronica anagalis-aquatica*

5839 water starwort
f callitriche
d Wasserstern
l *Callitriche*

5840 water-starwort family
f callitrichacées
d Wassersterngewächse; Kallitrichazeen
l *Callitrichaceae*

*** waterweed → 2022**

5841 waterwort
f élatine
d Tännel
l *Elatine*

5842 waterwort family
f élatinacées
d Tännelgewächse
l *Elatinaceae*

5843 wavy hairgrass
f canche flexueuse
d Drahtschmiele; geschlängelte Schmiele; geschlängelige Schmiele; Schlangenschmiele; Haferschmiele
l *Deschampsia flexuosa*

5844 wavy rhubarb
f rhubarbe ondulée
d gemeiner Rhabarber; krauser Rhabarber
l *Rheum undulatum; Rheum rhabarbarum*

5845 wax gourd
f bénincase

d Wachskürbis
l *Benincasa*

* **wax myrtle → 483**

5846 wax palm
f céroxylon des Andes
d Wachspalme der Anden
l *Ceroxylon andicola*

5847 wax plant
f hoya
d Porzellanblume; Wachsblume
l *Hoya*

5848 wax tree
f sumac faux-vernis
d Talgsumach
l *Toxicodendron succedaneum; Rhus succedanea*

5849 waxy corn
f maïs cireux
d Wachsmais
l *Zea mays* convar. *ceratina*

5850 waxy mushroom
d rötlicher Lacktrichterling
l *Laccaria laccata; Clitocybe laccata*

5851 wayaka yam bean
f pachyrrhize anguleuse
d Yamsbohne
l *Pachyrrhizus angulatus*

5852 wayfaring tree
f viorne cotonneuse
d Schlinge; wolliger Schneeball
l *Viburnum lantana*

5853 wayside cerastium
f céraiste à cinq étamines
d Sandhornkraut
l *Cerastium semidecandrum*

5854 wayside speedweel
d glänzender Ehrenpreis
l *Veronica polita*

5855 weaselsnout
f ortie jaune
d gelbe Goldnessel; gelbe Taubnessel
l *Galeobdolon luteum; Lamium galeobdolon*

5856 weaver's-broom
f genêt d'Espagne
d Binsenginster
l *Spartium*

5857 weaver's-broom; Spanish broom
f genêt jonciforme; spartier d'Espagne
d Pfriemenstrauch
l *Spartium junceum*

5858 wedge-leaf candollea
f candollée
l *Candollea cuneiformis*

5859 weeping beech
f hêtre pleureur
d Hängebuche; Trauerbuche
l *Fagus sylvatica* var. *pendula*

* **weeping birch → 2131**

5860 weeping forsythia
f forsythie à fleurs pendantes
d überhängende Forsythie
l *Forsythia suspensa*

* **weeping merulius → 2880**

5861 weigela
f diervillée
d Weigelie
l *Weigela*

5862 weld mignonette; dyer's rocket; dyer's weld
f gaude; herbe à jaunir; réséda des teinturiers
d Färberresede; Färberwau; Gelbkraut
l *Reseda luteola*

5863 Welsh onion
f ciboule
d Winterzwiebel
l *Allium fistulosum*

5864 Welsh poppy
f méconopside de Cambrie
d Mohnling; Keulenmohn; Scheinmohn
l *Meconopsis cambrica*

5865 Welwitschia
f welwitschie
d Welwitschie; Tumbo
l *Welwitschia*

5866 wendlandia
f wendlandie
l *Wendlandia*

5867 western bracken
f fougère aigle commune
d Adlerfarn
l *Pteridium aquilinum pubescens; Pteris aquilina pubescens*

* **western hemlock → 3996**

5868 western larch
f mélèze occidental
d abendländische Lärche
l *Larix occidentalis*

* **western yellow pine → 4259**

5869 West Indian gherkin; bur cucumber; anguria
f angurie; concombre des Antilles
d Arada-Gurke
l *Cucumis anguria*

5870 West Indies balsa; balsa
f ochrome
d Balsabaum; Hasenpfotenbaum
l *Ochroma lagopus; Ochroma pyramidale*

5871 West Indies mahogany
f acajou de Cuba; acajou mahagon; acajou à meubles; arbre d'acajou
d echtes Kuba-Mahagoni; echtes Mahagoni; westindischer Mahagonibaum
l *Swietenia mahagoni*

5872 wet-land deer vetch; large bird's-foot trefoil
f lotier des marais; lotier velu
d Sumpfhornklee; Sumpfschotenklee
l *Lotus uliginosus*

* **Weymouth pine → 1982**

5873 wheat
f blé
d Weizen
l *Triticum*

5874 wheat
f blé tendre; blé ordinaire
d Brotweizen; echter Weizen; gemeiner Weizen
l *Triticum aestivum; Triticum vulgare*

* **wheat grass → 1640**

5875 Wheeler sotol
f dasylire de Wheeler
d Wheelers Rauhlilie
l *Dasylirion wheeleri*

5876 wheel-stamen tree
f trochodendre
d Radbaum
l *Trochodendron*

5877 whiptree
f luhée
d Lühea
l *Luehea*

* **white alder → 5323**

5878 white ash
f frêne blanc (d'Amérique)
d Weissesche; amerikanische Esche
l *Fraxinus americana; Fraxinus acuminata; Fraxinus alba*

5879 whitebark pine
f pin blanc
d weissstengelige Kiefer
l *Pinus albicaulis*

5880 white beak sedge
d weisse Schnabelried; weisse Schnabelsimse
l *Rhynchospora alba*

5881 whitebeam mountain ash
f alisier blanc; alisier de Bourgogne
d Mehlbeere; Mehlbeerbaum
l *Sorbus aria*

5882 white bedstraw
f gaillet blanc
d gemeines Labkraut; Grasstern; weisses Waldstroh; Wiesenlabkraut
l *Galium mollugo*

* **white beet → 5398**

5883 white bryony
f bryone blanche
d weisse Zaunrübe
l *Bryonia alba*

5884 white butterbur
f pétasite blanc
d weisse Pestwurz
l *Petasites albus*

5885 white cabbage
f chou blanc
d Weisskohl; Weisskraut; Kappes
l *Brassica oleracea* var. *capitata alba*

5886 white canary tree; white Chinese olive
f canari blanc; canari de Cochinchine
d chinesische weisse Olive; weisser Kanarienbaum
l *Canarium album*

5887 white cedar false cypress
d Kugelzypresse; weisse Zeder;

Zederzypresse
l *Chamaecyparis thyoides*

* **white Chinese olive → 5886**

5888 white cinquefoil
f potentille blanche
d weisses Fingerkraut
l *Potentilla alba*

5889 white clover
f trèfle rampant; trèfle blanc
d Kriechklee; weisser Wiesenklee; Weissklee
l *Trifolium repens*

5890 white dammar pine
f agathis blanc
d Dammarabaum
l *Agathis alba; Dammara alba*

5891 white dead nettle
f lamier blanc; ortie blanche
d weisse Taubnessel
l *Lamium album*

5892 white durra
f sorgho penché; riz égyptien
l *Sorghum cernuum*

5893 white Egyptian lotus
f lotier d'Egypte
d ägyptische Lotosblume
l *Nymphaea lotus*

5894 white false hellebore
f vératre blanc
d weisser Germer
l *Veratrum album*

5895 white fir
f sapin concolore
d Silbertanne
l *Abies concolor*

5896 white fringe tree
f arbre de neige
d Giftesche; virginische Schneeblume
l *Chionanthus virginica*

5897 white-leaf Japanese magnolia
f magnolia du Japon
d rote Magnolie
l *Magnolia obovata; Magnolia hypoleuca*

5898 white lily-turf
f ophiopogon jaburan
d Jaburan-Schlangenbart
l *Ophiopogon jaburan; Mondo jaburan*

5899 white lupine
f lupin blanc
d weisse Lupine; Weisslupine
l *Lupinus albus*

5900 white mignonette
f réséda blanc
d weisse Reseda; weisser Wau
l *Reseda alba*

* **white mistletoe → 2113**

5901 white-moon petunia; night-flowering petunia
f pétunia blanc
d weisse Petunie
l *Petunia nyctanginiflora; Petunia axillaris*

5902 white mulberry
f mûrier blanc
d weisser Maulbeerbaum
l *Morus alba*

5903 white mullein
d mehlige Königskerze
l *Verbascum lychnitis*

5904 white mustard
f moutarde blanche
d weisser Senf; englischer Senf
l *Sinapis alba; Brassica hirta; Brassica alba*

5905 white oak
f chêne blanc
d weisse Eiche; Weisseiche
l *Quercus alba*

5906 white poplar
f peuplier blanc; ypréau
d Silberpappel; Weisspappel
l *Populus alba*

5907 white quebracho
f quebracho
d Quebrachobaum
l *Aspidosperma*

5908 white sandalwood
f santal blanc
d Sandelholzbaum
l *Santalum album*

5909 white sapote
f sapote blanc

d weisse Sapote
l *Casimiroa edulis*

5910 white Spanish broom
f genêt blanc d'Espagne; genêt du Portugal
l *Cytisus multiflorus*

5911 whitespot betony
f épiaire des bois; ortie puante
d Waldziest
l *Stachys sylvatica*

5912 whitespot giant arum
f amorphophallus campanulé
d glockenförmiger Amorphophallus
l *Amorphophallus campanulatus*

5913 white spruce; Canadian spruce
f épicéa blanc; épicéa du Canada; sapin de Normandie; sapinette blanche
d Schimmelfichte; Weissfichte
l *Picea alba; Picea glauca; Picea canadensis*

5914 white-stem pondweed; long pondweed
f potamot allongé
d langstieliges Laichkraut
l *Potamogeton praelongus*

5915 white stonecrop
f orpin blanc
d weisse Fetthenne; weisser Mauerpfeffer
l *Sedum album*

5916 white swallowwort
f dompte-venin
d Hundstod; Schwalbenschwanz; Giftwurzel
l *Cynanchum vincetoxicum; Vincetoxicum officinale*

5917 white sweet clover; Bokhara clover
f mélilot blanc
d Bucharaklee; weisser Steinklee; weisser Honigklee; Hanfklee; Kabulklee; Pferdeklee; Riesenklee
l *Melilotus albus*

5918 whitetop
f hyménophyse
d Pfeilkresse
l *Hymenophysa; Cardaria*

5919 whitetop sedge
f dichromène
l *Dichromena*

5920 white truffle
d Weisstrüffel
l *Choeromyces*

5921 white vine spinach; Malabar nightshade
f épinard de Malabar
d indischer Spinat; Malabarspinat
l *Basella alba*

* **white water lily → 2132**

5922 white willow
f saule blanc
d Silberweide; Weissweide
l *Salix alba*

5923 whitish russula
f russule sans lait
d Milchlingstäubling; milchlingsähnlicher Weisstäubling
l *Russula delica*

* **whitish sedge → 4970**

5924 whitlowwort
f illécèbre verticillé
d quirliges Knorpelkraut
l *Illecebrum verticillatum*

* **whitlowwort → 1897, 3788**

5925 whorled pedicularis
f pédiculaire verticillée
d quirlblättriges Läusekraut
l *Pedicularis verticillata*

5926 whorled plantain
d Sandwegerich
l *Plantago indica*

5927 whorled Solomon's-seal
f sceau de Salomon verticillé
d quirlblättriges Salomonssiegel; quirlblättrige Weisswurz
l *Polygonatum verticillatum*

5928 whortleberry willow
f saule faux-myrte
d Heidelbeerweide; Moorweide
l *Salix myrtilloides*

5929 Wichura rose
f rosier de Wichura
d Wichuras Rose
l *Rosa wichuraiana*

5930 widdringtonia
f widdringtonia

d Widdringtonia
l *Widdringtonia*

5931 wide-leaved sea lavender
f statice à larges feuilles
l *Limonium latifolium; Statice latifolia*

5932 wide-world parnassia
f parnassie des marais
d Sumpfherzblatt
l *Parnassia palustris*

5933 widgeon weed
f ruppie
d Salde; Ruppie
l *Ruppia*

5934 widgeon weed
d Meerfaden; Meersalde; Schnabelsalde; Strandsalde; Wasserriemen
l *Ruppia maritima*

5935 wigandia
f wigandie
l *Wigandia*

* **wild angelica → 6029**

5936 wild basil savory
d Borstenquendel; Wirbeldost
l *Clinopodium vulgare; Satureia vulgaris*

5937 wild bean
f strophostyle
l *Strophostyles*

5938 wild bugloss
f lycopside
d Krummhals
l *Lycopsis*

5939 wild bugloss
f lycopside des champs
d Ackerkrummhals
l *Lycopsis arvensis*

5940 wild cabbage
f chou
d Gemüsekohl; Küchenkohl; Gartenkohl
l *Brassica oleracea*

5941 wild calla; water arum
f calla des marais
d Drachenwurz; Schweinekraut; Schweinsohr; Sumpfschlangenkraut; Sumpfschlangenwurz
l *Calla palustris*

* **wild camomile → 2452**

5942 wild carrot
f carotte sauvage
d Gelbrübe; gelbe Rübe; gemeine Mohrrübe; wilde Möhre
l *Daucus carota*

5943 wild celery
f céleri
d echte Sellerie; gemeine Sellerie; Sellerie; Eppich
l *Apium graveolens*

5944 wild celery
f vallisnérie
d Wasserschraube
l *Vallisneria*

5945 wild coffee
f psychotria
d Brechstrauch
l *Psychotria*

5946 wild garden parsnip
d Waldpastinak
l *Pastinaca sativa* ssp. *sylvestris*

5947 wild ginger; asarabacca
f asaret
d Haselwurz
l *Asarum*

* **wild hydrangea → 5039**

5948 wild indigo
f baptisie
d Färberhülse; Wildindigo
l *Baptisia*

5949 wild ixora
f isertie
l *Isertia*

* **wild mallow → 2797**

* **wild marjoram → 1484**

5950 wild myrobalan plum
f prunier myrobalan
d Kirschpflaume; Myrobalane
l *Prunus cerasifera divaricata; Prunus divaricata*

5951 wild oat
f folle avoine; avéneron
d Flughafer; Wildhafer; Windhafer
l *Avena fatua*

5952 wild ochra; wild okra
f malachre
d Weichmalve
l *Malachra*

5953 wild olive
f olivier sauvage
d wilder Ölbaum
l *Olea oleaster*

5954 wild pansy
f pensée
d Stiefmütterchen; dreifarbiges Veilchen
l *Viola tricolor*

5955 wild radish
f ravenelle
d Hederich
l *Raphanus raphanistrum*

5956 wild rice
f zizanie
d Wasserreis
l *Zizania*

5957 wild rye; lyme grass
f élyme
d Haargerste; Strandroggen
l *Elymus*

5958 wild senna
f casse du Maryland
l *Cassia marilandica; Cassia medsgeri*

* **wild service tree → 1128**

* **wild strawberry → 2127**

5959 wild sweet crab apple
f pommier odorant
d wohlriechender Zierapfel
l *Malus coronaria*

5960 wild tobacco
f acniste
l *Acnistus*

5961 wild wheat
d wilder Emmer
l *Triticum dicoccoides*

5962 wildwood buttercup
d Waldhahnenfuss
l *Ranunculus nemorosus*

5963 Willmott winter hazel
f corylopsis de Miss Willmott
d Willmotts Scheinhasel
l *Corylopsis willmottiae*

5964 willow; osier
f saule
d Weide
l *Salix*

5965 willow family
f salicacées
d Weidengewächse; Salikazeen
l *Salicaceae*

5966 willow fire-wheel tree
f sténocarpe à feuilles de saule
d Weidenschmalfrucht
l *Stenocarpus salignus*

5967 willow-leaf red quebracho
f quebracho rouge; quebracho coloré
d roter Quebrachobaum
l *Schinopsis balansae*

5968 willow-leaved aster
f aster à feuilles de saule
d weidenblättrige Aster; Weidenaster
l *Aster salicifolia*

5969 willow-leaved eucalyptus; black peppermint
f eucalyptus à feuilles de saule
d Pfefferminzbaum; Wangara
l *Eucalyptus salicifolia*

5970 willow-leaved inula
f inule à feuilles de saule
d Weidenalant; weidenblättriger Alant
l *Inula salicana*

5971 willow-leaved pear
f poirier à feuilles de saule
d weidenblättriger Birnbaum
l *Pyrus salicifolia*

5972 willow-leaved spirea; Aaron's-beard
f spirée à feuilles de saule
d Weidenspierstrauch; weidenblättriger Spierstrauch
l *Spiraea salicifolia*

5973 willow lettuce
f laitue à feuilles de saule
d Weidenlattich
l *Lactuca saligna*

5974 willow oak
f chêne à feuilles de saule
d Weideneiche
l *Quercus phellos*

5975 willowweed
f épilobe
d Weidenröschen
l *Chamaenerium; Epilobium*

5976 Wilson barberry
f vinettier de Lady Wilson
d Wilsons Berberitze
l *Berberis wilsonae*

5977 Wilson deutzia
f deutzie de Wilson
d Wilsons Deutzie
l *Deutzia wilsoni*

*** Wilson fir → 5978**

5978 Wilson Sakhalin fir; Wilson fir
f sapin de Wilson
d Wilsons Tanne
l *Abies wilsoni; Abies sachalinensis nemorensis*

5979 Wilson spirea
f spirée de Wilson
d Wilsons Spierstrauch
l *Spiraea wilsoni*

5980 Wilson spruce
f épicéa de Wilson
d Wilsons Fichte
l *Picea wilsoni*

5981 wind bent grass; wind grass; corn grass
f agrostide des champs
d Ackerschmiele; Ackerstraussgras; Ackerhafer; Ackerwindhalm; Schlinggras; Taugras; Windhalm
l *Apera spica-venti; Agrostis spica-venti*

5982 windmill grass; finger grass
f chloris
d Fingergras
l *Chloris*

5983 windmill palm
f trachycarpe
d Hanfpalme
l *Trachycarpus*

5984 wineberry
f aristotélie
d chinesischer Jasmin
l *Aristotelia*

*** wineberry → 3784**

5985 wine mauritia; murity palm
f mauritier
d Buruti; Wein-Mauritiuspalme
l *Mauritia vinifera*

*** wine palm → 3537**

5986 wine yeast
f levain
d Weinhefe
l *Saccharomyces ellipsoideus*

5987 wing cactus
d Flügelkaktus
l *Pterocactus*

5988 wing celtis
f ptéroceltis
d Flügelzürgel
l *Pteroceltis*

5989 winged elm
f orme ailé
l *Ulmus alata*

5990 winged euonymus
f fusain ailé
d Flügelspindel; geflügelter Spindelbaum; Korkflügelspindelbaum; Korkspindel
l *Euonymus alatus*

5991 winged everlasting
f immortelle des sables
d Sandimmortelle; Papierknöpfchen
l *Ammobium alatum*

5992 winged lophira
f lophire pyramidal
l *Lophira alata*

*** winged pea → 5197**

5993 winged yam
f igname ailée; grande igname
d geflügelte Yamswurzel
l *Dioscorea alata*

5994 winghead
f ptérocéphale
d Flügelkopf
l *Pterocephalus*

5995 wing-kelp; murlin; kombu
d Flügeltang
l *Alaria*

5996 wing nut
f ptérocarya

d Flügelnuss
l *Pterocarya*

5997 wing-seeded spurry
f spergule à cinq étamines
d fünfmänniger Spark
l *Spergula pentandra*

5998 winter aconite
f éranthe
d Winterling
l *Eranthis*

5999 winter cress
f barbarée; herbe de Sainte-Barbe
d Barbarakraut; Bärbelkraut; Barbenkraut; Winterkresse
l *Barbarea*

6000 winter currant; blood currant; red-flowering currant
f groseillier sanguin
d blutrote Johannisbeere; Blutjohannisbeere; rotblühende Johannisbeere
l *Ribes sanguineum*

6001 winter fat
f eurotia
d Hornmelde
l *Eurotia*

6002 winter grape
f vigne d'Espagne; cépage américain
d Winterrebe
l *Vitis berlandieri*

6003 wintergreen
f gaulthérie
d nordamerikanisches Wintergrün; Steinbeere
l *Gaultheria*

* **wintergreen → 4377**

* **wintergreen family → 4378**

6004 winter hazel
f corylopsis
d Scheinhasel
l *Corylopsis*

6005 winter rape
f navet; chou-navet
d Raps; Reps(kohl)
l *Brassica napus*

6006 winter savory
f sarriette montagnarde
d Bergbohnenkraut; Winterbohnenkraut
l *Satureia montana*

6007 Winter's bark (drimys)
f drimys de Winter
d Winters Gewürzrindenbaum
l *Drimys winteri*

6008 winter squash
f courge commune
d Riesenkürbis
l *Cucurbita maxima*

6009 winter sweet
f chimonanthe précoce
d früher Gewürzstrauch
l *Chimonanthus praecox*

* **winter vetch → 2693**

6010 wire vine
f muhlenbeckie
d Scheinknöterich
l *Muehlenbeckia*

6011 wistaria
f glycine
d Glyzine
l *Wistaria; Wisteria*

* **witch grass → 4026**

6012 witch hazel
f noisetier de sorcière; hamamélide
d Hexennuss; Zaubernuss; Zauberstrauch
l *Hamamelis*

6013 witch-hazel family
f hamamélidacées
d Hamamelisgewächse; Hamamelidazeen; Amberbäume; Balsambäume
l *Hamamelidaceae*

6014 woad
f pastel
d Waid
l *Isatis*

6015 woad-waxen
f genêt
d Ginster
l *Genista*

6016 wolfberry; desert thorn; squawbush; boxthorn
f lyciet

d Bocksdorn; Teufelszwirn
l *Lycium*

6017 wolffia
f wolffia
d Entenlinse; Zwerglinse
l *Wolffia*

6018 wolfsbane monkshood
f aconit étrangle-loup
d Wolfseisenhut
l *Aconitum lycoctonum; Aconitum vulparia*

6019 wonder violet
f violette étonnante
d Wunderveilchen
l *Viola mirabilis*

* **wood agaric → 2326**

6020 wood apple
f féronie
d Elefantenapfelbaum
l *Feronia*

6021 wood bedstraw
f gaillet des forêts
d Waldlabkraut
l *Galium silvaticum*

* **wood betony → 3383**

6022 woodbine (honeysuckle)
f chèvrefeuille des bois
d Waldgeissblatt; deutsches Geissblatt; nördliches Geissblatt
l *Lonicera periclymenum*

* **wood blewit → 214**

6023 wood bluegrass
f pâturin des bois
d Hainrispengras
l *Poa nemoralis*

6024 wood crane's-bill
f géranium des bois
d Waldstorchschnabel
l *Geranium sylvaticum*

6025 wood cudweed
f gnaphale des forêts
d Waldruhrkraut
l *Gnaphalium sylvaticum*

6026 wood fern
f dryoptéris
d Wurmfarn
l *Dryopteris*

6027 wood figwort
d knotige Braunwurz; Feigwarzenkraut; Knotenbraunwurz
l *Scrophularia nodosa*

6028 wood germander
f germandrée sauvage
d Salbeigamander; Waldgamander
l *Teucrium scordonia*

6029 woodland angelica; wild angelica
f angélique des bois; angélique sauvage
d Waldangelika; Waldbrustwurz; Waldengelwurz; wilde Brustwurz
l *Angelica sylvestris*

6030 woodland beaked chervil
f anthrisque sauvage
d Waldkerbel; wilder Kerbel; Wiesenkerbel
l *Anthriscus sylvestris*

6031 woodland bulrush
f scirpe des bois
d Waldbinse; Waldsimse
l *Scirpus sylvaticus*

6032 woodland European grape
d Waldweinrebe
l *Vitis vinifera sylvestris*

6033 woodland forget-me-not
f myosotis des bois
d Bergvergissmeinnicht; Waldvergissmeinnicht
l *Myosotis sylvatica*

6034 woodland groundsel
f séneçon des bois
d Waldgreiskraut; Waldkreuzkraut
l *Senecio sylvaticus*

6035 wood melick
d einblütiges Perlgras; zartes Perlgras
l *Melica uniflora*

6036 wood nettle
f laportée
d Brennpflanze
l *Laportea*

6037 wood nymph; one-flowered pyrola
f pyrole uniflore
d Moosauge; einblütiges Wintergrün
l *Moneses uniflora; Pyrola uniflora*

6038 wood pimpernel
f lysimaque de bois
d Hainfelberich
l *Lysimachia nemorum*

6039 wood reed
f cinna
l *Cinna*

6040 woodruff
f aspérule
d Waldmeister; Färbermeier
l *Asperula*

6041 wood rush
f luzule
d Hainsimse; Marbel
l *Luzula*

* **wood sanicle → 2121**

6042 wood sedge
f laîche des bois
d Waldsegge; Waldriedgras
l *Carex sylvatica*

6043 woodsia
f woodsia
d Wimperfarn
l *Woodsia*

* **wood sorrel → 3985**

6044 wood-sorrel oxalis
f surelle-petite-oseille; alléluia
d Waldsauerklee
l *Oxalis acetosella*

6045 wood starwort
f stellaire des bois
d Waldsternmiere; Hainmiere
l *Stellaria nemorum*

6046 wood vetch
f grande vesce des montagnes
d Waldwicke
l *Vicia sylvatica*

* **woody aloe → 299**

6047 woolly betony; woolly woundwort
f épiaire laineuse
d Samtblatt
l *Stachys lanata*

6048 woolly draba
f drave tomenteuse
d filziges Felsenblümchen
l *Draba tomentosa*

6049 woolly lactarius; woolly milk cap
f lactaire à toison
d Birkenmilchling; zottiger Birkenreizker
l *Lactarius torminosus*

6050 woolly-leaf mock orange
f seringa du Népal
l *Philadelphus tomentosus; Philadelphus nepalensis*

* **woolly milk cap → 6049**

6051 woolly-pod vetch
f vesce à fruit velu
d rauhsamige Vogelwicke
l *Vicia dasycarpa*

6052 woolly terminalia
f sain
l *Terminalia tomentosa*

6053 woolly-twiged willow
f saule à rameaux velus
d filzästige Weide; langblättrige Weide
l *Salix dasyclados*

6054 woolly willow
d wollige Weide
l *Salix lanata*

* **woolly woundwort → 6047**

6055 wool mullein
d grosse Königskerze; grossblütige Königskerze; grossblumiges Wollkraut
l *Verbascum thapsiforme*

* **worm moss → 1623**

6056 wormseed goosefoot; wormseed tea; Mexican tea
f thé du Mexique; ansérine ambroisie
d mexikanisches Teekraut; mexikanischer Tee
l *Chenopodium ambrosioides*

* **wormwood → 4640**

* **woundwort → 550, 1389, 3159, 3227, 3987**

6057 wrinkled pholiota; gypsy
f pholiote aux chèvres; pholiote ridée
d Reifpilz; Runzelschüppling; Zigeuner
l *Pholiota caperata; Rozites caperata*

6058 wrinkle-leaf rockrose
f ciste crépu

d krause Zistrose
l *Cistus crispus*

6059 wulfenia
f wulfénie
d Wulfenie; Kuhtritt
l *Wulfenia*

* **wych elm → 4771**

6060 xylocarpus
f xylocarpe
d Granatenbaum
l *Xylocarpus*

6061 xylosma; manzanillo
f xylosme
l *Xylosma*

Y

6062 yam
f igname
d Yamswurzel
l *Dioscorea*

6063 yam bean
f pachyrrhize
d Yamsbohne
l *Pachyrrhizus*

6064 yam family
f dioscoréacées
d Dioskoreazeen; Yamspflanzen; Schmeerwurzgewächse; Yamswurzelgewächse
l *Dioscoreaceae*

* **yampee → 1748**

6065 Yangtao actinidia
f actinidie de Chine
d chinesische Aktinidie; chinesischer Strahlengriffel
l *Actinidia chinensis*

6066 yard long cowpea; asparagus bean
d Spargelbohne
l *Dolichos sesquipedalis; Vigna sesquipedalis*

6067 yarrow
f achillée
d Schafgarbe
l *Achillea*

6068 yaupon
f thé des Apalaches; apalachine
l *Ilex vomitoria*

6069 yeast; saccharomyces
f levure; saccharomyces
d echter Hefepilz; echte Hefe
l *Saccharomyces*

6070 yeast fungi
f saccharomycétacées
d Hefepilze; Sprosspilze
l *Saccharomycetaceae*

6071 Yeddo spruce
f épicéa de Yezo
d Ajanfichte; Yedofichte
l *Picea jezoensis; Picea ajanensis*

6072 yellow alpine skullcap
d Hopfenhelmkraut
l *Scutellaria lupulina; Scutellaria alpina* var. *lupulina*

6073 yellow-and-blue forget-me-not
f myosotis versicolore
d buntes Vergissmeinnicht
l *Myosotis versicolor*

6074 yellow avens
d steife Nelkenwurz
l *Geum strictum*

6075 yellow bedstraw
f gaillet vrai; gaillet jaune
d echtes Labkraut; gelbes Waldstroh
l *Galium verum*

* **yellow-berried nightshade → 6103**

6076 yellow birch
f bouleau jaune
d Gelbbirke; gelbe Birke
l *Betula lutea*

* **yellow-brown boletus → 820**

6077 yellow buckeye
f pavier jaune; pavier fauve
l *Aesculus octandra; Aesculus flava; Pavia flava*

6078 yellow centaurea
d Sommerflockenblume
l *Centaurea solstitialis*

6079 yellow cephalaria
f céphalaire des Alpes
d Alpenschuppenkopf
l *Cephalaria alpina*

6080 yellow corydalis
f corydale jaune
d gelber Lerchensporn
l *Corydalis lutea*

6081 yellow-cracked boletus; submentose mushroom
f bolet subtomenteux; cèpe mou
d filziger Röhrling; Ziegenlippe
l *Boletus subtomentosus*

* **yellow day lily → 2334**

6082 yellow-devil hawkweed
d reichblütiges Habichtskraut
l *Hieracium floribundum*

6083 yellow everlasting
f immortelle des sables

d Sandstrohblume
l *Helichrysum arenarium*

6084 yellow-eyed grass; sword plant
f xyride
d Degenbinse; Degenkraut
l *Xyris*

6085 yellow field cress
d Waldkresse; wilde Sumpfkresse
l *Rorippa silvestris; Nasturtium silvestre*

6086 yellow flag iris
f iris faux-acore
d gelbe Schwertlilie; Wasserschwertlilie; Teichlilie; Gilgenwurzel
l *Iris pseudoacorus*

6087 yellow flax
f reinwardtia
l *Reinwardtia*

6088 yellow-flowered gourd; egg squash
f coloquinelle
d Eierkürbis
l *Cucurbita pepo* var. *ovifera*

* **yellow-flowered restharrow → 5048**

6089 yellow foxglove
f digitale à grandes fleurs
d grossblütiger Fingerhut; blassgelber Fingerhut
l *Digitalis grandiflora; Digitalis ambigua*

6090 yellow foxtail
f sétaire glauque
d graugrüne Borstenhirse; gelbhaariger Fennich; fuchsgelbes Fennichgras
l *Setaria glauca*

6091 yellow gentian
f gentiane jaune; grande gentiane
d Amarellkraut; Bitterwurz(el); gelber Enzian; grosser Enzian
l *Gentiana lutea*

* **yellow granadilla → 5815**

6092 yellow-green silene
d grünliches Leimkraut
l *Silene chlorantha*

6093 yellow hawthorn
f épine à fruits jaunes
d gelber Weissdorn
l *Crataegus flava*

6094 yellow-heart prickly ash
f satiné jaune; bois satiné d'Amérique
d westindischer Seidenholzbaum
l *Zanthoxylum flavum*

6095 yellow horn
f xanthocère
d Gelbhorn
l *Xanthoceras*

6096 yellow horn poppy
f glaucie jaune
d gelber Hornmohn
l *Glaucium flavum*

6097 yellow marsh saxifrage
d Moorsteinbrech
l *Saxifraga hirculus*

6098 yellow meadow rue
f pigamon jaune
d gelbe Wiesenraute
l *Thalictrum flavum*

6099 yellow Mexican water lily
f nénuphar jaune
d gelbe Seerose; mexikanische Seerose
l *Nymphaea flava; Nymphaea mexicana*

6100 yellow mignonette
f réséda à fleurs jaunes; réséda sauvage
d gelbe Reseda
l *Reseda lutea*

6101 yellow mombin
f mombin jaune
d Mombinpflaume; Schweinspflaume
l *Spondias lutea; Spondias mombin*

6102 yellow mountain saxifrage
d Borstensteinbrech; Fetthennensteinbrech
l *Saxifraga aizoides; Leptasea aizoides*

6103 yellow nightshade; yellow-berried nightshade
f morelle jaune
d zottiger Nachtschatten
l *Solanum luteum; Solanum villosum*

* **yellow oat grass → 6109**

6104 yellow pea vine
f gesse jaune; orobe jaune

d gelbe Platterbse
l *Lathyrus luteus*

* **yellow poplar → 5636**

6105 yellowroot
f xanthorrhize
d Gelbwurz
l *Xanthorrhiza; Zanthorrhiza*

6106 yellow sedge
f laîche jaunâtre; carex jaune
d gelbes Riedgras; gelbe Segge
l *Carex flava*

6107 yellow-stem white willow
f osier jaune; osier vitellin; verdoisis; amarinié
d Dotterweide
l *Salix vitellina; Salix alba vitellina*

6108 yellow sweet clower
f mélilot officinal
d gelber Steinklee; echter Steinklee; gebräuchlicher Steinklee; echter Honigklee; Ackersteinklee; Melilotenklee
l *Melilotus officinalis*

* **yellow trefoil → 638**

6109 yellow trisetum; yellow oat grass; golden oat gras
f avoine jaunâtre
d Goldhafer
l *Trisetum flavescens; Avena flavescens*

6110 yellow unicorn plant
d Gemsenhörner
l *Ibicella lutea; Martynia lutea*

6111 yellow vetch
f vesce jaune
d gelbe Wicke
l *Vicia lutea*

6112 yellow whitlowwort
f badianier à petites fleurs
d kleinblütiger Sternanis
l *Illicium parviflorum*

6113 yellow wild indigo
f baptisie des teinturiers
d gelbe Färberhülse
l *Baptisia tinctoria*

6114 yellowwood
f cladraste
d Gelbholz
l *Cladrastis*

6115 yellow wood anemone
f anémone fausse-renoncule
d gelbe Osterblume; gelbes Windröschen
l *Anemone ranunculoides*

6116 yellowwort
f chlorette; chlore
d Bitterling
l *Chlora; Blackstonia*

* **yerba maté → 4048**

6117 yew
f if
d Eibe(nbaum)
l *Taxus*

6118 yew family
f taxacées
d Eibengewächse
l *Taxaceae*

6119 ylang-ylang; ilang-ilang
f ylang-ylang
d Ilang-Ilang; Ylang-Ylang
l *Cananga odorata*

6120 yucca; Adam's-needle
f yucca
d Palmlilie; Adamsnadel; Bajonettbaum
l *Yucca*

6121 Yugoslav globe thistle
f échinope du Bannat
l *Echinops bannaticus*

6122 yulan magnolia
f magnolia yulan
d Lilienmagnolie
l *Magnolia denudata; Magnolia conspicua; Magnolia yulan*

Z

6123 Zabel spirea
f spirée de Zabel
d Zabels Spierstrauch
l *Spiraea zabeliana*

6124 Zanzibar oil vine
d Talerkürbis
l *Telfairia pedata*

6125 zedoary turmeric
f zédoaire
d Zitwer
l *Curcuma zedoaria*

6126 zephyr lily
f zéphyranthe
d Zephyrblume
l *Zephyranthes*

6127 Zeyher mock orange
f seringa de Zeyher
d Zeyhers Pfeifenstrauch
l *Philadelphus zeyheri*

6128 zigzag clover
f trèfle intermédiaire
d mittlerer Klee
l *Trifolium medium*

6129 zinnia
f zinnia
d Zinnie
l *Zinnia*

6130 ziziphora
f ziziphore
d Zizifer
l *Ziziphora*

6131 zygomycetes
f zygomycètes
d Jochpilze; Zygomyzeten
l *Zygomycetes*

FRANÇAIS

abaca 2
abélie 3
abobra 1659
abobra à feuilles laciniées 1660
abrétier 3784
abricotier 1350
abricotier commun 1350
abricotier d'Alexandrie 4336
abricotier de Briançon 773
abricotier de Mandchourie 3457
abricotier de Saint-Domingue 3453
abricotier des Antilles 3453
abricotier de Sibérie 4891
abricotier du Japon 3015
abricotier noir 4336
abricotier pourpré 4336
abrome 4
abronie 5
absinthe 1564
absinthe commune 1564
absinthe de mer 3496
absinthe des rivages 489
absinthe officinale 1564
absinthe petite 3496
absinthe pontique 4548
absinthe romaine 4548
abutilon 7
abutilon hybride 2898
abutilon ordinaire 1217
acacia 10
acacia à bois noir 655
acacia à catechu 1056
acacia à gomme 2662
acacia à longues épines 103
acacia à longues feuilles 5409
acacia blanchâtre 4957
acacia commun 636
acacia d'Arabie 399
acacia de Constantinople 4948
acacia de Farnèse 5351
acacia de Sibérie 3329, 4617
acacia du Sénégal 2662
acacia noir 2637
acacia rose 4555
acajou à meubles 5871
acajou à pommes 1378
acajou cédrel 1234
acajou d'Afrique 2985
acajou de Cuba 5871
acajou du Sénégal 4821
acajou femelle 1234
acajou mahagon 5871
acanthacées 14
acanthe 12
acanthe épineuse 5151
acanthe molle 5077
acanthe sauvage 4770
acantholimon 4306
acanthopanax 11
accroche-abeille 780
acéna 4854
acéracées 3482
ache des montagnes 2418
achillée 6067
achillée millefeuille 1565
achillée musquée 3769
achillée sternutatoire 5054
achimène 15
achyranthe 1108
acnide 5518
acniste 5960
aconit 3660
aconit anthora 4373
aconit étrangle-loup 6018
aconit napel 17
aconit panaché 3465
acore 5365
acore graminé 3073
acore odorant 1918
acore vrai 1918
acrocome 18
acronychie 19
acronychie d'Australie 480
acrostiche 20
acrostique 20
actée 441
actée en épi 616
actinidiacées 22
actinidie 21
actinidie à feuilles dentées 748
actinidie de Chine 6065
actinidie denticulée 748
actinidie kolomikta 3176
actinidie polygame 4968
actinomyces 23
adénandre 29
adénanthère 493
adénophore 3198
adénophore à feuilles de lis 3302
adinandre 30
adlumie 32
adonide 33
adonide d'automne 4164
adonide d'été 5317
adonide printanière 5179
adoxacées 3684
adoxe 3765
adoxe muscatelle 3766
æchmée 35
æschynanthe 466
æschynomène 3109
afzélie 53
afzélie d'Afrique 39
agalancé 1875
agapanthe 54
agaric 57
agaricacées 2468
agaricales 58
agaric à odeur de farine 4226
agaric couleur de miel 2838
agaric cultivé 2419
agaric des bois 2326
agaric des bruyères 2870
agaric des champs 1476
agaric des forêts 2326
agaric des jachères 2870
agaric visqueux 2516
agathis 1795
agathis blanc 5890
agave 59
agave d'Amérique 1092
agave faux-fourcroya 2781
agave sisal 4976
agave vert noirâtre 4328
agérate 60
agérate du Mexique 3594
agératoire 60
aglaie 61
aglaonème 62
agneau de Scythie 4784
agnos 1120
agriophylle 64
agripaume 1475
agropyre 1640
agrostemma 1610
agrostide 534
agrostide blanche 4469
agrostide commun 1335
agrostide des champs 5981
agrostide des chiens 5708
agrostide des marais 1665
agrostide vulgaire 1335
agrume 1246
agrumes 1247
aigremoine 63
aigremoine commune 1348
aigremoine eupatoire 1348
aigremoine odorante 2355
aiguille 91
aiguille de berger 4861
ail 2432
ailante 65
ailante glanduleux 5584
ail à toupet 5450
ail blanc 2432
ail commun 2432
ail de cerf 3368
ail des champs 2579, 4284
ail des chiens 5450
ail de Schubert 4762
ail des ours 4415
ail des vignes 2223
ail doré 3303
ail rocambole 2474
aïphane 4601
airelle 692
airelle à corymbes 2796
airelle des marais 717
airelle du Canada 969
airelle myrtille 3784
airelle noire 717
airelle ponctuée 1645
airelle rouge 1645
aïzoacées 1033
ajonc 2566
ajonc d'Europe 1426
ajonc nain 1942
akébie 73
akée 74
akée d'Afrique 75
alavert 3642
albizzie 79
albizzie anthelminthique 3757
alcanna 97
alchémille 3202
alchémille commune 1454
alchémille des Alpes 3709
alchémille des champs 2226
alchémille vulgaire 1454
alchornée 1225
aldrovande 83
aldrovande vésiculeuse 5793
alectryon 84
alerce 2267

alerce de Patagonie 4071
alerce de Tasmanie 5448
alétris 5221
alévrite 992, 5640
alfa 2062
algues 92
algues brunes 815
algues rouges 4433
algues vertes 2621
alguette 1504
aliboufier 5056
aliboufier benzoïn 5315
aliboufier du Sumatra 5315
aliboufier officinal 1915
alisier blanc 5881
alisier de bois 1128
alisier de Bourgogne 5881
alisier de Laponie 3902
alisier torminal 1128
alisier tranchant 1128
alismacées 95
alisme plantain d'eau 211
alkanna 97
alléluia 6044
alliaire 2433
alliez 609
alocasie 108
aloès 109
aloès à gorge de perdrix 3132
aloès corne de bélier 299
aloès faux-dragonnier 4392
aloès féroce 1003
aloès officinal 5601
aloès panache 3132
aloès soccotrin 5076
aloès vrai 5601
aloès vulgaire 3568
alonzoa 3529
alopécure des Alpes 133
alphitonie 5578
alphitonie élevée 5579
alpinia 158
alpiste 984
alpiste des Canaries 985
alpiste roseau 4473
alstonie 165
alstrémère 166
alternanthère 3111
althée 169, 3507
alysse 172
alysse à calice persistant 4007
alysse épineuse 5152
alysse maritime 5352
alysse montagnarde 3693
alysse odorante 5352
alysse saxatile 2551
amadouvier 5529
amande de terre 1230
amandier 104
amandier à coque tendre 2354
amandier à fleurs du Japon 105
amandier amer 597
amandier de Chine 2311
amandier nain de Géorgie 4611
amandier nain de Russie 4611
amanitacées 173
amanite à étui 4853
amanite bulbeuse 1820
amanite de César 918
amanite engaînée 4853
amanite panthère 4032
amanite phalloïde 1820
amanite printanière 5180
amanite rougeâtre 713
amanite tue-mouches 2319
amanite vireuse 1842
amarantacées 175
amarante 174
amarante à fleurs en queue 3387
amarante crête-de-coq 1410
amarante paniculée 5005
amarante queue-de-renard 3387
amarante réfléchie 4464
amarantine globuleuse 1421
amarinié 6107
amaryllidacées 177
amaryllide 176
amaryllis 176
amaryllis à bandes 448
amaryllis belladone 526
amaryllis de Guernsey 2658
amaryllis de Joséphine 2469
amaryllis jaune 2158
amaryllis pourpre 4742
ambrette 3761
ambroisie 4400
ambroisie à feuilles d'armoise 1513
amélanchier 4832
amélanchier à feuilles rondes 1949
amélanchier commun 2428
amélanchier du Canada 4838
amélanchier du Japon 342
amidonnier 2029
ammi 72, 216
ammobium 217
ammophile 487
amome des Indes 1419
amoora 218
amorphe 2171
amorphe arbustif 2948
amorphophallus 220
amorphophallus campanulé 5912
amorphophallus de Rivière 1848
amourette 643
ampélopsis 221
amsinckie 2217
amsonie 223
anabase 233
anacamptide 234
anacardiacées 1043
anacardier 1042, 1378
anacycle 2460
anagallide 4185
anagallide des champs 4747
anamirte 2264
ananas 235, 4191
ananas cultivé 4191
anaphalide 4094
anaphalide nacrée 1488
anchiétée 236
anchuse 840
ancolie 1341
ancolie commune 2087
ancolie de jardins 2087
ancolie de Skinner 4985
ancolie du Canada 188
andrachné 239
andréacées 240
andromède 241
andromède à feuilles de polium 723
andromède d'Europe 723
andromède en arbre 5095
androsace 4527
androsace de Suisse 5401
androselle 4527
anémone 242
anémone à feuilles de narcisse 3796
anémone couronnée 4273
anémone des Alpes 112
anémone des bois 2134
anémone des fleuristes 4273
anémone des prés 3554
anémone du Japon 3013
anémone éclatante 2273
anémone fausse-renoncule 6115
anémone printanière 5721
anémone pulsatille 2116, 5177
anémone sauvage 5061
aneth 1399, 1857
aneth doux 1411
aneth odorant 1399
angelin 246, 907
angélique 245, 4044
angélique des bois 6029
angélique en arbre d'Amérique 1849
angélique en arbre de Chine 1167
angélique en arbre du Japon 3016
angélique officinale 2406
angélique palustre 714
angélique sauvage 6029
angioptère 5730
angiospermes 248
angréacum 746
angurie 5869
anhalonie 3344
anis 253
anis étoilé vrai 5607
anisochèle 255
anisochile 255
anisoptère 3590
anneslée 257
anonacées 1752
anone 1751
anone chérimolier 1132
anone hérissée 5093
anone muriquée 5093
anone réticulée 849
ansérine 2561
ansérine ambroisie 6056
ansérine blanche 3212
ansérine des murs 3837
antennaire 4363
antennaire des Carpathes 1028
antennaire dioïque 1510
anthémis 963
anthémis des champs 2220
anthémis des teinturiers 2528
anthémis noble 4546
anthéric 269
anthistirie 3131
antholise 3415
anthosperme 182
anthrisque 494
anthrisque commun 856

anthrisque sauvage 6030
anthure 270
anthyllide 271
anthyllide vulnéraire 3159
antidesme 566, 1164
antoinette 2262
aouara d'Afrique 46
aouret 3642
apalachine 6068
aphananthe 274
aphélandre 275
aphélandre orangée 3936
aphyllophorales 276
apio 277
apios 4282
apocyn 1872
apocynacées 1873
apocyn chanvrin 2775
apocyn gobe-mouches 5173
apocyn vénitien 4291
aponogéton 5807
aponogétonacées 278
apophylle 279
aporoselle 280
aposeris 281
aptandre 290
apulée 291
aquilaire 1958
aquilégie 1341
arabette 4516
arabette de Gérard 2450
arabette du Caucase 143, 5774
arabette hérissée 2688
aracées 334
arachide 4087, 4088
araliacées 2495
aralie 295
aralie du Japon 3034, 5675
aralie épineuse 1849
araucarie 297
araucarie d'Australie 855
araucarie de Bidwill 855
araucarie de Cunningham 1730
araucarie du Brésil 4049
araucarie du Chili 3658
araucarie élevée 3877
araucariées 298
araujie 658
arbousier 3418
arbousier busserole 500
arbousier commun 5276
arbousier des Pyrénées 5276
arbre à baume 3085
arbre à beurre 468, 898, 899
arbre à bourre 3134
arbre à chapelet 4553
arbre à chapelets 1160
arbre à chou 246, 907
arbre à cire 5104
arbre à coca 2883
arbre à copahu 1583
arbre à coton 3440
arbre à encens 2937
arbre à gale 1501
arbre à gomme 399
arbre à grives 2114
arbre à gutta 3443
arbre à guttapercha 3443
arbre à kapok 3134
arbre à kola 5293
arbre à la pluie 4404
arbre à laque 533, 3050
arbre à liège de l'Amour 226
arbre à maté 4048
arbre à mâture 2930
arbre à melon 4035
arbre à myrrhe 1477
arbre à noix vomique 3898
arbre à pain 769, 2998
arbre à perruque 1527, 5033
arbre à poison 1501
arbre à résine 3275
arbre à résine du Pérou 940
arbre à savonnette 5102
arbre à soie 79
arbre à teck 1544
arbre à tek 1544
arbre à thé 1543
arbre au poivre 1120, 3296
arbre aux cloches d'argent 4950
arbre aux concombres 1726
arbre aux fraises 5276
arbre aux perles 1529
arbre aux quarante écus 2493
arbre à vernis 3050
arbre d'acajou 5871
arbre d'argent 4964
arbre de fer 2966
arbre de Gordon 2493
arbre de jamba 865
arbre de Judas 3114
arbre de Judée 3114
arbre de Kala 2356
arbre de mango 1464
arbre de mastic 3275
arbre de Moïse 2261
arbre de mort 3455
arbre d'encens 875
arbre de neige 2379, 5896
arbre de Paradis 1973
arbre de rata 2964
arbre des escargots 1439
arbre de Shisham 4977
arbre des pagodes du Japon 3059
arbre des voyageurs 3409
arbre de vie 1973
arbre sacré 45
arbre triste 3851
arbrisseau ambre 182
arbuste aux papillons 892
arceuthobium 1947
archangélique 245, 2406
archegoniates 301
archimycètes 302
archytée 303
arctotide 312
ardisie 5117
ardisie du Japon 3018
arec 313
arénaire 4692
areng 2558
arenga 5305
aréquier 313
aréthuse 314
argalon 1227
argasse 1523
argémone 4302
argentine 5064
argousier 4787
argousier faux-nerprun 1523
argyreia 339
argyrolobe 4962
argythamne 4953
arisème 2999
aristide 5499
aristoloche 1932
aristoloche à grandes fleurs 4104
aristoloche clématite 592
aristoloche de Mandchourie 3460
aristoloche de Virginie en arbre 1404
aristoloche siphon 1404
aristolochiacées 593
aristotélie 5984
armérie 5502
armérie maritime 1545
armillaire de miel 2838
armoise 4640
armoise à balais 3966
armoise amère 1564
armoise annuelle 5396
armoise commune 3747
armoise maritime 3496
armoise petit-dragon 5445
armoise romaine 4548
arnébie 322
arnica 323
arnica des montagnes 3694
arnigo 2685
arnique 323
arnoséride 3213
arnoséris 3213
aronie 1224
aronie rouge 4446
aronque 2519
arracacia 277, 325
arrête-bœuf 962, 4484, 5492
arrhénanthère 3905
arroche 4663
arroche des jardins 2421
arroche de Tartarie 5457
arroche-fraise 675
arrow-root de la Jamaïque 541
arrow-root des Antilles 541
arrow-root des Indes orientales 1983
arrow-root de Tahiti 2244
artabotrys 5422
artichaut commun 331
artichaut de Jérusalem 3094
artichaut d'hiver 3094
artocarpe à feuilles entières 2998
artocarpe incisé 769
arum 333
arundinaire 994
arundinelle 335
asaret 5947
asaret d'Europe 2133
asclépiadacées 3621
asclépiade 3620
asclépiade à ouate 1469
asclépiade de Curaçao 676
asclépiade de Syrie 1469
asclépiade incarnate 5345
asclépiade tubéreuse 894

ascomycètes 337
ascophylle 4539
asiminier 1751, 4077
asiminier trilobé 1485
aspérelle 740
asperge 350
asperge à feuilles aiguës 4848
asperge commune 1351
asperge de Sprenger 5178
asperge plumeuse 2204
aspergette 478
aspergille 351
aspérule 6040
aspérule des sables 5201
aspérule odorante 5395
aspérule tinctoriale 1957
asphodèle 352
asphodèle blanc 757
asphodéline 3003
aspidie 354
aspidistra 353
asplénie 5161
asplénie vert 2634
assa fétida 336
astélie 357
aster 358
aster à feuilles de saule 5968
aster amelle 2973
aster de la Nouvelle-Angleterre 3839
aster de la Nouvelle-Belgique 3842
aster des Alpes 113
aster de Sibérie 4892
aster des Pyrénées 4370
astérées 359
aster linière 2102
aster maritime 5597
aster œil-de-Christ 2973
astilbe 360
astilbe de Chine 1169
astragale 3353
astragale à gomme-adragante 5568
astragale des Alpes 137
astragale du Danemark 4350
astragale glycyphylle 3619
astragale pois chiche 3701
astrance 3531
astrocarpe 361
astrocaryum 5626
astrone 5234
astrophyte 5213
atalantie 362
atalantie d'Inde 2927
athamante 3715
atractyle 1864
atragène 125
atraphace 2522
attalée 365
aubépine altaïque 167
aubépine à un style 4973
aubépine d'Arnold 324
aubépine d'Ellwanger 2017
aubépine du Canada 972
aubépine épineuse 2038
aubépine monogyne 4973
aubergine 2414
aubifoin 1613
aubour 3192
aubriétie 366
aucuba 367
aucuba du Japon 3019
aulne 81
aune 81
aune à feuilles en cœur 2971
aune blanc 5122
aune blanchâtre 5122
aune commun 2072
aune d'Orégon 4432
aune du Caucase 1066
aune du Japon 3012
aunée 2955
aunée britannique 784
aunée ensiforme 5407
aunée officinale 2010
aune glutineux 2072
aune noir 2072
aune rouge 4432
aune rugueux 2732
aune vert d'Europe 2105
auriculariales 369
auricule 368
auricule de jardins 368
aurone 3923
aurone des champs 4641
aurone femelle 1776
aurone sauvage 4641
autobasidiomycètes 381
avelinier 2098, 2246
avéneron 5951
avicennie 386
avocatier 183
avodiré 41
avoine 1481, 3904
avoine à fourrage 3333
avoine courte 3333
avoine d'Algérie 4457
avoine des prés 3553
avoine d'Orient 4927
avoine élevée 5427
avoine jaunâtre 6109
avoine laineuse 1553
avoine nerveuse 3374
avoine nue 3789
avoine pubescente 2687
avoine stérile 252
avoine strigeuse 3374
axyris 388
azalée 389, 4490
azalée couchée 114
azalée de l'Inde 2933
azalée du Pont 4264
azalée pontique 4264
azara à petites feuilles 752
azarole 390
azarolier 390
azérolier 390
azobé 2005
azolle 391

baccaurée 404
bacchante de Virginie 1974
baccharide 405
bacope 5811
bactris 5154
badamier 5600
badenier 254
badiane 254, 3014
badianier 3014
badianier à petites fleurs 6112
badianier du Cambodge 959
baguenaudier 662
baguenaudier commun 1362
baguenaudier d'Orient 3950
balanophoracées 412
balata 1355
balisier 996
balisier comestible 1993
ballote 425
ballote fétide 614
ballote noire 614
balsaminacées 433
balsamine 5051
balsamine à petites fleurs 5023
balsamine des bois 5560
balsamine des jardins 2407
balsamine glanduleuse 4596
balsamite 1632
bambou 437
bambou commun 5491
bambou doré 2527
bambou noir 615
bananier 439, 1358
bananier à fibres 2
bananier commun 4219
bananier d'Abyssinie 8
bananier de Chine 1936
bananier du Japon 3020
bananier du paradis 4219
bananier japonais 3020
bananier textile 2
banian 738
banian d'Inde 2929
banksia 442
banyan 444
baobab 445
baphier 968
baptisie 5948
baptisie des teinturiers 6113
baquois 4779
barbadine 2475
barbarée 610, 5999
barbarée printanière 1966
barbeau 1613
barbe de bouc 2519, 3559
barbe de bouc commune 5410
barbe de Jupiter 1527, 3119
barbe de moine 1870
barbon 706
bardane 859
bardane commune 2601
bardane tomenteuse 1636
bardanette 2459, 5246
barkhaisie 2725
barosme 827
bartonie 459
bartschia 460
basellacées 3414
baselle 5736
basidiolichens 461
basidiomycètes 462
basilic 463
basilic commun 5354
basilic sauvage 464
bassia 468
batate 5380

bâton de Jacob 1449
bâton de St.-Jacques 1449
bâton de St.-Joseph 4648
batrachosperme 479
bauhinier 481
bauhinier cotonneux 4654
baumier 711
baumier de Giléad 432
baumier de Jamaïque 430
baumier du Canada 432
baumier du Pérou 429, 4153
baumier du Tolu 5541
bdellium d'Afrique 44
bec-de-grue 91
bec-de-grue colombin 3369
bec-de-grue disséqué 1757
bec-de-grue mauvin 5026
bec-de-grue pigeon 3369
beckmannie 5015
bédille 2100
bégonia 523
bégoniacées 524
bégonia rex 355
bégonia tuberculeux hybrides 5621
beilschmiedie 5016
belamcanda de Chine 620
bel indien 408
belladone 525
belle-de-jour 1941
belle-de-nuit 1415, 2341
belvédère 531
bénincase 5845
benjoin 5127
benoîte 385
benoîte aquatique 5791
benoîte d'eau 5791
benoîte de rivage 5791
benoîte urbaine 1354
berbéridacées 453
berbéride 2075
berce 1649
berce brancursine 2825
berce de Sibérie 4898
berce pubescente 1069
berce spondyle 2825
berchémie 5336
bergamotier 536
bergénie 537
bergénie à feuilles charnues 3254
bergénie à feuilles cordiformes 2736
bergie 538
béringène 2414
berle 5828
berle à feuilles étroites 3811
berlinia 539
bermudienne 698
bermudienne à feuilles étroites 1364
bertéroa 2160
bétel 549
bétoine 550
bétoine des montagnards 3694
bétoine officinale 1361
bette 1360
bette à cardes 1116
betterave 520
betterave à sucre 5298
betterave fourragère 2322, 3471
betterave potagère 1360, 2408, 4438
betterave rouge 2408, 4438
betterave sauvage 4785
betterave sucrière 5298
bétulacées 579
bibacier 3375
bicorne 1846
bicorne odorante 5363
bident 521
bident penché 3866
bident triparti 857
bifora 555
bigardier 5092
bignone 567
bignone à vrilles 1702
bignone orangée 1702
bignoniacées 5612
bignonier 567
bilimbi 576
billbergia 68
bistorte 2079
bixacées 612
blanchard velouté 1553
blanchette 2088
blé 5873
blechnum 4726
blechnum commun 1821
blé compact 1288
blé d'Abyssinie 2064
blé de Jérusalem 2029
blé de Pologne 4247
blé de Turquie 3434
blé dur 1930
blé locular 2004
blé miracle 1288
blé noir 832, 1370
blé ordinaire 5874
blé poulard 4286
blé rameux 756
blé tendre 5874
blétille 672
blette 675
bl(e)uet 1613
bl(e)uet vivace 3697
bocconie 4230
boerhaavie 5131
bois à balles 203
bois à cassave 1849
bois à lardoir 2096
bois carré 2096
bois-cuir 3259
bois d'angelin 246
bois d'argan 315
bois d'ébène 42
bois de Cam 4869
bois de campêche 3359
bois de Cayan 3968
bois de chat 5049
bois de chien 2309
bois de Chine 5640
bois de corail 4676
bois de corail dur 4869
bois de corail tendre 1902
bois de dhau 3207
bois de fer 2966
bois de fer de Birmanie 865
bois de gaïac 1460, 3293
bois de guitare 2218
bois de la Sainte Croix 2113
bois de liège des Antilles 3316
bois de lièvre 2529
bois de nagas 2966
bois dentelle 3319
bois de pouine 683
bois de rose 474, 648, 4570
bois de Sainte Lucie 3426
bois de Saint-Martin 3009
bois de sappan 4702
bois de Sibucoa 4702
bois de Surinam 5338
bois de vie 1460
bois d'oreille 2194
boisduvalie 5144
bois franc 2039
bois gentil 2194
bois jaune 1527, 1912
bois jaune du Gabon 3984
bois joli 2194
bois pial 3312
bois puant 2078, 3575
bois rouge de Kodar 4679
bois rouge des Andamans 237
bois rouge d'Inde 4544
bois sanguin 683
bois satiné d'Amérique 6094
bois satiné d'Inde 1107
bois-serpent 272
bois sucré 1459
bolet 727
bolet à chair jaune 4448
bolet amer 598
bolet bai 485
bolet blafard 3396
bolet comestible 1093
bolet de pin 2574
bolet des bouviers 4846
bolet élégant 2011
bolet granulé 2574
bolet jaune 820, 2011
bolet marron 1140
bolet moucheté 5700
bolet noble 1093
bolet orangé 3940
bolet raboteux 582
bolet roux 3940
bolet rude 582
bolet rugueux 582
bolet satan 679
bolet subtomenteux 6081
bolet tacheté 5700
boltonie 728
bomarée 729
bombacacées 4944
bombacées 4944
bombax 730
bonduc 3847
bon-Henri 2559
bonnet carré 2096
bonnet de prêtre 2096
borasse 4019
borraginacées 735
bosée 2544
boswellie 2363
boswellie de Carter 553
botryche 2577
botryche de Virginie 4427
boucage 4186

bougainvillier 742
bouillon-blanc 2276
bouleau 578
bouleau à canots 4037
bouleau à feuilles de peuplier 2594
bouleau à papier 4037
bouleau bleu 5357
bouleau de Chine 1171
bouleau de Daourie 1783
bouleau de Forrest 2333
bouleau de Maximowicz 3650
bouleau de Middendorff 3612
bouleau de Potanin 4280
bouleau d'Erman 2057
bouleau de Schmidt 4758
bouleau d'Europe 2131
bouleau du Japon 3076
bouleau gris 2594
bouleau jaune 6076
bouleau merisier 5357
bouleau nain 1935, 3390
bouleau noir 4504
bouleau pubescent 4323
bouleau verruqueux 2131
boulette commune 1423
bouquet parfait 5393
bourdaine 2513
bourdaine marine 1523
bourgène 2513
bourrache 734
bourrache officinale 1366
bourreau des arbres 185
bourrérie 5286
bourse à pasteur 4862
bourse de berger 4862
boussingaultie 3413
bouteloue 2572
bouton d'or 5425
bouton noir 525
bouvardie 747
boviste plombé 3244
boykinie 753
brachyélytre 4880
brachypode 2164
brachypode des bois 4996
brachypode penné 3033
brahée 5358
brasénie 5835
brassie 5132
bravoa à fleurs géminées 3604
brésillet de Fernambouc 4296
brésillet des Indes 4702
brésine 4466
breslinge 2635
bridélie 776
brize 4134, 4381
broccoli asperge 800
brodiée 801
brome 802
brome de Madrid 5109
brome des champs 2219
brome de Schrader 4483
brome des prés 3546, 5034
brome des seigles 1138
brome des toits 1123
brome du Japon 3024
brome inerme 5034
broméliacées 4192
bromélie 804
brome madrilène 5109
brome mou 5078
brome stérile 4287
brosime 770
brouillard 1433
broussonétie 4039
browallie 814
bruckenthalie 5142
brugnonier 3818
brunelle 4816
brunelle à grandes fleurs 559
brunelle commune 1525
brunfelsie 4405
brunsvigie de l'Impératrice Joséphine 2469
bruyère 2742, 2744
bruyère à balais 546
bruyère arborescente 5581
bruyère caminet 1701
bruyère cendrée 5668
bruyère clarin 1701
bruyère commune 4772
bruyère de la Méditerranée 594
bruyère des neiges 5186
bruyère en arbre 5581
bruyère tétragone 1701
bryone 824
bryone blanche 5883
bryone dioïque 4439
bryophytes 825
buddléia 892
buddléia changeant 3941
buddléia de David 3941
buddléia du Japon 3026
buddléia du Père David 3941
budlège 892
bugle 838
bugle de Genève 2442
bugle rampant 1030
buglosse 840
buglosse de Barrelier 1959
buglosse d'Italie 2975
buglosse officinale 1371
bugrane 962, 4484
buis 749
buis bénit 1368
buis commun 1368
buis de Chine 1161
buis de Mahon 5108
buis du Japon 3054
buis piquant 885
buisson ardent 2261
buisson ardent à feuilles étroites 3801
buisson ardent écarlate 4743
buis toujours vert 1368
bulbocode 3557
bulnésie 851
buniade 1617
bunya bunya 855
buphthalme 3986
buplèvre 5493
buplèvre à feuilles en faux 2156
buplèvre à feuilles rondes 4589
buplèvre ligneux 4888
buplèvre menu 4998
bursaire 874
burséracées 876
busserole 3480
busserole alpine 141
busserole officinale 500
butée 886
butomacées 2315
butome 2313
buxacées 751
byrsonime 905
byssus rouge 4537

cabombe 2184
cacaoyer 912
cachryde 913
cactacées 915
cactier 914
cade 4298
cadie 916
caféier 1318
caféier d'Arabie 292
caféier de Libéria 3287
cailcedrat du Sénégal 4821
caille-lait 510
cajeputier 920, 3579
calaba à fruits ronds 2943
calaba commun 1104
calabure 3007
caladion 925
calamagrostide 4475
calamagrostide épigeios 1131
calamagrostide terrestre 1131
calamagrostis 4475
calament 4720
calament des Alpes 148
calament officinal 927
calandrinie 4529
calathéa 928
calcéolaire 5011
calcéolaire à feuilles entières 878
calcéolaire du Mexique 3595
calcéolaire hybride herbacée 5760
calcéolaire pantoufle 5760
calebassier 924
calebassier à feuilles longues 1373
calendule 929
caliméride 3526
calla 949
calla des marais 5941
calla d'Ethiopie 1374
callianthème 951
callicarpe 2370
calligon 952
callitrichacées 5840
callitriche 5839
callitriche d'automne 384
callitriche printanier 5802
callitris 1777
callitris articulé 296
callune 2744
calochorte 3494
calonyction 3670
calophylle 508
calotropis 953
calycanthacées 5278
calycanthe 5386
calycanthe fertile 4011
calycanthe multiflore 1540
calypso 956

calypso bulbeuse 957
calystégie 2509
camarine 1706
camarine noir 627
camash 958
camash comestible 1375
camassia 958
camélée 5194
camélia 960
camélia du Japon 1376
camélia sasanqua 4711
caméline 572, 2168
camérisier à balais 2099
caminet 1701
camomille 963
camomille des teinturiers 2528
camomille ordinaire 2452
camomille puante 3542
camomille romaine 4546
camomille vraie 2452
campanulacées 529
campanule 528
campanule à bouquets 1797
campanule à feuilles de pêcher 4083
campanule à feuilles d'ortie 1643
campanule à feuilles rondes 690
campanule à grandes fleurs 424
campanule à larges feuilles 2599
campanule alpine 116
campanule barbue 502
campanule carillon 999
campanule des Carpathes 1026
campanule de Scheuchzer 4756
campanule de Sibérie 4894
campanule gantée 1643
campanule pyramidale 1157
campanule raiponce 4414
camphrée 965
camphrier 966
camphrier de Bornéo 1367
campsis 5611
camptosore 5763
canaigre 981
canari 989
canari blanc 5886
canari commun 3085
canari de Cochinchine 5886
canari noir 623
canavalie 2997
canche 2677
canche bleue 3674
canche caryophyllée 4959
canche cespiteuse 5628
canche flexueuse 5843
canche précoce 1963
candollée 5858
canelle 995
canjerana 991
canna 996
cannabinacées 2777
cannabine faux-chanvre 471
cannacées 997
canne aromatique 4574
canne à sucre 5300
canneberge 1658
canneberge à gros fruits 3224
canneberge commune 5019
canneberge ponctuée 1645
canne de jonc 1380
canne de Provence 2482
canne d'Inde 2934
cannelier 1238
cannelier casse 1047
cannelier de Ceylan 1105
cannelier-giroflée 1275
cantaloup 998
cantharellacées 1001
caoutchouc des jardins 2944
caoutchoutier de Céara 1081
capillaire 3430
capillaire cunéiforme 1828
capillaire de Montpellier 5099
capillaire tendre 2182
capparidacées 1009
câprier 1007
câprier commun 1377
caprifiguier 1412
caprifoliacées 2842
capron 2722
capuce de moine 17
capucine 3812
capucine des Canaries 986
capucine tubéreuse 5617
caragana 4095
caragana arborescent 4910
caragana argenté 4913
carambolier bilimbi 576
carambolier vrai 1012
carape 1657
carape de Guyane 2660
cardamine 600
cardamine à petites fleurs 5021
cardamine des prés 1723
cardamine hérissée 4111
cardamome 1014
cardère 5469
cardère sauvage 5717
cardère velue 2692
cardiaire 1475
cardiaque 1475
cardiosperme 2741
cardon d'Espagne 5112
cardon sauvage 1015
carex 4810
carex dioïque 1860
carex élongé 2023
carex hérissé 2690
carex jaune 6106
carex vulpin 2348
caricacées 4034
carline 1017
carline acaule 5035
carludovice 1018
carludovique 1018
carotte 2410
carotte de Philadelphie 5651
carotte sauvage 5942
caroubier commun 1023
carouge à miel 1439
carpésium 1029
carrote 1035
carthame 4636
carthame des teinturiers 4637
carthame officinal 4637
carvi 1013
carya aquatique 5809
caryer 2795
caryer amer 605
caryer blanc 4841
caryer cotonneux 3640
caryer de Chine 1186
caryer glabre 4179
caryer muscade 3896
caryer pacanier 4100
caryocar nucifère 4725
caryophyllacées 4198
caryopteris 689
caryote 2265
caryote brûlante 5540
cascarillier 1041
casque de Jupiter 17
casse 4823
casse du Maryland 5958
casse fistuleuse 2548
cassine 1048
cassiopée 1049
cassiopée-mousse 307
cassiopée tétragone 2260
cassissier 2081
castanopsis 2145
castillèje 4003
castillier 3882
castilloa 2665
casuarina 516
casuarina à feuilles de prêle 2876
casuarinacées 517
casuarina de Cunningham 1731
cataire 1060, 3830
cataire d'Ukraine 5676
catalpe 1053
catalpe commun 5097
catalpe du Japon 1175
catalpe élégant 3879
catanance 1737
catananche 1737
catasète 3659
catha 294
cattleya 1065
cattleya jaune citron 5634
caucalier 2166
caulophylle 1322
cavalam 2733
céanothe 1080
cécropie 4331
cédrat 1243
cédratier 1243
cèdre 1082
cèdre à encens 936
cèdre bâtard 1218
cèdre blanc 1973
cèdre blanc de Californie 936
cèdre bleu 688
cèdre d'Atlas 364
cèdre de Chypre 1778
cèdre de Cuba 1234
cèdre de l'Himalaya 1836
cèdre d'encens de Californie 936
cèdre des Bermudes 543
cèdre de Singapour 866
cèdre de Virginie 1980
cèdre du Liban 1083
cédrèle 1084
cédrèle de Chine 1205
cédrèle odorante 1234

cèdre puant 2303
cèdre rouge 1980
célastracées 5203
célastre 608
célastre grimpant 185
céleri 1086, 5943
céleri à côtes 2411
célosie 1307
célosie argentée 2192
celsie 1088
centaurée 1089
centaurée ambrette 4355, 5388
centaurée chausse-trappe 5232
centaurée de montagne 3697
centaurée d'Europe orientale 1067
centaurée jacée 817
centaurée noire 625
centaurée scabieuse 4734
centenille 1109, 2179
centranthe 1091
centranthe rouge 3119
centrolobe 4275
cépage américain 6002
cèpe 1093
cèpe annulaire 820
cèpe à pied rouge 4448
cèpe bai 485
cèpe blafard 3396
cèpe châtain 1140
cèpe de Bordeaux 1093
cèpe des bouviers 4846
cèpe du diable 679
cèpe jaune 820
cèpe jaune des pins 2574
cèpe jaune du mélèze 2011
cèpe marron 1140
cèpe mou 6081
cèpe orangé 3940
cèpe rude 582
cèpe satan 679
céphalaire 1094
céphalaire des Alpes 6079
céphalaire de Syrie 5417
céphalaire de Tartarie 5453
céphalante 902
céphalanthère 4163
céphalotaxe 4232
céphalotaxe de Fortune 1196
céphalotaxus 4232
céphalote 1096
céraiste 1145
céraiste à cinq étamines 5853
céraiste aquatique 5795
céraiste cotonneux 5064
céraiste de Bieberstein 5461
céraiste des Alpes 122
céraiste des champs 5230
céraiste tomenteux 5064
cératiole 4686
cératocarpe 1097
cératophyllacées 2862
cératophylle 2860
cératoptéris 5803
cératozamie 2854
cercidophylle du Japon 3048
cercocarpe 3712
cercospora 1099
cerfeuil 1137, 4658
cerfeuil bâtard 1648
cerfeuil bulbeux 5651
cerfeuil des fous 1648
cerfeuil sauvage 856
cerise carée 446
cerise de Antilles 446
cerise du Brésil 2653
cerisier 1133
cerisier à feuilles dentées en scie 3951
cerisier à grappes 2078
cerisier carré 4206
cerisier de Cayenne 4206
cerisier de la Chine 3398
cerisier de Sainte Lucie 3426
cerisier des sables 547, 4683
cerisier de Tahiti 3977
cerisier de Virginie 1384
cerisier doux 1011
cerisier du Japon 1174
cerisier du Mexique 1011
cerisier du Portugal 4277
cerisier griottier 3679
cerisier griottier nain 2644
cerisier Mahaleb 3426
cerisier noir 626, 1384
cerisier tardif 626
céropégie 1102
céroxylon des Andes 5846
césalpiniacées 4824
césalpinie 758
césalpinie hérissée 4296
cestreau 3098
cétérach 4737
cétraire 1103
cétraire d'Islande 2916
chaenomeles 2312
chalef 2006
chalef à feuilles étroites 4616
chalef argenté 4951
chalef multiflore 1135
chamédaphné 3257
chamédorée 1112
champignon 3758
champignon de couche 2419
champignon de fumier 2419
champignon de Paris 2419
champignon noir 2858
champignon polonais 1093
chanterelle 1000
chanterelle comestible 1115
chanterelle commune 1115
chanvre 2772
chanvre commun 2773
chanvre de Calcutta 4591
chanvre de Deccan 3146
chanvre de Manille 2
chanvre de Virginie 5518
chanvre du Bengale 5331
chara 5267
charagne 5267
charançonne 1738
char de Vénus 17
chardon 782, 5487
chardon à bonnetier 2390
chardon acanthoïde 13
chardon à foulon 2390, 5717
chardon argenté 669
chardon aux ânes 4770
chardon béni 670
chardon béni commun 671
chardon crépu 1740
chardon de Kerner 3155
chardon d'Espagne 1015
chardon étoilé 5232
chardon-Marie 669
chardonnette 1638
chardon onoporde 4770
chardon penché 3759
chardon-Roland 5047
charme 2108, 2852
charme à feuilles en cœur 2738
charme blanc 2108
charme commun 2108
charme d'Amérique 198
charme de la Caroline 198
charme d'Europe 2108
charme d'Orient 3956
charme du Caucase 1073
charme du Japon 3045
charme houblon 2849
charophytes 1119
châtaigne d'eau 5794
châtaignier 1139
châtaignier chincapin 99
châtaignier commun 2086
châtaignier d'Amérique 187
châtaignier de Saint-Domingue 1732
châtaignier du Brésil 765
châtaignier du Japon 3027
chayote comestible 1122
chéilanthe 3323
chélidoine 1085
chélidoine grande éclaire 2606
chêne 3899
chêne à écailles dorées 1002
chêne à feuilles de châtaignier 1141
chêne à feuilles de houx 505
chêne à feuilles de laurier 3233
chêne à feuilles de saule 5974
chêne à feuilles en lyre 3982
chêne à galles 85
chêne a glands sessile 1928
chêne à grappes 2041
chêne à gros fruits 870
chêne à lattes 4865
chêne aquatique 5827
chêne à trouchets 1928
chêne bicolore 5347
chêne blanc 5905
chêne blanc aquatique 3982
chêne buisson 505
chêne châtaignier 1142
chêne châtaignier des marais 5343
chêne chevelu 2128
chêne cocciné 4746
chêne commun 2041
chêne commun sessile 1928
chêne d'Angleterre 2041
chêne de Bourgogne 2128
chêne de Caroline 3342
chêne de Hongrie 2979
chêne de Mirbeck 987
chêne de Mongolie 3653
chêne denté 1786
chêne de Pannonie 2979

chêne des marais 4200
chêne d'Espagne 2814, 5101
chêne des Pyrénées 4374
chêne de Turquie 2128
chêne du Liban 3261
chêne du Pont 4265
chêne du Portugal 4278
chêne écarlate 4746
chêne étoilé 4279
chêne faux-liège 2814
chêne femelle 1928
chêne hongrois 2979
chêne kermès 3153
chêne-liège 1603
chêne lombard 2128
chêne macédonien 3403
chêne noir (d'Amérique) 5827
chêne occidental 554
chêne pédonculé 2041
chêne prin 5343
chêne pubescent 4324
chêne quercitron 644
chêne rouge 5101
chêne rouge d'Amérique 3883
chêne rouvre 1928, 2041
chêne sessile 1928
chêne vélani 5697
chêne velu 2128
chêne vert 2832
chenille 4767
chenillette 4767
chénopode blanc 3212
chénopode bon-Henri 2559
chénopode quinoa 4390
chénopodiacées 2562
chérimbelier 3977
chérimolier 1132
chérimolier du Pérou 1132
chervi(s) 4986
cheveu de Vénus 5099
cheveux du roi 5576
chèvrefeuille 2841
chèvrefeuille à baie noire 631
chèvrefeuille à balais 2099
chèvrefeuille à fruits bleus 5356
chèvrefeuille alpestre 161
chèvrefeuille des bois 6022
chèvrefeuille des haies 2099
chèvrefeuille des jardins 5370
chèvrefeuille de Tartarie 5455
chèvrefeuille d'Europe 2099
chèvrefeuille du Japon 3042
chèvrefeuille toujours vert 5613
chicoracées 1232
chicorée 1147
chicorée endive 2032
chicorée sauvage 1381
chicot 1319
chicot dioïque 3151
chicot du Canada 3151
chiendent 1640, 1879
chiendent à crête 1684
chiendent allongé 5432
chiendent chevelu 5252
chiendent désertique 5210
chiendent de Smith 707
chiendent intermédiaire 2952
chiendent rampant 4379
chigomier 1344
chimaphile 4202
chimaphile en ombelle 1496
chimonanthe précoce 6009
chiocoque 3617
chironie 5229
chloranthacées 1221
chloranthe 1220
chlore 6116
chlorelle 1222
chlore perfoliée 1567
chlorette 6116
chloris 5982
chlorophora 2396
chlorophora élevé 2963
chlorophora tinctorial 1912
chlorophycées 2621
chocho 1122
choin 724
chondre 2960
chondre crépu 2961
chondrille 4983
chou 906, 5940
chou blanc 5885
chou broccoli 800
chou cabus 2734
chou champêtre 585
chou de Bruxelles 823
chou de Chine 1806, 4005
chou de Milan 4722
chou de Savoie 4722
chou de Shanton 4158
chou de Suède 4626
chou-fleur 1079
chou frisé pommé 4722
chou marin 1326, 1387
chou marin de Tartarie 5452
chou-navet 4626, 6005
chou palmiste 909
chou pommé 2734
chou pommé frisé 4722
chou-rave 3175
chou rouge 4444
chou vert 3128
chroococcacées 1228
chrysanthème 1229
chrysanthème à corymbe 3569
chrysanthème à grandes feuilles 5441
chrysanthème d'automne 3687
chrysanthème des Alpes 123
chrysanthème des blés 1609
chrysanthème de Sibérie 3179
chrysanthème des jardins 1710
chrysanthème des moissons 1609
chrysanthème grande-marguerite 3987
chrysanthème leucanthème 3987
chrysanthème-matricaire 2210
chrysobalane 1315
chrysobalane icaquier 2914
chrysogone 2549
chrysopogon 4750
chusquéa 1231
ciboule 5863
ciboulette 1219
cicerole 1144
cichoracées 1232
cicutaire 5808
cicutaire aquatique 2130
cierge 1100
cierge à grandes fleurs 4384
cierge géant 4645
ciguë aquatique 2130
ciguë vireuse 2130
cimicaire 837
cimifuge 837
cinéraire 1237
cinéraire hybride 1385
cinéraire maritime 4958
cinna 6039
circée 2031
circée des Alpes 124
circée parisienne 4055
cirier 483
cirier de Californie 948
cirier de Louisiane 5104
cirolier 4487
cirolier commun 2698
cirolier du Brésil 410
cirse 5487
cirse acaule 5238
cirse des champs 975
cirse lancéolé 850
cisse 5577
cistacées 4532
ciste 4531
ciste à feuilles de laurier 3235
ciste à feuilles de peuplier 4271
ciste à feuilles de sauge 4670
ciste à gomme 2664
ciste crépu 6058
ciste hérissé 2689
citharexylon 2218
citron 3268
citronnier 1243, 3268
citronnier à trois folioles 5592
citronnier trifolié 5592
citrouille 4330
civette 1219
cladanthe 1248
cladion 4727
cladonie 1249
cladonie rangifère 4479
cladraste 6114
cladraste à bois jaune 213
clandestine écailleuse 5549
clarkie 1253
clausène 5779
clavaire 2520
clavaire crêpue 1683
clavaire dorée 2531
clavaire en forme de chou-fleur 4468
clavaire jaune 4013
clavalier 4295
clavalier à feuilles de frêne 1506
clavalier de Bunge 2282
clavariacées 1588
clayton(i)e 5181
clayton(i)e de Cuba 3627
clématite 1260
clématite à feuilles simples 5080
clématite à panicules 5353
clématite à vrilles 5723
clématite azurée 3297
clématite bleue 2976

clématite brûlante 4227
clématite de Corée 3180
clématite de Jackman 3000
clématite des bois 5571
clématite des haies 5571
clématite de Sibérie 4896
clématite des montagnes 244
clématite de Soungari 5330
clématite de Virginie 5756
clématite d'Orient 3952
clématite dressée 2646
clématite du Père Armand 319
clématite fausse-vigne 2976
clématite flammèle 4227
clématite laineuse 3858
clématite odorante 4227
clématite tangoute 2532
clématite viorne 3255
cléome 5128
clérodendron 2510
cléthra 5323
cléthracées 1261
clianthe 4059
clinopode 926
clintonie 490
clitocybe 1270
clitocybe en entonnoir 2395
clitocybe nébuleux 1274
clitocybe odorant 5385
clitocybe orange 2165
clitopile petite prune 4226
clitoria 4178
clivie 3126
clivie vermillon 4745
clochette d'hiver 1530
clusiacées 2669
clusier 1290
clypéole 1297
cobée 1301
cobée grimpante 4337
cocculus 5044
cochléaria 4783
cochléaria d'Angleterre 2043
cochléaria officinal 1522
cochlosperme 4860
coco de mer 1887
cocotier 1314
cocotier des Maldives 1887
cocotier du Chili 5419
cocrête 4430
cocriste 4430
codium 1316
codonopside 340
cœlogyne 1317
cœur de bœuf 849
cœur-de-Marie 1363
cognassier 4389
cognassier à fruit en gourde 1414
cognassier commun 1512
cognassier de Chine 1181
cognassier de Maule 3038
cognassier du Bengale 408
cognassier du Japon 2312, 3038
coïx larme 31
colchique 382
colchique d'automne 1353
coléanthe 3741
coleus 1325
coleus scutellaire 1386
coliole 1325
collétie 1330
collétie épineuse 1769
collinsie 1331
collinsonia 2863
collomie 1332
collybie 1333
collybie à pied velouté 5709
collybie butyracée 901
colocase 2013
colocase des anciens 1806
colocasie 2013
colombe rouge 2028
coloquinelle 6088
coloquinte commune 1334
colubrina 1340
columbine 1341
colurie 1281
colza 1343, 4417
comaret 3502
combrétacées 3776
commélinacées 5135
comméline 1814
commiphore 3779
compagnon blanc 2140
composées 1570
comptonie 5364
concombre à petites pyramides 2855
concombre commun 1724
concombre d'âne 5202
concombre d'attrape 5202
concombre des Antilles 5869
concombre métulifère 2855
condalie 1571
condurango 1572
conifères 1579
coniosélin 2770
conjuguées 1580
conopholide 990
conringia 2710
consoude 1345
consoude du Caucase 1068
consoude rugueuse 4297
consoude tubéreuse 5619
convolvulacées 3683
conyze 1239
cooperie 4402
copaïer 1583
copaïer officinal 1582
copalme d'Amérique 210
copalme d'Orient 3965
copayer 1583
copernicie 1584
copernicie à cire 1021
coprinacées 1586
coprin chevelu 4843
coprin noir 2951
coprin noir d'encre 2950
coptide 2556
coquelicot 1616
coquelourde 243
coquelourde des jardins 967, 4558
coqueret 2645
coqueret alkékenge 5275
coqueret comestible 1894
coqueret du Pérou 4156
coqueret pubescent 1894
coracan 4399
coralline 1591
corbeille d'argent 143, 2144
corbeille d'or 2551
cordyline 1899
cordyline à fleurs terminales 1402
cordyline australe 2470
coréopsis 1596
coréopsis élégant 4212
corête du Japon 3049
corette potagère 4283
coriandre 1597, 1598
coriariacées 1600
corisperme 5517
corisperme à feuilles d'hysope 2911
cormier 4833
cornacées 1882
corneille en bouquets 5629
cornifle 2860
cornouiller à fruits blancs 5454
cornouiller à grandes fleurs 2309
cornouiller blanc 5454
cornouiller de Sibérie 4901
cornouiller de Suède 1937
cornouiller femelle 683
cornouiller mâle 1611
cornouiller osier 4458
cornouiller sanguin 683
cornouiller stolonifère 4458
cornuelle 5794
cornutie 1621
coronille 1712
coronille arbisseau 4766
coronille bigarrée 1713
coronille changeante 1713
coronille des jardins 4766
coronope 5783
corossolier 1751
corossolier à fruit hérissé 5297
corrigiole 1622
corrigiole littorale 5271
corroyère à feuilles de myrte 3782
cortinaire violet 5739
cortuse 506
corydale 1628
corydale jaune 6080
corydale noble 4897
corylopsis 6004
corylopsis de Miss Willmott 5963
corylopsis en épis 5147
corylopsis pauciflore 891
corynéphore 1284
corynocarpe 3135
corypha 1629
cosmos 1630
cosmos bipenné 1392
coste 5158
cotonéastre 1633
cotonéastre à feuilles entières 2089
cotonéastre à fruits noirs 618
cotonéastre à grappes 4437
cotonéastre cotonneux 774
cotonéastre de Dammer 501
cotonéastre de Zabel 1134
cotonéastre horizontal 4515
cotonéastre tomenteux 774
cotonnier 1635
cotonnier américain 5685

cotonnier d'Asie 3283
cotonnier d'Egypte 4790
cotonnière 2318
cotonnier en arbre 349
cotule 759
cotylédon 1639
couche-couche 1748
coudonier 1512
coudrier 2098, 2246
coulemelle 4051
couleuvrée 4439
courge 2568, 4330
courge commune 6008
courge de Siam 3437
courge musquée 1747
couronne des blés 1391
couronne impériale 1711
cousso 3190
crambe 1326
crambe maritime 1387
cranson 2871
crapaudine 2967
crassulacées 3970
crassule 1661
craterelle corne d'abondance 2858
crépide 2725
crépide bisannuelle 4577
crépide des toits 3803
cresson alénois 2412
cresson d'eau 5796
cresson de fontaine 5796
cresson de Para 4045
cresson des prés 1723
cresson d'Inde 1478
cresson officinal 5796
crête-de-coq 1410, 1683
crête de paon 4676
crételle 1877
crételle commune 1681
crételle des prés 1681
crine 1696
crinole 1696
crinole à longues feuilles 2707
crithme 4674
crithme marin 4675
crocanthème 2387
crocus 1699
crocus d'Heufell 2790
crocus printanier 1395
croisette 1704
croisier 1349
croix de Jérusalem 3451
croix de Malte 3451, 4332
croix de St.-André 4647
croix de St.-Jacques 526
crosne du Japon 332
crotalaire 4425, 5331
crotalaire intermédiaire 5000
crotalaire lancéolée 3215
crotalaire remarquable 4884
crotalaire striée 5283
croton 1705
croton révulsif 4334
croton tillé 4334
crozophore 5653
crucianelle 1703
crucifères 908
crupine 1714
crypside 4294
cryptocoryne 1716
cryptogames 1717
cryptogramme 4512
cryptomérie 1718
cryptomérie du Japon 3029
cubèbe 1722
cucas enroulé du Japon 4643
cucubale 656
cucurbitacées 2569
cumin 1728
cumin de Malte 1729
cunile 5265
cunninghamia 1163
cunninghamia de Chine 1383
cupressacées 1775
cupulifères 514
curcuma 1551, 5648
cuscutacées 1871
cuscute 1870
cuscute de Gronovius 2641
cuscute d'Europe 2607
cuscute du thym 1400, 2287
cuscute du trèfle 1280
cuscute mineure 1280
cuscute odorante 1153
cuvière 1762
cyananthe 1763
cyanophycées 701
cyathée 5580
cycadacées 1765
cycade 1764
cycas 1764
cyclamen 1766
cyclamen de Perse 2990
cyclamen d'Europe 2092
cyclanthacées 1767
cyclanthère 1768
cyclophore 2197
cycnoche 5348
cycnoque 5348
cymbalaire 465, 3147
cynanque 5340
cynoglosse 2879
cynoglosse printanière 1672
cynomoir 1771
cynomorion 1771
cynosure à crêtes 1681
cypelle 1772
cypéracées 4811
cyphomandre bétacé 5589
cyprès 1773
cyprès à fruits de pois 4724
cyprès à gros fruits 3665
cyprès chauve 414
cyprès commun 2978
cyprès de Cachemire 3137
cyprès de Californie 2571
cyprès de Gowen 2571
cyprès de la Louisiane 414
cyprès de Lambert 3665
cyprès de l'Arizona 317
cyprès de Lawson 3243
cyprès de Lusitanie 3597
cyprès de MacNab 3405
cyprès de Monterrey 699
cyprès de Montezuma 3667
cyprès de Nutka 3874
cyprès des cimetières 3727
cyprès du Sawara 4724
cyprès funèbre 3727
cyprès pleureur 3727
cyprès pyramidal 4367
cyprès toruleux 551
cyprès toujours vert 2978
cypripède 3203
cyrtanthe 1780
cyrtostachys 4795
cystoptéris 657
cystoptéris fragile 786
cystoptéris montagnard 3696
cytinet 1781
cytise 809, 3192
cytise à grappes 2529
cytise aubour 2529
cytise blanc 4276
cytise capité 557
cytise commun 2529
cytise d'Autriche 375
cytise d'Ecosse 4773
cytise des Alpes 4773
cytise en épis 5137
cytise faux ébénier 2529, 4773
cytise pourpré 4339

dacryde 2894
dactyle 1309
dactyle pelotonné 3943
dagussa 4399
dahlia 1782
dahlia commun 2413
dalbergie 4570
damasonie 1793
damasonie commune 5219
dame-d'onze-heures 1535
dammara 1795
danthonie 1798
daphné 1800
daphné bois-gentil 2194
daphné bois-joli 2194
daphné camélée 4559
daphné du Caucase 1070
daphné lauréole 5192
daphnoïdées 3608
darlingtonie 941
dasylire 5090
dasylire chevelu 822
dasylire de Wheeler 5875
datisque 1810
dattier 1807
dattier des Canaries 983
dattier de Zanzibar 4820
dattier fruitier 1808
dattier sauvage 5302
datura 1811
dauphinelle 3226
dauphinelle consoude 2330
dauphinelle des blés 2330
dauphinelle des champs 2330
dauphinelle des jardins 4524
dauphinelle de Tauride 2305
dauphinelle élevée 518
davallie 2711
deinbollie 1826
dendrobe 1829
dendrocalamus 2467

denhamie 1830
dentaire 5548
dentaire bulbifère 1592
dent-de-chien 1878
dent-de-lion 2723
dentelaire 4225
derris 3100
desmidiées 1841
desmodie 5516
deutzie 1843
deutzie à feuilles crénelées 2397
deutzie à grandes fleurs 1961
deutzie de Schneider 4759
deutzie de Siebold 4932
deutzie de Wilson 5977
deutzie grêle 4995
deutzie scabre 2397
diale 1851
diale de Guinée 5715
diapensie 1852
diarrhène 495
diascie 5665
diatomées 1853
dicentre 668
dichapétale 1854
dichromène 5919
dicotylédones 1855
dicrane 2331
dictame 1866
dictame blanc 2435
dictame commun 2435
didisque 3194
dieffenbachie 5631
diélytre 668
dièrame 2016
dierville 880
diervillée 5861
digitale 2346
digitale à grandes fleurs 6089
digitale jaune 5280
digitale laineuse 2614
digitale pourprée 1416
digitaria 2254
dillénie 1858
dimorphothèque 1005
dioclée 1295
dionée 5718
dioscoréacées 6064
diosme 1861
diphyllée 5678
diplotaxe 5775
diplotaxe à feuilles ténues 5007
diplotaxe des murailles 5257
dipsacacées 5471
diptérocarpacées 1862
diptérocarpe 2668
diptérocarpe à baume de gurjan 1432
diptérocarpe élevé 2035
dircé 3259
dircé des marais 363
discostigme 1863
dividivi 1867
dodartie 1869
dodonée 2850
dodonée fébrifuge 1251
dolique 1883
dolique de Floride 1823
dolique d'Egypte 2897
dombéya 1885
dompte-venin 5916
doradille 5161
doradille noire 651
doradille septentrionale 2329
dorême 5316
dorême ammoniac 731
dorine à feuilles alternes 168
doronic 3276
doronic d'Autriche 379
doronic du Caucase 1075
dorsténie 5559
doryanthe 5119
dorycnie 982
double feuille 895
douce-amère 604
doucette 2088
douglas 1888
douglas à gros cônes 556
douglas bleu 4541
douglas d'Orégon 1401
douglas du Colorado 4541
douglas du Japon 3031
douglas vert 1401
doum 1890
doumier 1890
dracocéphale 1901
dracocéphale à fleurs de thym 5510
dracocéphale d'Autriche 376
dracocéphale de Moldavie 3645
dracocéphale de Ruysch 4629
draconcule 5254
dragone 5445
dragonnier 1898
dragonnier balsamique 2357
dragonnier commun 1900
dragonnier des Canaries 1900
drave 1897
drave alpine 4519
drave de Norvège 3888
drave des glaciers 2498
drave de Sibérie 4902
drave printanière 5183
drave tomenteuse 6048
drimys 1903
drimys de Winter 6007
droséracées 5326
drosère 5325
drosère à feuilles rondes 4588
drosère d'Angleterre 2044
drosophyllum 1908
dryade 1920
dryade à 8 pétales 3726
drymaire 4808
dryobalanops 736
dryoptéris 6026
dryoptéris à crêtes 1685
dryoptéris commun 3901
dryoptéris de Braun 760
dryoptéris spinuleux 5546
duboisie 1606, 1922
duchesnée 3643
duradeu 3642
durante 4992
durio 1927
dyckie 1953
dysoxyle 4106
ébénacées 1986
ébène 42
ébénier 1987
échalotte 4844
échevérie 1988
échinacée 1989
échinocactus 1990
échinocéréus 1991
échinocyste 3639
échinodore 862
échinopanax 1847
échinope 2508
échinope boule azurée 5027
échinope de Russie 4615
échinope du Bannat 6121
échinophore 4300
échinopsis 2749
échitès 4718
éclaire 2606
éclairette 2241
edelweiss 1992
edelweiss alpin 1406
edgeworthie 4038
edgeworthie à papier 3958
égilope 2517
égilops 2517
églantier 1875
églantier odorant 5359
églantier rouge 5359
égopode 2570
égopode commune 596
eichhornie 5810
élaeocarpus 2008
élais de Guinée 46
élaphomyce 2717
élatinacées 5842
élatine 5841
élatine à trois étamines 5501
éléagnacées 3926
éléocarpe 2008
éléocarpe d'Amérique 189
éléocarpe de Nouvelle-Zélande 2809
éléocharis 5145
éléodendre 2177
éléodendre du Cap 1004
éléphantope 2014
éleusine 2015, 4399
éleusine d'Inde 2563
ellébore 2766
ellébore d'Orient 3272
ellébore noir 1226
ellébore vert 2628
élodée 2022
élodée d'Argentine 1832
élodée du Canada 976
elsholtzie 2024
élyme 5957
élyme des sables 2094
embélie 2025
embothrie 2027
émilie 5451
empétracées 1707
empètre noir 627
encrier 2951
endive 2032
endivie 2032
engrain 2004
énocarpe 403

entandrophragme 4698
entérolobe 1967
entomophthorées 2047
épacride 2048
épeautre 5124
éperon de chevalier 2330
épéru 5764
épervière 2726
épervière à toison 4842
épervière de Savoie 4723
épervière des murs 5769
épervière des prés 2224
épervière ombellée 3804
épervière orangée 3942
éphédracées 3107
éphèdre d'Amérique 192
éphèdre de Californie 934
éphèdre de Gérard 2449
éphèdre du Canada 192
éphèdre vert 2625
éphémère de Rio 5780
éphémère de Virginie 5752
épiaire 550
épiaire alpine 117
épiaire de Germanie 3731
épiaire des bois 5911
épiaire des champs 2230
épiaire des marais 3501
épiaire laineuse 6047
épicéa 5189
épicéa bicolore 80
épicéa blanc 5913
épicéa commun 3886
épicéa de Balfour 417
épicéa de Brewer 772
épicéa de Corée 3187
épicéa de Finlande 2256
épicéa de Glehn 4657
épicéa de l'Himalaya 2806
épicéa d'Engelmann 2034
épicéa de Sargent 4709
épicéa de Schrenk 4761
épicéa de Serbie 4830
épicéa de Sibérie 4915
épicéa de Sitka 4979
épicéa de Wilson 5980
épicéa de Yezo 6071
épicéa d'Orient 3964
épicéa du Canada 5913
épicéa du Caucase 3964
épicéa du Colorado 1338
épicéa du Japon 5523
épicéa noir 653
épicéa pungent 1338
épicéa rouge 4465
épicéa spinuleux 4942
épi d'eau 4262
épigée 5569
épilobe 5975
épilobe à petites fleurs 5024
épilobe de Hornemann 2857
épilobe de Lamy 3214
épilobe des Alpes 156
épilobe des marais 3522
épilobe des montagnes 3723
épilobe en épi 2262
épilobe hérissé 2695
épimède 2049
épinard 4304, 5148
épinard de Malabar 5921
épinard de muraille 5773
épinard de Nouvelle-Zélande 3844
épinard fraise 5273
épinard malabre 5153
épinard oseille 4072
épinard piquant 5153
épinard sauvage 2559
épine 2729
épine à fruits jaunes 6093
épine à fruits noirs 2082
épine blanche 2038
épine de Heldreich 2761
épine de Lambert 3209
épine de rat 885
épine de Robeson 4508
épine d'Espagne 390
épine d'Orient 4960
épine du Christ 1227, 1439
épine écarlate 5485
épine ergot de coq 1313
épine fleurie 2038
épine jaune 5112
épine marante 1523
épine noire 2082, 5014
épine rouge 4452
épine toujours verte 2039
épinette à la bière 653
épinette noire 653
épine-vinette 452
épine-vinette à feuilles de buis 3421
épine-vinette commune 2075
épine-vinette de Corée 3178
épipactide 2050
épiphylle 3247
éponge végétable 5564
équisétacées 2877
érable 3481
érable à feuilles d'aubépine 2730
érable à feuilles de charme 2853
érable à feuilles de frêne 750
érable à feuilles rondes 5735
érable à grandes feuilles 562
érable à nervures roussâtres 4470
érable à sucre 5304
érable blanc 4966
érable champêtre 2756
érable coloré 4004
érable de Cappodoce 1327
érable de Crète 3957
érable de Heldreich 418
érable de la Colchide 1327
érable de l'Orégon 562
érable de Mandchourie 3464
érable de Montpellier 3669
érable de Norvège 3885
érable de Pennsylvanie 5284
érable de Siebold 4934
érable de Tartarie 5456
érable de Trautvetter 4442
érable du Japon 2391, 3056
érable faux platane 4217
érable gris 4036
érable jaspé 5284
érable négundo 750
érable Nikko 3856
érable noir 637
érable palme 3056
érable plane 3885
érable rouge 4455
érable sycomore 4217
éragrostide 3385
éranthe 5998
érechtite 869
érémostachys 1839
érémure 1838
érémure élégant 3435
ergot 1258, 2053
érianthe 4228
éricacées 2745
érigéron 2292
érigéron âcre 602
érigéron du Canada 2878
ériochloé 1735
ériogone 2054
ériophoron 1637
ériophylle 2055
éritric 2056
érodium 2789
érophile 2188
ers 609
ers tétrasperme 2343
ers velu 5531
ervilier 609
érysimon 3772
érysiphacées 4289
érythrée centaurée 1909
érythrine 1587
érythrine à corail 1390
érythrine crête-de-coq 1311
érythrone 2190
érythroxylacées 1303
érythroxylum 1304
escallone 2060
escariole 4299
escarole 4299
eschscholtzie 2554
eschscholtzie de Californie 943
escourgeon 4981
esparcette 1519, 4646
estragon 5445
éthionème 5261
éthuse 36
eucalyptus 2065
eucalyptus à feuilles de saule 5969
eucalyptus à feuilles d'osier 4493
eucalyptus commun 5447
eucalyptus de Gunn 1233
eucephalartos 3125
eucharis 2066
euclidium 2067
eugénier 2068
eugénier à fleurs solitaires 4206
eugénier du Brésil 761
eupatoire 732
eupatoire à feuilles de chanvre 2776
eupatoire chanvrine 2776
euphorbe 2070
euphorbe des vignes 4159, 5328
euphorbe épurge 1008
euphorbe petit cyprès 1774
euphorbe polychrome 1750
euphorbe réveille-matin 5328
euphorbe superbe 1500
euphorbiacées 5191

euphraise 2152
euphraise officinale 1910
eurotia 6001
eurycome 2137
euterpe 2138
évax 2139
évernie 2150
évernie des pruniers 5204
évodie 2151
évodie du Japon 3032
exochorde 4090
exochorde de Korolkow 5645

fabagelle 498
fabiane imbriquée 4154
fabrecoulier 2106
fagacées 514
fanabrégié 2106
farouche 1694
farsétia 2185
fatsie 2187, 3034
faucillère 2155
fausse-giroflée 5573
fausse guimauve 1217
fausse oronge 2319
fausse roquette 4526
faux-acacia 636, 3354
faux-capillaire 3431
faux coqueret 285
faux-corête 635
faux cyprès 2167
faux cyprès obtus 2812
faux cyprès pisifère 4724
faux ébénier 3192
faux indigo 2171, 2948
faux indigotier 2948
faux-jalap 1415
faux-mélèze 2539
faux-mélèze de Kaempfer 3388
faux-millet 5032
faux nénuphar 2297
faux-pistachier 2083
faux-poivrier 4125
faux-poivrier à feuilles de térébinthe 766
faux-poivrier commun 940
faux quinquina 5324
faux sapin 3886
faux-senné 1362
faux tremble 4380
fédia corne d'abondance 52
feijoa 2195
félicie 2196
fendlera 2199
fenouil 1411, 2200
fenouil amer 1411
fenouil bâtard 1399
fenouil de Florence 2300
fenouil des chevaux 2866
fenouil marin 4675
fenouil puant 1399
fenugrec 2203
fer-à-cheval 2874
féronie 6020
féronie géant 2012
féronier 2012
férulage 2207
férule 2471
férule as(s)a fétida 336
férule commune 1418
férule persique 336
fétuque 2208
fétuque améthyste 215
fétuque bleue 215
fétuque des moutons 4858
fétuque des prés 3547
fétuque élevée 3547
fétuque faux-roseau 4474
fétuque géante 2472
fétuque hétérophylle 4839
fétuque ovine 4858
fétuque raide 5251
fétuque roseau 4474
fétuque rouge 1676, 4450
fève de cheval 2864
fève d'Egypte 2811
fève de marais 788
fève de tonca 1933, 5545
féverole 2864
févier 2840
févier à grosses épines 573
févier à miel 1439
févier à trois pointes 1439
févier d'Amérique 1439
févier d'Argentine 219
févier de Calabar 1816
févier de Caspienne 1044
févier de Chine 1187
févier du Japon 3041
févier tonka 1933
ficaire 2241, 4181
ficoïde 3592
ficoïde glaciale 2917
figuier 2240
figuier commun 1412
figuier d'Adam 4219
figuier de Barbarie 2936
figuier de Bengale 444
figuier des banians 738
figuier des pagodes 738
figuier d'Inde 4301
figuier élastique 2944
figuier grimpant 1267
figuier sycomore 5408
filage 2318
filao à feuilles de prêle 2876
filao dressé 1299
filaria 3642
filaria à feuilles étroites 3808
filaria à larges feuilles 5586
filipendule 3562
filipendule à 6 pétales 1907
fiorin blanche 4469
fistuline hépatique 515
fitzroya 2267
flacourtiacées 2941
flacourtie 2270
flacourtie de Cochinchine 4025
flacourtie de Madagascar 4413
flacourtie de Sonde 4605
flambé 2458
flavérie 2285
flèche d'eau 3924
fléchière 327, 3924
fléole 5527
fléole des Alpes 154
fléole des prés 5528
fléole des sables 4690
fleur d'Adonis 33
fleur de coucou 4398
fleur de la jalousie 4464
fleur de l'amour 4464
fleur de la nuit 3850
fleur de la passion 697
fleur des dames 1436
fleur d'une heure 2316
flindersie 2294
flindersie d'Australie 426
flourensie 5444
flouve 5724
flouve odorante 5389
flûteau 5829
foie de bœuf 515
foin du Tibet 2731
foin précoce 1963
folle avoine 252, 5951
fontanésie 2323
fontinale 5823
fontumia à caoutchouc 3205
forstéronie 2335
forsythie 2337
forsythie à fleurs pendantes 5860
fougère aigle 26
fougère aigle commune 5867
fougère en arbre 4945
fougère femelle 3349
fougère femelle commune 3199
fougère mâle 3445
fougère royale 4593
fougères 2206
fourcrée gigantesque 4209
fragon épineux 885
fraisier 5272
fraisier ananas 4195
fraisier caperonnier 2722
fraisier commun 2127
fraisier des bois 2127
fraisier des Indes 2932
fraisier de Virginie 5754
fraisier du Chili 1150
fraisier vert 2635
framboisier 4420, 4462
framboisier d'Amérique 624
framboisier des marais 5342
framboisier du Canada 2359
framboisier du Nord 3880
framboisier jaune 1273
frangipanier 2360
frankénia 2361
frankéniacées 2362
frankénie 2361
franquène 2361
fraxinelle commune 2435
fréesie 2365
frêne 338
frêne à bois jaune 2526
frêne à feuilles simples 4972
frêne à fleurs 2308
frêne à manne 2308
frêne à petites feuilles 4583
frêne blanc (d'Amérique) 5878
frêne bleu 687
frêne commun 2073
frêne de Caroline 1024

frêne de Mandchourie 3458
frêne de Pennsylvanie 4434
frêne des marais 1024
frêne de Syrie 5416
frêne d'Europe 2073
frêne dimorphe 93
frêne doré 2526
frêne élevé 2073
frêne noir 613
frêne oxyphylle 3798
frêne rouge 4434
frêne vert 2622
frésillon 1508
fritillaire 2380
fritillaire damier 1126
fritillaire du Kamtschatka 3130
fromental 5427
fuchsia 2389
fuchsia commun 3422
fumariacées 2393
fumeterre 2392
fumeterre officinale 1911
funaire 2394
funkie 4221
funkie à feuilles lancéolées 3060
funkie lancéolée 3060
funtumia 4946
fusain 2069
fusain ailé 5990
fusain à larges feuilles 793
fusain d'Amérique 805
fusain d'Europe 2096
fusain du Japon 2146
fusain nain 1939
fusain noir 1981
fusain verruqueux 5787
fusain vert 2146
fustet 5033
fustet commun 1527

gaertnère 873
gagée 2399
gaïac 3293
gaïac blanc 1460
gaïac officinal 1460
gaillarde 2400
gaillarde vivace 1491
gaillet blanc 5882
gaillet des forêts 6021
gaillet des marais 3500
gaillet gratteron 1055
gaillet jaune 6075
gaillet vrai 6075
gaînier 1098
gaînier commun 3114
gaînier de Chine 1199
gaînier du Canada 1979
galane 5655
galanga de l'Inde 2609
galantine 1530
galantine d'hiver 1530
galax 2401
galé 5366
galéga 2521
galéolope tétrahit 781
galéope 2778
galéope à feuilles étroites 3805
galéope bigarré 3846
galéope commun 4453
galéope pubescent 1895
galéopse 2778
galéopside à feuilles étroites 3805
galinsoga 4386
galipée 2403
galipée officinale 251
gallinette menotte 2531
galtonia 5319
gambier 532
gandasuli 2491
garance 1463, 3410
garance tinctoriale 1463
garcinie 2405
gardénie 2416
garde robe 1776
garrye 4947
gastérie 2436
gastérie verruqueuse 3990
gastéromycètes 2437
gastridie 3863
gastrolobe 4240
gattilier 3296
gattlier 3296
gaude 5862
gaulthérie 6003
gaulthérie couchée 1125
gaura 2438
gaylussacie 2884
gazanie 2440
gazon du Parnasse 4058
géastre 1969
gélatine de terre 5225
gelidium 55
gelsémie 2441
genêt 6015
genêt à balais 4769
genêt allemand 2463
genêt blanc d'Espagne 5910
genêt d'Angleterre 3823
genêt d'Espagne 5856
genêt des teinturiers 1563
genêt du Portugal 5910
genêt jonciforme 5857
genévrier 3118
genévrier à l'encens 2926
genévrier à prunes 4231
genévrier a rameaux arqués 2801
genévrier commun 1453
genévrier d'Afrique 43
genévrier de Californie 937
genévrier de Chine 4369
genévrier de Formose 2332
genévrier de Grèce 2618
genévrier de la Vallée d'Andorre 238
genévrier de Phénicie 4168
genévrier de Sargent 4707
genévrier des Bermudes 543
genévrier des montagnes 3703
genévrier des rivages 4874
genévrier de Syrie 5418
genévrier de Virginie 1980
genévrier de Wallich 5771
genévrier d'Occident 4938
genévrier du Mexique 3599
genévrier écailleux 4974
genévrier élevé 43
genévrier flasque 3598
genévrier gercé 100
genévrier oxycèdre 4298
genévrier piquant 4298
genévrier prostré 3039
genévrier rigide 3820
genévrier sabine 4719
génipa 2444
génipa des Antilles 2443
gentianacées 2446
gentiane 2445
gentiane à feuilles d'asclépiade 3622
gentiane amarelle 260
gentiane-asclépiade 3622
gentiane à tiges courtes 1291
gentiane champêtre 3550
gentiane ciliée 2684
gentiane croisette 1700
gentiane de Bavière 482
gentiane de neige 5063
gentiane des champs 3550
gentiane des Pyrénées 4371
gentiane germanique 2457
gentiane jaune 6091
gentiane pourprée 4346
gentiane printanière 5185
géoglosse 1970
géonoma 4840
géraniacées 2448
géranium 2447
géranium à feuilles molles 1892
géranium à feuilles rondes 4584
géranium à grosses racines 570
géranium brun 1931
géranium découpé 1757
géranium des bois 6024
géranium des collines 5686
géranium des horticulteurs 2873
géranium de Sibérie 4905
géranium des marais 718
géranium des prés 3551
géranium des Pyrénées 4372
géranium herbe-à-Robert 2787
géranium nain 5026
géranium pourpré 4342
géranium rosat 4568
géranium sanguin 680
gerbera 2451
germaine en buisson 1312
germandrée 2453
germandrée de montagne 3705
germandrée d'Orient 3954
germandrée femelle 1758
germandrée petit chêne 1113
germandrée sauvage 6028
gesneria 2464
gesneriacées 2465
gesse 4098
gesse à feuilles rondes 4587
gesse à larges feuilles 4133
gesse chiche 2279
gesse cultivée 2587
gesse des bois 2278
gesse des marais 3510
gesse des montagnes 3716
gesse des prés 3555
gesse de Tanger 5438
gesse domestique 2587
gesse du printemps 607

gesse hérissée 4578
gesse jaune 6104
gesse noire 645
gesse odorante 5378
giclet 5202
gilie 2487
gillénie 2172
gingembre 1419, 2489
ginkgo 2492
ginkgoacées 3432
ginkgo bilobé 2493
ginsang 346
ginseng 346, 2494
ginseng à cinq feuilles 195
ginseng à trois feuilles 1940
giraumon(t) 4330, 5705
giroflée 5258, 5768
giroflée commune 1556
giroflée des dames 1794
giroflée des jardins 263
giroflée des murailles 1556
giroflée jaune 1556
giroflier 1283
girole 1000
glaïeul 2499
glaïeul bleu 2458
glaïeul commun 1420
glaïeul de Gand 771
glaïeul des marais 5500
glaïeul des moissons 1612
gland de terre 2649
glauce 4796
glaucie cornue 652
glaucie jaune 6096
glaucière 2859
glaucion 2859
gléchome 2647
gleicheniacées 2502
globulaire 2506
globulaire à feuilles en cœur 2737
globulariacées 2505
gloire de neige 2512
glouteron 859, 2601
gloxinie 1424, 2515
glu de chêne 515
gluttier 1204, 4699
glycérie 3476
glycérie aquatique 806
glycérie flottante 5820
glycine 6011
glycine à folioles nombreuses 3365
glycine à longues grappes 3078
glycine d'Amérique 212
glycine de Chine 1211
glycine du Japon 3078
glycine en arbre 212
glycine floribonde 3078
glycine tubéreuse 207
glyptostrobe 1162
gnaphale 1727
gnaphale blanc jaunâtre 3092
gnaphale des forêts 6025
gnavelle 262
gnétacées 3107
gnète 3105
godétie 2524
goémon 4538
gombaud 3921
gombo 3921
gombo chanvre 3146
gommart 875
gommier 2065
gommier bleu 5447
gomphide glutineux 2516
gomphie 2557
gomphrène 2503
goodyère 4428
goodyère rampante 1675
gordonie 2564
gouet 333
gouet à feuilles maculées 3378
gouet commun 3378
gouet d'Italie 2972
gouet maculé 3378
gouet serpentaire 1536
gourbet 4475
gourde bouteille 922
gourde calebasse 922
gourdin d'Hercule 1849
goyave 1431
goyavier 1431, 2656
goyavier à feuilles de laurier 1631
grâce-de-Dieu 1913
graminées 2581
grand ajonc 1426
grand aspidistra 1352
grand corossolier 5297
grand dragonnier 1900
grande ammi 2603
grande astrance 2611
grande aunée 2010
grande bacury 411
grande berce 2825
grande brize 569
grande camomille 2210
grande capucine 1478
grande centaurée 1090
grande chélidoine 2606
grande ciguë 4241
grande consoude 1389
grande cuscute 2607
grande éclaire 2606
grande gentiane 6091
grande igname 5993
grande marguerite 1542
grande mauve 2797
grande ortie 574
grande patience 4258
grande pensée jaune 2136
grande pervenche 563
grande pimprenelle 2409, 2608
grande souchette 790
grande vesce des montagnes 6046
grand galanga 2609
grand mélilot 5426
grand millet 5087
grand orpin 2612
grand persil de montagne 3715
grand plantain 4502
grand rosage 4557
grand roseau 2482
grand soleil 1539
grapefruit 2578
grassette 900
gratiole 2754
gratiole officinale 1913
gratteron 1055
grémil 1428, 2640
grémil des champs 1614
grémil pourpre et bleu 1669
grenadier 4255
grenadille 4070
grenadille bleue 697
grenadille vineuse 2475
greuvier 2639
greuvier à feuilles de tilleul 3317
grévillée 2638
grindélie 2663
grindélie robuste 4873
griottier 3483, 5091
grisaille 2597
grisard 2597
grisset 1523
gros blé 4286
gros chiendent 542
gros daraden 5586
groseillier 1745, 2560
groseillier à grappes 3882
groseillier à maquereau 2103
groseillier d'Amérique 447
groseillier de Mandchourie 3459
groseillier de Meyer 3607
groseillier des Alpes 128
groseillier des Barbados 447
groseillier des buffles 1276
groseillier des Carpathes 1027
groseillier des Indiens 2935
groseillier des rochers 4520
groseillier doré 2535
groseillier d'Orient 3953
groseillier du Colorado 1336
groseillier du Japon 3030
groseillier fétide 4991
groseillier noir d'Amérique 186
groseillier noir d'Europe 2081
groseillier rouge 3882
groseillier rouge des marais 208
groseillier sanguin 6000
guaréa 3768
guayule 2657, 4065
guettarde 5714
guettardie 5714
gui 3631
gui blanc 2113
gui commun 2113
guimauve 169
guimauve officinale 3507
guizotie 3849
gunnère 2667
gunnère chilienne 1148
gunnère du Chili 1148
guttifères 2669
guzmannia 2670
gymnadénie 4481
gymnandre 3206
gymnospermes 2671
gynandropsis 5129
gynérion 4021
gynérion argenté 4818
gynure 5713
gypsophile 1433, 2672
gypsophile des murailles 1749
gypsophile de Steven 5243
gypsophile paniculée 397

gypsophile rampante 1670
gypsophile visqueuse 5248
gyrole rouge 3940
gyromitre comestible 1994
gyroselle 4871

habénaria 2673
hacquétie 2675
halénie 5193
halésie 4950
halidryde 4797
halimodendron 4667
halodule 4870
halorage 4786
haloxylon 4729
hamamélidacées 6013
hamamélide 6012
hamamélide de Chine 1212
hamamélide de Virginie 1562
hamamélide du Japon 3079
hamamélis printanier 5727
hamamélis velouté 1212
hamélie 2699
hamiltonie 2700
haricot 497
haricot à feuilles angulaires 34
haricot à œil noir 1059
haricot commun 3158
haricot d'Angola 3755
haricot de Lima 3309, 4940
haricot de Madagascar 4940
haricot des Indes 4748
haricot d'Espagne 4748
haricot dolique 1393
haricot écarlat 4748
haricot-fleur 4748
haricot-mungo 3754
harmale 2714
haworthie 2728
hédéome 2178
hédychie 2491
hédyscèpe 5680
hélénie 5052
héléocharis 5145
hélianthe annuel 1539
hélianthème 5334
hélicodicéros 5666
héliconie 2763
hélictère 4781
héliopside 2764
héliotrope 2765
héliotrope du Pérou 1436
héliptère 5333
hellébore 2766
hellébore fétide 507
hellébore orientale 3272
hélonie 5346
helvellacées 2768
helvelle 2767
helvelle crépue 1697
helvelle lacuneuse 3196
helxine 402
hémanthe 678
hémanthe de Catherine Saunders 3138
hémérocalle 1815
hémérocalle de Middendorff 3613
hémérocalle de Thunberg 5504
hémérocalle fauve 2334
hémérocalle graminée 2585
hémérocalle jaune 2334
henné 2782
hépatique 2783
hépatique des jardins 1437
hépatique trilobée 3343
herbe à chapelets 31
herbe à éléphant 3793
herbe à éternuer 5054
herbe à jaunir 1563, 5862
herbe à la bosse 2628
herbe à la flèche 3924
herbe à lait 1470
herbe à la manne 5820
herbe à l'esquinancie 5201
herbe à manne 3476
herbe à ouate 1469
herbe à plumets 4020
herbe à rosarie 31
herbe à setons 2628
herbe au pauvre homme 1913
herbe au scorbut 1522
herbe aux bisons 836
herbe aux chantres 2757
herbe aux charpentiers 1430
herbe aux chats 1060
herbe aux cure-dents 5547
herbe aux écus 3651
herbe aux goutteux 596
herbe aux gueux 5571
herbe aux mites 3690
herbe aux perles 1428
herbe aux plumets 2115
herbe aux turquoises 1944
herbe aux verrues 2606
herbe aux vipères 5744
herbe à 9 chemises 3368
herbe citron 3270
herbe de Bahia 409
herbe de Dallis 1789
herbe de feu 1226
herbe de grâce 1516
herbe de Guinée 2661
herbe de Harding 1761
herbe de Sainte-Barbe 610, 5999
herbe de Sainte Claire 2606
herbe de Saint Fiacre 1436
herbe de Saint-Guillaume 63
herbe de Saint Jacques 4401
herbe de Saint Jean 1537
herbe de Saint Philippe 1956
herbe des charpentiers 4401
herbe des Pampas 4020
herbe de Vasey 5703
herbe du Saint-Esprit 2406
herbe du Turc 1372
herbe du vent 243
herbe royale 5354
herbe sacrée 2427, 2910
herbe Saint-Roch 1413
herbe triste 1415, 2341
herminie 3763
herniaire 877
herniaire glabre 1372
herniaire velue 2682
herse 954
hétéranthère 3743
hêtre 513
hêtre antarctique 268
hêtre blanc 2077
hêtre commun 2077
hêtre d'Amérique 184
hêtre de bois 2077
hêtre d'Engler 2036
hêtre de Siebold 4931
hêtre d'Orient 3949
hêtre du Japon 3022
hêtre pleureur 5859
hêtre pourpré 4345
heuchère 171
hévéa 2791, 4050
hiba 2161
hibbertie 904
hibiscus 2794
hickory 2795
hièble 3570
hiérochloé 5367
himanthoglosse 3346
hippéastre 176
hippocastanacées 2865
hippocratée 2813
hippocrépide 2874
hirnéole oreille de Judas 3101
hoférie 2822
holostée 3730
homale 2833
hopée 3588
hopée odorante 5486
hortensia 2902
hortensia commun 565
hortensia des jardins 565
hortensia de Virginie 5039
hortensia en arbre 5039
hortensia en panicule 4028
hortensia grimpant 1269
hottonie 2193
hottonie des marais 2097
houblon 2847
houblon commun 1440
houblon du Japon 3043
houlque 5710
houlque d'Alep 3103
houlque laineuse 1553
houlque molle 2461
houlque odorante 5368
houque 5710
houstonie 710
houx 2826
houx commun 2039
houx d'Amérique 196
houx d'Angleterre 2039
houx hérisson 2751
houx verticillé 1560
hovène 4406
hovène sucré 3066
howée 4828
hoya 5847
hugonie 2885
humée 2886
humée élégante 2340
hunnemannie 2534
hutchinsie 2895
hydnacées 2901
hydne 2900
hydne coralloïde 1589

hydne imbriqué 2922
hydne sinué 2752
hydnocarpe 1121
hydnophyte 2899
hydrangelle arborescente 5039
hydrangelle des jardins 565
hydraste 2546
hydraste du Canada 2547
hydrocharidacées 2383
hydrocharide 2381
hydrocléis 5831
hydrocotyle 4114
hydrocotyle commune 1489
hydrodicte 5826
hydrophyllacées 5814
hydrophylle 5813
hygrophile 5233
hygrophoracées 2905
hygrophore 2904
hygrophore conique 1577
hyménocallide 5130
hyménomycètes 4180
hyménophyllée 2248
hyménophyllées 2249
hyménophyse 5918
hypécoon 2906
hypéricacées 4652
hypholome 2907
hyphomycètes 2245
hypne 2908
hysope 2909
hysope des garigues 1515
hysope officinale 2910

ibéride 993
ibéride blanche 4523
icaquier 1315
if 6117
if à prunes 4232
if commun 2045
if d'Angleterre 2045
if de Californie 3998
if de Chine 1215
if du Canada 980
igname 6062
igname ailée 5993
igname bulbifère 69
igname comestible 1214
igname de Chine 1241
igname du Brésil 1748
ijal 1130
ilicacées 2827
illécèbre 2919
illécèbre verticillé 5924
immortelle 2149
immortelle à bractées 5279
immortelle bleue 4793
immortelle des Alpes 1406
immortelle des sables 5991, 6083
impérate 4713
impératoire 3532
incarvillée 2924
indigo 5603
indigo bâtard 2171, 2948
indigotier 2947, 5603
inga 2949
inule à feuilles de saule 5970
inule britannique 784
inule conyze 1239
inule de Bretagne 784
iochrome 5738
ipécacuana annelé 2957
ipomée 3682
ipomée pourpre 1474
irésine 677
iriartée 5253
iridacées 2959
iris 2958
iris à feuilles de gramen 2582
iris Armes de France 2458
iris à tige sans feuilles 5268
iris d'Allemagne 2458
iris de Florence 2301
iris de Grossheim 2642
iris de Koempfer 4393
iris de Perse 4148
iris de Sibérie 4907
iris d'Espagne 5111
iris de Suse 3728
iris deuil 3728
iris d'Ibérie 2913
iris du Caucase 1074
iris du Taurus 1943
iris faux-acore 6086
iris japonais 3047
iris pâle 5371
iris panaché 2889
iris pygmée 1943
iris tigré 620
ischnosiphon 5532
isertie 5949
isoétacées 4388
isoète 4387
isolome 2969
isopyre 2970
isotome 4887
itéa 5387
itéa de Virginie 5755
ivraie 4631
ivraie à fleurs nombreuses 2980
ivraie d'Italie 2980
ivraie énivrante 1804
ivraie multiflore 2980
ivraie vivace 4135
ivraie vraie 1804
ixia 2993
ixie 2993
ixiolire 2994
ixiolire montagnard 5415
ixora 2995

jacaranda 2996
jacaranda à feuilles de mimosa 4849
jacaranda du Brésil 764
jacée 817
jacinthe 2896
jacinthe commune 1445
jacinthe d'eau 1557
jacksonie 3002
jacobée 4401
jacquier 2998
jacquinie 5211
jalap vrai 3005
jalousie 5393
jambo longue 3010
jambose de Malaque 3912
jambosier 4556
jamélonier 3010
jamésie 3011
jamésie américaine 1264
jaquier noir 634
jarilla 5172
jarousse 4098
jarrah 3081
jasione 3082
jasione des montagnes 4857
jasione vivace 4736
jasmin 3083
jasmin à grandes fleurs 1052
jasmin bâtard 5376
jasmin commun 1450
jasmin d'Arabie 293
jasmin de Virginie 1550
jasmin du Cap 3439
jasmin officinal 1450
jeffersonie 5662
jonc 4608
jonc à coton 1637
jonc commun 1517
jonc d'eau 2600
jonc de crapaud 5535
jonc des Alpes 146
jonc des chaisiers 2600
jonc des jardiniers 1517
jonc des tonneliers 2600
jonc fleuri 2314
jonc glauque 1517
jonchée 3674
joncinelle 4201
jonc maritime 4800
jonc odorant 1918
jonésie 4704
jonquille 3110, 3795
jonquine 852
joubarbe 2882
joubarbe à toile d'araignée 5133
joubarbe des toits 2779
juanulloa 3112
jubée 3113
juglandacées 5778
jujubier 3115
jujubier commun 1452
jujubier de Jamaïque 1321
jujubier grimpant 76
julienne 4522
julienne des dames 1794
juncaginacées 3116
juncasées 4609
jurinéa 3120
jusquiame 2823
jusquiame blanche 4586
jusquiame noire 633
jussiée 5832
justicie 3121
jute 3122, 4591

kadsoura 3123
kadsoura du Japon 4744
kæmpférie 4485
kalmie 3129
kalmie à feuilles étroites 3210
kalmie à larges feuilles 3710
kalmie glauque 720
kapokier 3134

kapoquier 3134
kava 3140
kawa 3140
kayée 3142
kéaki 3080
kennédie 3148
kennedye 3148
kentie 3149
kernéra 3154
kerrie 3156
kerrie du Japon 3049
kételéerie 3157
kételéerie de Fortune 2338
kételéerie du P. David 1812
ketmie 1185, 2794
ketmie à feuilles de tilleul 3316
ketmie comestible 3921
ketmie de Syrie 4886
ketmie élevée 3008
kévazingo 826
kickxie 2317
kigélie 4716
kirengeshoma 3163
kleinhovie 3165
klugie 3166
knautia 3167
kniphofie 5551
kobrésie 1302
kochie 5318
kochie à balais 531
koélérie 3173
koélérie à crète 4292
koélérie glauque 702
koelreutérie 2555
koelreutérie paniculée 4027
koenigie 3174
kok-saghyz 4613
kolatier 1323
kosteletzkye 4664
kouème 3917
kousso 3190
kudzu 3188
kudzu du Japon 5505
kumquat 3189
kumquat à fruits oblongs 3980
kumquat à fruits ronds 3528
kunzée 3698

labiées 3629
laboulbéniacées 3191
lachenalie tricolore 5591
lachnanthe 4463
lactaire 3195
lactaire à lait abondant 3938
lactaire à toison 6049
lactaire délicieux 1827
lactaire douceâtre 5372
lactaire grosse poire 3938
lactaire orangé doré 3938
lactaire poivré 4123
lagénaire 2567
lagerstrémie 2580
lagerstrémie des Indes 1394
lagerstrémie d'Inde 1394
lagoséride 3204
lagotis 3206
lagunaire 5307
lagure 2713
laîche 4810
laîche aquatique 5834
laîche blanchâtre 4970
laîche capillaire 2678
laîche compacte 2348
laîche des bois 6042
laîche des montagnes 3718
laîche des rivages 5282
laîche des sables 4688
laîche digitée 2252
laîche dioïque 1860
laîche distante 1865
laîche élevée 4107, 5656
laîche en gazon 5630
laîche fangeuse 3744
laîche glauque 1020
laîche jaunâtre 6106
laîche noirâtre 650
laîche pâle 4010
laîche paniculée 4030
laîche précoce 5725
laîche pulicaire 2293
laîche velue 2690
laîche vésiculeuse 674
laiteron 1531, 5105
laiteron des champs 2236
laiteron des marais 3517
laiteron épineux 4305
laitier 4250
laitue 3281
laitue à feuilles de saule 5973
laitue cultivée 2417
laitue de mer 1524
laitue lombarde 4545
laitue pommée 2735
laitue romaine 4545
laitue sauvage 4299
laitue vireuse 603
laitue vivace 4132
lallemantie 3208
lamarckie 2550
lamier 1818
lamier amplexicaule 2780
lamier blanc 5891
lamier maculé 5166
lamier pourpré 4343
laminaire 667
laminaire à sucre 5299
laminaire du Japon 4807
laminaire palmé 4806
laminariacées 3145
lampourde 1306
lampsane 3860
lampsane commune 1479
lamy 899
lance du Christ 839, 2085
landolphie 2666
langue de bœuf 515
langue de châtaignier 515
langue de chien 2879
langue de chien commune 1444
langue de serpent 1346
lanse 3217
lanse de Sonde 1886
lantana 3218
lapagerie 1152
lapagerie rose 4445
lapeyrousie 2169
laportée 6036
lardizabale 3223
larmes de Job 31
larmille 31
larrée 1680
laser 3227
lasie 3228
latanier 3229
lathrée 3230
lathrée écailleuse 5549
lauracées 3232
lauréole des bois 5192
lauréole femelle 2194
lauréole mâle 5192
laurier 3231
laurier alexandrin 88
laurier Bourbon 4436
laurier-cerise 1456, 3238
laurier cheval 88
laurier commun 2615
laurier d'Alexandrie 88
laurier d'Apollon 2615
laurier de Californie 938
laurier de l'Inde 3412
laurier de Saint-Antoine 2262
laurier des Iroquois 4712
laurier du Portugal 4277
laurier-rose 3925
laurier-rose d'Europe 1482
laurier-tin 3237
lavande 3239
lavande commune 5604
lavande de mer 4793
lavande femelle 5604
lavande spic 795
lavande vraie 5604
lavanèse 1425
lavatère 5582
lavatère de Thuringe 5507
lavatère en arbre 5716
layie 5519
lécanore 3262
lécanore comestible 3477
lecokie 3263
lécythide 3657
lède 3265
lédon 3265
lédon des marais 1719
leersia 1753
légumineuses 3267
leitnérie 1605
lemnacées 1925
lentibulariacées 665
lenticule 1924
lentille 1457, 3273
lentille d'Auvergne 3930
lentille d'eau 1924
lentille d'Espagne 2587
lentille tétrasperme 2343
lentillon 1457
lentin joli 4739
lentin squameux 4739
lentinus 3274
léonotis 3320
léontice 3322
léonure 1475, 3689
lépiote 3278
lépiote élevée 4051

leptochloé 5171
leptosperme 5472
leptosyne 1596
lespédézie 3279
lespédézie de Chine 1189
lespédézie de Corée 3183
lespédézie du Japon 1458
lespédézie soyeuse 1189
lespédézie stipulacée 3183
leucanthème des Alpes 123
leucanthème vulgaire 3987
leucocrine 5226
leucophylle 4961
levain 5986
levure 6069
levure de bière 1359
lewisie 3285
lézardelle 3347
liane à bœuf 1265, 2046
liane aux voyageurs 3486
liane-réglisse 3090
liardier 647
liatride 2439
libanotide 3286
libertie 3288
libidibi 1867
libocèdre 2925
libocèdre décurrent 936
libocèdre de Nouvelle-Zélande 4229
libocèdre du Chili 1149
licanie 3289
lichen 3290
lichen crin-de-cheval 2868
lichen de la manne 3477
lichen de prunellier 5204
lichen des rennes 4479
lichen d'Islande 2916
licuala 3292
lierre 2988
lierre commun 2040
lierre d'Angleterre 2040
lierre de bois 2040
lierre de Crimée 1691
lierre de Perse 1324
lierre des murailles 3147
lierre de terre 1429
lierre d'Europe 2040
lierre du Caucase 1324
lierre grimpant 2040
lierre japonais 3028
lierre terrestre 1429
ligulaire 2542
ligustique 3294
lilas 3295
lilas commun 1461
lilas de Bretschneider 2686
lilas de Hongrie 2890
lilas de l'Amour 228
lilas de l'Himalaya 2803
lilas de Perse 4149
lilas de Rouen 1190
lilas d'été 892
lilas japonais 3075
lilas penché 3867
lilas varin 1190
liliacées 3301
lime acide 3311
limette acide 3311
limettier 3311
limettier hérissé 3538
limnanthe 3548
limnochare 5711
limodore 2589
limonelle 3311
limonier 3268
limoselle 3745
lin 2286
linacées 2288
lin à feuilles étroites 3802
lin à fibre longue 2214
lin à graines 3915
linaigrette 1637
linaire 5533
linaire à feuilles de genêt 811
linaire alpestre 155
linaire commun 887
linaire de Dalmatie 1791
linaire de Macédonie 3404
linaire d'Italie 2984
linaire du Canada 3922
lin cultivée 2214
lin d'Autriche 378
lin de la Nouvelle-Zélande 3843
lindélofie 3313
lindernia 2179
lin jaune d'or 2538
linnée 5660
lin purgatif 4335
lin vivace 4129
liparis 5658
lippia 3324
liquidambar 5369
liquidambar d'Amérique 210
liriope 3308
lis 3300
lis à longues fleurs 1972
lis blanc 3417
lis bulbifère 842
lis couleur isabelle 3791
lis de Carniole 1022
lis de Chalcédoine 1111
lis de Guernsey 2658
lis de Ledebour 3264
lis de Malabar 3436
lis de St.-Bruno 4649
lis de St.-Jacques 393
lis des étangs 2132
lis doré du Japon 2525
lis du Japon 3053
liseron 2509
liseron des champs 2100
liseron des haies 2748
liseron pourpre 1474
liseron scammonée 4741
liseron soldanelle 4804
liseron tricolore 1941
lis gigantesque 2478
lis japonais 3053
lis jaune 842, 2334, 3269
lis martagon 3524
lis orange 3939
lis rose des Egyptiens 2811
lis royal 4476
listère 5657
lis tigré 5522
litchi 3397
litchi de Chine 3398
litchi ponceau 3398
lithosperme des champs 1614
littonia 3340
littorelle 4879
livèche 3384
livèche officinale 2418
livistone 2183
livistone biroo 3087
lloïdie 159
loasa 1155
lobéliacées 3351
lobélie 3350
lobélie de Dortmann 5819
locular 2004
loganiacées 3358
loiseleurie couchée 114
loiseleurie des Alpes 114
lonchitis 3708
lonchocarpe 3216
lophanthe 2476
lophire 3372
lophire à grandes feuilles 2005
lophire pyramidal 5992
loranthacées 3632
loranthe 3377
loropétale 3380
lotier 1824
lotier corniculé 588
lotier d'Egypte 5893
lotier des marais 5872
lotier rouge 5197
lotier velu 5872
louffa 5564
louffa cylindrique 5290
lucée 446
ludwigia 4812
luhée 5877
lunaire 2834
lunaire annuelle 1884
lunaire vivace 4130
lupin 3395
lupin blanc 5899
lupin bleu 703
lupin changeant 5096
lupin hérissé 2084
lupin jaune 2135
lupin muant 5096
lupin polyphylle 5789
lupin vivace 5327
lupuline 638
luzerne 3565
luzerne arborescente 5583
luzerne bigarrée 472
luzerne commune 90
luzerne cultivée 90
luzerne en arbre 5583
luzerne en faux 4921
luzerne hérissée 933
luzerne houblon 638
luzerne maculée 5170
luzerne orbiculaire 903
luzerne tachetée 5170
luzerne tropicale 3279
luzule 6041
luzule de Forster 2336
luzule des champs 2239
luzule des Sudètes 5296

luzule de Wahlenberg 5762
luzule en épi 5141
lychnide de Chalcédoine 3451
lychnide des prés 4398
lychnide du soir 2140
lychnide visqueuse 1250
lyciet 6016
lyciet à feuilles d'halimus 3535
lyciet commun 3535
lyciet de Barbarie 450
lyciet de Chine 1213
lyciet de Russie 4612
lycope 839
lycope d'Europe 2085
lycoperdacées 4326
lycopode 1285
lycopode des Alpes 127
lycopode en massue 4606
lycopodiacées 1286
lycopside 5938
lycopside des champs 5939
lygodium 1266
lysimaque 3370
lysimaque commune 2540
lysimaque de bois 6038
lysimaque ponctuée 5169
lythracées 3371
lythraire 3400
lythraire à feuilles d'hysope 2912

maceron 87
mâche 2088
mâche noire 5492
maclure 3971
maclure à feuilles d'oranger 3972
maclure épineux 3972
mâcre 5794
macrotomia 4314
macrozamia 3406
madie 1151, 3416
magnolia acuminé 1726
magnolia à fleurs de lis 3304
magnolia à grandes fleurs 5098
magnoliacées 3424
magnolia de Campbell 964
magnolia de Soulange 4715
magnolia de Virginie 5355
magnolia du Japon 5897
magnolia étoilé 5227
magnolia parasol 5679
magnolia yulan 6122
magnolier 3423
mahonia 3429
mahonia à feuilles de houx 3947
mahonia de Fortune 1192
mahonia du Japon 3055
mahot rouge à petites feuilles 744
maianthème 492
maianthème à deux feuilles 5671
maïs 3434
maïs cireux 5849
maïs corné 2295
maïs denté 1835
maïs farineux 2307
maïs perlé 4268
maïs sucré 5301
maïs tendre 2307
maïs vêtu 4233
malachre 5952
malaxide 25
malaxide des marais 3250
malcolmie 3444
malherbe 2210
malope 3449
malpighier 446, 3450
malvacées 3448
malvastrum 2173
mamillaire 3454
mammée 3452
mandarinier 3162, 3468
mandragore 3469
manettie 3470
manglier 200, 3475
mangoustan du Malabar 3474
mangoustanier 3474
manguier 1464, 3472
manguier fétide 180
manioc 1046
manioc amer 1379
manioc comestible 1379
manioc doux 67, 5360
manisure 3108
manne de Perse 961
manne de Pologne 5820
manne sanguinale 2683
manne terrestre 2683
manteau royal 1341
marante 329, 541
maranthacées 330
marasme 3484
marasme d'Oréade 2154
marattiacées 3485
marchantie 1465, 3487
margose 434
margousier azadirachta 3492
margousier azedarach 1160
marguerite des champs 3987
marjolaine 3497, 3967, 5375
marmottier 773
marron d'eau 5794
marronnier 828
marronnier à fruit turbiné 3046
marronnier blanc 1443
marronnier commun 1443
marronnier de Californie 931
marronnier de Chine 1188
marronnier d'Inde 1443
marrube 2816
marrube aquatique 2085
marrube blanc 1442
marrube commun 1442
marsault 2523
marsdénie 3499
marsilée 4128
marsiliacées 3523
martyniacées 5683
martynie 3527
massette 1062, 1380
maté 4048
matricaire 3541
matricaire inodore 4751
matthiole 5258
matthiole blanchâtre 1538
maurandie 3536
mauritier 3537, 5985
mauve 3447
mauve à feuilles rondes 1945
mauve alcée 2830
mauve crépue 1743
mauve de Mauritanie 3576
mauve des Juifs 4283
mauve frisée 1743
mauve musquée 2830, 3760
mauve royale 2788
mauve sauvage 2797
mayaque 4267
mayenne 2414
maytène 3540
maytène à feuilles de houx 2831
maytène du Chili 1154
méconopside 708
méconopside de Cambrie 5864
médaille de Judas 1884
médéola 3564
médicinier 449, 3838
médinilla 3567
mélampyre 1655
mélampyre violet 696
mélanconiales 3580
mélastomacées 3581
mélèze 3222
mélèze d'Amérique 1977
mélèze de l'Himalaya 2802
mélèze de Sibérie 4908
mélèze d'Europe 2110
mélèze du Japon 3051
mélèze du Sikkim 2802
mélèze occidental 5868
méliacées 3428
mélianthe 2839
mélicoque 2837
mélicyte touffu 2808
mélie 1159
mélilot 5362
mélilot à petite fleur 267
mélilot blanc 5917
mélilot officinal 6108
mélinet 2843
mélinis à petite fleur 3644
mélique 3582
mélique de Crimée 1693
mélique de Transylvanie 5570
mélisse 427
mélisse de Constantinople 341
mélisse des Moluques 341
mélisse officinale 1356
mélitte 3583
mélitte à feuilles de mélisse 469
melloco 5622
mélocactus 3584
melon 3762
melon d'eau 5821
mélongène 2414
melon serpent 5046
mélothrie 3586
ménispermacées 3672
ménisperme 3671
ménisperme de Daourie 347
ménisperme du Canada 1472
menthe 3628
menthe à feuilles rondes 283
menthe aquatique 5822
menthe citronnée 535
menthe des champs 2229

menthe poivrée 4122
menthe pouliot 4113
menthe sauvage 2869
menthe verte 5120
mentzélie 3587
ményanthe 716
ményanthe trifolié 1365
menziésie 4989
méon 415
mercuriale 3589
mercuriale annuelle 2785
mercuriale des bois 1876
mérianelle 3415
merisier 3544
merisier à grappes 2078
mertensia 691
mérue 3420
merveille du Pérou 1415, 2341
mésembryanthème 3592
méthonique 2511
métroxylon 4644
mibore 3609
michélie 3610
micocoulier 2674
micocoulier austral 2106
micocoulier de Chine 1183
micocoulier de Corée 3182
micocoulier de Provence 2106
micocoulier de Tournefort 3955
micocoulier de Virginie 1434
micocoulier d'Orient 2928
micocoulier du Caucase 1072
micocoulier occidental 1434
micromérie 3611
micromérie de Grèce 2619
mignonette 1468
mikanie 3616
mikanie grimpante 1268
milla 3602
millepertuis 4651
millepertuis à grandes fleurs 1
millepertuis commun 1537
millepertuis des montagnes 3721
millet 4315
millet à chandelles 4092
millet commun 4315
millet des oiseaux 2352
millet diffus 201
millet perle 4092
millet pied-de-coq 456
millet sauvage 3624
miltonia 4031
mimosa 3625
mimosa blanchâtre 4957
mimosa blanchissant 4957
mimosacées 3626
mimosa de Bailey 1300
mimule 3655
mimule jaune 2541
mimule musqué 3764
mimusops 848
minette 638
minette dorée 638
mirabelle 3630
miroir de Vénus 5719
miscanthe 4956
misotte 96
mitchelle 4066
mitelle 3634
mitraire 3635
mnie 3638
mnium 3638
mohwa 2920
moisissure 3691
molène 3752
molène aux teignes 3690
molène de Phénicie 4351
molène fausse-phlomide 1255
molène noire 641
molène pyramidale 4368
molène violette 4351
molinie 3646
molinie bleue 3674
mollugine 1032
molucelle 3647
mombin jaune 6101
momordique à feuilles de vigne 434
momordique balsamine 431
momordique baumier 431
monarde 512
monnaie du Pape 1884
monnoyère 2232
monocotylédonées 3661
monodora 3662
monstéra 3663
montie 2938
moracées 3750
morée 3676
morelle 3853
morelle faux-piment 3095
morelle grimpante 604
morelle jaune 6103
morelle noir 643
morène 2382
moricaude 1994
morille 3678
morille à tête ronde 1473
morille comestible 1473
morille conique 1578
morille des pins 1683
morille fauve 1473
morille fétide 5256
morille vulgaire 1473
morillon 1994
morine 3681
moringa 2872
mors de grenouille 2382
mors du diable 3561
mo(u)caya 3737
moureiller 905
mouron 4185
mouron bleu 705
mouron des champs 4747
mouron des oiseaux 1146
mouron mignon 722
mousse d'azur 2056
mousse de Ceylan 56
mousse de Corse 1623
mousse de Jaffna 56
mousse des jardinières 2908
mousse d'Irlande 2961
mousse perlée 2961
mousseron d'automne 2154
mousses 5605
moutarde 3771
moutarde à feuilles 2939
moutarde blanche 5904
moutarde brune 2939
moutarde du Loiret 2939
moutarde noire 642
moutarde sauvage 1118
moyenne stellaire 1146
mucoracées 3739
mucune 1644, 3740
muflier 5050
muflier des jardins 1528
muflier oronce 1619
muguet 3305
muguet de mai 3306
muguet du Japon 1944
muhlenbeckie 6010
muhlenbergie 3748
mûrier 3749
mûrier à papier 1486
mûrier blanc 5902
mûrier de Chine 1486
mûrier de Renard 2080
mûrier des haies 2080
mûrier noir 640
mûrier rouge 4456
murraye 3084
musacées 440
muscadier 3894
muscadier commun 1480
muscari 2579
muscari à grappe 5214
muscari à toupet 5450
muscari botryoïde 1427
muscari chevelu 5450
myagre 3773
mycène 3774
myopore 3775
myosotis 2327
myosotis à grandes fleurs 5399
myosotis des Açores 392
myosotis des Alpes 132
myosotis des bois 6033
myosotis des champs 2222
myosotis des marais 5602
myosotis gazonnant 5627
myosotis palustre 5602
myosotis versicolore 6073
myricacées 484
myricaire 2181
myricaire d'Allemagne 2456
myricaire de Daourie 1784
myriophylle 4060
myriophylle à épis 5138
myriophylle brésilien 763
myriophylle verticillé 974
myrique 483
myristicacées 3895
myrocarpe 3778
myrocarpe à feuilles denses 762
myrrhis odorant 5361
myrsinacées 3780
myrtacées 3783
myrte 3781
myrte commun 5606
myrte d'Australie 370
myrte épineux 885
myrtille 3784
myrtille à feuilles persistantes 1645
myrtille rouge 1645

myxomycètes 5006

naïadacées 3787
nandine 3790
nandine fruitière 4634
narcisse à bouquet 4248
narcisse de Constantinople 4248
narcisse des jardins 4237
narcisse des poètes 4237
narcisse faux-narcisse 1397
nard raide 3533
narthèce 715
navet 6005
navet du diable 4439
navet jaune 4626
navet oléifère 1343
navette 3817
nayade 3786
néflier 3577
néflier commun 3578
néflier du Japon 3376
neillie 3824
neillie de Chine 1193
neillie du Tibet 5514
nélombo 3381
nélombo d'Orient 2811
nélumbo 3381
némésie 3825
ne-me-touchez-pas 5560
némophile 3826
ne-m'oublie(z)-pas 2327
nénufar 5817
nénuphar 5817
nénuphar blanc 2132
nénuphar jaune 6099
néottie 590
népenthacées 3829
népenthe 3828
népète chataire 1060
néphrolépis 5405
néphrome 3831
neptunia 4825
nérier à feuilles de laurier 1482
nériette 2262
nérine 3832
nerprun 830
nerprun à feuilles d'aune 82
nerprun alaterne 2974
nerprun alpin 121
nerprun bourdaine 2513
nerprun de Californie 932
nerprun de Chine 1173
nerprun de Pursh 1040
nerprun des Alpes 121
nerprun des rochers 4513
nerprun des teinturiers 4145
nerprun du Japon 3025
nerprun du Liban 3260
nerprun purgatif 1369
nerprun tinctorial 1173
nertère humble 491
neslie paniculée 422
néviusie 5068
nez coupé 2083
nicandre 284
nicandre physalide 285
nidulariées 589
nielle des blés 1391
nierembergie 1733
nigelle 2201
nigelle de Crète 2415
nigelle de Damas 3386
nipa 3859
nitelle 3861
nitraire 3862
nivéole 1530, 5062
nivéole d'été 5322
nivéole printanière 5188
noble épine 2038
noisetier 2098, 2246
noisetier commun 2098
noisetier d'Amérique 194
noisetier de Byzance 5646
noisetier de Californie 935
noisetier de Chine 1179
noisetier de Constantinople 5646
noisetier de Mandchourie 3456
noisetier de sorcière 1562, 6012
noisetier de Turquie 5646
noisetier du Japon 3035
noisetier du Queensland 4385
noisetier du Tibet 5513
noisetier franc 2473
noisetier tubuleux 2473
noix d'arec 548
noix muscade 1480
noix vomique 3898
nolane 3871
noline 3872
nonette jaune 2574
nonette voilée 820
nonnée 3873
nopal à cochenille 1305
nostoc 3889
nostocacées 2375
nostoc commune 5225
nothofagus 2163
nouret 3992
noyer 5777
noyer à balais 4179
noyer à beurre 897
noyer cendré 897
noyer commun 4151
noyer cordiforme 2281
noyer cultivé 4151
noyer de Chine 1208
noyer de Mandchourie 3466
noyer de Para 765
noyer de Siebold 4936
noyer des Indes 3438
noyer des Moluques 992
noyer des pourceaux 4179
noyer des rochers 5479
noyer du Brésil 765
noyer du Japon 2493
noyer noir d'Amérique 1975
nufar 1647
nufar jaune 2090
nuphar 1647
nyctanginacées 2342
nyctérinie 3852
nymphéacées 5818
nympheau 2297
nyssacées 5643

oca 3918
ochne 3907
ochrome 5870
ocotéa 3908
odontite 460
odontite rouge 4435
odontoglosse 3910
œdogonie 3911
œil-de-cheval 2955
œil-de-faisan 4237
œillet 4197
œillet à bouquets 1019
œillet à delta 3433
œillet de Chine 1165
œillet de poète 5393
œillet des bois 1837
œillet des Chartreux 1037
œillet des fleuristes 1019
œillet des prés 4398
œillet des sables 2255
œillet giroflée 1019
œillet mignardise 1129, 2588
œillet superbe 3298
œillet velu 1837
œnanthe 5799
œnanthe aquatique 2251
œnanthe fistuleuse 5625
œnanthe safranée 2771
œnothère 2141
oïdium 3914
oignon 2420, 3931
oignon patate 3753
oignon sous terre 3753
okoumé 3920
oldfieldie d'Afrique 51
oléacées 3928
oléaire 1788
olivier 3927
olivier commun 1483
olivier de Bohême 2006, 4616
olivier de Normandie 683
olivier sauvage 5953
ombellifères 1036
ombilic 3816
ombilicaire 4536
omphalode 3815
onagracées 2142
onagre 2141
onagre bisannuelle 1408
oncidier 3929
onocléa délicate 4826
ononis épineux 5492
onoporde 1638
onoporde acanthe 4770
onoporde à feuilles d'acanthe 4770
onoporde de Crimée 5462
onosma 3932
ophioglossacées 28
ophioglosse 27
ophioglosse commun 1346
ophiopogon 3307
ophiopogon jaburan 5898
ophrys des marais 3250
ophrys mouche 2320
opopanax 3934
orange amer 5092
oranger de Malabar 408
oranger des falaises 5428
oranger des Osages 3971

oranger des savetiers 5354
oranger doux 5377
orbignya 396
orcanète 1954
orcanette 1954
orchidacées 3945
orchidée 3944
orchis 3946
orchis militaire 5079
oreille d'âne 2235
oreille de noyer 3992
oreille-de-souris 5064
oreille d'ours 368
oréodoxe 4594
oréoptéris 3884
orge 454
orge à crinière 2350
orge à deux rangs 5672
orge à six rangs 4981
orge carrée 455
orge des rats 3729
origan 1484, 3967
orme 2018
orme à feuilles de charme 5040
orme à grappes 4521
orme ailé 5989
orme à larges feuilles 191
orme à liège 1602
orme à longues pédoncules 5174
orme aquatique 5801
orme à trois feuilles 1441
orme blanc 191, 4771
orme blanc d'Amérique 191
orme blanc d'Europe 4614
orme champêtre 5040
orme cilié 4614
orme d'Amérique 191
orme de Chine 1177
orme de Cornouailles 1615
orme de montagne 4771
orme de Samarie (à trois feuilles) 1441
orme de Sibérie 2021, 4903
orme des rochers 4521
orme de Thomas 4521
orme diffus 5174
orme étalé 4614
orme liège 1602
orme nain 4903
orme rouge 5040
orme roux 5012
orne de Chine 1168
orne d'Europe 2308
ornithogale 5228
ornithogale des Pyrénées 478
orobanchacées 813
orobanche 812
orobanche majeure 2604
orobanche mineure 1279
orobanche rameuse 2774
orobe 4099
orobe jaune 6104
oronce 2533
oronge ciguë blanche 5180
oronge ciguë verte 1820
oronge ciguë vireuse 1842
oronge vraie 918
orostachys 5263
oroxylon 2945
orpin 5264
orpin âcre 2553
orpin bâtard 5673
orpin blanc 5915
orpin d'Espagne 5114
orpin intermédiaire 2148
orpin pourpré 4354
orpin réfléchi 3089
orpin reprise 3341
orpin rosat 4567
orpin rose 4567
orpin rougeâtre 4467
orpin velu 2691
orseille 3325
ortie 3835
ortie à pilules 4547
ortie blanche 4410, 5891
ortie brûlante 1874
ortie de Chine 4410
ortie dioïque 574
ortie du Canada 977
ortie jaune 5855
ortie puante 5911
ortie rouge 4343
orvale 1254
oryzopsis 4499
oseille 1868
oseille à feuilles obtuses 601
oseille aquatique 5798
oseille commune 2429
oseille crépue 1741
oseille des brebis 4859
oseille ronde 2373
osier amandier 106
osier jaune 6107
osier pâle 106
osier rouge 4352
osier viminal 467
osier vitellin 6107
osmanthe 3973
osmonde 3974
osmonde royale 4593
osmorrhize 5381
osmundacées 3975
ostéomèle 733
ostrowskye 2468
ostryer 2849
ostryer à feuilles de charme 2107
ostryer de Virginie 197
ostryer du Japon 3044
ottélie 3978
ouratée 3979
owala 3984
owénie 2030
owénie à feuilles de cerisier 1136
oxalidacées 5088
oxydendre 5094
oxydendre en arbre 5095
oxylobe 3991
oxyrie 3719
oxytropide 1662
oxytropide de l'Oural 5690

pachycentrie 3994
pachyrrhize 6063
pachyrrhize anguleuse 5851
pachysandre 3995
padouk 47
padouk de Birmanie 864
pahudie 4002
pain d'oiseau 2553
paka 917
palafoxie 4006
palaque 3814
palétuvier des montagnes 1290
palétuvier noir 200
palissandre de l'Inde 1984
palissandre du Brésil 648
palissandre du Sénégal 42
paliure 4015
paliure épineux 1227
palmacées 4018
palmetto 909
palmier 4016
palmier à cire 1021
palmier à (l')huile 46, 3916
palmier à sucre 2558
palmier-canne 3737
palmier chanvre 2339
palmier dattier 1807
palmier de Chine 2339
palmier de la Guadeloupe 2654
palmier doum 2001
palmier éventail d'Australie 371
palmier éventail de Chine 1178
palmier éventail de Cochinchine 3442
palmier nain 1948, 3572
palmiers 4018
palmier sagou 1764
palomet 2629
pamplemoussier 4329
panais 2422, 4064
panarine 3788
pancratie 4023
pandanacées 4780
pandanées 4780
pang(i)e 2324, 4024
panic 4026
panic à tiges grêles 1561
panicaut 2058
panicaut champêtre 5047
panicaut des Alpes 709
panicaut des champs 5047
panicaut maritime 4789
panic capillaire 1561
panic érigé 5403
panic millet 4315
panic rameux 818
panic vert 2624
panis des marais 456
pantoufle 5011
panus 4033
papareh 434
papavéracées 4274
papayacées 4034
papayer 4035
paphinie 4040
papilionacées 4041
pappophore 4042
papyrier 1486
pâquerette 1787
pâquerette vivace 2037
paquerolle 470
parépon 4085, 5154

pariétaire 4105
pariétaire officinale 5773
parinare 4052
parisette 4053
parisette à quatre angles 2786
parisette à quatre feuilles 2786
parkie 3864
parkie du Soudan 5670
parméliacées 4057
parmélie 4056
parnassie 4058
parnassie des marais 5932
parochète 4847
paronyque 3788
parrotie 4061
parrotie de Perse 4150
parrye 4062
parsonsie 1736
pasanie 5439
pas-d'âne 1339
pas-d'âne commun 1388
paspale 4067
paspale d'Urville 5703
passerage 4126
passerage à fleurs denses 4293
passerage champêtre 2233
passerage de Virginie 5749
passifloracées 4069
passiflore 4068
passiflore bleue 697
passiflore comestible 4070
pastel 6014
pastel cultivé 1956
pastel des teinturiers 1956
pastenade 2422
pastenaque 2422
pastèque 5821
patate douce 5380
patchouli 2792
patchouli de Cablin 911
patience des Alpes 129
patience des jardins 4072
pâtisson 883
patrinie 4073
patte de lion 1406
pâturin 700
pâturin à feuilles étroites 3799
pâturin annuel 258
pâturin bulbeux 844
pâturin commun 4579
pâturin comprimé 970
pâturin de Balfour 416
pâturin des Alpes 118
pâturin des bois 6023
pâturin des marais 2345
pâturin des prés 3150
paturon blanc 1476
paullinie 2655, 4074
paulownia tomenteux 4595
paulownie 4075
paulownie impériale 4595
paumelle 5672
pavette 4076
pavier 828
pavier blanc 739
pavier fauve 6077
pavier glabre 3913
pavier jaune 6077
pavier rouge 4441
pavot 4272
pavot argémone 2153
pavot coquelicot 1616
pavot cornu 2859
pavot des Alpes 140
pavot des Pyrénées 4375
pavot de Tournefort 3960
pavot douteux 3367
pavot somnifère 3933
paxille 4078
paxille enroulé 2956
payéna 4079
pêcher 4082
pédaliacées 4101
pédérie 2211
pédiculaire 3383
pédiculaire à sceptre 1117
pédiculaire de Laponie 3219
pédiculaire des bois 3382
pédiculaire des marais 2117
pédiculaire des Sudètes 5295
pédiculaire verticillée 5925
pédilanthe 5010
pégane 4102
pégane harmale 2714
peigne de Vénus 4861
peireskia 4136
pélargonier 4103
pélargonium 4103
pélargonium à feuilles de lierre 2991
pélargonium à grandes fleurs 575
pélargonium à ombelle 4568
pélargonium malodorant 4566
pélargonium très odorant 3897
peltaire 4864
peltandre 326
peltigère 2648
peltiphylle 4731
peltogyne 4348
pénicillaire 4108
pénicille 4109
pénicille glauque 2632
pennisétum 4110
pensée 5954
pentace 5489
pentacme 4115
pentapère 4116
pentas 5216
pentstémon 4117
pentstémon de Hartweg 2718
pentzie 4855
pépéromie 4119
péplide 5833
pépon 4330
perce-muraille 5773
perce-neige 1530, 5060
pérille 4138
pérille faux-basil 1492
périploque 4949
périploque grec 2616
péristrophe 4139
perle des potagères 4986
pernettye 4141
pérowskie 4142
perruque du diable 1280
persée 4144
persicaire cuspidée 3036
persicaire de Sakhaline 4656
persicaire douce 5168
persicaire du Levant 4309
persil 4063
persil bâtard 4124
persil sauvage 856, 1417
pervenche 4140
pervenche mineure 1493
pesse 3488
pesse commune 3489
peste d'eau 976
pétalostémon 4290
pétasite 888
pétasite blanc 5884
pétasite officinal 4340
petit corail 4743
petit cyprès 2167
petite astrance 1946
petite bardane 1306
petite berle 545
petite brize 3335
petite capucine 882
petite ciguë 2325
petite douve 5121
petite lentille d'eau 1403
petite marguerite 2037
petite massette 3800
petite massue 2520
petite mauve 1945
petite oseille 4859
petit épeautre 2004
petite pervenche 1493
petite pimprenelle 867, 5018
petite vrillée 2100
petit gris d'automne 1859
petit houx 885
petit mélinet 3280
petit millet 5032
petit muguet 5395
petit-nénuphar commun 2378
petit œillet d'Inde 2369
petit poivre 1120
petit radis 5030
petit vesceron 5531
pétrée 4157
pétrocallide 4510
pétunia 4160
pétunia blanc 5901
pétunia hybride 1495
pétunia violet 5742
pétunie 1495
peucédan 2824
peuplier 4270
peuplier à feuilles de laurier 3234
peuplier baumier 5100
peuplier baumier de Giléad 428
peuplier baumier d'Occident 942
peuplier blanc 5906
peuplier commun 647
peuplier cotonnier noir 942
peuplier de Berlin 540
peuplier de Corée 3185
peuplier de Frémont 2366
peuplier de la Caroline 250, 1978
peuplier de Lombardie 3360
peuplier de l'Ontario 428
peuplier de Maximowicz 3062
peuplier de Sargent 4214

peuplier de Virginie 1978
peuplier d'Italie 3360
peuplier du Canada 979
peuplier du Missouri 5100
peuplier faux tremble 4380
peuplier gris(ard) 2597
peuplier noir 647
peuplier noir du Nouveau-Monde 1978
peuplier odoriférant 3654
peuplier pyramidal 3360
peuplier sombre 819
peuplier tremble 2074
peuplier tremble d'Europe 2074
peuplier tremble du Japon 4929
pézizacées 1734
phacélie 4161
phacélie à feuilles de tanaisie 5443
phalangère à fleur de lis 4648
phalaris 984
phallacées 4162
phalle puant 5256
phanérogames 2310
phasque 1968
phégopteris 3797
phellodendre de l'Amour 226
phellodendre-liège 1604
phéophycées 815
philodendron 4165
phippsie 2915
phlomide 3097
phlomide sous-ligneuse 1451
phlomide tubéreuse 5620
phlox 4166
phlox de Drummond 1919
phlox de Sibérie 4911
phlox paniculé 5320
pholiote aux chèvres 6057
pholiote changeante 1114
pholiote précoce 1964
pholiote ridée 6057
phormier 2215
photinie 4169
photinie de Davidson 1813
phragmite 4472
phragmite commun 1514
phryma 3373
phycomycètes 5824
phyllanthe 3249
phyllanthe officinal 2026
phyllanthe sour 3977
phyllocacte 3247
phylloclade 1087
phyllodoce 3706
phyllospadix 5337
phyllostachys 4170
physalis 2645
physocarpe 3857
physostégie 3321
physostigma 921
phytéléphas 2986
phytéléphas macrocarpe 1447
phytocrène 4171
phytolaccacées 4244
phytolaque 4243
phytolaque à 10 étamines 1502
phytophthore 4172
pia 2244
picramnie 599
picrasme 4383
picride 3989
picride échioïde 783
pied bleu 214
pied-d'alouette des champs 2330
pied-d'alouette vivace hybride 2305
pied-de-chat alpestre 142
pied-de-chat commun 1510
pied-de-chèvre 4733
pied-d'éléphant 2014, 5558
pied-de-lièvre 4394
pied-de-lion 1454
pied-de-pigeon 3369
pied-de-veau 3378
pied-d'oiseau commun 1526
piéride 4176
pigamon 3556
pigamon à feuilles d'ancolie 1342
pigamon à feuilles étroites 3806
pigamon jaune 6098
pigamon mineur 3392
pignet 2983
pignon d'Inde 449
pilée 1259
pilocarpe 4184
piloselle 3734
pilulaire 4183
pimélée 4498
piment 102, 4459
piment annuel 884
piment de la Jamaïque 102
piment giroflé 102
piment giroflier 102
pimprenelle 868
pimprenelle aquatique 808
pimprenelle sanguisorbe 5018
pin 4190
pin à amandes 1337
pin à aubier 4208
pin à bois lourd 4259
pinacées 4194
pin à écorce en dentelle 3193
pin à encens 3352
pin à gros cônes 1642
pin à longues feuilles 3366
pinanga 4187
pin à sucre 5306
pin ayacahuite 3606
pin blanc 1982, 5879
pin bon 2983
pin cembro 5402
pin cembro de Sibérie 4917
pin d'Alep 86
pin de Balkans 419
pin de Banks 3001
pin de Bordeaux 1296
pin de Boston 3366
pin de Calabre 1625
pin de Chilghoza 1156
pin de Chine 1195
pin de Corée 3184
pin de Corse 1625
pin de cowrie 3139
pin de Cuba 4993
pin de Genève 4775
pin de Gérard 1156
pin de Heldreich 2762
pin de Hudson 3001
pin de Jersey 5750
pin de Jérusalem 86
pin de J. Sabine 1856
pin de Lambert 5306
pin de l'île de Norfolk 3877
pin de Lord Weymouth 1982
pin de Macédoine 419
pin de Masson 3530
pin de montagne 5400
pin de Monterrey 3666
pin de Montezuma 3668
pin de Murray 3355
pin d'Engelmann 2033
pin de Parana 4049
pin de Russie 4775
pin des Canaries 988
pin de Sibérie 4917
pin des Landes 1296
pin des Pyrénées 2052
pin des rochers de Suisse 5402
pin des temples 3023
pin de Ténériffe 988
pin de Torrey 5557
pin de Virginie 5750
pin du Chili 3658
pin du Père Armand David 320
pin épineux 595
pin femelle 1156
pin jaune 4882
pin jaune du Nouveau-Monde 4259
pin maritime 1296
pin mugho 3746
pin Napoléon 3193
pin noir 380
pin noir d'Amérique 4208
pin noir du Japon 3023
pin palustre 3366
pin parasol 2983
pin pignon 2983
pin pinier 2983
pin pleureur de l'Himalaya 2807
pin pondéreux 4259
pin queue de Renard 2353
pin raide 4208
pin résineux 4208
pin rigide 4208
pin rouge d'Amérique 4460
pin rouge du Japon 3067
pin rouge du Nord 4775
pin sabine 1856
pin sauvage 4775
pin souple 3310
pin sylvestre 4775
pin vrillé 4876
pin Weymouth 1982
pipe allemande 1404
pipéracées 4121
piptadénie 4203
piptadénie commune 1497
piptadénie d'Afrique 48
piptanthe 4204
pissenlit 1796
pissenlit officinal 1398
pistachier 4205
pistachier commun 1498
pistachier d'Atlas 3724
pistachier lentisque 3275

pistachier térébinthe 5476
pistie 5816
pithécolobium 272
pittone 5562
pittosporacées 5537
pittospore 4210
pittospore de Chine 5538
pittospore ondulé 3937
pittospore tobira 5538
pivoine 4118
pivoine en arbre 5585
pivoine mâle 1590
pivoine officinale 1490
plagianthe 5659
plagiobotryde 4269
plagiochile 4211
planère 5800
plantaginacées 4220
plantain 4218
plantain d'Asie 348
plantain d'eau 2593
plantain des Alpes 162
plantain des montagnes 3717
plantain lancéolé 829
plantain moyen 5379
plantain psyllium 2290
plante au compas 4571
plantes à spores 1717
plaqueminier 4152
plaqueminier commun 1494
plaqueminier d'Europe 1809
plaqueminier de Virginie 1494
plaqueminier d'Italie 1809
plaqueminier d'Orient 1809
plaqueminier du Levant 1809
plaqueminier faux-lotier 1809
plaqueminier kaki 3127
platanacées 4216
platane 4215
platane à feuilles d'érable 3361
platane de Linden 3006
platane de Londres 3361
platane d'Espagne 5113
platane d'Occident 205
platane d'Orient 3959
platane du Mexique 3601
platanthéra 4403
platycérium 5205
platycérium à corne d'élan 1534
platycodone 423
platycodone à grandes fleurs 424
platymiscie 3402
platystémone 4222
plectranthe 5190
pleurogyne 2198
pleuropogon 4819
pleurote en coquille 3992
pleurote ostracé 3992
pluchée 4223
plumbaginacées 3246
plumeau 2097
plume d'autruche 3976
plutée couleur de cerf 2189
podagraire 596
podocarpacées 4236
podocarpe 4235
podocarpe montagnard 5449
podophylle 3539
podophylle à feuilles peltées 1845
podophylle américain 1845
podophylle indien 2804
podostémon 4506
pogonie 4238
poinsettie 4239
poireau 3266
poirée 3473
poirée à carde 5398
poirée à couper 1116
poirée ordinaire 1116
poirier 4089
poirier à feuilles d'amandier 107
poirier à feuilles de bouleau 580
poirier à feuilles de saule 5971
poirier amandier 107
poirier commun 1487
poirier de Chine 5694
poirier de Malaque 3912
poirier de neige 5066
poirier des sables 4687
pois 4080
pois carré 2946, 3498
pois chiche 1144
pois cultivé 2423
pois d'Angola 4177
pois de cœur 2741
pois de Jérusalem 3754
pois de pigeon 2231
pois de senteur 5378
pois de vache 1393, 1650
pois du Brésil 1393
pois du Cap 3309, 4940
pois gras 4578
pois sabre 1448
poivre d'eau 3511
poivre de la Jamaïque 102
poivre de murailles 2553
poivre d'Espagne 884
poivre d'Inde 884
poivre girofle 102
poivre sauvage 1120
poivrier 4120
poivrier du Japon 3063
poivrier du Pérou 4125
poivrier noir 646
poivron 102
polémoniacées 4167
polémonie 4246
polémonie bleue 2620
pollinia 5303
polycarpe 3479
polycnème 4249
polygalacées 3623
polygala commun 1470
polygala de Sibérie 4909
polygonacées 833
polymnie 3248
polypode 4251
polypodiacées 4252
polypogon 504
polyporacées 4253
polypore destructeur 4254
polypore du bouleau 583
polypore écailleux 1921
polypore en touffe 2385
polypore soufré 5312
polypore versicolore 1569
polystic 2828
polystic falciforme 2881
polytric 2676
pomacées 282
pomaderre d'Australie 5746
pomme cythère 178
pomme de cajou 1378
pomme de Jérusalem 431
pomme de merveille 431
pomme de Tahiti 3912
pomme de terre 4281
pomme de terre céleri 277
pomme épineuse 3102
pomme-liane 5815
pommier 288
pommier à baies 4899
pommier à bouquets 1180
pommier à feuilles de prunier 4091
pommier commun 1349
pommier de Sibérie 4899
pommier de Siebold 5554
pommier floribond 3037
pommier odorant 5959
pommier paradis 4046
pommier sauvage 1349
pontédériacées 4175
pontédérie 4174
populage 3508
populage des marais 1466
populage des marécages 1466
porcelle 1061
porcelle enracinée 5164
porphyre 3241
porreau 3266
portefeuille 3419
portulacacées 4359
portulacaire 4361
potamogétonacées 4263
potamot à feuilles obtuses 712
potamot allongé 5914
potamot de Fries 2376
potentille 1242
potentille ansérine 4969
potentille arbisseau 879
potentille argentée 4954
potentille blanche 5888
potentille de neige 5069
potentille de Norvège 3887
potentille des rochers 1263
potentille dorée 2530
potentille dressée 5555
potentille printanière 5182
potentille rampante 1667
poule de bois 3992
pourpier 4358
pourpier à grandes fleurs 1505
pourpier d'hiver 3627
pourpier potager 1509
pratelle 1476
prêle 2875
prêle des bourbiers 5344
prêle des champs 2225
prêle des marais 3506
prêle des prés 3552
prêle des tourneurs 4777
prêle d'hiver 4777
prénanthe 4429
prénanthe pourpré 4353

pressin 2007
primevère 4307
primevère à coupes 5550
primevère à grandes fleurs 2042
primevère de Chine 1197
primevère du Japon 3064
primevère élevée 3988
primevère farineuse 586
primevère officinale 1652
primevère vulgaire 2042
primulacées 4308
prinsépie 4310
pritchardia 4311
priva 3373
prosope 1467, 3593
protée 4318
protée mellifère 5308
prune de la Gallissonière 206
prunellier 5014
prunier 4224
prunier abricotier 289
prunier commun 2424
prunier damas 847
prunier d'Arménie 1350
prunier dattier 1809
prunier de l'Oklahoma 3919
prunier de Pennsylvanie 4188
prunier des Alpes 773
prunier des grèves 488
prunier de Watson 4684
prunier du Canada 978
prunier du Pacifique 3164
prunier du Portugal 4277
prunier épineux 5014
prunier japonais 3061
prunier myrobalan 3777, 5950
prunier sauvage 847
prunier trilobé 2311
psalliote des champs 1476
pseudobactérie 4319
psoralée 4782
psychotria 5945
ptélée 2851
ptélée trifolié 1441
ptéride 754
ptéridophytes 4321
ptéris crétoise 1686
ptérocarpe 3999
ptérocarya 5996
ptérocaryer à feuilles de frêne 1077
ptérocaryer à feuilles de sorbier 1077
ptérocaryer de Chine 1210
ptérocaryer de Rehder 4477
ptérocaryer du Japon 3077
ptéroceltis 5988
ptérocéphale 5994
ptéroxyle 5053
ptychosperma 4322
puègne blanche 683
puéraire 3188
puéraire hirsute 5505
puine blanc 1508
pulicaire 2291
pulmonaire 3393
pulmonaire à feuilles étroites 1651
pulmonaire à feuilles molles 3711
pulmonaire officinale 1462
pulsatille des prés 3554
pulsatille printanière 5721
purshie 3698
puschkinie 4362
putine 5586
pycnanthème 3714
pyrénomycètes 2277
pyrèthre 4376
pyrèthre d'Afrique 2460
pyrèthre de Dalmatie 1790
pyrèthre rose 2306
pyrolacées 4378
pyrole 4377
pyrole à feuilles rondes 2119
pyrole uniflore 6037
pyrole unilatérale 4926

quamoclit 5220
quapalier 5013
quassier 4382
quassier amer 5338
quassier de Jamaïque 3009
quatélé 3657
quebracho 5907
quebracho blanc 1559
quebracho coloré 5967
quebracho rouge 3379, 5967
quenouille 1062, 1380
queue-de-cheval 2225, 3506
queue-de-chien 2353
queue-de-rat 2225
quillai 5070
quillai savonneux 5071
quillaya 5070
quinoa 4390
quinquina 1236
quisqualis 4391

racine jaune 2410
râcle 4682
radiole 2289
radiole à feuilles de lin 101
radis 2425, 4395
rafflésia 3664
rafflésiacées 4397
raifort sauvage 2871
raiponce 3636
raiponce en épi 5139
raiponce orbiculaire 420
raisin d'Amérique 1502
raisin de mer 3106
raisin d'ours 500
raisinier 4788
raisin maritime 3104
rajania 5623
ramalina 4407
ramboutan 4409
ramie 2174
ramondie 4411, 4412
ramtil 2063
raoulia 4416
râpette 2459, 3419
raphia 4396
rapistre 4418
raquette à aiguillon unique 1507
raquette commune 1507
raquette figuier d'Inde 2936
ratoncule 3736
rauwolfia 1844
rave 585, 5650
ravenala 5572
ravenelle 5955
ray-grass anglais 4135
ray-grass d'Italie 2980
redoul 1599
régalisse 1459
réglisse 1459, 3291
réglisse de bois 1503
rehmannie 4478
reineckie 4480
reine-claude 2627
reine des prés 2112, 5160
reine-marguerite 1158
reine-marguerite commune 1382
reinwardtia 6087
renonculacées 890
renoncule 889
renoncule âcre 5425
renoncule à feuilles d'aconit 16
renoncule alpestre 3677
renoncule aquatique 5797
renoncule bulbeuse 841
renoncule de neige 5059
renoncule des champs 1608
renoncule des glaciers 2497
renoncule des montagnes 3699
renoncule d'Illyrie 2921
renoncule d'Orient 4146
renoncule flammette 5121
renoncule flammule 5121
renoncule-langue 5544
renoncule rampante 1666
renoncule scélérate 673
renoncule tête d'or 2552
renouée 3172
renouée à feuilles de patience 1739
renouée bistorte 2079
renouée brûlante 3511
renouée de Boukhara 726
renouée des Alpes 131
renouée des buissons 2755
renouée de Siebold 3036
renouée des oiseaux 4317
renouée de Turkestan 726
renouée liseron 1926
renouée persicaire 5168
renouée vivipare 5759
requiénie 2820
réséda 3614
réséda à fleurs jaunes 6100
réséda blanc 5900
résédacées 3615
réséda des teinturiers 5862
réséda odorant 1468
réséda sauvage 6100
restiole 4552
réveille-matin 5328
rhamnacées 831
rhapidophylle 3821
rhapis 3201
rhapontic 4486
rhéédie 4487
rhéxia 3545
rhéxie 3545
rhinanthe 4430
rhinanthe crête de coq 1308

rhipsalide 4488
rhizocton(i)e 4489
rhododendron 4490
rhododendron en arbre 5587
rhododendron hirsute 2430
rhododendron pontique 4264
rhodomyrte 4565
rhodopaxille nu 214
rhodophycées 4433
rhodotype 3099
rhodotype faux kerria 635
rhodyménie 4791
rhopalostylis 4491
rhubarbe 4492
rhubarbe anglaise 2426
rhubarbe de l'Himalaya 2805
rhubarbe des moines 129
rhubarbe du paysan 4159
rhubarbe officinale 3566
rhubarbe ondulée 5844
rhubarbe palmée 5089
rhubarbe rhapontic 2426
rhynchospore 496
riccie 4494
richarde 950
richarde d'Afrique 1374
ricin 1050
ricin commun 1051
ricinelle 1585
ricinodendre d'Afrique 3770
rivinie 4575
riz 4495, 4496
riz d'eau 266
riz du Canada 266
riz égyptien 5892
robinier 3354
robinier du Nouveau-Mexique 3841
robinier faux-acacia 636
robinier hispide 4555
robinier rose 4555
robinier visqueux 1252
rocelle 3325, 4509
rocouyer 256
roëlla 4542
roemérie 344
rogérie 4543
roi des palmiers 1721, 4594
romanzowie 3633
romarin 4563
romarin officinal 4564
ronce 619
ronce à feuilles d'orme 2020
ronce à fruits bleus 2093
ronce arctique 305
ronce bleuâtre 2093
ronce bleue 2093
ronce commune 2080
ronce d'Autriche 374
ronce des haies 2080
ronce des rochers 5260
ronce d'Occident 624
ronce du Canada 2359, 5490
ronce élégante 4660
ronce faux mûrier 1273
ronce-framboise 3357
ronce frutescente 2080
ronce odorante 2359
rondelétie 4549
rondelette 4549
roquette 4525
roquette de mer 4799
roquette d'Orient 5574
rosacées 4560
rosage 4490
rosage de Laponie 3220
rosage des Alpes 4530
rosage en arbre 5587
rosage rouillé 4530
rose à cent feuilles 910
rose alpestre 145
rose à petites fleurs 5025
roseau 1514, 2481
roseau à balais 1514
roseau à quenouille 2482
roseau commun 1514
roseau des bois 1131
roseau des étangs 1380
rose de Chine 1185
rose de Daourie 1785
rose de France 2371
rose de Harrison 2715
rose de Jéricho 3091
rose de Noël 1226
rose des Alpes 4530
rose des haies 1875
rose de sorcière 1875
rose d'Inde 394
rose du Nil 2811
roselle 4562
rose mousseuse 3685
rose pommier 287
rose rouge 2371
rose-trémière 2829
rose velue 287
rosier 4554
rosier à feuilles rouges 4454
rosier à odeur de thé 5468
rosier Banks 443
rosier bijou 4580
rosier blanc 1634
rosier cannelle 1240
rosier centfeuilles 910
rosier centfeuilles pompon 4256
rosier d'Albert 78
rosier d'amour 4256
rosier de Banks 443
rosier de Begger 522
rosier de Belgique 1792
rosier de Bengale 2143
rosier de Bengale pompon 4580
rosier d'Ecosse 4776
rosier de Damas 1792
rosier de l'Ile Bourbon 745
rosier de l'Inde 1200
rosier de Macartney 3401
rosier de mai 1240
rosier de Miss Lawrence 4580
rosier de Nutka 3875
rosier de pomme 287
rosier de Provence 910
rosier de Provins 2371
rosier des Alpes 145
rosier des champs 2234
rosier des chiens 1875
rosier des haies 2583
rosier des quatre saisons 1792
rosier de Virginie 5751
rosier de Wichura 5929
rosier du Japon 4602
rosier du Saint Sacrement 1240
rosier églantine 5359
rosier mousseux 3685
rosier moussu 3685
rosier multiflore 3068
rosier musqué 3767
rosier petit Saint François 861
rosier pimprenelle 4776
rosier pompon 4580
rosier rouillé 5359
rosier rugueux 4602
rosier thé 5468
rosier tomenteux 5543
rosier toujours fleuri 2143
rosier toujours vert 2147
rossolis 5325
rossolis à feuilles rondes 4588
rosulaire 5262
rotale 4573
rotang 4424, 4574
rotang sang-dragon 5314
roxburghie 4592
ruban d'eau 871
rubiacées 3411
rudbeckie 1573
rudbeckie hérissée 629
rudbeckie laciniée 1755
rudbeckie pourprée 4344
rue 4597
rue de chèvre 1425
rue de muraille 5776
rue des prés 136
rue fétide 1516
ruellia 3478
rue sauvage 2714
ruines-de-Rome 3147
ruizia 4604
ruppie 5933
ruprechtie 5745
russulacées 4622
russule 4621
russule bleue et jaune 686
russule charbonnière 686
russule cyanoxanthe 686
russule émétique 2028
russule fétide 2209
russule noircissante 628
russule sans lait 5923
russule verdoyante 2629
rutacées 4599
ruyschie 4628

sabadille 4633
sabal 4017
sabbatie 4561
sabine 4719
sabinier 4719
sablier 4681
sabline 4692
sabline à feuilles de serpolet 5511
sabline des Baléares 1626
sabot de Notre-Dame 3203
sabot de Vénus 2109, 3203
saccharomyces 6069
saccharomycétacées 6070

safran bâtard 4637
safran d'Allemagne 4637
safran d'automne 4638
safran des Indes 1551
safran des prés 1353
safran du Gâtinais 4638
safran officinal 4638
safran printanier 1395
sagesse des chirurgiens 2296, 5442
sagine 4093
sagine couchée 4313
sagine des horticulteurs 1624
sagine noueuse 3171
sagittaire 327
sagou du Japon 4643
sagoutier 5041
sain 6052
sainegrain 711
sainfoin 1519, 5390
sainfoin à fleur violette 90
sainfoin cultivé 1519
sainfoin de prés 1519
sainfoin des Alpes 2368
sainfoin d'Espagne 5310
sainfoin d'Italie 5310
sainfoin oscillant 5474
salade 2417
salicacées 5965
salicaire 4349
salicaire à feuilles d'hysope 2912
salicaire effilée 5782
salicorne 2500, 3514
saliquier 1736
salpiglossis 5624
salpiglossis des jardins 4738
salsepareille 2623
salsepareille squine 1166
salsifis 4662
salsifis à feuilles de poireau 5706
salsifis noir 649
salvadora 4669
salviniacées 4672
salvinie 4671
samandure 4673
samandure d'Inde 3848
sambéquier 2009
samole 807
samole de Valerand 808
sandersonie 4685
sandoric 4695
sandoric d'Inde 4696
sanguenite 3496
sanguinaire 681
sanguinaire du Canada 682
sanguisorbe 2409
sanicle 4693
sanicle d'Europe 2121
sansevière 4694
sansevière cylindrique 2918
sansevière de Ceylan 1106
santal 4677
santalacées 4678
santal blanc 5908
santal du Venezuela 430
santal rouge des Antilles 1902
santal rouge des Indes 863
santoline 3240
santoline petit cyprès 1776
sanve 1118
sanvitalie 4697
sapin 2257
sapin à aiguilles 3462
sapin à grandes bractées 779
sapin à liège 1601
sapin (à) queue de tigre 5523
sapin argenté 4955
sapin baumier 432
sapin baumier du Sud des Etats-Unis 2364
sapin bleu 1338
sapin ciguë 973
sapin concolore 5895
sapindacées 5073
sapin d'Algérie 94
sapin d'Arizona 1601
sapin d'Asie Mineure 1235
sapin de Cilicie 1235
sapin de Corée 3181
sapin de Crimée 3876
sapin de Douglas 1401, 1888
sapin de Faxon 2191
sapin de Fraser 2364
sapin de Grèce 2617
sapin de Kawakami 3141
sapin de l'Arizona 2034
sapin de l'Himalaya 2800
sapin de Lorraine 4955
sapin de Low 3997
sapin de Nikko 3855
sapin de Nordmann 3876
sapin de Normandie 4955, 5913
sapin de Pindrow 4189
sapin de Sakhaline 4655
sapin de Sibérie 4904
sapin d'Espagne 5110
sapin de Vancouver 2573
sapin de Veitch 5707
sapin de Webb 2800
sapin de Wilson 5978
sapin double 653
sapin du Canada 973
sapin du Caucase 3876
sapin du Japon 3649
sapin du Sikkin 4942
sapin écailleux 2271
sapinette blanche 5913
sapinette de Sibérie 4915
sapinette d'Orient 3964
sapinette noire 653
sapinette rouge 4465
sapin géant 2573
sapin Momi 3649
sapin noble 3865
sapin parasol du Japon 5682
sapin pectiné 4955
sapin rouge de Californie 4451
sapin sacré du Mexique 4635
saponaire 5074
saponaire des vaches 1653
saponaire officinale 743
saponaire savonière 743
sapotacées 4701
sapote blanc 5909
sapotier 4700
sapotil(l)ier 4700
saprolégniées 4703
saraca 4704
sargasse 4705
sarracéniacées 204
sarracénie 4207
sarracénie pourpre 1499
sarrasin 832, 1370
sarrasin de Tartarie 5446
sarrette 4728
sarrette des teinturiers 1955
sarriette 4721
sarriette de jardins 5321
sarriette des champs 5187
sarriette montagnarde 6006
sassafras 4712
sassafras officinal 1520
satiné jaune 6094
satyre fétide 5256
satyre puant 5256
saucissonnier 4716
sauge 4639
sauge à grandes fleurs 560
sauge des bois 5743
sauge de Sibthorp 4919
sauge des prés 3558
sauge éclatante (du Brésil) 4749
sauge fausse-verveine 5720
sauge glutineuse 5249
sauge officinale 2427
sauge sclarée 1254
sauge verticillée 3299
saule 5964
saule à feuilles aiguës 4850
saule à feuilles d'amandier 106
saule à feuilles de laurier 1801
saule à feuilles de pêcher 4084
saule à feuilles rondes 3253
saule à grandes feuilles 564
saule à oreillettes 4581
saule à rameaux velus 6053
saule arctique 311
saule baumier 435
saule blanc 5922
saule cendré 2598
saule d'Amérique 654
saule daphné 1801
saule de Babylone 400
saule d'Egypte 2002
saule de Hooker 2846
saule de Laponie 3221
saule de Richardson 4501
saule de Russie 4620
saule des bancs de sable 4680
saule de Silésie 4943
saule de Sitka 4980
saule de vanniers 467
saule du désert 1840
saule émoussé 3891
saule faux-myrte 3785, 5928
saule fragile 787
saule glauque 2596
saule gris 2598
saule laurier 3236
saule marsault 2523
saule marsaux 2523
saule nain 4366
saule noir 1801
saule noircissant 693
saule pentandrique 3236

saule pleureur 400
saule pleureur du Japon 5508
saule pourpré 4352
saule rampant 1679
saule réticulé 3833
saule rouge 4352
saule vert-de-mer 2596
saule viminale 467
sauromate 3345
saururacées 3348
saussurée 4717
saussurée des Alpes 147
savonnier 5072
savonnier des Antilles 5102
savonnier du Sénégal 4822
saxaoul 4729
saxifragacées 4732
saxifrage 4730
saxifrage à feuilles opposées 5663
saxifrage à feuilles rondes 791
saxifrage aizoon 71
saxifrage des ombrages 3362
saxifrage dorée 2545
saxifrage étoilée 5231
saxifrage granulée 3560
saxifrage sarmenteuse 5277
scabieuse 4735
scabieuse des champs 2235
scabieuse des jardins 5383
scabieuse gorge de pigeon 1891
scabieuse succise 3561
scammonée du Levant 4741
sceau de Salomon 5082
sceau de Salomon à grandes feuilles 799
sceau de Salomon multiflore 2071
sceau de Salomon officinal 1916
sceau de Salomon verticillé 5927
scheuchzeria 4754
scheuchzeriacées 4755
schime 2659
schime à feuilles entières 1802
schinopsis 4461
schizandre 3425
schizandre de Chine 1191
schizanthe 893
schizéacées 1742
schizocode 2377
schizomycètes 2266
schizophragme 2903
schizostyle 1695
schlumbergérie 1656
schmidélie 4757
schotie 3124
sciadopitys 5681
scille 5200
scille à deux feuilles 5664
scille aimable 5223
scille d'automne 383
scille de Sibérie 4916
scille d'Italie 2982
scille maritime 4798, 4878
scille officinale 4878
scindapse 2989
scirpe 852
scirpe de lacs 2600
scirpe des bois 6031
scirpe gazonnant 1822
scirpe maritime 4801
sclarée 1254
scléranthe 3168
sclérie 4431
sclérochloa 2701
sclérodermatacées 2706
scléroderme 1405, 2705
sclérotinia 4764
scolopendre officinale 2716
scolyme 3993
scolyme d'Espagne 5112
scopiure 4767
scopolie 4765
scopolie de Carniole 2122
scorsonère 4831
scrofulaire 2242
scrofulaire aquatique 5804
scrofulaire printanière 5184
scrofulariacées 2243
scutellaire 4987
scutellaire d'Orient 3962
sébestier 1594
sécurigère 2720
sécurinéga 4809
seigle 4630
seiglin 1138
sélaginellacées 4815
sélaginelle 4814
sélaginelle de Martens 3525
sélin 1138
sémul 3440
séné 4823
séneçon 2651
séneçon à feuilles de roquette 3809
séneçon cinéraire 4958
séneçon de Fuchs 2388
séneçon de Jacob 4401
séneçon des bois 6034
séneçon en arbre 1974
séneçon visqueux 5247
séneçon vulgaire 1430
séné d'Egypte 89
sénegré 2203
séné Tinnevelly 1575
sénevé 1118
sensitive pudique 4827
séquoia 4829
séquoia à feuilles d'if 945
séquoia géant 2483
séquoia gigantesque 2483
séquoia toujours vert 945
seringa 3641
seringa à bouquets 5376
seringa à feuilles en cœur 2739
seringa à grandes feuilles 2819
seringa à grandes fleurs 571
seringa à petites feuilles 3328
seringa commun 5376
seringa de Coulter 1641
seringa de Falconer 2157
seringa de Gordon 2565
seringa de Schrenk 4760
seringa de Zeyher 6127
seringa du Caucase 1078
seringa du Japon 4714
seringa du Népal 6050
seringa inodore 4752
seringa mexicain 3600
seringa verruqueux 5788
serjania 5335
serpentaire commune 1536
serpolet 3688
serradelle 587, 1526
sésame 3961, 4834
sésame d'Allemagne 572
sesbanie 4835
séséli 4836
seslérie 3675
seslérie bleue 704
sétaire 780
sétaire d'Italie 2352
sétaire glauque 6090
sétaire vert 2624
sétaire verticillée 2844
shepherdia 834
shérardie 2227
shérardie des champs 2228
shorée 4872
shorée robuste 4661
shortie 4881
sibbaldia 4889
sibérienne 4918
sicyote 858
sicyote anguleux 5765
sida 4925
sidalcée 1127
sidérite 2967
sidérodendron 4928
sidéroxylon 3117
siégesbeckie 4653
siéglingie 4937
siéversie 4941
silène 1054
silène à bouquets 5394
silène acaule 3686
silène à cinq plaies 2372
silène à fruits pendants 1905
silène conique 1576
silène d'Autriche 149
silène de France 2372
silène de Ruprecht 4607
silène de Sibérie 4895
silène de Tartarie 5459
silène d'Italie 2981
silène enflé 663
silène noctiflore 3850
silène penché 3869
simarubacées 66
simarube 4971
simarube amer 3968
siphonanthe 5615
siphonia 4975
siphonode 2987
siphonode d'Australie 372
sison 2835
sisymbre 4978
skimmie 4984
skimmie du Japon 3069
smilacées 5031
smilacine 5081
soja 5106
solanacées 3854
soldanelle 111
soldanelle alpestre 2496
soldanelle montagnarde 2602
soleil à feuilles argentées 4963

soleil miniature 1725
solidage 2543
solorina 5083
sophora 4001
sophora du Japon 3059
sorbaire 2180
sorbaire à feuilles de sorbier 5691
sorbaire de Lindley 3318
sorbier 3695
sorbier à feuilles d'aune 1831
sorbier de bois 1128
sorbier de l'Amour 231
sorbier de Laponie 3902
sorbier des montagnes d'Amérique 202
sorbier des montagnes d'Europe 2114
sorbier des oiseleurs 2114
sorbier discolore 5058
sorbier domestique 4833
sorbier du Japon 3058
sorbier finlandais 3902
sorbier intermédiaire 5350
sorgho 5086
sorgho à balais 810
sorgho d'Afrique 5087
sorgho d'Alep 3103
sorgho douro 1929
sorgho du Soudan 5294
sorgho menu 5294
sorgho penché 5892
sorgho sucré 5085, 5309
souchet 2280, 4364
souchet à feuilles alternes 5677
souchet à papier 4043
souchet comestible 1230
souchet odorant 2402
souchet rond 3893
souci 929
souci d'eau 1466
souci des jardins 4285
souci officinal 4285
soude 4618
soude kali 1518
soya 5106
spadiciflores 5107
sparassis crêpu 1683
sparaxide 5781
sparganiacées 872
sparganier 871
spargoule 5196
spargoule des champs 1620
sparmannie 5115
sparmannie d'Afrique 49
sparte 2062
spartier d'Espagne 5857
spartine 1593
spergulaire 4689
spergule 5196
spergule à cinq étamines 5997
sphaeriales 5126
sphagnacées 4096
sphaigne 5125
sphénogyne 5692
spigélie 5136
spilanthe(s) 5162
spiranthe 3197
spiranthe d'automne 1455
spirée 5160
spirée à feuilles de bouleau 581
spirée à feuilles de germandrée 2455
spirée à feuilles de prunier 775
spirée à feuilles de saule 5972
spirée à feuilles larges 797
spirée alpestre 151
spirée blanche d'Amérique 3807
spirée chinoise 1201
spirée crénelée 5067
spirée de Billiard 577
spirée de Douglas 1889
spirée de Rosthorn 4572
spirée de Sargent 4708
spirée de Thunberg 5506
spirée de Virginie 5753
spirée de Wilson 5979
spirée de Zabel 6123
spirée du Japon 3070
spirée intermédiaire 3963
spirée tomenteuse 2703
spirodèle 1923
spirogyre 4260
spondias 3648
sporobole 1906
stangéria 2205
stapélie 1034
staphyléacées 660
staphylier 659
staphylier imparipenné 2083
staphylier penné 2083
statice à larges feuilles 5931
statice commun 3573
statice de Gmelin 4914
statice de Tartarie 5458
statice sinué 3890
stauntonie 5237
stauntonie à 6 folioles 3071
stellaire 5235
stellaire des bois 6045
stellaire graminée 3338
stellaire holostée 1971
sténanthe 5239
sténocarpe 2263
sténocarpe à feuilles de saule 5966
sténogyne 5240
stéphanandre 5241
stéphanandre de Tanaka 5437
stéphanandre incisée 1754
sterculiacées 1223
sterculie à feuilles d'érable 2274
sterculie des rochers 457
sterculie(r) 741
sterculier à feuilles de platane 1194
sternbergie 5242
stévie 5244
sticta 5250
stipe 3819
stipe plumeuse 2115
stokésie 5259
stramoine 1811
stramoine commune 3102
stramoine en arbre 2304
stramoine metel 2810
stramoine odorante 247
stranvésie 5270
stranvésie du Père David 1202
stratiote 5836
stratiote faux-aloès 5837
strélitzie 584
streptocarpe 1006
streptope 5669
streptope embrassant 1257
strombosie 5285
strophanthus 5287
stropharia 5288
strophostyle 5937
strychnos ignatier 4650
stryphnodendre 170
stryphnodendre barbatimao 451
stuartie 5245
stuartie monodelphe 5430
stylidier 5289
styracacées 5269
styrax officinal 1915
suaéda 4813
subulaire 387
subulaire aquatique 5792
sucepin 2940
sumac 5313
sumac à bois glabre 5042
sumac à laque 3050
sumac amarante 5206
sumac à 5 folioles 2268
sumac brillant 2275
sumac copal 2275
sumac de Chine 1203
sumac des corroyeurs 4920
sumac de teinturiers 1527
sumac faux-vernis 5848
sumac fustet 1527
sumac odorant 2358
sumac thézéra 2268
sumac vénéneux 1501, 4242
sumac vernis 3050
sureau 2009
sureau à grappes 2120
sureau arborescent 5588
sureau commun 2095
sureau d'Amérique 190
sureau d'eau 2091
sureau du Canada 190
sureau du Pérou 4155
sureau noir (d'Europe) 2095
sureau rouge 2120
sureau rouge d'Europe 2120
surelle 3985
surelle corniculée 1673
surelle oca 3918
surelle-petite-oseille 6044
surette 2429, 3977
swainsonie 5339
swertie 5397
swertie vivace 119
swiéténie 3427
sympétales 5413
symphonie 5414
symphonie du Gabon 2398
symphorine 5057
symphorine à feuilles rondes 2935
symphorine à fruits blancs 1529
symphorine commune 1529
symplocacées 5374
symplocarpe 4990
symplocos 5373

tabac 5536
tabac commun 1548
tabac indien 2942
tabac rustique 395
tabébuia 5614
tabernémontane 1654
tacca 5421
taccacées 4173
tagète 3493
talauma 5423
talipot 5424
tama 899
tamaricacées 5436
tamarinier 5433, 5434
tamaris 5435
tamaris africain 50
tamaris à quatre étamines 2344
tamaris à 5 étamines 2269
tamaris commun 2374
tamaris d'Afrique 50
tamaris de France 2374
tamaris de Kashgar 3136
tamaris de Russie 3909
tamier 622
tanaisie 5440
tanaisie vulgaire 1542
taro 1806
tarriétie 5266
tarriétie de Java 3088
taxacées 6118
taxodiacées 5463
teck 5465
técoma 5610
teff 5473
tek 1544
télékia élégant 2740
télèphe 3969
telfairie 3917
téosinte 5475
téraspic d'été 2504
térébinthe 5476
terre-noix 2724
testudinaire 5558
tête de méduse 2838
tête de mort 2479
tétracère 5477
tétragonolobe 1824
tétrapanax 4500
thalassie 5654
thapsie 1817
thé 1543
théacées 5464
thé de brousse 500
thé de Java 558
thé de Jersey 3840
thé de l'abbé Galois 1177
thé de montagne 1125
thé des Apalaches 6068
thé de Terre-Neuve 1125
thé d'Europe 1917
thé du Mexique 6056
théier 1543, 5467
théléphoracées 3256
thélesperme 2636
thélygonacées 1770
thélygone 5481
thélymitre 5482
thélypteris 3504
théméda 3131
thériaque d'Angleterre 1113
thermopside 5483
thé rouge 1484
thésion 476
thladianthe 5616
thladianthe douteuse 3467
thlaspi 4112
thlaspi blanc 4523
thlaspi des jardiniers 2504
thlaspi violet 2504
thlaspi vivace 2144
thouinier 5496
thrinace 5480
thrinax 5480
thunbergie 1272
thuya 300
thuya à feuilles en herminette 2793
thuya américain 1973
thuya articulé 296
thuya de Chine 3948
thuya de Corée 3177
thuya de Lobb 2484
thuya d'Occident 1973
thuya d'Orient 3948
thuya du Canada 1973
thuya du Japon 3017
thuya géant 2484
thuya géant de Californie 2484
thuya gigantesque 2484
thym 5509
thym aux chats 1064
thym commun 1546
thyméléacées 3608
thymélée 5116
thym serpolet 3688
thym vulgaire 1546
tiarelle 2321
tibourbou 5515
tigridie 5520
tigridie oeil de paon 1547
tikoki 5524
tiliacées 3315
tillandsia 5525
tillée 4365
tillée aquatique 1511
tillétiacées 5526
tilleul 3314
tilleul à grandes feuilles d'Europe 561
tilleul à petites feuilles 3327
tilleul argenté 4965
tilleul argenté pleureur 4967
tilleul commun 2111
tilleul d'Afrique 6
tilleul d'Amérique 199
tilleul d'Amérique pleureur 4967
tilleul d'Asie 1692
tilleul dasystyla 1076
tilleul de Crimée 1076
tilleul de Hollande 561, 2111
tilleul de l'Amour 229
tilleul de Mandchourie 3463
toddalie 5539
toddalie d'Asie 343
tofieldie 2162
tomate 1549, 5542
topinambour 3094
toque 4987
toque d'Orient 3962
toque élevée 5429
tordyle 2719
torénie 5553
torilis 2759
tormentille 5555
torreya 5556
torreya à feuilles d'if 2303
torreya de Californie 947
torreya de Chine 1206
torreya du Japon 3074
torreya élevé 1206
toulicie 5561
toupélo 5642
tourette 4517
tourette glabre 5566
tournesol 1539, 5329
toute-épice 2415
tout-épice(s) 102
tozzie 5567
trachélie 5503
trachélosperme 5224
trachycarpe 5983
trachylobe 4421
tradescantie 5134
tragus 860
traînasse 4317
tramète bossu 2485
tréculie 5575
trèfle 1278
trèfle aggloméré 1292
trèfle anglais 1694
trèfle bâtard 164
trèfle blanc 5889
trèfle bleu 711
trèfle commun 4447
trèfle d'Alexandrie 2000
trèfle d'eau 1365
trèfle de Dalmatie 2888
trèfle de Pannonie 2888
trèfle de Perse 4147
trèfle des Alpes 126
trèfle des champs 3225, 4394
trèfle des montagnes 3702
trèfle des prés 4447
trèfle de Virginie 1441
trèfle douteux 5292
trèfle du Japon 1458
trèfle élégant 4883
trèfle faux-lupin 473
trèfle filiforme 5497
trèfle hybride 164
trèfle incarnat 1694
trèfle intermédiaire 6128
trèfle jaune des sables 3159
trèfle musque 711
trèfle odorant 711
trèfle oscillant 5474
trèfle porte-fraise 5274
trèfle rampant 5889
trèfle renversé 4147
trèfle rouge 4447
trèfle rougeâtre 2351
trèfle souterrain 5291
trèfle violet 4447
tremble du Japon 4929
tremblette 4134

tribule 954
tribule terrestre 4332
trichilie 611
tricholome équestre 2051
tricholome nu 214
tricholome prétentieux 1859
trichomane 2247
trichosanthe 5045
trichosanthe serpentin 1996
tricrate 5
tricyrtis 5534
trientale 5218
triglochin 4234
triglochin des marais 328
triglochin maritime 4877
trigonelle 711, 5593
trillie 5594
trinie 5595
triodie 5596
triostée 2867
tripe de roche 4535
tripsac 2404
trisétaire 2176
tristanie à feuilles en bouquets 778
tritome 5551
tritonie 5599
trochodendre 5876
troène 4312
troène brillant 2514
troène commun 1508
troène de Californie 944
troène de Chine 1198
troène de l'Amour 232
troène des bois 1508
troène du Japon 3065
trolle 2507
trolle d'Asie 4906
trolle de Chine 1182
trolle d'Europe 1422
trompette de la mort 2858
trompette des morts 2858
tropéolacées 3813
truffe 5608
truffe de St. Jean 4619
truffe d'été 4619
truffe du Périgord 4137
tsuga 2769
tsuga commun 973
tsuga de Chine 1184
tsuga de Mertens 3707
tsuga de Siebold 4933
tsuga diversifolié 3040
tsuga du Canada 973
tsuga du Japon 3040, 4933
tsuga du Nouveau-Monde 3996
tubérales 5609
tubéreuse 5618
tubéreuse bleue 40
tue-chien 1353
tulbaghie 5632
tulipe 5633
tulipe de Biberstein 552
tulipe d'Eichler 2003
tulipe de Mormons 3494
tulipe sauvage 2302
tulipier 5635
tulipier de Chine 1207
tulipier de Virginie 5636
tulipier du gabon 527
tunique 5641
turnère 5649
turquette glabre 1372
turraée 5212
tussilage pas-d'âne 1388
typha à feuilles larges 1380
typhacées 1063

uapaca 5674
ulluque 5622
ulmacées 2019
ulve 4794
upas antiar 5684
urédinées 4623
uréra 4778
uriam d'Assam 3086
urosperme 4856
urticacées 3836
usnée 5693
usnée barbue 503
ustilaginacées 854
utriculaire 664
uvette 3104, 3106
uvulaire 530

vachotte 3938
vacoua 1521, 4779
valérianacées 5696
valériane 5695
valériane de montagne 3722
valériane des jardins 3119
valériane dioïque 3520
valérianelle 1618
valériane officinale 1552
vallisnérie 5944
vallisnérie spiralée 5159
vanda 5698
vanillier 3605, 5699
varaire noir 630
varech 4538
varech alimentaire 407
varech denté 1038
varech vésiculeux 666
vatérie 5704
vatérie de l'Inde 4196
vauchérie 2626
veilleuse 1353
vératre 2170
vératre blanc 5894
vératre noir 630
vératre vert 193
verbénacées 5729
verbésine 1709
verdiau 4352, 4471
verdoisis 6107
verek 2662
verge d'or à feuilles de graminée 2586
verge d'or commune 2101
verge d'or du Canada 971
verge d'or tardive 3892
vergerette 2292
vergerette âcre 602
vergerette alpine 130
vergerette annuelle 259
vergerette du Canada 2878
vernis du Japon 5584
vernonie 2965
véronique 5123
véronique à feuilles de lierre 2992
véronique à feuilles de serpolet 5512
véronique agreste 2237
véronique à longues feuilles 1289
véronique aquatique 509
véronique beccabunga 509
véronique bleu améthyste 475
véronique de Bachofen 406
véronique des Alpes 150
véronique des champs 1532
véronique des montagnes 3720
véronique de Tournefort 5563
véronique digitée 2253
véronique en épi 5140
véronique germandrée 2454, 2892
véronique mâle 1532, 1917
véronique officinale 1917
véronique printanière 5726
véronique rustique 2237
véronique sous-ligneuse 4534
verrucaire 5785
verveine 5728
verveine-citronelle 3271
verveine des Indes 3270
verveine en arbre 3271
verveine officinale 2129
vesce 5731
vesce à feuilles de pois 4097
vesce à feuilles étroites 3810
vesce à fruit velu 6051
vesce à grosses graines 568
vesce commune 1554
vesce craque 591
vesce cultivée 1554
vesce de Narbonne 3794
vesce de Pannonie 2893
vesce de printemps 1554
vesce de Russie 2693
vesce des buissons 2462
vesce des haies 2760
vesce du Bengale 4356
vesce grêle 5004
vesce jaune 6111
vesce pisiforme 4097
vesce pourpre noire 4356
vesce velue 2693
vésicaire 661
vesse-de-loup 4325
vesse-de-loup ciselée 1328
vesse-de-loup géante 2479
vesse-de-loup perlée 5784
vétiver 5732
victoria (regia) 5830
vigne 2104, 2575
vigne à fruits bleus 224
vigne américaine 4503
vigne de battures 4503
vigne de Judée 604
vigne de l'Amour 227
vigne d'Espagne 6002
vigne épineuse 777
vigne folle 1664
vigne foxée 2386
vigne Isabelle 2347
vigne musquée 3756
vigne odorante 2386

vigne rouge 1057
vigne-vierge 1664
vigne-vierge à trois becs 3028
vigne-vierge à 5 folioles 5748
vigne-vierge du Japon 3028
vigne-vierge grimpante 5748
villarésie 5734
vinettier 452, 2075
vinettier à feuilles de buis 3421
vinettier d'Asie 345
vinettier de Bretschneider 4338
vinettier de Darwin 1805
vinettier de Hooker 2845
vinettier de Lady Wilson 5976
vinettier de l'Amour 225
vinettier de Sargent 4706
vinettier de Sibérie 4893
vinettier de Siebold 4930
vinettier de Thunberg 3021
vinettier du Canada 98
vinettier subverruqueux 5786
violacées 5740
violette 5737
violette biflore 5661
violette cornue 2856
violette des champs 2238
violette des chiens 1880
violette des sables 4691
violette du Cap 1347
violette élevée 5431
violette étonnante 6019
violette odorante 5392
violier jaune 1556
viorne 5733
viorne à feuilles de prunier 632
viorne à manchettes 3792
viorne cotonneuse 5852
viorne de Carles 3186
viorne de Siebold 4935
viorne obier 2091
viorne odorante 5391
vipérine 5744
vipérine à feuilles de plantain 4357
vipérine commune 1555
virgilier 5747
virgilier à bois jaune 213
viscaire alpine 306
vismie 5758
vitacées 2576
voandzou 1574
volubilis 1474
volubilis purgatif 3005
vornouiller 1881
vrillée 2100
vulnéraire 3159
vulpin 2349
vulpin des prés 3549
vulpin genouillé 5805

wachendorfie 5761
wahlenbergie 4511
waldsteinie 458
wallénie 5767
wallichie 5770
washingtonia 5790
weldénie 5226
wellingtonia géant 2483
welwitschie 5865
wendlandie 5866
widdringtonia 5930
widdringtonia de Schwarz 4763
wigandie 5935
wolffia 6017
woodsia 6043
woodwardie 1110
wulfénie 6059

xanthium 1306
xanthocère 6095
xanthocère à feuilles de sorbier 4868
xanthorrhée 2591
xanthorrhize 6105
xanthosome 3441
xanthoxyle 4295
xéranthème 2923
xéranthème annuel 1446
ximénie d'Amérique 5428
xylocarpe 6060
xylosme 6061
xyride 6084

yèble 3570
ylang-ylang 6119
ypréau 5906
yucca 6120
yucca à feuilles d'aloés 110
yucca filamenteux 24
yucca magnifique 3692
yucca-pied d'éléphant 846

zamier 1581
zannichellie 4266
zantédesquie 950
zantédesquie éthiopienne 1374
zauschnérie 2259
zédoaire 6125
zéphyranthe 6126
zinnia 6129
zinnia élégant 1568
zinzibéracées 2490
zizanie 5956
ziziphore 6130
zostéracées 1998
zostère 1997
zostère varech 1407
zoydie 3242
zoysie 3242
zoysie du Japon 3052
zygadène 1819
zygomycètes 6131
zygophyllacées 499

DEUTSCH

Aalbaum 2841
Aalbeere 2081
Aalohrenbaum 2095
Aaronsstab 333
Aasblume 1034, 3378
Aashornbaum 2095
Aaspflanze 1034
Abbisskraut 3561
Abelia 3
Abelie 3
Abendblume 2993
abendländische Lärche 5868
abendländische Platane 205
abendländischer Lebensbaum 1973
Aberraute 3923
abessinische Banane 8
abessinischer Weizen 2064
abessinisches Liebesgras 5473
Abrandkraut 3923
Abronie 5
Absinth 1564
Abura 6
Abutilon 7
Acajubaum 1378
Achillenkraut 1565
Achimenes 15
achtblättrige Silberwurz 3726
Ackerbilsenkraut 633
Ackerbohne 788, 3794
Ackerborstendolde 5176
Ackerbrombeere 2093
Ackerchampignon 2870
Ackerdistel 975
Ackerehrenpreis 2237
Ackereichel 2649
Ackererbse 2231
Ackerfelberich 2540
Ackerfrauenmantel 2226
Ackerfuchsschwanz 4464
Ackerfuchsschwanz(gras) 3735
Ackergänsedistel 2236
Ackergauchheil 4747
Ackerglockenblume 4414
Ackergoldklee 2848
Ackergünsel 2650
Ackerhafer 5981
Ackerhahnenfuss 1608
Ackerhellerkraut 2232
Ackerhohlzahn 4453
Ackerhoniggras 2461
Ackerhornkraut 5230
Ackerhundskamille 2220
Ackerklee 4394
Ackerklettenkerbel 5176
Ackerknäuel 262
Ackerknautie 2235
Ackerknöterich 1620
Ackerkohl 1118, 2710
Ackerkratzdistel 975
Ackerkrone 1391
Ackerkrummhals 5939
Ackerlinse 1457
Ackerlöwenmaul 1619
Ackermäuseschwänzchen 5530
Ackermennig 1348
Ackerminze 2229
Ackermohn 1616
Ackernachtnelke 3850
Ackernuss 2649
Ackernüsschen 422
Ackerplatterbse 2587
Ackerrittersporn 2330
Ackerröschen 2272
Ackerrose 2234, 2583
Ackerröte 2227, 2228
Ackersalat 3211
Ackersaudistel 2236
Ackerschachtelhalm 2225
Ackerschmalwand 3733
Ackerschmiele 5981
Ackerschötendotter 5573
Ackerschöterich 5573
Ackersenf 1118
Ackerskabiose 2235
Ackerspark 1620
Ackerspörgel 1620
Ackersteinklee 6108
Ackersteinsame 1614
Ackerstiefmütterchen 2238
Ackerstraussgras 5981
Ackertäschelkraut 2232
Ackertaubnessel 2780
Ackertrespe 2219
Ackerveilchen 2238
Ackervergissmeinnicht 2222
Ackerweizen 1655
Ackerwicke 1554
Ackerwinde 2100
Ackerwindhalm 5981
Ackerwitwenblume 2235
Ackerziest 2230
Acrocomie 18
Adamsnadel 6120
Adansons Sabal 1948
Adlerbaum 1958
Adlerfarn 26, 5867
Adlerholzbaum 1958
Adlumie 32
Adonisröschen 33
Adsukibohne 34
Adventsstern 1500
Adzukibohne 34
Affenblume 3655
Affenbrotbaum 445
Affenknabenkraut 3656
Affentreppen 481
Affodill 352
afrikanische Gurke 2855
afrikanische Ölpalme 46
afrikanischer Hanf 4694
afrikanischer Myrrhenstrauch 44
afrikanische Schmucklilie 40
afrikanische Sparmannie 49
afrikanische Tamariske 50
afrikanische Wicke 5438
afrikanische Zimmerlinde 49
Afterkamille 963
Afterkreuzkraut 869
Agarikazeen 2488
Agave 59
Ageratum 60
Agrume 1246
Agrumen 1247
ägyptische Lotosblume 5893
ägyptischer Feigenbaum 5408
ägyptischer Kümmel 72
ägyptischer Schotendorn 399
ägyptische Sesbanie 4081
ägyptische Weide 2002
ägyptische Zehrwurzel 1806, 2013
Ahlbaum 2078
Ahlbeere 2081
Ahlbeerstrauch 2081
Ahlkirsche 2078, 2099
Ahorn 3481
ahornblättrige Platane 3361
Ahorngewächse 3482
ährenblütiges Tausendblatt 5138
Ährenchristophskraut 616
ährengrasähnliche Trespe 4483
Ährenheide 5142
Ährenlilie 715
Ährensimse 5141
Ährentausendblatt 5138
ährige Felsenbirne 1949
ährige Hainsimse 5141
ährige Johannisbeere 3882
ähriger Ehrenpreis 5140
ähriger Erdbeerspinat 675
ähriger Gänsefuss 675
ähriger Goldhafer 5146
ähriger Grannenhafer 5146
ährige Scheinhasel 5147
ähriges Christophskraut 616
ähriges Tausendblatt 5138
ährige Teufelskralle 5139
Ahuaca 183
Ailanthus 65
Aizoazeen 1033
Ajanfichte 6071
Akanthazeen 14
Akanthus 12
Akanthusgewächse 14
Akazie 10
Akebie 73
Akelei 1341
akeleiblättrige Wiesenraute 1342
Akeleiwiesenraute 1342
Aki-Baum 75
Aktinidie 21
Aktinidie Kolomikta 3176
Aktinidien 22
Alant 2955
Alantbeere 2081
Albeere 2081
Albergine 2414
Alberts Blumenspiere 5645
Alberts Rose 78
Albizzie 79
Alchemistenkraut 1454
Alcocks Fichte 80
Aldrovandie 83
Aleppobartgras 3103
Aleppofichte 86
Aleppohirse 3103
Aleppokiefer 86
Alexandriner Klee 2000
alexandrische Aprikose 4336
alexandrischer Lorbeer 88
Alfagras 2062
Algarobas-Linse 3930
Algarova 1023
Algen 92
Algenpilze 5824

Alhandel 1334
Alhenna 2782
Alhornbaum 2095
Alkanna 1954
Alkannawurzel 97, 1954
Alkekengi 5275
Alleepappel 3360
Allermannsharnisch 3368
Allerseelenaster 3687
Alligatorapfel 4257
Alligatorbirne 183
Almenkraut 604
Almenrausch 4530
Almenraute 604
Almer 2078
Aloe 109
aloeblättrige Palmlilie 110
Aloekrebsschere 5837
Alokasie 108
Alonsoblume 3529
Alpbalsam 4530
Alpenampfer 129
Alpenanemone 112
Alpenaster 113
Alpenaurikel 368
Alpenbärentraube 141, 4320
Alpenbärlapp 127
Alpenbarts(ch)ie 115
Alpenberufkraut 130
Alpenbinse 146
Alpenbirke 1935
Alpenbohnenbaum 4773
Alpendistel 709
Alpenehrenpreis 150
Alpenerle 2105
Alpenfelsenblümchen 4519
Alpenfettkraut 3700
Alpenfrauenmantel 3709
Alpenfuchsschwanz 133
Alpengänsekresse 143
Alpengeissblatt 161
Alpengelbling 4890
Alpenglöckchen 111, 506
Alpenglockenblume 116
Alpengoldregen 4773
Alpenhabichtskraut 135
Alpenhahnenfuss 3677
Alpenheiderich 5186
Alpenhelm 460
Alpenhexenkraut 124
Alpenhornkraut 122
Alpenjohannisbeere 128
Alpenkatzenpfötchen 142
Alpenklee 126, 4347
Alpenknöterich 131
Alpenkresse 143
Alpenkreuzdorn 121
Alpenleimkraut 149, 3686
Alpenleinkraut 155
Alpenlieschgras 154
Alpenmannsschild 144
Alpenmariengras 153
Alpenmastkraut 310
Alpenmohn 140
Alpenpechnelke 306
Alpenquendel 148
Alpenrachen 5567
Alpenrausch 4490
Alpenrebe 125
Alpenrispengras 118
Alpenrose 145, 389, 4490
Alpensadebaum 4719
Alpenschärfling 147
Alpenscharte 147, 4717
Alpenschmiele 134
Alpenschön 951
Alpenschuppenkopf 6079
Alpensonnenröschen 152
Alpenspierstrauch 151
Alpenspindelbaum 793
Alpensteineibe 5449
Alpenstern 1946
Alpensüssklee 2368
Alpentragant 137
Alpentroddelblume 2496
Alpenveilchen 1766
Alpenvergissmeinnicht 132
Alpenwaldrebe 125
Alpenwegerich 162
Alpenwiesenraute 136
Alpenwimperfarn 157
Alpenwucherblume 123
Alpenziest 117
Alpinie 158
Alpkraut 732
Alpranke 604
Alraunpflanze 3469
Alraunwurzel 3469
Alsike 164
Alsikeklee 164
Alstroemerie 166
Altaifichte 4915
Altaiweissdorn 167
Altaiwindröschen 2962
Amarant 174
Amarantgewächse 175
Amaranthazeen 175
Amarantine 2503
Amarellkraut 6091
Amaryllidazeen 177
Amaryllis 176
Amaryllisgewächse 177
Amazonenmandel 765
Ambarihanf 3146
Ambatsch 179
Amberbaum 182, 5369
Amberbäume 6013
Amberkraut 1064
Amberstrauch 182
Ambla-Baum 2026
Ambla-Blattblume 2026
Ambraflockenblume 181
Ambrosienkraut 4400
Ameisenbaum 4331
amerikanische Agave 1092
amerikanische "Blueberry" 2796
amerikanische Buche 184
amerikanische Eberesche 202
amerikanische Esche 5878
amerikanische Färbereiche 644
amerikanische Grauesche 4434
amerikanische Hasel 194
amerikanische Jamesie 1264
amerikanische Johannisbeere 186
amerikanische Kastanie 187
amerikanische Lärche 1977
amerikanische Ölfrucht 189
amerikanische Pflaume 206
amerikanischer Amberbaum 210
amerikanischer Geweihbaum 3151
amerikanische Rotfichte 4465
amerikanischer Silberbaum 4951
amerikanischer Spindelstrauch 805
amerikanischer Storaxbaum 210
amerikanischer Trompetenbaum 5097
amerikanischer Zürgelbaum 1434
amerikanische Säckelblume 3840
amerikanische Seide 2641
amerikanisches Fussblatt 1845
amerikanisches Gelbholz 213
amerikanisches Meerträubchen 192
amerikanische Sumpfzypresse 414
amerikanische Terpentinkiefer 3352
amerikanische Ulme 191
amerikanische Zitterpappel 4380
Amethystschwingel 215
Amla 2026
Ammelkorn 2029
Ammer 2029
Ammoniakpflanze 731, 5316
Amorphophallus 220
Amourettengras 4381
Ampelidazeen 2576
Ampfer 1868
Ampferknöterich 1739
Ampferwurzel 4258
Amselbeere 1369
Amselkraut 1342
Amsinckie 2217
Amsonie 223
Amurberberitze 225
Amureberesche 231
Amurflieder 228
Amurkorkbaum 226
Amurlinde 229
Amurrainweide 232
Amurrebe 227
Anakardiazeen 1043
Ananas 235, 4191
Ananaserdbeere 4195
Ananasgewächse 4192
Ananaskirsche 4156
Anassa 235
Andertanne 297
Andiroba 2660
Andorn 2816
Andreäazeen 240
Andreaskraut 4647
Anemone 242
Angelika 245
Angelikabaum 1167, 1849
Angelika(wurzel) 2406
Angergras 700
Angerling 1476, 2870
Angertrespe 2681
Angiospermen 248
Angolaerbse 1574
Angusturabaum 251
Anis 253
Anischampignon 2870
Anisegerling 2870
Anissägeblättling 4739
Anistrichterling 5385

Annattostrauch 256
Anonazeen 1752
Anone 1751
Anonengewächse 1752
ansehnliches Mannagras 806
antarktische Scheinbuche 268
Antialkoholikerpilz 2950
Antillenakazie 5351
Antillenkassie 5351
Antoniuskraut 2262
Apfelbaum 288, 1349
Apfelbeere 1224
Apfelbeerstrauch 1224
apfelfrüchtiger Osagedorn 3972
Apfelquitte 286
Apfelrose 287
Apfelsine 5377
Aponogetonazeen 278
Apothekerrose 2371
Apothekerstockmalve 3507
Apozynazeen 1873
Aprikose 1350
Aprikosenbaum 1350
Aprikosenpflaume 289
Aprikose von St. Domingo 3453
Aprilblume 2134
Aquacate 183
Aquifoliazeen 2827
arabische Akazie 399
arabische Luzerne 5170
arabische Mahaleb 3426
arabischer Jasmin 293
arabischer Kaffeebaum 292
arabischer Schneckenklee 5170
Arada-Gurke 5869
Araliazeen 2495
Aralie 295
Araliengewächse 2495
Araukarie 297
Araukariengewächse 298
Arazeen 334
Arbuse 5821
Archegoniaten 301
Archers Fitzroya 5448
Archimyzeten 302
Archispermen 2671
Ardisie 5117
Areka 313
Arekanusspalme 548
Arekapalme 313, 548
Arethuse 314
Arganbaum 315, 4150
Arganiabaum 315
Arizonakiefer 318
Arizonatanne 1601
Arizonazypresse 317
arktisches Rispengras 304
arktische Weide 311
Armands Kiefer 320
Armands Waldrebe 319
armblütiges Riedgras 2212
armblütiges Sumpfried 2213
armblütige Sumpfbinse 2213
armblütige Sumpfsimse 2213
Armleuchteralgen 1119
Armleuchtergewächs 5267
Armleuchtergewächse 1119
Armutblume 1253
Arnebie 322
Arnika 323
Arnolds Weissdorn 324
Aroideen 334
Arongewächse 334
Aronstab 333
Aronstabgewächse 334
Aronwurzel 333
Arracacha 325
Artischocke 331
Artistolochiazeen 593
Arzneilöffelkraut 1522
Aschenblume 1237
Aschenkannenkraut 2225
Aschenkraut 1237
Aschenpflanze 1237
aschgrauer Walnussbaum 897
aschgraue Weide 2598
Aschkraut 1237
Aschlauch 3266
Aschweide 2598
asiatische Berberitze 345
asiatische Felsenbirne 342
asiatischer Wegerich 348
asiatische Trollblume 4906
askalonische Zwiebel 4844
Asklepia 1558
Asklepiadazeen 3621
Askomyzeten 337
Aspe 2074
Aspergillus 351
Asperiofoliazeen 735
Aspidistra 353, 1352
Assam-Tee 356
Astblume 1248
Aster 358
Astflechte 4407
ästige Sommerwurz 2774
Astilbe 360
astlose Graslilie 4648
Astmoos 2908
Asvatha 738
äthiopische Zantedeschie 1374
Atlasbeerbaum 1128
Atlasblume 2524, 4130
Atlaszeder 364
Ätna-Berberitze 37
Atragene 125
Attich 3570
Aubergine 2414
Aubrietie 366
Auferstehungspflanze 3285
aufgeblasenes Leimkraut 663
aufrechte Osterluzei 592
aufrechter Igelsame 2126
aufrechtes Fingerkraut 5555
aufrechtes Glaskraut 5773
aufrechte Studentenblume 394
aufrechte Trespe 3546
aufrechte Waldrebe 2646
Augenbohne 1393
Augenpappel 2830
Augentrost 2152
Augentrostgras 1971
Aukuba 367
Aurikel 368
ausdauernde Jasione 4736
ausdauernder Knäuel 4131
ausdauernder Lein 4129
ausdauerndes Bingelkraut 1876
ausdauerndes Gänseblümchen 2037
ausdauerndes Knäuelkraut 4131
ausdauerndes Silberblatt 4130
ausdauernde Swertie 119
ausgebreitete Kuhschelle 5177
ausgebreitete Melde 2186
ausgebreitete Studentenblume 2369
ausgedehnte Segge 3364
Austerbaum 200
Austernpilz 3992
Austernschwamm 3992
Austernseitling 3992
Austrahlheide 2048
australische Cordyline 2470
australische Keulenlilie 2470
australischer Gummibaum 4957
australische Seideneiche 4945
australische Silberakazie 4957
Autobasidiomyzeten 381
Avocatobaum 183
Avocatobirne 183, 4144
Avogatenbaum 183
Awapfeffer 3140
Azalee 389, 4490
Azalie 389
Azarol-Hagedorn 390
Azerazeen 3482
Azerineen 3482

Baboen 398
Babul 399
Bachbunge 509
Bachbungen-Ehrenpreis 509
Bachehrenpreis 509
Bachnelkenwurz 5791
Bachofens Ehrenpreis 406
Bachpilze 2437
Bachsteinbrech 120
Bachweidenröschen 5024
Backenklee 982
Badekraut 2418
Baileys Akazie 1300
Bajonettbaum 6120
Bajonettpflanze 2918
Bajree 4092
Bakope 5811
Balam 4079
Balanophorazeen 412
Balata 1355
Balatabaum 1355
Bälbaum 408
Baldgreis 1430
Baldrian 5695
Baldriangewächse 5696
Baldschuanknöterich 726
Balfours Fichte 417
Balfours Kiefer 2353
Balfours Rispengras 416
Balgmais 4233
Ballonblume 423
Ballonrebe 2741
Ballote 425
Balsabaum 5870
Balsamapfel 431
Balsambaum 430, 5552
Balsambäume 6013

Balsamdrachenbaum 2357
Balsamfeige 1290
Balsamgewächse 1043
Balsamgurke 431
Balsaminazeen 433
Balsamine 5051
Balsaminengewächse 433
Balsampappel 5100
Balsamstrauch 5552
Balsamtanne 432
baltische Binse 436
Bambus 437, 1231
Bambusgras 437
Bambuspalme 3408, 4396
Bambusrohr 437
Banane 439, 1358, 4219
Bananengewächse 440
Bandalgen 1841
Bandblume 2542
Bandflechte 2150, 4407
Bandgras 3905
Bandweide 467
Baniane 738
Banks Kiefer 3001
Banks Rose 443
Banyan 738
Banyanbaum 444
Baobab 445
Barbados-Aloe 3568
Barbados-Kirsche 3450
Barbados-Stachelbeere 447
Barbarakraut 610, 5999
Bärbelkraut 5999
Barbenkraut 5999
Bärenklau 12, 1649
Bär(en)lauch 4415
Bärenohr 312
Bärenschote 3353, 3619
Bärentapengewächse 14
Bärentraube 3480
Bärenwurzel 415
Barg-Lang 2580
Bärlapp 1285
Bärlappgewächse 1286
Bärlapplauch 4415
Bärohr 368
Barreliers Ochsenzunge 1959
Bärsanikel 506
Bartblume 689
Bartfaden 1088, 4117
Bartflechte 503, 5693
Bartflechten-Tillandsie 5576
Bartgerste 706
Bartgras 706, 1985
Barthafer 5003
bärtige Glockenblume 502
Bartkoralle 1589
Bartmoos 503, 1968
Bartnelke 5393
Bartsaturei 3611
Bartschie 460
Bartsie 460
Bartweide 4352
Bartweizen 1930
Bärwinde 2509
Bärwurz 415
Bärwurzel 415
Baselbeere 2038
Basellazeen 3414
Baselle 5736
Basellgewächse 3414
Basellkartoffel 3413
Basidienpilze 462
basidientragende Pilze 462
Basidiolichenen 461
Basidiomyzeten 462
Basilie 463
Basilienkraut 463
Basilikenkraut 463
Basilikum 463, 5354
Bastardabutilon 2898
Bastardbegonie 5075
Bastardeberesche 3902
Bastardfetthenne 2148
Bastardfuchsschwanz 5005
Bastardindigo 2171
Bastardklee 164
Bastardlorbeer 3237
Bastardmahagoni 4698
Bastardmehlbeere 3902
Bastardpantoffelblume 5760
Bastardpetunie 1495
Bastardrittersporn 2305
Bastardrömerie 5741
Bastardsafran 4637
Bastardtaubnessel 1756
Bastpilz 2838
Bastulme 4614
Batate 5380
Batatopflaume 4413
Batokpflaume 4413
Bauchhülse 4240
Bauernkirsche 3544
Bauernkresse 2232
Bauernrose 1490
Bauernschminke 1614
Bauernsenf 2232
Bauerntabak 395
baumartige Aloe 299
baumartige Hortensie 5039
baumartige Pfingstrose 5585
baumartiger Erbsenstrauch 4910
baumartiger Schneckenklee 5583
Bäumchenweide 3339
Baum der Reisenden 3409, 5572
Baumerika 5581
Baumhasel 5646
Baumheide 5581
Baumholder 2095
Baumlieb 4165
Baummalve 5582
Baummörder 608
Baumschlinge 4949
Baumtomate 5589
Baumweichsel 5091
Baumwollbaum 730, 3134
Baumwolle 1635
Baumwucherer 1829
Baumwürger 608
bayrischer Enzian 482
Becherblume 868, 5018
Becherfarn 5580
Becherflechte 1249
Becherfrüchtler 514
Becherglocke 3198
Becherkätzchen 4947
Becherling 2767
Becherpflanze 4571
Becherpilze 1734
Becherprimel 5550
Becherschwertel 1772
Beckmannsgras 209, 5015
Bedecktsamer 248
Bedecktsamige 248
Beerenapfelbaum 4899
Beereneibe 2045
Beerennelke 656
Beerenölweide 1135
Beerentang 4705
Beerentaubenkropf 544
beerentragender Hühnerbiss 544
Beerentraube 3425
Beerenzierapfel 4899
Beerenzwetsche 1315
Beerkraut 1348
Beermelde 675
Beetkohl 1116
Beggers Rose 522
Begoniazeen 524
Begonie 523
Begoniengewächse 524
begranntes Ruchgras 265
behaarte Alpenrose 2430
behaarte Hainsimse 2696
behaarte Karde 2692
behaarte Platterbse 4578
behaarter Ginster 2685
behaarter Günsel 2442
behaarter Mauerpfeffer 2691
behaarter Odermennig 2679
behaarte Segge 2690
behaartes Liebesgras 2931
behaartes Riedgras 2690
behaartes Schamkraut 4111
behaartes Weidenröschen 2695
behaarte Wicke 5531
behaartfrüchtige Platterbse 4578
Beifuss 4640
beifussblättriges Traubenkraut 1513
beilblättriger Lebensbaum 2793
Beilwicke 2720
Beilwurz 525
Beinbrech 715
Beinholz 1508, 2099
Beinwell 1345
Beinwurz(el) 1345
Beisskohl 1116
Belbaum 408
Belladonna 525
Benediktenkarde 670
Benediktenkraut 385, 670
bengalischer Hanf 5331
Bennussbaum 2872
Benthalm 3674
Benzoë-Storaxbaum 5315
Benzoinbaum 5127
berauschender Kälberkropf 1648
Berberidazeen 453
Berberitze 452
Berberitzengewächse 453
Berbesbeere 2075
Berchemie 5336
bereifte Brombeere 2093
Bergahorn 4217

Bergamottbaum 536
Bergamotte 536
Bergamottminze 535
Bergangelika 295
Bergaster 2973
Bergbaldrian 3722
Bergblasenfarn 3696
Bergbohnenkraut 6006
Bergehrenpreis 3720
Bergeiche 1142
Bergenie 537
Bergerle 2105
Bergfarn 3884
Bergfenchel 4836
Bergflachs 476
Bergflockenblume 3697
Berggamander 3705
Berghaarstrang 3715
Berghahnenfuss 3699
Berghähnlein 3796
Berghartheu 3721
Bergheilkraut 4857
Bergheilwurz 3713
Berghellerkraut 139
Bergholunder 2120
Berghopfen 1442
Bergixlilie 5415
Bergjasione 4857
Bergjohanniskraut 3721
Bergkaffee 292
Bergkaiserkrone 4997
Bergkiefer 5400
Bergklee 3702
Bergknoblauch 2474
Bergknöterich 131
Bergkornblume 3697
Bergkronenwicke 3704
Berglauch 1934, 3144
Berglinde 3327
Berglorbeer 3129
Berglungenkraut 3711
Bergmahagoni 1097, 3712
Bergminze 3714, 4720, 4721
Bergmispel 2089
Bergpalme 1112
Bergpetersilie 3715
Bergplatterbse 3716
Bergriedgras 3718
Bergrispengras 796
Bergrose 145
Bergrüster 4771
Bergsandknöpfchen 4857
Bergsegge 3718
Bergsilge 3715
Bergsteinkraut 3693
Bergsteinsame 1669
Bergtroddelblume 2602
Bergulme 4771
Bergvergissmeinnicht 6033
Bergwaldhyazinthe 2605
Bergwaldrebe 244
Bergwegerich 3717
Bergweidenröschen 3723
Bergwohlverleih 323, 3694
Bergwundklee 160
Bergziest 5687
Berle 545, 5209
Berliner Pappel 540
Bermudagras 542
Bermudazeder 543
Bernardinerkraut 671
Bertram 4376
Bertramsgarbe 5054
Bertram(wurz) 2210
Bertramwurzel 2460
Beruf(s)kraut 2292
Beschreikraut 2292
Besenbeifuss 3966
Besenginster 809, 4769
Besengras 1593
Besenheide 2744, 4772
Besenhirse 810
Besenkraut 531, 2744, 5318
Besenradmelde 531
Besenrauke 2296, 5442
Besenried 3674
Besensommerzypresse 531
Besensorghum 810
Besing 3784
betäubender Kälberkropf 1648
Bete 1116
Betel 549
Betelnusspalme 313, 548
Betelpalme 313, 548
Betelpfeffer 549
Betulazeen 579
Beutelbaum 874
Biberbaum 5355
Biberklee 1365
Biberkraut 1909
Bibernellrose 4776
Bick(el)beere 3784
Biebersteins Hornkraut 5461
Biebersteins Tulpe 552
biegsames Nixkraut 5002
Bienenbalsam 512
Bienenbrot 4161
Bienenorchis 519
Bienenragwurz 519
Bienensang 1818, 3583
Bierhefe 1359
Bigardie 5092
Bignoniazeen 5612
Bignoniengewächse 5612
Billbergie 68
Billiards Spierstrauch 577
Bilsenkraut 2823
Bindendivie 2032
Bindesalat 4545
Binetsch 4304
Bingelkraut 3589
Binkel 1288
Binse 4608
Binsengewächse 4609
Binsenginster 5856
Binsenkaktus 4488
Binsenknorpellattich 4610
Binsenlilie 698
Binsenseide 1637
Biota 300
Birke 578
birkenartige Gewächse 579
birkenblättriger Birnbaum 580
birkenblättriger Spierstrauch 581
Birkenerle 2105
Birkengewächse 579
Birkenhainbuche 2108
Birkenmilchling 6049
Birkenpilz 582
Birkenporling 583
Birkenröhrling 582
Birkenwurz 2207
Birnbaum 4089
Birnbäumchen 4926
Birnentang 2477
Birnentraube 5422
Birnkraut 4377
Birnmoose 5605
Bisamblume 181
Bisamdistel 3759
Bisamflockenblume 4355, 5388
Bisamhyazinthe 3277
Bisamklee 711
Bisamkraut 3765
Bisamkürbis 1747
Bisamstrauch 3761
Bischofskappe 3634
Bischofsmütze 2049
Bitterdistel 671
bittere Cassave 1379
bittere Juka 1379
Bitterenzian 260
bitterer Maniok 1379
Bitteresche 4971
Bittereschengewächse 66
bittere Schleifenblume 4523
bittere Springgurke 434
Bitterholz 4382, 4383
Bitterholzgewächse 66
Bitterklee 716, 1365
Bitterkraut 3989
Bitterling 6116
Bittermandelbaum 104, 597
Bittermaniok 1379
Bitternuss 605
Bitterorange 5092
Bitterschopf 109
Bittersüss 604
Bitterwurz(el) 3285, 4561, 6091
Bitterzitrulle 1334
Bixazeen 612
Blasenalge 2626
Blasenbaum 1362, 2555
Blasenbinse 4754
Blasenesche 2555
Blasenfarn 657
Blasenkirsche 2645
Blasennuss 2083
Blasenriedgras 674
Blasenschötchen 661
Blasenschote 662
Blasensegge 674
Blasensenne 662
Blasenspiere 3857
Blasenstengel 4539
Blasenstrauch 662, 2083
Blasentang 666
blasige Wasserfalle 5793
blasse Schwertlilie 5371
blasses Weidenröschen 4012
blassgelber Fingerhut 6089
Blattähre 2527, 4170
Blattang 667
Blattblume 3249

Blatteibe 1087
Blätterpilz 57, 2419
Blätterpilze 2488
Blätterschwamm 57
Blätterschwämme 2488
Blattkaktus 3247
Blattkeimer 1855
Blattkohl 3128
blattlose Platterbse 2592
Blaualgen 701
Blaublume 1763
Blaudolde 3194
blaue Akazie 3078
blaue Atlaszeder 688
blaue Clematis 3297
blaue Douglasfichte 4541
blaue Dürwurz 602
blaue Giftbeere 285
blaue Heckenkirsche 5356
blaue Kugeldistel 1423
blaue Lupine 703
blaue Luzerne 90
blaue Malve 3576
blaue Osterblume 3554
blauer Bockshornklee 711
blauer Eisenhut 17
blauer Gauchheil 705
blauer Gummibaum 5447
blauer Heinrich 1555
blauer Helm 2094
Blauerlattich 4132
blauer Natternkopf 1555
Blauesche 687
blaues Fettkraut 3700
blaues Geissblatt 5356
blaues Kopfgras 704
blaues Lungenkraut 1651
blaues Pfeifengras 3674
blaues Sperrkraut 2620
blaue Swertie 119
blaue Waldrebe 2976, 5080
Blaufichte 695, 1338
Blauglöckchen 1763
Blauglockenbaum 4075, 4595
Blaugras 704, 3675
blaugraue Kammschmiele 702
Blaugrünalgen 701
blaugrüne Algen 701
blaugrüne Binse 2704
blaugrüne Quecke 2952
blaugrüne Rose 4454
blaugrüne Zypresse 3597
Blauheide 3706
Blauholz(baum) 3359
Blaukissen 366
Blauklee 4847
Blaukraut 4444
Blauregen 3078
blauroter Steinsame 1669
Blauspiere 4918
Blaustern 5200
Blautanne 695
Blauweide 1687
Blauzederwacholder 4974
bleiche Segge 4010
bleiches Riedgras 4010
bleiches Wintergrün 2633
Bleichmoos 5125
Bleichsellerie 2411
bleigrauer Zwergbovist 3244
Bleistift-Wacholder 1980
Bleiwurz 4225
Bleiwurzgewächse 3246
Blende 832
Bletilla 672
Blumenbinse 2313, 4754
Blumenbinsengewächse 4755
Blumenesche 2308
Blumenhartriegel 2309
Blumenkohl 1079
Blumenkresse 3812
Blumenlieschgewächse 2315
Blumenrohr 996
Blumenrohrgewächse 997
Blumensimse 4754
Blumenspiere 4090
Blumes Buntlippe 1386
Blutampfer 684
Blutauge 3502
Blutbaum 3359
Blutblume 678
Blutbuche 2077, 4345
Blutdorn 4452
Blütenpflanzen 2310
Blütenschweif 270
Blütentange 4433
Blutfennich 2683
Blutfingergras 2683
Blutfingerhirse 2683
Bluthirse 1985, 2683
Blutholzbaum 3359
Blutjohannisbeere 6000
Blutklee 1694
Blutkraut 680, 681, 2079, 2409, 5018
Blutkrautknöterich 2079
Blutnelke 1037
Blutpilz 679, 3940
blutrote Fingerhirse 2683
blutrote Johannisbeere 6000
blutroter Storchschnabel 680
blutrote Zwergmispel 2089
Blutschuppe 1781
Blutstorchschnabel 680
Blutströpfchen 1037, 4164
Blutweiderich 3400, 4349
Blutwurz 681, 5555
Boberelle 5275
Bocksbart 4662
Bocksbeere 2081, 2093
Bocksdorn 450, 6016
Bockshauhechel 2518
Bockshorn 4662
Bockshornbaum 1023
Bockshornklee 5593
Bocksknöterich 2522
Bocksorchis 3346
Bockspetersilie 4733
Bocksweizen 2522
bodenfrüchtiger Klee 5291
Bodenkohlrabi 4626
Bodenlorbeer 5569
Boerhaavie 5131
Bogenflieder 3867
bogengrannige Trespe 3024
Bogenhanf 4694
Bogenlilie 1780
Bohne 497
Bohnenbaum 809, 3192, 3680
Bohnenkaper 498
Bohnenkraut 4721, 5321
Bohnenstrauch 809
Bohnenwinde 1883
Bolle 2420
Bologneserfenchel 2300
Boltonie 728
Bomarea 729
Bombakazeen 4944
Bombayhanf 3146, 5331
Bombazeen 4944
Bootfarn 5730
Borassuspalme 4019
Boretsch 1366
Bornkresse 5796
Bornmüllers Tanne 737
Borraginazeen 735
Borretsch 734, 1366
Borretschgewächse 735
Borstendolde 2759
Borstenfedergras 4110
Borstengras 5499
Borstenhirse 780, 2352
Borstenkarde 2692
Borstenkraut 2130
Borstenquendel 5936
Borstenschwingel 4839
Borstensteinbrech 6102
borstige Robinie 4555
borstiges Rapünzchen 2977
Böskraut 1817
Bottenbinse 4666
Bougainvillea 742
Bourbonrose 745
Boussingaultie 3413
Bouvardia 747
Bovist(stäubling) 4325
Boykinie 753
Brachdistel 5047
Brachpilz 1476
Brachsenkraut 4387
Brachsenkräuter 4388
Brachsenkrautgewächse 4388
Brainders Weissdorn 3845
Brandbaum 299
Brandknabenkraut 1803
Brandkraut 3097
Brandpilze 854
Brandschopf 1307
Brasilettoholz 4296
brasilianische Araukarie 4049
brasilianische Doppeldreizackpalme 767
brasilianische Futterhirse 1789
brasilianische Kirschmyrte 761
brasilianischer Pfefferbaum 766
brasilianischer Quassiabaum 5338
brasilianisches Tausendblatt 763
brasilianische Wachspalme 1021
Brasilienholz 758
Brasilnussbaum 765
Brätling 3938
Brauereihefe 1359
Braunalgen 815
Braunbeere 2081
braune Faltenmorchel 1994

braune Haselwurz 2133
Braunelle 4816
brauner Dachpilz 2189
brauner Dosten 1484
brauner Habichtsschwamm 2922
brauner Storchschnabel 1931
brauner Streifenfarn 3431
brauner Wulstblätterschwamm 4853
braune Schnabelsimse 816
braunes Schnabelried 816
braungelber Frauenschuh 2109
Braunheil 4816
Brauns Schildfarn 760
braun werdende Weide 77
Braunwurz 2242
Braunwurzgewächse 2243
Braut im Haar 3386
Brautspiere 2431
Brechnuss 3838
Brechnussbaum 3898
Brechnussgewächse 3358
Brechstrauch 5945
Brechveilchen 2957
Breiapfelbaum 4700
Breislauch 1219
Breitblattkresse 2118
breitblättrige Fetthenne 3341
breitblättrige Glockenblume 2599
breitblättrige Lorbeerrose 3710
breitblättrige Ölweide 4951
breitblättrige Platterbse 4133
breitblättriger Ehrenpreis 2892
breitblättriger Holzritterling 790
breitblättriger Lavendel 795
breitblättriger Pfeifenstrauch 2819
breitblättriger Rohrkolben 1380
breitblättriger Rübling 790
breitblättriger Spierstrauch 797
breitblättriger Spindelbaum 793
breitblättrige Sitter 792
breitblättriges Knabenkraut 798
breitblättrige Steinlinde 5586
breitblättrige Sumpfwurz 792
breitblättrige Weisswurz 799
breite Haftdolde 789
Breitfaden 4222
Breitkölbchen 4403
Breitlauch 3266
Breitwegerich 4502
brennende Adonisblume 2272
brennende Liebe 3451
brennender Hahnenfuss 5121
brennendes Teufelsauge 2272
brennende Waldrebe 4227
Brennessel 3835
Brennesselgewächse 3836
Brennhahnenfuss 5121
Brennhülse 1823
Brennkraut 1614, 2646, 5425
Brennpalme 2265, 5540
Brennpflanze 6036
Brennwinde 1155
Bresling 2635
Bretschneiders Berberitze 4338
Brettspielblume 1126
Brewers Fichte 772
Briançon-Aprikose 773
Brisslauch 1219
Broccoli 800
Brockenbirke 1935
Brockenmoos 2916
Brodiaea 801
Brokkoli 800
Brombeere 619
Brombeerstrauch 619
Bromeliazeen 4192
Bromelie 804
Bromelien 4192
Bronzenblatt 2401
Brotfarnpalme 3125
Brotfruchtbaum 2998
Brotnussbaum 770
Brotpalme 3125
Brottraube 5422
Brotweizen 5874
Browallie 814
Bruchheil 3545
Bruchkraut 877
Bruchweide 787
Brunfelsie 4405
Brunnenkresse 5796
Brüsseler Kohl 823
Brustbeerbaum 1594
Brustbeere 1594
Brustwarzendistel 3454
Brustwurz 245
Brustwurzel 245, 2406
Bryonie 824
Bryophyten 825
Bubiköpfchen 402
Buccostrauch 827
Bucharaklee 5917
Buche 513
Buchenfarn 3797
Buchengewächse 514
Buchenpilz 3992, 5529
Buchenschwamm 5529
Buchesche 2108
Buchs 749
Buchsbaum 749
Buchsbaumgewächse 751
Büchsenkraut 2179
buchtige Königskerze 3571
Buchweizen 832, 1370
Buckelbeere 2884
buckelige Wasserlinse 5404
Buckeltramete 2485
Buffalogras 836
Büffelbeere 834, 4952
Büffelgras 836
Büffelklette 835
Bügelholz 294
Bukettnarzisse 4248
Bukkostrauch 827
Bulbophyllum 843
Bundblume 470
Bungekiefer 3193
Buntblatt 1325
bunte Aloe 3132
bunte Kronenwicke 1713
bunte Kronwicke 1713
bunte Peltsche 1713
bunter Eisenhut 3465
bunter Hohlzahn 3846
bunter Porling 1569
bunter Schachtelhalm 5701
bunter Steinbrech 5663
bunte Schwertlilie 2889
buntes Krönel 1713
buntes Lungenkraut 1462
buntes Vergissmeinnicht 6073
bunte Wolfsmilch 1750
Buntlippe 1325
Buntnessel 1325
Buntwurz 925
Bunya-Bunya-Baum 855
Bürgelkraut 4358
Burgunderrose 861
Burgunderrübe 3471
burgundische Eiche 2128
Burserazeen 876
Bürstengras 504
Bürstenkraut 4637
Bürstenmoos 1435
Buruti 5985
Bürzel 954
Bürzeldorn 954
Burzel(kraut) 1509, 4358
Buschbohne 881
Büschelblume 2476, 4161
Büschelglockenblume 1797
Büschelkraut 4161, 5516
Büschelnelke 5393
Büschelrose 3068
Büschelschön 4161
Buschgoldregen 809
Buschklee 3279
Buschveilchen 2134
Buschweide 787
Buschwindröschen 2134
Butterbaum 898, 899, 4852
Butterblume 889, 1398, 1466
Butternussbaum 897, 4725
Butterpilz 820
Butterraps 572
Butterröhrling 820
Butterrübling 901
Buttersame 572
Butterwurzel 3700
Bux 749
Buxazeen 751

Caesalpinie 758
Calabar-Bohne 1816
Calandrinie 4529
Camassie 958
Camote 5380
Campbells Magnolie 964
Canaigrewurzel 981
Caryotapalme 2265
Cassave 1046, 1379
Catawbarebe 2347
Catechu-Akazie 1056
Catechu-Palme 548
Catjangbohne 1059
Cattleya 1065
Cayenne-Pfeffer 884
Ceara-Kautschukbaum 1081
Cedrat 1243
Cedrat-Zitrone 1243
Cedrelabaum 1234
Cedrelaholzbaum 1234
Celosie 1307
Celsie 1088

Centifolie 910
Cerealien 2581
Cerkospore 1099
Ceylon-Bogenhanf 1106
ceylonischer Zimtbaum 1105
Ceylon-Moos 56
Ceylon-Sansevieria 1106
Ceylon-Zimtbaum 1105
Champak 5423
Champignonartige 2488
Chayote 1122
Chenopodeen 2562
Chenopodiazeen 2562
Cherimoya 1132
Chicasapflaume 1143
Chicot 1319
Chileerdbeere 1150
Chilefichte 3658
chilenische Araukarie 3658
chilenische Flusszeder 1149
chilenische Seide 1153
Chilepalme 3113
Chinabaum 1236
Chinagras 2174, 4410
Chinakohl 4005
Chinanessel 2174
Chinarindenbaum 1236
Chinastechwinde 1166
Chinazimtbaum 1047
Chinesennelke 1165
chinesische Aktinidie 6065
chinesische Aralie 1167
chinesische Astilbe 1169
chinesische Banane 1936
chinesische Birke 1171
chinesische Birne 4687
chinesische Dattelpflaume 3127
chinesische Eibe 1215
chinesische Esche 1168
chinesische Flügelnuss 1210
chinesische Gartenaster 1382
chinesische Glyzinie 1211
chinesische Goldlärche 3388
chinesische Hasel 1179
chinesische Hemlocktanne 1184
chinesische Hickorynuss 1186
chinesische Jute 1217
chinesische Kiefer 1195
chinesische Livistone 1178
chinesische Nelke 1165
chinesische Pappel 1058
chinesische Primel 1197
chinesische Rainweide 1198
chinesischer Blauregen 1211
chinesischer Bocksdorn 1213
chinesischer Eibisch 1185
chinesischer Flieder 1190
chinesischer Götterbaum 5584
chinesischer Hanf 1217
chinesischer Jasmin 5984
chinesischer Klebsame 5538
chinesischer Kohl 4005
chinesischer Liguster 1198
chinesische Rose 1185
chinesische Rosskastanie 1188
chinesischer Pfeffer 2282
chinesischer Quittenbaum 1181
chinesischer Rhabarber 3566
chinesischer Sadebaum 4369
chinesischer Schotenbaum 1187
chinesischer Spierstrauch 1201
chinesischer Strahlengriffel 6065
chinesischer Sumach 1203
chinesischer Surenbaum 1205
chinesischer Talgbaum 1204
chinesischer Tee 1543
chinesischer Tulpenbaum 1207
chinesischer Wacholder 4369
chinesischer Walnussbaum 1208
chinesischer Zimtbaum 1047
chinesischer Zürgelbaum 1183
chinesische Scheinquitte 1181
chinesische Schlüsselblume 1197
chinesische schwarze Olive 623
chinesisches Judasbaumblatt 1199
chinesisches Spaltkölbchen 1191
chinesisches Zuckerrohr 5085
chinesische Traubenspiere 1193
chinesische Trollblume 1182
chinesische weisse Olive 5886
chinesische Wistarie 1211
chinesische Yamswurzel 1241
chinesische Zelkowe 1216
chinesische Zwittertanne 1383
chinesische Zypresse 551
Chironie 5229
Chloranthazeen 1221
Choleradistel 3326, 5155
Christianswurz 2628
Christophskraut 441
Christrose 1226, 2766
Christusakazie 1439
Christusdorn 1227, 1439
Christ(us)palme 1051
Christustränengras 31
Chrysantheme 1229
Chrysanthemum 1229
Chufa 1230
Claytonie 5181
Clematis 1260
Clintonia 490
Colanussbaum 1323
Coleus 1325
Colletie 1330
Collinsie 1331
Colorado-Johannisbeere 1336
Colorado-Sandlilie 5226
Columbo-Wurzel 955
Commeline 1814
Conringie 2710
Coquito 5419
Coulters Kiefer 1642
Coulters Pfeifenstrauch 1641
Crabbaum 1657
Cryptomerie 1718
Cubebafrucht 1722
Cunninghams Araukarie 1730
Cunninghams Keulenbaum 1731
Cyclanthera 1768
Cypergras 2280

Dachpalme 4017
Dachpippau 3803
Dachtrespe 1123
Dagussa 4399
Dahlie 1782
dahurische Birke 1783
dahurischer Mondsame 347
dahurische Rose 1785
dahurische Tamariske 1784
dalmatische Insektenblume 1790
dalmatisches Leinkraut 1791
Damasonie 1793
Damaszener Rose 1792
Damensattel 1499
Dammarabaum 1795, 5890
Dammaraharzbaum 477
Dammers Zwergmispel 501
dänischer Tragant 4350
Dari 5087
Darlingtonie 941
Darrgras 5367
Darwins Berberitze 1805
Dattelpalme 1807
Dattelpflaume 1809
Dattelpflaume(nbaum) 4152
Datura 1811
Dauerklee 90
Dauerknäuel 4131
Dauerlupine 5789
Daunengras 1637
Davallie 2711
Davidsons Glanzmispel 1813
Davids Stranvesia 1202
Davids Tanne 1812
Decksamer 248
Degenbinse 6084
Degenkraut 6084
Dekkanhanf 3146
Delebpalme 4019
Deodarzedar 1836
Desmidiazeen 1841
deutsche Hundszunge 2630
deutsche Luzerne 4921
deutsche Mispel 3578
deutsche Purgierwinde 2748
deutscher Bärenklau 2825
deutscher Blasenstrauch 1362
deutscher Buchsbaum 2096
deutscher Enzian 2457
deutscher gefleckter Ingwer 3378
deutscher Ginster 2463
deutscher Indigo 1956
deutscher Kicher 2587
deutscher Lorbeer 626
deutscher Tabak 1548
deutscher Ziest 3731
deutsche Schwertlilie 2458
deutsches Filzkraut 1396
deutsches Geissblatt 6022
deutsche Skammonie 2748
deutsches Weidelgras 4135
deutsche Tamariske 2456
deutsche Trüffel 4619
deutsche Waldrebe 5571
deutsche Zypresse 2456
Deutzie 1843
Diatomeen 1853
dichtährige Segge 4303
dichtblättrige Wasserpest 1832
dichtblütige Kresse 4293
dichtes Laichkraut 1833
Dickblatt 1661, 4365
Dickblattgewächse 3970

Dickblättler 2905
dickblättriger Schwarztäubling 628
dickblättrige Wickelwurz 3254
dickblättrige Wolfsmilch 4805
Dickblume 3714
Dickröhrling 727
Dickröschen 5262
Dickrübe 1360
dickschaliger Kartoffelbovist 1405
Dieffenbachie 5631
Dierville 880
Diffinbohne 3309
Dikotyledonen 1855
Dikotylen 1855
Dill 1857
Dillenie 1858
Dillfenchel 1399, 1857
Dillkraut 1857
Dingel 2589
Dinkel(weizen) 5124
Dioskoreazeen 6064
Dipsazeen 5471
Diptam 1866
Dipterokarpazeen 1862
Distel 782
Distelsommerwurz 5488
Dividivi 1867
Dodonäe 2850
Doldengewächse 1036
Doldenmilchstern 1535
Doldenrebe 221
doldentraubiger Milchstern 1535
Doldenwinterlieb 1496
doldige Blumenbinse 2314
doldiger Bauernsenf 2504
doldiger Milchstern 1535
doldige Schleifenblume 2504
doldige Schwanenblume 2314
doldiges Habichtskraut 3804
doldige Spurre 3004
doldiges Winterlieb 1496
doldige Wucherblume 3569
Dombeya 1885
Doompalme 2001
Doppelblatt 498
Doppelblattgewächse 499
Doppelblume 1260
Doppeldreizackpalme 5598
Doppelkappe 32
Doppelkokos 1887
Doppelmalve 1127
Doppelsame 5775
Doppelsporn 1363
Dorant 2816, 5050
Dornapfel 3102
Dornbaum 1439
Dornengras 4294
Dornenkirsche 4310
Dornfarn 5546
dorniger Akanthus 5151
dorniger Schotenbaum 1439
dornige Spitzklette 5155
Dornkronenbaum 2840
Dornmelde 468
Dornrose 1875
Dorn(schlehe) 5014
Dorsche 4626
Dorstenie 5559
Dost 3967
Dotter 2168
Dotterblume 3508
Dotterschwamm 1115
Dotterweide 6107
Douglasfichte 1401
Douglasie 1401, 1888
Douglas Spierstrauch 1889
Douglastanne 1401, 1888
Doumpalme 2001
Dourabaum 45, 3864
Dourg 5087
Dracaene 1898
Drachenbaum 1900
Drachenbaumaloe 4392
Drachenblutbaum 1900
Drachenblutpalme 5314
Drachenkopf 1901
Drachenlilie 1900
Drachenpalme 1900
Drachenwurz 949, 1536, 2079, 5254, 5941
Dragon 5445
Drahtschmiele 5843
Drehähre 2100
Drehfrucht 1006
Drehling 3992, 4139
Drehmoos 2394
Drehwurz 2100, 3197
Dreiblatt 716, 5594
dreiblättrige Lederblume 1441
dreiblättriger Fieberklee 1365
dreiblättrige Zitrone 5592
Dreiblattzitrone 5592
dreiblütiges Labkraut 5384
dreidornige Gleditschie 1439
dreifarbiges Veilchen 5954
dreifarbige Winde 1941
dreifingeriger Steinbrech 4600
Dreifingersteinbrech 4600
dreifurchige Wasserlinse 5217
dreigriffeliges Hornkraut 5236
Dreihöckerblume 5534
dreihörniges Labkraut 1607
dreikantige Simse 5590
dreilappige Mandel 2311
dreilappiger Ahorn 3669
dreilappiger Pappaubaum 1485
dreilappiger Pfirsichstrauch 2311
dreilappiges Traubenkraut 2480
dreimänniger Tännel 5501
dreimännige Weide 106
Dreimasterblume 5134
Dreimasterblume aus Rio de Janeiro 5780
dreispaltiges Labkraut 5017
dreispaltige Yamswurzel 1748
dreispitzige Jungfernrebe 3028
dreiteiliger Ehrenpreis 2253
dreiteiliger Zweizahn 857
Dreizack 4234
Dreizackgewächse 3116
Dreizackpflanzen 3116
Dreizahn 4937, 5596
Dresch-Lein 2214
Droserazeen 5326
Drossel 2105
Drosselbeere 2091, 2114
Drudenkraut 4606
Drummonds Flammenblume 1919
Drüsenbaum 493
Drüsenblume 493
Drüsenglocke 3198
Drüsenklee 4782
Drüsenstrauch 3838
drüsentragender Kampferbaum 3827
Drüsenträger 3198
drüsiger Götterbaum 5584
drüsiges Springkraut 4596
Dryade 1920
Dschungelgras 4713
Dshusgun 952
Dshusgunstrauch 952
Duffinbohne 4940
Duftblume 1850, 3973
Duftblüte 1850, 3973
duftender Mangobaum 853
duftende Schlüsselblume 1652
Duftheliotrop 1436
duftige Schmiele 2595
Duftodermennig 2355
Duftrebe 2386, 4503
Duftrose 4753
Duftschneeball 3186
Duftsteinrich 5352
Dufttrichter 2441
Dulldill 2823
Dullkraut 2823
Dumpalme 1890, 2001
Dünenhafer 2076, 2094
dunkelblättrige Pappel 819
dunkelgrüne Agave 4328
dunkelgrünes Weidenröschen 1952
dunkelvioletter Dickfuss 5739
dunkle Königskerze 641
dünnästiger Pippau 5038
Dünngras 5171
Dünnschwanz 4923
dünnwurzelige Segge 1595
durchlöchertes Hartheu 1537
durchwachsenblättrige Kresse 1256
durchwachsenblättriges Täschelkraut 5494
durchwachsene Kresse 1256
durchwachsenes Hellerkraut 5494
durchwachsenes Laichkraut 5495
Durianbaum 1927
Durra 1929, 5086
Durragras 5087
Durra(hirse) 5087
Dürrbehndel 2434
Dürrbein 2434
Dürreiche 1928
Dürrenstaude 1453
Durrha 5087
Dürrkraut 2292
Dürrlilitze 1611
Dürrwurz 1239
Duwock 3506
Dyckie 1953

Ebenbaum 1987
Ebenholz 42, 1987
Ebenholzgewächse 1986
Eberesche 2114, 3695

ebereschenblättrige Fiederspiere 5691
Ebereschenfiederspiere 5691
Ebereschenspiere 5084
Eberraute 3923
Eberreis 3923
Eberwurz 1017
Echinokaktus 1990
echte Akazie 4957
echte Alkanna 2782
echte Aloe 3568, 5601
echte Bärentraube 500
echte Betonie 1361
echte Brechwurz 2957
echte Brunnenkresse 5796
echte Dattelpalme 1808
echte Engelwurz 2406
echte Farne 2206
echte Galleiche 85
echte Goldrute 2101
echte Goldtraube 2535
echte Hagebuttenrose 287
echte Hauswurz 2779
echte Hefe 6069
echte Heidelbeere 3784
echte Hirse 4315
echte Hortensie 565
echte Hundszunge 1444
echte Kamille 2452
echte Kastanie 2086
echte Katzenminze 1060
echte Kicher 1144
echte Kokospalme 1314
echte Kölme 927
echte Königskerze 2276
echte Mandarine 3162
echte Mandel 104
echte Meerzwiebel 4878
echte Mielitz 806
echte Mispel 3578
echte Myrte 5606
echte Nelkenwurz 1354
echte Ochsenzunge 1371
echte Pfingstrose 1490
echte Pistacie 1498
echte Pistazie 1498
echte Quitte 1512
echter Baldrian 1552
echter Beinfuss 3747
echter Beinwell 1389
echter Brasilholzbaum 4296
echter Champignon 2419
echter Dost 1484
echter Ebenholzbaum 1987
echter Efeu 2040
echter Ehrenpreis 1917
echter Eibisch 3507
echter Eisenhut 17
echter Elsebeerbaum 1128
echter Erdbeerspinat 5273
echter Erdrauch 1911
echter Essigbaum 4920
echter Feigenbaum 1412
echter Feigenkaktus 2936
echter Fenchel 1411
echter Feuerschwamm 5529
echter Fuchsschwanz 3387
echter Gagel 5366
echter Gamander 1113
echter Geissbart 1425
echter Gelbling 1115
echter Giftsumach 1501
echter Ginkgobaum 2493
echter Haarstrang 1438
echter Hausschwamm 2880
echter Hefepilz 6069
echter Honigklee 6108
echter Jasmin 1450
echter Kaffeebaum 292
echter Kalmus 1918
echter Kampferbaum 966
echter Kapernstrauch 1377
echter Kerbel 4658
echter Kreuzdorn 1369
echter Kümmel 1013
echter Lanzenbaum 1886
echter Lavendel 5604
echter Lein 2214
echter Löwenzahn 1398
echter Mandelbaum 104
echter Mousseron 4226
echte Rockenbolle 2474
echter Ölbaum 1483
echte Rosskastanie 1443
echter Pastinak 2422
echter Perückenstrauch 1527
echter Pfeffermilchling 4123
echter Pockholz 1460
echter Quendel 927
echter Quittenbaum 1512
echter Reizker 1827
echter Rhabarber 3566
echter roter Sandelholzbaum 863
echter Rotholzbaum 4296
echter Safran 4638
echter Sassafras 1520
echter Schwarzkümmel 2415
echter Steinklee 6108
echter Steinsame 1428
echter Storaxbaum 1915
echter Tabak 1548
echter Waldmeister 5395
echter Weizen 5874
echter Ysop 2910
echter Zunderschwamm 5529
echte Sagopalme 5157
echte Salbei 2427
echtes Alpenglöckchen 2496
echtes Barbarakraut 610
echtes Basilienkraut 5354
echtes Benediktenkraut 671
echtes Berufkraut 602
echte Seidenpflanze 1469
echtes Eisenkraut 2129
echte Sellerie 5943
echtes Federgras 2115
echtes Geissblatt 5370
echtes Glasschmalz 3514
echtes Gnadenkraut 1913
echtes Herzgespann 1475
echtes Johanniskraut 1537
echtes Katzenkraut 1060
echtes Kuba-Mahagoni 5871
echtes Labkraut 6075
echtes Leinkraut 887
echtes Löffelkraut 1522
echtes Lungenkraut 1462
echtes Mädesüss 2112
echtes Mahagoni 5871
echtes Salomonssiegel 1916
echtes Seegras 1407
echtes Seifenkraut 743
echtes Springkraut 5560
echtes Steinkraut 2551
echtes Tännelkraut 4851
echtes Tausendgüldenkraut 1909
echte Steinbeere 5260
echtes Zuckerrohr 5300
echte Tigerblume 1547
echte Trüffel 5608
echte Vanille 3605
echte Waldrebe 5571
echte Weintraube 2104
echte Zypresse 2978
eckige Haargurke 5765
Edeldistel 2058
Edelesche 2073
Edelgamander 1113
Edelkamille 4546
Edelkastanie 1139, 2086
Edelmarguerite 4376
Edelpflaume 2627
Edelpilz 1093, 1476
Edelpilze 2488
Edelreizker 1827
Edelrose 1200
Edeltanne 3865
Edelwalnuss 4151
Edelweiss 1406, 1992
Edelwicke 5378
edler Lorbeer 2615
edler Walnussbaum 4151
Efeu 2988
efeublättrige Gunderrebe 1429
efeublättrige Pelargonie 2991
efeublättriger Ehrenpreis 2992
efeublättriges Leinkraut 3147
Efeuehrenpreis 2992
Efeugewächse 2495
Efeugundermann 1429
Efeutute 2989
Ehrenpreis 5123
Eibe(nbaum) 6117
eibenblättrige Torreye 2303
Eibengewächse 6118
Eibensequoie 945
Eibenzypresse 414
Eibisch 169, 2794
Eiche 3899
Eichelbecherling 4764
Eichenfarn 3797, 3901
Eichenmispel 3377
Eichenwald 4808
Eichenwirrling 3900
Eichenzwergknäueling 606
Eichhörnchenfuchsschwingel 803
Eichhornie 1557, 5810
Eichhornschwanzfederschwingel 803
Eichlers Tulpe 2003
Eichpilz 1093
Eidechsenschwanz 3347
Eidechsenschwanzpflanzen 3348
Eidechsenwurz 3345
Eierfrucht 2414, 3853

Eierkürbis 6088
Eierpflanze 2414
Eierschwamm 1115
eiförmiges Sumpfried 3983
eiförmige Sumpfbinse 3983
eiförmiges Zweiblatt 1999
Einbeere 4053
Einblatt 25
einblütige Erve 3930
einblütige Glockenblume 4756
einblütiges Perlgras 6035
einblütiges Wintergrün 6037
einblütige Wicke 3930
einfache Wiesenraute 5009
eingriffeliger Weissdorn 4973
einjährige Flammenblume 1919
einjähriger Beifuss 5396
einjähriger Feinstrahl 259
einjähriger Knäuel 262
einjähriger Mauerpfeffer 264
einjähriger Ziest 2758
einjähriges Bingelkraut 2785
einjähriges Knäuelkraut 262
einjähriges Madikraut 1151
einjährige Sonnenblume 1539
einjährige Spreublume 1446
einjähriges Rispengras 258
einkeimblättrige Pflanzen 3661
einköpfiges Berufkraut 309
Einkorn 2004
Einorchis 3763
Einsamähre 1839
einseitswendiges Wintergrün 4926
einstacheliger Feigenbaum 1507
Eisbeere 3617
Eisblume 2917
Eiseiche 1928
Eisenbaum 3117
Eisenbirke 4758
Eisenholz 4061
Eisenholzbaum 315, 2964, 2966, 3117, 4150
Eisenhut 3660
eisenhutblättriger Hahnenfuss 16
Eisenkraut 5728
Eisenkrautgewächse 5729
Eisenmassbaum 2964
Eisenveilchenbaum 5447
Eiskoralle 1589
Eiskraut 70, 2917, 3592
Eiskrautgewächse 1033
Eispflanze 2917
Eiszapfen 2425
Eläagnazeen 3926
Elefantenapfelbaum 2012, 6020
Elefantenfuss 2014, 5558
Elefantengras 3793
Elefantenlausbaum 1378
eleganter Röhrling 2011
Elemibaum 3085
Elend 5047
Eleusine 2015
Elfenbeindistel 5487
Elfenbeinpalme 1447, 2986
Elfenblume 2049
Eller 81
Ellerling 2904
elliptische Rose 4753
Ellwangers Weissdorn 2017
Elsbeerbaum 1128
Elsbeere 1128
Else 81
Emilie 5451
Emmer 2029
Emodi-Rhabarber 2805
Empetrazeen 1707
empfindlicher Krempling 2956
Endivie 2032
Engelkraut 323
Engelmannfichte 2034
Engelmannkiefer 2033
Engelsauge 710
Engelsblümchen 1510
Engelstrompete 247, 2304
Engelsüss 1503, 4251
Engelwurz 245, 2406
Englers Buche 2036
englische Erbse 5197
englische Mispel 2428
englischer Ginster 3823
englischer Senf 5904
englischer Sonnentau 2044
englischer Weizen 4286
englisches Leimkraut 2372
englisches Löffelkraut 2043
englisches Raigras 4135
Ensete 8
Entenflott 1924
Entenfuss 1845
Entengras 5820
Entengrün 1924
Entengrütze 1924
Entenlinse 1924, 6017
entferntährige Segge 1865, 4482
entferntähriges Riedgras 1865
Entomophthorazeen 2047
Enzian 2445
Enziangewächse 2446
Eppich 1086, 2241, 5943
Equisetazeen 2877
Erbse 4080
Erbsenbaum 4095
erbsenfrüchtige Lebensbaumzypresse 4724
Erbsenstrauch 4095, 4910
Erbsenwicke 4097
Erdapfel 3094
Erdbeerbaum 3418, 5276
Erdbeere 5272
Erdbeergewürzstrauch 1540
Erdbeerklee 5274
Erdbeerstrauch 1540
Erdbirne 207, 3094, 4282
Erdbohne 4087
Erdbürzeldorn 4332
Erdeichel 1907, 2649, 4087
Erderbse 1574
Erdkastanie 2724
Erdkohlrabi 4626
Erdkohlrübe 4626
Erdkratzdistel 5238
Erdmandel 1230, 2649, 4087
Erdmandelzypergras 1230
Erdnuss 2649, 4087
Erdnussplatterbse 2649
Erdpalme 4840
Erdrauch 2392
Erdrauchgewächse 2393
Erdraute 2392
Erdrose 385
Erdrübe 4626
Erdscheibe 1766
Erdschlüsselblume 2042
Erdstachelnuss 954
Erdstern 1969
Erdzunge 1970
erhabene Wechselscheide 4642
Erikazeen 2745
Erizinen 2745
Erle 81
erlenblättrige Eberesche 1831
erlenblättrige Eiche 1779
erlenblättriger Kreuzdorn 82
Erman's Birke 2057
Erve 609, 3273
Erysiphazeen 4289
Erythea 2059
Erythroxylazeen 1303
Erzengelwurz 2406
Escariol 2032
Esche 338, 2073
Eschenahorn 750
eschenblättrige Flügelnuss 1077
eschenblättriger Ahorn 750
eschenblättriger Diptam 2435
eschenblättriger Gelbhorn 4868
Eschenkastanie 2145
Escheverie 1988
Eschscholtzie 2554
Eseldistel 1638, 3759
Eselfeige 5408
Eselsbohne 5028
Eselsgurke 5202
Eselskerbel 1648
Eselsohr 3378
Eselwolfsmilch 3251
Eskallonie 2060
Eskariol 2032
Esparsette 4646
Esparto(gras) 2061, 2062
Espasel 1042
Espe 2074
Esper 1519
essbare Arracacha 277
essbare Canna 1993
essbare Linse 1457
essbare Platterbse 2587
essbarer Flügeltang 407
essbares Blumenrohr 1993
essbare Yamswurzel 1214
Essigbaum 5313
Essigdorn 2075
Essigrose 2371
Esskastanie 2086
Estragon 5445
Eugenie 2068
Eukalyptus 2065
Euphorbiazeen 5191
europäische Edeltanne 4955
europäische Hopfenbuche 2107
europäische Korallenwurz 1960
europäischer Siebenstern 2125
europäischer Spindelbaum 2096
europäischer Zürgelbaum 2106

europäisches Alpenveilchen 2092
europäisches Pfaffenhütchen 2096
europäische Trollblume 1422
europäische Zwergpalme 3572
ewiger Klee 90

Fabiana 4154
Fächerahorn 3056
Fächerbaum 2492
Fächerpalme 5424
Fackelbaum 3352, 4775
Fackelblume 3752
Fackeldistel 1100
Fackelkraut 3752
Fackellilie 5551
Fadenbinse 5498
Fadenfingerhirse 5037
fadenförmige Binse 5498
Fadenhirse 5037
Fadenklee 5292
Fadenröhre 4391
fädentragende Palmlilie 24
Fadenwicke 2343
fädige Palmlilie 24
Fagazeen 514
fahler Sturmhut 4373
Fahlweide 467
Fahnenwicke 1662
Fahnwicke 1662
Falconers Pfeifenstrauch 2157
Fallsame 1906
falsche Jalape 1415
falsche Kamille 2210, 2220, 4751
falscher Bastardindigo 2948
falscher Eierschwamm 2165
falscher Hederich 1118
falscher Jasmin 1078, 3641
falscher Pfeifenstrauch 5376
falscher Porst 723
falscher Safran 4637
falscher Wurmfarn 3199
falsche Trüffel 1405
Faltenlilie 159
Faltensamenpalme 4322
Färberblume 1563
Färberdistel 1955
Färberdorn 830, 4145
Färberflechte 3325
Färberginster 1563
Färberhülse 5948
Färberhundskamille 2528
Färbermeier 1957, 6040
Färbermeister 1957
Färberochsenzunge 1954
Färberresede 5862
Färberröte 1463, 3410
Färbersaflor 4637
Färberscharte 1955, 4728
Färberwaid 1956
Färberwau 5862
Färberwurzel 1463
farnartige Gewächse 4321
Farne 2206
Farneibe 1087
Farnkrautmännchen 3445
Farnkrautwurzel 3445
Farnmyrte 5364
Farnpalme 1764, 4643
Farnpalmen 1765
Farnpflanzen 4321
Farnstrauch 5364
Fasanenstrauch 1362
Fasel 1883, 3158
Faselbohne 2897
Faserlein 2214
Faserpalmlilie 24
Faserschirm 5595
Fasole 3158
Fatsie 2187
Faulbaum 830, 2078, 2513
Faulbeerbaum 2078
Faulschimmel 3914
Faulweide 3236
Faxons Tanne 2191
Federähre 4110
Federborstengras 4110
Federbuschflechte 2868
Federgras 3819, 4475
Federharzbaum 2791
federiges Pfriemengras 2115
Federmohn 4230
Federnelke 2588
Federspargel 2204
Feigbohne 3395
Feigenbaum 2240
Feigendistel 2936, 4301
Feigenkaktus 4301
Feigenwurzel 2241
Feigwarzenkraut 2241, 6027
Feigwurz 2241, 4181
Feijoa 2195
feinblättriges Hasenohr 4998
feinblättrige Wicke 755
Feinblattwicke 755
feines Hasenohr 4998
feingespitzter Knöterich 3036
Feingras 5171
Feinstrahl 2292
Feisterling 1683
Felberich 3370
Feldahorn 2756
Feldampfer 4859
Feldbeifuss 4641
Feldbohne 788
Feldedelpilz 1476
Feldegerling 1476
Feldehrenpreis 1532
Feldenzian 3550
Felder 3370
Felderbse 2231
Feldflohkraut 2878
Feldhysop 1515
Feldklee 2848
Feldkresse 2233, 2895
Feldlöwenmaul 1619
Feldmännertreu 5047
Feldmannstreu 5047
Feldmargerite 1609
Feldmohn 1616
Feldnelke 1037
Feldpappel 647
Feldraute 1516, 1911
Feldrittersporn 2330
Feldrose 1875, 2234
Feldrüster 5040
Feldsalat 1618, 2088
Feldschafgarbe 1565
Feldschwindling 2154
Feldsimse 2239
Feldspark 1620
Feldthymian 3688
Feldulme 5040
Feldveilchen 2238
Feldwegwarte 1381
Feldwicke 1554
Feldwinde 2100
Felicie 2196
Fellstreuling 2705
Felsenahorn 3669
Felsenbeifuss 4540
Felsenbirne 2199, 4832
Felsenblümchen 1897
Felsenehrenpreis 4534
Felsenfingerkraut 1263
Felsenjohannisbeere 4520, 4900
Felsenkirsche 3426
Felsenmastkraut 310
Felsenmoos 2961
Felsennelke 1129, 5641
Felsenpfeffer 3089
Felsenschleifenblume 4514
Felsensegge 4533
Felsensteinkraut 2551
Felsenstrauch 389
Felsenulme 4521
Felsenwalnuss 5479
Felslabkraut 2743
Fench 780
Fenchel 1411, 2200
Fenchelholzbaum 1520, 4144
Fendlere 2199
Fennich 780
Ferkelkraut 1061
Ferkelnuss 4179
Fernambukholzbaum 768, 4296
Feste 2725
Fetisch-Bohne 1448
Fettblatt 5811
Fettblume 1466
Fetthenne 5264
Fetthennensteinbrech 6102
Fettkörbchen 5713
Fettkraut 900
Fettkräuter 665
Fettmännchen 3211
Feuerahorn 230
Feuerbaum 2812
Feuerblume 1616
Feuerbohne 4748
Feuerdorn 2261, 3038
Feuerkolben 2999
Feuerkraut 2262
Feuerlilie 842
Feuermohn 1616
Feuerröschen 5317
feuerrotes Adonisröschen 2272
Feuersauerdorn 4659
Feuerzypresse 2812
Fichte 5189
Fichtenpilz 1093
Fichtenspargel 2940
Fieberbaum 4712, 5447
Fieberheilbaum 2065, 5447
Fieberklee 716

Fieberkraut 2210
Fieberrindenbaum 1236
Fieberstrauch 5127
Fieberweide 787
Fieberwurzel 2867, 5221
fiederblättriger Jaborandistrauch 5143
Fiederpolster 759
fiederspaltige Kranzspiere 1754
fiederspaltige Takka 2244
Fiederspiere 2180, 5084
Fiederzwenke 3033
Filipendelwurz 1907
Filz 780
filzästige Weide 6053
Filzgrünalge 1316
filzige Klette 1636
filzige Königskerze 1255
filziger Röhrling 6081
filziger Ziest 3731
filzige Segge 1896
filziges Felsenblümchen 6048
filziges Hornkraut 5064
filziges Riedgras 1896
filzige Zwergmispel 774, 4437
Filzklette 1636
Filzkönigskerze 1255
Filzkraut 2318, 3097
Filzrose 5543
Filzscharte 3120
Fingerehrenpreis 2253
Fingergras 5982
Fingerhirse 2015, 2254, 4399
Fingerhundszahn 542
Fingerhut 2346
Fingerkraut 1242
Fingerriedgras 2252
Fingersegge 2252
Fingertang 4806
Finkensame 422
finnische Fichte 2256
Fioringras 617
Firnisbaum 3050
Firnissumach 3050
Fischers Nelke 1165
Fischgras 2184
Fischweide 467
Fisole 497, 3158
Fitzroya 2267
flache Binse 4582
flaches Quellried 2284
flaches Rispengras 970
Flachs 2214, 2286
Flachsdotter 572
Flachsgewächse 2288
Flachslilie 2215
Flachsseide 2287
flachstengeliges Laichkraut 2283
Flacourtiengewächse 2941
Flader 2756
Fladnitzer Felsenblümchen 308
Flakourtiazeen 2941
Flamingoblume 270
Flammenblume 4166
flammendes Herz 668
Flaschenbaum 1751
Flaschenbaumgewächse 1752
Flaschenbovist 5784
Flaschenkürbis 922, 2567
Flaschenstäubling 5784
Flatterbinse 1517
Flattergras 3624
flatterige Waldhirse 201
Flatterrüster 4614, 5174
Flattertrespe 3024
Flatterulme 4614, 5174
Flaumeiche 4324
Flaumhafer 2687
flaumiges Steinrösel 4559
Flaumquecke 5252
Flaumwiesenhafer 2687
Flechtbinse 2600
Flechte 3290
Flechtgras 1870
Flechtstraussgras 4469
Flechtweide 467
Fleckblume 5162
Fleckenaron 2999
Fleckenklee 1425
Fleckschierling 4241
Fleischblume 4398
fleischfarbige Rosskastanie 4441
Fleischkraut 2871
Fleischlauch 3266
Fleischling 582
fleischrote Seidenpflanze 5345
Fleischschwamm 515
Flieder 3295
Fliederbaum 2095
Fliederspeer 892
Fliegenblätterling 2319
Fliegenkraut 3102
Fliegenorchis 2320
Fliegenpilz 2319
Fliegenragwurz 2320
Fliegenschwamm 2319
Flockenblume 1089
Flockenstreuling 2479
Flockenwurz 3120
Flohknöterich 5168
Flohkraut 2291, 2292, 5168
Flohriedgras 2293
Flohsegge 2293
Florentiner Habichtskraut 3161
florentinische Schwertlilie 2301
Florida-Zeder 543
Floriden 4433
Flügelbohne 2946
Flügeleiche 736
Flügelerbse 1824
Flügelfruchtbaum 3999
Flügelhartheu 5199
Flügelkaktus 3247, 5987
Flügelkopf 5994
Flügelnuss 5996
flügelsamiger Spörgel 3490
Flügelspindel 5990
Flügeltang 5995
Flügelzürgel 5988
Flughafer 5951
Flussampfer 5798
Flusszeder 2925
flutender Schwaden 5820
flutendes Froschkraut 2299
Föhre 4190
Fontanesie 2323
Forrests Birke 2333
Forsythie 2337
Fortunes Kopfeibe 1196
Fortunes Mahonie 1192
Fortunes Tanne 2338
Frangipani 2360
Frankeniazeen 2362
Fransenenzian 2684
Fransenschwertel 5781
Franzkraut 1348
Franzosengras 3905
Franzosenkraut 3326, 4386
französischer Ahorn 3669
französischer Lein 2367
französische Rose 2371
französischer Spinat 2429
französischer Sumach 3782
französisches Leimkraut 2372
französisches Raigras 3905, 5427
französische Tamariske 2374
Franzweizen 832
Frasers Tanne 2364
Frauenbiss 2454
Frauendistel 669
Frauendorn 1875
Frauenflachs 887
Frauenhaar 5099
Frauenhaar(farn) 3430
Frauenherz 1363
Frauenholz 2099
Frauenkilte 1794
Frauenmantel 3202
Frauenschuh 3203
Frauenspiegel 5719
Frauentäubling 686
Frauenviole 1794
Freesie 2365
Freisamer 2671
Freizell-Spaltalgen 1228
fremder Ehrenpreis 4360
Fremonts Pappel 2366
freudiggrüner Ahorn 1327
Fritillarie 2380
Froschbiss 2381
Froschbissgewächse 2383
Froschkraut 673
Froschlaichalge 479
Froschlöffel 5829
Froschlöffelgewächse 95
Froschpfeffer 673
fruchtbarer Gewürzstrauch 4011
fruchtbares Rispengras 2345
Fruchtbaum 3066
frühe Haferschmiele 1963
Früheiche 2041
früher Gewürzstrauch 6009
früher Schmielenhafer 1963
früher Schüppling 1964
frühes Riedgras 5725
Frühlingsadonisröschen 5179
Frühlingsbarbenkraut 1966
Frühlingsberberitze 5722
Frühlingsbraunwurz 5184
Frühlingsehrenpreis 5726
Frühlingsenzian 5185
Frühlingsfingerkraut 5182
Frühlingsgedenkemein 1672
Frühlingsheide 5186

Frühlingshungerblümchen 5183
Frühlingsknotenblume 5188
Frühlingsküchenschelle 5721
Frühlingskuschelle 5721
Frühlingsplatterbse 607
Frühlingsriedgras 5725
Frühlingssafran 1395
Frühlingsschlüsselblume 1652
Frühlingsteufelsauge 5179
Frühlingswasserstern 5802
Frühlingswulstling 5180
Frühlingszaubernuss 5727
Frühlorchel 1994
Frühmorchel 2767
Frühsegge 5725
Fuchsbeere 2080
Fuchsbohne 5483
fuchsgelbes Fennichgras 6090
Fuchsie 2389
Fuchsklee 2351
Fuchskreuzkraut 2388
Fuchsriedgras 2348
fuchsrotes Quellried 4443
Fuchsschwanzamarant 3387
Fuchsschwanzgewächse 175
Fuchsschwanz(gras) 174, 2349
Fuchssegge 2348
Fuchstraube 2347, 2786
Fukus 4538
Fumariazeen 2393
fünfblättrige Jungfernrebe 5748
fünfblättriger Sumach 2268
Fünfling 5216
fünfmänniger Spark 5997
Funkie 4221
Fussblatt 1845, 3539
Futtererbse 2231
Futterrübe 2322, 3471
Futterwassermelone 1245
Futterwicke 1554

gabelästiges Leimkraut 2328
Gabelfarn 2501
gabeliges Leimkraut 2328
Gabelzahn 1709
Gabelzahnmoos 2331
Gagel 483
Gagelgewächse 484
Gagelstrauch 483, 5366
Gagelsträucher 484
Gaillardie 2400
Galanga 2609
Galgant-Alpinie 2609
Galipee 2403
Gallenpilz 598
Gallenröhrling 598
Gallerttang 373
gallische Tamariske 2374
Gamagras 2404
Gamander 2453
Gamanderehrenpreis 2454
Gambohanf 3146
Gänseblümchen 1787, 2037
Gänseblume 1787
Gänsedistel 5105
Gänsefingerkraut 4969
Gänsefuss 2561
Gänsefussgewächse 2562
Gänsegarbe 4516
Gänsekohl 4516, 4969
Gänsekraut 4516
Gänsekresse 4516
Gänsekressling 3733
Gänsemiere 1146
Gänsepappel 3447
Gänserich 4969
Gänsestrenzel 596
ganzblättrige Pantoffelblume 878
Garcinia 2405
Gardenie 2416
Gartenampfer 4072
Gartenanemone 4273
Gartenapfelbaum 1349
Gartenaster 1158
Gartenbalsamine 2407
Gartenbingelkraut 2785
Gartenbohne 3158
Gartenboretsch 1366
Gartendahlie 2413
Gartendill 1399
Gartenedelpilz 2419
Gartenengelwurz 2406
Gartenerbse 2423
Gartenfuchsschwanz 3387
Gartengänsedistel 1531
Gartengänsekresse 5774
Gartenglockenblume 999
Gartengurke 1724
Gartenhirnkraut 5354
Gartenhyazinthe 1445
Gartenkerbel 4658
Gartenkohl 5940
Gartenkoriander 1598
Gartenkresse 2412
Gartenkürbis 4330
Gartenlattich 2417, 3281
Gartenlavendel 5604
Gartenleimkraut 5394
Gartenlevkoje 1538
Gartenliebstöckel 2418
Gartenlöwenmaul 1528
Gartenmelde 2421
Gartenmelisse 1356
Gartenmohn 3933
Gartenmöhre 2410
Gartenmontbretie 5599
Gartennandine 4634
Gartennelke 1019
Gartenpastinak 2422
Gartenpetersilie 1417
Gartenpetunie 1495
Gartenpfefferkraut 5321
Gartenpfingstrose 1490
Gartenpimpinelle 5018
Gartenplatterbse 4133
Gartenquendel 5321
Gartenrapunzel 1408
Gartenrauke 4525
Gartenraute 1516
Gartenresede 1468
Gartenrettich 2425
Gartenrhabarber 2426
Gartenringelblume 4285
Gartenrittersporn 2305, 4524
Gartenrosmarin 4564
Gartensalat 2417, 3281
Gartensalbei 2427
Gartenschierling 2325
Gartenschwarzwurzel 649
Gartensiegwurz 1420
Gartenspargel 1351
Gartenspinat 4304
Gartenspringkraut 2407
Gartenstrohblume 5279
Gartenthymian 1546
Gartenwicke 5378
Gartenwolfsmilch 4159
Gartenzinnie 1568
Gartenzwiebel 2420
Gartenzypresse 1776
Gaspeldorn 1426
Gasterie 2436
Gasteromyzeten 2437
Gauchblume 600
Gauchheil 4185
gauchheilblättriges Weidenröschen 156
Gauchheilweidenröschen 156
Gauchraden 4398
Gaugau 2966
Gauklerblume 3655
Gaylussacie 2884
Gazanie 2440
gebauter Hanf 2773
gebauter Koriander 1598
gebauter Lein 2214
gebauter Pastinak 2422
Gebirgshexenkraut 124
Gebirgsklee 4347
Gebirgsrose 145
Gebirgssteinkraut 2551
gebräuchliche Ochsenzunge 1371
gebräuchlicher Baldrian 1552
gebräuchlicher Steinklee 6108
gebräuchlicher Storaxbaum 1915
gebräuchliches Eisenkraut 2129
gebräuchliches Lungenkraut 1462
gebuckelter Trichterling 2395
gebuckelte Tramete 2485
Gedenkenmein 3815
Geduldampfer 4072
gefelderter Grüntäubling 2629
gefiederte Klappernuss 2083
gefiederter Rautenfarn 794
gefiederte Zwenke 3033
gefingerte Segge 2252
gefleckte Flockenblume 5165
gefleckte Gauklerblume 1471, 2541
gefleckter Aronstab 3378
gefleckter Schierling 4241
geflecktes Knabenkraut 5163
gefleckte Taubnessel 5166
gefleckte Wolfsmilch 5167
geflügelter Spindelbaum 5990
geflügelte Yamswurzel 5993
gefranster Enzian 2684
gegenblättriger Steinbrech 5663
gegenblättriges Milzkraut 3935
gegenständiges Milzkraut 3935
gehäuftblättrige Spitzenblume 1355
gehörnter Sauerklee 1673
Geigenholz 2218
Geigenholzbaum 2218
Geissbart 2519

Geissbaum 2073
Geissblatt 2841
Geissblattgewächse 2842
Geisselgras 4431
Geisselkaktus 4488
Geisselsegge 4431
Geissfuss 596
Geissklee 809, 1425
Geissraute 2521
gekielter Feldsalat 3143
gekielter Lauch 3144
gekieltes Rapünzchen 3143
geknäuelte Glockenblume 1797
geknäuelter Klee 1292
geknäueltes Hornkraut 3732
geknieter Fuchsschwanz 5805
gekinetes Fuchsschwanzgras 5805
gekörnte Hirschtrüffel 3399
gekrümmte Hainsimse 1746
gekrümmter Dünnschwanz 4924
gelappte Kranzspiere 5437
gelappter Schildfarn 3349
Gelbbirke 6076
gelbbräunlicher Trichterling 2395
Gelbdolde 87
gelbe Alpenrose 4264
gelbe Birke 6076
gelbe Färberhülse 6113
gelbe Goldnessel 5855
gelbe Hauhechel 5048
gelbe Junkerlilie 1449
gelbe Lupine 2135
gelbe Luzerne 4921
gelbe Märzblume 1397
gelbe Narzisse 1397
gelbe Nixblume 2090
gelbe Osterblume 6115
gelbe Platterbse 6104
gelber Ackerklee 3225
gelber Blasenstrauch 1362
gelber Bockshornklee 2203
gelber Enzian 6091
gelbe Reseda 6100
gelber Fingerhut 5280
gelber Fingerstrauch 879
gelber Günsel 2650
gelber Hanf 471
gelber Hartriegel 1611
gelber Honigklee 588
gelber Hornklee 588
gelber Hornmohn 6096
gelber Hornstrauch 1611
gelber Ingwer 1551
gelber Jakobsstab 1397
gelber Klee 3159
gelber Lein 2538
gelber Lerchensporn 6080
gelbe Rose 374
gelber Portulak 1509
gelber Seidelbast 5192
gelber Stachelpilz 2752
gelber Steinklee 6108
gelbe Rübe 2410, 5942
gelber Weissdorn 6093
gelber Wundklee 3159
gelber Ziegenbart 2531, 4013
gelbes Benediktenkraut 671
gelbe Schönmalve 1217
gelbe Schwertlilie 6086
gelbe Seerose 2090, 6099
gelbe Segge 6106
gelbes Flachskraut 887
gelbe Skabiose 1663
gelbes Liebstöckel 2418
gelbes Löwenmaul 887
gelbes Riedgras 6106
gelbe Sternbergie 2158
gelbes Veilchen 2136
gelbes Waldstroh 6075
gelbes Windröschen 6115
gelbe Taglilie 2334, 3269
gelbe Taubnessel 5855
gelbe Teichrose 2090
gelbe Tripmadam 3089
gelbe Viole 1556
gelbe Vogelwicke 2587
gelbe Wicke 6111
gelbe Wiesenraute 6098
gelbe Wiesenwicke 2587
gelbe Zeitlose 1652
gelbfilziger Spierstrauch 2703
gelbhaariger Fennich 6090
Gelbharzbaum 2591
Gelbholz 1527, 4295, 6114
Gelbholzsumach 1527
Gelbhorn 6095
Gelbkiefer 4259
Gelbklee 638
Gelbkraut 5862
gelblicher Bockshornklee 2203
gelblichweisser Klee 5311
gelblichweisses Ruhrkraut 3092
Gelbling 4889, 4890
Gelblupine 2135
Gelbrübe 5942
Gelbstern 2399
Gelbveigelein 1556
Gelbviole 1556
Gelbweiderich 3370
Gelbwurz 5648, 6105
Gelbwurzel 1551, 2546
Gelenkblume 3321
Gelenkwurz 1916
Gelidium 55
gemeine Akelei 2087
gemeine Bärentraube 500
gemeine Berberitze 2075
gemeine Birne 1487
gemeine Blasenkirsche 5275
gemeine Brunnenkresse 5796
gemeine Buche 2077
gemeine Dattelpalme 1808
gemeine Erdbeere 2127
gemeine Erdnuss 4088
gemeine Erle 2072
gemeine Feldbohne 5028
gemeine Felsenbirne 2428
gemeine Fichte 3886
gemeine Flockenblume 817
gemeine Gallertalge 5225
gemeine Gänsedistel 1531
gemeine Glockenheide 1701
gemeine Goldrute 2101
gemeine Graukresse 2818
gemeine Haferschmiele 4959
gemeine Hainbinse 2239
gemeine Hainbuche 2108
gemeine Hasel(nuss) 2098, 2246
gemeine Haselwurz 2133
gemeine Hopfenbuche 2107
gemeine Hühnerhirse 456
gemeine Hundszunge 1444
gemeine Hyazinthe 1445
gemeine Jujube 1452
gemeine Jungfernrebe 5484
gemeine Kichererbse 1144
gemeine Kiefer 4775
gemeine Kokospalme 1314
gemeine Kratzbeere 2093
gemeine Kreuzblume 1470
gemeine Kuhblume 1398
gemeine Lärche 2110
gemeine Lederblume 1441
gemeine Mahonie 3947
gemeine Melde 2186
gemeine Mispel 3578
gemeine Mohrrübe 5942
gemeine Morchel 1473
gemeine Myrte 5606
gemeine Nachtviole 1794
gemeine Ochsenzunge 1371
gemeine Osterluzei 592
gemeine Passionsblume 697
gemeine Pestwurz 4340
gemeine Quecke 4379
gemeine Rainwurz 1508
gemeiner Alpenstrauch 128
gemeiner Andorn 1442
gemeiner Aronstab 3378
gemeiner Augentrost 1910
gemeine Rauke 4525
gemeiner Baldrian 1552
gemeiner Bambus 1357
gemeiner Bärenklau 2825
gemeiner Bastardindigo 2948
gemeiner Beifuss 5396
gemeiner Beinwell 1389
gemeiner Blasenstrauch 1362
gemeiner Bocksdorn 3535
gemeiner Bohnenbaum 2529
gemeiner Boretsch 1366
(gemeiner) Brotbaum 769
(gemeiner) Brotfruchtbaum 769
gemeiner Daun 781
gemeiner Dill 1399
gemeiner Dosten 1484
gemeiner Drachenbaum 1900
gemeiner Efeu 2040
gemeiner Ehrenpreis 1917
gemeiner Erdrauch 1911
gemeiner Feigenbaum 1412
gemeiner Feldsalat 3211
gemeiner Fenchel 1411
gemeiner Feuerdorn 4743
gemeiner Flieder 1461
gemeiner Frauenmantel 1454
gemeiner Froschlöffel 211
gemeiner Gamander 1113
gemeiner Gilbweiderich 2540
gemeiner Goldregen 2529
gemeiner grosser Erbsenstrauch 4910
gemeiner Hafer 1481
gemeiner Hanf 2773

gemeiner Hartbovist 1405
gemeiner Hartriegel 683
gemeiner Hohlzahn 781
gemeiner Hopfen 1440
gemeiner Hornbaum 2108
gemeiner Hornklee 588
gemeiner Huflattich 1388
gemeiner Judasbaum 3114
gemeiner Kalmus 1918
gemeiner Kammgras 1681
gemeiner Kanarienbaum 3085
gemeiner Kapernstrauch 1377
gemeiner Kellerhals 2194
gemeiner Kerbel 856
gemeiner Klettenkerbel 5688
gemeiner Krähenfuss 1678
gemeiner Kreuzdorn 1369
gemeiner Kugelamarant 1421
gemeiner Kümmel 1013
gemeiner Leberpilz 515
gemeiner Liguster 1508
gemeiner Löwenzahn 1398
gemeiner Manglebaum 200
gemeiner Mauerpfeffer 2553
gemeiner Natternkopf 1555
gemeine Robinie 636
gemeine Rosskastanie 187, 1443
gemeiner Pastinak 2422
gemeiner Perückenstrauch 1527
gemeiner Portulak 1509
gemeiner Quittenbaum 1512
gemeiner Rainkohl 1479
gemeiner Reiherschnabel 91
gemeiner Reis 4496
gemeiner Rhabarber 5844
gemeiner Rosmarin 4564
gemeiner roter Kopfklee 4447
gemeiner Sadebaum 4719
gemeiner Sandbüchsenbaum 4681
(gemeiner) Sandhafer 2076
(gemeiner) Sanikel 2121
gemeiner Schneeball 2091
gemeiner Seidelbast 2194
gemeiner Spinat 4304
gemeiner Spindelstrauch 2096
gemeiner Spitzkiel 4213
gemeiner Stachelbeerstrauch 2103
gemeiner Stechapfel 3102
gemeiner Stechdorn 1227
gemeiner Tabak 1548
gemeiner Taxbaum 2045
gemeiner Teufelsabbiss 3561
gemeiner Trompetenbaum 5097
gemeiner Tulpenbaum 5636
gemeine Rübe 1360
gemeiner Wacholder 1453
gemeiner Walnussbaum 4151
gemeiner Wasserdost 2776
gemeiner Weissdorn 2038
gemeiner Weizen 5874
gemeiner Wolfstrapp 2085
gemeiner Wunderbaum 1051
gemeiner Ysop 2910
gemeines Angergras 3150
gemeines Bambusrohr 1357
gemeines Barbarakraut 610
gemeines Bilsenkraut 633
gemeine Schafgarbe 1565
gemeine Schmiele 5628
gemeine Schneebeere 1529
gemeine Sellerie 5943
gemeines Fettkraut 3700
gemeines Greiskraut 1430
gemeines Heidekraut 4772
gemeines Hexenkraut 4055
gemeines Hornblatt 2861
gemeines Hornkraut 2221
gemeine Sichelmöhre 1409
gemeine Siegwurz 1420
gemeines Immergrün 1493
gemeines Katzenpfötchen 1510
gemeines Knäuelgras 3943
gemeines Knaulgras 3943
gemeines Kreuzkraut 1430
gemeines Labkraut 5882
gemeines Lauchkraut 2433
gemeines Leinkraut 887
gemeines Meerträubchen 3106
gemeines Moschuskraut 3766
gemeine Sommerwurz 1282
gemeine Sonnenblume 1539
gemeine Sophienrauke 2296
gemeines Pfeilkraut 3924
gemeines Rispengras 4579
gemeines Ruchgras 5389
gemeines Rutenkraut 1418
gemeines Schilfrohr 1514
gemeines Schneeglöckchen 1530
gemeines Seifenkraut 743
gemeines Sonnenröschen 1515
gemeines Steckenkraut 1418
gemeines Straussgras 1335
gemeines Sumpfried 719, 1533
gemeines Süssholz 1459
gemeine Stechpalme 2039
gemeines Teichrohr 1514
gemeine Sumpfkresse 721
gemeines Zittergras 4134
gemeines Zymbelkraut 3147
gemeine Tamariske 2374
gemeine Tollkirsche 525
gemeine Traubenkirsche 2078
gemeine Waldrebe 5571
gemeine Wasseraloe 5837
gemeine Wegwarte 1381
gemeine Wicke 1554
gemeine Wiesenkresse 1723
gemeine Wucherblume 3987
gemeine Zwiebel 2420
Gemsblume 3694
Gemsenhörner 6110
Gemsheide 114
Gemshorn 1846
Gemskresse 2895
Gemswurz 3276
Gemüseampfer 4072
Gemüseartischocke 331, 1015
Gemüsekohl 5940
Gemüselauch 4284
Gemüsepappel 4283
Gemüseplatterbse 2587
Gemüsespargel 1351
Gemüsespinat 4304
Genipabaum 2444
Genipbaum 2443, 2444
Genisarobaum 4404
Genter Gladiole 771
Genter Siegwurz 771
Gentianazeen 2446
Georgine 1782
Geraniazeen 2448
Geranie 2447
Gerards Kiefer 1156
Gerards Meerträubchen 2449
Gerbera 2451
Gerbermyrte 1599
Gerberstrauch 1599, 4920
Gerberstrauchgewächse 1600
Gerbersumach 4920
Germer 2170
Gersch 596
Gerste 454
Gerstendinkel 2029
Gerstentrespe 1138
Gerstenweizen 1930
geruchlose Kamille 4751
geruchlose Primel 3988
gesägtblättrige Kirsche 3951
geschlängelige Schmiele 5843
geschlängelte Schmiele 5843
Geschwulstkraut 2612
Gesneriazeen 2465
Gesnerie 2464
Gesneriengewächse 2465
gestirnte Pflaume 1012
gestreckter Ehrenpreis 2709
gestreckter Felsenstrauch 114
gestreifter Klee 3169
gestutzte Weide 3891
getäfelter Stäubling 1328
Getreidekraut 832, 4390
getüpfeltes Hartheu 1537
Gevatterblume 3987
Geweihbaum 1319
Geweihfarn 1534, 5205
Geweihstuppe 1249
Gewitterblume 5242
gewöhnliche Agave 1092
gewöhnliche Eseldistel 4770
gewöhnliche Fuchsie 3422
gewöhnliche Kratzdistel 850
gewöhnliche Kreuzblume 1470
gewöhnliche Ochsenzunge 1371
gewöhnliche Pestwurz 4340
gewöhnliche Primel 2042
gewöhnlicher Beinwell 1389
gewöhnlicher Hohlzahn 781
gewöhnlicher Igelsame 2126
gewöhnlicher Odermennig 1348
gewöhnlicher Schotenklee 588
gewöhnliches Ferkelkraut 5164
gewöhnliches Leinkraut 887
gewöhnliches Lungenkraut 1462
gewöhnliches Rispengras 4579
gewöhnliche Waldrebe 5571
Gewürzampfer 2373
Gewürzkerbel 4658
Gewürzkoriander 1598
Gewürzlilie 4485
Gewürzmüllen 1120
Gewürznelkenbaum 1283
Gewürzrinde 4823
Gewürzrindenbaum 1903
Gewürzstrauch 5386

Gewürzstrauchgewächse 5278
gezähnelter Schneckenklee 933
gezähnter Honigklee 1834
gezähnter Steinklee 1834
Gichtbeere 2081
Gichtkraut 1913, 2787
Gichtmorchel 5256
Gichtrose 1490
Gichtrübe 824
Gichtschwamm 5256
Giersch 596, 2570
Giesskannenschimmel 351
Giftbaum 5684
Giftbeere 284
Gifteiche 1501
Giftesche 5896
Gifthahnenfuss 673
Giftheil 4373
giftiger Efeu 1501
giftiger Merk 2130
giftiger Wasserschierling 2130
Giftknippe 2655, 4074
Giftlattich 603
Giftlorchel 1994
Giftranunkel 673
Giftsalat 603
Giftsumach 4242
Giftwasserschierling 2130
Giftwicke 1713
Giftwurzel 5916
Gilbkraut 1563, 2606
Gilbweiderich 3370
Gilgen 2458, 2494
Gilgenkraut 4285
Gilgenwurzel 6086
Gilia 2487
Gilke(nkraut) 4285
Gillenie 2172
Gingellikraut 2063, 3849
Ginkgoazeen 3432
Ginkgo(baum) 2492
Ginkgogewächse 3432
Ginseng 195, 346, 2494
Ginster 6015
ginsterblättriges Leinkraut 811
Gipskraut 2672
Gladiole 2499
Glanzbaum 61
glänzender Ehrenpreis 5854
glänzende Wiesenraute 3806
Glanzgras 984
Glanzkölbchen 275
Glanzkraut 3356, 5658
Glanzmispel 4169
Glanzrauke 3363
Glanzstrauch 4498
Glaskirsche 5091
Glaskraut 4105
Glasschmalz 2500
Glasweizen 1930
Glattblattaster 3842
glattblättriger Sumach 5042
glattblättrige Rüster 5040
glatte Endivie 4545
glattes Bruchkraut 1372
glattes Tausendkorn 1372
Glatthafer 3905
Gleditschie 2840
Glehns Fichte 4657
Gleicheniazeen 2502
Gleichsaum 2969
Gleichzipfel 4887
Gleisse 36
Gletscherfelsenblümchen 2498
Gletscherhahnenfuss 2497
Gliederkaktus 3247
Gliederkraut 2967
Gliedkraut 1565
Globulariazeen 2505
Glöckel 506
Glockenblume 528
Glockenblumengewächse 529
glockenförmiger Amorphophallus 5912
Glockenheide 340, 2742
Glockenrebe 1301, 4337
Glockenwinde 3871
Glotzblume 1422
Gloxinie 2515
Glyzine 6011
Gmelins Strandflieder 4914
Gnadenkraut 2754
Gnemonbaum 5149
Gnetazeen 3107
Goabohne 2946
Goldampfer 2536
Goldapfel 1549
Goldaster 3687
Goldaugenschmetterlingsstrauch 3941
Goldbaldrian 4073
Goldball 1755
Goldbandlilie 2525
Goldbartgras 1688, 4750
Goldblatt 367, 3019, 5211
Goldblume 1229, 4285
Golddistel 3993
Goldesche 2526
Goldfaden 2556
Goldfelberich 2540
Goldfingerkraut 2530
Goldfliederkamille 4376
Goldflöckchen 2337
Goldfruchtpalme 896
goldgelbe Aphelandre 3936
goldgelbe Johannisbeere 2535
goldgelbe Koralle 2531, 4013
goldgelber Lärchenröhrling 2011
Goldhaar(aster) 2102
Goldhafer 6109
Goldhafer(gras) 2176
Goldhahnenfuss 2552
Goldholunder 2537
Goldholzsumach 1527
Goldjohannisbeere 2535
Goldkeule 2533
Goldkiefer 4259
Goldklee 2848
Goldknöpfchen 5425
Goldkolben 2542
Goldköpfchen 3326
Goldkörbchen 2551
Goldkraut 1430
Goldlack 1556, 5768
Goldlärche 2539
Goldlauch 3303
Goldmilz 168, 2545
Goldmohn 2554
Goldnarbe 3441
Goldorange 3019
Goldpflaume 178
Goldranunkel 2507
Goldregen 3192
Goldröhrling 2011
Goldröschen 3049, 3156
Goldrute 2543
Goldrutenbaum 2544
goldschuppige Eiche 1002
Goldsteinbrech 168
Goldtanne 4451
Goldtraube 2535
Goldwaldrebe 2532
Goldwurz 2606
Goldwurzel 352
Gommer 4247
Gordonie 2564
Gordons Pfeifenstrauch 2565
Gordonstrauch 2564
Götterbaum 65
Götterblume 4871
Götterduft 1861
Götterpflaume 4152
Gottesauge 4143
Gottesgnadenkraut 2754
Gottesurteilbohne 921, 1816
Gottvergess 425
Götzenbaum 738
Götzenholz 4716
Gowens Zypresse 2571
Gräberweide 400
Grabzypresse 3923
Gramineen 2581
Granatapfel(baum) 4255
granatapfelblättrige Malpighie 446
Granatenbaum 6060
Gränke 241, 723
Grannenhafer 2176
Grannenhirse 4499, 5032
grannenlose Trespe 5034
Grapefruit 2578
Grasartige 2581
grasartiger Kalmus 3073
grasartiges Laichkraut 5702
Grasbaum 2591
grasblättrige Eiche 939
grasblättrige Goldrute 2586
grasblättriger Froschlöffel 2593
grasblättrige Schwertlilie 2582
grasblättriges Sandkraut 2584
Grasblume 1019, 5502
Gräser 2581
Graslauch 1219, 2474
Graslilie 269
Grasmiere 3338
Grasnelke 1019, 5502
Grasnelkengewächse 3246
Grasplatterbse 2592
Grasschwertel 698
Grasschwertlilie 2582
Grasstern 5882
Grassternmiere 3338
Grasweizen 1930
Graubirke 4037
graue Heide 5668

grauer Ahorn 4036
grauer Faltentintling 2950
Grauerle 5122
grauer Ritterling 1859
grauer Schötendorn 2817
grauer Schöterich 2817
grauer Walnussbaum 897
graue Segge 4970
graues Hartgras 2702
graues Riedgras 4970
graues Silbergras 2595
graues Sonnenröschen 2821
graue Weide 2007, 2598
graugrüne Borstenhirse 6090
graugrüne Quecke 2952
graugrünes Schillergras 702
graugrünes Weidenröschen 3214
Graukappe 1274
Graukopf 1274
Graukresse 2160
Graumelde 4663
Graunuss 897
Graupappel 2597
Grauweide 2598
Green 2871
Greiskraut 2651
Grensel 2646
Grensing 2646
Gretchen im Busch 3386
Gretchen im Grünen 3386
Gretel im Busch 3386
Grevillie 2638
Grewie 2639
Gricken 832
griechische Baumschlinge 2616
griechische Micromerie 2619
griechischer Baldrian 2620
griechisches Heu 2203
griechische Tanne 2617
griechische Zeder 4298
Griffelblume 5289
Grindelie 2663
Grindkraut 1430
grossblättrige Deutzie 1961
grossblättrige Linde 561
grossblättrige Pfingstrose 1590
grossblättriger Ahorn 562
grossblättrige Salbei 560
grossblättrige Weide 564
grossblättrige Wucherblume 5441
grossblumige Cinerarie 1385
grossblumige Pelargonie 575
grossblumige Pfingstrose 1490
grossblumiges Wollkraut 6055
grossblütige Brunelle 559
grossblütige Königskerze 6055
grossblütige Magnolie 5098
grossblütige Pfeifenblume 4104
grossblütiger Enzian 1291
grossblütiger Fingerhut 6089
grossblütiger Pfeifenstrauch 571
grossblütige Wicke 4885
grosse Ammei 2603
grosse Bibernelle 2608
grosse Braunelle 559
grosse Brennessel 574
grosse Brunelle 559
grosse Eberwurz 5035
grosse Fetthenne 3341
grosse Gänseblume 3987
grosse Hasel(nuss) 2473
grosse Hundsrose 2234
grosse Kapuzinerkresse 1478
grosse Klette 2601
grosse Knorpelmöhre 2603
grosse Kokardenblume 1491
grosse Königskerze 6055
grosse Maiblume 1916
grosse Massliebe 3987
grosse Moosbeere 3224
grosse Nessel 574
grosser Baldrian 1552
grosser Ehrenpreis 2892
grosser Enzian 6091
grosser Eppich 2418
grosser Galgant 2609
grosser Gelbfuss 2516
grosser Hahnenfuss 5544
grosser Heinrich 2010
grosser Knorpellattich 4610
grosser Krallenklee 1526
grosser Krümmling 4610
grosser Odermennig 2355
grosser Rosenapfel 3912
grosser Sauerampfer 2429
grosser Schachtelhalm 2610
grosser Schirmling 4051
grosser Schirmpilz 4051
grosser Schmierling 2516
grosser Tiekbaum 1544
grosser Tintenpilz 4843
grosser Vogelfuss 1526
grosser Wegerich 4502
grosser Wiesenknopf 2409
grosses Alpenglöckchen 2602
grosse Schlüsselblume 3988
grosses Flohkraut 1413
grosses Hexenkraut 4055
grosses Immergrün 563
grosses Liebesgras 5255
grosses Löwenmaul 1528
grosse Sommerwurz 2604
grosses Riedgras 4107
grosses Rindsauge 2740
grosses Schöllkraut 2606
grosses Straussgras 617
grosses Tausendgüldenkraut 1909
grosse Sternblume 2611
grosse Sterndolde 2611
grosse Sternmiere 1971
grosses Vergissmeinnicht 1672
grosses Windröschen 5061
grosses Wintergrün 2119
grosses Zittergras 569
grosses Zweiblatt 1999
grosse Telekie 2740
grosse Torreye 1206
grosse Waldhyazinthe 2605
grosse Wallwurz 1389
grosse Wucherblume 3987
grosse Zirneiche 2128
grossfrüchtige Elfenbeinpalme 1447
grossfrüchtiger Wacholder 4231
grossfrüchtige Steinnusspalme 1447
grossfrüchtige Zypresse 3665
Grossheims Schwertlilie 2642
Grosstanne 2573
grosswurzeliger Storchschnabel 570
Grubenlorchel 3196
grubiger Schleimrübling 4550
Grünalgen 2621
grünblütiges Wintergrün 2633
Grundbirne 3094
Grundeichel 4087
Grundfeste 2725, 3803
Grundflockenbaum 4734
Grundheil 1917
Grundkraut 4735
grüne Borstenhirse 2624
grüne Christwurz 2628
grüne Hohlzunge 2384
Grüneiche 2832
grüne Minze 5120
grünender Pfeffermilchling 4123
grüne Nieswurz 2628
grüner Anistrichterling 5385
grüner Fennich 2624
grüner Germer 193
grüner Giftwulstling 1820
grüner Knäuel 262
grüner Knollenblätterpilz 1820
grüner Kolbenschimmel 2631
Grünerle 2105
grüner Pinselschimmel 2632
grüner Pippau 5038
grüner Streifenfarn 2634
Grünesche 2622
grüne Sitter 792
grünes Meerträubchen 2625
grünfester Pippau 3803
grünliche Kuckucksblume 2605
grünlicher Täubling 2629
grünliches Leimkraut 6092
grünliches Wintergrün 2633
grünliche Waldhyazinthe 2605
Grünling 2051
grünschuppiger Täubling 2629
Grünwattle 2637
grusinische Mandel 4611
Grützschwaden 5820
Guajabenbaum 2656
Guajakbaum 1460, 3293
Guajara 2914
guatemalische Palmlilie 846
Guava 1431
Guavenbaum 2656
Guayabe 1431
Guayave 1431, 2656
Guayule 2657, 4065
Guernseylilie 2658
Guettarde 5714
Guinea-Bogenhanf 5382
Guinea-Gras 2661
guineische Palme 46
Gummiakazie 2662
Gummiapfel 508
Gummiarabikum-Baum 2662
Gummibaum 2944
Gummitraganth 5568
Gummiträger 2669
Gummiwurz(el) 3934
Gunderkraut 2647
Gundermann 2647
Gunderrebe 2647

Gunnere 2667
Günsel 838
Gurgemei 1551
Gurke 1724
Gurkenbaum 576, 1726
Gurkendill 1857
Gurkengewächse 2569
Gurkenkraut 1366, 1399, 1857
Gurkenmagnolie 1726
Gürtelkraut 4606
guter Heinrich 2559
Guttaperchabaum 3443, 3814
Guttiferen 2669
Guzmannie 2670
Gymnospermen 2671
Gynanderpflanze 5129

Haarananas 5525
Haarbirke 4323
Haarblume 1996, 5045
Haarfarn 2247
haarfrüchtige Balsampappel 942
haarfrüchtige Pappel 942
Haargerste 5957
Haargurke 858
Haarhainsimse 2696
Haarhirse 1561
haarige Lupine 2084
haarige Platterbse 4578
haariges Bruchkraut 2682
haariges Schamkraut 4111
haariges Tausendkorn 2682
Haarkraut 101
Haarmoos 2676
Haarnixe 2184
Haarschotengras 2572
haarstengelige Segge 2678
Haarstrang 1907, 2824
Habenaria 2673
Habichtskraut 2726
habichtskrautähnliches Bitterkraut 2727
Habichtskrautbitterich 2727
Habichtskrautbitterkraut 2727
Habichtspilz 2922
Hackenkopf 952
Hacquetie 2675
Hafer 3904
Haferdistel 975
Haferkirsche 3544
Haferpflaume 847
Haferschlehe 847
Haferschmiele 2677, 5843
Haferwurzel 5706
Haftdolde 2166
Hagebuche 2108
Hagebutte 1875
Hagedorn 1875
Hagerose 287
Hahnendorn 1313
Hahnenfuss 889
Hahnenfussgewächse 890
Hahnenfusspflanzen 890
Hahnenkamm 1307, 1410, 2192, 2368, 2531, 4013
hahnenkammförmiger Schildfarn 1685
Hahnenklötenwurzel 1353
Hahnenkopf 5390
Hahnensporn 5190
Hainampfer 684
Hainbirke 5357
Hainblume 3826
Hainbuche 2852
Hainbuchenahorn 2853
Hainfelberich 6038
Hainkresse 2450
Hainmarbel 2652
Hainmiere 6045
Hainrispengras 6023
Hainsalbei 5743
Hainsimse 6041
Hainveilchen 4507
Hainwachtelweizen 696
Hainwicke 2462
Hakenlilie 1696
hakige Spitzklette 486
Halbgräser 4811
Halbrohr 994
Halde 4266
Halfa 2061
Hallimasch 2838
Halsbandschwindling 1329
Halskraut 1525, 5503
Hamamelidazeen 6013
Hamamelisgewächse 6013
Hamelie 2699
Hammelmöhre 4064
Hammerstrauch 3098
Handaralie 3034
Händelwurz 4481
handförmiger Riementang 4806
handlappiger Rhabarber 5089
Handpilz 2520
Hanf 2772
hanfartige Hundswolle 2775
Hanfgewächse 2777
Hanfhundsgift 2775
Hanfklee 5917
Hanfkraut 887
Hanfnessel 781, 2778
Hanfpalme 2339, 3572, 5983
Hanfpappel 2797
Hanfweide 467
Hanfwürger 2774
Hängebaum 400
Hängebirke 2131
Hängebuche 5859
Hängeflieder 3867
hängende Sandbirke 2131
hängende Segge 4107
hängendes Herz 668
hängendes Leimkraut 1905
hängende Weissbirke 2131
Hängesilberlinde 4967
Hängeweide 400
Harfenstrauch 5190
Harmelkraut 4102
Harmelraute 4102
Harrisons Rose 2715
Hartbovist 2705
Hartboviste 2706
Hartgras 2701
Hartheu 4651
Hartheugewächse 2669, 4652
Hartholz 4809
Hartmais 2295
Hartriegel 683, 1881, 4312
Hartriegelgewächse 1882
Hartwegs Bartfaden 2718
Hartweizen 1930
Harzaffodil 2591
Harzeiche 1925
Harzföhre 4460
Harzkiefer 4460
Harzklee 4782
Harzlabkraut 2743
Haselerle 2732
Haselnuss 2098
Haselwurz 5947
Hasenbovist 1328
Hasenbrot 2239
Hasenklee 1519, 4394
Hasenlattich 4429
Hasenohr 3206, 5493
Hasenpappel 3447
Hasenpfötchen 1510
Hasenpfotenbaum 5870
Hasenriedgras 2712
Hasenröhrling 1140
Hasenschwanzgras 2713
Hasensegge 2712
Hasenstaubpilz 1328
Hauhechel 4484, 5492
Hauseberesche 4833
Hauslaub 2882
Hauswurz 2882
Hauszwiebel 2420
Hautfarn 2247, 2248
Hautfarne 2249
Hautpilze 4180
Haworthie 2728
Hechskraut 1744
hechtgraue Weide 2007
Hechtrose 4454
Heckdorn 2038
Heckenberberitze 2075
Heckenkerbel 856
Heckenkirsche 2841
Heckenknöterich 2755
Heckenrose 1627, 1875
Heckensame 2566
Heckenwicke 2462
Hedeoma 2178
Hederich 5955
Hefepilze 6070
Hegeweide 467
Heide 2742
Heidegewächse 2745
Heidegrütze 832
Heidegünsel 1030, 2442
Heidekorn 832
Heidekraut 2742, 2744
Heidekräuter 2745
Heidekrautgewächse 2745
Heidekuhschelle 5177
Heidelbeere 692
Heidelbeerweide 5928
Heidelberger Schlosskraut 465
Heidenelke 3433
Heidenkorn 832, 1370
Heiderling 2870, 4621
Heideröschen 1875, 4559
Heidewacholder 1453

heidnisches Wundkraut 2101
Heil aller Schäden 4693
Heilandstee 1348
Heildistel 670
Heilglöckchen 506
Heiligenblume 3240
Heiligenkraut 1776, 3240
Heiligenpflanze 3240
heiliger Feigenbaum 738
Heiliggeistwurzel 2406
Heilkraut 2825, 4693
Heilwurz 3286, 3934, 4836
Heilwurzel 3507
Heilziest 1361
Heimsaat 1332
Heinrich 2955
Heister 2108
Heldreichs Ahorn 418
Heldreichs Kiefer 2762
Heldreichs Weissdorn 2761
Helenakraut 2010
Helenenkraut 2010
Helikonie 2763
Heliotrope 2765
Hellerkraut 2232, 4112
helles Bilsenkraut 4586
Helmblume 3946
Helmbohne 2897
Helmknabenkraut 5079
Helmkraut 4987
Helmling 3774
Helonie 5346
Helvellazeen 2768
Helxine 402
Hemlocktanne 2769
Henequen 2781
Henequen-Agave 2781
Henna 2782
Hennastrauch 2782
Herbstadonisröschen 4164
Herbstblatte 1274
Herbstblaustern 383
Herbstdrehähre 1455
Herbstdrehwurz 1455
Herbsterle 2732
Herbstfeuerröschen 4164
Herbstflieder 3941, 5320
Herbstlorchel 1697
Herbstlöwenzahn 2159
Herbstmusseron 2154
Herbstrose 1353
Herbstrübe 5650
Herbstsafran 4638
Herbststernblume 3842
Herbsttrompete 2858
Herbstwasserstern 384
Herbstzeitlose 1353
Herkuleskraut 1649
Herlitze 1611
Herlitzenbaum 1611
Herrenpilz 1093, 1476
Herrnkümmel 2835
Herzblatt 4058
herzblättrige Erle 2971
herzblättrige Hainbuche 2738
herzblättrige Kugelblume 2737
herzblättriger Pfeifenstrauch 2739
herzblättrige Wickelwurz 2736
Herzblume 668
Herzenkelch 2066
herzförmiger Walnussbaum 2281
herzförmige Walnuss 2281
Herzgespann 3689
Herzlilie 4221
Herznussbaum 1042
Herzsame 2741
Herzzweiblatt 3881
Heteranthera 3743
Heuffels Krokus 2790
Heuschreckenbaum 3354
Heusenkraut 4812
Hevea 2791
Hexenkraut 441, 1537, 2031
Hexennuss 6012
Hexenpilz 3396
Hexenröhrling 3396
Hexenschwamm 3396
Hexenzwirn 3535
Hibalebensbaum 2161, 2793
Hibbertia 904
Hickory(baum) 2795
Hickorynussbaum 2795, 4100
Himalaja-Flieder 2803
Himalaja-Weymouthskiefer 2807
Himalaja-Zeder 1836
Himbeere 4420, 4462
Himbeererdbeere 5754
Himbeerstrauch 4420
Himmelfahrtsblümchen 1510
Himmelsauge 1814
Himmelsaugengewächse 5135
Himmelsbaum 5584
Himmelsbrand 3690
Himmelsgerste 5124
Himmelsherold 2056
Himmelsleiter 2620, 4246
Himmelsleitergewächse 4167
Himmelslilie 2458
Himmelsschlüssel(chen) 4307
Himmelstau 5820
Hinoki der Japaner 2812
Hiobsträne 31
Hiobstränengras 31
Hippokastanazeen 2865
Hirnkraut 463
Hirschbeere 2080
Hirschdorn 1369
Hirschheil 3286
Hirschholunder 2120
Hirschklee 2776
Hirschkohl 1462
Hirschkolbensumach 5206
Hirschling 1827, 2922
Hirschmangold 1462
Hirschsprung 1622
Hirschstreuling 2717
Hirschtrüffel 2717
Hirschwurz 3738
Hirschzunge 2716, 2922
Hirse 4026, 4315
Hirsegras 3624
Hirsenklee 638
Hirsepilz 5700
Hirtengras 5527
Hirtentäschel(kraut) 4862
hochwüchsiges Veilchen 5431
Höckerblume 5534
Höckerkelch 1736
Hofraute 1516
Hofrun 1516
hohe Brombeere 2798
hohe Deutzie 2397
hohe Nephrolepis 1541
hohe Primel 3988
Höherahorn 5712
hoher Glatthafer 5427
hoher Götterbaum 5584
hoher Honigklee 5426
hoher Rittersporn 518
hoher Steinklee 5426
hoher Wacholder 2618
hoher Wiesenhafer 5427
hohe Schlüsselblume 3988
hohes Fingerkraut 5555
hohes Habichtskraut 3160
hohes Helmkraut 5429
hohe Sommerwurz 2604
hohe Zypresse 551
Hohldotter 3773
hohle Pferdesaat 5625
Hohlnarbe 1317
Hohlsame 555
Hohlweide 2523
Hohlwurz 1628
Hohlzahn 2778
Holder 2009
Holderbaum 2009
holländische Linde 2111
holländischer Tabak 1548
Holland-Linde 2111
Holler 2009
Hollunder 2009
Hölperchen 1645
Holquahitl 4022
Holunder 2009
Holunderbaldrian 4569
holunderblättriger Baldrian 4569
Holzapfelbaum 1349
Holzbirnbaum 1487
Holzbirne 1487
Holzkirsche 3544
Holzölbaum 5640
Holzschwamm 2880, 4254
Hondurasrinde 599
Honigblume 3167
honiggelber Ringling 2838
Honiggras 5710
Honigklee 5362
Honigkraut 427
Honigpalme 3113
Honigpilz 2838
Honigspringkraut 4596
Honigstrauch 2839
Hooker Berberitze 2845
Hookers Weide 2846
Hopfen 1440, 2847
Hopfenbaum 1441, 2849
Hopfenbuche 2849
Hopfenhainbuche 2107
Hopfenhelmkraut 6072
Hopfenklee 638
Hopfenluzerne 638
Hopfenseide 2607
Hopfenstrauch 1441

Hornbaum 2852
Hornblatt 2860
Hornblattgewächse 2862
Hornbusch 4686
Hornemanns Weidenröschen 2857
Hornfarn 5803
Hornklee 1824
Hornkraut 1145, 2860
Hornkümmel 2330
Hornlabkraut 1607
Hornmelde 6001
Hornmohn 2859
Hornpalmfarn 2854
Hornsauerklee 1673
Hornstrauch 1881
Hornstrauchgewächse 1882
Hornveilchen 2856
Hornwurz 5666
Hortensie 2902
Houstonie 710
Hovenie 4406
Hufeisenklee 2874
Huflattich 1339
Hügelerdbeere 2635
Hügelmädesüss 1907
Hügelmeier 5201
Hügelmeister 5201
Hügelstorchschnabel 5686
Hügelvergissmeinnicht 1962
Hühnerbiss 656
Hühnerdarm 1146
Hühnerfuss 706
Hühnerfussgras 1985
Hühnerhirse 1310
Hühnerkraut 1146
Hühnerpolei 3688
Hühnertod 633, 643
Hülsdorn 2039
Hülse 2039
Hülsenfruchtgewächse 3267
Hülsenfrüchtige 3267
Hülsenfrüchtler 3267
Hülsengewächse 3267
Hulst 2039
Humea 2886
Hundeblume 1398, 1796
hundertblättrige Rose 910
hundertjährige Aloe 1092
Hundsbaum 2096
Hundsbeere 2099
Hundsdorn 1875
Hundsgift 1872
Hundsgiftgewächse 1873
Hundsgras 1640, 3943
Hundshirse 542
Hundskamille 963
Hundskerbel 856
Hundskirsche 2099
Hundskohl 2785, 5481
Hundskohlgewächse 1770
Hundskolben 1771
Hundskraut 633
Hundslauch 2223
Hundslevkoje 3444
Hundsnelke 743
Hundspetersilie 36, 2325
Hundsrauke 4526
Hundsrose 1875, 2234
Hundsrute 1771
Hundsstraussgras 5708
Hundstod 5916
Hundsveilchen 1880
Hundswolle 1872
Hundswürger 5340
Hundswurz 234
Hundszahn 1878
Hundszahngras 542, 1879
Hundszunge 2879
Hungerblümchen 2188
Hungerblume 1897, 2188
Hungermoos 2916
Hunnemannie 2534
Hutmorchel 1473
Hutmorchelpilze 2768
Hutpilze 58
Hyazinthe 2896
Hydrocharideen 2383
Hydrocharitazeen 2383
Hydrophyllazeen 5814
Hymenomyzeteen 4180
Hymenophylazeen 2249
Hyphomyzeteen 2245

iberische Schwertlilie 2913
Igelföhre 1296
Igelgurke 3639
Igelkaktus 1990, 2749
Igelkerzenkaktus 1991
Igelkolben 871
Igelkolbengewächse 872
Igelkopf 871, 1989
Igelkopfgewächse 872
Igelkraftwurz 1847
Igelkraut 1354
Igellock 2860
Igelpolster 4306
Igelsame 5246
Igelsäulenkaktus 1991
Igelschlauch 862
Igelschwamm 1589
Igelskolben 871, 3102
Ignatiusbohne 4650
Iguape-Nussbaum 992
Ikakopflaumenbaum 2914
Ilang-Ilang 6119
Ilizineen 2827
illyrischer Hahnenfuss 2921
Ilme 2018
Immenblatt 3583
immerblühende Begonie 4143
immerblühendes Schiefblatt 4143
Immergrün 70, 2882, 2988, 4140
immergrüne Bärentraube 500
immergrüne Eiche 2832
Immergrüneiche 2832
immergrüner Bauernsenf 2144
immergrüner Buchsbaum 1368
immergrüner Kreuzdorn 2974
immergrüne Rose 2147
immergrüner Schneeball 3237
immergrüne Sequoie 945
immergrünes Geissblatt 5613
immergrüne Zypresse 2978
Immergrüngewächse 1873
Immerschön 2149
Immerschönkraut 4772
Indigopflanze 2947
Indigostrauch 2947, 5603
indische Alpenrose 2933
indische Azalee 2933
indische Eleusine 2563
indische Erdbeere 2932
indische Feige 1452, 2929, 2936
indische Kartoffel 5380
indische Lotusblume 2811
indischer Brotbaum 2998
indischer Entenfuss 2804
indischer Flachs 4591
indischer Flieder 3492
indischer Hanf 2775
indischer Korallenbaum 4676
indischer Lotus 2811
indischer Mangobaum 1464
indische Rose 1200
indischer Sandoribaum 4696
indischer Spinat 5921
indischer Zedrachbaum 1160
(indischer) Zibetbaum 1927
indisches Blumenrohr 2934
indische Seerose 3381
indisches Fussblatt 2804
indische Waldpalme 5302
indische Wucherblume 3687
Inga 2949
Ingber 2489
Ingwer 1419, 2489
Ingwergewächse 2490
Ingwertanne 3243
Inkalilie 166
Inkarnatklee 1694
Inkarvillea 2924
Insektenpulverpflanze 1790
Ipekakuanha 2957
Iresine 677
Iridazeen 2959
Irideen 2959
Iris 2958
irländisches Moos 2961
Isabellaholz 4144
Isabellarebe 2347
isländisches Moos 2916
isländische Sumpfkresse 721
Isoetazeen 4388
Isoeteen 4388
Isop 2909
Isorabaum 4781
italienische Hirse 2352
italienische Kiefer 2052
italienische Maiblume 2071
italienische Meerzwiebel 2982
italienische Mispel 390
italienische Narzisse 3110
italienische Ochsenzunge 2975
italienische Pappel 3360
italienischer Aronstab 2972
italienischer Blaustern 2982
italienischer Hahnenkopf 5310
italienisches Geissblatt 5370
italienisches Leimkraut 2981
italienisches Leinkraut 2984
italienische Spitzklette 486
italienisches Raygras 2980
italienisches Rohr 2482
italienische Steinkiefer 2983

italienische Waldrebe 2976
Iva 3769
Ixlilie 2994
Ixora 2995

Jaborandistrauch 4184
Jaburan-Schlangenbart 5898
Jacaranda 2996
Jacarandaholz 764
Jackbohne 1448
Jackfruchtbaum 2998
Jackmans Waldrebe 3000
Jacksonie 3002
Jaffnamoos 56
Jägergrün 1285
Jägerkraut 951
Jägermütze 1499
Jagua 2443
Jahrhundertpflanze 59
jähriges Angergras 258
Jakaranda 2996
Jakobs(greis)kraut 4401
Jakobskreuzkraut 4401
Jakobsleiter 2620
Jakobslilie 393
Jalapawinde 3005
Jamaika-Pfeffer 102
Jambea 865
Jambolanabaum 3010
Jamesie 3011
Japanhopfen 3043
japanische Aprikose 3015
japanische Ardisie 3018
japanische Artischocke 332
japanische Aukube 3019
japanische Azalee 1170
japanische Banane 3020
japanische Berberitze 3021
japanische Bergangelika 5675
japanische Bergulme 3461
japanische Birke 3076
japanische Buche 3022
japanische Cryptomerie 3029
japanische Douglasfichte 3031
japanische Eberesche 3058
japanische Erle 3012
japanische Faserbanane 3020
japanische Flügelnuss 3077
japanische Gleditschie 3041
japanische Glyzine 3078
japanische Hainbuche 3045
japanische Hirse 3057
japanische Hopfenbuche 3044
japanische Johannisbeere 3030
japanische Kamelie 1376
japanische Kartoffel 332
japanische Kastanie 3027
japanische Kirsche 1174
japanische Koniogramme 438
japanische Kryptomerie 3029
japanische Lärche 3051
japanische Lilie 3053
japanische Mahonie 3055
japanische Mispel 3376
japanische Mohrenhirse 3133
japanische Primel 3064
japanischer Ahorn 2391
japanischer Buchsbaum 3054
japanischer Christusdorn 3041
japanischer Flieder 3075
japanischer Hopfen 3043
japanischer Judasbaum 3048
japanischer Kampferbaum 966
japanischer Klee 1458
japanischer Kreuzdorn 3025
japanischer Kugelfaden 4744
japanischer Kumquat 3528
japanischer Lebensbaum 3017
japanischer Liguster 3065
japanische Rose 1376
japanische Rotkiefer 3067
japanischer Palmentang 4807
japanischer Pfeffer 3063
japanischer Quittenbaum 3038
japanischer Rosinenbaum 4406
japanischer Schlangenbart 1944
japanischer Schnurbaum 3059
japanischer Sommerflieder 3026
japanischer Spierstrauch 3070
japanischer Spindelstrauch 2146
japanischer Sternanis 3014
japanischer Trompetenbaum 1175
japanischer Zierapfel 5554
japanische Sagopalme 4643
japanische Scheinquitte 1414, 3038
japanische Schirmtanne 5682
japanische Schwarzkiefer 3023
japanisches Geissblatt 3042
japanische Skimmie 3069
japanische Sophore 3059
japanisches Windröschen 3013
japanische Trespe 3024
japanische Ulme 3461
japanische Zaubernuss 3079
japanische Zeder 3029
japanische Zelkowe 3080
japanische Zypresse 2793
Japanknollen 332
Japanmispel 3375
Japanzeder 3029
Jaqueira 2998
Jasione 3082
Jasmin 3083
Jasminglanz 2416
Jasmintrompete 5610, 5611
Jasminwurzel 2441
Javapalisander 1984
Javatee 558
Jeffersonie 5662
Jelängerjelieber 1461, 5370
Jerichorose 3091
Jerusalemartischocke 3094
Jerusalembohne 3754
Jerusalemkerze 1839
Jerusalemsblume 3451
Jerusalemskreuz 3451
Jesuitennuss 5794
Jesuitentee 4048
Jochalgen 1580
Jochbaum 2108
Jochblatt 498
Jochblümchen 1819
Jochpilze 6131
Johannisapfel 4046
Johannisbeere 1745
Johannisbeerstrauch 1745
Johannisblume 3694, 3987
Johannisbrotbaum 1023
Johannisbrotgewächse 4824
Johannisgürtel 4606
Johanniskraut 2612, 4651
Johanniskrautgewächse 4652
Johanniswedel 5410
Jonquille 3110
Josika-Flieder 2890
Juanulloa 3112
Juckbohne 1644, 3740
Juckfasel 1644
Judasbaum 1098
Judaslinde 1098
Judasohr 3101
Judenbart 5277
Judenbaum 1098
Judendorn 1452, 3115
Judenkirsche 2645, 3095, 5275
Judenohr 3101
Judenpappel 4283
Judenstern 1535
Juglandazeen 5778
Jungfer im Busch 3386
Jungfer im Grünen 3386
Jungfer in Haaren 3386
Jungferngras 1971
Jungfernhaar 5099
Jungfernherz 668
Jungfernrebe 221, 1664
Jungfernwein 1664
Junkaginazeen 3116
Junkazeen 4609
Junkerlilie 3003
Jupiterbart 2779
Jurianuss 765
Jussiea 5832
Justicie 3121
Jute 3122
Jutepflanze 4591

Kabulklee 5917
Kaffee 1318
Kaffeebaum 1318
Kaffeestrauch 1318
Kaffeetragant 5349
Kaffeewicke 5349
Kaffeewurzel 1230
Kaffernhirse 5087
Kaffernkorn 5087
kahlblättrige Jujube 1452
kahler Krempling 2956
kahler Sumach 5042
kahler Wimperfarn 5043
kahles Bruchkraut 1372
kahles Ferkelkraut 5036
kahle Weide 163
Kaimanstrauch 635, 3099
Kaiserbaum 4075
Kaisereiche 1786
Kaiserkrone 1711, 2380
kaiserliche Paulownie 4595
Kaiserling 918
Kaiserpaulownie 4595
Kaiserschwamm 918
Kaiserwurzel 3532
Kajeputbaum 920, 3579
Kajubaum 1378

Kakaobaum 912
Kakaomalve 4
Kakibaum 3127
Kakipflaume 3127, 4152
Kaktazeen 915
Kakteen 915
Kaktus 914
Kaktusgewächse 915
Kalababaum 508
kalabrische Kiefer 1625
Kaladie 925
Kälberkropf 1137
Kalebasbaum 924
Kalebasse 922, 2567
Kalebassenbaum 924, 1373
Kalebassenmuskat 3662
Kalebassenmuskatnuss 923
Kalendel 929
kalifornische Flusszeder 936
kalifornische Hemlockstanne 3996
kalifornische Kiefer 3310
kalifornischer Gagelstrauch 948
kalifornischer Hasel(nuss)strauch 935
kalifornischer Kreuzdorn 932
kalifornischer Lorbeer 938
kalifornischer Mohn 943
kalifornische Rosskastanie 931
kalifornischer Wacholder 937
kalifornischer Walnussbaum 930
kalifornisches Meerträubchen 934
kalifornische Torreye 947
Kalinkenbeerstrauch 2091
Kalisalzkraut 1518
Kalla 949
Kallitrichazeen 5840
Kalmie 3129
Kalmückennuss 4611
Kalmus 5365
Kalo 2013
Kalykanthazeen 5278
Kalzeolarie 5011
Kamassie 958
Kameldorn 961
Kamel(l)ie 960
Kamille 963, 3541
Kamillenrautenfarn 794
Kammarante 1720
Kammblume 2476
kammförmiges Laichkraut 2202
Kammgras 1877
Kamminze 1682, 2024
Kammlaichkraut 2202
Kammquecke 1684
Kammwurmfarn 1685
Kampanulazeen 529
Kampeschebaum 3359
Kampferkraut 965
Kampferlorbeer 966
Kampferölbaum 736
Kämpfers Iris 4393
Kämpfers Lärche 3051
Kämpfers Schwertlilie 4393
Kanadapappel 1978
kanadische Akelei 188
kanadische Berberitze 98
kanadische Blutwurz 682
kanadische Brombeere 5490
kanadische Eibe 980
kanadische Felsenbirne 4838
kanadische Gelbwurzel 2547
kanadische Goldrute 971
kanadische Hemlockstanne 973
kanadische Nessel 977
kanadische Orangewurz(el) 2547
kanadische Pappel 979, 1978
kanadische Pflaume 978
kanadische Rebe 5748
kanadischer Holunder 190
kanadische Rispendolde 2836
kanadischer Judasbaum 1979
kanadischer Mondsame 1472
kanadischer Reis 266
(kanadischer) Schusserbaum 3151
kanadischer Weissdorn 972
kanadisches Berufkraut 2878
kanadisches Leinkraut 3922
kanadische Wasserpest 976
Kanarienbaum 989
Kanariengras 985
Kanarienhirse 985
Kanariennuss 989
kanarische Dattelpalme 983
kanarische Kiefer 988
Kanavalie 2997
Kaneelbaum 995
Känguruhbaum 516
Kannazeen 997
Kannenkraut 2225
Kannenpflanze 3828
Kannenstaude 3828
Kannenstrauch 3828
Kannenstrauchgewächse 3829
Kannenträger 3828
Kanonenbaum 4331
Kanonenblume 1259
Kanonierblume 1259
Kantalupe 998
Kantenhartheu 5198
Kantenheide 3706
Kantenlauch 249
Kantharelle 1000
Kantharellen 1001
kantiger Lauch 249
kantiger Spindelstrauch 2096
Kapmaiblume 2365
Kapokbaum 3134
Kapokwollbaum 3134
Kapparidazeen 1009
Kappenstendel 956
Kap(p)erngewächs 1009
Kap(p)ernstrauch 1007
Kap(p)ernsträucher 1009
Kappes 2734, 5885
Kapraute 29
Kaprifoliazeen 2842
Kapringelblume 1005
Kapschotendorn 103
Kapselgeissblatt 880
Kapstachelbeere 4156
Kapuzinerbart 1381
Kapuzinerkraut 3386
Kapuzinerkresse 3812
Kapuzinerkressengewächse 3813
Kapuzinerpilz 582
karaibische Kiefer 4993
Karakabaum 3135
Karambole 1012
Kardamompflanze 1014
Karde 1015, 5469
Kardendistel 3681, 5469
Kardengewächse 5471
Kardifiol 1079
Kardinalblume 3350
Kardobenediktendistel 671
Kardone 1015
Karfiol 1079
Karikazeen 4034
Karl-Ludwigspalme 1018
Karlsdistel 5035
Karlsszepter-Läusekraut 1117
Karnaubapalme 1021
Karobe 1023
Karobenbaum 1023
Karolina-Pappel 250
karolinische Hainbuche 198
Karotte 2410
Karpatenglockenblume 1026
Karpatenjohannisbeere 1027
Karpatenkatzenpfötchen 1028
Karpatenspierstrauch 3963
Kartäusernelke 1037
Kartoffel 4281
Kartoffelbovist 2705
Kartoffelnzwiebel 3753
Kartoffelpilz 4172
Kartoffelrose 4602
Karve 1013
Karwe 1013
Karyophyllazeen 4198
Kaschmirzypresse 3137
Kaschubaum 1042, 1378
Käseklee 711
Käsepappel 3447
Käsepappelgewächse 3448
Kaskarillenstrauch 1041
kaspische Lotosblume 1045
kaspische Weide 4850
Kassawe 1379
Kassawestrauch 1379
Kassie 4823
Kastanie 1139
Kastanienbaum 1139
kastanienblättrige Eiche 1141
Kastanieneiche 1141
Kastanienpilz 485
Kastillea 4003
Kastorbohne 1051
Kasuarinazeen 517
Kasuarine 516
Kasuarinengewächse 517
Katalpe 1053
Katalpenbaum 1053
Katappenbaum 5600
Katharinas Blutblume 3138
Kathstrauch 294, 1657
Katsurabaum 3048
Katzengras 1309
Katzenkerbel 1911
Katzenklee 4394
Katzenkraut 1064, 1309, 3830, 4122
Katzenminze 3830
Katzenpfötchen 4094, 4363
Kauchin 3162

kaukasische Erle 1066
kaukasische Fetthenne 5673
kaukasische Gänsekresse 5774
kaukasische Gemswurz 1075
kaukasische Hainbuche 1073
kaukasische Insektenblume 2306
kaukasische Linde 1076
kaukasischer Beinwell 1068
kaukasischer Pfeifenstrauch 1078
kaukasischer Seidelbast 1070
kaukasischer Zürgelbaum 1072
kaukasische Schwertlilie 1074
kaukasische Zelkowe 2021
Kaukasusfetthenne 5673
Kaukasusfichte 3964
Kaukasustanne 3876
Kaukasusvergissmeinnicht 821
Kaupfeffer 549
Kaurifichte 1795, 3139
Kautschukbaum 2665
Kautschuklöwenzahn 4613
Kawakami-Tanne 3141
Kawapfeffer 3140
Kawastrauch 3140
kegeliger Saftling 1577
kegeliger Wulstling 1842
kegeliges Leimkraut 1576
Kegelleimkraut 1576
keilblättrige Rose 4753
keilförmiger Buschklee 1189
keilförmiger Lappenfarn 1828
Kelchsteinkraut 4007
Kellerhals 1800
Kennedie 3148
Kentia 3149
kerbblättriger Spierstrauch 5067
Kerbel 494
Kerbelkraut 494
Kerbelrübe 5651
Kerguelenkohl 3152
Kermesbeere 1502, 4243
Kermesbeerengewächse 4244
Kermeseiche 3153
Kermespflanzen 4244
Kerners Distel 3155
Kernobstgehölze 282
Kernpilze 2277
Kerrie 3156
Kerzenbeerstrauch 5104
Kerzenhirse 4092
Kerzenkaktus 1100
Kerzennussbaum 992
Keste 2086
Kestenbaum 2086
Keteleerie 3157
Kettenblume 1796
Kettenfarn 1110
Keuladerfarn 3323
Keule 2520
Keulenbärlapp 4606
Keulenbaum 516, 1899
keulenfüssiger Trichterling 1287
Keulenlilie 1899
Keulenmohn 708, 5864
Keulenpalme 1581, 3406
Keulenpilz 2520
Keulenpilze 1588
Keulenschmiele 2595
Keulenschwamm 2520
keusche Sinnpflanze 4827
Keuschlamm 3296
Kicher 4098
Kicherling 2587
Kichertragant 3701
Kiebitzblume 2380
Kiebitzei 1126
Kiefer 4190
Kieferngewächse 4194
Kiefer von Bordeaux 1296
Kielkrone 953
Kiem 1013
Kienporst 1719
Kieselalgen 1853
Kigelie 4716
Kikar 399
Kirschbaum 1133
Kirsche 1133
Kirschlorbeer 1456, 3238
Kirschmyrte 2068
Kirschpflaume 3777, 5950
kirschroter Speitäubling 2028
Kirschtanne 446
Kirschwurzelkraut 3715
Kitschbaum 2078
Kit(t)upalme 5540
Kladonie 1249
Klammerstrauch 4718
Klanglein 1825
Klapper 2232
Klapperhülse 4425
Klapperkopf 3903
Klappernuss 659
Klappernussgewächse 660
Klapperschlangenkraut 971
Klapperschote 4425
Klappertopf 4426, 4430
Klarinettenrohr 2482
Klatschmohn 1616
Klebbaum 4699
Klebe 2607
Klebeiche 1928
Kleberklee 1519
Klebgras 4682
Klebkraut 1055, 5249
Kleblabkraut 1055
Klebnelke 1054, 1250
Klebraden 1250
klebrige Erle 2072
klebrige Nachtnelke 5757
klebriger Gänsefuss 3096
klebrige Robinie 1252
klebrige Salbei 5249
klebriges Greiskraut 5247
klebriges Kreuzkraut 5247
klebriges Leimkraut 5757
klebriges Schleierkraut 5248
Klebsame 4210
Klebsamengewächse 5537
Klee 1278
Kleebaum 3192
Kleeblatt-Lederstrauch 1441
Kleefarn 4128
Kleefarne 3523
Kleefarngewächse 3523
Kleeseide 1280, 1400
Kleestrauch 1441, 2529, 2851
Kleeteufel 1279
Kleetod 1279
Kleewürger 1279
Kleiderbaum 205
kleinblättrige Königskerze 2276
kleinblättrige Linde 3327
kleinblättriger Efeu 2040
kleinblättriger Pfeifenstrauch 3328
kleinblättriger Sommerflieder 3941
kleinblättrige Steinlinde 3327
kleinblättrige Wiesenraute 3392
kleinblütiger Klee 5470
kleinblütige Rose 5025
kleinblütiger Steinklee 267
kleinblütiger Sternanis 6112
kleinblütiger Storchschnabel 5026
kleinblütiges Knopfkraut 3326
kleinblütiges Schaumkraut 5021
kleinblütiges Springkraut 5023
kleinblütiges Weidenröschen 5024
kleine Bibernelle 4733
kleine Bisamhyazinthe 1427
kleine Braunelle 1525
kleine Brennessel 1874
kleine Brunelle 1525
kleine Kamelie 4711
kleine Kapuzinerkresse 882
kleine Klauenschote 3337
kleine Klette 5020
kleine Küchenschelle 3554
kleine Luzerne 3332
kleine Malve 1945
kleine Moosbeere 5019
kleine Nessel 1874
kleine Quecke 4469
kleiner Ampfer 4859
kleiner Baldrian 3520
kleiner Blutegerling 2326
kleiner Dorant 1619
kleiner Erbsenstrauch 4617
kleiner Feigenkaktus 1507
kleiner Klee 5292
kleiner Knöterich 4999
kleiner Lämmersalat 5029
kleiner Reis von Peru 4390
kleiner Sauerampfer 4859
kleiner Storchschnabel 5026
kleiner Vogelfuss 3337
kleiner Waldegerling 2326
kleiner Wiesenknopf 5018
kleine Sabal 1948
kleines Filzkraut 4994
kleines Glanzgras 3336
kleines Habichtskraut 3734
kleines Hexenkraut 124
kleines Immergrün 1493
kleines Laichkraut 401
kleines Liebesgras 3331
kleines Mädesüss 1907
kleines Nagelkraut 5018
kleine Sommerwurz 1279
kleines Schlammkraut 5825
kleines Springkraut 5023
kleines Sumpfried 1950
kleine Sumpfbinse 1950
kleine Sumpfsimse 1950
kleines Wintergrün 5065
kleines Zittergras 3335

kleines Zweiblatt 3881
kleine Taglilie 2585
kleine Taubnessel 4343
kleine Wachsblume 3280
kleine Waldhyazinthe 895
kleine Wasserlinse 1403
kleine Wiesenraute 3392
kleine Wolfsmilch 1951
kleinfrüchtiger Dotter 3334
kleinfrüchtiger Leindotter 3334
Kleingriffel 25
kleinköpfiger Pippau 5038
Kleinling 1109
kleinster Igelkolben 3252
kleinster Schneckenklee 3332
Klematis 1260
Kleopatranadel 1838
Klethragewächse 1261
Klethrazeen 1261
Klette 859
klettenartiger Igelsame 2126
Kletteneiche 870
Klettengras 860
Klettenkerbel 494, 2759
Klettenkraut 859
Klettenlabkraut 1055
Kletterfarn 1266
Kletterfeige 1267
Klettergras 860
Kletterhortensie 1269
Kletterkerbel 494
kletternde Trompetenblume 1550
Klettertrompete 5611
Klettertulpe 4337
Klimme 5577
Klimmstrauch 532
Klimm-Ylang-Ylang 5422
Klingelblume 4511
Klistierkraut 2785
Klivie 3126, 4745
Klopfklette 1306
Klosterbeere 2560
Klusie 1290
Kluster 2113
Knabenkraut 3946
Knabenkrautgewächse 3945
Knackbeere 2635
Knackelbeere 2635
Knackmandelbaum 2354
Knackweide 787
Knäuel 3168
Knäuelampfer 1293
Knäuelglockenblume 1797
Knäuelgras 1309
Knäuelhornkraut 3732
Knäueling 4033
Knäuelsimse 1293
Knaul 3168
Knaulblume 1332
Knaulgras 1309
Knautie 3167
Knebel 4093
Knickbeere 2127
Knicker 4317
Knoblauch 2432
Knoblauchgamander 5806
Knoblauchhederich 2433
Knoblauchpilz 2434
Knoblauchrauke 2433
Knoblauchschwamm 2434
Knoblauchschwindling 2434
Knochenbrech 715
Knochenholz 2099
Knöllchenknöterich 5759
Knöllchensteinbrech 3560
Knollenbaselle 5622
Knollenbegonie 5621
Knollenblätterpilze 173
Knollenbrandkraut 5620
Knollenerbse 4282
Knollenhahnenfuss 841
Knollenkapuzinerkresse 5617
Knollenkerbel 5651
Knollenknöterich 5759
Knollensauerklee 3918
Knollenwinde 5380
Knollenyams 69
Knollenziest 332
Knollenzwiebel 2420
knollige Platterbse 2649
knolliger Beinwell 5619
knolliger Dingel 2590
knolliger Hahnenfuss 841
knolliges Rispengras 844
knollige Wallwurz 5619
Knopfkraut 4386
Knorpelblume 2919
Knorpelkraut 2919, 4249
Knorpellattich 4983
Knorpelmöhre 216
Knorpeltang 2960
Knotenblume 5062
Knotenbraunwurz 6027
knotenfrüchtige Wicke 609
Knotenfuss 5669
Knotenkraut 3096
Knotenmastkraut 3171
Knotentintling 2950
Knöterich 3172
Knöterichgewächse 833
knotige Braunwurz 6027
knotiger Beinwell 5619
knotiger Klettenkerbel 3170
knötiges Mastkraut 3171
Kobäe 1301
Kobobaum 4939
Kobresie 1302
Köcherbaum 4392
Koenigie 3174
Kohl 906
Kohlbaum 246, 907
Kohlgänsedistel 1531
Kohllauch 4284
Kohlmalve 1743
Kohlpalme 2138, 4594
Kohlpappel 1743
Kohlportulak 1509
Kohlrabi 3175
Kohlraps 1343
Kohlrübe 4626
Kohlsaat 4417
Köhr 1013
Kokainstrauch 2883
Kokardenblume 2400
Kokastrauch 1304, 2883
Kokkelstrauch 5044
Kokospalme 1314
Kokospflaume 2914
Kok-Saghys 4613
Kola 1323
Kolabaum 1323, 5293
Kolbenbärlapp 4606
Kolbenfaden 62
Kolbenhirse 2352
Kolbenpalmen 1767
Kolbenschimmel 351
Kolbenschosser 412
Kolbenschwamm 1970
Kolbenweizen 1288
Kolibritrompete 2259
Kölle 4721, 5321
Kollinsonie 2863
Koloquinte 1334
Koloquinthe 1334
Koloradofichte 1338
Koloradosandlilie 5226
Komfrey 1345
Kommafarn 1685
Kommelinazeen 5135
Kommeline 1814
Kompasspflanze 4571
Kompositen 1570
Kondurango 1572
Koniferen 1579
Königin der Nacht 4384
Königsbegonie 355
Königsdattelpalme 983
Königsfarn 3974, 4593
Königsfliegenpilz 4032
Königsgranadille 2475
Königskerze 3752
Königslilie 4476
Königspalme 4594
Königspalme des Antillen 1721
Königsrose 4118
Konjugaten 1580
Konvolvulazeen 3683
Kopaivabalsambaum 1583
Kopaivabaum 1583
Kopalbaum 5704
Kopalfichte 1795
Kopernicie 1584
Kopfbinse 1010
Kopfblume 902
Köpfchenblütler 1570
Köpfchenknöterich 4199
Köpfchenschimmel 3691
Kopfeibe 4232
Kopfgras 3675
kopfiger Gänsefuss 675
kopfiger Goldregen 557
Kopfklee 1278
Kopfkohl 2734
Kopflauch 421
Kopforchis 4163
Kopfried 724
Kopfsalat 2735
Kopfsimse 724
Kopfständel 4163
Kopoubohne 3188, 5505
Korakan 4399
Korallenbaum 1390, 1587, 3095
Korallenbeere 491, 2935
Korallenbohne 1587

Korallenflechte 2968
Korallenholunder 2120
Korallenkaktus 4488
Korallenkirsche 3095, 5275
Korallenmoos 1623
Korallenschwamm 1589
Korallenstachelbart 1589
Korallenstrauch 1311, 1587
Korallenwurz 1591
Korbblütengewächse 1570
Korbblütler 359, 1570
Korbblütler mit Zungenblüten 1232
Körbchenblütler 1570
Korbeiche 5343
Korbmarante 928
Korbweide 467, 4352
Kordie 1594
Kordyline 1899
koreanische Berberitze 3178
koreanische Doppelblume 3180
koreanische Fichte 3187
koreanische Kiefer 3184
koreanische Pappel 3185
koreanischer Klee 3183
koreanischer Lebensbaum 3177
koreanischer Zürgelbaum 3182
koreanische Tanne 3181
Koriander 1597
Koriariazeen 1600
Korkbaum 1603, 1604
Korkeiche 554, 1603
Korkflügelspindelbaum 5990
Korkholz 1605
Korkschwamm 3274
Korkspindel 5990
Korkulme 1602
Kornazeen 1882
Kornblume 1089, 1613
Körnchenröhrling 2574
Kornelkirsche 1611
Kornelkirschgewächse 1882
Körnersteinbrech 3560
körniger Steinbrech 3560
Kornmohn 1616
Kornnelke 1391
Kornrade 1391, 1610
Korntrespe 1138
Kornwicke 1554
Kornwinde 2100
korsikanisches Wurmmoos 1623
korsische Kiefer 1625
Kosobaum 3190
Kostbeere 3882
Kostwurz 5158
Krachmandelbaum 2354
Kraftwurz(el) 2494
Kragenblume 1029
Krähenaugenbaum 3898
Krähenbeere 1706
Krähenbeerengewächse 1707
Krähenfuss 2330, 5783
Krähenfusswegerich 1708
Krainer Hahnenfuss 2921
Krainer Lilie 1022
Krampfdistel 4770
Krampfkraut 2112
Kränbeere 1645
Kranzblume 2491
Kranzlauch 5632
Kranzlichtnelke 4558
Kranzspiere 5241
Krapp 1463, 3410
Krassulazeen 3970
Kratzdistel 5487
Krausbeere 2560
krausblättriger Ampfer 1741
Krausdistel 5047
krause Distel 1740
krause Eiche 4845
krause Malve 1294, 1743
krause Minze 1698
Krauseminze 1698
krauser Ampfer 1741
Krauserampfer 1741
krauser Rhabarber 5844
krauser Ziegenbart 1683
krauses Laichkraut 1744
krause Zistrose 6058
Kraut 2734
krautiger Backenklee 2784
Krautweide 4366
Krebsblume 1705
Krebsdistel 1638
Krebskraut 5653
Krebsschere 5836
Krebswurz 2079
Kreisblume 2460
Kreisflechte 4535
kreisförmige Weide 4590
Krempling 4078
Kren 2871
Kreosotbusch 1680
Kresse 4126
kretischer Saumfarn 1686
Kreutzblümchen 4250
Kreuzbaum 1051, 2756
Kreuzbeere 1369
Kreuzblatt 1703
kreuzblättrige Wolfsmilch 1008
Kreuzblümchengewächse 3623
Kreuzblume 4250
Kreuzblütler 908
Kreuzdorn 830
Kreuzdorngewächse 831
Kreuzenzian 1700
Kreuzgras 2015
Kreuzkraut 1237, 2651
Kreuzkümmel 1728, 1729
Kreuzlabkraut 1704
Kreuzrebe 1702
Kreuzstrauch 405
Kreuztännel 5501
Kreuzwurm 2651
Krieche 847
kriechende Hauhechel 962
kriechender Günsel 1030
kriechender Hahnenfuss 1666
kriechende Rose 2234
kriechende Scheinbeere 1125
kriechendes Felsenröschen 114
kriechendes Fingerkraut 1667
kriechendes Schleierkraut 1670
kriechende Weide 1679
Kriechenpflaume 847
Kriechgünsel 1030
Kriechhauhechel 962
Kriechkiefer 3072
Kriechklee 5889
Kriechwacholder 238, 1671
Krimbibernelle 1689
Krimefeu 1691
Krimeseldistel 5462
Krimlinde 1692
Krimlotwurz 1690
Krimpbohne 2997
Krimperlgras 1693
Krimschleifenblume 5460
Krimschwertlilie 1943
Krokus 1699
Krönel 1712
Kronenanemone 4273
Kronenblume 2380
Kronenlichtnelke 4558
kronloses Mastkraut 273
Kronsbeere 1645
Kronwicke 1712
Krönzel 2560
Krösling 2154
Krötenbinse 5535
Krötenkraut 3096
Krötenlilie 5534
Krötenmelde 3102
Krötenschwamm 4032
Krugblatt 4207
Krugfarn 2711
Krugpflanze 3828
Krukenbaum 3657
Krullfarn 2248
Krummähre 4795
Krummhals 5938
Krummholzkiefer 3746, 5400
Krümmling 4983
Kruppbohne 881
Kruziferen 908
Kryptogamen 1717
Kryptomerie 1718
kubanisches Burzelkraut 3627
Kuba-Spinat 3627
Kuba-Zeder 1234
Kubebenpfeffer 1722
Kuchenbaum 3048
Küchenkohl 5940
Küchenschwindling 2434
Küchenzwiebel 2420
Kuckucksblume 3946, 4398, 4403
Kuckucksblütler 3945
Kuckucksnelke 4398
Kudzu 3188
Kugelamarant 2503
Kugelbinse 4201
Kugelblume 2506
Kugelblumengewächse 2505
Kugeldistel 2508
Kugelfaden 3123
kugelige Teufelskralle 420
Kugelkaktus 3454
Kugelkernpilze 5126
Kugelpilzartige 5126
Kugelschötchen 3154
Kugelzypresse 5887
Kuhbaum 1646
Kuhblume 1466, 1796
Kuhbohne 1650
Kuherbse 1393

Kuhkraut 1653
Kuhmaul 2516
Kuhmilchbaum 1646
Kuhpilz 4846
Kuhröhrling 4846
Kuhschelle 243
Kuhtritt 6059
Kukurbitazeen 2569
Kümmel 1013
Kümmelmyrte 3010
Kümmelsilge 4817
Kümmerlingskraut 1399, 1546
Kumquat 3189
Kundebohne 1393
Kunigundenkraut 2776
Kupferblatt 1585
Kupuliferen 514
Kürbis 2568
Kürbisbaum 924, 1373
Kürbisgewächse 2569
kurdistanische Pflaume 321
kurzblättrige Eibe 3998
Kurzhafer 3333
Küstenkiefer 4876
Küstensequoie 945, 4829
Kutte 1633

Labiaten 3629
Labkraut 510
Labkrautgewächse 3411
Lablabbohne 2897
Labmeister 511
Laboulbeniazeen 3191
lackartiger Schotendotter 5573
Lackmusflechte 3325, 4509
Lackmuskraut 5653
Lackviole 1556
Lagenarie 2567
Lagerströmie 2580
Laichkrautgewächse 4263
Lakritze 1459
Lallemantie 3208
Lamarckie 2550
Lambertshaselnuss 2473
Lambertskiefer 5306
Lambertsnuss 2473
Lambertsweissdorn 3209
Lambertszypresse 3665
Laminariazeen 3145
Lämmerklee 638
Lämmersalat 3213
Lampionblume 2645
Landreitgras 1131
Landrohrgras 1131
Landschilfgras 1131
Landtabak 1548
langährige Quecke 5432
langblättrige Akazie 5409
langblättrige Rauke 3363
langblättriger Ehrenpreis 1289
langblättriger Sonnentau 2044
langblättrige Weide 4680, 6053
langblütige Lilie 1972
Langbohne 1393
langdorniger Orangenbaum 3538
lange Kurkume 1551
langes Zypergras 2402
Langfaden 1344
Langfadengewächse 3776
langfüssige Weide 1298
langhaariges Habichtskraut 3734
langkapselige Jute 4283
langstieliges Laichkraut 5914
Lanzenbaum 3217
lanzenblättrige Funkie 3060
lanzenblättrige Kratzdistel 850
Lanzenfarn 3708
Lanzenrosette 35
Lanzenschildfarn 3708
Lanzettkratzdistel 850
Lapagerie 1152
Lapeyrousie 2169
Lappenbeere 4788
Lappenblume 2906
lappländisches Läusekraut 3219
lappländische Weide 3221
Lärche(nbaum) 3222
Lärchenschwamm 4333
Lardizabala 3223
Larvenblume 3655
Laserkraut 3227
Lasie 3228
Latanie 3229
Lattich 3281
Laubheide 5323
Laubholzschüppling 1114
Laubkaktus 4136
Laublatsche 2105
Laubmoose 5605
Lauch 3931
Lauchgamander 5806
Lauchschwamm 2434, 3484
Laugenblume 759
Laurazeen 3232
Laurinen 3232
Läusekraut 1719, 3383
Läusesamen 4633
Lavendel 3239
Lavendelheide 241, 723
Lavendelweide 2007
Lawsons Lebensbaumzypresse 3243
Lawsons Scheinzypresse 3243
Lawsons Zypresse 3243
Laya 5519
Lebensbaum 300
Lebensbaumzypresse 2167
Leberbalsam 60
Leberblümchen 1437, 2783, 3343
Leberdistel 4299
Leberklette 63
Leberklettenkraut 1348
Leberkraut 2133, 3487
Lebermoos 2783
Leberstockkraut 2418
Leberwurstbaum 4716
Ledebours Lilie 3264
Lederbaum 1599, 2851
Lederblatt 3257
lederblättrige Rose 3258
Lederholz 363, 3259
Lederorangenbaum 4329
Lederstrauch 2851
Leguminosen 3267
Lehne 3885
Leichkraut 4262
Leiereiche 3982
leierförmige Eiche 3982
Leierholz 2218
Leimkraut 1054
Leimmistel 2113
Leimsaat 1332
Lein 2286
Leinbaum 3885
Leinblatt 476
Leindotter 572, 2168
Leingewächse 2288
Leinkraut 5533
Leinlolch 2708
Leinseide 2287
Leinwandbaum 3319
Leistenpilze 1001
Leistling 1000
Lemnazeen 1925
Lemongras 3270
Lentibulariazeen 665
Leopardblume 620
Lerchensporn 1628
Lespedezie 3279
Leuchte 2327
Leuchterbaum 200, 1102
Leuchterblume 1102
Levkoje 5258
Lewisie 3285
Libanoneiche 3261
Libanonzeder 1083
Liberiakaffee 3287
Libertie 3288
Lichtbaum 200
Lichtblume 382, 3557
Lichterbaum 200
Lichtmyrte 483
Lichtnelke 967
Lichtnussbaum 992
Liebesapfel 1549
Liebesbaum 1098
Liebesblume 54
Liebesgras 3385, 4381
Liebkraut 510
Liebstock 2418
Liebstöckel 3384
liegende Azalee 114
liegender Dreizahn 2746
liegender Ehrenpreis 2709
liegender Klee 2848, 3225
liegender Krähenfuss 1678
liegendes Fingerkraut 1031
liegendes Hartheu 1677
liegendes Mastkraut 4313
liegendes Scharfkraut 2459
liegendes Schlangenäuglein 2459
Lieschgras 5527
Lieschkolben 1062
Lieschrohr 1062
Ligularie 2542
Liguster 4312
Likualapalme 3292
Lila 1461
Lilak 1461
Liliazeen 3301
Lilie 3300
Liliengewächse 3301
Lilienmagnolie 3304, 6122
Lilienschweif 1838
Lilienschwertel 3308

Liliensimse 2162
Liliputrose 4580
Limabohne 3309
Limbaholz 38
Limettenbaum 3311
Limettenzitrone 3311
Limone(nbaum) 3268
Lindelofie 3313
Linde(nbaum) 3314
lindenblättrige Grewie 3317
lindenblättriger Eibisch 3316
Lindengewächse 3315
Lindens Platane 3006
Lindleys Fiederspiere 3318
Linse 3273
Linsenbaum 809, 1362
Linsenerve 3273
Linsenwicke 609, 3930
Lippenblütler 3629
Lippenfarn 3323
Lippie 3324
Litschibaum 3397, 3398
Littonie 3340
Livistone 2183
Loase 1155
Lobelie 3350
Lobeliengewächse 3351
Lochblume 3655
Löcherpilz 1093
Löcherpilze 4253
Löcherschwämme 4253
Löffelkraut 4783
Loganbeere 3357
Logangewächse 3358
Loganiazeen 3358
Lolch 4631
lolchartige Segge 4632
Lonchocarpus 3216
Lonizere 2841
Lontaropalme 4019
Loosbaum 2510
Loranthazeen 3632
Lorbeer(baum) 3231
lorbeerblättrige Granadille 5815
lorbeerblättrige Grenadille 5815
lorbeerblättrige Pappel 3234
lorbeerblättrige Passionsblume 5815
lorbeerblättriger Flachsbaum 566
lorbeerblättrige Zistrose 3235
Lorbeereiche 3233
Lorbeergewächse 3232
Lorbeerkirsche 1456
Lorbeerkraut 5192
Lorbeerpappel 3234
Lorbeerrose 1482, 3129, 3925
Lorbeerschneeball 3237
Lorbeerseidelbast 5192
Lorbeerweide 3236
Lorchel 2767
Lorcheln 2768
Lord Howe Palme 4828
Losbaum 2510
Lotuspflaume 1809
Lotwurz 3932
Löwenklau 12
Löwenmaul 5050
Löwenohr 3320
Löwenschwanz 3320
Löwenzahn 1796, 2723
Lows Tanne 3997
Ludwigie 4812
Luffa-Gurke 5290
Lühea 5877
Lungenflechte 2916, 3394, 5250
Lungenkraut 3393
Lungenmoos 2916, 3394
Lungenwurz 691
Lupine 3395
Lupinenklee 473
Luzerne 90, 3565
Lykopodiazeen 1286
Lythrazeen 3371

Maasliebe 1787
Macahuba 3737
Macartneys Rose 3401
Macawbaum 3737
Machorka 395
MacNabs Zypresse 3405
Macoya 3737
Macrozamie 3406
Madagaskarbohne 1448, 3309, 4940
madagaskarische Kartoffel 2721
Madagaskarpflaume 4413
madagassische Muskatnuss 3407
madagassische Sagopalme 3408
Mädchenauge 1596
Mädchenhaarbaum 2492
Mädchensalat 3211
Madeiralorbeer 3412
Mädesüss 3562
Madie 1151, 3416
Madrashanf 5331
Mafureirabaum 611
Maggikraut 3384
Magistranzwurzel 3532
Magnoliazeen 3424
Magnolie 3423
Magnoliengewächse 3424
Maguah 1092
Maguey 1092
Mahagonibaum 3427
Mahalebkirsche 3426
Mähnengerste 2350
Mahonie 3429
Maiapfel 1845, 3539
Maiblümchen 3306
Maiblume 1398, 1796, 3306
Maiblumenbaum 4950
Maiglöckchen 3305, 3306
Maiglöckchenbaum 4950
Maililie 3306
Mairan 5375
Mairose 1240
Mais 3434
Majoran 3497
makedonische Eiche 3403
makedonisches Leinkraut 3404
malabarischer Rosenapfel 3912
Malabarjasmin 1052
Malabarspinat 5153, 5921
maledivische Nuss 1887
Malope 3449
Malpigie 3450
Maltesekreuz 3451
Malvazeen 3448
Malve 3447
Malvengewächse 3448
Mamaobaum 4035
Mammeibaum 3452
Mammiapfel 3453
Mammillarie 3454
Mammutbaum 2483, 4829
Mandarine 3468
Mandelbaum 104
Mandelbäumchen 2311
mandelblättriger Birnbaum 107
Mandelweide 106, 4084
Mänderle 4000
Mandioka 1046
mandschurische Esche 3458
mandschurische Hasel 3456
mandschurische Johannisbeere 3459
mandschurische Linde 3463
mandschurische Pfeifenblume 3460
mandschurischer Ahorn 3464
mandschurischer Aprikosenbaum 3457
mandschurischer Strahlengriffel 3176
mandschurischer Walnussbaum 3466
Mandua 4399
Mandubi-Bohne 4087
Mandubi-Erbse 1574
Manettie 3470
Manglebaum 3475
Mangobaum 3472
Mangold 1116, 3473
Mangostana 3474
Mangostanbaum 3474
Mangostane 3474
Mangotanne 3472
Mangrovebaum 200, 386, 3475
Mangrovefarn 20
Manilahanf 2
Maniok 1046
Maniokstrauch 3838
Mannaesche 2308
Mannaflechte 3262
Mannagras 3476
Mannaschwaden 5820
Männertreu 1672, 2058, 2454, 4789
männliches Farnkraut 3445
männliches Knabenkraut 3446
Mannsbart 706
Mannsblut 1537
Mannsknabenkraut 3446
Mannsschield(kraut) 4527
Mannstreu 2058, 4789
Manschenill(en)baum 3455
Manzanillabaum 3455
Marante 329
Marattiazeen 3485
Marbel 6041
Marettig 2871
Margaretenblume 1787, 3987
Margerite 3987
Marguerite 1787, 3987
Marienblatt 1632, 3987
Marienblume 3987
Mariendistel 669
Marienflachs 887
Marienglocke 999
Mariengras 984, 1620, 5367

Marienkraut 3694
Marienmantel 1454
Mariennessel 1442
Marienröschen 2140
Marienschlüssel 1652
Marienträngras 31
Marienveilchen 999
Mariettenveilchen 999
Marille 1350
Marjoran 5375
Markbaum 179
Markbinse 4608
Markerbse 3498
Maronenbaum 2086
Maronenpilz 485.
Maronenröhrling 485
Maronpflaume 4413
Marschbohne 788
Marsdenie 3499
Marsiliazeen 3523
Martagonlilie 3524
Martens Moosfarn 3525
Martynie 3527
Martyniengewächse 5683
Marua 4399
Marupa 4047
Märzbecher 5188
Märzblümchen 3343
Märzblume 5060, 5188
Märzveilchen 5392
Märzwurzel 1354
Maskenblume 3655
Masseller 2756
Masserle 2756
Mas(s)holder 2756
Masslieb(chen) 1787
Massons Kiefer 3530
Mastixbaum 3275
Mastixpistazie 3275
Mastixstrauch 3275
Mastkraut 4093
Matebaum 4048
Matepflanze 4048
Matestrauch 4048
Mauerandorn 1442
Mauerdoppelsame 5257
Mauerfelsenblümchen 5766
Mauergipskraut 1749
Mauerhabichtskraut 5769
Mauerkraut 1749
Mauerlattich 5772
Mauernelkenwurz 1354
Mauerpfeffer 5264
Mauerpippau 3803
Mauerrampe 5257
Mauerrauke 5776
Mauerraute 3788, 5161
Mauerstreifenfarn 5776
Maulbeerbaum 3749
Maulbeerfeige(nbaum) 5408
Maulbeergewächse 3750
Maulwurfskraut 1008
Maurandie 3536
Mauritiushanf 4209
Mauritiuspalme 3537
Mäusefras 3775
Mäusefuchsschwanz 3735
Mäusefuchsschwingel 4423
Mäusegerste 3729
Mäuseohr 2327
Mäuseschwanz 3736
Mäuseschwänzchen 3736
Mäuseschwanzfederschwingel 4423
Mäuseschwanzfuchsschwingel 4423
Maximowiczs Birke 3650
Maximowiczs Pappel 3062
Medeola 3564
Medinilla 3567
Meeraloe 5836
Meerampfer 2536
Meerbrachdistel 4789
Meereiche 4797
Meerfaden 5934
Meerfenchel 4674, 4675
Meerkohl 1326, 1387
Meerkohlwinde 4804
Meerkokos 1887
Meerkraut 2500
Meerlattich 1524
Meerlavendel 4793
Meerlinse 1924
Meernixenkraut 5156
Meerrettich 2871
Meersaite 4792
Meersalat 1524, 4794
Meersalde 5934
Meersenf 4799
Meersimse 4801
Meerstrandbinse 4800
Meerstranddreizack 4877
Meerstrandkiefer 1296
Meerstrandstreu 4789
Meerstrandwegerich 2123
Meerstrandwinde 4804
Meerträubchen 3104
Meerträubchengewächse 3107
Meertraube 4788
Meerträubel 3104, 3105
Meerwermut 1776, 3496
Meerwurzeldistel 4789
Meerzwiebel 4798
Mehlbanane 4219
Mehlbeerbaum 5881
Mehlbeere 5881
mehlige Königskerze 5903
Mehlpilz 4226
Mehlprimel 586
Mehlschlüsselblume 586
Mehlschmergel 4390
Mehltaupilze 4289
mehrjähriges Gänseblümchen 2037
mehrzeilige Gerste 455
Meisterwurz(el) 3532
Mekkabalsam 3563
Mekkasennes 1575
Melanganapfel 2414
Melankonialen 3580
Melastomazeen 3581
Melde 2561, 4663
meldenblättriger Teufelszwirn 3535
Meliazeen 3428
Melilotenklee 5362, 6108
Melisse 427
Melissenimmenblatt 469
Melonenbaum 4035
melonenblättriger Nachtschatten 3585
Melonengranadille 2475
Melone(ngurke) 3762
Melonenkaktus 3584
Melonenkürbis 883, 5705
Melothrie 3586
Meluchia 4283
Mengel 388
Menispermazeen 3672
Mentzelie 3587
Menziesie 4989
Mercuriuskraut 2785
Merk 5828
Merkurialkraut 2785
Mesquitebaum 1467, 3593
Mesquitestrauch 3593
Mesquitobaum 1467
mexikanische Lilie 526
mexikanische Nopalpflanze 1305
mexikanische Pantoffelblume 3595
mexikanische Platane 3601
mexikanischer Klee 3596
mexikanischer Leberbalsam 3594
mexikanischer Pfeifenstrauch 3600
mexikanischer Sadebaum 3599
mexikanischer Tee 6056
mexikanische Seerose 6099
mexikanisches Läusekraut 1914
mexikanisches Teekraut 6056
mexikanische Sumpfzypresse 3667
mexikanische Teosinte 3603
Mexikostern 3602
Meyers Johannisbeere 3607
Michelie 3610
Micromerie 3611
Middendorffs Birke 3612
Middendorffs Taglilie 3613
Mikanie 3616
Milchbaum 1646, 3618
Milchblätterschwamm 3195
Milchblume 2401, 4250
Milchbrätling 3938
Milchdistel 669, 2236
Milchkraut 3860, 4796
Milchlattich 3281
Milchling 3195
milchlingsähnlicher Weisstäubling 5923
Milchlingstäubling 5923
Milchreizker 3938
Milchschwamm 3938
Milchstern 5228
Milchwicke 1176
Miltonie 4031
Milzfarn 5161
Milzkraut 2545
Mimosazeen 3626
Mimose 3625
mimosenblättrige Jacaranda 4849
Mimosengewächse 3626
Minze 3628
Mirabelle 3630
Mirbecks Eiche 987
Mirobalanenbaum 2026
Mispel 3577
Mistel 3631
Mistelgewächse 3632

Mistschwamm 2950, 2951
Mitsumata 3958
Mittagsblume 3592
Mittagsblumengewächse 1033
Mittelmeerhafer 4457
Mittelmeerheide 594
Mittelmeertrespe 5109
mittlerer Klee 6128
mittlerer Wegerich 5379
mittleres Fingerkraut 1893
mittleres Wintergrün 2953
mittleres Zittergras 4134
Mohar(hirse) 2352
Mohn 4272
Mohngewächse 4274
Mohnling 708, 5864
Möhre 1035
Mohrenhaftdolde 2753
Mohrenhirse 1929, 5086, 5087
Mohrmeddhu 553
Mohrrübe 2410
Molchschwanz 3347
Molt(e)beere 1273
molukkische Zuckerpalme 2558
Mombinpflaume 3048, 6101
Momitanne 3649
Monarde 512
Monatsklee 90
Monatsrettich 2425
Monatsrose 1792, 3767
Mönchskolben 3873
Mönchskraut 3873
Mönchspfeffer 1120, 3296
Mönchsrhabarber 4072
Mönchsrharbarber 129
Mönchswurz 3694
Mondbohne 4940
Mondklee 90
Mondkraut 4130
Mondraute 2577
Mondraute(nfarn) 3673
Mondsame 3671
Mondsamengewächse 3672
Mondviole 2834
Mondwinde 3670
mongolische Eiche 3653
Monke 3082
Monokotyledonen 3661
Monokotylen 3661
Monstera 1101, 3663
Monterreykiefer 3666
Montezumakiefer 3668
Moorbeere 717
Moorbeerengewächse 2745
Moorbirke 4323
Moorblasenbinse 3515
Moorglocke 4511
Moorheide 3706
Moorheidelbeere 717
Moorkratzdistel 3519
Moorlabkraut 5341
Moorreitgras 5008
Moorsimse 496
Moorsteinbrech 6097
Moorweide 5928
Moosauge 6037
Moosbeere 692, 1658
Moosblümchen 4365
Moosblume 1466
Moosdistel 1531
Moose 825
Moosfarn 4814
Moosfarngewächse 4815
Moosglöckchen 5660
Moosheide 307, 1049, 3706
Mooskraut 4814
Mooskrautgewächse 4815
Moospflanzen 825
Moosrose 3685
Moosweidenröschen 2857
Moräe 3676
Morastbirke 1935
Morazeen 3750
Morchel 3678
Morchelpilze 2768
morgenländische Fichte 3964
morgenländische Platane 3959
morgenländischer Ackerkohl 5574
morgenländischer Knöterich 4309
morgenländischer Lebensbaum 3948
morgenländischer Mohn 3960
Morgenröschen 943, 5394
Morindafichte 2806
Morine 3681
Mormonentulpe 3494
Moschusblümchen 3766
Moschusblume 4355, 5388
Moschuserdbeere 2722
Moschusgauklerblume 3764
Moschuskraut 3764, 3765
Moschuskrautgewächse 3684
Moschuskürbis 1747
Moschusmalve 3760
Moschusrose 3767
Moschusschafgarbe 3769
Mosestränengras 31
Moskitobaum 1467
Mostgummibaum 1233
Mostjöcken 1645
Mostpalme 403
Mottenkönig 5190
Mottenkraut 1719
Mottenpflanze 5190
Mousseron 2434
Muhlenbergia 3748
Mukorazeen 3739
Mulga 3751
Müllen 1120, 3296
Mult(e)beere 1273
Mumebaum 3015
Mummel 1647
Mungbohne 3754
Mungobohne 3755
Münze 3628
Münzfelberich 3651
Murrayakiefer 3355
Musazeen 440
Muscadinerebe 3756
Muschelblümchen 2970
Muschelblume 3647, 5816
Muschelpilz 3992
Muskatellerkraut 1254
Muskatellersalbei 1254
Muskatnussbaum 1480, 3662, 3894
Muskatsalbei 1254
Musseron 2434
Mutterkamille 2210
Mutterkorn 1258, 2053
Mutterkraut 1356, 2210, 3747, 4065, 4796
Mutterwurz 415, 714, 1411, 3294, 3694
Mützenstrauch 3635
Myrikazeen 484
Myristikazeen 3895
Myrizeen 484
Myrobalane 3777, 5950
Myrobalanenbaum 1124
Myrrahbaum 9
Myrrhenkerbel 5361
Myrrhenkraut 87
Myrrhenstrauch 3779
Myrsinengewächse 3780
Myrtazeen 3783
Myrte 3781
Myrtenblatt 3564
Myrtengewächse 3783
Myrtenheide 3579
Myrtenweide 3785
Myxomyzeten 5006

Nabelkraut 1489, 1639, 3816
Nabelnuss 3815
Nabelwurz 1639
Nachmittagsblume 3592
Nachtblume 293
Nachtkerze 2141
Nachtkerzengewächse 2142
Nachtleimkraut 3850
Nachtlichtnelke 3850
Nachtschatten 604, 3853
Nachtschattengewächse 3854
Nachtviole 4522
Nachtzauke 895
nackte Jungfer 1353
nackte Jungfrau 5060
nackter Blätterschwamm 214
nackter Ritterling 214
Nackthafer 3789
Nacktsamer 2671
nacktstengelige Schwertlilie 5268
Nadeleiche 4200
Nadelhölzer 1579
Nadelkerbel 4861
Nadelpalme 3821
Nadelsumpfried 3822
Nadelsumpfsimse 3822
Nagami Kumquat 3980
Nagasbaum 2966
Nagelbeere 3907
Nagelblume 1461
Nägelchenbaum 1461
Nägelein 1461
Nägeleinspilz 2154
Nagelkraut 3479, 3788
Najadazeen 3787
Nanas 235
Nandine 3790
Nankinglilie 3791
Napfkraut 3873
Narbonner Wicke 3794
Nardensame 2415
Narzisse 3795
Narzissengewächse 177
Natterblümchen 1470

Nattergras 649
Natter(n)farn 27
Natter(n)farne 28
Natter(n)farngewächse 28
Natter(n)kopf 5744
natter(n)kopfähnliches Bitterkraut 783
Natter(n)kopfbitterkraut 783
Natter(n)wurz 1346, 2079
Natter(n)zunge 27
nebelgrauer Trichterling 1274
Nebelkappe 1274
Negerhirse 4092, 5086
Neillie 3824
Nektarine(npfirsich) 3818
Nelke 4197
Nelkengewächse 4198
Nelkenhafer(schmiele) 4959
Nelkenpfeffer 102
Nelkenpfefferbaum 102
Nelkenpflanzen 4198
Nelkenschmielenhafer 4959
Nelkenschwamm 2154
Nelkenschwindling 2154
Nelkensommerwurz 1282
Nelkenwurz 385, 5136
Nelkenzimtbaum 1275
Nemesie 3825
Nepenthazeen 3829
Nephrolepis 5405
Neptunie 4825
Nerine 3832
Nessel 3835
Nesselblatt 1585, 3529
nesselblättrige Glockenblume 1643
Nesselgewächse 3836
Nesselglockenblume 1643
Nesselröschen 3049
Nesselschirm 2667
Nesselseide 2607
Nestpilze 589
Nestwurz 590
netzaderige Weide 3833
Netzan(n)one 849
Netzblatt 1675, 4428
netzförmiger Flaschenbaum 849
Netzmelone 3834
Netzschwertel 2499
netzstieliger Hexenröhrling 3396
Netzweide 3833
neubelgische Aster 3842
neumexikanische Robinie 3841
neuseeländische Flachslilie 3843
neuseeländischer Dammarabaum 3139
neuseeländischer Flachs 2215, 3843
neuseeländischer Spinat 3844
Niaulibaum 920
Nichtblätterpilze 276
Nichtlamellenpilze 276
nickende Distel 3759
nickende Küchenschelle 3554
nickender Blaustern 4916
nickender Milchstern 3870
nickender Steinbrech 1904
nickender Zweizahn 3866
nickendes Leimkraut 3869
nickendes Perlgras 3868
nickendes Wintergrün 4926
Nickwurz 4429
niedergestreckte Zwergmispel 4515
niederliegende Brombeere 3880
niederliegendes Johanniskraut 1677
niederliegendes Mastkraut 4313
niedrige Binse 845
niedriger Bambus 2643, 3389
niedriger Erbsenstrauch 4617
niedrige Schwarzwurzel 725
niedrige Zwergpalme 3572
Nierembergie 1733
Nierenbaum 1042
Nierenflechte 3831
Nierenschild 3831
Nierentee 558
Niesholz 5053
Nieswurz 2170, 2766
Nigersaat 2063
Nikanpalme 4491
Nikkoahorn 3856
Nikkotanne 3855
Nimbaum 3492
Nipapalme 3859
Nipponahorn 4004
Nissegras 3863
Nittabaum 45
Nixkraut 3786
Nixkrautgewächse 3787
Nolimetangere 5560
Nolina 3872
nordamerikanische Hortensie 5039
nordamerikanische Platane 205
nordamerikanischer Reis 266
nordamerikanisches Wintergrün 6003
nordische Alpenrose 3220
nordische Brombeere 305, 1273
nordische Erle 2072
nordische Himbeere 305
nordische Johannisbeere 3882
nordische Käsemalve 1945
nordische Malve 1945
nordischer Drachenkopf 4629
nordisches Felsenblümchen 4519
nordisches Katzenpfötchen 142
nordisches Labkraut 3878
nordische Zwergbirke 1935
nordkarolinische Hickory 1025
nördlicher Streifenfarn 2329
nördliches Geissblatt 6022
Nordmannstanne 3876
Norfolktanne 3877
Norne 957
norwegisches Felsenblümchen 3888
norwegisches Fingerkraut 3887
Nostok 3889
Nostokazeen 2375
nubische Tanne 94
Nuck 2063
Numbo-Knollen 2721
Nüsschensalat 3211
Nussdolde 913
Nusseibe 3074, 5556
Nussgras 3893
Nusskiefer 1856
Nusspalme 2265
nusstragende Stinkeibe 3074
nusstragende Torreye 3074
Nutkalebensbaumzypresse 3874
Nutkarose 3875
Nutkascheinzypresse 3874
Nutkazypresse 3874
Nyatoh 3814
Nymphäazeen 5818
Nyssazeen 5643

obergärige Bierhefe 1359
Oberkohlrübe 3175
Oberrübe 3175
Obst-Banane 1358
Ochsenauge 3986
Ochsenherz 1751
Ochsenherzapfel 849
Ochsenpilz 4846
Ochsenzunge 515, 840
Odermennig 63, 1348
offenblütige Waldrebe 3297
Öhrchenweide 4581
Öhrenhellerkraut 5494
Ohrlappenpilzartige 369
Ohrlappenpilze 369
Ohrpilz 3101
Ohrweide 4581
Okra 3921
Okwabaum 5575
Ölbaum 1483, 3927
ölbaumartige Gewächse 3928
Ölbaumgewächse 3928
Ölbohne 5106
Öldotter 2168
Oleander 1482, 3925
Olearie 1788
Oleaster 2006
Oleazeen 3928
Ölfrucht 2008
ölgebendes Madikraut 1151
Olive 1483
Ölkamellie 2799
Öllein 3915
Ölnussbaum 897
Ölpalme 46, 3916
Ölraps 1343
Ölrauke 4525
Ölstrauch 2177
Ölweide 2006
Ölweidengewächse 3926
Omorikafichte 4830
Oncidium 3929
Ontariopappel 428
Ophioglossazeen 28
Opuntie 4301
Orange 1246, 5377
Orangenbaum 3019
Orangenraute 3084
orangerotes Habichtskraut 3942
oranger Rauhfuss 3940
Orangewurzel 2546
Orchidazeen 3945
Orchidee 3944
Orchideen(gewächse) 3945
Orchis 234
orientalische Braunwurz 5804
orientalische Buche 3949
orientalische Flockenblume 1067
orientalische Hainbuche 3956

orientalische Johannisbeere 3953
orientalische Platane 3959
orientalischer Amberbaum 3965
orientalischer Blasenstrauch 3950
orientalischer Gamander 3954
orientalischer Hornbaum 3956
orientalischer Lebensbaum 3948
orientalischer Mohn 3960
orientalischer Sesam 3961
orientalischer Weissdorn 4960
orientalisches Helmkraut 3962
orientalische Waldrebe 3952
Orleangewächse 612
Orleansstrauch 256
Orobanchazeen 813
Orseilleflechte 3325
Osagedorn 3971
Osageorange 3971
Osmundazeen 3975
Osterglocke 1397
Osterluzei 1932
Osterluzeigewächse 593
Osterpalme 400
österreichische Eiche 2128
österreichische Gemswurz 379
österreichische Kresse 377
österreichischer Drachenkopf 376
österreichischer Geissklee 375
österreichischer Lein 378
österreichische Schwarzkiefer 380
österreichische Sumpfkresse 377
ostindische Hanfrose 3146
ostindischer Hanf 5331
ostindischer Kampferbaum 736
ostindischer Rosenholzbaum 1984
ostindischer Rotholzbaum 4702
Ostlinde 3327
Ostritzwurzel 1438
Ostrowskie 2468
Ostseerohr 1131
Ottelie 3978
Otternzunge 2079
ovalblättrige Weide 3981
ovaler Kumquat 3980
Oxalidazeen 5088

Pachysandra 3995
Palafoxie 4006, 5244
Palisanderbaum 648
Palisanderholz 2996
Palme 4016
Palmen 4018
Palm(en)farn 1764
Palmenfarne 1765
Palmettopalme 909
Palmfarn 1581
Palmlilie 6120
Palmweide 2523
Palmyrapalme 4019
Pampasgras 4020, 4021, 4818
Panamapalme 1018
Pandanazeen 4780
Pandane 4779
Pandang 4779
Pangibaum 2324
Pankrazlilie 4023
pannonischer Klee 2888
pannonische Wicke 2893
Pantherblume 620
Pantherpilz 4032
Pantherschwamm 4032
Pantoffelbaum 1603
Pantoffelblume 5011
Pantoffelstrauch 5010
Papageiblume 2400
Papageienblatt 3111
Papau(baum) 4077
Papaverazeen 4274
Papayabaum 4035
Papayazeen 4034
Papierbirke 4037
Papierblume 1488, 2923
Papierknöpfchen 217, 5991
papierliefernde Edgeworthia 3958
Papiermaulbeerbaum 1486, 4039
Papierstaude 4043
Papilionaten 4041
Papilionazeen 4041
Pappel 4270
pappelblättrige Birke 2594
pappelblättrige Zistrose 4271
Pappelfeigenbaum 738
Pappelrose 2788
Pappelweide 647
Paprika 884, 4459
Papyrusstaude 4043
Paradiesapfel 1549, 4046
Paradiesfeige 439, 1358, 4219
Paradieslilie 4649
Paradiesvogelblume 584
Paraguayteepflanze 4048
Parakautschukbaum 2791, 4050
Parakresse 4045
Paranakiefer 4049
Paranussbaum 765
Parasolpilz 3278
Pariser Anis 2300
Pariser Labkraut 4054
Parklinde 2111
Parmeliazeen 4057
Parrotie 4061
Parrye 4062
Parzenkraut 2130
Passiflorazeen 4069
Passionsblume 4068
Passionsblumengewächse 4069
Pasternak 4064
Pastinak(e) 4064
patagonische Fitzroya 4071
Paternosterbaum 1160, 2083
Paternostererbse 4553
Patienzkraut 4072
Patschuli(pflanze) 2792
Paullinie 4074
Paulownie 4075
Pechbaum 3886
Pechkiefer 4208
Pechnelke 1250
Pechpalme 5154
Pechsame 5538
Pechtanne 3886
Pedaliazeen 4101
Peireskie 4136
Peitschenaffodil 3003
Peitschenkaktus 4422
Pekanbaum 4100
Pekannussbaum 4100
Pekingkohl 4158
Pelargonie 4103
Pellefarn 1262
Pel(l)uschke 2231
Pelzfarn 1271
pennsylvanischer Ahorn 5284
Pentapere 4116
Peperomie 4119
perennierende Lupine 5327
Perigord-Trüffel 4137
Perille 1492, 4138
Peristrophe 4139
Perlbusch 4090
Perlenbeerstrauch 2370
Perlenschwamm 713
Perlfarn 4826
Perlgras 3582, 4110
Perlhirse 4092
Perlkraut 4110
Perlmoos 2961
Perlpfötchen 4094
Perlpilz 713
Perlzwiebel 2474, 3266
Pernambukholzbaum 4296
Pernettya 4141
Perovskie 4142
Perowskie 4142
persische Akazie 4948
persische Erdscheibe 2990
persische Insektenblume 2306
persische Parrotie 4150
persischer Ehrenpreis 5563
persischer Flieder 4149
persischer Hahnenfuss 4146
persischer Klee 4147
persisches Alpenveilchen 2990
persische Schwertlilie 4148
peruanischer Balsambaum 429
peruanischer Holunder 4155
peruanischer Pfefferbaum 940
Perubalsambaum 429, 4153
Perückenbaum 1527
Perückenstrauch 5033
peruvianische Sonnenwende 1436
Pestilenzwurz 888
Pestwurz 888
Petersbart 4941
Petersilie 1417, 4063
Petersschlüssel 1652
Petersstrauch 1529
Peträa 4157
Petunie 4160
Pfaffenbeere 2081
Pfaffenhütchen 2069
Pfaffenhütlein 2069
Pfaffenkraut 1398
Pfaffenpfötchen 2069
Pfaffenröhrlein 1398
Pfaffenröslein 2069
Pfahlrohr 2481
Pfannengras 4067
Pfauenlilie 1547
Pfeffer 4120
Pfefferbaum 4125
Pfeffergewächse 4121
Pfefferknöterich 3511
Pfefferkraut 2118, 4119, 4721, 5321

Pfefferkuchenbaum 2001
Pfefferling 1115
Pfefferminzbaum 5969
Pfefferminze 4122
Pfeifenbinse 3646
Pfeifengras 3646
Pfeifenholz 2523
Pfeifenstrauch 1461, 3641
Pfeilblatt 108
Pfeilkraut 327
Pfeilkresse 4127, 5918
Pfeilwurz 329, 541, 3924
Pfeilwurzgewächse 330
Pfennigkraut 2232, 3651, 4112
Pferdebinse 2600
Pferdebohne 2864
Pferdefleischholzbaum 1355
Pferdefuss 1388
Pferdeklee 5917
Pferdeminze 512
Pferdenzahnmais 1835
Pferdeschwanz 2225
Pferdesessel 2866
Pfingstblume 4118
Pfingstnelke 1129
Pfingstrose 4118
Pfirsich(baum) 4082
pfirsichblättrige Glockenblume 4083
Pfirsichpalme 4085, 5154
Pflaume 2424
Pflaumenaprikose 4336
Pflaumenbandflechte 5204
Pflaume(nbaum) 4224
pflaumenblättriger Apfelbaum 4091
pflaumenblättriger Schneeball 632
pflaumenblättriger Spierstrauch 775
Pflaumenpfeffer 1220
Pflaumenpilz 4226
pflaumfrüchtiger Wacholder 5418
Pflückerbse 2423
Pfriemengras 3819
Pfriemenkresse 387
Pfriemenstrauch 5857
Phazelie 4161
Pheespalme 3543
Phillostachys 4170
Photinie 4169
Phykomyzeteen 5824
Phyllokaktus 3247
Physostegie 3321
Phytolakkazeen 4244
Pieris 4176
Pillenbaum 5128
Pillenbrennessel 4547
Pillenfarn 1674, 4183
Pillenkraut 4183
Pillenriedgras 4182
Pillensegge 4182
Pilz 3758
Pilzalgen 5824
Pimentbaum 102
Pimpernuss 659, 1498
Pimpernussgewächse 660
Pimpinelle 4186
Pinangpalme 4187
Pinchots Wacholder 4440
Pindowapalme 365
Pindrows Tanne 4189
Pinheiro 4049
Pinie 2983
Pinksnelke 2588
Pinselgras 4092
Pinselhirse 4092
Pinselschimmel 4109
Pipal 738
Piperazeen 4121
Pippau 2725
Piptadenie 4203
Pirolazeen 4378
Pisang 439
Pisangfeige 439
Pistazie 4205
Pistie 5816
Pitangabaum 4206
Pitjungbaum 2324
Pittosporazeen 5537
Plantaginazeen 4220
Platanazeen 4216
Platane 4215
platanenblättriger Stinkbaum 1194
Platanengewächse 4216
Platterbse 4098, 4099
Platthalmrispe 970
Plossobaum 533
Plumeria 2360
Pogostemon 911
Polarbirke 1935
Polarweide 4245
Poleigränke 723
Polei(minze) 4113
Polemoniazeen 4167
polnischer Weizen 4247
Polyantherose 3068
Polygalazeen 3623
Polygaleen 3623
Polygonazeen 833
Polymnie 3248
polynesische Steinnusspalme 4644
Polyporazeen 4253
Pomaquinte 1334
Pomazeen 282
Pomeen 282
Pomelie 2578
Pomeranze(nbaum) 5092
Pompelmus 4329
Pontederiazeen 4175
Pontederie 4174
Pontederiengewächse 4175
pontische Alpenrose 4264
pontische Azalee 4264
pontische Eiche 4265
pontische Hasel 1071
pontischer Weissdorn 390
Porlinge 4253
Porree 3266
Porsch 1719
Porst 1719, 3265
portugiesische Eiche 4278
portugiesischer Kirschlorbeer 4277
portugiesische Zeder 2926
Portulacarie 4361
Portulak 4358
Portulakazeen 4359
Portulakgewächse 4359
Portulakröschen 1505
Porzellanblümchen 3362
Porzellanblume 5847
Potanins Birke 4280
Prachtaloe 3692
Prachtglocke 2468
prächtige Himbeere 4660
Prachtkerze 2438
Prachtlilie 2511
Prachtnelke 3298
Prachtsalbei 4749
Prachtscharte 2439
Prachtspiere 360
Prachtstrauch 2027
Prachtzierapfel 1180
Prärieklee 4290
Prärielilie 958
Präriemalve 1127
Preiselbeere 1645
Preisslauch 3266
Presshefe 1359
Priesterbaum 2510
Primel 4307
Primelgewächse 4308
Primelstrauch 5569
Primulazeen 4308
Pritchardie 4311
Proboszidea 1846
Prophetenblume 4314
Proteusgewächs 4318
Provencerrose 2371
Provinzrose 2371
Prunelle 4816
Prunkbohne 4748
Prunkspiere 4090
Prunkwinde 3682, 5220
Pseudobakterie 4319
Pteridophyten 4321
Puffbohne 788
Puffmais 4268
Pulai 165
Pulqueagave 4328
Pulverholz 2513
Pumpelrose 4118
Punktfarn 2828
Punktfelberich 5169
punktierter Gilbweiderich 5169
Punschpflanze 3271
Purgierbeere 1369
Purgierdolde 1817
Purgierkroton 4334
Purgierlein 4335
Purgiermoos 2916
Purgiernuss 449, 3838
Purgierstrauch 5116
Purgierwinde 3005, 4741
Purpuralgen 4433
Purpurenzian 4346
Purpurfetthenne 4354
Purpurfichte 4341
Purpurfuchsie 3422
Purpurginster 4339
Purpurglöckchen 171
Purpurgranadille 4070
Purpurhasenlattich 4353
Purpurkissen 366
Purpurklee 2351
Purpurknabenkraut 3200
purpurner Enzian 4346
purpurner Fingerhut 1416

purpurner Storchschnabel 4342
purpurroter Fingerhut 1416
purpurrotes Knabenkraut 3200
purpurrote Sommerwurz 694
Purpurskabiose 5383
Purpurtang 3241
Purpurtanne 1039
Purpurtrichterwinde 1474
Purpurweide 4352
Purpurwicke 4356
Purpurwinde 1474
Purshs Kreuzdorn 1040
Puschkinie 4362
pyramidenförmige Pappel 3360
Pyramidenkönigskerze 4368
Pyramidenpappel 3360
Pyrenäen-Eiche 4374
Pyrenäen-Milchstern 478
Pyrenäen-Stiefmütterchen 2856
pyrenäische Aster 4370
pyrenäische Eiche 4374
pyrenäischer Enzian 4371
pyrenäischer Mohn 4375
pyrenäischer Storchschnabel 4372
Pyrenomyzeteen 2277

Quantelbeerbaum 2428
Quassiaholzbaum 4382
Quebrachobaum 4461, 5907
Quecke 1640
Queckentrespe 5034
Quellenbaum 3409
Queller 2500
Quellkraut 2938
Quellmoos 5823
Quendel 464, 926, 5509
quendelblättriger Ehrenpreis 5512
quendelblättriges Sandkraut 5511
Quendelehrenpreis 5512
Quendelsandkraut 5511
Quendelseide 1400
Quetschblume 3467, 5616
Quetsche 2114
Quetschgurke 5616
Quicke 2114
Quillajabaum 5070, 5071
Quirinkraut 1388
quirlblättrige Schirmtanne 5682
quirlblättriges Läusekraut 5925
quirlblättriges Salomonssiegel 5927
quirlblättrige Weisswurz 5927
quirlblütige Borstenhirse 2844
quirliger Tannenwedel 3489
quirlige Salbei 3299
quirliges Knorpelkraut 5924
quirliges Tausendblatt 974
Quirlsalbei 3299
Quitschbeere 2114
Quitsche 2114
Quitschstrauch 2114
Quitte(nbaum) 4389
Quittenmispel 1633

Rabinschen 3211
Rachenblütler 2243
Rachenlilie 3415
Rachenschwertel 3415
Radbaum 5876
radblättrige Seekanne 2378
Rade 1391, 1610
Radendistel 5047
Radi 4399
Radies 2425
Radies(chen) 5030
Radmelde 5318
Raffiabastpalme 3408
Raffiaweinpalme 3408
Rafflesiazeen 4397
Ragi 4399
Ragwurz 3946
Rahmapfel 849, 5297
Raigras 4631
Rainbeere 1369
Rainfarn 1542, 5440
rainfarnblättrige Büschelblume 5443
Rainkohl 3860
Rainsalat 3860
Rainweide 4312
Rajanie 5623
Raketenblume 5551
Rambutan 4327, 4409
Ramie 4410
Ramondie 4411
Ramsel 4250
Ramtilkraut 2063
Randspärkling 3490
Rangapfel 4068
Range 2607
Rangoonbohne 3309, 4940
Ranunkel 889
Ranunkelstrauch 3156
Ranunkulazeen 890
Ranunkulgewächse 890
Raoulie 4416
Raps 4417, 6005
Rapsdotter 4418
Rapünzchen 1618, 3211
Rapunze 3211
Rapunzelglockenblume 4414
Rasamalabaum 4419
Rasenbinse 845
Raseneiche 2041
Rasenriedgras 5630
Rasenschmiele 5628
Rasensimse 1822
Rasenspiere 4528
Rasensteinbrech 946
Rasenstrauch 1335
Rasenwurz 2823
rasige Haarsimse 1822
rasiger Steinbrech 946
rasige Segge 5630
rasiges Haargras 1822
rasiges Hornkraut 2221
rasiges Vergissmeinnicht 5627
Raspelstrauch 5477
Rasselblume 1737
Rasselstendel 3659
Rauhafer 3374
Rauhblattaster 3839
Rauhblattgewächse 735
Rauhblättrige 735
rauhblättrige Alpenrose 2430
rauhblättrige Rose 3491
rauhe Deutzie 2397
rauhe Gänsedistel 4305
rauhe Gänsekresse 2688
rauhe Nelke 1837
rauher Beinwell 4297
rauher Eibisch 2815
rauher Klee 4576
rauher Wiesenhafer 3553
rauhes Hornblatt 2861
rauhe Simse 5420
rauhes Knäuelgras 3943
rauhes Lieschgras 785
rauhes Rispengras 4579
rauhes Veilchen 2694
rauhes Vergissmeinnicht 1962
rauhe Wallwurz 4297
rauhhaarige Alpenrose 2430
rauhhaarige Gänsekresse 2688
rauhhaariger Fuchsschwanz 4464
rauhhaariger Sonnenhut 629
rauhhaarige Sommerzypresse 2680
rauhhaariges Schamkraut 4111
rauhhaarige Wicke 5531
Rauhhülse 4421
Rauhkarde 2390
Rauhlilie 5090
rauhsamige Vogelwicke 6051
Rauhschopf 5090
Rauhweizen 4286
Rauke 4978
raukenblättriges Greiskraut 3809
raukenblättriges Kreuzkraut 3809
Raukenkohl 4525
Raukensenf 4978, 5442
Raupenähre 5015
Raupenkraut 4767
Rauschbaum 1120
Rauschbeere 717, 1706
Rauschpfeffer 3140
Raute 4597
Rautenanemone 4598
Rautenfarn 2577
Rautengewächse 4599
Rauwolfia 1844
Rebe 2575
Rebendolde 5799
Rebengewächse 2576
Rebhunbeere 4066
Rechenblume 5373
Redwood 4829
Regenbaum 4404
rehbrauner Dachpilz 2189
Rehders Flügelnuss 4477
Rehfellchen 2922
Rehmannie 4478
Rehpilz 2922
reichblütiger Zierapfel 3037
reichblütiges Habichtskraut 6082
Reifpilz 6057
Reifweide 1801
Reiherschnabel 2789
Reineckie 4480
Reineclaude 2627
Reis 4495
Reismelde 4390
Reispapierbaum 4500
Reisquecke 1753, 4497
Reitgras 4475
Reneklode 2627
Rentierflechte 4479

Rentiermoos 4479
Reps(kohl) 6005
Reseda 3614
Resedagewächse 3615
Resedazeen 3615
Rettich 2425, 4395
Rexbegonie 355
Rhabarber 4492
Rhamnazeen 831
Rhapontik 2426
Rheindorn 1523
Rhizoctonia 4489
Rhododendron 4490
Rhodophyzeen 4433
Ribisel 3882
Ribitzel 3882
Richardsons Weide 4501
Ried 1514
Riedgräser 4811
Riedgrasgewächse 4811
Riemenblatt 3126
Riemenblume 3377, 3380
Riemenblumengewächse 3632
Riemenblüte 3380
Riemenlippe 2673
Riementang 667
Riemenzunge 3346
Riesenbambus 2467
Riesenblume 3664
Riesenbohne 2997
Riesenbovist 2479
Riesenerdbeere 2722
Riesenfenchel 2471
Riesenhülse 1265, 2046
Riesenhyazinthe 5319
Riesenkaktus 4645
Riesenklee 5917
Riesenknöterich 3036
Riesenkürbis 6008
Riesenlebensbaum 2484
Riesenlorbeer 5098
Riesenorange 4329
Riesenroggen 4247
Riesenschilf 2482
Riesenschirmling 4051
Riesenschwingel 2472
Riesensequoie 2483
Riesensüssgras 806
riesige Lilie 2478
riesige Palmlilie 846
Riffelbeere 1645
Rindenpilze 3256
Rindsauge 3986
Rindswurz 633
Ringblume 2460
Ringelblume 929
Ringelrose 929
ringloser Wulstling 4853
Rippenfarn 1821, 4726
Rippenmangold 5398
Rispal 2916
Rispe 700
rispenblütige Waldrebe 5353
Rispenfarn 3974
Rispengipskraut 397
Rispengras 700
Rispenhafer 1481
Rispenhirse 4315
Rispenstrauch 2181
rispige Flockenblume 5165
rispige Hortensie 4028
rispiger Amarant 3387
rispiger Eisenhut 4029
rispige Segge 4030
rispiges Riedgras 4030
rispige Waldrebe 5353
Ritschling 1827
Rittersporn 3226
Ritterstern 176, 526
Riviers Amorphophallus 1848
Rivine 4575
Rivinie 4575
Rizinus 1051
Robertskraut 2787
Robesons Weissdorn 4508
Robinie 3354
Rocambole 2474
Rodel 3478
Roggen 4630
Roggenrose 1391
Roggentrespe 1138
Rohr 1514, 2481
rohrartige Marante 541
rohrartige Pfeilwurz 541
rohrartiger Fuchsschwanz 1668
rohrartiger Schwingel 4474
rohrartiges Fuchsschwanzgras 1668
rohrartiges Reitgras 1131
rohrartiges Silbergras 1320
Röhrenkassie 2548
Röhrenrosskümmel 5625
Röhrenschwamm 727
Rohrfuchsschwanz 1668
Rohrglanzgras 4473
Rohrgras 994
röhrige Nebendolde 5625
röhrige Rebendolde 5625
Rohrkolben 1062
Rohrkolbengewächse 1063
Rohrkolbenhirse 4092
Rohrkümmel 1729
Röhrling 727
Rohrpalme 4424
Rohrpfeilwurz 541
Rohrpompe 1062
Rohrschilf 4472
Rohrschwingel 4474
Rollfarn 4512
Romanzoffie 3633
Römerie 344
römische Bete 5398
römische Hundskamille 4546
römische Kamille 4546
römischer Beifuss 4548
römischer Kohl 5398
römischer Koriander 2415
römischer Kümmel 1729
römischer Salat 4545
römische Salbei 1254
römische Wicke 3794
Rondeletie 4549
rosablättriger Helmling 4086
rosablättriger Krempling 4226
rosarotes Weidenröschen 4012
Rosazeen 4560
Rose 4554
Rosellahanf 4562
Rosenapfel 287, 4556
Rosenapfel(baum) 1858
Rosenbaum 4490
Rosenbinse 845
Roseneibisch 1185
Rosengeranium 4566
Rosengewächse 4560
Rosenginster 4339
Rosenholz 474, 4570
Rosenklee 1694
Rosenkohl 823
Rosenkranzbaum 3059
Rosenkranzpappel 1978
Rosenlapagerie 4445
Rosenlorbeer 1482
Rosenmalve 2830
Rosenmelde 5639
Rosenmyrte 4565
Rosenpappel 2830, 3921
rosenrote Robinie 4555
Rosenwirsing 823
Rosenwurz 4567
Rose von Sharon 1
rosiges Weidenröschen 4012
Rosmarienweide 5387
Rosmarin 4563
Rosmarinheide 241, 723
Rossfenchel 2251, 2866
Rosshuf 1388
Rosskastanie 828
Rosskastanienbaum 828
Rosskastaniengewächse 2865
Rosskümmel 1438, 4124
Rossmalve 2797
Rossminze 2869
Rosspappel 2797
rostblättrige Alpenrose 4530
rostbraune Kopfsimse 4624
rostbraunes Kopfried 4624
rostfarbiges Kopfried 4624
Rosthorns Spierstrauch 4572
Röstling 1827
Rostpilze 4623
rostrote Alpenrose 4530
Rotalgen 4433
Rotangpalme 4424, 4574
rotbeeriger Wacholder 4298
rotbeerige Zaunrübe 4439
rotblättrige Rose 4454
rotblühende Johannisbeere 6000
rotblühende Pavie 4441
rotbrauner Frauenschuh 2109
Rotbuche 2077
Röte 3410
rote Akazie 4555
rote Erle 2072
rote Eskallonie 4449
rote Fetthenne 4354
Rötegewächse 3411
rote Heckenkirsche 2099
rote Heidelbeere 1645
rote Hühnerwurz 680
Roteibe 2045
Roteiche 3883, 5101
rote Immortelle 1421
rote Johannisbeere 3882
rote Magnolie 5897

rote Mangrove 200
rötender Wulstling 713
rote Ochsenzunge 1954
rote Pestwurz 4340
roter Ahorn 4455
roter Beinwell 1389
roter Blasenstrauch 3950
roter Fingerhut 1416
roter Fliegenpilz 2319
roter Fuchsschwanz 3387
roter Gauchheil 4747
roter Hartriegel 683
roter Holunder 2120
roter Hornmohn 652
roter Klee 4447
Roterle 2072, 4432
roter Maulbeerbaum 4456
rote Rosskastanie 4441
roter Pfeffer 884
roter Quebrachobaum 3379, 5967
roter Sandelholzbaum 4676, 4679
roter Schwingel 4450
roter Steinbrech 5663
rote Rübe 2408, 4438
roter Widerton 3431
roter Zahntrost 4435
rote Salbei 4749
Rotesche 4434
rote Schneebeere 2935
rote Seifenwurzel 743
rotes Fingergras 2683
rotes Frauenhaar 3431
rote Spargelbohne 5197
rote Spornblume 3119
rotes Quellried 4443
rotes Straussgras 1335
rote Taubnessel 4343
rote Traubenkirsche 1384
rote Zaunrübe 4439
rote Zeder 1980
rotfrüchtige Kutte 2089
rotfrüchtiger Apfelbeerstrauch 4446
rotfrüchtiger Wacholder 4168
rotfrüchtige Zaunrübe 4439
Rotfüsschen 4448
Rotfussröhrling 4448
rotgelbe Taglilie 2334
Rothäutchen 3940
Rothautröhrling 3940
Rotholz 1304
Rotholzbaum 768
Rothölzer 1303
Rotkappe 3940
Rotklee 1694, 4447
Rotkohl 4444
Rotkopf 3940
Rotkraut 4444
rötlicher Lacktrichterling 5850
rötliche Schuppenwurz 5549
rötliches Dickblatt 4467
rötliches Fingerkraut 4837
rotnerviger Ahorn 4470
rotrandige Narzisse 1397
Rotrüster 5040
Rotschwingel 1676, 4450
Rotulme 5012, 5040
Rotwurzel 3840
Roucoustrauch 256
Rouenflieder 1190
Rübe 520
Rübendistel 4486
Rübenkerbel 5651
Rübenkohl 585
Rubiazeen 3411
Rübkohl 3175
Rübling 1333
Rübsen 585, 3817
Ruchgras 5724
Rudbeckie 1573
Ruellie 3478
Ruhmesblume 4059
Ruhm(es)krone 2511
Ruhrkraut 1727
Rührmichnichtan 5560
Ruhrrindenbaum 4971
Ruhrwurz 5555
Rujastrauch 1527
Rüllsaat 2168
rumelische Weymoutskiefer 419
rundblättrige Esche 4583
rundblättrige Felsenbirne 2428
rundblättrige Glockenblume 690
rundblättrige Malve 1945
rundblättrige Minze 283
rundblättrige Platterbse 4587
rundblättriger Ahorn 5735
rundblättriger Sonnentau 4588
rundblättriger Steinbrech 791
rundblättriger Storchschnabel 4584
rundblättriges Hasenohr 4589
rundblättriges Wintergrün 2119
rundblättrige Weide 3253
rundblättrige Wucherblume 2887
rundes Hasenohr 4589
runde Teufelskralle 420
rundkapselige Jute 4591
rundköpfiger Lauch 421
Rundmorchel 1473
Rundpflaume 2627
Runzelbart 314
Runzelerle 2732
runzeliger Rapsdotter 4603
runzeliger Windsbock 4603
Runzelrose 4602
Runzelschüppling 6057
Rupfsalat 1759
Ruppie 5933
Ruprechtskraut 2787
Ruprechts Leimkraut 4607
Ruprechtsstorchschnabel 2787
Rüssellilie 1738
Rüsselschwertel 698
russische Flockenblume 4627
russischer Bocksdorn 4612
russischer Klee 3159
russischer Senf 2939
russisches Fingerkraut 1893
russisches Salzkraut 1518
russische Tamariske 3909
russische Weide 4620
Russkopf 1859
Rüster 2018
rüsterblättrige Brombeere 2020
Rutazeen 4599
rutenförmige Hirse 5403
Rutenkaktus 4488
Rutenmelde 2186
Rutenmorchel 5256
Rutenpalme 3201
Rutenpilze 4162
Rutenweiderich 5782
Rutenwolfsmilch 3251

Saarbaum 647
Saarbuche 647
Saarvogelfuss 1526
Saaterbse 2423
Saatesparsette 1519
Saatgerste 455
Saathafer 1481
Saathanf 2773
Saatkuhnelke 1653
Saatlein 2214
Saatleindotter 572
Saatlinse 1457
Saatmohn 3367
Saatplatterbse 2587
Saatrübe 5650
Saatsiegwurz 1612
Saatwicke 1554
Saatwucherblume 1609
Sabadill(a) 1914, 4633
Sabal(palme) 909, 4017
Sabinerbaum 4719
sabinischer Wacholder 4719
Sachalinknöterich 4656
Sachalintanne 4655
Sachsenstern 1535
Säckelblume 1080, 2049
Sadewacholder 4719
Saflor 4636
Safranholzbaum 1048
Safrankrokus 4638
Safranlilie 3939
Safranwurz 5648
Saftling 2904
Saftschopf 1092
Sagobaum 1764, 5041
Sagobäume 1765
Sagopalme 1764, 5041
Sagwirenpalme 2558
Salai 2937
Salamanderbaum 566
Salat 2417, 3281
Salatrapünzchen 3211
Salatrübe 4438
Salbaum 4872
Salbei 4639
salbeiblättrige Zistrose 4670
Salbeigamander 6028
Salde 5933
Sale 2523
Salikazeen 5965
Salomonsnuss 1887
Salomonssiegel 5082
Salomons Wundernuss 1887
Salpeterstrauch 3862
Salviniazeen 4672
Salweide 2523
Salzbaum 4667
Salzbinse 4234, 4666
Salzbunge 807
Salzhornkraut 3514
Salzkraut 2500, 4618

Salzmelde 4813, 5208
Salzschwaden 96
Salzsimse 5420
Salzspärkling 4665
Salzsteppenstrauch 4729
Salzstrauch 4667
Salzwermut 3496
Samaunbaum 2324
Sammetblume 3493
Sammetgras 2713
Samtbaum 226
Samtblatt 4398, 6047
Samtblume 174, 3493
Samtfussrübling 5709
Samtfusswinterpilz 5709
Samtgras 2713
Samtkraut 4262
Samtpalme 3229
Samtpappel 7, 4925
Samtröschen 2037
Samttrespe 5078
Sandalgewächse 4678
Sandarakbaum 296, 1777
Sandarakzypresse 296, 1777
Sandbirke 2131
Sandbirne 4687
Sanddorn 1523, 4787
Sandelholzbaum 4677, 5908
Sandelholzgewächse 4678
Sanderbse 2693
Sandesparsette 2891
Sandferkelkraut 5036
Sandginster 2685
Sandglöckchen 4857
Sandgras 487
Sandhaargras 2094
Sandhafer 3374
Sandhalm 487, 2076
Sandheide 4772
Sandhornkraut 5853
Sandimmortelle 217, 5991
Sandkirsche 4683
Sandknöpfchen 3082
Sandkraut 4692
Sandlieschgras 4690
Sandluzerne 472
Sandmohn 2153
Sandnelke 2255
Sandoribaum 4695
Sandpilz 5700
Sandreitgras 1131
Sandriedgras 4688
Sandroggen 2094
Sandrohr 487, 2076
Sandröhrling 5700
Sandschaumkresse 4518
Sandschilf 2076
Sandschwingel 1676
Sandsegge 4688
Sandsilbergras 2595
Sandstrohblume 6083
Sandveilchen 4691
Sandverbene 5
Sandwegerich 5926
Sandwicke 2693
Sanikel 4693
Sankt-Peter-Stab 2101
Sansibarerbse 3755
Santalazeen 4678
Santelgewächse 4678
Sanvitalie 4697
Saphirbeere 5373
Sapindazeen 5073
Sapindengewächse 5073
Sapindusfichte 3964
Sapotazeen 4701
Sapotengewächse 4701
Sapotillbaum 4700
Sappanholzbaum 4702
Saprolegniazeen 4703
Saranahlilie 3130
Sarepta-Senf 2939
Sargents Berberitze 4706
Sargents Fichte 4709
Sargents Pappel 4214
Sargents Spierstrauch 4708
Sargents Wacholder 4707
Saribupalme 3087
Sarmentazeen 2576
Sarrazeniazeen 204
Sassafras 4712
Sassafraslorbeer 1520
Satanpilz 679
Saturei 4721
Saubohne 788
Saudistel 1531, 5105
Sauerampfer 2429
Sauerapfel 5093
Sauerbaum 5094
Sauerdorn 452, 2075
Sauerdörner 453
Sauerdorngewächse 453
sauere Sobbe 5093
Sauergräser 4811
Sauerkirschbaum 5091
Sauerkirsche 3483, 5091
Sauerklee 3985
Sauerkleegewächse 5088
Säuerling 138, 3719
Sauersack 5093
Sauer(weichsel)kirsche 3679
Sauerzitrone 3268
Saufenchel 1438
Saukraut 643, 2418
Sa(u)lbaum 4661
Säulenblume 5289
Säulenkaktus 1100, 1991
Säulenschirmpalme 5680
Saumfarn 754
Saumnarbe 2198
Saumpilz 2907
saure Blattblume 3977
saure Limette 3311
Saururazeen 3348
Saustockkraut 2418
Savoyer Habichtskraut 4723
Savoyerkohl 4722
Sawarascheinzypresse 4724
Sawarazypresse 4724
Saxaul 4729
Saxifragazeen 4732
Schabenkönigskerze 3690
Schabenkraut 3096, 3690
Schabziegerklee 711
Schachblume 1126
Schachbrettblume 1126
Schachtelhalm 2875
Schachtelhalmgewächse 2877
Schachtelhalmmeerträubchen 3652
Schafbeere 3792
Schafchampignon 2870
Schafegerling 2870
Schafgarbe 6067
Schaflinse 1713
Schafpilz 582
Schafschwingel 4858
Schafskabiose 4857
Schafsnase 2516
schaftlose Primel 2042
Schalmeienrohr 2482
Schalotte 4844
Schamblume 466, 4178
schamhafte Sinnpflanze 4827
Schampflanze 3109
Schantungkohl 4158
Scharbockskraut 1522, 2241, 4181
scharfer Blasenzug 4227
scharfer Hahnenfuss 5425
scharfer Mauerpfeffer 2553
scharfes Berufkraut 602
scharfe Wolfsmilch 3251
Scharfhülse 4421
Scharfkraut 3419
Schärfling 4717
Scharlach 1254
Scharlachbeere 4243
Scharlachdorn 1976, 5485
Scharlacheiche 4746
Scharlachlilie 1111
Scharlachquitte 3038
scharlachrote Salbei 4749
Scharte 2439, 4728
Schattenblume 492, 5081
Schattenpalme 5424
Schaumblüte 2321
Schaumkraut 600
Scheelkorn 4076
Scheibenflechte 3262
Scheibenklee 903
Scheibenkraut 4864
Scheibenschötchen 4864
Scheidengras 3741
Scheidenpilz 4853
Scheidenschwamm 4853
Scheidenstreifling 4853
Scheinakazie 636, 3354
Scheinbuche 2163
Scheineibe 4232
Scheineller 5323
Scheinerle 5323
Scheinglocke 3163
Scheingräser 4811
Scheinhasel 6004
Scheinkamelie 5245
Scheinkastanie 2145
Scheinkerrie 635, 3099
Scheinknöterich 6010
Scheinlärche 2539
Scheinmalve 2173
Scheinmohn 708, 5864
Scheinmyrte 2264
Scheinquitte 2312
Scheinwanzenblume 1596
Scheinziest 550

Scheinzypresse 2167
Schellbeere 1273
Schellenblume 528, 3302
Schellkraut 1085
Schellwurz 1085
Scherte 2130
Scheuchzeriazeen 3116, 4755
Scheuchzerie 4754
Scheuchzers Glockenblume 4756
Scheuerkraut 2225
Schibaum 898
Schicksalbaum 2510
Schiebikenstrauch 2095
Schiefblatt 523
Schiefteller 15
schierlingsblättriger Reiherschnabel 91
Schierlingsreiherschnabel 91
Schierlingssilge 2770
Schierlingstanne 973, 2769
Schildampfer 2373
Schildblatt 4731
Schildblume 353, 1352, 5655
Schildehrenpreis 3518
Schildfarn 354, 2828
Schildflechte 2648, 4056
Schildklee 2891
Schildkraut 172, 1297, 5655
Schildkrötenpflanze 5558
Schildnarbe 1352
Schildträger 4987
Schilf 852, 2481
Schilfpalme 4574, 5480
Schilf(rohr) 4472
Schillergras 3173
Schimmelfichte 5913
Schimmelkraut 2318
Schimmelweide 1801
Schimmerbaum 4318
Schindeleiche 4865
Schinussbaum 4852
Schirmakazie 79
Schirmblatt 5678
Schirmfichte 5682
Schirmmagnolie 5679
Schirmpalme 1629, 3572
Schirmpflanzen 1036
Schirmpilz 3278
Schirmtanne 5681
Schizäazeen 1742
Schizomyzeteen 2266
schlaffes Rispengras 3725
Schlafkraut 633
Schlafmohn 3933
Schlammkraut 3745
Schlammriedgras 3744
Schlammschachtelhalm 5344
Schlammsegge 3744
Schlammwasserstern 4261
Schlangenbart 3307
Schlangenbeere 1164
Schlangenfackeldistel 1100
Schlangengras 649
Schlangenhaargurke 5045
Schlangenholzbaum 3898
Schlangenkaktus 4422
Schlangenknoblauch 2474
Schlangenkraut 949, 5655
Schlangenlauch 2474
Schlangenlilie 1900
Schlangenmelone 5046
Schlangenmoos 4606
Schlangenschmiele 5843
Schlangenwurz 949
schlanke Hainsimse 2336
schlanke Simse 2336
schlanke Wicke 5004
Schlauchpflanze 4207
Schlauchpilze 337
Schlauchschwämme 337
Schleh(busch) 5014
Schlehdorn 5014
Schlehe(ndorn) 5014
Schleiergipskraut 397
Schleiergras 261
Schleierkraut 397, 2672
Schleifenblume 993
Schleimalge 3889
Schleimapfelbaum 408
Schleimkraut 5835
Schleimling 3889
Schleimpilze 5006
Schleimsame 1332
Schlenke 1796
schlesische Weide 4943
Schleudersamengras 1906
Schlingbohne 1883
Schlinge 5852
Schlinggras 5981
schlitzblättrige Karde 1760
schlitzblättriger Sonnenhut 1755
schlitzblättriger Storchschnabel 1757
schlitzblättriger Wegerich 1708
Schlitzblattwegerich 1708
Schlitzblume 893
Schlüpfsame 1714
Schlüsselblume 4307
Schlüsselblumengewächse 4308
Schmalblattlungenkraut 1651
schmalblättrige Hainsimse 2652
schmalblättrige Kurkume 1983
schmalblättrige Lorbeerrose 3210
schmalblättrige Lupine 703
schmalblättrige Ölweide 4616
schmalblättriger Doppelsame 5007
schmalblättriger Feuerdorn 3801
schmalblättriger Hohlzahn 3805
schmalblättriger Klee 2250
schmalblättriger Lein 3802
schmalblättriger Merk 3811
schmalblättriger Quendel 3688
schmalblättriger Rohrkolben 3800
schmalblättriger Rüsselschwertel 1364
schmalblättrige Simse 2652
schmalblättriges Lungenkraut 1651
schmalblättriges Rispengras 3799
schmalblättrige Steinlinde 3808
schmalblättrige Wicke 3810
Schmalblattwicke 3810
schmaler Doppelsame 5007
schmaler Merk 545
schmaler Streifenfarn 2329
schmale Wiesenraute 3806
Schmalfrucht 2263
Schmalkopfmohn 3367
Schmalwand 3733
Schmalz 2168
Schmalzblume 1466
Schmalzmelde 4802
Schmeerwurz 622
Schmeerwurzgewächse 6064
Schmelzkraut 52
Schmergel 2561
Schmerling 2574
Schmerwurz 622
Schmerzwurzel 3341
Schmetterlingsblütler 4041
Schmetterlingsporling 1569
Schmetterlingsstrauch 892
Schmiele 2677
Schminkbeere 4243
Schminkbohne 3158
Schminkwurz 1954
Schmirgel 1466
Schmuckkörbchen 1392, 1630
Schmucklilie 54
Schmuckmalve 7, 1127
Schmucktanne 297
Schmuckzypresse 1777
Schnabelnachtschatten 835
Schnabelried 496
Schnabelsalde 5934
Schnabelschötchen 2067
Schnabelsimse 496
Schnalle 1616
Schneckenblume 2460
Schneckenklee 3565
Schneckensame 4860
Schneeball(strauch) 5733
Schneebaum 2379
Schneebeere 2379, 3617, 5057
Schneebirne 5066
Schnee-Enzian 5063
Schneefingerkraut 5069
Schneeflocke 2144
Schneeflockenbaum 2379
Schneeflockenblume 2379
Schneeflockenstrauch 2379
Schneeglanz 2512
Schneeglöckchen 1530, 5060, 5188
Schneeglöckchenbaum 4950
Schneeglocke 5068
Schneehahnenfuss 5059
Schneelilie 5188
Schneepilz 1859
Schneerose 1226, 2766
Schneespierstrauch 2431
Schneesteinbrech 5055
Schneestolz 2512
Schneetröpfchen 1530
Schneeweide 2596
Schneide(gras) 4727
Schneiders Deutzie 4759
Schnellefeu 1268
Schnittkohl 4626
Schnittlauch 1219
Schnittmangold 1116
Schnittsalat 1759
Schnurbaum 4001
Schokoladenbaum 912
Schöllkraut 1085
Schöllwurz 2606
Schönbeere 2370

Schönblatt 508
schöne Meerzwiebel 5223
schöne Palmlilie 3692
schöner Ampfer 2216
schöner Blaustern 5223
Schönhäutchen 5130
Schönmalve 7
Schönminze 4720
Schönmütze 2065
Schopffackeldistel 3584
schopfige Bisamhyazinthe 5450
Schopfkerzenkaktus 1095
Schopflilie 822
Schopfpalme 18
Schopftintenpilz 4843
Schopftintling 4843
Schopfträubel 5450
Schotenbaum 2840
Schotendorn 10, 1439
Schotendotter 3772
Schotenklee 1824
Schotenkresse 3733
Schotenpfeffer 884
Schöterich 3772
Schotie 3124
schottische Mutterwurz 4774
schottische Zaunrose 5359
Schotts Erdpalme 316
Schraubenalge 4260
Schraubenbaum 4779, 4781
Schraubenbaumgewächse 4780
Schraubenbohne 3593
Schraubenorchis 3197
Schraubenpalme 1521, 4779
Schraubenpalmengewächse 4780
Schraubenstengel 5159
Schraubenvallisnerie 5159
Schrencks Pfeifenstrauch 4760
Schriftfarn 4737
Schuberts Lauch 4762
Schuppeneibe 2894
Schuppenflechte 1103
Schuppenheide 1049
Schuppenkopf 1094
Schuppenried 1302
Schuppenrindenhickory 4841
Schuppenschwanz 4923
Schuppentanne 4049
Schuppenwurz 3230
Schuppenzeder 2925
schuppiger Porling 1921
schuppiger Sägeblättling 4739
schuppiger Schwarzfussporling 1921
Schüsselflechte 4056
Schusserbaum 1319
Schusterpalme 353, 367
Schuttbingelkraut 2785
Schwabenkorn 5124
Schwaden 3476
Schwalbenkraut 2606
Schwalbenschwanz 3620, 5916
Schwalbenschwanzgewächse 3621
Schwalbenwurz 5340
Schwalbenwurzenzian 3622
Schwammbeere 1121
Schwammgurke 5290, 5564
Schwammkürbis 5564
Schwanblume 5348
Schwanenblume 2132, 2313
Schwanenblumengewächse 2315
Schwanenkraut 2314
Schwanzblume 270
Schwanzsame 4856
Schwarzahorn 637
Schwarzbirke 4504
Schwarzbuche 2849
Schwarzdorn 5014
schwarze Akazie 2637
schwarze Aprikose 4336
schwarze Ballote 614
schwarze Christblume 1226
schwarze Christwurz 1226
schwarze Erbse 3794
schwarze Flockenblume 625
schwarze Heckenkirsche 631
schwarze Himbeere 624
Schwarzeiche 5827
schwarze Johannisbeere 2081
schwarze Königskerze 641
schwarze Kopfsimse 621
schwarze Krähenbeere 627
schwarze Linde 199
schwärzender Saftling 1577
schwarze Nieswurz 630, 1226
schwarze Platterbse 645
schwarzer Andorn 425, 614
schwarzer Geissklee 5137
schwarzer Goldregen 5137
schwarzer Gottvergess 614
schwarzer Holunder 2095
schwarzer Kanarienbaum 623
schwarzer Koriander 2415
schwarzer Kreuzkümmel 2415
Schwarzerle 2072
schwarzer Maulbeerbaum 640
schwarzer Nachtschatten 643
schwarzer Pfeffer 646
schwarzer Senf 642
schwarzer Streifenfarn 651
schwarzer Walnussbaum 1975
schwarzer Weissdorn 2082
schwarzes Bilsenkraut 633
Schwarzesche 613
schwarze Segge 650
schwarzes Geissblatt 631
schwarzes Kopfried 621
schwarze Teufelskralle 639
schwarze Tollkirsche 525
schwarze Walnuss 1975
schwarze Weide 654
schwarze Zwergmispel 618
schwarzfaseriger Ritterling 1859
schwarzfrüchtiges Christophskraut 616
Schwarzfüllhorn 2858
Schwarzholzakazie 655
Schwarzkiefer 380, 653
Schwarzkohl 642
Schwarzkraut 616
Schwarzkümmel 2201, 2415
schwärzliche Marbel 5296
Schwarzlinde 199
Schwarzmundgewächse 3581
Schwarznessel 425, 614, 4138
Schwarznuss 1975
Schwarzpappel 647
Schwarzs Widdringtonia 4763
Schwarzweide 654
schwarzwerdender Geissklee 5137
schwarzwerdende Weide 693
Schwarzwurz 1345, 1389
Schwarzwurzel 649, 4831
Schwedenklee 164
schwedische Eberesche 5350
schwedische Luzerne 4921
schwedischer Drachenkopf 4629
schwedischer Hartriegel 1937
schwedischer Kaffee 5349
Schwefelblume 270
schwefelgelbe Koralle 2531, 4013
Schwefelkopf 2907
Schwefelporling 5312
Schwefelwurz 3738
Schweifblume 270
Schweinekraut 5941
Schweinewurz 949
Schweinrüssel 698
Schweinsgummibaum 5414
Schweinsohr 5941
Schweinspflaume 6101
Schweizer Klee 1519
Schweizer Mangold 5398
Schweizer Mannsschield 5401
Schwerins Steinapfel 1172
schwertblättriger Alant 5407
Schwertbohne 1448, 5406
Schwertel 2499
Schwertlilie 698, 2958
Schwertliliengewächse 2959
Schwimmblatt 4671
Schwimmblattgewächse 4672
schwimmender Froschbiss 2382
schwimmendes Froschkraut 2299
schwimmendes Laichkraut 2298
schwimmende Wassernuss 5794
Schwimmfarn 4671
Schwimmfarne 4672
Schwimmfarngewächse 4672
Schwindelkorn 1598, 1804
Schwindling 3484
Schwingel 2208
Schwingelgras 2208
Schwingelschilf 4505
Sebadille 1914
sechsblättrige Stauntonia 3071
sechsmänniger Tännel 4982
Sechszack 4877
sechszeilige Gerste 4981
Seebeere 4786
Seedorn 1523
Seefenchel 4675
Seeföhre 1296
Seegras 1997
Seegrasgewächse 1998
seegrüne Quecke 2952
seegrüne Weide 2596
Seehaide 2361
Seeigelkaktus 2749
Seekanne 2297
Seekiefer 1296
Seekohl 1326
Seekokos 1887
Seekreuzdorn 1523
Seelenholz 2099

Seelilie 2132
Seemangold 4785
Seemannstreu 4789
Seerose 5817
Seerosengewächse 5818
Seesimse 2600
Seetang 4538
Seetraube 4788
Segge 4810
seggenartige Kobresie 1016
Seide 1870
Seidelbast 1800
Seidelbastgewächse 3608
Seidenakazie 4948
Seideneiche 2638
Seidengewächse 1871
Seidengras 4228
Seidenkautschuk 4946
Seidenpflanze 3620
Seidenpflanzengewächse 3621
Seifenbaum 5070, 5071, 5072, 5102
Seifenbaumgewächse 5073
Seifenblume 5072
Seifenkraut 5074
Seifennussbaum 5072
Seilgras 4552
Selaginellazeen 4815
selbstkletternde Jungfernrebe 5748
Sellerie 1086, 5943
Semmelpilz 2752, 5700
Semmelstachelpilz 2752
Semmelstoppelpilz 2752
Senegal-Akazie 2662
Senf 3771
Sennesblätterstrauch 89
Sennesstrauch 4823
Sensendüwel 265
serbische Fichte 4830
Serjania 5335
Serradella 587, 1526
Sesam 3961, 4834
Sesbanie 4835
Seschellennuss 1887
Sesel 4836
Seychellennuss 1887
Seychellenpalme 1887
Shortie 4881
Sibbaldie 4889
sibirische Aster 4892
sibirische Berberitze 4893
sibirische Doppelblume 4896
sibirische Fichte 4915
sibirische Flammenblume 4911
sibirische Glockenblume 4894
sibirische Hundswolle 4291
sibirische Kiefer 4917
sibirische Kreuzblume 4909
sibirische Lärche 4908
sibirische Meerzwiebel 4916
sibirischer Apfelbaum 4899
sibirischer Aprikosenbaum 4891
sibirischer Bärenklau 4898
sibirischer Blaustern 4916
sibirischer Hartriegel 4901
sibirischer Lerchensporn 4897
sibirischer Storchschnabel 4905
sibirischer Wacholder 3703
sibirische Schwertlilie 4907
sibirisches Felsenblümchen 4902
sibirisches Futterkraut 4898
sibirisches Korn 4247
sibirisches Leimkraut 4895
sibirische Tanne 4904
sibirische Wucherblume 3179
Sibthorps Salbei 4919
sichelförmiger Schildfarn 2881
Sichelhasenohr 2156
Sichelklee 4921
Sichelluzerne 4921
Sichelmöhre 2155
Sichelpunktfarn 2881
Sichelschildfarn 2881
Sicheltanne 1718
Sichelwolfsmilch 4922
siebenbürgisches Perlgras 5570
Siebenstern 5218
Siebenundsiebzigerlei 711
Siebenzeit 2203
Siebolds Ahorn 4934
Siebolds Berberitze 4930
Siebolds Buche 4931
Siebolds Deutzie 4932
Siebolds Hasel 3035
Siebolds Hemlocktanne 4933
Siebolds Pappel 4929
Siebolds Schneeball 4935
Siebolds Walnussbaum 4936
Siegesbeckie 4653
Siegmarskraut 2830
Siegmarswurz 2830
Siegwurz 2499, 3368
Sievabohne 4940
Silau 4124
Silberahorn 4966
Silberbaum 920, 2492, 4964
Silberbäume 3926
Silberblatt 2834
silberblättrige Sonnenblume 4963
Silberdistel 669, 1017
Silberfingerkraut 4954
Silbergras 1284, 2550, 4713
Silberhafer 3333
Silberhülse 4962
silberiger Brandschopf 2192
Silberkappe 4962
Silberkerze 837
Silberkraut 339
Silberlinde 4965
Silbermarbel 2652
Silberpappel 5906
Silbersalzstrauch 4913
Silberscharte 3120
Silbertanne 5895
Silberweide 4951, 5922
Silberwein 4968
Silberwurz 1920
Silberzeder 364
Simarubazeen 66
Simse 852
Simsenlilie 2162
Sinau 1454, 3202
Sinfin 90
Singrün 4140
Sinnpflanze 3109, 3625
Sisalagave 4976
Sisalhanf 4976
Sissoo 4977
Sitkafichte 4979
Sitkaweide 4980
Sitkazypresse 3874
Sitter 2050
sizilianischer Sumach 4920
Skabiose 4735
Skabiosenflockenbaum 4734
skandinavische Johannisbeere 3882
Skariol 4299
Skimmie 4984
Skinners Akelei 4985
Skopolie 4765
Skorbutkraut 1522
Skordienkraut 5806
Skorzonere 4831
Skrofelkraut 2242
Skrofulariazeen 2243
skytisches Lamm 4784
Sockelblume 2049
Sodakraut 1518
Sode 4813
Sojabohne 5106
Sokotraaloe 5076
Solanazeen 3854
Soldatenkraut 3534
Sommeradonisröschen 5317
Sommeraster 1158
Sommerblutströpfchen 5317
Sommerefeu 1268
Sommereiche 2041
Sommerendivie 4545
Sommerfetthenne 264
Sommerflieder 892
Sommerflockenblume 6078
Sommergerste 5672
Sommerhyazinthe 5319
Sommerknäuel 262
Sommerknotenblume 5322
Sommerlevkoje 263
Sommerlinde 561
Sommerlingsstrauch 892
Sommermajoran 5375
Sommermalve 3449
Sommerpappel 2788
Sommerraps 3817
Sommerrispengras 258
Sommerrübsen 585
Sommertamariske 2269
Sommertrüffel 4619
Sommerwurz 812, 990
Sommerwurzgewächse 813
Sommerziest 2758
Sommerzwiebel 2420
Sommerzypresse 5318
Sondrio 3275
Sonnenauge 2764
Sonnenbaum 2812
Sonnenblume 5329
Sonnenbraut 5052
Sonnendistel 5035
Sonnenflügel 5333
Sonnengold 2149
Sonnenhut 1573
Sonnenkraut 5325
Sonnenröschen 2387, 5334
Sonnenrose 5329
Sonnenstern 2740

Sonnentau 5325
Sonnentaugewächse 5326
Sonnenwende 1381, 2765
Sonnenwolfsmilch 5328
Sonnenzypresse 2812
Sophienkraut 5442
Sophienrauke 2296
Sophore 4001
Sorgho 5086
Sorghum 5086
Soskonblume 2049
Soulanges Magnolie 4715
Spadizifloren 5107
Spaltalgen 701
Spaltastfarne 1742
Spaltblume 893
Spaltglocke 2377
Spaltgriffel 1695
Spalthortensie 2903
Spaltkölbchen 3425
Spaltpilze 2266
spanische Artischocke 1015
spanische Eiche 2814
spanische Eier 2414
spanische Erbse 2587
spanische Fetthenne 5114
spanische Golddistel 5112
spanische Iris 5111
spanische Linse 2587, 3930
spanische Platane 5113
spanischer Flieder 1461
spanischer Klee 2891
spanischer Lauch 3266
spanischer Pfeffer 884, 4459
spanischer Wacholder 4298
spanische Schwarzwurzel 649
spanische Schwertlilie 5111
spanisches Rohr 4424
spanisches Süssholz 1459
spanische Tanne 5110
spanische Wicke 5378
spanische Zeder 2926, 4298
Sparganiazeen 872
Spargel 350, 1351
Spargelbohne 1824, 6066
Spargelerbse 1824
Spargelklee 5197
Spargelkohl 800
Spark 5196
Spärkling 4689
Sparmannie 5115
sparriger Schüppling 4740
Spars 350
Spartgras 2061
spätblühende Goldrute 3892
spätblühende Traubenkirsche 626
Spatelfaden 4668
spätes Rispengras 2345
Spätrispe 2345
Spatzenzunge 5116
Spatzenzungengewächse 3608
Speckmelde 2785
Speerblume 5119
Speierling 4833
Speiselorchel 1994
Speisemöhre 2410
Speisemorchel 1473
Speiserhabarber 2426
Speiserübe 1360, 2408
Speisetomate 1549
Speisetrüffel 5608
Spelling 847
Spelt(weizen) 5124
Spelz 5124
Spelzmais 4233
Sperbe 4833
Sperberkraut 2409, 5018
Spergel 5196
sperrige Binse 2747
sperriger Schotendotter 5175
sperriger Schöterich 5175
Sperrkraut 4246
Sperrkrautgewächse 4167
Sphagnazeen 4096
Sphagnum 5125
spiegelndes Laichkraut 4866
Spierstaude 3562
Spierstrauch 5160
spiessblättrige Weide 2697
Spiesseiche 4200
spiessförmige Weide 2697
Spiessleinkraut 4851
Spiessmelde 5118
Spiesstanne 1163
Spiessweide 2697
Spilling 847
Spinat 5148
Spindelbaum 2069
Spindelbaumgewächse 5203
Spindelkraut 1864
Spindelstrauch 2069
Spindelsträucher 5203
Spinnendistel 671
Spinnenpflanze 5128
Spiralhülse 272
Spitzahorn 3885
spitzblättrige Esche 3798
spitzblättriger Ahorn 3885
spitzblättriger Strahlengriffel 748
spitzblättrige Schwarzwurzel 3574
spitzblättrige Weide 4850
Spitzenblume 848, 5117
spitzes Pfeilkraut 3924
spitzes Silberblatt 4130
spitzhütiger Knollenblätterpilz 1842
Spitzklee 3702
Spitzklette 1306
Spitzmorchel 1578
Spitzorchis 234
Spitzwegerich 829
Sporenpflanzen 1717
Sporenständerpilze 462
Spörgel 4689, 5196
Spornblume 1091
Spottnussbaum 3640
spreizende Melde 2186
Spreizschöterich 5175
Sprekelie 393
Sprengers Spargel 5178
Spreublume 1108, 2923
Springkraut 5051
Springkrautgewächse 433
Spritzgurke 5202
Sprödblätterpilze 4622
sprossende Felsennelke 3330
sprossender Bärlapp 2954
Sprossenkohl 823
Sprossenschlinke 3911
Sprosspilze 6070
Spurre 3730
Ssant 399
Stabwurz 3923
Stachelaloe 1003
Stachelan(n)one 5093
Stachelbeere 2560
Stachelbeerstrauch 2560
Stacheldistel 13
Stacheldolde 4300
Stachelfarn 3349
Stachelginster 2566
Stachelgras 860, 4682
Stachelgurke 1122, 3639
Stachelhirse 456
stachelige Nachtkerze 5022
stachelige Rose 4776
stach(e)liges Kammgras 2750
stachelige Spitzklette 5155
Stachelkeil 11
Stachelkiefer 595
Stachelkraftwurz 11
Stachellattich 4299
Stachelmohn 4302
Stachelnelke 4306
Stachelnuss 1811
Stachelnüsschen 4854
Stachelpanax 11
Stachelphoenix 5150
Stachelpilze 2901
Stachelrose 4776
Stachelsalad 4299
Stachelsandnelke 4306
Stachelschwamm 2900
Stachelsegge 4303
stachelspitziges Laichkraut 2376
stachliger Flaschenbaum 5093
stachliger Mäusedorn 885
Stammblatt 1322
Ständerpilze 462
Stangenspargel 1351
Stangerie 2205
Stapelie 1034
Staphyleazeen 660
starke Eiche 2041
Stärkemais 2307
starrer Haarstrang 3738
Starrgras 5251
staubfädige Palmlilie 24
staubfädige Yucca 24
Stäubling 4325
Stäublinge 4326
Staubpilz 4325
Staubpilze 4326
Staubschwamm 4325
Staudenaster 3526
Staudenflammenblume 5320
Staudengerste 5672
Staudenkohl 3128
Staudenkraut 2262
Staudenphlox 5320
Staudensonnenblume 5052
Staudentrompete 2924
Staudenwicke 4133
Stauntonia 5237
Stechapfel 1811

Stechdorn 1369, 2039, 4015, 5014
Stecheiche 2039, 2832
stechender Hohlzahn 781
Stechfichte 1338
Stechginster 1426, 2566
Stechhülse 2826
Stechhülsengewächse 2827
Stechnelke 4306
Stechpalme 2826
stechpalmenartige Gewächse 2827
Stechpalmengewächse 2827
Stechtanne 3157
Stechwinde 2623
Stechwindengewächse 5031
Steckenkraut 336, 2471
Steckenpalme 3201
steife Nelkenwurz 6074
steifer Lauch 5281
steifer Sauerklee 1566
steifes Borstengras 3533
steife Segge 5656
steifes Riedgras 5656
steife Waldrebe 2646
steife Wolfsmilch 5689
Steifgras 5251
steifhaariger Schneckenklee 933
Steinapfel 733
Steinbeere 619, 1645, 2786, 5260, 6003
Steinbibernell 4733
Steinbrech 4730
Steinbrechbibernelle 4733
Steinbrechgewächse 4732
Steineibe 4235
Steineibengewächse 4236
Steineiche 1928, 2832
Steinerdbeere 3643
Steinesche 2073
Steinfeder 3431
Steinfingerkraut 1263
Steinflachs 2115
Steingünsel 2543
Steinhanf 1810
Steinhirse 1614
Steinklee 638, 5362
Steinkraut 172
Steinkresse 168
Steinlabkraut 2743
Steinlinde 3642
Steinlorbeer 3237
Steinmais 2295
Steinminze 4720
Steinmispel 2089
Steinmorchel 1994
Steinnusspalme 2986
Steinpfeffer 2553
Steinpilz 727, 1093
Steinquendel 464, 4720, 5187
Steinquitte 1633
Steinröschen 4559
Steinsame 2640
Steinschmückel 4510
Steinstorchschnabel 3369
Steintäschel 5261
Steinweichsel 3426
Steinwurz 1348
Steinwurzelkraut 1348
Stelzenpalme 5253
Stendelwurz 4403
stengellose Distel 5238
stengellose Kratzdistel 5238
stengellose Primel 2042
stengellose Schlüsselblume 2042
stengelloses Leimkraut 3686
Stengelmangold 5398
Stengelsellerie 2411
stengelumfassender Knotenfuss 1257
stengelumfassende Taubnessel 2780
Steppenflachs 2214
Steppenlilie 3435
Steppenraute 2714
Steppenwindröschen 5061
Sterkuliazeen 1223
Sterkuliengewächse 1223
Sternanis 254, 5607
Sternanispflanze 3014
Sternapfel 5211
Sternapfelbaum 919
Sternbalsam 3852
Sternbergie 5242
Sternblume 358, 1146
Sternblume von Neu-England 3839
Sterndolde 3531
Sternflockenblume 5232
Sternjasmin 5224
Sternkaktus 5213
Sternkiefer 1296
Sternkraut 5235
Sternmagnolie 5227
Sternmiere 5235
Sternmoos 3638, 4093
Sternnelke 3433
Sternnusspalme 5626
Sternriedgras 5222
Sternsteinbrech 5231
Sternwinde 5220
Sternwurz 5263
Stevens Gipskraut 5243
Stevie 5244
Stewartie 5245
Stichsalat 1759
Stichwurz 824
Stiefmütterchen 5954
Stielblütengras 4956
Stieleiche 2041
Stielfaden 4506
Stielmangold 5398
Stielporlinge 4253
Stieltännel 4982
Stinkandorn 614
Stinkasant 336
Stinkbaum 741, 3575
stinkende Hauhechel 2518
stinkende Hundskamille 3542
stinkende Nierwurz 507
stinkender Mangobaum 180
stinkendes Wanzenkraut 4988
stinkende Taubnessel 614
Stinkesche 2151
Stinkhauhechel 2518
Stinkholz 3908
Stinkhundskamille 3542
Stinkknackbeere 2211
Stinkkohl 281, 4990
Stinkmorchel 5256
Stinksalat 281
Stinktäubling 2209
Stinkteufel 604
Stinkwacholder 4719
Stockgerste 4981
Stockmalve 169, 2829
Stockmorchel 1994, 2767
Stockrose 2829
Stockschüppling 1114
Stockschwamm 1114
Stockschwämmchen 1114
Stokesie 5259
Stoppelpilz 2752, 2900
Stoppelpilze 2901
Stoppelrübe 5650
Storaxbaum 5056
Storchschnabel 2447
Storchschnabelgewächse 2448
Strahlengriffel 21
Strahlengriffelgewächse 22
Strahlenpalme 3292
Strahlenpilze 23
strahllose Kamille 4193
Strandampfer 2536
Strandaster 5597
Strandbinse 4800
Strandbohne 1448
Stranddreizack 4877
Strandflieder 3573, 4793
Strandgerste 2094
Strandhafer 487, 2094
Strandhirschsprung 5271
Strandkasuarine 2876
Strandkiefer 1296
Strandkohl 1326
Strandkreuzkraut 4958
Strandling 2124, 4879
Strandmännertreu 4789
Strandmannstreu 4789
Strandmelde 4875
Strandnelke 3573
Strandpflaume 488
Strandplatterbse 3495
Strandroggen 2094, 5957
Strandrübe 1360
Strandsalde 5934
Strandsalzschwaden 4803
Strandsandhalm 2076
Strandsimse 4801
Strandsode 4802, 4813
Strandwegerich 2123
Strandwinde 4804
Stranvesie 5270
Stränze 1946
strauchartige Päonie 5585
strauchartiger Spargel 4848
Strauchbohne 4177
Straucherbse 4177
Strauchhyazinthe 5214
strauchige Cordyline 1402
strauchige Keulenlilie 1402
strauchige Pfingstrose 5585
strauchiger Ehrenpreis 4534
strauchiger Fingerstrauch 879
strauchiges Filzkraut 1451
strauchiges Fingerkraut 879
Strauchkirsche 2644
Strauchmalve 5582
Strauchmandel 4611
Strauchpalme 3572

Strauchpäonie 5585
Strauchpappel 5582
Strauchpfingstrose 5585
Strauchplatane 3361
Strauchportulak 4361
Strauchsonnenwende 5562
Straussfarn 3976, 4826
Straussfelberich 5629
Straussgras 534
Strausskirsche 3095
Streichkraut 1810
Streifenfarn 5161
Streitkolbenbaum 516
Strichkraut 1810
Strobe 1982
Strobeldorn 331
Strohblume 217, 2149, 2923
Strophant 5287
Stubenefeu 1268
Studentenblume 3493, 5393
stumpfblättriger Ampfer 601
stumpfblättriges Laichkraut 712
stumpfer Rhabarber 2426
stumpfes Silberblatt 1884
Stundenblume 2316
Stundeneibisch 2316
sturmblättriger Hahnenfuss 16
Sturmhut 3660
Sturmhut-Hahnenfuss 16
Stutzblattkaktus 3247
Stutzweide 3891
Styrakazeen 5269
Styraxgewächse 5269
Suarinussbaum 4725
Sudangras 5294
Südeiche 5439
Sudetenhainsimse 5296
Sudetenläusekraut 5295
südeuropäische Tamariske 2374
südliche Livistone 371
südliche Mehlbeere 1995
südlicher Abbiss 5103
südlicher Wimperfarn 4625
südlicher Zürgelbaum 2106
südliches Geissblatt 5370
Südseemyrte 5472
Sugizeder 3029
Sumachgewächse 1043
Sumpfbaldrian 3520
Sumpfbärlapp 3503
Sumpfbeere 1658
Sumpfbinse 2600, 3906, 5145
Sumpfblasenbinse 3515
Sumpfblume 3548
Sumpfblumenbinse 3515
Sumpfblutauge 3502
Sumpfbrombeere 1273
Sumpfbrustwurz 714
Sumpfdotterblume 1466
Sumpfdreiblatt 1365
Sumpfdreizack 328
Sumpfeibisch 3761
Sumpfeiche 2876, 4200
Sumpfengelwurz 714
Sumpffarn 3504
Sumpfgänsedistel 3517
Sumpfgarbe 5054
Sumpfgras 4294
Sumpfgreiskraut 4009
Sumpfhaarstrang 3509
Sumpfheide 1701
Sumpfherzblatt 5932
Sumpfhimbeere 1273
Sumpfholunder 5324
Sumpfhornklee 5872
Sumpfkiefer 3366
Sumpfkirsche 2078
Sumpfklee 716
Sumpfknabenkraut 3093
Sumpfkratzdistel 3519
Sumpfkreuzkraut 4009
Sumpfkrug 5332
Sumpflabkraut 3500
Sumpfläusekraut 2117
Sumpflieb 5711
Sumpfliliensimse 4768
Sumpflöffelchen 3512
Sumpfmädesüss 2112
Sumpfmoos 5125
Sumpfpippau 3505
Sumpfplatterbse 3510
Sumpfporst 1719
Sumpfprimel 2097
Sumpfquendel 5833
Sumpfried 5145
Sumpfriedgras 3516
Sumpfrispengras 2345
Sumpfrodel 2117
Sumpfrohrkraut 3391
Sumpfruhrkraut 3391
Sumpfsassafras 5355
Sumpfschachtelhalm 3506
Sumpfschlangenkraut 5941
Sumpfschlangenwurz 5941
Sumpfschnabel 3548
Sumpfschotenklee 5872
Sumpfsegge 3516
Sumpfsiegwurz 5500
Sumpfsimsenlilie 4768
Sumpfstern 5397
Sumpfstorchschnabel 718
Sumpfstraussgras 1665
Sumpfteichfaden 1504
Sumpfveilchen 3521
Sumpfvergissmeinnicht 5602
Sumpfwasserfeder 2097
Sumpfwasserstern 5802
Sumpfweidenröschen 3522
Sumpfwurz 2050
Sumpfzeder 414
Sumpfziest 3501
Sumpfzypressengewächse 5463
Sunn-Hanf 5331
Suppenpilz 2154
Surenbaum 1084
Surinam-Giftbaum 2820
Süssack 5297
Süssdolde 5361
süsse Balsampflaume 178
süsse Cassave 67
süsse Kartoffel 5380
süsse Mombinapflaume 178
süsser Maniok 5360
süsser Tragant 3619
Süssgras 3476
Süssholz 3291
Süssholztragant 3619
Süsskartoffel 5380
Süsskirsche(nbaum) 3544
Süssklee 1519, 5390
süsslicher Milchling 5372
Süssling 2752, 3938
Süssmaniok 67
Süssschnee 2491
Süsswurzel 4986
Swainsonie 5339
Swietenie 3427
Sykomore 205, 5408
Sympetalen 5413
Symplokazeen 5374
Synanthereen 1570
Syringe 3295
syrische Esche 5416
syrische Raute 2714
syrischer Eibisch 4886
syrischer Schuppenkopf 5417
syrische Seidenpflanze 1469

Tabak 5536
Tabakwürger 2774
Tacca 5421
Tafelbanane 2763
Taf(gras) 5473
Taglilie 1815
Taguabaum 1447
Tahitipfeilwurz 2244
Takkazeen 4173
Talerkürbis 3917, 6124
Talgbaum 899, 4196
Talgsumach 5848
Tallipotbaum 5424
Tallipotpalme 5424
Tamarikazeen 5436
Tamarinde 5433, 5434
Tamariske 2181, 5435
Tamariskgewächse 5436
Tang 4538
Tange 815
tangerische Platterbse 5438
Tanne 2257
Tännel 5841
Tännelgewächse 5842
Tännelkraut 2317
Tannenbärlapp 2258
Tannenklee 271, 3159
Tannenwedel 3488
Tarant 5397
Taro 1806, 2013
Tartsenflechte 2916
Täschelkraut 2232, 4112
Tatarengeissblatt 5455
Tatarenleimkraut 5459
Tatarenmelde 5457
tatarische Heckenkirsche 5455
tatarische Melde 5457
tatarischer Ahorn 5456
tatarischer Buchweizen 5446
tatarischer Hartriegel 4901
tatarischer Meerkohl 5452
tatarischer Schuppenkopf 5453
tatarischer Strandflieder 5458
tatarisches Heidekorn 5446
tatarisches Leimkraut 5459
Taubenerbse 4177

Taubenkerbel 1911
Taubenkropf 656, 663
Taubenskabiose 1891
Taubenstorchschnabel 3369
taube Trespe 4287
Taublatt 1908
Täubling 4621
Taubnessel 1818
Tauglocke 4542
Taugras 5981
Taumelkälberkropf 1648
Taumelkerbel 1648
Taumellolch 1804
Tausendblatt 4060
Tausendgüldenkraut 1090
Tausendkorn 877
Tausendschön 1787, 2037, 3387
Taxodiazeen 5463
Taxodie 414
Tazette 4248
Teakbaum 1544, 5465
Teebaum 5467
Teegewächse 5464
Teerose 5468
Teestrauch 1543, 5467
Teestrauchgewächse 5464
Teichampfer 5798
Teichbinse 2600
Teichfaden 1504, 4266
Teichkalmus 1918
Teichkolben 1062
Teichlilie 6086
Teichlinse 1923, 1924
Teichrose 1647, 2132
Teichrosengewächse 5818
Teichschachtelhalm 5344
Teichschaumkraut 5021
Teichschilf 1514
Teichsimse 1924
Teichwasserstern 4261
Telabun 4399
Telegraphenpflanze 5474, 5516
Telekie 3986
Teosinte 5475
Teparybohne 5478
Teppichmargerite 2055
Terpentinpistazie 5476, 5647
Teufelsapfel 3102
Teufelsauge 33
Teufelsbart 112
Teufelsbaum 4014
Teufelsdreck 336
Teufelshaar 2607
Teufelskralle 3636
Teufelspuppe 5275
Teufelswurz 633
Teufelszwirn 1870, 2607, 6016
Teufelszwirngewächse 1871
Theriakwurzel 2406
Thrinaxpalme 5480
Thuja 300
Thuje 300
Thunbergie 1272
Thunbergs Berberitze 3021
Thunbergs Pueraria 5505
Thunbergs Spierstrauch 5506
Thunbergs Taglilie 5504
Thüringer Fingerkraut 4837
Thüringer Strauchpappel 5507
thüringische Strauchmalve 5507
Thymeläazeen 3608
Thymian 5509
thymianblättriger Drachenkopf 5510
thymianblättriger Ehrenpreis 5512
Thymianseide 1400
tibetanisches Heu 2731
tibetische Hasel 5513
tibetische Traubenspiere 5514
Tiekbaum 1544, 5465
Tienschan-Fichte 4761
Tigerblume 5520
Tigergras 5521
Tigerlilie 5520, 5522
Tigerschwanzfichte 5523
Tiglibaum 4334
Tiliazeen 3315
Tillandsie 5525
Tilletiazeen 5526
Timotheegras 5528
Tintenbeere 3426
Tintenbeerstrauch 1508
Tintenblätterpilz 2950, 2951
Tintenkirsche 3426
Tintenpilze 1586
Tintenschwamm 2950, 2951
Tintling 2951
Titanenwurz 2466
Tocusso 4399
Tofieldie 2162
Tollgerste 1804
Tollkerbel 4241
Tollkraut 3102, 4765
Tollrübe 2122
Tomate 1549, 5542
Tonkabaum 1933, 5545
Topffruchtbaum 3657
Topfgreiskraut 1385
Topinambur 3094
Torenie 5553
Torfbeere 1273, 5019
Torfgränke 3257
Torfmoos 5125
Torfmoose 4096
Torfmyrte 4141
Torreys Kiefer 5557
Totenblume 3493, 4285
Totentrompete 2858
Tournefortie 5562
Tourneforts Ehrenpreis 5563
Tourneforts Zürgelbaum 3955
Tradeskantie 5134
Tradeskantie aus Rio de Janeiro 5780
Tragant 3353
tränendes Herz 1363
Träneneibe 2894
Tränenkiefer 2807
Tränenschwamm 2880
Tränenweide 400
Trapp 3322
Traube 2575
Träubel 2579
Traubenbaum 4788
traubenblütiger Steinbrech 71
traubenblütiges Klettengras 5207
Traubeneiche 1928
Traubenfarn 3974
Traubengamander 1758
Traubenhafer 1798
Traubenholunder 2120
Traubenkirsche 1384
Traubenkraut 3096, 4400
Traubenschmergel 3096
Traubenspiere 3824
Traubensteinbrech 71
Traubentrespe 413
Traubenwucherblume 3569
traubige Schneebeere 1529
traubige Trespe 413
Trauerbaum 3851
Trauerbirke 2131
Trauerblume 1852
Trauerbuche 5859
Traueriris 3728
Trauerkilte 1794
trauernde Viole 1794
Trauerschwertlilie 3728
Trauersegge 650
Trauerviole 1794
Trauerweide 400, 5508
Trauerzypresse 3727
Träuschling 5288
Trautvetters Ahorn 4442
Trebol 3402
Tremse 1613
Trespe 802
Trichterkelch 3647
Trichterlilie 4023, 4649
Trichterling 1270
Trichtermelisse 3647
Trichterschwertel 2016
Trichterwinde 1941
Trifttragant 4350
Trinie 5595
Tritome 5551
Trockenblume 2149
Troddelblume 111
Trollblume 2507
Trolle 2507
Trompetenbaum 1053, 4331
Trompetenbaumgewächse 5612
Trompetenblatt 1499, 4207
Trompetenblume 567, 2924, 5611
Trompetenblütler 5612
Trompetennarzisse 1397
Trompetenwinde 5610
Trompetenzunge 4738, 5624
Tropensporn 746
Tropenwurz 108
tropfender Faltenschwamm 2880
trübes Fingerkraut 4837
Trüffel 5608
Trüffelartige 5609
Trüffeln 5609
Trüffelpilze 5609
trügerischer Lauch 1934
Trunkelbeere 717
Trüschling 1476
Tschirimajabaum 1132
Tuberaeen 5609
Tuberose 5618
Tuchkarde 2390
Tulbaghie 5632
Tulpe 5633

Tulpenbaum 5635
Tumbo 5865
Tungbaum 992
Tung(öl)baum 5640
Tupelobaum 5642
Tüpfelfarn 1503, 4251
Tüpfelfarngewächse 4252
Tüpfelhartheu 1537
Turbenkapsel 2321
Türkenbund(lilie) 3524
Türkenmohn 3960
turkestanische Eberesche 5644
türkische Baumhasel 5646
türkische Bohne 4748
türkische Hasel 5646
türkischer Drachenkopf 3645
türkischer Flieder 1461
türkischer Klee 2891
türkischer Pfeffer 884
türkischer Rhabarber 3566
türkischer Schwarzkümmel 3386
türkischer Tabak 1548
türkischer Weizen 3434
Turmeris 1551
Turm(gänse)kresse 5565
Turmkraut 4517, 5566
Turnera 5649
Turnip 5650
Turra 2185
Turräe 5212
Typhazeen 1063

überhängende Forsythie 5860
übersehenes Reitgras 5008
Ufererdröschen 5791
Uferhirschsprung 5271
Uferriedgras 5282
Ufersegge 5282
Ufersumpfkresse 222
Uferweide 2007
Ufer(wein)rebe 2386, 4503
Uferwolfstrapp 2085
Uferzaunwinde 2748
Ukkanübaum 300
ukrainische Katzenminze 5676
Ulmazeen 2019
Ulme 2018
ulmenblättrige Spiräe 2455
Ulmengewächse 2019
Ulmenspierstrauch 2455
Umbelliferen 1036
unbegrannte Trespe 5034
unbewehrte Trespe 5034
unechte Akazie 636
unechter Bärenklau 2825
unechter Ehrenpreis 475
unechter Muskat 2175
unechte Robinie 636
unechtes Leinkraut 4585
unechtes Tännelkraut 4585
ungarische Eiche 2979
ungarische Rauke 5638
ungarischer Enzian 1799
ungarischer Flieder 2890
ungarischer Klee 2888
ungarische Silberlinde 4965
ungarische Wicke 2893
untergetauchte Wasserlinse 5217
unterirdischer Klee 5291
Unterkohlrabi 4626
Upasbaum 5684
Uralfahnenwicke 5690
Urdbohne 3755
Uredineen 4623
Urpilze 302
Ursinie 5692
Urtikazeen 3836
Urtizeen 3836
Usambara-Veilchen 1347
Ustilaginazeen 854

Valerianazeen 5696
Vanda 5698
Vanille 5699
Vanillenheliotrop 1436
Vanillenstrauch 1436
Vaucherie 2626
Veilchen 5737
Veilchenalge 4537
Veilchengewächse 5740
Veilchenholz 2486
Veilchenmoos 4537
Veilchensteinflechte 4537
Veilchenstrauch 5738
Veilchentabak 395
Veitchs Tanne 5707
Venusfinger 1444
Venusfliegenfalle 5718
Venushaar 3430, 5099
Venuskamm 4861
Venusnabel 3816
Venusschuh 2109, 3203
Venusspiegel 5719
veränderliche Buddleie 3941
veränderliche Lupine 5096
Verbesine 1709
Vereinblütler 1570
Verek 2662
Vergissmeinnicht 2056, 2327
verkehrtkegelförmige Schlüsselblume 5550
verlängerte Segge 2023
verlängertes Riedgras 2023
Vermeinkraut 476
Vernonie 2965
verschiedenblättrige Hemlocktanne 3040
verschiedenblättriger Schwingel 4839
verschiedenfarbige Dahlie 2413
Versteckblume 1715
Verwachsenkronblättrige 5413
verwechselte Trespe 2681
Vetiverwurzel 5732
Vexiergurke 5202
Vexiernelke 4558
Viehbohne 788
Viehgras 700, 3476
vielblättrige Lupine 5789
vielblütige Maiblume 2071
vielblütige Rose 3068
vielblütige Salbei 3299
vielblütiges Salomonssiegel 2071
vielblütige Weisswurz 2071
vielfarbige Aloe 3132
vielfarbige Wolfsmilch 1750
vielteiliger Rautenfarn 794
vielzählige Lupine 5789
vielzeilige Gerste 455
vierblättrige Einbeere 2786
vierkantige Heide 2260
Vierlingskraut 4073
viermännige Tamariske 2344
viersamige Wicke 2343
Vieruhrblume 1415, 2341
Viktoria 5830
violette Königskerze 4351
violette Petunie 5742
violetter Dickfuss 5739
violetter Natternkopf 4357
violetter Ritterling 214
violetter Weiderich 2912
violette Sommerwurz 694
Virgilie 5747
virginische Dattelpflaume 1494
virginische Dreimasterblume 5752
virginische Eiche 3342
virginische Erdbeere 5754
virginische Hopfenbuche 197
virginische Kiefer 5750
virginische Kresse 5749
virginische Pappel 1978
virginischer Hanf 5518
virginischer Jasmin 1550
virginischer Moorstrauch 5755
virginische Rose 5751
virginischer Rautenfarn 4427
virginischer Sadebaum 1980
virginischer Spierstrauch 5753
virginischer Tabak 1548
virginischer Wacholder 1980
virginische Schneeblume 5896
virginische Sumpfzypresse 414
virginische Traubenkirsche 1384
virginische Waldrebe 5756
Virgintabak 1548
Vismie 5758
Vitazeen 2576
Vogelbeerbaum 2114, 3695
Vogelbeere 2114
Vogelfuss(klee) 587
Vogelhirse 2352
Vogelkirsche 3544
Vogelknöterich 4317
Vogelkolbenhirse 2352
Vogelkraut 1146
Vogelmiere 1146
Vogelmilch 5228
Vogelnest 590
Vogelseide 2607
Vogelwicke 591
Volkamerie 2510
Volkmannie 2510
voreilender Schüppling 1964

Wachendorfie 5761
Wacholder 3118
Wacholdermistel 1947
Wacholderschwamm 1827
Wachsbaum 5104
Wachsblätterpilze 2905
Wachsblume 1558, 2843, 5847
Wachsgagel 5104
Wachsjambuse 3010
Wachskürbis 5845

Wachsmais 5849
Wachsmyrte 483
Wachspalme der Anden 5846
wachsstieliger Trichterling 4867
Wachsstrauch 5366
Wachtelweizen 1655
Wahlenbergie 4511
Wahlenbergs Simse 5762
Waid 6014
Walch 2517
Waldakelei 2087
Waldanemone 5061
Waldangelika 6029
Waldbingelkraut 1876
Waldbinse 6031
Waldbrustwurz 6029
Waldchampignon 2326
Walddattelpalme 5302
Waldedelpilz 2326
Waldehrenpreis 1917
Waldengelwurz 6029
Walderbse 4099
Walderdbeere 2127
Waldfrauenfarn 3199
Waldgamander 6028
Waldgeissbart 5410
Waldgeissblatt 6022
Waldgerste 2094
Waldgreiskraut 6034
Waldhabichtskraut 5769
Waldhahnenfuss 5962
Waldhainsimse 2613
Waldhasel(nuss)strauch 2098, 2246
Waldheidelbeere 3784
Waldhirse 3624
Waldholunder 2120
Waldhyazinthe 4403
Waldkerbel 6030
Waldkiefer 4775
Waldkirsche 3544
Waldklee 4347
Waldklettenkerbel 5688
Waldkresse 6085
Waldkreuzkraut 6034
Waldlabkraut 6021
Waldläusekraut 3382
Waldlilie 5594
Waldmalve 2797
Waldmarbel 2613
Waldmeister 6040
Waldpastinak 5946
Waldplatterbse 2278
Waldporling 4254
Waldrebe 1260
Waldriedgras 6042
Waldrispengras 796
Waldroggen 2094
Waldruhrkraut 6025
Waldsauerklee 6044
Waldschachtelbaum 5411
Waldsegge 6042
Waldsimse 2613, 6031
Waldsteinia 458
Waldsternmiere 6045
Waldstorchschnabel 6024
Waldveilchen 5412
Waldvergissmeinnicht 6033
Waldvögelein 4163
Waldweidenröschen 2262
Waldweinrebe 6032
Waldwicke 6046
Waldwindröschen 5061
Waldziest 5911
Waldzwenke 4996
Walkerdistel 2390
Wallabarholzbaum 5764
Wallichs Wacholder 5771
Walloneneiche 5697
Wallwurz 1345
Walnussbaum 5777
Walnussbaumgewächse 5778
Walpurgiskraut 3673
Wampibaum 5779
Wandelklee 5474, 5516
Wandelröschen 3218
wanderndes Blatt 5763
Wandnessel 4105
Wangara 5969
Wanzenbeere 2081
Wanzenblume 1596
Wanzendill 1598
Wanzenkraut 837, 1598, 1719
Wanzenrose 374
Wanzensame 5517
Warzenberberitze 5786
Warzenbirke 2131
Warzenflechte 5785
Warzenkaktus 3454
Warzenkraut 2606
Warzenpfeifenstrauch 5788
Warzensame 2636
Warzenspindelbaum 5787
Warzenstäubling 5784
warzige Gasterie 3990
warzige Hirschtrüffel 3399
warziger Spindelbaum 5787
Washingtonie 5790
Wasserähre 5807
Wasseraloe 5836
Wasserblatt 5813
Wasserblattgewächse 5814
Wasserbrotwurzel 1806, 2013
Wasserdarm 5795
Wasserdickblatt 1511
Wasserdost 732
Wasserehrenpreis 5838
Wassereiche 5827
Wasserfalle 83
Wasserfarn 391
Wasserfeder 2193
Wasserfenchel 2251, 5799
Wasserfreund 5233
Wassergreiskraut 3513
Wasserhaarnixe 2184
Wasserhahnenfuss 5797
Wasserhalmgewächse 665
Wasserhanf 2776
Wasserhickory 5809
Wasserhyazinthe 1557, 4174, 5810
Wasserkastanie 5794
Wasserkelch 1716
Wasserkerbel 2251
Wasserklee 1365
Wasserknöterich 5812
Wasserkraut 2546
Wasserkresse 222
Wasserkreuzkraut 3513
Wasserkrug 1499
Wasserlasche 4494
Wasserliesch 2314
Wasserlieschgewächse 2315
Wasserlilie 2132
Wasserlimone 5815
Wasserlinse 1924
Wasserlinsen 1925
Wasserlinsengewächse 1925
Wasserlobelie 5819
Wassermelone 5821
Wassermerk 545, 5828
Wasserminze 5822
Wassermohn 5831
Wassernabel 1489, 4114
Wassernetzalge 5826
Wassernusspalme 4644
Wasserpest 2022
Wasserpfefferknöterich 3511
Wasserpfriemenkresse 5792
Wasserprimel 2193
Wasserrebendolde 2251
Wasserreis 266, 5956
Wasserriemen 5934
Wasserrose 2132, 5817
Wasserrosengewächse 5818
Wasserrübe 5650
Wasserschierling 5808
Wasserschild 5835
Wasserschlauch 664
Wasserschlauchgewächse 665
Wasserschlüssel 5831
Wasserschneeball 2091
Wasserschraube 5944
Wasserschwaden 806
Wasserschwertlilie 6086
Wassersegge 5834
Wassersenf 2776
Wasserstern 5839
Wassersterngewächse 5840
Wassersumpfkresse 222
Wassersüssgras 806
Wasserulme 5800
Wasserwegerich 211
Wasserweichkraut 5795
Wau 3614
Waugewächse 3615
Weberdistel 2390
Weberkarde 2390
wechselblättriges Milzkraut 168
wechselblättriges Zypergras 5677
Wegdistel 850
Wegdorn 830
Weg(e)breit 4218
Wegedistel 13
Wegekraut 510
Wegerauke 2757
Wegerich 2290, 4218
Wegerichfroschlöffel 211
Wegerichgewächse 4220
Weg(e)tritt 4218
Wegwarte 1147
wehrlose Trespe 5034
weicher Akanthus 5077
weicher Hohlzahn 1895
weiche Rose 685
weicher Storchschnabel 1892

weiches Honiggras 2461
weiche Trespe 5078
weiche Zaubernuss 1212
weichhaarige Eiche 4324
weichhaariger Hohlzahn 1895
weichhaarige Rose 287
weichhaariger Stechapfel 2810
Weichkraut 1032
Weichmais 2307
Weichmalve 5952
Weichorchis 25
Weichselbaum 3544
Weichsel(kirsche) 3426, 3483
Weichwurz 25, 3250
Weide 5964
Weidenalant 5970
Weidenaster 5968
weidenblättrige Aster 5968
weidenblättriger Alant 5970
weidenblättriger Birnbaum 5971
weidenblättriger Spierstrauch 5972
Weidendorn 1523
Weideneiche 2041, 5974
Weidengewächse 5965
Weidenlattich 5973
Weidenröschen 5975
Weidensanddorn 1523
Weidenschmalfrucht 5966
Weidenseide 2641
Weidenspierstrauch 5972
Weidenwegerich 5379
Weiderich 3400
Weiderichgewächse 3371
Weigelie 5861
Weihnachtskaktus 3247
Weihnachtsrose 1226, 3091
Weihnachtsstern 1500, 4239
Weihrauchbaum 553, 2363, 2926
Weihrauchkiefer 3352
Weinberglauch 2223
Weinhefe 5986
Weinkermesbeere 1502
Wein-Mauritiuspalme 5985
Weinpalme 18, 3537, 4019
Weinraute 15[illegible]6
Weinrebe 2575
Weinrebengewächse 2576
Weinrose 5359
Weinstock 2575
Weismutskiefer 1982
Weissbaum 920, 3579
Weissbecher 1733
Weissbeerenhartriegel 5454
Weissbirke 2131
Weissbuche 2108, 2852
Weissdorn 2729
Weissdornakazie 103
weissdornblättriger Ahorn 2730
weisse Aschwurz 2435
weisse Eiche 5905
weisse Fetthenne 5915
weisse Hickory 3640
Weisseiche 5905
weisse Lichtnelke 2140
weisse Lilie 3417
weisse Lupine 5899
weisse Melde 3212
weisse Mistel 2113
weisse Narzisse 4237
weisse Osterblume 2134
weisse Pestwurz 5884
weisse Petunie 5901
weisser Ackerkohl 5574
weisser Affodill 757
weisser Ahorn 4217
weisser Amarant 5637
weisser Diptam 2435
weisser Dorant 1442, 5054
weisse Reseda 5900
weisser Fuchsschwanz 5637
weisser Gänsefuss 3212
weisser Germer 5894
weisser Goldregen 4276
weisser Gummibaum 4493
weisser Hartriegel 4458
weisser Honigklee 5917
weisser Kanarienbaum 5886
weisser Knäuel 4131
weisser Knollenblätterpilz 5180
weisser Kümmel 1729
weisser Kürbis 1209
Weisserle 5122
weisser Mauerpfeffer 5915
weisser Maulbeerbaum 5902
weisse Rose 1634
weisse Rosskastanie 1443
weisser Schöterich 5574
weisser Senf 5904
weisser Sesam 3961
weisser Sisal 2781
weisser Spierstrauch 3807
weisser Steinklee 5917
weisser Teebaum 920
weisse Rübe 5650
weisser Wau 5900
weisser Wiesenklee 5889
weisser Windhalm 4469
weisse Sapote 5909
weisses Bilsenkraut 4586
Weissesche 5878
weisse Schnabelried 5880
weisse Schnabelsimse 5880
weisse Seerose 2132
weisses Fingerkraut 5888
weisses Quebracho 1559
weisses Straussgras 4469
weisses Veilchen 4008
weisses Waldstroh 5882
weisse Taubnessel 5891
weisse Teufelskralle 5139
weisse Wucherblume 3987
weisse Zaunrübe 5883
weisse Zeder 5887
Weissfaden 3245
Weissfichte 5913
weissgraue Segge 4970
weissgraues Felsenblümchen 5667
weissgraues Riedgras 4970
Weissgummibaum 875
Weisskiefer 1856
Weissklee 5889
Weisskohl 5885
Weisskraut 5885
weissliche Pappel 428
Weisslorbeer 5355
Weisslupine 5899
Weissmoos 5125
Weisspappel 5906
weissrandiger Fuchsschwanz 4316
Weissrüster 191
weissstengelige Kiefer 5879
Weisstanne 4955
Weissträffel 5920
Weissulme 191
Weissweide 5922
Weisswurz 5082
Weisszeder 2167
Weisszüngel 3282
Weizen 5873
Wellingtonie 2483
welsche Mispel 390
welsche Nuss 4151
welscher Fennich 2352
welscher Weizen 4286
welsches Weidengras 2980
Welschkohl 4722
Welschkorn 3434
Welschnussbaum 4151
Welwitschie 5865
Wendelähre 3197
Wendelbeere 2081
Wendelblumenklee 4147
wenigblütige Scheinhasel 891
wenigblütiges Sumpfried 2213
wenigblütige Weisswurz 1916
Wermut 1564
Wermutbeifuss 1564
westindischer Mahagonibaum 5871
westindischer Nierenbaum 1378
westindischer Seidenholzbaum 6094
westindische Stachelbeere 447
westliche Balsampappel 942
Wetterdistel 5035
Wettermoos 2394
Wetterrose 2830
Wetterrösel 2316
Weymouthskiefer 1982
Wheelers Rauhlilie 5875
Whites Widdringtonia 3637
Wichuras Rose 5929
Wicke 5731
Wickelwurz 537, 3254
Wicklinse 609
Widdringtonia 5930
Widerbart 5195
Widerstoss 3573, 4793
Widerton 2676
Wiegekraut 1564
Wielandskraut 3556
Wiesenalant 784
Wiesenaugentrost 1910
Wiesenbärenklau 2825
Wiesenbibernell 2409
Wiesenbocksbart 3559
Wiesenchampignon 1476
Wiesenegerling 1476
Wiesenfahnenwicke 4213
Wiesenfeste 4577
Wiesenflachs 887
Wiesenflockenblume 817
Wiesenfrauenmantel 1454
Wiesenfuchsschwanz(gras) 3549
Wiesengeld 3651
Wiesenglockenblume 4408

Wiesenhabichtskraut 2224
Wiesenhafer 3905
Wiesenhornklee 588
Wiesenkammgras 1681
Wiesenkerbel 1648, 6030
Wiesenklee 4447
Wiesenknöterich 2079
Wiesenkopf 867
Wiesenkresse 600
Wiesenkreuzblümchen 1470
Wiesenkreuzblume 1470
Wiesenkreuzkraut 4401
Wiesenkuhschelle 3554
Wiesenkümmel 1013
Wiesenlabkraut 5882
Wiesenlein 4335
Wiesenlieschgras 5528
Wiesenlolch 4135
Wiesenmangold 1365
Wiesenmargarite 3987
Wiesenpippau 4577
Wiesenplatterbse 2587, 3555
Wiesenraute 3556
Wiesenrispe 3150
Wiesenrispengras 3150
Wiesenruchgras 5389
Wiesensafran 1353
Wiesensalbei 3558
Wiesensauerampfer 2429
Wiesenschachtelhalm 3552
Wiesenschaumkraut 1723
Wiesenschwertlilie 4907
Wiesenschwingel 3547
Wiesenseide 1400
Wiesensilge 4124
Wiesenstorchschnabel 3551
Wiesenveilchen 2694
Wiesenvergissmeinnicht 5627
Wiesenwucherblume 3987
Wildbete 4785
Wilddotter 3334
wilde Blasenkirsche 5275
wilde Brustwurz 6029
wilde Eberesche 2114
wilde Erbse 2231
wilde Erdbeere 2127
wilde Karde 5717
wilde Malve 2797
wilde Möhre 5942
wilde Narde 2133
wilde Platterbse 2278
wilder Birnbaum 1487
wilder Dost 1484
wilder Emmer 5961
wilder Gamander 2454
wilder Jasmin 1078
wilder Kastanienbaum 828
wilder Kerbel 6030
wilder Lattich 4299
wilder Majoran 1484
wilder Mohn 1616
wilder Ölbaum 5953
wilde Rose 1875
wilder Pfefferstrauch 2194
wilder Reis 266, 740, 4497
wilder Rosmarin 1719
wilder Safran 1353, 4637
wilder Senf 1118
wilder Spinat 2421
wilder Wein 221, 1664, 5484, 5748
wilder Ysop 5321
wildes Barbarakraut 610
wildes Süssholz 1503, 3619
wilde Sumpfkresse 6085
wilde Tulpe 2302
wilde Zitrone 1845
Wildhafer 5951
Wildindigo 5948
Williams Igelkaktus 3591
Willmotts Scheinhasel 5963
Wilsons Berberitze 5976
Wilsons Deutzie 5977
Wilsons Fichte 5980
Wilsons Spierstrauch 5979
Wilsons Tanne 5978
Wimperfarn 6043
windende Osterluzei 1404
Windengewächse 3683
Windenglocke 340
Windenknöterich 1926
Windhafer 5951
Windhalm 534, 5981
Windröschen 242
Windröslein 242
Winkelriedgras 4482
Winteraster 3687
Winterblatt 4881
Winterbohnenkraut 6006
Wintereiche 1928
Winterendivie 2032
Wintergrün 4377
Wintergrüngewächse 4378
Winterkresse 5999
Winterlauch 3266
Winterlevkoje 1538
Winterlieb 4202
Winterlinde 3327
Winterling 5998
Wintermondviole 4130
Winterportulak 3627
Winterrebe 6002
Winterrose 1226
Winterschachtelhalm 4777
Winters Gewürzrindenbaum 6007
Winterspargel 649
Winterzwiebel 5863
Wirbeldost 5936
Wirsingkohl 4722
Wirtelborstenhirse 2844
Wirtelfennich 2844
Witwenblume 3167
wohlriechende Becherglocke 3302
wohlriechende Dracaena 2357
wohlriechende Himbeere 2359
wohlriechende Pappel 3654
wohlriechende Platterbse 5378
wohlriechende Resede 1468
wohlriechender Gewürzstrauch 1540
wohlriechender Pfeifenstrauch 1078, 5376
wohlriechender Zierapfel 5959
wohlriechendes Geissblatt 5370
wohlriechendes Gemshorn 5363
wohlriechendes Mariengras 5368
wohlriechendes Ruchgras 5389
wohlriechendes Veilchen 5392
Wohlverleih 323
Wolfsbeere 2786
Wolfsbohne 3395
Wolfseisenhut 6018
Wolfskirsche 525
Wolfsklee 638
Wolfsmilch 2070
Wolfsmilchgewächse 5191
Wolfstrapp 839
Wolfswut 525
Wollbaum 3134
Wollbäume 4944
Wollblume 271, 3159
Wolldistel 4770
Wollfruchtkaktus 3344
Wollgras 1637
wolliger Fingerhut 2614
wolliger Schneeball 5852
wolliges Honiggras 1553
wollige Weide 6054
wolliggestiefelter Schirmling 4863
Wollklee 3159
Wollklette 1636
Wollknöterich 2054
Wollkraut 3752
Wollmispel 3375
Wolltraube 3375
Wollziest 3731
Worbel 2127
Wruke 4626
Wucherblume 1229
Wulfenie 6059
Wulstlingartige 173
Wundblume 271
Wunderbalsamapfel 431
Wunderbaum 1050
Wunderblume 1415, 2341
Wunderblumengewächse 2342
Wunderstrauch 4391
Wunderveilchen 6019
Wundklee 271, 3159
Wundkraut 2612
Würger 812
Würgerenzian 3622
Wurmfarn 3445, 6026
Wurmkraut 1542
Wurmsamenkraut 3284
Wurstkraut 3688, 5375
Wurzelbaum 200
wurzellose Wasserlinse 4551
wurzelnde Jasmintrompete 1550
Wurzelpetersilie 5652
Wurzelrübling 4550
Würzsilge 2835
Wüstenpalme 110
Wüstenziest 1839
Wüterich 4241, 5808
Wutschierling 2130

Yamsbohne 5851, 6063
Yamspflanzen 6064
Yamswurzel 6062
Yamswurzelgewächse 6064
Yarraholz 3081
Yedofichte 6071
Ylang-Ylang 6119
Ysop 2909
ysopblättriger Wanzensame 2911

Ysopweiderich 2912

Zabels Spierstrauch 6123
Zabels Zwergmispel 1134
Zackenschötchen 1617
Zackenschote 1617
zahme Eberesche 4833
Zahnbürstenbaum 4669
Zahnlilie 1878, 2190
Zahnmais 1835
Zahnstocherkraut 5547
Zahntrost 460
Zahnwehholz 1506
Zahnwurz 600, 5548
Zahnzunge 3910
Zähpilze 4253
Zannichellie 4266
Zantedeschie 950
Zäpfchenkraut 530
Zapfenbäume 1579
Zapfenholz 2513
Zapfenpalme 1581
Zapfenträger 1579
zarte Binse 4288
zarter Gauchheil 722
zarter Lappenfarn 2182
zarter Lein 5001
zartes Hasenohr 4998
zartes Perlgras 6035
zartes Quellgras 806
zartes Scheidengras 3742
Zaserblume 3592
Zauberblume 4704
Zauberhasel 1562
Zaubernuss 6012
Zauberstrauch 6012
Zauke 3306
Zaungiersch 596
Zaunhopfen 1440
Zaunkirsche 2099
Zaunlattich 4299
Zaunlilie 269
Zaunrebe 1664
Zaunrose 1875
Zaunrübe 824
Zaunwicke 2760
Zaunwinde 2509, 2748
Zauschnerie 2259
Zebraholzbaum 4275
Zeder 1082
Zederbaum 1082
Zedertanne 3157
Zederwacholder 4298
Zederzypresse 5887
Zedrachbaum 1159
Zedrachgewächse 3428
Zedrele 1084
Zedrelenbaum 1234
Zedrelenholzbaum 1234
Zedrobaum 1084
Zehrkraut 1361
Zehrwurz 333
Zeichenwurz 2999
Zeiland 2194, 5192, 5194
Zeilfarn 20
Zeitlose 382
Zentifolie 910
Zephyrblume 6126
Zeratophyllazeen 2862
zerbrechlicher Blasenfarn 786
Zerreiche 2128
Zeyhers Pfeifenstrauch 6127
Zichorie 1147, 1381
Ziegenbart 5410
Ziegenklee 2203
Ziegenlippe 6081
Ziegenraute 1425
Ziegenweide 2523
Zierbelkiefer 5402
Zierbelnusskiefer 5402
Zierbelnusstanne 5402
zierliche Deutzie 4995
zierliches Hartheu 4912
zierliche Wicke 5004
Ziermalve 2788
Ziernessel 1325
Zierspark 3969
Zierwicke 4098
Ziest 550
Zigarrenbaum 5097
Zigeuner 6057
Zigeunerkorn 633
zilizische Tanne 1235
Zimmerahorn 7
Zimmerlinde 5115
Zimmerstrauch 1540
Zimmertanne 297, 3877
Zimtapfel 5297
Zimtbaum 1238
Zimtbrombeere 2359
Zimterdbeere 2722
Zimthimbeere 2359
Zimtkassie 1047
Zimtlorbeer 1238
Zimtrindenbaum 995
Zimtröschen 1078, 5376
Zimtrose 1240
Zinchone 1236
Zindelbast 5192
Zinerarie 1237
Zingiberazeen 2490
Zinnamom 1238
Zinnie 6129
Zinnkraut 2225
Zipfelkraut 5833
Zipolle 2420
Zirmet 2719
Zistazeen 4532
Ziströschen 4531
Zistrose 4531
Zistrosengewächse 4532
Zitronat 1243
Zitronat-Zitrone 1243
Zitrone 1246
Zitronellgras 1244
Zitrone(nbaum) 3268
Zitronengras 3270
Zitronenkraut 1356, 3271
Zitronenmelisse 1356
Zitronenpelargonie 3897
Zitronenpilz 5700
Zitronenstrauch 3324
Zitterespe 2074
Zittergras 4381
Zitterpappel 2074
Zitwer 6125
Zitwersamenkraut 3284
Zizifer 6130
Zonalpelargonie 2873
Zottelblume 716
Zottelwicke 2693
Zottenblume 716
zottige Fetthenne 2691
zottiger Birkenreizker 6049
zottiger Günsel 2442
zottiger Nachtschatten 6103
zottiges Habichtskraut 4842
zottiges Weidenröschen 2695
zottige Wicke 2693
Zuchtchampignon 2419
Zuckerahorn 5304
Zuckerapfel 5297
Zuckerbirke 5357
Zuckerbusch 5308
Zuckererbse 2423
Zuckergras 5303
Zuckerhirse 5085
Zuckerkiefer 5306
Zuckermais 5301
Zuckermelde 2421
Zuckermelone 3762
Zuckerpalme 5305
Zuckerriementang 5299
Zuckerrohr 5300
Zuckerrose 2371
Zuckerrübe 5298
Zuckerschotenbaum 1439
Zuckersorgho 5085
Zuckersorghum 5309
zuckersüsser Stärkemais 5215
Zuckertang 5299
Zuckerwurzel 4986
Zügelorchis 2673
Züngel 4481
Zungenblütler 1232
Zungenfarn 2716
Zungenhahnenfuss 5544
Zungenpilz 515
Zürgelbaum 2674
zusammengedrückte Binse 4582
zusammengedrücktes Rispengras 970
Zusammengesetztblütler 1570
Zweiblatt 5657
zweiblättrige Kuckucksblume 895
zweiblättrige Meerzwiebel 5664
zweiblättriger Blaustern 5664
zweiblättrige Schattenblume 5671
zweiblättrige Stendelwurz 895
zweiblättrige Waldhyazinthe 895
zweiblütiges Veilchen 5661
Zweifadengras 495
zweifarbige Eiche 5347
zweifarbiges Mädchenauge 4212
zweifarbige Weide 5466
Zweiflügelnussbaum 2668
zweigriffeliger Weissdorn 2038
Zweigweizen 756
zweihäusige Gichtrübe 4439
zweihäusige Segge 1860
zweihäusiges Katzenpfötchen 1510
zweihäusiges Riedgras 1860
zweihäusige Zaunrübe 4439
zweijährige Nachtkerze 1408

zweijähriger Pippau 4577
Zweikeimblättrige 1855
Zweikorn 2029
zweischäriger Klee 1965
zweisporiger Egerling 2419
zweiteiliges Leinkraut 1277
Zweizahn 521
zweizeilige Gerste 5672
Zwenke 2164
Zwergazalee 114
Zwergbambus 4710
Zwergbanane 1936
Zwergbirke 1935
Zwergeiche 505
Zwergfelsenröschen 114
Zwergfilzkraut 4994
Zwergflachs 2289
Zwerggras 3609
Zwergholunder 3570
Zwergigelkolben 3252
Zwergkastanie 99
Zwergkirschbaum 2644
Zwergkirsche 2644
Zwergkugelberberitze 3421
Zwerglein 101
Zwerglinse 4551, 6017
Zwerglorbeer 3257
Zwergmandel 4611
Zwergmaulbeere 1273
Zwergmäuseschwanz 5530
Zwergmispel 1633
Zwergölbaum 5194
Zwergpfeffer 4119
Zwergquitte 2089
Zwergrohrpalme 1112
Zwergruhrkraut 1938
Zwergschmiele 2677
Zwergschneckenklee 3332
Zwergschwertlilie 1943
Zwergspindelstrauch 1939
Zwergulme 4903
Zwergweichsel 2644
Zwergweizen 1288
Zwetsche(nbaum) 2424
Zwetschge 2424
Zwetschke 2424
Zwiebel 3931
Zwiebelfenchel 2300
Zwiebellauch 2420
zwiebeltragende Zahnwurz 1592
Zwiebelzahnwurz 1592
Zwieselbeere 3544
Zwillingsblatt 5662
Zwillingsblume 3604
Zyane 1613
Zyanophyzeen 701
Zygomyzeten 6131
Zygophyllazeen 499
Zykadazeen 1765
Zykadee 1764
Zykaden 1765
Zyklamen 1766
Zyklanthazeen 1767
zylindrischer Bogenhanf 2918
Zyperazeen 4811
Zypergras 2280
Zyperngras 4364
Zypresse 1773
zypressenartige Santolina 1776
Zypressenfichte 300
Zypressengewächse 1775
Zypressenheide 1049
Zypressenkraut 3240
Zypressenwolfsmilch 1774

LATINE

Abelia 3
Abelmoschus esculentus 3921
Abelmoschus moschatus 3761
Abies 2257
Abies alba 4955
Abies amabilis 1039
Abies arizonica 1601
Abies balsamea 432
Abies bornmuelleriana 737
Abies cephalonica 2617
Abies cilicica 1235
Abies concolor 5895
Abies concolor lowiana 3997
Abies faxoniana 2191
Abies firma 3649
Abies fraseri 2364
Abies grandis 2573
Abies holophylla 3462
Abies homolepis 3855
Abies kawakami 3141
Abies koreana 3181
Abies lasiocarpa arizonica 1601
Abies lowiana 3997
Abies magnifica 4451
Abies nobilis 3865
Abies nordmanniana 3876
Abies numidica 94
Abies pindrow 4189
Abies pinsapo 5110
Abies procera 3865
Abies religiosa 4635
Abies sachalinensis 4655
Abies sachalinensis nemorensis 5978
Abies sibirica 4904
Abies spectabilis 2800
Abies squamata 2271
Abies veitchi 5707
Abies venusta 779
Abies webbiana 2800
Abies wilsoni 5978
Abobra 1659
Abobra tenuifolia 1660
Abobra viridiflora 1660
Abroma 4
Abronia 5
Abrus 4553
Abrus praecatorius 3090
Abutilon 7
Abutilon avicennae 1217
Abutilon hybridum 2898
Abutilon theophrasti 1217
Acacia 10
Acacia aneura 3751
Acacia arabica 399
Acacia baileyana 1300
Acacia catechu 1056
Acacia dealbata 4957
Acacia decurrens 2637
Acacia farnesiana 5351
Acacia homalophylla 2486
Acacia horida 103
Acacia longifolia 5409
Acacia melanoxylon 655
Acacia senegal 2662
Acaena 4854
Acalypha 1585
Acanthaceae 14
Acantholimon 4306
Acanthopanax 11
Acanthophoenix 5150
Acanthus 12
Acanthus mollis 5077
Acanthus spinosus 5151
Acer 3481
Aceraceae 3482
Acer campestre 2756
Acer cappadocium 1327
Acer carpinifolium 2853
Acer circinatum 5735
Acer crataegifolium 2730
Acer creticum 3957
Acer dasycarpum 4966
Acer ginnala 230
Acer griseum 4036
Acer heldreichi 418
Acer japonicum 2391
Acer laetum 1327
Acer macrophyllum 562
Acer mandshuricum 3464
Acer monspessulanum 3669
Acer negundo 750
Acer nigrum 637
Acer nikoense 3856
Acer orientale 3957
Acer palmatum 3056
Acer pennsylvanicum 5284
Acer pictum 4004
Acer platanoides 3885
Acer pseudoplatanus 4217
Acer rubrum 4455
Acer rufinerve 4470
Acer saccharinum 4966
Acer saccharum 5304
Acer sieboldianum 4934
Acer tataricum 5456
Acer trautvetteri 4442
Acer velutinum 5712
Achillea 6067
Achillea millefolium 1565
Achillea moschata 3769
Achillea ptarmica 5054
Achimenes 15
Achras zapota 4700
Achyranthes 1108
Acinos 464
Acinos alpinus 148
Acnida cannabina 5518
Acnistus 5960
Aconitum 3660
Aconitum anthora 4373
Aconitum lycoctonum 6018
Aconitum napellus 17
Aconitum paniculatum 4029
Aconitum variegatum 3465
Aconitum vulparia 6018
Acorus 5365
Acorus calamus 1918
Acorus gramineus 3073
Acorus japonicus 3073
Acrocomia 18
Acrocomia sclerocarpa 3737
Acronychia 19
Acronychia baueri 480
Acrostichum 20
Actaea 441
Actaea spicata 616
Actinidia 21
Actinidia arguta 748
Actinidiaceae 22
Actinidia chinensis 6065
Actinidia kolomikta 3176
Actinidia polygama 4968
Actinomyces 23
Adansonia digitata 445
Adenandra 29
Adenanthera 493
Adenanthera pavonina 4676
Adenophora 3198
Adenophora lilifolia 3302
Adhathoda vasica 3438
Adiantum 3430
Adiantum capillus-veneris 5099
Adiantum cuneatum 1828
Adiantum tenerum 2182
Adinandra 30
Adlumia 32
Adonis 33
Adonis aestivalis 5317
Adonis autumnalis 4164
Adonis flammea 2272
Adonis vernalis 5179
Adoxa 3765
Adoxaceae 3684
Adoxa moschatellina 3766
Aechmea 35
Aegilops 2517
Aegle marmelos 408
Aegopodium 2570
Aegopodium podagraria 596
Aeschynanthus 466
Aeschynomene 3109
Aeschynomene elaphroxylon 179
Aesculaceae 2865
Aesculus 828
Aesculus californica 931
Aesculus chinensis 1188
Aesculus flava 6077
Aesculus glabra 3913
Aesculus hippocastanum 1443
Aesculus octandra 6077
Aesculus parviflora 739
Aesculus pavia 4441
Aesculus turbinata 3046
Aethionema 5261
Aethusa 36
Aethusa cynapium 2325
Afzelia 53
Afzelia africana 39
Agapanthus 54
Agapanthus africanus 40
Agapanthus umbellatus 40
Agaricaceae 2488
Agaricales 58
Agaricus 57
Agaricus arvensis 2870
Agaricus bisporus 2419
Agaricus campestris 1476
Agaricus cervinus 2189
Agaricus glutinosus 2516
Agaricus involutus 2956
Agaricus nebularis 1274
Agaricus nudus 214
Agaricus portentosus 1859
Agaricus silvaticus 2326

Agathis 1795
Agathis alba 5890
Agathis australis 3139
Agave 59
Agave americana 1092
Agave atrovirens 4328
Agave fourcroydes 2781
Agave sisalana 4976
Ageratum 60
Ageratum mexicanum 3594
Aglaia 61
Aglaonema 62
Agrimonia 63
Agrimonia eupatoria 1348
Agrimonia odorata 2355
Agrimonia pilosa 2679
Agriophyllum 64
Agrocybe praecox 1964
Agropyron 1640
Agropyron cristatum 1684
Agropyron desertorum 5210
Agropyron elongatum 5432
Agropyron intermedium 2952
Agropyron repens 4379
Agropyron smithi 707
Agropyron trichophorum 5252
Agropyrum 1640
Agrostemma 1610
Agrostemma githago 1391
Agrostis 534
Agrostis alba 4469
Agrostis canina 5708
Agrostis capillaris 1335
Agrostis gigantea 617
Agrostis palustris 1665
Agrostis spica-venti 5981
Agrostis tenuis 1335
Agrostis vulgaris 1335
Ailanthus 65
Ailanthus altissima 5584
Ailanthus glandulosa 5584
Aiphanes 4601
Aira 2677
Aira capillaris 261
Aira caryophyllea 4959
Aira praecox 1963
Aizoaceae 1033
Aizoon 70
Ajuga 838
Ajuga chamaepitys 2650
Ajuga genevensis 2442
Ajuga reptans 1030
Akebia 73
Alaria 5995
Alaria esculenta 407
Albizzia 79
Albizzia anthelmintica 3757
Albizzia julibrissin 4948
Albizzia odoratissima 2356
Alchemilla 3202
Alchemilla alpina 3709
Alchemilla arvensis 2226
Alchemilla vulgaris 1454
Alchornea 1225
Aldrovanda 83
Aldrovanda vesiculosa 5793
Alectoria jubata 2868
Alectorolophus 4426
Alectryon 84
Alectryon excelsum 5524
Aletris 5221
Aleurites fordii 5640
Aleurites moluccana 992
Algae 92
Alhagi camelorum 961
Alhagi pseudoalhagi 961
Alisma 5829
Alisma gramineum 2593
Alisma natans 2299
Alisma plantago-aquatica 211
Alismataceae 95
Alkanna 97
Alkanna tinctoria 1954
Alliaria officinalis 2433
Allium 3931
Allium angulosum 249
Allium ascalonicum 4844
Allium carinatum 3144
Allium cepa 2420
Allium cepa *var.* aggregatum 3753
Allium falax 1934
Allium fistulosum 5863
Allium moly 3303
Allium oleraceum 4284
Allium porrum 3266
Allium sativum 2432
Allium schoenoprasum 1219
Allium schuberti 4762
Allium scorodoprasum 2474
Allium sphaerocephalum 421
Allium strictum 5281
Allium ursinum 4415
Allium victorialis 3368
Allium vineale 2223
Alnus 81
Alnus cordata 2971
Alnus glutinosa 2072
Alnus incana 5122
Alnus japonica 3012
Alnus oregana 4432
Alnus oregona 4432
Alnus rubra 4432
Alnus rugosa 2732
Alnus subcordata 1066
Alnus viridis 2105
Alocasia 108
Aloe 109
Aloe arborescens 299
Aloe barbadensis 3568
Aloe dichotoma 4392
Aloe ferox 1003
Aloe soccotrina 5076
Aloe succotrina 5076
Aloe variegata 3132
Aloe vera 5601
Aloe vulgaris 3568
Alonsoa 3529
Alopecurus 2349
Alopecurus alpinus 133
Alopecurus geniculatus 5805
Alopecurus myosuroides 3735
Alopecurus pratensis 3549
Alopecurus ventricosus 1668
Aloysia citriodora 3271
Alphitonia 5578
Alphitonia excelsa 5579
Alpinia 158
Alpinia galanga 2609
Alsine 5235
Alstonia 165
Alstonia scholaris 4014
Alstroemeria 166
Alternanthera 3111
Althaea 169
Althaea hirsuta 2815
Althaea officinalis 3507
Althaea rosea 2829
Altingia excelsa 4419
Alyssum 172
Alyssum calycinum 4007
Alyssum maritimum 5352
Alyssum montanum 3693
Alyssum saxatile 2551
Alyssum spinosum 5152
Amanita caesarea 918
Amanitaceae 173
Amanita muscaria 2319
Amanita pantherina 4032
Amanita phalloides 1820
Amanita rubescens 713
Amanita vaginata 4853
Amanita verna 5180
Amanita virosa 1842
Amanitopsis vaginata 4853
Amaranthaceae 175
Amaranthus 174
Amaranthus albus 5637
Amaranthus blitoides 4316
Amaranthus caudatus 3387
Amaranthus graecizans 5637
Amaranthus hybridus 5005
Amaranthus paniculatus 3387
Amaranthus retroflexus 4464
Amaranthus spinosus 5153
Amaryllidaceae 177
Amaryllis 176
Amaryllis belladonna 526
Amaryllis vittata 448
Amberboa 181
Amberboa turanica 4355
Ambrosia 4400
Ambrosia artemisifolia 1513
Ambrosia trifida 2480
Amelanchier 4832
Amelanchier asiatica 342
Amelanchier canadensis 4838
Amelanchier ovalis 2428
Amelanchier rotundifolia 2428
Amelanchier spicata 1949
Ammi 216
Ammiaceae 1036
Ammi majus 2603
Ammi visnaga 5547
Ammobium 217
Ammobium alatum 5991
Ammophila 487
Ammophila arenaria 2076
Amoora 218
Amorpha 2171
Amorpha fruticosa 2948
Amorphophallus 220
Amorphophallus campanulatus 5912
Amorphophallus rivieri 1848
Amorphophallus titanum 2466

Ampelopsis 221
Ampelopsis brevipedunculata 224
Ampelopsis quinquefolia 5748
Amsinckia 2217
Amsonia 223
Amygdalus communis 104
Amygdalus communis *var.* amarus 597
Amygdalus communis *var.* fragilis 2354
Amygdalus georgica 4611
Amygdalus nana 4611
Amygdalus triloba 2311
Amyris 5552
Amyris balsamifera 430
Anabasis 233
Anacamptis 234
Anacardiaceae 1043
Anacardium 1042
Anacardium occidentale 1378
Anacharis canadensis 976
Anacyclus 2460
Anagallis 4185
Anagallis arvensis 4747
Anagallis coerulea 705
Anagallis tenella 722
Anagyris foetida 3575
Anamirta 2264
Ananas 235
Ananas comosus 4191
Ananas sativus 4191
Anaphalis 4094
Anaphalis margaritaceae 1488
Anastatica hierochuntica 3091
Anchietea 236
Anchusa 840
Anchusa azurea 2975
Anchusa barrelieri 1959
Anchusa italica 2975
Anchusa officinalis 1371
Andira 246
Andira inermis 907
Andrachne 239
Andreaeaceae 240
Andromeda 241
Andromeda polifolia 723
Andropogon 706
Andropogon halepensis 3103
Andropogon ischaemum 1985
Andropogon sorghum 5087
Androsace 4527
Androsace alpina 144
Androsace helvetica 5401
Anemone 242
Anemone alpina 112
Anemone altaica 2962
Anemone coronaria 4273
Anemone fulgens 2273
Anemone hepatica 1437
Anemone japonica 3013
Anemonella thalictroides 4598
Anemone narcissiflora 3796
Anemone nemorosa 2134
Anemone patens 5177
Anemone pratensis 3554
Anemone pulsatilla 2116
Anemone ranunculoides 6115
Anemone sylvestris 5061
Anemone vernalis 5721
Anethum 1857
Anethum graveolens 1399
Angelica 245
Angelica archangelica 2406
Angelica palustris 714
Angelica sylvestris 6029
Angiopteris 5730
Angiospermae 248
Angraecum 746
Anhalonium 3344
Anisochilus 255
Anisoptera 3590
Anneslea 257
Anneslia 257
Annona 1751
Annonaceae 1752
Annona cherimola 1132
Annona glabra 4257
Annona muricata 5093
Annona reticulata 849
Annona squamosa 5297
Antennaria 4363
Antennaria alpina 142
Antennaria carpathica 1028
Antennaria dioica 1510
Anthemis 963
Anthemis arvensis 2220
Anthemis cotula 3542
Anthemis nobilis 4546
Anthemis tinctoria 2528
Anthericum 269
Anthericum liliago 4648
Antholyza 3415
Anthophyta 2310
Anthospermum 182
Anthoxanthum 5724
Anthoxanthum aristatum 265
Anthoxanthum odoratum 5389
Anthriscus 494
Anthriscus cerefolium 4658
Anthriscus sylvestris 6030
Anthriscus vulgaris 856
Anthurium 270
Anthyllis 271
Anthyllis montana 160
Anthyllis vulneraria 3159
Antiaris toxicaria 5684
Antidesma 1164
Antidesma bunius 566
Antirrhinum 5050
Antirrhinum majus 1528
Antirrhinum orontium 1619
Apeiba tiburbu 5515
Apera spica-venti 5981
Aphananthe 274
Aphelandra 275
Aphelandra aurantiaca 3936
Aphyllophorales 276
Apios 4282
Apios americana 207
Apios tuberosa 207
Apium 1086
Apium graveolens 5943
Apium graveolens *var.* dulce 2411
Apocynaceae 1873
Apocynum 1872
Apocynum androsaemifolium 5173
Apocynum cannabinum 2775
Apocynum sibiricum 4291
Apocynum venetum 4291
Aponogeton 5807
Aponogetonaceae 278
Apophyllum 279
Aporocactus flagelliformis 4422
Aporosella 280
Aposeris 281
Aptandra 290
Apuleia 291
Aquifoliaceae 2827
Aquilaria 1958
Aquilegia 1341
Aquilegia canadensis 188
Aquilegia skinneri 4985
Aquilegia vulgaris 2087
Arabidopsis thaliana 3733
Arabis 4516
Arabis alpina 143
Arabis arenosa 4518
Arabis caucasica 5774
Arabis gerardi 2450
Arabis glabra 5566
Arabis hirsuta 2688
Arabis turrita 5565
Araceae 334
Arachis 4087
Arachis hypogaea 4088
Aralia 295
Araliaceae 2495
Aralia chinensis 1167
Aralia chinensis glabrescens 3016
Aralia cordata 5675
Aralia edulis 5675
Aralia elata 3016
Aralia japonica 3034
Aralia spinosa 1849
Araucaria 297
Araucaria angustifolia 4049
Araucaria araucana 3658
Araucaria bidwillii 855
Araucaria brasiliana 4049
Araucariaceae 298
Araucaria cunnighami 1730
Araucaria excelsa 3877
Araujia 658
Arbutus 3418
Arbutus unedo 5276
Arceuthobium 1947
Archangelica officinalis 2406
Archegoniatae 301
Archimycetes 302
Archytea 303
Arcterica 4176
Arctium 859
Arctium lappa 2601
Arctium minus 5020
Arctium tomentosum 1636
Arctostaphylos 3480
Arctostaphylos alpinus 141
Arctostaphylos uva-ursi 500
Arctotis 312
Arctous 4320
Arctous alpinus 141
Ardisia 5117
Ardisia japonica 3018
Areca 313

Areca catechu 548
Arenaria 4692
Arenaria balearica 1626
Arenaria graminifolia 2584
Arenaria serpyllifolia 5511
Arenga 5305
Arenga pinnata 2558
Arenga saccharifera 2558
Arethusa 314
Argania sideroxylon 315
Argemone 4302
Argyreia 339
Argyrolobium 4962
Argythamnia 4953
Ariocarpus 3344
Arisaema 2999
Aristida 5499
Aristolochia 1932
Aristolochiaceae 593
Aristolochia clematitis 592
Aristolochia durior 1404
Aristolochia grandiflora 4104
Aristolochia macrophylla 1404
Aristolochia mandshuriensis 3460
Aristolochia sipho 1404
Aristotelia 5984
Armeniaca dasycarpa 4336
Armeniaca mandshurica 3457
Armeniaca mume 3015
Armeniaca sibirica 4891
Armeniaca vulgaris 1350
Armeria 5502
Armeria maritima 1545
Armillaria mellea 2838
Armoracia lapathifolia 2871
Arnebia 322
Arnica 323
Arnica montana 3694
Arnoseris 3213
Arnoseris minima 5029
Aronia 1224
Aronia arbutifolia 4446
Arracacia 325
Arracacia esculenta 277
Arracacia xanthorrhiza 277
Arrhenatherum 3905
Arrhenatherum elatius 5427
Artabotrys 5422
Artemisia 4640
Artemisia abrotanum 3923
Artemisia absinthium 1564
Artemisia annua 5396
Artemisia campestris 4641
Artemisia cina 3284
Artemisia dracunculus 5445
Artemisia maritima 3496
Artemisia pontica 4548
Artemisia rupestris 4540
Artemisia scoparia 3966
Artemisia stellariana 489
Artemisia vulgaris 3747
Artocarpus altilis 769
Artocarpus heterophyllus 2998
Artocarpus incisus 769
Artocarpus integer 2998
Artocarpus integrifolius 2998
Artocarpus lakoocha 3207
Arum 333
Arum dracunculus 1536
Arum italicum 2972
Arum maculatum 3378
Aruncus 2519
Aruncus sylvestris 5410
Arundinaria 994
Arundinella 335
Arundo 2481
Arundo donax 2482
Asarum 5947
Asarum europaeum 2133
Asclepiadaceae 3621
Asclepias 3620
Asclepias curassavica 676
Asclepias incarnata 5345
Asclepias syriaca 1469
Asclepias tuberosa 894
Ascomycetes 337
Ascophyllum 4539
Ascyrum crux-andreae 4647
Ascyrum hypericoides 4647
Asimina 4077
Asimina triloba 1485
Asparagus 350
Asparagus acutifolius 4848
Asparagus officinalis 1351
Asparagus plumosus 2204
Asparagus sprengeri 5178
Asperella 740
Aspergillus 351
Aspergillus glaucus 2631
Asperugo 3419
Asperugo procumbens 2459
Asperula 6040
Asperula cynanchica 5201
Asperula galioides 511
Asperula odorata 5395
Asperula tinctoria 1957
Asphodeline 3003
Asphodeline lutea 1449
Asphodelus 352
Asphodelus albus 757
Aspidistra 353
Aspidistra elatior 1352
Aspidium 354
Aspidium filixmas 3445
Aspidosperma 5907
Aspidosperma quebracho-blanco 1559
Asplenium 5161
Asplenium adiantum-nigrum 651
Asplenium ruta-muraria 5776
Asplenium septentrionale 2329
Asplenium trichomanes 3431
Asplenium viride 2634
Astelia 357
Aster 358
Asteraceae 359
Aster alpinus 113
Aster amellus 2973
Aster novae-anglica 3839
Aster novi-belgi 3842
Aster pyrenaeus 4370
Aster salicifolia 5968
Aster sibiricus 4892
Aster tripolium 5597
Astilbe 360
Astilbe chinensis 1169
Astragalus 3353
Astragalus alpinus 137
Astragalus boeticus 5349
Astragalus cicer 3701
Astragalus danicus 4350
Astragalus glycyphyllus 3619
Astragalus gummifer 5568
Astragalus hypoglotis 4350
Astragalus sinicus 1176
Astrantia 3531
Astrantia major 2611
Astrantia minor 1946
Astrocarpus 361
Astrocaryum 5626
Astronium 5234
Astrophytum 5213
Atalantia 362
Atalantia monophylla 2927
Athyrium filixfemina 3199
Atractylis 1864
Atraphaxis 2522
Atriplex 4663
Atriplex hastata 5118
Atriplex hortensis 2421
Atriplex litoralis 4875
Atriplex patula 2186
Atriplex patula hastata 5118
Atriplex pedunculata 5208
Atriplex rosea 5639
Atriplex tatarica 5457
Atropa belladonna 525
Atropis 96
Attalea 365
Aubrietia 366
Aucoumea 3920
Aucuba 367
Aucuba japonica 3019
Aurantioideae 1247
Aureolaria 3903
Auricularia auricula-judae 3101
Auriculariales 369
Autobasidiomycetes 381
Avena 3904
Avena barbata 5003
Avena brevis 3333
Avena byzantina 4457
Avena fatua 5951
Avena flavescens 6109
Avena nuda 3789
Avena orientalis 4927
Avena pratensis 3553
Avena pubescens 2687
Avena sativa 1481
Avena sativa orientalis 4927
Avena sterilis 252
Avena strigosa 3374
Avenastrum pratensis 3553
Averrhoa bilimbi 576
Averrhoa carambola 1012
Avicennia 386
Axyris 388
Azadirachta indica 3492
Azalea 389
Azalea pontica 4264
Azara microphylla 752
Azolla 391

Baccaurea 404

Baccharis 405
Baccharis halimifolia 1974
Bacopa 5811
Bactris 5154
Balanophoraceae 412
Ballota 425
Ballota foetida 614
Ballota nigra 614
Balsaminaceae 433
Bambusa 437
Bambusa arundinacea 5491
Bambusa nigra 615
Bambusa vulgaris 1357
Banksia 442
Baphia 968
Baphia nitida 4869
Baptisia 5948
Baptisia tinctoria 6113
Barbarea 5999
Barbarea verna 1966
Barbarea vulgaris 610
Barkhausia 2725
Barosma 827
Barringtonia acutangula 1130
Bartonia 459
Bartsia 460
Bartsia alpina 115
Basella 5736
Basella alba 5921
Basellaceae 3414
Basella tuberosa 5622
Basidiolichenes 461
Basidiomycetes 462
Bassia 468
Bassia hirsuta 2680
Bassia latifolia 2920
Batis 4668
Batrachospermum 479
Bauhinia 481
Bauhinia tomentosa 4654
Beckmannia 5015
Beckmannia syzigachne 209
Begonia 523
Begoniaceae 524
Begonia rex 355
Begonia semperflorens 4143
Begonia socotrana 5075
Begonia tuberhybrida 5621
Beilschmiedia 5016
Belamcanda chinensis 620
Bellis 1787
Bellis perennis 2037
Bellium 470
Benincasa 5845
Benincasa cerifera 1209
Benzoin 5127
Berberidaceae 453
Berberis 452
Berberis aetnensis 37
Berberis aggregata 4659
Berberis amurensis 225
Berberis asiatica 345
Berberis bretschneideri 4338
Berberis buxifolia 3421
Berberis canadensis 98
Berberis darwini 1805
Berberis hookeri 2845
Berberis japonica 3021
Berberis koreana 3178
Berberis sargentiana 4706
Berberis sibirica 4893
Berberis sieboldi 4930
Berberis thunbergi 3021
Berberis vernae 5722
Berberis verruculosa 5786
Berberis vulgaris 2075
Berberis wilsonae 5976
Berchemia 5336
Berchemia scandens 76
Bergenia 537
Bergenia cordifolia 2736
Bergenia crassifolia 3254
Bergia 538
Berlinia 539
Berteroa 2160
Berteroa incana 2818
Bertholletia excelsa 765
Berula 545
Berula erecta 5209
Beta 520
Beta cicla 1116
Beta maritima 4785
Beta saccharifera 5298
Beta vulgaris 1360
Beta vulgaris hortensis 4438
Beta vulgaris *ssp.* cicla 3473
Beta vulgaris *ssp.* esculenta 2408
Beta vulgaris *ssp.* esculenta *var.* rubra 4438
Beta vulgaris *ssp.* rapacea 3471
Beta vulgaris *var.* cicla 5398
Beta vulgaris *var.* macrorrhiza 2322
Betonica 550
Betonica officinalis 1361
Betula 578
Betulaceae 579
Betula chinensis 1171
Betula dahurica 1783
Betula ermani 2057
Betula forresti 2333
Betula japonica 3076
Betula lenta 5357
Betula lutea 6076
Betula maximowiczii 3650
Betula middendorffi 3612
Betula nana 1935
Betula nigra 4504
Betula papyrifera 4037
Betula pendula 2131
Betula populifolia 2594
Betula potanini 4280
Betula pubescens 4323
Betula pumila 3390
Betula schmidti 4758
Betula verrucosa 2131
Bidens 521
Bidens cernua 3866
Bidens tripartita 857
Bifora 555
Bignonia 567
Bignonia capreolata 1702
Bignoniaceae 5612
Bilderdykia convolvulus 1926
Billbergia 68
Biota 300
Biota orientalis 3948
Bischofia javanica 3086
Bistorta officinalis 2079
Bistorta vivipara 5759
Bixaceae 612
Bixa orellana 256
Blackstonia 6116
Blechnum 4726
Blechnum spicant 1821
Bletilla 672
Blighia 74
Blighia sapida 75
Blysmus compressus 2284
Blysmus rufus 4443
Bocconia 4230
Boehmeria 2174
Boehmeria nivea 4410
Boerhaavia 5131
Boisduvalia 5144
Boletus 727
Boletus aurantiacus 3940
Boletus badius 485
Boletus bovinus 4846
Boletus castaneus 1140
Boletus chrysenteron 4448
Boletus edulis 1093
Boletus elegans 2011
Boletus felleus 598
Boletus granulatus 2574
Boletus grevillei 2011
Boletus luridus 3396
Boletus luteus 820
Boletus rufus 3940
Boletus satanus 679
Boletus scaber 582
Boletus subtomentosus 6081
Boletus variegatus 5700
Boletus versipellis 3940
Boltonia 728
Bomarea 729
Bombacaceae 4944
Bombax 730
Bombax malabaricum 3440
Borago 734
Borago officinalis 1366
Borassus flabellifer 4019
Bor(r)aginaceae 735
Bosea yervamora 2544
Boswellia 2363
Boswellia carteri 553
Boswellia serrata 2937
Botrychium 2577
Botrychium lunaria 3673
Botrychium multifidum 794
Botrychium virginianum 4427
Bougainvillea 742
Bourreria 5286
Boussingaultia 3413
Bouteloua 2572
Bouvardia 747
Bovista plumbea 3244
Boykinia 753
Brachychiton acerifolium 2274
Brachychiton rupestris 457
Brachyelytrum 4880
Brachypodium 2164
Brachypodium pinnatum 3033
Brachypodium sylvaticum 4996
Brahea dulcis 5358

Brasenia 5835
Brassia 5132
Brassica 906
Brassica alba 5904
Brassica campestris 585
Brassicaceae 908
Brassica chinensis 4005
Brassica hirta 5904
Brassica juncea 2939
Brassica kaber 1118
Brassica napobrassica 4626
Brassica napus 6005
Brassica napus *var.* napus 4417
Brassica napus *var.* oleifera 1343
Brassica nigra 642
Brassica oleracea 5940
Brassica oleracea *var.* acephala 3128
Brassica oleracea *var.* botrytis 1079
Brassica oleracea *var.* capitata 2734
Brassica oleracea *var.* capitata alba 5885
Brassica oleracea *var.* capitata rubra 4444
Brassica oleracea *var.* cymosa 800
Brassica oleracea *var.* gemmifera 823
Brassica oleracea *var.* gongylodes 3175
Brassica oleracea *var.* sabauda 4722
Brassica pekinensis 4158
Brassica rapa 5650
Brassica rapa *var.* oleifera 3817
Bravoa geminiflora 3604
Bridelia 776
Briza 4381
Briza maxima 569
Briza media 4134
Briza minor 3335
Brodiaea 801
Bromelia 804
Bromeliaceae 4192
Bromus 802
Bromus arvensis 2219
Bromus catharticus 4483
Bromus commutatus 2681
Bromus erectus 3546
Bromus inermis 5034
Bromus japonicus 3024
Bromus madritensis 5109
Bromus mollis 5078
Bromus racemosus 413
Bromus secalinus 1138
Bromus sterilis 4287
Bromus tectorum 1123
Bromus unioloides 4483
Brosimum 770
Brosimum aubleti 5049
Brosimum galactodendron 1646
Brosimum utile 1646
Broussonetia 4039
Broussonetia papyrifera 1486
Browallia 814
Bruckenthalia 5142
Brunfelsia 4405
Brunnera 821
Brunsvigia gigantea 2469
Brunsvigia josephinae 2469
Bryales 5605
Bryonia 824
Bryonia alba 5883
Bryonia dioica 4439
Bryophyta 825
Buchloe dactyloides 836
Buddleia 892
Buddleia davidi 3941
Buddleia japonica 3026
Buddleia variabilis 3941
Bulbocodium 3557
Bulbophyllum 843
Bulboschoenus maritimus 4801
Bulnesia 851
Bunias 1617
Bunium 2724
Buphthalmum 3986
Buphthalmum speciosum 2740
Bupleurum 5493
Bupleurum falcatum 2156
Bupleurum fruticosum 4888
Bupleurum rotundifolium 4589
Bupleurum tenuissimum 4998
Bursaria 874
Bursera 875
Burseraceae 876
Butea 886
Butea frondosa 533
Butea monosperma 533
Butomaceae 2315
Butomus 2313
Butomus umbellatus 2314
Butyrospermum 898
Butyrospermum parkii 4852
Buxaceae 751
Buxus 749
Buxus balearica 5108
Buxus japonica 3054
Buxus microphylla japonica 3054
Buxus sempervirens 1368
Byrsonima 905

Cabomba 2184
Cabralea canjerana 991
Cabralea congerana 991
Cachrys 913
Cactaceae 915
Cactus 914
Cadia 916
Caesalpinia 758
Caesalpinia bonducella 3847
Caesalpinia brasiliensis 768
Caesalpiniaceae 4824
Caesalpinia coriaria 1867
Caesalpinia crista 3847
Caesalpinia echinata 4296
Caesalpinia sappan 4702
Cajanus cajan 4177
Cajanus indicus 4177
Cakile 4799
Caladium 925
Calamagrostis 4475
Calamagrostis epigeios 1131
Calamagrostis neglecta 5008
Calamintha 926, 4720
Calamintha acinos 5187
Calamintha alpina 148
Calamintha officinalis 927
Calamus 4424
Calamus rotang 4574
Calandrinia 4529
Calathea 928
Calceolaria 5011
Calceolaria herbeohybrida 5760
Calceolaria integrifolia 878
Calceolaria mexicana 3595
Calendula 929
Calendula officinalis 4285
Calimeris 3526
Calla 949
Calla palustris 5941
Callianthemum 951
Callicarpa 2370
Calligonum 952
Callistephus 1158
Callistephus chinensis 1382
Callitrichaceae 5840
Callitriche 5839
Callitriche autumnalis 384
Callitriche stagnalis 4261
Callitriche verna 5802
Callitris 1777
Callitris quadrivalvis 296
Calluna 2744
Calluna vulgaris 4772
Calochortus 3494
Calonyction 3670
Calophyllum 508
Calophyllum calaba 1104
Calophyllum inophyllum 2943
Calopogon 2589
Calopogon pulchellus 2590
Calotropis 953
Caltha 3508
Caltha palustris 1466
Calycanthaceae 5278
Calycanthus 5386
Calycanthus fertilis 4011
Calycanthus floridus 1540
Calypso 956
Calypso bulbosa 957
Calystegia 2509
Calystegia sepium 2748
Calystegia soldanella 4804
Camassia 958
Camassia esculenta 1375
Camelina 2168
Camelina microcarpa 3334
Camelina sativa 572
Camellia 960
Camellia drupifera 2799
Camellia japonica 1376
Camellia oleifera 2799
Camellia sasanqua 4711
Camellia sinensis 1543
Campanula 528
Campanula alpina 116
Campanula barbata 502
Campanula carpatica 1026
Campanulaceae 529
Campanula glomerata 1797
Campanula grandiflora 424
Campanula latifolia 2599
Campanula medium 999
Campanula patula 4408
Campanula persicifolia 4083
Campanula pyramidalis 1157

Campanula rapunculus 4414
Campanula rotundifolia 690
Campanula scheuchzeri 4756
Campanula sibirica 4894
Campanula trachelium 1643
Camphora officinarum 966
Camphorosma 965
Campsis 5611
Campsis radicans 1550
Camptosorus 5763
Cananga odorata 6119
Canarium 989
Canarium album 5886
Canarium commune 3085
Canarium nigrum 623
Canarium pimela 623
Canavalia 2997
Canavalia ensiformis 1448
Canavalia gladiata 5406
Candollea cuneiformis 5858
Canella 995
Canna 996
Cannabinaceae 2777
Cannabis 2772
Cannabis sativa 2773
Cannaceae 997
Canna edulis 1993
Canna indica 2934
Cantharellaceae 1001
Cantharellus 1000
Cantharellus aurantiacus 2165
Cantharellus cibarius 1115
Capparidaceae 1009
Capparis 1007
Capparis spinosa 1377
Caprifoliaceae 2842
Capsella bursa pastoris 4862
Capsicum 4459
Capsicum annuum 884
Capsicum frutescens 884
Caragana 4095
Caragana arborescens 4910
Caragana frutescens 4617
Caragana frutex 4617
Caragana microphylla 3329
Carapa 1657
Carapa guianensis 2660
Cardamine 600
Cardamine hirsuta 4111
Cardamine parviflora 5021
Cardamine pratensis 1723
Cardaminopsis arenosa 4518
Cardaria 5918
Cardaria draba 4127
Cardiospermum 2741
Carduus 782
Carduus acanthoides 13
Carduus crispus 1740
Carduus kerneri 3155
Carduus marianus 669
Carduus nutans 3759
Carex 4810
Carex acutiformis 3516
Carex aquatilis 5834
Carex arenaria 4688
Carex atrata 650
Carex caespitosa 5630
Carex canescens 4970
Carex capillaris 2678
Carex chordorrhiza 1595
Carex digitata 2252
Carex dioica 1860
Carex distans 1865
Carex elata 5656
Carex elongata 2023
Carex extensa 3364
Carex flava 6106
Carex glauca 1020
Carex hirta 2690
Carex leporina 2712
Carex limosa 3744
Carex loliacea 4632
Carex montana 3718
Carex pallescens 4010
Carex paniculata 4030
Carex pauciflora 2212
Carex pendula 4107
Carex pilosa 2690
Carex pilulifera 4182
Carex praecox 5725
Carex pulicaris 2293
Carex remota 4482
Carex riparia 5282
Carex rupestris 4533
Carex spicata 4303
Carex stellulata 5222
Carex sylvatica 6042
Carex tomentosa 1896
Carex vesicaria 674
Carex vulpina 2348
Caricaceae 4034
Carica papaya 4035
Carlina 1017
Carlina acaulis 5035
Carludovica 1018
Carnegiea gigantea 4645
Carpesium 1029
Carpinus 2852
Carpinus americana 198
Carpinus betulus 2108
Carpinus caroliniana 198
Carpinus caucasica 1073
Carpinus cordata 2738
Carpinus japonica 3045
Carpinus orientalis 3956
Carthamus 4636
Carthamus tinctorius 4637
Carum carvi 1013
Carum copticum 72
Carya 2795
Carya alba 3640
Carya aquatica 5809
Carya blanc 4841
Carya carolinae-septentrionalis 1025
Carya cordiformis 605
Carya glabra 4179
Carya illinoensis 4100
Cary amara 605
Carya myristicaeformis 3896
Carya ovata 4841
Carya pecan 4100
Carya porcina 4179
Carya sinensis 1186
Carya tomentosa 3640
Caryocar nucifera 4725
Caryophyllaceae 4198
Caryopteris 689
Caryota 2265
Caryota urens 5540
Casimiroa edulis 5909
Cassia 4823
Cassia acutifolia 89
Cassia angustifolia 1575
Cassia fistula 2548
Cassia marilandica 5958
Cassia medsgeri 5958
Cassine 1048
Cassiope 1049
Cassiope hypnoides 307
Cassiope tetragona 2260
Castanea 1139
Castanea crenata 3027
Castanea dentata 187
Castanea japonica 3027
Castanea pumila 99
Castanea sativa 2086
Castanopsis 2145
Castanospermum australe 3680
Castilla 2665
Castilla elastica 4022
Castilleja 4003
Castilloa 2665
Casuarina 516
Casuarinaceae 517
Casuarina cunninghamiana 1731
Casuarina equisetifolia 2876
Casuarina stricta 1299
Catabrosa aquatica 806
Catalpa 1053
Catalpa bignonioides 5097
Catalpa ovata 1175
Catalpa speciosa 3879
Catananche 1737
Catasetum 3659
Catha edulis 294
Cattleya 1065
Cattleya citrina 5634
Caucalis 2166
Caucalis daucoides 2753
Caulophyllum 1322
Ceanothus 1080
Ceanothus americanus 3840
Cecropia 4331
Cedrela 1084
Cedrela odorata 1234
Cedrela sinensis 1205
Cedrela toona 866
Cedrus 1082
Cedrus atlantica 364
Cedrus atlantica glauca 688
Cedrus brevifolia 1778
Cedrus deodara 1836
Cedrus libani 1083
Ceiba pentandra 3134
Celastraceae 5203
Celastrus 608
Celastrus scandes 185
Celosia 1307
Celosia argentea 2192
Celosia cristata 1410
Celsia 1088
Celtis 2674
Celtis australis 2106
Celtis caucasica 1072

Celtis koraiensis 3182
Celtis occidentalis 1434
Celtis sinensis 1183
Celtis tourneforti 3955
Cenchrus 4682
Centaurea 1089
Centaurea calcitrapa 5232
Centaurea cyanus 1613
Centaurea jacea 817
Centaurea maculosa 5165
Centaurea montana 3697
Centaurea moschata 5388
Centaurea nigra 625
Centaurea orientalis 1067
Centaurea ruthenica 4627
Centaurea scabiosa 4734
Centaurea solstitialis 6078
Centaurium 1090
Centaurium umbellatum 1909
Centranthus 1091
Centranthus ruber 3119
Centrolobium 4275
Centunculus 1109
Cephaelis ipecacuanha 2957
Cephalanthera 4163
Cephalanthus 902
Cephalaria 1094
Cephalaria alpina 6079
Cephalaria syriaca 5417
Cephalaria tatarica 5453
Cephalocereus 1095
Cephalotaxus 4232
Cephalotaxus fortunei 1196
Cephalotus 1096
Cerastium 1145
Cerastium alpinum 122
Cerastium arvense 5230
Cerastium biebersteini 5461
Cerastium caespitosum 2221
Cerastium cerastioides 5236
Cerastium glomeratum 3732
Cerastium semidecandrum 5853
Cerastium tomentosum 5064
Cerasus 1133
Cerasus avium 3544
Cerasus besseyi 547
Cerasus glandulosa 105
Cerasus pumila 4683
Ceratiola 4686
Ceratocarpus 1097
Ceratonia siliqua 1023
Ceratophyllaceae 2862
Ceratophyllum 2860
Ceratophyllum demersum 2861
Ceratopteris 5803
Ceratozamia 2854
Cercidiphyllum japonicum 3048
Cercis 1098
Cercis canadensis 1979
Cercis chinensis 1199
Cercis japonica 1199
Cercis siliquastrum 3114
Cercocarpus 3712
Cercospora 1099
Cereus 1100
Cereus flagelliformis 4422
Cereus giganteus 4645
Cereus grandiflorus 4384
Cerinthe 2843
Cerinthe minor 3280
Ceropegia 1102
Ceroxylon andicola 5846
Cestrum 3098
Ceterach 4737
Cetraria 1103
Cetraria islandica 2916
Chaenomeles 2312
Chaenomeles japonica 3038
Chaenomeles lagenaria 1414
Chaenomeles maulei 3038
Chaenomeles sinensis 1181
Chaerophyllum 1137
Chaerophyllum bulbosum 5651
Chaerophyllum temulum 1648
Chamaecyparis 2167
Chamaecyparis lawsoniana 3243
Chamaecyparis nootkatensis 3874
Chamaecyparis obtusa 2812
Chamaecyparis pisifera 4724
Chamaecyparis thyoides 5887
Chamaedaphne 3257
Chamaedorea 1112
Chamaenerium 5975
Chamaenerium angustifolium 2262
Chamaepericlymenum suecicum 1937
Chamaepeuce 5487
Chamaerops humilis 3572
Chara 5267
Charophyta 1119
Chayota edulis 1122
Cheilanthes 3323
Cheiranthus 5768
Cheiranthus cheiri 1556
Chelidonium 1085
Chelidonium majus 2606
Chelone 5655
Chenopodiaceae 2562
Chenopodium 2561
Chenopodium album 3212
Chenopodium ambrosioides 6056
Chenopodium bonus-henricus 2559
Chenopodium botrys 3096
Chenopodium capitatum 675
Chenopodium foliosum 5273
Chenopodium murale 3837
Chenopodium quinoa 4390
Chilopsis linearis 1840
Chimaphila 4202
Chimaphila umbellata 1496
Chimonanthus praecox 6009
Chiococca 3617
Chionanthus 2379
Chionanthus virginica 5896
Chionodoxa 2512
Chironia 5229
Chlora 6116
Chloranthaceae 1221
Chloranthus 1220
Chlora perfoliata 1567
Chlorella 1222
Chloris 5982
Chlorophora 2396
Chlorophora excelsa 2963
Chlorophora tinctorial 1912
Chlorophyceae 2621
Chloroxylon swietenia 1107
Choeromyces 5920
Chondrilla 4983
Chondrilla juncea 4610
Chondrus 2960
Chondrus crispus 2961
Chorda filum 4792
Chroococcaceae 1228
Chrozophora 5653
Chrysallidocarpus 896
Chrysanthemum 1229
Chrysanthemum alpinum 123
Chrysanthemum balsamita 1632
Chrysanthemum cinerariaefolium 1790
Chrysanthemum coccineum 2306
Chrysanthemum coronarium 1710
Chrysanthemum corymbosum 3569
Chrysanthemum indicum 3687
Chrysanthemum inodorum 4751
Chrysanthemum koreanum 3179
Chrysanthemum leucanthemum 3987
Chrysanthemum macrophyllum 5441
Chrysanthemum majus 1632
Chrysanthemum parthenium 2210
Chrysanthemum roseum 2306
Chrysanthemum rotundifolium 2887
Chrysanthemum segetum 1609
Chrysanthemum sibiricum 3179
Chrysobalanus 1315
Chrysobalanus icaco 2914
Chrysogonum 2549
Chrysophyllum 5211
Chrysophyllum cainito 919
Chrysopogon 4750
Chrysopogon gryllus 1688
Chrysosplenium 2545
Chrysosplenium alternifolium 168
Chrysosplenium oppositifolium 3935
Chukrasia tabularis 1218
Chusquea 1231
Cibotium barometz 4784
Cicer arietinum 1144
Cichoriaceae 1232
Cichorium 1147
Cichorium intybus 1381
Cicuta 5808
Cicuta virosa 2130
Cimicifuga 837
Cimicifuga foetida 4988
Cinchona 1236
Cineraria 1237
Cineraria cruenta 1385
Cineraria maritima 4958
Cinna 6039
Cinnamomum 1238
Cinnamomum camphora 966
Cinnamomum cassia 1047
Cinnamomum glanduliferum 3827
Cinnamomum zeylanicum 1105
Circaea 2031
Circaea alpina 124
Circaea lutetiana 4055
Cirsium 5487
Cirsium acaule 5238
Cirsium arvense 975
Cirsium lanceolatum 850
Cirsium palustre 3519

Cissus 5577
Cissus gongylodes 3486
Cistaceae 4532
Cistus 4531
Cistus crispus 6058
Cistus hirsutus 2689
Cistus ladaniferus 2664
Cistus laurifolius 3235
Cistus populifolius 4271
Cistus salvifolius 4670
Citharexylum 2218
Citrullus colocynthis 1334
Citrullus vulgaris 5821
Citrullus vulgaris citroides 1245
Citrus 1246
Citrus amara 5092
Citrus aurantifolia 3311
Citrus aurantium 5092
Citrus bergamia 536
Citrus bigardia 5092
Citrus decumana 4329
Citrus grandis 4329
Citrus hystrix 3538
Citrus limetta 3311
Citrus limon 3268
Citrus limonia 3268
Citrus margarita 3980
Citrus maxima 4329
Citrus medica 1243
Citrus nobilis 3162
Citrus nobilis deliciosa 3468
Citrus paradisi 2578
Citrus reticulata 3468
Citrus sinensis 5377
Citrus trifoliata 5592
Citrus vulgaris 5092
Cladanthus 1248
Cladium 4727
Cladonia 1249
Cladonia rangiferina 4479
Cladrastis 6114
Cladrastis lutea 213
Clarkia 1253
Clausena 5779
Clavaria 2520
Clavaria aurea 2531
Clavaria botrytis 4468
Clavariaceae 1588
Clavaria flava 4013·
Claviceps 1258
Claviceps purpurea 2053
Claytonia 5181
Claytonia perfoliata 3627
Clematis 1260
Clematis alpina 125
Clematis alpina *var.* sibirica 4896
Clematis armandi 319
Clematis cirrhosa 5723
Clematis flammula 4227
Clematis integrifolia 5080
Clematis jackmani 3000
Clematis koreana 3180
Clematis lanuginosa 3858
Clematis montana 244
Clematis orientalis 3952
Clematis paniculata 5353
Clematis patens 3297
Clematis recta 2646
Clematis songarica 5330
Clematis tangutica 2532
Clematis viorna 3255
Clematis virginiana 5756
Clematis vitalba 5571
Clematis viticella 2976
Cleome 5128
Clerodendron 2510
Clethra 5323
Clethraceae 1261
Clianthus 4059
Clinopodium 926
Clinopodium vulgare 5936
Clintonia 490
Clitocybe 1270
Clitocybe aurantiaca 2165
Clitocybe candicans 4867
Clitocybe clavipes 1287
Clitocybe infundibuliformis 2395
Clitocybe laccata 5850
Clitocybe nebularis 1274
Clitocybe odora 5385
Clitopilus prunulus 4226
Clitoria 4178
Clivia 3126
Clivia miniata 4745
Clusia 1290
Clusiaceae 2669
Clypeola 1297
Cneorum 5194
Cnicus 670
Cnicus benedictus 671
Cobaea 1301
Cobaea scandens 4337
Cobresia 1302
Cobresia caricina 1016
Coccoloba 4788
Cocculus 5044
Cochlearia 4783
Cochlearia anglica 2043
Cochlearia officinalis 1522
Cochlospermum 4860
Cochorium endivia 2032
Cocos nucifera 1314
Codium 1316
Codonopsis 340
Coelococcus 4644
Coeloglossum viride 2384
Coelogyne 1317
Coffea 1318
Coffea arabica 292
Coffea liberica 3287
Coix lacrima jobi 31
Cola 1323
Cola acuminata 5293
Colchicum 382
Colchicum autumnale 1353
Coleanthus 3741
Coleanthus subtilis 3742
Coleus 1325
Coleus blumei 1386
Coleus parviflorus 2721
Coleus scutellarioides 1386
Colletia 1330
Colletia spinosa 1769
Collinsia 1331
Collinsonia 2863
Collomia 1332
Collybia 1333
Collybia butyracea 901
Collybia platyphylla 790
Collybia radicata 4550
Collybia velutipes 5709
Colocasia 2013
Colocasia antiquorum 1806
Colocasia esculenta 1806
Colubrina 1340
Coluria 1281
Colutea 662
Colutea arborescens 1362
Colutea orientalis 3950
Comarum palustre 3502
Combretaceae 3776
Combretum 1344
Commelina 1814
Commelinaceae 5135
Commiphora 3779
Commiphora abyssinica 9
Commiphora africana 44
Commiphora myrrha 1477
Commiphora opobalsamum 3563
Compositae 1570
Comptonia 5364
Condalia 1571
Coniferae 1579
Coniogramme japonica 438
Conioselinum 2770
Conium maculatum 4241
Conjugatae 1580
Conopholis 990
Conringia 2710
Conringia orientalis 5574
Convallaria 3305
Convallaria majalis 3306
Convolvulaceae 3683
Convolvulus 2509
Convolvulus arvensis 2100
Convolvulus scammonia 4741
Convolvulus sepium 2748
Convolvulus soldanella 4804
Convolvulus tricolor 1941
Cooperia 4402
Copaifera 1583
Copaifera copallifera 4939
Copaifera officinalis 1582
Copernicia 1584
Copernicia cerifera 1021
Coprinaceae 1586
Coprinus 2951
Coprinus atramentarius 2950
Coprinus comatus 4843
Coptis 2556
Corallorrhiza 1591
·Corallorrhiza trifida 1960
Corchorus 3122
Corchorus capsularis 4591
Corchorus olitorius 4283
Cordia 1594
Cordyline 1899
Cordyline australis 2470
Cordyline terminalis 1402
Coreopsis 1596
Coreopsis tinctoria 4212
Coriandrum 1597
Coriandrum sativum 1598
Coriaria 1599

Coriariaceae 1600
Coriaria myrtifolia 3782
Corispermum 5517
Corispermum hyssopifolium 2911
Cornaceae 1882
Cornus 1881
Cornus alba 5454
Cornus alba sibirica 4901
Cornus florida 2309
Cornus mas 1611
Cornus sanguinea 683
Cornus sibirica 4901
Cornus stolonifera 4458
Cornus suecica 1937
Cornutia 1621
Coronaria 967
Coronaria coriacea 4558
Coronilla 1712
Coronilla emerus 4766
Coronilla montana 3704
Coronilla varia 1713
Coronopus 5783
Coronopus procumbens 1678
Corrigiola 1622
Corrigiola littoralis 5271
Cortaderia 4020
Cortaderia selloana 4818
Cortinarius violaceus 5739
Cortusa 506
Corydalis 1628
Corydalis lutea 6080
Corydalis nobilis 4897
Corydalis sibirica 4897
Corylopsis 6004
Corylopsis pauciflora 891
Corylopsis spicata 5147
Corylopsis willmottiae 5963
Corylus 2246
Corylus americana 194
Corylus avellana 2098
Corylus avellana pontica 1071
Corylus californica 935
Corylus chinensis 1179
Corylus colurna 5646
Corylus mandshurica 3456
Corylus maxima 2473
Corylus pontica 1071
Corylus sieboldiana 3035
Corylus sieboldiana mandshurica 3456
Corylus tibetica 5513
Corynephorus 1284
Corynephorus canescens 2595
Corynocarpus 3135
Corypha 1629
Corypha umbraculifera 5424
Cosmos 1630
Cosmos bipinnatus 1392
Costus 5158
Cotinus 5033
Cotinus coggygria 1527
Cotoneaster 1633
Cotoneaster dammeri 501
Cotoneaster horizontalis 4515
Cotoneaster integerrima 2089
Cotoneaster melanocarpa 618
Cotoneaster racemiflora 4437
Cotoneaster tomentosa 774
Cotoneaster zabeli 1134
Cotula 759
Cotyledon 1639
Covillea 1680
Crambe 1326
Crambe maritima 1387
Crambe tatarica 5452
Crassocephalum 5713
Crassula 1661
Crassulaceae 3970
Crataegus 2729
Crataegus altaica 167
Crataegus arnoldiana 324
Crataegus azarolus 390
Crataegus brainderi asperifolia 3845
Crataegus canadensis 972
Crataegus chrysocarpa phoenicea 1976
Crataegus coccinea 5485
Crataegus crusgalli 1313
Crataegus ellwangeriana 2017
Crataegus flava 6093
Crataegus heldreichi 2761
Crataegus lambertiana 3209
Crataegus monogyna 4973
Crataegus nigra 2082
Crataegus orientalis 4960
Crataegus oxyacantha 2038
Crataegus pontica 390
Crataegus robesoniana 4508
Crataegus rotundifolia 1976
Crataegus sanguinea 4452
Craterellus cornucopioides 2858
Crepis 2725
Crepis biennis 4577
Crepis capillaris 5038
Crepis paludosa 3505
Crepis tectorum 3803
Crepis virens 5038
Crescentia 924
Crescentia cajete 1373
Crinum 1696
Crinum longifolium 2707
Crithmum 4674
Crithmum maritimum 4675
Crocanthemum 2387
Crocus 1699
Crocus heuffelianus 2790
Crocus sativus 4638
Crocus vernus 1395
Crotalaria 4425
Crotalaria intermedia 5000
Crotalaria juncea 5331
Crotalaria lanceolata 3215
Crotalaria mucronata 5283
Crotalaria spectabilis 4884
Crotalaria striata 5283
Croton 1705
Croton eleuteria 1041
Croton tiglium 4334
Crucianella 1703
Cruciferae 908
Crupina 1714
Crypsis 4294
Cryptanthus 1715
Cryptocoryne 1716
Cryptogamae 1717
Cryptogramma 4512
Cryptomeria 1718
Cryptomeria japonica 3029
Cryptotaenia canadensis 2836
Ctenanthe 1720
Cucubalus 656
Cucubalus baccifer 544
Cucumis anguria 5869
Cucumis melo 3762
Cucumis melo flexuosus 5046
Cucumis melo *var.* cantalupensis 998
Cucumis melo *var.* reticulatus 3834
Cucumis metuliferus 2855
Cucumis sativus 1724
Cucurbita 2568
Cucurbitaceae 2569
Cucurbita ficifolia 3437
Cucurbita maxima 6008
Cucurbita melanosperma 3437
Cucurbita moschata 1747
Cucurbita pepo 4330
Cucurbita pepo *var.* giraumontia 5705
Cucurbita pepo *var.* melopepo 883
Cucurbita pepo *var.* ovifera 6088
Cuminum 1728
Cuminum cyminum 1729
Cunila 5265
Cunninghamia 1163
Cunninghamia sinensis 1383
Cunnunghamia lanceolata 1383
Cupania 1732
Cupania sapida 75
Cuphae 1736
Cupressaceae 1775
Cupressus 1773
Cupressus arizonica 317
Cupressus californica 2571
Cupressus cashmeriana 3137
Cupressus ducloxiana 551
Cupressus funebris 3727
Cupressus goveniana 2571
Cupressus goveniana *var.* glauca 699
Cupressus lusitanica 3597
Cupressus macnabiana 3405
Cupressus macrocarpa 3665
Cupressus sempervirens 2978
Cupressus sempervirens *var.* pyramidalis 4367
Cupressus torulosa 551
Curculigo 1738
Curcuma 5648
Curcuma angustifolia 1983
Curcuma longa 1551
Curcuma zedoaria 6125
Cuscuta 1870
Cuscutaceae 1871
Cuscuta epilinum 2287
Cuscuta epithymum 1400
Cuscuta europaea 2607
Cuscuta gronovi 2641
Cuscuta suaveolens 1153
Cuscuta trifolii 1280
Cuviera 1762
Cyananthus 1763
Cyanophyceae 701
Cyathea 5580
Cycadaceae 1765

Cycas 1764
Cycas revoluta 4643
Cyclamen 1766
Cyclamen europaeum 2092
Cyclamen indicum 2990
Cyclamen persicum 2990
Cyclanthaceae 1767
Cyclanthera 1768
Cyclophorus 2197
Cycnoches 5348
Cydonia 4389
Cydonia maulei 3038
Cydonia oblonga *var.* maliformis 286
Cydonia sinensis 1181
Cydonia vulgaris 1512
Cymbalaria 465
Cymbalaria muralis 3147
Cymbopogon citratus 3270
Cymbopogon nardus 1244
Cynanchum 5340
Cynanchum vincetoxicum 5916
Cynara cardunculus 1015
Cynara scolymus 331
Cynocrambaceae 1770
Cynodon 1879
Cynodon dactylon 542
Cynoglossum 2879
Cynoglossum germanicum 2630
Cynoglossum officinale 1444
Cynomorium 1771
Cynosurus 1877
Cynosurus cristatus 1681
Cynosurus echinatus 2750
Cynoxylon floridum 2309
Cypella 1772
Cyperaceae 4811
Cyperus 2280
Cyperus alternifolius 5677
Cyperus esculentus 1230
Cyperus longus 2402
Cyperus papyrus 4043
Cyperus rotundus 3893
Cyphomandra betacea 5589
Cypripedium 3203
Cypripedium calceolus 2109
Cyrtanthus 1780
Cyrtonium falcatum 2881
Cyrtostachys 4795
Cystopteris 657
Cystopteris fragilis 786
Cystopteris montana 3696
Cytinus 1781
Cytisus 809
Cytisus albus 4276
Cytisus alpinus 4773
Cytisus austriacus 375
Cytisus capitatus 557
Cytisus laburnum 2529
Cytisus multiflorus 5910
Cytisus nigricans 5137
Cytisus purpureus 4339
Cytisus scorparius 4769
Cytisus supinus 557

Dacrydium 2894
Dactylis 1309
Dactylis glomerata 3943
Daedalea quercina 3900
Daemonorops draco 5314
Dahlia 1782
Dahlia variabilis 2413
Dalbergia 4570
Dalbergia latifolia 1984
Dalbergia melanoxylon 42
Dalbergia nigra 648
Dalbergia sissoo 4977
Damasonium 1793
Damasonium stellatum 5219
Dammara alba 5890
Danae racemosa 88
Danthonia 1798
Daphne 1800
Daphne caucasica 1070
Daphne cneorum 4559
Daphne laureola 5192
Daphne mezereum 2194
Darlingtonia 941
Dasiphora fruticosa 879
Dasylirion 5090
Dasylirion acrotrichum 822
Dasylirion wheeleri 5875
Dasystoma 3903
Datisca 1810
Datisca cannabina 471
Datura 1811
Datura arborea 2304
Datura metel 2810
Datura stramonium 3102
Datura suaveolens 247
Datura tatula 3102
Daucus 1035
Daucus carota 5942
Daucus carota *var.* sativus 2410
Davallia 2711
Deinbollia 1826
Delphinium 3226
Delphinium ajacis 4524
Delphinium consolida 2330
Delphinium cultorum 2305
Delphinium elatum 518
Delphinium hybridum 2305
Dendrobium 1829
Dendrocalamus 2467
Denhamia 1830
Dentaria 5548
Dentaria bulbifera 1592
Derris 3100
Deschampsia 2677
Deschampsia alpina 134
Deschampsia caespitosa 5628
Deschampsia flexuosa 5843
Descurainia 5442
Descurainia sophia 2296
Desmidiaceae 1841
Desmodium 5516
Desmodium gyrans 5474
Deutzia 1843
Deutzia gracilis 4995
Deutzia grandiflora 1961
Deutzia scabra 2397
Deutzia schneideriana 4759
Deutzia sieboldiana 4932
Deutzia wilsoni 5977
Dialium 1851
Dialium guineense 5715
Dianthus 4197
Dianthus arenarius 2255
Dianthus armeria 1837
Dianthus barbatus 5393
Dianthus caesius 1129
Dianthus carthusianorum 1037
Dianthus caryophyllus 1019
Dianthus chinensis 1165
Dianthus deltoides 3433
Dianthus fischeri 1165
Dianthus plumarius 2588
Dianthus superbus 3298
Diapensia 1852
Diarrhena 495
Diascia 5665
Diatomeae 1853
Dicentra 668
Dicentra spectabilis 1363
Dichapetalium 1854
Dichromena 5919
Dicorynia paraensis 4044
Dicotyledoneae 1855
Dicranum 2331
Dictamnus 1866
Dictamnus albus 2435
Dicypellium caryophyllatum 1275
Didelotia africana 826
Didiscus 3194
Dieffenbachia 5631
Dielytra 668
Dierama 2016
Diervilla 880
Digitalis 2346
Digitalis ambigua 6089
Digitalis grandiflora 6089
Digitalis lanata 2614
Digitalis lutea 5280
Digitalis purpurea 1416
Digitaria 2254
Digitaria ischaemum 5037
Digitaria linearis 5037
Digitaria sanguinalis 2683
Dillenia 1858
Dimorphotheca 1005
Dioclea 1295
Dionaea 5718
Dioscorea 6062
Dioscorea alata 5993
Dioscorea batatas 1241
Dioscorea bulbifera 69
Dioscoreaceae 6064
Dioscorea esculenta 1214
Dioscorea trifida 1748
Diosma 1861
Diospyros 4152
Diospyros chinensis 3127
Diospyros ebenum 1987
Diospyros kaki 3127
Diospyros lotus 1809
Diospyros virginiana 1494
Diphylleia 5678
Diplotaxis 5775
Diplotaxis muralis 5257
Diplotaxis tenuifolia 5007
Dipsacaceae 5471
Dipsacus 5469
Dipsacus fullonum 2390
Dipsacus laciniatus 1760
Dipsacus pilosus 2692

Dipsacus sylvestris 5717
Diptercarpaceae 1862
Dipterocarpus 2668
Dipterocarpus tuberculatus 2035
Dipterocarpus turbinatus 1432
Dipteryx 5545
Dipteryx odorata 1933
Dirca 3259
Dirca palustris 363
Discostigma 1863
Dodartia 1869
Dodecatheon 4871
Dodonaea 2850
Dodonaea viscosa 1251
Dolichos 1883
Dolichos japonicus 5505
Dolichos lablab 2897
Dolichos sesquipedalis 6066
Dolichos sinensis 1393
Dombeya 1885
Dombeya punctata 744
Dorema 5316
Dorema ammoniacum 731
Doronicum 3276
Doronicum austriacum 379
Doronicum caucasicum 1075
Dorstenia 5559
Doryanthes 5119
Dorycnium 982
Dorycnium herbaceum 2784
Draba 1897
Draba alpina 4519
Draba fladnizensis 308
Draba glacialis 2498
Draba incana 5667
Draba muralis 5766
Draba norvegica 3888
Draba sibirica 4902
Draba tomentosa 6048
Draba verna 5183
Dracaena 1898
Dracaena australis 2470
Dracaena draco 1900
Dracaena fragrans 2357
Dracocephalum 1901
Dracocephalum argunense 4629
Dracocephalum austriacum 376
Dracocephalum moldavicum 3645
Dracocephalum ruyschianum 4629
Dracocephalum thymiflorum 5510
Dracunculus 5254
Dracunculus vulgaris 1536
Drimys 1903
Drimys winteri 6007
Drosera 5325
Drosera anglica 2044
Droseraceae 5326
Drosera rotundifolia 4588
Drosophyllum 1908
Dryas 1920
Dryas octopetala 3726
Drymaria 4808
Dryobalanops 736
Dryobalanops aromatica 1367
Dryopteris 6026
Dryopteris cristata 1685
Dryopteris filixmas 3445
Dryopteris linnaeana 3901
Dryopteris oreopteris 3884
Dryopteris phegopteris 3797
Dryopteris spinulosa 5546
Dryopteris thelypteris 3504
Duboisia 1922
Duboisia myoporoides 1606
Duchesnea 3643
Duchesnea indica 2932
Duranta 4992
Durio 1927
Dyckia 1953
Dysoxylum 4106

Ebenaceae 1986
Ecballium elaterium 5202
Echeveria 1988
Echinacea 1989
Echinacea purpurea 4344
Echinocactus 1990
Echinocereus 1991
Echinochloa 1310
Echinochloa crusgali 456
Echinochloa crusgali frumentacea 3057
Echinochloa frumentacea 3057
Echinocystis 3639
Echinodorus 862
Echinopanax 1847
Echinophora 4300
Echinops 2508
Echinops bannaticus 6121
Echinops exaltatus 4615
Echinopsilon hirsutum 2680
Echinopsis 2749
Echinops ritro 5027
Echinops sphaerocephalus 1423
Echites 4718
Echium 5744
Echium plantagineum 4357
Echium violaceum 4357
Echium vulgare 1555
Edgeworthia 4038
Edgeworthia papyrifera 3958
Eichhornia 5810
Eichhornia crassipes 1557
Elaeagnaceae 3926
Elaeagnus 2006
Elaeagnus angustifolia 4616
Elaeagnus argentea 4951
Elaeagnus commutata 4951
Elaeagnus multiflora 1135
Elaeis 3916
Elaeis guineensis 46
Elaeocarpus 2008
Elaeocarpus americanus 189
Elaeocarpus dentatus 2809
Elaeodendron 2177
Elaeodendron capense 1004
Elaphomyces 2717
Elaphomyces cervinus 3399
Elatinaceae 5842
Elatine 5841
Elatine hexandra 4982
Elatine triandra 5501
Eleocharis 5145
Eleocharis acicularis 3822
Eleocharis eupalustris 719
Eleocharis ovata 3983
Eleocharis palustris 1533
Eleocharis parvula 1950
Eleocharis pauciflora 2213
Elephantopus 2014
Elettaria cardamomum 1014
Eleusine 2015
Eleusine coracana 4399
Eleusine indica 2563
Elisma natans 2299
Elodea 2022
Elodea canadensis 976
Elodea densa 1832
Elsholtzia 2024
Elsholtzia cristata 1682
Elymus 5957
Elymus arenarius 2094
Embelia 2025
Embothrium 2027
Emilia 5451
Empetraceae 1707
Empetrum 1706
Empetrum nigrum 627
Entada 2046
Entada phaseoloides 1265
Entada scandens 1265
Entandrophragma 4698
Enterolobium 1967
Entomophthoraceae 2047
Epacris 2048
Eperua 5764
Ephedra 3104
Ephedra americana 192
Ephedra californica 934
Ephedraceae 3107
Ephedra distachya 3106
Ephedra equisetina 3652
Ephedra gerardiana 2449
Ephedra viridis 2625
Epigaea 5569
Epilobium 5975
Epilobium alpinum 156
Epilobium angustifolium 2262
Epilobium hirsutum 2695
Epilobium hornemanni 2857
Epilobium lamyi 3214
Epilobium montanum 3723
Epilobium obscurum 1952
Epilobium palustre 3522
Epilobium parviflorum 5024
Epilobium roseum 4012
Epimedium 2049
Epipactis 2050
Epipactis latifolia 792
Epiphyllum 3247
Epipogon aphyllum 5195
Equisetaceae 2877
Equisetum 2875
Equisetum arvense 2225
Equisetum hyemale 4777
Equisetum limosum 5344
Equisetum majus 2610
Equisetum palustre 3506
Equisetum pratense 3552
Equisetum sylvaticum 5411
Equisetum variegatum 5701
Eragrostis 3385
Eragrostis abyssinica 5473
Eragrostis cilianensis 5255

Eragrostis major 5255
Eragrostis megastachya 5255
Eragrostis minor 3331
Eragrostis pilosa 2931
Eragrostis poaeoides 3331
Eranthis 5998
Erechtites 869
Eremostachys 1839
Eremurus 1838
Eremurus spectabilis 3435
Erianthus 4228
Erica 2742
Erica arborea 5581
Erica carnea 5186
Ericaceae 2745
Erica cinerea 5668
Erica mediterranea 594
Erica scoparia 546
Erica tetralix 1701
Erigeron 2292
Erigeron acris 602
Erigeron alpinus 130
Erigeron annuus 259
Erigeron canadensis 2878
Erigeron uniflorus 309
Eriobotrya 3375
Eriobotrya japonica 3376
Eriocaulon 4201
Eriochloa 1735
Eriogonum 2054
Eriophorum 1637
Eriophyllum 2055
Eritrichium 2056
Erodium 2789
Erodium cicutarium 91
Erophila 2188
Eruca sativa 4525
Erucastrum 4526
Ervum lens 1457
Eryngium 2058
Eryngium alpinum 709
Eryngium campestre 5047
Eryngium maritimum 4789
Erysimum 3772
Erysimum canescens 2817
Erysimum cheiranthoides 5573
Erysimum officinale 2757
Erysimum repandum 5175
Erysiphaceae 4289
Erythea 2059
Erythea edulis 2654
Erythraea centaurium 1909
Erythrina 1587
Erythrina corallodendron 1390
Erythrina crista-gali 1311
Erythronium 2190
Erythronium dens canis 1878
Erythroxylaceae 1303
Erythroxylum 1304
Erythroxylum coca 2883
Escallonia 2060
Escallonia rubra 4449
Eschscholtzia 2554
Eschscholtzia californica 943
Eucalyptus 2065
Eucalyptus globulosus 5447
Eucalyptus gunni 1233
Eucalyptus marginata 3081
Eucalyptus salicifolia 5969
Eucalyptus viminalis 4493
Eucephalartos 3125
Eucharidium 1253
Eucharis 2066
Eucheuma speciosum 373
Euchlaena 5475
Euchlaena mexicana 3603
Euclidium 2067
Eugenia 2068
Eugenia brasiliensis 761
Eugenia caryophyllata 1283
Eugenia cumini 3010
Eugenia dombeyi 2653
Eugenia malaccensis 3912
Eugenia myrtifolia 370
Eugenia paniculata australis 370
Eugenia uniflora 4206
Euonymus 2069
Euonymus alatus 5990
Euonymus americanus 805
Euonymus atropurpureus 1981
Euonymus europaeus 2096
Euonymus japonicus 2146
Euonymus latifolius 793
Euonymus nanus 1939
Euonymus verrucosa 5787
Eupatorium 732
Eupatorium cannabinum 2776
Euphorbia 2070
Euphorbiaceae 5191
Euphorbia cyparissias 1774
Euphorbia epithymoides 1750
Euphorbia esula 3251
Euphorbia exigua 1951
Euphorbia falcata 4922
Euphorbia helioscopia 5328
Euphorbia lathyrus 1008
Euphorbia maculata 5167
Euphorbia paralias 4805
Euphorbia peplus 4159
Euphorbia polychroma 1750
Euphorbia pulcherrima 1500
Euphorbia stricta 5689
Euphorbia virgata 3251
Euphrasia 2152
Euphrasia officinalis 1910
Eupritchardia 4311
Eurotia 6001
Eurycoma 2137
Euterpe 2138
Evax 2139
Evernia 2150
Evernia prunastri 5204
Evodia 2151
Evodia glauca 3032
Exochorda 4090
Exochorda alberti 5645
Exochorda korolkowi 5645
Exogonium purga 3005

Fabiana imbricata 4154
Fagaceae 514
Fagopyrum 832
Fagopyrum esculentum 1370
Fagopyrum sagittatum 1370
Fagopyrum tataricum 5446
Fagus 513
Fagus americana 184
Fagus engleriana 2036
Fagus grandifolia 184
Fagus japonica 3022
Fagus orientalis 3949
Fagus sieboldi 4931
Fagus sylvatica 2077
Fagus sylvatica *var.* atropunica 4345
Fagus sylvatica *var.* pendula 5859
Falcaria 2155
Falcaria sioides 1409
Falcaria vulgaris 1409
Farsetia 2185
Fatsia 2187
Fatsia japonica 3034
Fedia cornucopiae 52
Feijoa sellowiana 2195
Felicia 2196
Fendlera 2199
Feronia 6020
Feronia elephantum 2012
Ferula 2471
Ferula assa foetida 336
Ferula communis 1418
Ferulago 2207
Festuca 2208
Festuca amethystina 215
Festuca arenaria 1676
Festuca arundinacea 4474
Festuca bromoides 803
Festuca elatior 3547
Festuca elatior arundinacea 4474
Festuca gigantea 2472
Festuca heterophylla 4839
Festuca myuros 4423
Festuca ovina 4858
Festuca pratensis 3547
Festuca rubra 4450
Festuca rubra heterophylla 4839
Festuca rubra *var.* arenaria 1676
Ficaria 4181
Ficaria verna 2241
Ficus 2240
Ficus bengalensis 444
Ficus carica 1412
Ficus elastica 2944
Ficus indica 2929
Ficus pumila 1267
Ficus religiosa 738
Ficus sycomorus 5408
Filago 2318
Filago germanica 1396
Filago minima 4994
Filicineae 2206
Filipendula 3562
Filipendula hexapetala 1907
Filipendula ulmaria 2112
Firmanis simplex 1194
Fistulina hepatica 515
Fitzroya 2267
Fitzroya archeri 5448
Fitzroya cupressoides 4071
Fitzroya patagonica 4071
Flacourtia 2270
Flacourtia cataphracta 4025
Flacourtiaceae 2941
Flacourtia indica 4413
Flacourtia jangomas 4025

Flacourtia ramontschi 4413
Flacourtia rukam 4605
Flaveria 2285
Flindersia 2294
Flindersia australis 426
Flourensia 5444
Fluminea 4505
Foeniculum 2200
Foeniculum dulce 2300
Foeniculum vulgare 1411
Fomes fomentarius 5529
Fomes officinalis 4333
Fontanesia 2323
Fontinalis 5823
Forsteronia 2335
Forsythia 2337
Forsythia suspensa 5860
Fortunella 3189
Fortunella japonica 3528
Fortunella margarita 3980
Fragaria 5272
Fragaria ananassa 4195
Fragaria chilensis 1150
Fragaria indica 2932
Fragaria moschata 2722
Fragaria vesca 2127
Fragaria virginiana 5754
Fragaria viridis 2635
Frankenia 2361
Frankeniaceae 2362
Franseria 873
Fraxinus 338
Fraxinus acuminata 5878
Fraxinus alba 5878
Fraxinus americana 5878
Fraxinus angustifolia 3798
Fraxinus anomala 4972
Fraxinus caroliniana 1024
Fraxinus chinensis 1168
Fraxinus dimorpha 93
Fraxinus excelsior 2073
Fraxinus excelsior *var.* aurea 2526
Fraxinus mandshurica 3458
Fraxinus nigra 613
Fraxinus ornus 2308
Fraxinus parvifolia 4583
Fraxinus pennsylvanica 4434
Fraxinus pennsylvanica lanceolata 2622
Fraxinus potamophila 5644
Fraxinus quadrangulata 687
Fraxinus rotundifolia 4583
Fraxinus syriaca 5416
Fraxinus viridis 2622
Freesia 2365
Fritillaria 2380
Fritillaria camtschatcensis 3130
Fritillaria imperialis 1711
Fritillaria meleagris 1126
Fritillaria montana 4997
Fritillaria tenella 4997
Fuchsia 2389
Fuchsia magellanica 3422
Fucus 4538
Fucus serratus 1038
Fucus vesiculosus 666
Fumaria 2392
Fumariaceae 2393
Fumaria officinalis 1911
Funaria 2394
Fungus 3758
Funkia 4221
Funkia lancifolia 3060
Funtumia 4946
Funtumia elastica 3205
Furcraea gigantea 4209

Gaertneria 873
Gagea 2399
Gaillardia 2400
Gaillardia aristata 1491
Gaillardia grandiflora 1491
Galanthus 5060
Galanthus nivalis 1530
Galax 2401
Galega 2521
Galega officinalis 1425
Galeobdolon luteum 5855
Galeopsis 2778
Galeopsis angustifolia 3805
Galeopsis ladanum 4453
Galeopsis pubescens 1895
Galeopsis speciosa 3846
Galeopsis tetrahit 781
Galeopsis versicolor 3846
Galinsoga 4386
Galinsoga parviflora 3326
Galipea 2403
Galipea officinalis 251
Galium 510
Galium aparine 1055
Galium boreale 3878
Galium cruciatum 1704
Galium hercynicum 2743
Galium mollugo 5882
Galium palustris 3500
Galium parisiense 4054
Galium saxatile 2743
Galium silvaticum 6021
Galium tricorne 1607
Galium trifidum 5017
Galium triflorum 5384
Galium uliginosum 5341
Galium verum 6075
Galtonia 5319
Garcinia 2405
Garcinia mangostana 3474
Gardenia 2416
Garrya 4947
Gasteria 2436
Gasteria verrucosa 3990
Gasteromycetes 2437
Gastridium 3863
Gastrolobium 4240
Gaultheria 6003
Gaultheria procumbens 1125
Gaura 2438
Gaylussacia 2884
Gazania 2440
Geaster 1969
Gelidium 55
Gelsemium 2441
Genipa 2444
Genipa americana 2443
Genista 6015
Genista anglica 3823
Genista germanica 2463
Genista pilosa 2685
Genista scoparia 4769
Genista tinctoria 1563
Gentiana 2445
Gentiana amarella 260
Gentiana asclepiadea 3622
Gentiana bavarica 482
Gentiana campestris 3550
Gentianaceae 2446
Gentiana ciliata 2684
Gentiana clusii 1291
Gentiana cruciata 1700
Gentiana germanica 2457
Gentiana lutea 6091
Gentiana nivalis 5063
Gentiana pannonica 1799
Gentiana purpurea 4346
Gentiana pyrenaica 4371
Gentiana verna 5185
Geoglossum 1970
Geonoma 4840
Geonoma schottiana 316
Geraniaceae 2448
Geranium 2447
Geranium collinum 5686
Geranium columbinum 3369
Geranium dissectum 1757
Geranium macrorrhizum 570
Geranium molle 1892
Geranium palustre 718
Geranium phaeum 1931
Geranium pratense 3551
Geranium purpureum 4342
Geranium pusillum 5026
Geranium pyrenaicum 4372
Geranium robertianum 2787
Geranium rotundifolium 4584
Geranium sanguineum 680
Geranium sibiricum 4905
Geranium sylvaticum 6024
Gerbera 2451
Gesneria 2464
Gesneriaceae 2465
Geum 385
Geum rivale 5791
Geum strictum 6074
Geum urbanum 1354
Gilia 2487
Gillenia 2172
Ginkgo 2492
Ginkgoaceae 3432
Ginkgo biloba 2493
Gladiolus 2499
Gladiolus communis 1420
Gladiolus gandavensis 771
Gladiolus paluster 5500
Gladiolus segetum 1612
Glaucium 2859
Glaucium corniculatum 652
Glaucium flavum 6096
Glaux 4796
Glec(h)oma 2647
Glec(h)oma hederacea 1429
Gledits(ch)ia 2840
Gledits(ch)ia amorphoides 219
Gledits(ch)ia caspica 1044
Gledits(ch)ia japonica 3041

Gledits(ch)ia macrantha 573
Gledits(ch)ia sinensis 1187
Gledits(ch)ia triacanthos 1439
Gleichenia 2501
Gleicheniaceae 2502
Globularia 2506
Globulariaceae 2505
Globularia cordifolia 2737
Gloriosa 2511
Gloriosa superba 3436
Gloxinia 2515
Gloxinia speciosa 1424
Glyceria 3476
Glyceria aquatica 806
Glyceria fluitans 5820
Glycine hispida 5106
Glycine max 5106
Glycine soja 5106
Glycyrrhiza 3291
Glycyrrhiza glabra 1459
Glyptostrobus 1162
Gnaphalium 1727
Gnaphalium luteo-album 3092
Gnaphalium supinum 1938
Gnaphalium sylvaticum 6025
Gnaphalium uliginosum 3391
Gnetaceae 3107
Gnetum 3105
Gnetum gnemon 5149
Godetia 2524
Gomphia 2557
Gomphidius glutinosus 2516
Gomphrena 2503
Gomphrena globosa 1421
Goniolimon 4793
Goniolimon tataricum 5458
Goodyera 4428
Goodyera repens 1675
Gordonia 2564
Gossypium 1635
Gossypium arboreum 349
Gossypium barbadense 4790
Gossypium herbaceum 3283
Gossypium hirsutum 5685
Gracilaria helminthochorton 1623
Gracilaria lichenoides 56
Gramineae 2581
Gratiola 2754
Gratiola officinalis 1913
Grevillea 2638
Grevillea robusta 4945
Grewia 2639
Grewia tiliaefolia 3317
Grindelia 2663
Grindelia robusta 4873
Grossularia 2560
Guaiacum 3293
Guaiacum officinale 1460
Guarea 3768
Guarea trichilioides 203
Guettarda 5714
Guilielma gasipaes 4085
Guilielma speciosa 4085
Guizotia 3849
Guizotia abyssinica 2063
Gunnera 2667
Gunnera chilensis 1148
Guttiferae 2669
Guzmannia 2670
Gymnadenia 4481
Gymnocladus 1319
Gymnocladus canadensis 3151
Gymnocladus dioicus 3151
Gymnospermae 2671
Gynandropsis 5129
Gynerium 4021
Gynerium argenteum 4818
Gynura 5713
Gypsophila 2672
Gypsophila elegans 1433
Gypsophila muralis 1749
Gypsophila paniculata 397
Gypsophila repens 1670
Gypsophila stevenii 5243
Gypsophila viscosa 5248
Gyromitra esculenta 1994
Gyrophora 4535
Gyroporus castaneus 1140

Habenaria 2673
Habenaria viridis 2384
Hacquetia 2675
Haemanthus 678
Haemanthus katherinae 3138
Haematoxylon campechianum 3359
Hagenia abyssinica 3190
Halenia 5193
Halesia 4950
Halidrys 4797
Halimodendron 4667
Halimodendron argenteum 4913
Halimodendron halodendron 4913
Halodule 4870
Haloragis 4786
Haloxylon 4729
Hamamelidaceae 6013
Hamamelis 6012
Hamamelis japonica 3079
Hamamelis mollis 1212
Hamamelis vernalis 5727
Hamamelis virginiana 1562
Hamelia 2699
Hamiltonia 2700
Haworthia 2728
Hedeoma 2178
Hedera 2988
Hedera colchica 1324
Hedera helix 2040
Hedera taurica 1691
Hedychium 2491
Hedysarum 5390
Hedysarum coronaria 5310
Hedysarum hedysaroides 2368
Hedysarum obscurum 2368
Hedyscepe 5680
Helenium 5052
Heliamphora 5332
Helianthemum 5334
Helianthemum alpestre 152
Helianthemum canum 2821
Helianthemum nummularium 1515
Helianthus 5329
Helianthus annuus 1539
Helianthus argophyllus 4963
Helianthus cucumerifolius 1725
Helianthus debilis 1725
Helianthus tuberosus 3094
Helichrysum 2149
Helichrysum arenarium 6083
Helichrysum bracteatum 5279
Helicodiceros 5666
Heliconia 2763
Helicteres 4781
Helictotrichon pratense 3553
Heliopsis 2764
Heliotropium 2765
Heliotropium peruvianum 1436
Helipterum 5333
Helleborus 2766
Helleborus foetidus 507
Helleborus niger 1226
Helleborus orientalis 3272
Helleborus viridis 2628
Helonias 5346
Helvella 2767
Helvellaceae 2768
Helvella crispa 1697
Helvella esculenta 1994
Helvella lacunosa 3196
Helxine 402
Hemerocallis 1815
Hemerocallis flava 3269
Hemerocallis fulva 2334
Hemerocallis middendorffi 3613
Hemerocallis minor 2585
Hemerocallis thunbergi 5504
Hepatica 2783
Hepatica nobilis 3343
Hepatica triloba 3343
Heracleum 1649
Heracleum pubescens 1069
Heracleum sibiricum 4898
Heracleum sphondylium 2825
Heracleum wilhelmi 1069
Hermineria elaphroxylon 179
Herminium 3763
Herniaria 877
Herniaria glabra 1372
Herniaria hirsuta 2682
Hesperis 4522
Hesperis matronalis 1794
Heteranthera 3743
Heterospathe elata 4642
Heuchera 171
Hevea 2791
Hevea brasiliensis 4050
Hibbertia 904
Hibiscus 2794
Hibiscus abelmoschus 3761
Hibiscus cannabinus 3146
Hibiscus elatus 3008
Hibiscus esculentus 3921
Hibiscus rosa-sinensis 1185
Hibiscus sabdariffa 4562
Hibiscus syriacus 4886
Hibiscus tiliaceus 3316
Hibiscus tiliaceus elatus 3008
Hibiscus trionum 2316
Hieracium 2726
Hieracium alpinum 135
Hieracium aurantiacum 3942
Hieracium florentinum 3161
Hieracium floribundum 6082
Hieracium murorum 5769

Hieracium pilosella 3734
Hieracium praealatum 3160
Hieracium pratense 2224
Hieracium sabaudum 4723
Hieracium umbellatum 3804
Hieracium villosum 4842
Hierochloe 5367
Hierochloe alpina 153
Hierochloe odorata 5368
Himant(h)oglossum 3346
Hippeastrum 176
Hippeastrum vittatum 448
Hippocastanaceae 2865
Hippocratea 2813
Hippocrepis 2874
Hippomane mancinella 3455
Hippomarathrum 2866
Hippophae 4787
Hippophae rhamnoides 1523
Hippuris 3488
Hippuris vulgaris 3489
Hirneola auricula-judae 3101
Hoferia 2822
Holcus 5710
Holcus lanatus 1553
Holcus mollis 2461
Holcus sorghum sudanensis 5294
Holcus sudanensis 5294
Holosteum 3730
Holosteum umbellatum 3004
Homalium 2833
Hopea 3588
Hopea odorata 5486
Hordeum 454
Hordeum distichon 5672
Hordeum hexastichon 4981
Hordeum jubatum 2350
Hordeum murinum 3729
Hordeum vulgare 455
Hosta 4221
Hosta japonica 3060
Hosta lancifolia 3060
Hottonia 2193
Hottonia palustris 2097
Houstonia 710
Hovenia 4406
Hovenia dulcis 3066
Howea 4828
Hoya 5847
Hoya carnosa 1558
Hugonia 2885
Humea 2886
Humea elegans 2340
Humulus 2847
Humulus japonicus 3043
Humulus lupulus 1440
Hunnemannia 2534
Hura crepitans 4681
Hutchinsia 2895
Hyacinthus 2896
Hyacinthus botryoides 1427
Hyacinthus orientalis 1445
Hydnaceae 2901
Hydnocarpus 1121
Hydnophytum 2899
Hydnum 2900
Hydnum coralloides 1589
Hydnum imbricatum 2922
Hydnum repandum 2752
Hydrangea 2902
Hydrangea arborescens 5039
Hydrangea macrophylla 565
Hydrangea paniculata 4028
Hydrangea petiolaris 1269
Hydrastis 2546
Hydrastis canadensis 2547
Hydrocharis 2381
Hydrocharis morsus-ranae 2382
Hydrocharitaceae 2383
Hydrocleis 5831
Hydrocotyle 4114
Hydrocotyle vulgaris 1489
Hydrodictyon 5826
Hydrophyllaceae 5814
Hydrophyllum 5813
Hydrosme rivieri 1848
Hygrophila 5233
Hygrophoraceae 2905
Hygrophorus 2904
Hygrophorus conicus 1577
Hymenocallis 5130
Hymenomycetes 4180
Hymenophyllaceae 2249
Hymenophyllum 2248
Hymenophysa 5918
Hyoscyamus 2823
Hyoscyamus albus 4586
Hyoscyamus niger 633
Hypecoum 2906
Hypericaceae 4652
Hypericum 4651
Hypericum acutum 5199
Hypericum calycinum 1
Hypericum elegans 4912
Hypericum humifusum 1677
Hypericum montanum 3721
Hypericum perforatum 1537
Hypericum quadrangulum 5198
Hyphaene 1890
Hyphaene thebaica 2001
Hypholoma 2907
Hyphomycetes 2245
Hypnum 2908
Hypochoeris 1061
Hypochoeris glabra 5036
Hypochoeris radicata 5164
Hyssopus 2909
Hyssopus officinalis 2910

Iberis 993
Iberis amara 4523
Iberis saxatilis 4514
Iberis sempervirens 2144
Iberis taurica 5460
Iberis umbellata 2504
Ibicella lutea 6110
Ilex 2826
Ilex aquifolium 2039
Ilex aquifolium *var.* ferox 2751
Ilex opaca 196
Ilex paraguariensis 4048
Ilex verticillata 1560
Ilex vomitoria 6068
Ilicaceae 2827
Illecebrum 2919
Illecebrum verticillatum 5924
Illicium 254
Illicium anisatum 3014
Illicium cambodianum 959
Illicium parviflorum 6112
Illicium verum 5607
Impatiens 5051
Impatiens balsamina 2407
Impatiens noli tangere 5560
Impatiens parviflora 5023
Impatiens roylei 4596
Imperata 4713
Imperata cylindrica 1320
Imperatoria ostruthium 3532
Incarvillea 2924
Indigofera 2947
Indigofera tinctoria 5603
Inga 2949
Inodes palmetto 909
Inula 2955
Inula britannica 784
Inula conyza 1239
Inula ensifolia 5407
Inula helenium 2010
Inula salicana 5970
Iochroma 5738
Ipomoea 3682
Ipomoea batatas 5380
Ipomoea jalapa 3005
Ipomoea purpurea 1474
Iresine 677
Iriartea 5253
Iridaceae 2959
Iris 2958
Iris aphylla 5268
Iris caucasica 1074
Iris florentina 2301
Iris germanica 2458
Iris graminea 2582
Iris grossheimi 2642
Iris hispanica 5111
Iris iberica 2913
Iris kaempferi 3047
Iris laevigata 4393
Iris pallida 5371
Iris persica 4148
Iris pseudoacorus 6086
Iris pumila 1943
Iris sibirica 4907
Iris susiana 3728
Iris taurica 1943
Iris variegata 2889
Iris xiphium 5111
Isatis 6014
Isatis tinctoria 1956
Ischnosiphon 5532
Isertia 5949
Isidium 2968
Isnardia palustris 3512
Isoetaceae 4388
Isoetes 4387
Isoloma 2969
Isopyrum 2970
Isotoma 4887
Itea 5387
Itea virginica 5755
Ithyphallus impudicus 5256
Iva 5324
Ixia 2993

Ixiolirion 2994
Ixiolirion montanum 5415
Ixocomus bovinus 4846
Ixora 2995

Jacaranda 2996
Jacaranda acutifolia 4849
Jacaranda brasiliana 764
Jacaranda mimosaefolia 4849
Jacaranda mimosifolia 4849
Jacaranda ovalifolia 4849
Jacksonia 3002
Jamesia 3011
Jamesia americana 1264
Jasione 3082
Jasione montana 4857
Jasione perenis 4736
Jasminum 3083
Jasminum grandiflorum 1052
Jasminum officinale 1450
Jasminum sambac 293
Jatropha 3838
Jatropha curcas 449
Jatrorrhiza palmata 955
Jeffersonia 5662
Juanulloa 3112
Jubaea 3113
Jubaea spectabilis 5419
Juglandaceae 5778
Juglans 5777
Juglans californica 930
Juglans cinerea 897
Juglans cordiformis 2281
Juglans mandshurica 3466
Juglans nigra 1975
Juglans regia 4151
Juglans rupestris 5479
Juglans sieboldiana 4936
Juglans sieboldiana cordiformis 2281
Juglans sinensis 1208
Juncaceae 4609
Juncaginaceae 3116
Juncus 4608
Juncus alpinus 146
Juncus balticus 436
Juncus bufonius 5535
Juncus bulbosus 845
Juncus capitatus 1010
Juncus compressus 4582
Juncus effusus 1517
Juncus filiformis 5498
Juncus gerardi 4666
Juncus inflexus 2704
Juncus maritimus 4800
Juncus obtusiflorus 3906
Juncus squarrosus 2747
Juncus tenuis 4288
Juniperus 3118
Juniperus barbadensis 543
Juniperus bermudiana 543
Juniperus californica 937
Juniperus chinensis 4369
Juniperus communis 1453
Juniperus communis saxatilis 3703
Juniperus conferta 4874
Juniperus drupacea 5418
Juniperus excelsa 2618
Juniperus flaccida 3598
Juniperus formosana 2332
Juniperus horizontalis 1671
Juniperus horizontalis plumosa 238
Juniperus litoralis 4874
Juniperus macrocarpa 4231
Juniperus mexicana 3599
Juniperus montana 3703
Juniperus occidentalis 4938
Juniperus oxycedrus 4298
Juniperus pachyphloea 100
Juniperus phoenicea 4168
Juniperus pinchoti 4440
Juniperus procera 43
Juniperus procumbens 3039
Juniperus prostrata 1671
Juniperus recurva 2801
Juniperus rigida 3820
Juniperus sabina 4719
Juniperus sargentii 4707
Juniperus sibirica 3703
Juniperus squamata 4974
Juniperus thurifera 2926
Juniperus virginiana 1980
Juniperus wallichiana 5771
Jurinea 3120
Jussiaea 5832
Justicia 3121

Kadsura 3123
Kadsura japonica 4744
Kaempferia 4485
Kalmia 3129
Kalmia angustifolia 3210
Kalmia glauca 720
Kalmia latifolia 3710
Kalmia polifolia 720
Kayea 3142
Kennedia 3148
Kennedya 3148
Kentia 3149
Kernera 3154
Kerria 3156
Kerria japonica 3049
Keteleeria 3157
Keteleeria davidiana 1812
Keteleeria fortunei 2338
Khaya ivorensis 2985
Khaya senegalensis 4821
Kickxia 2317
Kickxia elatine 4851
Kickxia spuria 4585
Kigelia 4716
Kirengeshoma 3163
Kleinhovia 3165
Klugia 3166
Knautia 3167
Knautia arvensis 2235
Kniphofia 5551
Kobresia 1302
Kochia 5318
Kochia scoparia 531
Koeleria 3173
Koeleria cristata 4292
Koeleria glauca 702
Koelreuteria 2555
Koelreuteria paniculata 4027
Koenigia 3174
Kosteletzkya 4664
Kunzea 3698

Labiatae 3629
Laboulbeniaceae 3191
Laburnum 3192
Laburnum alpinum 4773
Laburnum anagyroides 2529
Laccaria laccata 5850
Lachenalia tricolor 5591
Lachnanthes 4463
Lactarius 3195
Lactarius deliciosus 1827
Lactarius piperatus 4123
Lactarius subdulcis 5372
Lactarius torminosus 6049
Lactarius volemus 3938
Lactuca 3281
Lactuca muralis 5772
Lactuca perennis 4132
Lactuca saligna 5973
Lactuca sativa 2417
Lactuca sativa *var.* capitata 2735
Lactuca sativa *var.* romana 4545
Lactuca sativa *var.* secalina 1759
Lactuca scariola 4299
Lactuca serriola 4299
Lactuca virosa 603
Lagenaria 2567
Lagenaria siceraria 922
Lagenaria vulgaris 922
Lagerstroemia 2580
Lagerstroemia indica 1394
Lagetta lintearia 3319
Lagoseris 3204
Lagotis 3206
Lagunaria 5307
Lagurus 2713
Lallemantia 3208
Lamarckia 2550
Laminaria 667
Laminariaceae 3145
Laminaria digitata 4806
Laminaria japonica 4807
Laminaria saccharina 5299
Lamium 1818
Lamium album 5891
Lamium amplexicaule 2780
Lamium galeobdolon 5855
Lamium hybridum 1756
Lamium maculatum 5166
Lamium purpureum 4343
La(m)psana 3860
La(m)psana communis 1479
Landolphia 2666
Lansium 3217
Lansium domesticum 1886
Lantana 3218
Lapageria 1152
Lapageria rosea 4445
Lapeirousia 2169
Lapeyrousia 2169
Laportea 6036
Laportea canadensis 977
Lappa major 2601
Lappula 5246
Lappula echinata 2126
Lardizabala 3223
Larix 3222

Larix americana 1977
Larix decidua 2110
Larix europaea 2110
Larix griffithi 2802
Larix japonica 3051
Larix kaempferi 3051
Larix laricina 1977
Larix leptolepis 3051
Larix occidentalis 5868
Larix sibirica 4908
Larrea 1680
Larrea divaricata 5172
Laserpitium 3227
Lasia 3228
Latania 3229
Lathraea 3230
Lathraea squamaria 5549
Lathyrus 4098
Lathyrus cicera 2279
Lathyrus hirsutus 4578
Lathyrus japonicus 3495
Lathyrus latifolius 4133
Lathyrus luteus 6104
Lathyrus maritimus 3495
Lathyrus montanus 3716
Lathyrus niger 645
Lathyrus nissolia 2592
Lathyrus odoratus 5378
Lathyrus palustris 3510
Lathyrus pratensis 3555
Lathyrus rotundifolius 4587
Lathyrus sativus 2587
Lathyrus sylvestris 2278
Lathyrus tingitanus 5438
Lathyrus tuberosus 2649
Lathyrus vernus 607
Lauraceae 3232
Laurocerasus 3238
Laurocerasus lusitanica 4277
Laurus 3231
Laurus nobilis 2615
Lavandula 3239
Lavandula latifolia 795
Lavandula officinalis 5604
Lavandula spica 5604
Lavandula vera 5604
Lavatera 5582
Lavatera arborea 5716
Lavatera thuringiaca 5507
Lavatera trimestris 2788
Lawsonia inermis 2782
Layia 5519
Lecanora 3262
Lecanora esculenta 3477
Leccinum aurantiacum 3940
Leccinum scabrum 582
Lecoc(k)ia 3263
Lecythis 3657
Ledum 3265
Ledum palustre 1719
Leersia 1753
Leersia oryzoides 4497
Legouzia 5719
Leguminosae 3267
Leitneria 1605
Lemna 1924
Lemna arrhiza 4551
Lemnaceae 1925
Lemna gibba 5404
Lemna minor 1403
Lemna trisulca 5217
Lens 3273
Lens culinaris 1457
Lentibulariaceae 665
Lentinus 3274
Lentinus lepideus 4739
Lentinus squamosus 4739
Leonotis 3320
Leontice 3322
Leontodon 2723
Leontodon autumnalis 2159
Leontopodium 1992
Leontopodium alpinum 1406
Leonurus 3689
Leonurus cardiaca 1475
Leopoldia 3277
Lepidium 4126
Lepidium apetalum 4293
Lepidium campestre 2233
Lepidium densiflorum 4293
Lepidium draba 4127
Lepidium latifolium 2118
Lepidium perfoliatum 1256
Lepidium sativum 2412
Lepidium virginicum 5749
Lepiota 3278
Lepiota clypeolaria 4863
Lepiota procera 4051
Lepista nuda 214
Leptasea aizoides 6102
Leptochloa 5171
Leptospermum 5472
Leptosyne 1596
Lespedeza 3279
Lespedeza cuneata 1189
Lespedeza sericea 1189
Lespedeza stipulacea 3183
Lespedeza striata 1458
Lesquerella 661
Leucadendron argenteum 4964
Leucaena 3245
Leucanthemum alpinum 123
Leucanthemum vulgare 3987
Leucocrinum 5226
Leucojum 5062
Leucojum aestivum 5322
Leucojum vernum 5188
Leucophyllum 4961
Leucorchis 3282
Levisticum 3384
Levisticum officinale 2418
Lewisia 3285
Liatris 2439
Libanotis 3286
Libanotis montana 3713
Libertia 3288
Libocedrus 2925
Libocedrus chilensis 1149
Libocedrus decurrens 936
Libocedrus doniana 4229
Libocedrus plumosa 4229
Licania 3289
Lichen 3290
Licuala 3292
Ligularia 2542
Ligusticum 3294
Ligusticum scoticum 4774
Ligustrum 4312
Ligustrum amurense 232
Ligustrum japonicum 3065
Ligustrum lucidum 2514
Ligustrum ovalifolium 944
Ligustrum sinense 1198
Ligustrum vulgare 1508
Liliaceae 3301
Lilium 3300
Lilium auratum 2525
Lilium bulbiferum 842
Lilium bulbiferum croceum 3939
Lilium candidum 3417
Lilium carniolicum 1022
Lilium chalcedonicum 1111
Lilium croceum 3939
Lilium giganteum 2478
Lilium japonicum 3053
Lilium ledebouri 3264
Lilium longiflorum 1972
Lilium martagon 3524
Lilium regale 4476
Lilium testaceum 3791
Lilium tigrinum 5522
Limnanthemum nymphaeoides 2378
Limnanthes 3548
Limnocharis 5711
Limodorum 2589
Limodorum tuberosum 2590
Limonium 4793
Limonium latifolium 5931
Limonium sinuatum 3890
Limonium vulgare 3573
Limosella 3745
Limosella aquatica 5825
Linaceae 2288
Linaria 5533
Linaria alpina 155
Linaria bipartita 1277
Linaria canadensis 3922
Linaria cymbalaria 3147
Linaria dalmatica 1791
Linaria elatine 4851
Linaria genistifolia 811
Linaria italica 2984
Linaria macedonica 3404
Linaria vulgaris 887
Lindelofia 3313
Lindera 5127
Lindernia 2179
Linnaea 5660
Linosyris vulgaris 2102
Linum 2286
Linum angustifolium 3802
Linum austriacum 378
Linum catharticum 4335
Linum flavum 2538
Linum gallicum 2367
Linum perenne 4129
Linum tenuifolium 5001
Linum usitatissimum 2214
Linum usitatissimum crepitans 1825
Linum usitatissimum *f.* brevimulticaula 3915
Liparis 5658
Liparis loeseli 3356
Lippia 3324

Lippia citriodora 3271
Liquidambar 5369
Liquidambar altingia 4419
Liquidambar orientalis 3965
Liquidambar styraciflua 210
Liriodendron 5635
Liriodendron chinense 1207
Liriodendron tulipifera 5636
Liriope 3308
Listera 5657
Listera cordata 3881
Listera ovata 1999
Litchi 3397
Litchi chinensis 3398
Lithocarpus 5439
Lithospermum 2640
Lithospermum arvense 1614
Lithospermum officinale 1428
Lithospermum purpureo-caeruleum 1669
Littonia 3340
Littorella 4879
Littorella uniflora 2124
Livistona 2183
Livistona australis 371
Livistona chinensis 1178
Livistona cochinchinensis 3442
Livistona rotundifolia 3087
Lloydia 159
Loasa 1155
Lobaria pulmonaria 3394
Lobelia 3350
Lobeliaceae 3351
Lobelia dortmanna 5819
Lobelia inflata 2942
Lobularia maritima 5352
Lodoicea maldivica 1887
Lodoicea seychellarum 1887
Loganiaceae 3358
Loiseleuria procumbens 114
Lolium 4631
Lolium italicum 2980
Lolium multiflorum 2980
Lolium perenne 4135
Lolium remotum 2708
Lolium temulentum 1804
Lonchocarpus 3216
Lonicera 2841
Lonicera alpigena 161
Lonicera caprifolium 5370
Lonicera coerulea 5356
Lonicera japonica 3042
Lonicera nigra 631
Lonicera periclymenum 6022
Lonicera sempervirens 5613
Lonicera tatarica 5455
Lonicera xylosteum 2099
Lophanthus 2476
Lophira 3372
Lophira alata 5992
Lophira alata procera 2005
Lophira procera 2005
Lophophora williamsi 3591
Loranthaceae 3632
Loranthus 3377
Loropetalum 3380
Lotus 1824
Lotus corniculatus 588
Lotus tetragonolobus 5197
Lotus uliginosus 5872
Ludwigia 4812
Ludwigia palustris 3512
Luehea 5877
Luffa 5564
Luffa cylindrica 5290
Lunaria 2834
Lunaria annua 1884
Lunaria rediviva 4130
Lupinus 3395
Lupinus albus 5899
Lupinus angustifolius 703
Lupinus hirsutus 2084
Lupinus luteus 2135
Lupinus mutabilis 5096
Lupinus perennis 5327
Lupinus polyphyllus 5789
Luzula 6041
Luzula arcuata 1746
Luzula campestris 2239
Luzula forsteri 2336
Luzula nemorosa 2652
Luzula pilosa 2696
Luzula spicata 5141
Luzula sudetica 5296
Luzula sylvatica 2613
Luzula wahlenbergi 5762
Lychnis 967
Lychnis alba 2140
Lychnis alpina 306
Lychnis chalcedonica 3451
Lychnis coronaria 4558
Lychnis flos cuculi 4398
Lychnis viscaria 1250
Lycium 6016
Lycium barbarum 450
Lycium chinense 1213
Lycium halimifolium 3535
Lycium ruthenicum 4612
Lycoperdaceae 4326
Lycoperdon 4325
Lycoperdon caelatum 1328
Lycoperdon giganteum 2479
Lycoperdon perlatum 5784
Lycopersicon 5542
Lycopersicon esculentum 1549
Lycopodiaceae 1286
Lycopodium 1285
Lycopodium alpinum 127
Lycopodium annotinum 2954
Lycopodium clavatum 4606
Lycopodium inundatum 3503
Lycopodium selago 2258
Lycopsis 5938
Lycopsis arvensis 5939
Lycopus 839
Lycopus europaeus 2085
Lygeum spartum 2061
Lygodium 1266
Lysimachia 3370
Lysimachia nemorum 6038
Lysimachia nummularia 3651
Lysimachia punctata 5169
Lysimachia thyrsiflora 5629
Lysimachia vulgaris 2540
Lythraceae 3371
Lythrum 3400
Lythrum hyssopifolia 2912
Lythrum salicaria 4349
Lythrum virgatum 5782

Macadamia ternifolia 4385
Macleaya 4230
Maclura 3971
Maclura aurantiaca 3972
Maclura pomifera 3972
Macrocystis pyrifera 2477
Macrotomia 4314
Macrozamia 3406
Madhuca indica 2920
Madia 3416
Madia sativa 1151
Maerua 3420
Magnolia 3423
Magnolia acuminata 1726
Magnolia campbelli 964
Magnoliaceae 3424
Magnolia conspicua 6122
Magnolia denudata 6122
Magnolia glauca 5355
Magnolia grandiflora 5098
Magnolia hypoleuca 5897
Magnolia liliflora 3304
Magnolia obovata 5897
Magnolia soulangeana 4715
Magnolia stellata 5227
Magnolia tripetala 5679
Magnolia virginiana 5355
Magnolia yulan 6122
Mahonia 3429
Mahonia aquifolium 3947
Mahonia fortunei 1192
Mahonia japonica 3055
Maianthemum 492
Maianthemum bifolium 5671
Majorana 3497
Majorana hortensis 5375
Malaceae 282
Malachium aquaticum 5795
Malachra 5952
Malaxis 25
Malaxis paludosa 3250
Malco(l)mia 3444
Malope 3449
Malpighia 3450
Malpighia punicifolia 446
Malus 288
Malus baccata 4899
Malus coronaria 5959
Malus floribunda 3037
Malus prunifolia 4091
Malus pumila 1349
Malus sieboldi 5554
Malus spectabilis 1180
Malus sylvestris 1349
Malus sylvestris *var.* paradisiaca 4046
Malva 3447
Malva alcea 2830
Malvaceae 3448
Malva crispa 1743
Malva mauritania 3576
Malva moschata 3760
Malva pusilla 1945
Malva rotundifolia 1945

Malvastrum 2173
Malva sylvestris 2797
Malva verticillata 1294
Mammea 3452
Mammea americana 3453
Mammillaria 3454
Mandragora 3469
Manettia 3470
Mangifera 3472
Mangifera foetida 180
Mangifera indica 1464
Mangifera odorata 853
Manihot 1046
Manihot aipi 67
Manihot dulcis 5360
Manihot esculenta 1379
Manihot glaziovi 1081
Manihot utilissima 1379
Manilkara bidentata 1355
Manisuris 3108
Maranta 329
Maranta arundinacea 541
Maranthaceae 330
Marasmius 3484
Marasmius oreades 2154
Marasmius rotula 1329
Marasmius scorodonius 2434
Marattiaceae 3485
Marchantia 3487
Marchantia polymorpha 1465
Mariscus 4727
Marrubium 2816
Marrubium vulgare 1442
Marsdenia 3499
Marsdenia cundurango 1572
Marsilea 4128
Marsiliaceae 3523
Martynia 3527
Martyniaceae 5683
Martynia fragrans 5363
Martynia lutea 6110
Matricaria 3541
Matricaria chamomilla 2452
Matricaria discoidea 4193
Matricaria inodora 4751
Matricaria matricaroides 4193
Matricaria suaveolens 4193
Matteuccia struthiopteris 3976
Matthiola 5258
Matthiola annua 263
Matthiola incana 1538
Maurandia 3536
Mauritia 3537
Mauritia vinifera 5985
Mayaca 4267
Maytenus 3540
Maytenus boaria 1154
Maytenus ilicifolia 2831
Meconopsis 708
Meconopsis cambrica 5864
Medeola 3564
Medicago 3565
Medicago arabica 5170
Medicago arborea 5583
Medicago hispida 933
Medicago lupulina 638
Medicago maculata 5170
Medicago minima 3332
Medicago orbicularis 903
Medicago sativa 90
Medicago sativa falcata 4921
Medicago varia 472
Medinilla 3567
Megotigea 5666
Melaleuca 3579
Melaleuca leucodendron 920
Melampyrum 1655
Melampyrum nemorosum 696
Melanconiales 3580
Melastomaceae 3581
Melia 1159
Melia azadirachta 3492
Melia azedarach 1160
Meliaceae 3428
Melianthus 2839
Melica 3582
Melica nutans 3868
Melica taurica 1693
Melica transilvanica 5570
Melica uniflora 6035
Melicocca 2837
Melicytus ramiflorus 2808
Melilotus 5362
Melilotus albus 5917
Melilotus altissimus 5426
Melilotus dentata 1834
Melilotus indica 267
Melilotus officinalis 6108
Melinis minutiflora 3644
Melissa 427
Melissa officinalis 1356
Melittis 3583
Melittis melissophyllum 469
Melocactus 3584
Melothria 3586
Menispermaceae 3672
Menispermum 3671
Menispermum canadense 1472
Menispermum dahuricum 347
Mentha 3628
Mentha aquatica 5822
Mentha arvensis 2229
Mentha citrata 535
Mentha crispa 1698
Mentha longifolia 2869
Mentha piperita 4122
Mentha pulegium 4113
Mentha rotundifolia 283
Mentha spicata 5120
Mentha viridis 5120
Mentzelia 3587
Menyanthes 716
Menyanthes trifoliata 1365
Menziesia 4989
Mercurialis 3589
Mercurialis annua 2785
Mercurialis perennis 1876
Mertensia 691
Merulius lacrimans 2880
Mesembryanthemum 3592
Mesembryanthemum crystallinum 2917
Mespilus 3577
Mespilus germanica 3578
Mesua ferrea 2966
Metrosiderosis 2964
Metroxylon 4644
Metroxylon rumphii 5157
Metroxylon sagu 5041
Meum 415
Mibora 3609
Michelia 3610
Micromelus 3695
Micromeria 3611
Micromeria graeca 2619
Microstylis 25
Mikania 3616
Mikania scandens 1268
Milium 3624
Milium effusum 201
Milla 3602
Miltonia 4031
Mimosa 3625
Mimosaceae 3626
Mimosa pudica 4827
Mimulus 3655
Mimulus guttatus 1471
Mimulus luteus 2541
Mimulus moschatus 3764
Mimusops 848
Mimusops balata 1355
Mirabilis 2341
Mirabilis jalapa 1415
Miscanthus 4956
Mitchella 4066
Mitella 3634
Mitragyna macrophylla 6
Mitragyna stipulosa 6
Mitraria 3635
Mnium 3638
Molinia 3646
Molinia caerulea 3674
Mollugo 1032
Moluc(c)ella 3647
Molucella laevis 341
Momordica balsamina 431
Momordica charantia 434
Monarda 512
Mondo 3307
Mondo jaburan 5898
Moneses uniflora 6037
Monocotyledones 3661
Monodora 3662
Monodora myristica 923
Monotropa 2940
Monstera 3663
Monstera deliciosa 1101
Montbretia 5599
Montia 2938
Montia perfoliata 3627
Moraceae 3750
Moraea 3676
Morchella 3678
Morchella conica 1578
Morchella esculenta 1473
Morina 3681
Moringa oleifera 2872
Morus 3749
Morus alba 5902
Morus nigra 640
Morus rubra 4456
Mucor 3691
Mucoraceae 3739
Mucuna 3740

Mucuna deeringiana 1823
Mucuna pruriens 1644
Muehlenbeckia 6010
Muehlenbergia 3748
Mulgedium 3281
Muntingia calabura 3007
Murraya 3084
Murraya exotica 1161
Musa 439
Musa basjoo 3020
Musaceae 440
Musa chinensis 1936
Musa ensete 8
Musa nana 1936
Musa paradisiaca 4219
Musa sapientum 1358
Musa textilis 2
Muscari 2579
Muscari botryoides 1427
Muscari comosum 5450
Muscari racemosum 5214
Myagrum 3773
Mycelis muralis 5772
Mycena 3774
Mycena galericulata 4086
Myoporum 3775
Myosotis 2327
Myosotis alpestris 132
Myosotis arvensis 2222
Myosotis azorica 392
Myosotis caespitosa 5627
Myosotis collina 1962
Myosotis dissitiflora 5399
Myosotis palustris 5602
Myosotis scorpioides 5602
Myosotis sylvatica 6033
Myosotis versicolor 6073
Myosurus 3736
Myosurus minimus 5530
Myrica 483
Myrica californica 948
Myricaceae 484
Myrica cerifera 5104
Myrica gale 5366
Myricaria 2181
Myricaria dahurica 1784
Myricaria germanica 2456
Myriophyllum 4060
Myriophyllum brasiliense 763
Myriophyllum spicatum 5138
Myriophyllum verticillatum 974
Myristica 3894
Myristicaceae 3895
Myristica fatua 2175
Myristica fragrans 1480
Myrocarpus 3778
Myrocarpus frondosus 762
Myroxylon 429
Myroxylon balsamum 5541
Myroxylon pereirae 4153
Myroxylon toluifera 5541
Myrrhis odorata 5361
Myrsinaceae 3780
Myrtaceae 3783
Myrtus 3781
Myrtus communis 5606
Myxomycetes 5006
Naiadaceae 3787
Naias 3786
Naias flexilis 5002
Naias marina 5156
Nandina 3790
Nandina domestica 4634
Nannorhops 3543
Narcissus 3795
Narcissus jonquilla 3110
Narcissus poeticus 4237
Narcissus pseudonarcissus 1397
Narcissus tazetta 4248
Nardus stricta 3533
Narthecium 715
Nasturtium officinale 5796
Nasturtium silvestre 6085
Naumburgia thyrsiflora 5629
Neillia 3824
Neillia sinensis 1193
Neillia thibetica 5514
Nelumbium 3381
Nelumbium caspicum 1045
Nelumbium nelumbo 2811
Nelumbium nuciferum 2811
Nelumbium speciosum 2811
Nelumbo 3381
Nemesia 3825
Nemophila 3826
Neottia 590
Nepenthaceae 3829
Nepenthes 3828
Nepeta 3830
Nepeta cataria 1060
Nepeta ucrainica 5676
Nephelium lappaceum 4409
Nephelium litchi 3398
Nephelium mutabile 4327
Nephrolepis 5405
Nephrolepis exaltata 1541
Nephroma 3831
Neptunia 4825
Nerine 3832
Nerium 3925
Nerium oleander 1482
Nerrine sarniensis 2658
Nertera depressa 491
Nertera granadensis 491
Neslia paniculata 422
Neviusia 5068
Nicandra 284
Nicandra physaloides 285
Nicotiana 5536
Nicotiana rustica 395
Nicotiana tabacum 1548
Nidulariaceae 589
Nierembergia 1733
Nigella 2201
Nigella damascena 3386
Nigella sativa 2415
Nipa 3859
Nitella 3861
Nitraria 3862
Nolana 3871
Nolina 3872
Nonnea 3873
Nopal coccinellifera 1305
Nopal cochenillifer 1305
Norta irio 3363
Nostoc 3889
Nostocaceae 2375
Nostoc commune 5225
Nothofagus 2163
Nothofagus antarctica 268
Notholaena 1271
Nuphar 1647
Nuphar luteum 2090
Nyctanginaceae 2342
Nyctanthes 3851
Nycterinia 3852
Nymphaea 5817
Nymphaea alba 2132
Nymphaeaceae 5818
Nymphaea flava 6099
Nymphaea lotus 5893
Nymphaea mexicana 6099
Nymphoides 2297
Nymphoides peltatum 2378
Nyssa 5642
Nyssaceae 5643

Ochna 3907
Ochroma lagopus 5870
Ochroma pyramidale 5870
Ocimum 463
Ocimum basilicum 5354
Ocotea 3908
Odontites 460
Odontites rubra 4435
Odontoglossum 3910
Oedogonium 3911
Oenanthe 5799
Oenanthe aquatica 2251
Oenanthe crocata 2771
Oenanthe fistulosa 5625
Oenanthe phellandrium 2251
Oenocarpus 403
Oenothera 2141
Oenothera biennis 1408
Oenothera muricata 5022
Oenothera parviflora 5022
Oidium 3914
Oldfieldia africana 51
Olea 3927
Oleaceae 3928
Olea europaea 1483
Olea oleaster 5953
Olearia 1788
Omphalodes 3815
Omphalodes verna 1672
Onagra 2141
Onagra biennis 1408
Onagraceae 2142
Oncidium 3929
Onobrychis 4646
Onobrychis arenaria 2891
Onobrychis sativa 1519
Onobrychis viciaefolia 1519
Onoclea sensibilis 4826
Ononis 4484
Ononis arvensis 2518
Ononis hircina 2518
Ononis natrix 5048
Ononis repens 962
Ononis spinosa 5492
Onopordon 1638
Onopordum 1638

Onopordum acanthium 4770
Onopordum tauricum 5462
Onosma 3932
Onosma tauricum 1690
Ophioglossaceae 28
Ophioglossum 27
Ophioglossum vulgatum 1346
Ophiopogon 3307
Ophiopogon jaburan 5898
Ophiopogon japonicus 1944
Ophrys apifera 519
Ophrys muscifera 2320
Oplopanax 1847
Opopanax 3934
Opuntia 4301
Opuntia ficus-indica 2936
Opuntia monocantha 1507
Opuntia vulgaris 1507
Orbignya oleifera 396
Orchidaceae 3945
Orchis 3944, 3946
Orchis latifolia 798
Orchis maculata 5163
Orchis mascula 3446
Orchis militaris 5079
Orchis palustris 3093
Orchis purpurea 3200
Orchis simia 3656
Orchis ustulata 1803
Oreodoxa 4594
Oreodoxa regia 1721
Origanum 3967
Origanum majorana 5375
Origanum vulgare 1484
Ornithogalum 5228
Ornithogalum nutans 3870
Ornithogalum pyrenaicum 478
Ornithogalum umbellatum 1535
Ornithopus 587
Ornithopus perpusillus 3337
Ornithopus sativus 1526
Orobanchaceae 813
Orobanche 812
Orobanche caryophyllacea 1282
Orobanche major 2604
Orobanche minor 1279
Orobanche purpurea 694
Orobanche ramosa 2774
Orobanche reticulata 5488
Orobus 4099
Orobus vernus 607
Orontium 2533
Orostachys 5263
Oroxylum 2945
Orthosiphon stamineus 558
Oryza 4495
Oryza sativa 4496
Oryzopsis 4499
Oryzopsis miliacea 5032
Osmanthus 3973
Osmanthus americanus 1850
Osmorrhiza 5381
Osmunda 3974
Osmundaceae 3975
Osmunda regalis 4593
Osteomeles 733
Osteomeles schwerinae 1172
Ostericum palustre 714
Ostrowskia 2468
Ostrya 2849
Ostrya carpinifolia 2107
Ostrya japonica 3044
Ostrya virginiana 197
Ottelia 3978
Ouratea 3979
Owenia 2030
Owenia cerasifolia 1136
Oxalidaceae 5088
Oxalis 3985
Oxalis acetosella 6044
Oxalis corniculata 1673
Oxalis crenata 3918
Oxalis stricta 1566
Oxalis tuberosa 3918
Oxycoccus 1658
Oxycoccus palustris 5019
Oxydendron 5094
Oxydendrum arboreum 5095
Oxylobus 3991
Oxyria 3719
Oxyria digyna 138
Oxytropis 1662
Oxytropis campestris 4213
Oxytropis uralensis 5690

Pachycentria 3994
Pachyrrhizus 6063
Pachyrrhizus angulatus 5851
Pachysandra 3995
Padus racemosa 2078
Padus serotina 626
Padus virginiana 1384
Paederia 2211
Paederota 4000
Paeonia 4118
Paeonia arborea 5585
Paeonia corallina 1590
Paeonia officinalis 1490
Paeonia suffruticosa 5585
Pahudia 4002
Palafoxia 4006
Palaquium 3814
Palaquium gutta 3443
Paliurus 4015
Paliurus spina-christi 1227
Palmaceae 4018
Palmae 4016
Panax 2494
Panax ginseng 346
Panax quinquefolium 195
Panax schinseng 346
Panax trifolium 1940
Pancratium 4023
Pandanaceae 4780
Pandanus 4779
Pandanus utilis 1521
Panellus stypticus 606
Pangium 4024
Pangium edule 2324
Panicum 4026
Panicum capillare 1561
Panicum maximum 2661
Panicum miliaceum 4315
Panicum ramosum 818
Panicum virgatum 5403
Panus 4033
Panus stypticus 606
Papaver 4272
Papaveraceae 4274
Papaver alpinum 140
Papaver argemone 2153
Papaver dubium 3367
Papaver orientale 3960
Papaver pyrenaicum 4375
Papaver rhoeas 1616
Papaver somniferum 3933
Papayaceae 4034
Paphinia 4040
Papilionaceae 4041
Pappophorum 4042
Papyrus antiquorum 4043
Paradisea 4649
Parietaria 4105
Parietaria officinalis 5773
Parinarium 4052
Paris 4053
Paris quadrifolia 2786
Parkia 3864
Parkia africana 45
Parkia biglobosa 5670
Parmelia 4056
Parmeliaceae 4057
Parnassia 4058
Parnassia palustris 5932
Parochetus 4847
Paronychia 3788
Parrotia 4061
Parrotia persica 4150
Parrya 4062
Parsonsia 1736
Parthenium 4065
Parthenium argentatum 2657
Parthenocissus 1664
Parthenocissus inserta 5484
Parthenocissus quinquefolia 5748
Parthenocissus tricuspidata 3028
Pasania 5439
Paspalum 4067
Paspalum dilatatum 1789
Paspalum notatum 409
Paspalum urvillei 5703
Passiflora 4068
Passiflora caerulea 697
Passifloraceae 4069
Passiflora edulis 4070
Passiflora laurifolia 5815
Passiflora macrocarpa 2475
Passiflora quadrangularis 2475
Pastinaca 4064
Pastinaca sativa 2422
Pastinaca sativa *ssp.* sylvestris 5946
Patrinia 4073
Paullinia 4074
Paullinia cupana 2655
Paulownia 4075
Paulownia tomentosa 4595
Pavetta 4076
Pavia 828
Pavia flava 6077
Paxillus 4078
Paxillus involutus 2956
Payena 4079
Pedaliaceae 4101
Pedicularis 3383

Pedicularis lapponica 3219
Pedicularis palustris 2117
Pedicularis sceptrum carolinum 1117
Pedicularis silvatica 3382
Pedicularis sudetica 5295
Pedicularis verticillata 5925
Pedilanthus 5010
Peganum 4102
Peganum harmala 2714
Peireskia 4136
Pelargonium 4103
Pelargonium capitatum 4568
Pelargonium grandiflorum 575
Pelargonium graveolens 4566
Pelargonium hederaefolium 2991
Pelargonium odoratissimum 3897
Pelargonium roseum 4566
Pelargonium zonale 2873
Pellaea 1262
Peltandra 326
Peltaria 4864
Peltigera canina 2648
Peltiphyllum 4731
Peltogyne 4348
Penicillaria 4108
Penicillium 4109
Penicillium glaucum 2632
Pennisetum 4110
Pennisetum glaucum 4092
Pennisetum purpureum 3793
Pennisetum spicatum 4092
Pennisetum typhoideum 4092
Pentace 5489
Pentaclethra macrophylla 3984
Pentacme 4115
Pentadesma butyracea 899
Pentapera 4116
Pentas 5216
Pen(t)stemon 4117
Pentstemon hartwegi 2718
Pentzia 4855
Peperomia 4119
Peplis 5833
Peranium 4428
Pereskia aculeata 447
Perilla 4138
Perilla frutescens 1492
Perilla ocymoides 1492
Periploca 4949
Periploca graeca 2616
Peristrophe 4139
Pernettya 4141
Perovskia 4142
Perowskia 4142
Persea 4144
Persea americana 183
Persea borbonia 4436
Persea indica 3412
Persicaria lapathifolia 1739
Persicaria orientalis 4309
Persica vulgaris 4082
Petalostemon 4290
Petalostemum 4290
Petasites 888
Petasites albus 5884
Petasites officinalis 4340
Petrea 4157
Petrocallis 4510
Petrophytum 4528
Petroselinum 4063
Petroselinum crispum latifolium 1417
Petroselinum crispum *var.* radicosum 5652
Petroselinum hortense 1417
Petroselinum sativum 1417
Petunia 4160
Petunia axillaris 5901
Petunia hybrida 1495
Petunia nyctanginiflora 5901
Petunia violaceae 5742
Peucedanum 2824
Peucedanum cervaria 3738
Peucedanum officinale 1438
Peucedanum oreoselinum 3715
Peucedanum ostruthium 3532
Peucedanum palustre 3509
Pezizaceae 1734
Phacelia 4161
Phacelia tanacetifolia 5443
Phaeophyceae 815
Phalaris 984
Phalaris arundinacea 4473
Phalaris canariensis 985
Phalaris minor 3336
Phalaris tuberosa *var.* stenoptera 1761
Phallaceae 4162
Phallus impudicus 5256
Phanerogamae 2310
Phascum 1968
Phaseolus 497
Phaseolus acutifolius 5478
Phaseolus angularis 34
Phaseolus aureus 3754
Phaseolus coccineus 4748
Phaseolus limensis 3309
Phaseolus lunatus 4940
Phaseolus lunatus *var.* macrocarpus 3309
Phaseolus multiflorus 4748
Phaseolus mungo 3755
Phaseolus vulgaris 3158
Phaseolus vulgaris humilis 881
Phellodendron 1604
Phellodendron amurense 226
Philadelphus 3641
Philadelphus caucasicus 1078
Philadelphus cordifolius 2739
Philadelphus coronarius 5376
Philadelphus coulteri 1641
Philadelphus falconeri 2157
Philadelphus gordonianus 2565
Philadelphus grandiflorus 571
Philadelphus inodorus 4752
Philadelphus latifolius 2819
Philadelphus mexicanus 3600
Philadelphus microphyllus 3328
Philadelphus nepalensis 6050
Philadelphus pubescens 2819
Philadelphus satsumanus 4714
Philadelphus satsumi 4714
Philadelphus schrenkii 4760
Philadelphus tomentosus 6050
Philadelphus verrucosus 5788
Philadelphus zeyheri 6127
Phillyrea 3642
Phillyrea angustifolia 3808
Phillyrea latifolia 5586
Philodendron 4165
Philodendron pertusum 1101
Phippsia 2915
Phleum 5527
Phleum alpinum 154
Phleum arenarium 4690
Phleum paniculatum 785
Phleum pratense 5528
Phlomis 3097
Phlomis fruticosa 1451
Phlomis tuberosa 5620
Phlox 4166
Phlox drummondi 1919
Phlox paniculata 5320
Phlox sibirica 4911
Phoenix 1807
Phoenix canariensis 983
Phoenix dactylifera 1808
Phoenix reclinata 4820
Phoenix sylvestris 5302
Pholiota caperata 6057
Pholiota mutabilis 1114
Pholiota praecox 1964
Pholiota squarrosa 4740
Pholiurus 4923
Pholiurus incurvus 4924
Phormium 2215
Phormium tenax 3843
Photinia 4169
Photinia davidsoniae 1813
Phragmites 4472
Phragmites communis 1514
Phryma 3373
Phycomycetes 5824
Phyllanthus 3249
Phyllanthus emblica 2026
Phyllantus acidus 3977
Phyllitis scolopendrium 2716
Phyllocactus 3247
Phyllocladus 1087
Phyllodoce 3706
Phyllospadix 5337
Phyllostachys 4170
Phyllostachys aurea 2527
Physalis 2645
Physalis alkekengi 5275
Physalis francheti 5275
Physalis peruviana 4156
Physalis pubescens 1894
Physocarpus 3857
Physostegia 3321
Physostigma 921
Physostigma venonosum 1816
Phytelephas 2986
Phytelephas macrocarpa 1447
Phyteuma 3636
Phyteuma nigrum 639
Phyteuma orbiculare 420
Phyteuma spicatum 5139
Phytocrenum 4171
Phytolacca 4243
Phytolacca americana 1502
Phytolaccaceae 4244
Phytolacca decandra 1502
Phytophthora 4172

Picea 5189
Picea abies 3886
Picea ajanensis 6071
Picea alba 5913
Picea alcockiana 80
Picea balfouriana 417
Picea bicolor 80
Picea brachytyla 4709
Picea breweriana 772
Picea canadensis 5913
Picea engelmanni 2034
Picea excelsa 3886
Picea fennica 2256
Picea glauca 5913
Picea glehni 4657
Picea jezoensis 6071
Picea koraiensis 3187
Picea koyamai 3187
Picea likiangensis balfouriana 417
Picea mariana 653
Picea morinda 2806
Picea nigra 653
Picea obovata 4915
Picea omorika 4830
Picea orientalis 3964
Picea polita 5523
Picea pungens 1338
Picea pungens *var.* glauca 695
Picea purpurea 4341
Picea rubens 4465
Picea rubra 4465
Picea sargentiana 4709
Picea schrenkiana 4761
Picea sitchensis 4979
Picea smithiana 2806
Picea spinulosa 4942
Picea wilsoni 5980
Picraena excelsa 3009
Picramnia 599
Picrasma 4383
Picrasma excelsa 3009
Picris 3989
Picris echioides 783
Picris hieracioides 2727
Pieris 4176
Pilea 1259
Pilocarpus 4184
Pilocarpus pennatifolius 5143
Pilularia 4183
Pilularia globulifera 1674
Pimelea 4498
Pimenta officinalis 102
Pimpinella 4186
Pimpinella anisum 253
Pimpinella magna 2608
Pimpinella saxifraga 4733
Pimpinella taurica 1689
Pinaceae 4194
Pinanga 4187
Pinguicula 900
Pinguicula alpina 3700
Pinus 4190
Pinus albicaulis 5879
Pinus arizonica 318
Pinus armandi 320
Pinus ayacahuite 3606
Pinus balfouriana 2353
Pinus banksiana 3001
Pinus brutius 2052
Pinus bungeana 3193
Pinus canariensis 988
Pinus caribaea 4993
Pinus cembra 5402
Pinus cembra sibirica 4917
Pinus cembroides edulis 1337
Pinus contorta 4876
Pinus contorta latifolia 3355
Pinus coulteri 1642
Pinus densiflora 3067
Pinus echinata 4882
Pinus edulis 1337
Pinus ellioti 4993
Pinus engelmanni 2033
Pinus excelsa 2807
Pinus flexilis 3310
Pinus gerardiana 1156
Pinus griffithi 2807
Pinus halepensis 86
Pinus halepensis brutia 2052
Pinus heldreichi 2762
Pinus koraiensis 3184
Pinus lambertiana 5306
Pinus laricio 1625
Pinus massoniana 3530
Pinus mitis 4882
Pinus montana 5400
Pinus montana mughus 3746
Pinus montezumae 3668
Pinus mughus 3746
Pinus mugo 5400
Pinus muricata 595
Pinus murrayana 3355
Pinus nigra 380
Pinus nigra poirentiana 1625
Pinus palustris 3366
Pinus peuce 419
Pinus pinaster 1296
Pinus pinea 2983
Pinus ponderosa 4259
Pinus ponderosa arizonica 318
Pinus pumila 3072
Pinus radiata 3666
Pinus resinosa 4460
Pinus rigida 4208
Pinus sabiniana 1856
Pinus sibirica 4917
Pinus sinensis 1195
Pinus strobus 1982
Pinus sylvestris 4775
Pinus tabulaeformis 1195
Pinus taeda 3352
Pinus thunbergi 3023
Pinus torreyana 5557
Pinus virginiana 5750
Piper 4120
Piperaceae 4121
Piper angustifolium 3534
Piper betle 549
Piper cubeba 1722
Piper methysticum 3140
Piper nigrum 646
Piptadenia 4203
Piptadenia africana 48
Piptadenia communis 1497
Piptanthus 4204
Piratinera guianensis 5049
Pirola 4377
Pirolaceae 4378
Pistacia 4205
Pistacia atlantica 3724
Pistacia lentiscus 3275
Pistacia mutica 5647
Pistacia terebinthus 5476
Pistacia vera 1498
Pistia 5816
Pisum 4080
Pisum arvense 2231
Pisum sativum 2423
Pisum sativum arvense 2231
Pisum sativum *var.* medullare 3498
Pithecolobium 272
Pithecolobium saman 4404
Pittosporaceae 5537
Pittosporum 4210
Pittosporum tobira 5538
Pittosporum undulatum 3937
Plagianthus 5659
Plagiobothrys 4269
Plagiochila 4211
Planera 5800
Planera aquatica 5801
Plantaginaceae 4220
Plantago 4218
Plantago alpina 162
Plantago asiatica 348
Plantago atrata 3717
Plantago coronopus 1708
Plantago indica 5926
Plantago lanceolata 829
Plantago major 4502
Plantago maritima 2123
Plantago media 5379
Plantago montana 3717
Plantago psyllium 2290
Platanaceae 4216
Platanthera 4403
Platanthera bifolia 895
Platanthera chlorantha 2605
Platanus 4215
Platanus acerifolia 3361
Platanus acerifolia hispanica 5113
Platanus hispanica 5113
Platanus lindeniana 3006
Platanus mexicana 3601
Platanus occidentalis 205
Platanus orientalis 3959
Platonia insignis 411
Platycerium 5205
Platycerium alcicorne 1534
Platycerium bifurcatum 1534
Platycodon 423
Platycodon grandiflorum 424
Platymiscium 3402
Platystemon 4222
Plectranthus 5190
Plectranthus fruticosus 1312
Pleioblastus humilis 3389
Pleioblastus pumilus 2643
Pleurogyna 2198
Pleuropogon 4819
Pleurotus ostreatus 3992
Pluchea 4223
Plumbaginaceae 3246
Plumbago 4225

Plum(i)er(i)a 2360
Pluteus cervinus 2189
Poa 700
Poa alpina 118
Poa angustifolia 3799
Poa annua 258
Poa arctica 304
Poa balfouri 416
Poa bulbosa 844
Poa chaixi 796
Poa compressa 970
Poa laxa 3725
Poa nemoralis 6023
Poa palustris 2345
Poa pratensis 3150
Poa trivialis 4579
Podocarpaceae 4236
Podocarpus 4235
Podocarpus alpinus 5449
Podophyllum 3539
Podophyllum emodi 2804
Podophyllum peltatum 1845
Podostemon 4506
Pogonia 4238
Pogostemon cablin 911
Pogostemon heyneanus 2792
Pogostemon patchouly 2792
Poinsettia 4239
Polemoniaceae 4167
Polemonium 4246
Polemonium caeruleum 2620
Polianthes 5618
Pollinia 5303
Polyalthia longifolia 2930
Polycarpon 3479
Polycnemum 4249
Polygala 4250
Polygalaceae 3623
Polygala sibirica 4909
Polygala vulgaris 1470
Polygonaceae 833
Polygonatum 5082
Polygonatum latifolium 799
Polygonatum multiflorum 2071
Polygonatum officinale 1916
Polygonatum verticillatum 5927
Polygonum 3172
Polygonum acivulare 4317
Polygonum alpinum 131
Polygonum amphibium 5812
Polygonum baldschuanicum 726
Polygonum bistorta 2079
Polygonum capitatum 4199
Polygonum convolvulus 1926
Polygonum cuspidatum 3036
Polygonum dumetorum 2755
Polygonum hydropiper 3511
Polygonum lapathifolium 1739
Polygonum minus 4999
Polygonum orientale 4309
Polygonum persicaria 5168
Polygonum sachalinense 4656
Polygonum viviparum 5759
Polymnia 3248
Polypodiaceae 4252
Polypodium 4251
Polypodium vulgare 1503
Polypogon 504
Polyporaceae 4253
Polyporus betulinus 583
Polyporus destructor 4254
Polyporus frondosus 2385
Polyporus officinalis 4333
Polyporus squamosus 1921
Polyporus sulphureus 5312
Polystichum 2828
Polystichum aculeatum 3349
Polystichum brauni 760
Polystichum falcatum 2881
Polystichum lobatum 3349
Polystichum lonchitis 3708
Polytrichum 2676
Polytrichum commune 1435
Pomaceae 282
Pomaderris apetala 5746
Poncirus trifoliata 5592
Pontederia 4174
Pontederiaceae 4175
Populus 4270
Populus alba 5906
Populus angulata 250
Populus balsamifera 5100
Populus berlinensis 540
Populus canadensis 979
Populus candicans 428
Populus canescens 2597
Populus cathayana 1058
Populus deltoides 1978
Populus deltoides missouriensis 5100
Populus fremonti 2366
Populus koreana 3185
Populus laurifolia 3234
Populus maximowiczi 3062
Populus nigra 647
Populus nigra *var.* italica 3360
Populus nigra *var.* pyramidalis 3360
Populus sargenti 4214
Populus sieboldi 4929
Populus suaveolens 3654
Populus tremula 2074
Populus tremuloides 4380
Populus trichocarpa 942
Populus tristis 819
Porphyra 3241
Porteranthus 2172
Portulaca 4358
Portulacaceae 4359
Portulaca grandiflora 1505
Portulaca oleracea 1509
Portulacaria afra 4361
Potamogeton 4262
Potamogetonaceae 4263
Potamogeton crispus 1744
Potamogeton densus 1833
Potamogeton friesi 2376
Potamogeton heterophyllus 5702
Potamogeton lucens 4866
Potamogeton natans 2298
Potamogeton obtusifolius 712
Potamogeton pectinatus 2202
Potamogeton perfoliatus 5495
Potamogeton praelongus 5914
Potamogeton pusillus 401
Potamogeton zosteriformis 2283
Potentilla 1242
Potentilla alba 5888
Potentilla anserina 4969
Potentilla argentea 4954
Potentilla aurea 2530
Potentilla erecta 5555
Potentilla fruticosa 879
Potentilla heptaphylla 4837
Potentilla intermedia 1893
Potentilla nivalis 5069
Potentilla nivea 5069
Potentilla norvegica 3887
Potentilla palustris 3502
Potentilla reptans 1667
Potentilla rupestris 1263
Potentilla silvestris 5555
Potentilla supina 1031
Potentilla tormentilla 5555
Potentilla verna 5182
Poterium 868
Poterium sanguisorba 5018
Prangos pabularia 2731
Prenanthes 4429
Prenanthes purpurea 4353
Primula 4307
Primula auricula 368
Primulaceae 4308
Primula elatior 3988
Primula farinosa 586
Primula japonica 3064
Primula obconica 5550
Primula officinalis 1652
Primula sinensis 1197
Primula veris 1652
Primula vulgaris 2042
Pringlea antiscorbutica 3152
Prinsepia 4310
Pritchardia 4311
Proboscidea 1846
Proboscidea fragrans 5363
Prosopis 3593
Prosopis chilensis 1467
Prosopis juliflora 1467
Protea 4318
Protea mellifera 5308
Prunella 4816
Prunella grandiflora 559
Prunella vulgaris 1525
Prunus 4224
Prunus americana 206
Prunus amygdalus 104
Prunus amygdalus *var.* fragilis 2354
Prunus angustifolia 1143
Prunus angustifolia *var.* watsoni 4684
Prunus armeniaca 1350
Prunus avium 3544
Prunus brigantina 773
Prunus cerasifera 3777
Prunus cerasifera divaricata 5950
Prunus cerasus 5091
Prunus cerasus austera 3679
Prunus cerasus marasca 3483
Prunus chicasa 1143
Prunus claudiana 2627
Prunus communis 2424
Prunus curdica 321
Prunus dasycarpa 4336
Prunus divaricata 5950
Prunus domestica 2424

Prunus fruticosa 2644
Prunus glandulosa 105
Prunus gracilis 3919
Prunus insititia 847
Prunus insititia syriaca 3630
Prunus italica 2627
Prunus japonica 1174
Prunus laurocerasus 1456
Prunus lusitanica 4277
Prunus mahaleb 3426
Prunus mandshurica 3457
Prunus maritima 488
Prunus mume 3015
Prunus nana 4611
Prunus nigra 978
Prunus padus 2078
Prunus pennsylvanica 4188
Prunus persica 4082
Prunus persica *var.* nectarina 3818
Prunus pumila 4683
Prunus salicina 3061
Prunus serotina 626
Prunus serotina salicifolia 1011
Prunus serrulata 3951
Prunus sibirica 4891
Prunus simoni 289
Prunus spinosa 5014
Prunus subcordata 3164
Prunus tenella 4611
Prunus triloba 2311
Prunus virginiana 1384
Prunus watsoni 4684
Psalliota arvensis 2870
Psalliota bispora 2419
Psalliota campestris 1476
Psalliota silvatica 2326
Pseudobacterium 4319
Pseudolarix 2539
Pseudolarix kaempferi 3388
Pseudotsuga 1888
Pseudotsuga glauca 4541
Pseudotsuga japonica 3031
Pseudotsuga macrocarpa 556
Pseudotsuga taxifolia 1401
Psidium 2656
Psidium friedrichsthalianum 1631
Psidium guajava 1431
Psidium laurifolium 1631
Psophocarpus tetragonolobus 2946
Psoralea 4782
Psychotria 5945
Ptaeroxylon 5053
Ptelea 2851
Ptelea trifoliata 1441
Pteretis nodulosa 3976
Pteridium aquilinum 26
Pteridium aquilinum pubescens 5867
Pteridophyta 4321
Pteris 754
Pteris aquilina pubescens 5867
Pteris cretica 1686
Pterocactus 5987
Pterocarpus 3999
Pterocarpus dalbergioides 237
Pterocarpus draco 1902
Pterocarpus indicus 863
Pterocarpus macrocarpus 864
Pterocarpus santalinus 4679
Pterocarpus soyauxi 47
Pterocarya 5996
Pterocarya caucasica 1077
Pterocarya fraxinifolia 1077
Pterocarya rehderiana 4477
Pterocarya rhoifolia 3077
Pterocarya sorbifolia 1077
Pterocarya stenoptera 1210
Pteroceltis 5988
Pterocephalus 5994
Ptychosperma 4322
Puccinellia maritima 4803
Pueraria 3188
Pueraria hirsuta 5505
Pueraria thunbergiana 5505
Pulicaria 2291
Pulicaria dysenterica 1413
Pulmonaria 3393
Pulmonaria angustifolia 1651
Pulmonaria montana 3711
Pulmonaria officinalis 1462
Pulsatilla 243
Pulsatilla patens 5177
Pulsatilla pratensis 3554
Pulsatilla vernalis 5721
Pulsatilla vulgaris 2116
Punica granatum 4255
Puschkinia 4362
Pycnanthemum 3714
Pycreus 4364
Pyracantha 2261
Pyracantha angustifolia 3801
Pyracantha coccinea 4743
Pyrenomycetes 2277
Pyrethrum 4376
Pyrethrum alpinum 123
Pyrethrum carneum 2306
Pyrethrum cinerariaefolium 1790
Pyrethrum corymbosum 3569
Pyrethrum macrophyllum 5441
Pyrethrum parthenium 2210
Pyrola 4377
Pyrolaceae 4378
Pyrola chlorantha 2633
Pyrola media 2953
Pyrola minor 5065
Pyrola rotundifolia 2119
Pyrola secunda 4926
Pyrola uniflora 6037
Pyrus 4089
Pyrus amygdaliformis 107
Pyrus betulifolia 580
Pyrus communis 1487
Pyrus nivalis 5066
Pyrus prunifolia 4091
Pyrus pyrifolia 4687
Pyrus salicifolia 5971
Pyrus ussuriensis 5694

Quamassia 958
Quamoclit 5220
Quassia 4382
Quassia amara 5338
Quebrachia 4461
Quebrachia lorentzi 3379
Quercus 3899
Quercus aegilops 5697
Quercus aenifolia 1779
Quercus agrifolia 939
Quercus alba 5905
Quercus bicolor 5347
Quercus borealis 3883
Quercus canariensis 987
Quercus castaneaefolia 1141
Quercus cerris 2128
Quercus chrysolepis 1002
Quercus coccifera 3153
Quercus coccinea 4746
Quercus crispula 4845
Quercus dentata 1786
Quercus frainetto 2979
Quercus graeca 5697
Quercus hispanica 2814
Quercus ilex 2832
Quercus ilicifolia 505
Quercus imbricaria 4865
Quercus infectoria 85
Quercus laurifolia 3233
Quercus libani 3261
Quercus lusitanica 4278
Quercus lyrata 3982
Quercus macedonica 3403
Quercus macrocarpa 870
Quercus macrolepis 5697
Quercus marilandica 634
Quercus michauxi 5343
Quercus mirbecki 987
Quercus mongolica 3653
Quercus mongolica grosseserrata 4845
Quercus montana 1142
Quercus nigra 5827
Quercus occidentalis 554
Quercus palustris 4200
Quercus petraea 1928
Quercus phellos 5974
Quercus pontica 4265
Quercus prinus 5343
Quercus pubescens 4324
Quercus pyrenaica 4374
Quercus robur 2041
Quercus rubra 5101
Quercus sessiliflora 1928
Quercus stellata 4279
Quercus suber 1603
Quercus suber occidentalis 554
Quercus tinctoria 644
Quercus trojana 3403
Quercus velutina 644
Quercus virens 3342
Quercus virginiana 3342
Quillaja 5070
Quillaja saponaria 5071
Quisqualis 4391

Radiola 2289
Radiola linoides 101
Rafflesia 3664
Rafflesiaceae 4397
Rajania 5623
Ramalina 4407
Ramaria aurea 2531
Ramischia secunda 4926
Ramonda 4411
Ramondia 4412
Randia dumetorum 3439

Ranunculaceae 890
Ranunculus 889
Ranunculus acer 5425
Ranunculus aconitifolius 16
Ranunculus acris 5425
Ranunculus alpestris 3677
Ranunculus aquatilis 5797
Ranunculus arvensis 1608
Ranunculus asiaticus 4146
Ranunculus auricomus 2552
Ranunculus bulbosus 841
Ranunculus ficaria 2241
Ranunculus flammula 5121
Ranunculus glacialis 2497
Ranunculus illyricus 2921
Ranunculus lingua 5544
Ranunculus montanus 3699
Ranunculus nemorosus 5962
Ranunculus nivalis 5059
Ranunculus repens 1666
Ranunculus sceleratus 673
Raoulia 4416
Raphanus 4395
Raphanus raphanistrum 5955
Raphanus sativus 2425
Raphanus sativus *var.* sativus 5030
Raphia 4396
Raphia pedunculata 3408
Raphia ruffia 3408
Rapistrum 4418
Rapistrum rugosum 4603
Rauwolfia 1844
Ravenala 5572
Ravenala madagascariensis 3409
Ravensara aromatica 3407
Razoumowskya 1947
Rehmannia 4478
Reineckia 4480
Reinwardtia 6087
Reseda 3614
Reseda alba 5900
Resedaceae 3615
Reseda lutea 6100
Reseda luteola 5862
Reseda odorata 1468
Restio 4552
Rhamnaceae 831
Rhamnus 830
Rhamnus alaternus 2974
Rhamnus alnifolia 82
Rhamnus alpina 121
Rhamnus alpinus 121
Rhamnus californica 932
Rhamnus cathartica 1369
Rhamnus frangula 2513
Rhamnus infectoria 4145
Rhamnus infectorius 4145
Rhamnus japonica 3025
Rhamnus libanotica 3260
Rhamnus purshiana 1040
Rhamnus saxatilis 4513
Rhamnus utilis 1173
Rhaphis gryllus 1688
Rhapidophyllum 3821
Rhapis 3201
Rhaponticum 4486
Rheedia 4487
Rheedia brasiliensis 410
Rheedia laterifolia 2698
Rheum 4492
Rheum emodi 2805
Rheum officinale 3566
Rheum palmatum 5089
Rheum rhabarbarum 5844
Rheum rhaponticum 2426
Rheum undulatum 5844
Rhexia 3545
Rhinanthus 4430
Rhinanthus crista-galli 1308
Rhipsalis 4488
Rhizoctonia 4489
Rhizophora 3475
Rhizophora mangle 200
Rhodiola rosea 4567
Rhododendron 389, 4490
Rhododendron arboreum 5587
Rhododendron ferrugineum 4530
Rhododendron flavum 4264
Rhododendron hirsutum 2430
Rhododendron indicum 2933
Rhododendron lapponicum 3220
Rhododendron luteum 4264
Rhododendron maximum 4557
Rhododendron molle 1170
Rhodomyrtus 4565
Rhodophyceae 4433
Rhodotypos 3099
Rhodotypos kerrioides 635
Rhodotypos scandens 635
Rhodymenia 4791
Rhopalostylis 4491
Rhus 5313
Rhus aromatica 2358
Rhus chinensis 1203
Rhus copallina 2275
Rhus coriaria 4920
Rhus cotinus 1527
Rhus glabra 5042
Rhus hirta 5206
Rhus pentaphylla 2268
Rhus succedanea 5848
Rhus toxicodendron 1501
Rhus typhina 5206
Rhus vernicifera 3050
Rhus vernix 4242
Rhynchospora 496
Rhynchospora alba 5880
Rhynchospora fusca 816
Ribes 1745
Ribes albinervium 208
Ribes alpinum 128
Ribes americanum 186
Ribes aureum 2535
Ribes carpathicum 1027
Ribes coloradense 1336
Ribes diacanthum 4900
Ribes glandulosum 4991
Ribes grossularia 2103
Ribes japonicum 3030
Ribes longiflorum 1276
Ribes mandshuricum 3459
Ribes meyeri 3607
Ribes nigrum 2081
Ribes odoratum 1276
Ribes orientale 3953
Ribes petraeum 4520
Ribes petraeum carpathicum 1027
Ribes prostratum 4991
Ribes reclinatum 2103
Ribes rubrum 3882
Ribes sanguineum 6000
Ribes saxatilis 4900
Ribes scandicum 3882
Ribes spicatum 3882
Ribes sylvestre 3882
Ribes triste 208
Ribes uva-crispa 2103
Ribes vulgare 3882
Riccia 4494
Richardia 950
Richardia scabra 3596
Richardsonia scabra 3596
Ricinodendron africanus 3770
Ricinus 1050
Ricinus communis 1051
Rivina 4575
Robinia 3354
Robinia hispida 4555
Robinia neomexicana 3841
Robinia pseudoacacia 636
Robinia viscosa 1252
Roccella 4509
Roccella tinctoria 3325
Roella 4542
Roemeria 344
Roemeria hybrida 5741
Rogeria 4543
Romanzoffia 3633
Rondeletia 4549
Rorippa amphibia 222
Rorippa armoracia 2871
Rorippa austriaca 377
Rorippa nasturtium-aquaticum 5796
Rorippa palustris 721
Rorippa silvestris 6085
Rosa 4554
Rosa agrestis 2583
Rosa alba 1634
Rosa alberti 78
Rosa alpina 145
Rosa arvensis 2234
Rosa banksiae 443
Rosa beggeriana 522
Rosa borboniana 745
Rosa canina 1875
Rosa caryophyllacea 4753
Rosaceae 4560
Rosa centifolia 910
Rosa centifolia *var.* muscosa 3685
Rosa centifolia *var.* parvifolia 861
Rosa centifolia *var.* pomponia 4256
Rosa chinensis 1200
Rosa chinensis *var.* minima 4580
Rosa chinensis *var.* semperflorens 2143
Rosa cinnamomea 1240
Rosa coriifolia 3258
Rosa corymbifera 1627
Rosa dahurica 1785
Rosa damascena 1792
Rosa eglanteria 5359
Rosa elliptica 4753
Rosa foetida 374
Rosa gallica 2371

Rosa glauca 4454
Rosa harrisoni 2715
Rosa illinoensis 4776
Rosa indica 1200
Rosa indica *var.* vulgaris 1200
Rosa inodora 4753
Rosa jundzilli 3491
Rosa macartnea 3401
Rosa marginata 3491
Rosa micrantha 5025
Rosa mollis 685
Rosa moschata 3767
Rosa multiflora 3068
Rosa muscosa 3685
Rosa nutkana 3875
Rosa odorata 5468
Rosa pendulina 145
Rosa pomifera 287
Rosa rubrifolia 4454
Rosa rugosa 4602
Rosa sempervirens 2147
Rosa spinosissima 4776
Rosa tomentosa 5543
Rosa villosa 287
Rosa virginiana 5751
Rosa wichuraiana 5929
Rosmarinus 4563
Rosmarinus officinalis 4564
Rosularia 5262
Rotala 4573
Roxburghia 4592
Rozites caperata 6057
Rubia 3410
Rubiaceae 3411
Rubia tinctorum 1463
Rubus 619, 4420
Rubus arcticus 305
Rubus caesius 2093
Rubus canadensis 5490
Rubus chamaemorus 1273
Rubus flagellaris 3880
Rubus fruticosus 2080
Rubus hispidus 5342
Rubus humifusus 3880
Rubus idaeus 4462
Rubus loganobaccus 3357
Rubus occidentalis 624
Rubus odoratus 2359
Rubus procerus 2798
Rubus saxatilis 5260
Rubus spectabilis 4660
Rubus ulmifolius 2020
Rudbeckia 1573
Rudbeckia hirta 629
Rudbeckia laciniata 1755
Rudbeckia purpurea 4344
Ruellia 3478
Ruizia 4604
Rumex 1868
Rumex acetosa 2429
Rumex acetosella 4859
Rumex alpinus 129
Rumex aquaticus 4258
Rumex conglomeratus 1293
Rumex crispus 1741
Rumex hydrolapathum 5798
Rumex hymenosepalus 981
Rumex maritimus 2536
Rumex obtusifolius 601
Rumex patientia 4072
Rumex persicarioides 2536
Rumex pulcher 2216
Rumex sanguineus 684
Rumex scutatus 2373
Ruppia 5933
Ruppia maritima 5934
Ruprechtia 5745
Ruscus aculeatus 885
Russula 4621
Russulaceae 4622
Russula cyanoxantha 686
Russula delica 5923
Russula emetica 2028
Russula foetens 2209
Russula nigricans 628
Russula virescens 2629
Ruta 4597
Rutaceae 4599
Ruta graveolens 1516
Ruyschia 4628

Sabadilla 4633
Sabadilla officinalis 1914
Sabal 4017
Sabal minor 1948
Sabal palmetto 909
Sabbatia 4561
Saccharomyces 6069
Saccharomyces cerevisiae 1359
Saccharomyces ellipsoideus 5986
Saccharomycetaceae 6070
Saccharum officinarum 5300
Sagina 4093
Sagina apetala 273
Sagina nodosa 3171
Sagina procumbens 4313
Sagina saginoides 310
Sagina subulata 1624
Sagittaria 327
Sagittaria japonica 3924
Sagittaria sagittifolia 3924
Sagittaria sinensis 3924
Saguerus 5305
Saintpaulia ionantha 1347
Salicaceae 5965
Salicornia 2500
Salicornia herbacea 3514
Salisburia 2492
Salix 5964
Salix acutifolia 4850
Salix aegyptiaca 2002
Salix alba 5922
Salix alba calva 1687
Salix alba vitellina 6107
Salix amygdalina 106
Salix amygdaloides 4084
Salix arbusculoides 3339
Salix arctica 311
Salix aurita 4581
Salix babylonica 400
Salix caprea 2523
Salix cinerea 2598
Salix coerulea 1687
Salix daphnoides 1801
Salix dasyclados 6053
Salix elaeagnus 2007
Salix elegantissima 5508
Salix fragilis 787
Salix fuscescens 77
Salix glabra 163
Salix glauca 2596
Salix grandifolia 564
Salix hastata 2697
Salix herbacea 4366
Salix hookeriana 2846
Salix incana 2007
Salix interior 4680
Salix lanata 6054
Salix lapponum 3221
Salix longifolia 4680
Salix longipes 1298
Salix myrsinifolia 693
Salix myrsinites 3785
Salix myrtilloides 5928
Salix nigra 654
Salix nigricans 693
Salix orbicularis 4590
Salix ovalifolia 3981
Salix pentandra 3236
Salix phylicifolia 5466
Salix polaris 4245
Salix purpurea 4352
Salix pyrolifolia 435
Salix repens 1679
Salix reticulata 3833
Salix reticulata subrotunda 4590
Salix retusa 3891
Salix richardsoni 4501
Salix rossica 4620
Salix rotundifolia 3253
Salix rubra 4471
Salix silesiaca 4943
Salix sitchensis 4980
Salix triandra 106
Salix viminalis 467
Salix vitellina 6107
Salmalia malabarica 3440
Salpiglossis 5624
Salpiglossis sinuata 4738
Salsola 4618
Salsola kali 1518
Salvadora 4669
Salvia 4639
Salvia glutinosa 5249
Salvia grandiflora 560
Salvia nemorosa 5743
Salvia officinalis 2427
Salvia pratensis 3558
Salvia sclarea 1254
Salvia sibthorpi 4919
Salvia splendens 4749
Salvia verbenacea 5720
Salvia verticillata 3299
Salvinia 4671
Salviniaceae 4672
Samandera 4673
Samandera indica 3848
Samandura 4673
Samandura indica 3848
Samanea saman 4404
Sambucus 2009
Sambucus arborescens 5588
Sambucus canadensis 190
Sambucus ebulus 3570

Sambucus nigra 2095
Sambucus nigra *var.* aurea 2537
Sambucus peruviana 4155
Sambucus pubens arborescens 5588
Sambucus racemosa 2120
Samolus 807
Samolus valerandi 808
Sandersonia 4685
Sandoricum 4695
Sandoricum indicum 4696
Sandoricum koetjape 4696
Sanguinaria 681
Sanguinaria canadensis 682
Sanguisorba 867
Sanguisorba minor 5018
Sanguisorba officinalis 2409
Sanicula 4693
Sanicula europaea 2121
Sansevieria 4694
Sansevieria cylindrica 2918
Sansevieria guineensis 5382
Sansevieria thyrsiflora 5382
Sansevieria zeylanica 1106
Santalaceae 4678
Santalum 4677
Santalum album 5908
Santolina 3240
Santolina chamaecyparissus 1776
Sanvitalia 4697
Sapindaceae 5073
Sapindus 5072
Sapindus saponaria 5102
Sapindus senegalensis 4822
Sapium 4699
Sapium sebiferum 1204
Saponaria 5074
Saponaria officinalis 743
Saponaria vaccaria 1653
Sapotaceae 4701
Saprolegniaceae 4703
Saraca 4704
Sarcodon imbricatus 2922
Sargassum 4705
Sarracenia 4207
Sarraceniaceae 204
Sarracenia purpurea 1499
Sasa 4710
Sassafras 4712
Sassafras albidum 1520
Sassafras officinale 1520
Satureia 4721
Satureia acinos 5187
Satureia alpina 148
Satureia calamintha 927
Satureia hortensis 5321
Satureia montana 6006
Satureia vulgaris 5936
Satureja 4721
Sauromatum 3345
Saururaceae 3348
Saururus 3347
Saussurea 4717
Saussurea alpina 147
Saxifraga 4730
Saxifraga aizoides 6102
Saxifraga aizoon 71
Saxifraga caespitosa 946
Saxifragaceae 4732
Saxifraga cernua 1904
Saxifraga granulata 3560
Saxifraga hirculus 6097
Saxifraga nivalis 5055
Saxifraga oppositifolia 5663
Saxifraga rivularis 120
Saxifraga sarmentosa 5277
Saxifraga stellaris 5231
Saxifraga tridactylis 4600
Saxifraga umbrosa 3362
Saxifrage rotundifolia 791
Scabiosa 4735
Scabiosa arvensis 2235
Scabiosa atropurpurea 5383
Scabiosa australis 5103
Scabiosa columbaria 1891
Scabiosa ochroleuca 1663
Scabiosa succisa 3561
Scandix pecten-veneris 4861
Scheuchzeria 4754
Scheuchzeriaceae 4755
Scheuchzeria palustris 3515
Schima 2659
Schima wallichi 1802
Schinopsis 4461
Schinopsis balansae 5967
Schinopsis lorentzi 3379
Schinus 4125
Schinus molle 940
Schinus terebinthifolius 766
Schizaeaceae 1742
Schizandra 3425
Schizandra chinensis 1191
Schizanthus 893
Schizocodon 2377
Schizomycetes 2266
Schizophragma 2903
Schizostylis 1695
Schlumbergera 1656
Schmidelia 4757
Schoenocaulon 4633
Schoenocaulon officinale 1914
Schoenus 724
Schoenus ferrugineus 4624
Schoenus nigricans 621
Schotia 3124
Sciadopitys 5681
Sciadopitys verticillata 5682
Scilla 5200
Scilla amoena 5223
Scilla autumnalis 383
Scilla bifolia 5664
Scilla italica 2982
Scilla sibirica 4916
Scindapsus 2989
Scirpus 852
Scirpus caespitosus 1822
Scirpus lacustris 2600
Scirpus maritimus 4801
Scirpus rufus 4443
Scirpus sylvaticus 6031
Scirpus tabernaemontani 5420
Scirpus triqueter 5590
Scleranthus 3168
Scleranthus annuus 262
Scleranthus perennis 4131
Scleria 4431
Sclerochloa 2701
Sclerochloa dura 2702
Scleroderma 2705
Sclerodermataceae 2706
Scleroderma vulgare 1405
Scleropoa 5251
Sclerotinia 4764
Scolochloa 4505
Scolopendrium vulgare 2716
Scolymus 3993
Scolymus hispanicus 5112
Scopolia 4765
Scopolia carniolica 2122
Scorpiurus 4767
Scorzonera 4831
Scorzonera hispanica 649
Scorzonera humilis 725
Scorzonera laciniata 3574
Scrophularia 2242
Scrophularia aquatica 5804
Scrophulariaceae 2243
Scrophularia nodosa 6027
Scrophularia vernalis 5184
Scutellaria 4987
Scutellaria alpina *var.* lupulina 6072
Scutellaria altissima 5429
Scutellaria lupulina 6072
Scutellaria orientalis 3962
Secale cereale 4630
Sechium edule 1122
Securigera 2720
Securinega 4809
Sedum 5264
Sedum acre 2553
Sedum album 5915
Sedum annuum 264
Sedum hispanicum 5114
Sedum hybridum 2148
Sedum maximum 2612
Sedum purpureum 4354
Sedum reflexum 3089
Sedum roseum 4567
Sedum rubens 4467
Sedum spurium 5673
Sedum telephium 3341
Sedum villosum 2691
Selaginella 4814
Selaginellaceae 4815
Selaginella martensi 3525
Selenicereus grandiflorus 4384
Selinum 4817
Sempervivum 2882
Sempervivum arachnoideum 5133
Sempervivum tectorum 2779
Senecio 2651
Senecio aquaticus 3513
Senecio cineraria 4958
Senecio cruentus 1385
Senecio erucifolius 3809
Senecio fuchsi 2388
Senecio jacobaea 4401
Senecio paludosus 4009
Senecio sylvaticus 6034
Senecio viscosus 5247
Senecio vulgaris 1430
Sequoia 4829
Sequoia gigantea 2483
Sequoia sempervirens 945
Serjania 5335

Serpula lacrimans 2880
Serratula 4728
Serratula tinctoria 1955
Sesamum 4834
Sesamum indicum 3961
Sesbania 4835
Sesbania aegyptiaca 4081
Seseli 4836
Seseli libanotis 3713
Sesleria 3675
Sesleria coerulea 704
Setaria 780
Setaria glauca 6090
Setaria italica 2352
Setaria verticillata 2844
Setaria viridis 2624
Shepherdia 834
Shepherdia argentea 4952
Sherardia 2227
Sherardia arvensis 2228
Shorea 4872
Shorea robusta 4661
Shorea wisneri 477
Shortia 4881
Sibbaldia 4889
Sibbaldia procumbens 4890
Sibiraea 4918
Sicyos 858
Sicyos angulatus 5765
Sida 4925
Sidalcea 1127
Sideritis 2967
Siderodendron 4928
Sideroxylon 3117
Siegesbeckia 4653
Sieglingia 4937
Sieglingia decumbens 2746
Sieversia 4941
Silaus 4124
Silene 1054
Silene acaulis 3686
Silene alpestris 149
Silene anglica 2372
Silene armeria 5394
Silene chlorantha 6092
Silene conica 1576
Silene cucubalus 663
Silene dichotoma 2328
Silene gallica 2372
Silene ïnflata 663
Silene italica 2981
Silene noctiflora 3850
Silene nutans 3869
Silene pendula 1905
Silene ruprechti 4607
Silene sibirica 4895
Silene tatarica 5459
Silene venosa 663
Silene viscosa 5757
Silphium 4571
Silybum marianum 669
Simaruba 4971
Simaruba amara 3968
Simarubaceae 66
Simaruba glauca 4047
Sinapis 3771
Sinapis alba 5904
Sinapis arvensis 1118
Sinapis juncea 2939
Sinningia 2515
Sinningia speciosa 1424
Siphonanthus 5615
Siphonia 4975
Siphonodon 2987
Siphonodon australe 372
Sison 2835
Sisymbrium 4978
Sisymbrium altissimum 5638
Sisymbrium irio 3363
Sisymbrium officinale 2757
Sisyrinchium 698
Sisyrinchium angustifolium 1364
Sium 5828
Sium angustifolium 3811
Sium sisarum 4986
Skimmia 4984
Skimmia japonica 3069
Sloanea 5013
Smilacaceae 5031
Smilacina 5081
Smilax 2623
Smilax china 1166
Smyrnium 87
Soja max 5106
Solanaceae 3854
Solanum 3853
Solanum dulcamara 604
Solanum heterodoxum 3585
Solanum luteum 6103
Solanum melongena 2414
Solanum nigrum 643
Solanum pseudocapsicum 3095
Solanum rostratum 835
Solanum tuberosum 4281
Solanum villosum 6103
Soldanella 111
Soldanella alpina 2496
Soldanella montana 2602
Solidago 2543
Solidago canadensis 971
Solidago graminifolia 2586
Solidago serotina 3892
Solidago virgaurea 2101
Solorina 5083
Sonchus 5105
Sonchus arvensis 2236
Sonchus asper 4305
Sonchus oleraceus 1531
Sonchus paluster 3517
Sophora 4001
Sophora japonica 3059
Sorbaria 2180
Sorbaria lindleyana 3318
Sorbaria sorbifolia 5691
Sorbaria tomentosa 3318
Sorbaronia 5084
Sorbus 3695
Sorbus alnifolia 1831
Sorbus americana 202
Sorbus amurensis 231
Sorbus aria 5881
Sorbus aucuparia 2114
Sorbus discolor 5058
Sorbus domestica 4833
Sorbus hybrida 3902
Sorbus intermedia 5350
Sorbus japonica 3058
Sorbus mougeotti 1995
Sorbus sçandica 5350
Sorbus torminalis 1128
Sorghum 5086
Sorghum cernuum 5892
Sorghum durra 1929
Sorghum halepense 3103
Sorghum saccharatum 5309
Sorghum technicum 810
Sorghum vulgare 5087
Sorghum vulgare durra 1929
Sorghum vulgare nervosum 3133
Sorghum vulgare saccharatum 5085
Sorghum vulgare sudanense 5294
Sorghum vulgare technicum 810
Soymida febrifuga 4544
Spadiciflorae 5107
Sparassis crispa 1683
Sparaxis 5781
Sparganiaceae 872
Sparganium 871
Sparganium minimum 3252
Sparmannia 5115
Sparmannia africana 49
Spartina 1593
Spartium 5856
Spartium junceum 5857
Spartium scoparium 4769
Spathodea campanulata 527
Specularia 5719
Spergula 5196
Spergula arvensis 1620
Spergula pentandra 5997
Spergularia 4689
Spergularia marginata 3490
Spergularia marina 4665
Spergularia salina 4665
Sphaeriales 5126
Sphagnaceae 4096
Sphagnum 5125
Sphenogyne 5692
Spigelia 5136
Spilanthes 5162
Spilanthes oleracea 4045
Spinacia 5148
Spinacia oleracea 4304
Spiraea 5160
Spiraea alba 3807
Spiraea alpina 151
Spiraea arguta 2431
Spiraea aruncus 5410
Spiraea betulifolia 581
Spiraea billardii 577
Spiraea callosa 3070
Spiraea chamaedryfolia 2455
Spiraea chinensis 1201
Spiraea crenata 5067
Spiraea douglasii 1889
Spiraea filipendula 1907
Spiraea japonica 3070
Spiraea latifolia 797
Spiraea media 3963
Spiraea prunifolia 775
Spiraea rosthornii 4572
Spiraea salicifolia 5972
Spiraea sargentiana 4708
Spiraea sorbifolia 5691

Spiraea thunbergi 5506
Spiraea tomentosa 2703
Spiraea virginiana 5753
Spiraea wilsoni 5979
Spiraea zabeliana 6123
Spiranthes 3197
Spiranthes autumnalis 1455
Spirodela 1923
Spirogyra 4260
Spondias 3648
Spondias cytherea 178
Spondias dulcis 178
Spondias lutea 6101
Spondias mombin 6101
Sporobolus 1906
Sprekelia 393
Stachys 550
Stachys alpina 117
Stachys annua 2758
Stachys arvensis 2230
Stachys germanica 3731
Stachys lanata 6047
Stachys officinalis 1361
Stachys palustris 3501
Stachys recta 5687
Stachys sieboldi 332
Stachys sylvatica 5911
Stangeria 2205
Stapelia 1034
Staphylea 659
Staphyleaceae 660
Staphylea pinnata 2083
Statice gmelini 4914
Statice latifolia 5931
Stauntonia 5237
Stauntonia hexaphylla 3071
Stellaria 5235
Stellaria graminea 3338
Stellaria holostea 1971
Stellaria media 1146
Stellaria nemorum 6045
Stenactis 2292
Stenanthium 5239
Stenocarpus 2263
Stenocarpus salignus 5966
Stenogynum 5240
Stephanandra 5241
Stephanandra flexuosa 1754
Stephanandra incisa 1754
Stephanandra tanakae 5437
Stercularia platanifolia 1194
Sterculia 741
Sterculia acerifolia 2274
Sterculiaceae 1223
Sterculia foetida 2733
Sterculia repestris 457
Sternbergia 5242
Sternbergia lutea 2158
Stevia 5244
Stewartia 5245
Stewartia monadelpha 5430
Sticta 5250
Sticta pulmonaria 3394
Stipa 3819
Stipa pennata 2115
Stipa tenacissima 2062
Stizolobium pruritum 1644
Stokesia 5259
Stranvaesia 5270
Stranvaesia davidiana 1202
Stratiotes 5836
Stratiotes aloides 5837
Strelitzia 584
Streptocarpus 1006
Streptopus 5669
Streptopus amplexifolius 1257
Strombosia 5285
Strophanthus 5287
Stropharia 5288
Strophostyles 5937
Strychnos ignati 4650
Strychnos nux vomica 3898
Stryphnodendron 170
Stryphnodendron barbatimao 451
Stuartia 5245
Stylidium 5289
Styracaceae 5269
Styrax 5056
Styrax benzoin 5315
Styrax officinalis 1915
Suaeda 4813
Suaeda maritima 4802
Subularia 387
Subularia aquatica 5792
Succisa australis 5103
Succisa pratensis 3561
Suillus bovinus 4846
Suillus grevillei 2011
Suillus luteus 820
Suillus variegatus 5700
Swainsona 5339
Swertia 5397
Swertia perennis 119
Swietenia 3427
Swietenia febrifuga 4544
Swietenia mahagoni 5871
Sympetalae 5413
Symphonia 5414
Symphonia gabonensis 2398
Symphoricarpos 5057
Symphoricarpos albus 1529
Symphoricarpos orbiculatus 2935
Symphoricarpos racemosus 1529
Symphoricarpus 5057
Symphytum 1345
Symphytum asperum 4297
Symphytum caucasicum 1068
Symphytum officinale 1389
Symphytum tuberosum 5619
Symplocaceae 5374
Symplocarpus 4990
Symplocos 5373
Synadenium 3618
Syndesmon thalictroides 4598
Synoum glandulosum 474
Syringa 3295
Syringa amurensis 228
Syringa amurensis japonica 3075
Syringa chinensis 1190
Syringa emodi 2803
Syringa japonica 3075
Syringa josikaea 2890
Syringa persica 4149
Syringa pubescens 2686
Syringa reflexa 3867
Syringa villosa 2686
Syringa vulgaris 1461
Syzygium aromaticum 1283
Syzygium cumini 3010
Syzygium jambos 4556
Syzygium malaccense 3912

Tabebuia 5614
Tabernaemontana utilis 1654
Tacca 5421
Taccaceae 4173
Tacca pinnatifida 2244
Tagetes 3493
Tagetes erecta 394
Tagetes patula 2369
Talauma 5423
Tamaricaceae 5436
Tamarindus 5433
Tamarindus indica 5434
Tamarix 5435
Tamarix africana 50
Tamarix gallica 2374
Tamarix hispida 3136
Tamarix odessana 3909
Tamarix pentandra 2269
Tamarix tetrandra 2344
Tamus 622
Tanacetum 5440
Tanacetum vulgare 1542
Taraxacum 1796
Taraxacum kok-saghyz 4613
Taraxacum officinale 1398
Tarrietia 5266
Tarrietia javanica 3088
Taxaceae 6118
Taxodiaceae 5463
Taxodium distichum 414
Taxodium mucronatum 3667
Taxus 6117
Taxus baccata 2045
Taxus brevifolia 3998
Taxus canadensis 980
Taxus chinensis 1215
Tecoma 5610
Tecoma radicans 1550
Tectona 5465
Tectona grandis 1544
Telekia 3986
Telekia speciosa 2740
Telephium 3969
Telfairia 3917
Telfairia pedata 6124
Tephrosia 2820
Terminalia catappa 5600
Terminaliaceae 3776
Terminalia chebula 1124
Terminalia superba 38
Terminalia tomentosa 6052
Testudinaria 5558
Tetracera 5477
Tetraclinis articulata 296
Tetragonia expansa 3844
Tetragonolobus 1824
Tetragonolobus purpureus 5197
Tetrapanax 4500
Teucrium 2453
Teucrium botrys 1758
Teucrium chamaedrys 1113
Teucrium marum 1064

Teucrium montanum 3705
Teucrium orientale 3954
Teucrium scordium 5806
Teucrium scordonia 6028
Thalassia 5654
Thalictrum 3556
Thalictrum alpinum 136
Thalictrum angustifolium 3806
Thalictrum aquilegifolium 1342
Thalictrum flavum 6098
Thalictrum minus 3392
Thalictrum simplex 5009
Thapsia 1817
Thea 5467
Theaceae 5464
Thea sinensis 1543
Thea sinensis *var.* assamica 356
Thelephoraceae 3256
Thelesperma 2636
Thelygonaceae 1770
Thelygonum 5481
Thelymitra 5482
Themeda 3131
Theobroma cacao 912
Thermopsis 5483
Thesium 476
Thladiantha 5616
Thladiantha dubia 3467
Thlaspi 4112
Thlaspi arvense 2232
Thlaspi montanum 139
Thlaspi perfoliatum 5494
Thouinia 5496
Thrinax 5480
Thujopsis 2161
Thujopsis dolabrata 2793
Thunbergia 1272
Thuya 300
Thuya gigantea 2484
Thuya japonica 3017
Thuya koraiensis 3177
Thuya occidentalis 1973
Thuya orientalis 3948
Thuya plicata 2484
Thuya standishi 3017
Thymelaea 5116
Thymelaeaceae 3608
Thymus 5509
Thymus serpyllum 3688
Thymus vulgaris 1546
Thysanolaena 5521
Tiarella 2321
Tigridia 5520
Tigridia pavonia 1547
Tilia 3314
Tilia americana 199
Tilia amurensis 229
Tilia argentea 4965
Tilia caucasica 1076
Tiliaceae 3315
Tilia cordata 3327
Tilia dasystyla 1076
Tilia euchlora 1692
Tilia europaea 2111
Tilia mandshurica 3463
Tilia parvifolia 3327
Tilia petiolaris 4967
Tilia platyphyllos 561
Tilia silvestris 3327
Tilia tomentosa 4965
Tilia vulgaris 2111
Tillaea 4365
Tillaea aquatica 1511
Tillandsia 5525
Tillandsia usneoides 5576
Tilletiaceae 5526
Toddalia 5539
Toddalia asiatica 343
Tofieldia 2162
Tofieldia palustris 4768
Toluifera pereirae 4153
Toona ciliata 866
Toona sinensis 1205
Tordylium 2719
Torenia 5553
Torilis 2759
Torilis arvensis 5176
Torilis japonica 5688
Torilis nodosa 3170
Torreya 5556
Torreya californica 947
Torreya grandis 1206
Torreya nucifera 3074
Torreya taxifolia 2303
Toulicia 5561
Tournefortia 5562
Toxicodendron radicans 1501
Toxicodendron succedaneum 5848
Toxicodendron vernicifluum 3050
Toxicodendron vernix 4242
Tozzia 5567
Trachelium 5503
Trachelospermum 5224
Trachycarpus 5983
Trachycarpus excelsus 2339
Trachycarpus fortunei 2339
Trachylobium 4421
Trachymene 3194
Tradescantia 5134
Tradescantia fluminensis 5780
Tradescantia montana 5752
Tradescantia virginiana 5752
Tragopogon 4662
Tragopogon porrifolius 5706
Tragopogon pratensis 3559
Tragus 860
Tragus racemosus 5207
Trametes gibbosa 2485
Trametes kalchbrenneri 2485
Trametes quercina 3900
Trametes versicolor 1569
Trapa natans 5794
Treculia 5575
Trema orientalis 2928
Trentepohlia iolithus 4537
Tribulus 954
Tribulus terrestris 4332
Trichilia 611
Tricholoma equestre 2051
Tricholoma nudum 214
Tricholoma portentosum 1859
Trichomanes 2247
Trichophorum caespitosum 1822
Trichosanthes 5045
Trichosanthes anguina 1996
Tricyrtis 5534
Trientalis 5218
Trientalis europaea 2125
Trifolium 1278
Trifolium agrarium 2848
Trifolium alexandrinum 2000
Trifolium alpestre 4347
Trifolium alpinum 126
Trifolium angustifolium 2250
Trifolium arvense 4394
Trifolium campestre 3225
Trifolium dubium 5292
Trifolium elegans 4883
Trifolium filiforme 5497
Trifolium fragiferum 5274
Trifolium glomeratum 1292
Trifolium hybridum 164
Trifolium incarnatum 1694
Trifolium lupinaster 473
Trifolium medium 6128
Trifolium montanum 3702
Trifolium ochroleucum 5311
Trifolium pannonicum 2888
Trifolium parviflorum 5470
Trifolium pratense 4447
Trifolium repens 5889
Trifolium resupinatum 4147
Trifolium rubens 2351
Trifolium sativum *var.* praecox 1965
Trifolium scabrum 4576
Trifolium striatum 3169
Trifolium subterraneum 5291
Triglochin 4234
Triglochin maritima 4877
Triglochin palustris 328
Trigonella 5593
Trigonella coerulea 711
Trigonella foenum-graecum 2203
Trillium 5594
Trinia 5595
Triodia 5596
Triosteum 2867
Tripsacum 2404
Trisetum 2176
Trisetum flavescens 6109
Trisetum spicatum 5146
Tristania conferta 778
Trithrinax 5598
Trithrinax brasiliensis 767
Triticum 5873
Triticum abyssinicum 2064
Triticum aestivum 5874
Triticum compactum 1288
Triticum compositum 756
Triticum dicoccoides 5961
Triticum dicoccum 2029
Triticum durum 1930
Triticum monococcum 2004
Triticum polonicum 4247
Triticum spelta 5124
Triticum turgidum 4286
Triticum vavilovianum 756
Triticum vulgare 5874
Tritoma 5551
Tritonia 5599
Trochodendron 5876
Trollius 2507
Trollius asiaticus 4906
Trollius chinensis 1182

Trollius europaeus 1422
Tropaeolaceae 3813
Tropaeolum 3812
Tropaeolum majus 1478
Tropaeolum minus 882
Tropaeolum peregrinum 986
Tropaeolum tuberosum 5617
Tsuga 2769
Tsuga canadensis 973
Tsuga chinensis 1184
Tsuga diversifolia 3040
Tsuga heterophylla 3996
Tsuga mertensiana 3707
Tsuga sieboldi 4933
Tuber 5608
Tuber aestivum 4619
Tuberales 5609
Tuber melanospermum 4137
Tulbaghia 5632
Tulipa 5633
Tulipa bibersteiniana 552
Tulipa eichleri 2003
Tulipa sylvestris 2302
Tunica 5641
Tunica prolifera 3330
Turgenia latifolia 789
Turnera 5649
Turraea 5212
Turraeanthus africanus 41
Turritis 4517
Turritis glabra 5566
Tussilago 1339
Tussilago farfara 1388
Tylopilus felleus 598
Typha 1062
Typha angustifolia 3800
Typhaceae 1063
Typha latifolia 1380

Uapaca 5674
Ulex 2566
Ulex europaeus 1426
Ulex nanus 1942
Ullucus tuberosus 5622
Ulmaceae 2019
Ulmus 2018
Ulmus alata 5989
Ulmus americana 191
Ulmus campestris 5040
Ulmus carpinifolia 5040
Ulmus carpinifolia suberosa 1602
Ulmus carpinifolia *var.* cornubiensis 1615
Ulmus effusa 5174
Ulmus foliacea 5040
Ulmus fulva 5012
Ulmus glabra 4771
Ulmus laciniata 3461
Ulmus laevis 4614
Ulmus montana 4771
Ulmus parvifolia 1177
Ulmus procera 5040
Ulmus pumila 4903
Ulmus racemosa 4521
Ulmus scabra 4771
Ulmus sibirica 4903
Ulmus suberosa 1602
Ulmus thomasi 4521
Ulva 4794
Ulva lactuca 1524
Umbelliferae 1036
Umbellularia californica 938
Umbilicaria 4536
Umbilicus 3816
Uncaria gambir 532
Uredinaceae 4623
Urena lobata 917
Urera 4778
Urginea 4798
Urginea maritima 4878
Urospermum 4856
Ursinia 5692
Urtica 3835
Urticaceae 3836
Urtica dioica 574
Urtica pilulifera 4547
Urtica urens 1874
Urtica utilis 4410
Usnea 5693
Usnea barbata 503
Ustilaginaceae 854
Utricularia 664
Utriculariaceae 665
Uvularia 530

Vaccinium 692
Vaccinium canadense 969
Vaccinium corymbosum 2796
Vaccinium macrocarpum 3224
Vaccinium myrtillus 3784
Vaccinium oxycoccos 5019
Vaccinium uliginosum 717
Vaccinium vitis-idaea 1645
Vagnera 5081
Valeriana 5695
Valerianaceae 5696
Valeriana dioica 3520
Valeriana excelsa 4569
Valeriana montana 3722
Valeriana officinalis 1552
Valeriana palustris 3520
Valerianella 1618
Valerianella carinata 3143
Valerianella dichotoma 3211
Valerianella eriocarpa 2977
Valerianella locusta 3211
Valerianella olitoria 2088
Vallisneria 5944
Vallisneria spiralis 5159
Vallota purpurea 4742
Vallota speciosa 4742
Vanda 5698
Vanilla 5699
Vanilla fragrans 3605
Vanilla planifolia 3605
Vateria 5704
Vateria indica 4196
Vaucheria 2626
Veratrum 2170
Veratrum album 5894
Veratrum nigrum 630
Veratrum viride 193
Verbascum 3752
Verbascum blattaria 3690
Verbascum lychnitis 5903
Verbascum nigrum 641
Verbascum phlomoides 1255
Verbascum phoeniceum 4351
Verbascum pyramidale 4368
Verbascum sinuatum 3571
Verbascum thapsiforme 6055
Verbascum thapsus 2276
Verbena 5728
Verbenaceae 5729
Verbena citriodora 3271
Verbena officinalis 2129
Verbesina 1709
Vernonia 2965
Veronica 5123
Veronica agrestis 2237
Veronica alpina 150
Veronica anagalis-aquatica 5838
Veronica arvensis 1532
Veronica bachofeni 406
Veronica beccabunga 509
Veronica chamaedrys 2454
Veronica fructiculosa 4534
Veronica hederaefolia 2992
Veronica latifolia 2892
Veronica longifolia 1289
Veronica montana 3720
Veronica officinalis 1917
Veronica peregrina 4360
Veronica persica 5563
Veronica polita 5854
Veronica prostrata 2709
Veronica scutellata 3518
Veronica serpyllifolia 5512
Veronica spicata 5140
Veronica spuria 475
Veronica tourneforti 5563
Veronica triphyllos 2253
Veronica verna 5726
Verrucaria 5785
Vesicaria 661
Vetiveria 5732
Viburnum 5733
Viburnum carlesi 3186
Viburnum lantana 5852
Viburnum lentago 3792
Viburnum odoratissimum 5391
Viburnum opulus 2091
Viburnum prunifolium 632
Viburnum pyrifolium 632
Viburnum sieboldi 4935
Viburnum tinus 3237
Vicia 5731
Vicia angustifolia 3810
Vicia articulata 3930
Vicia atropurpurea 4356
Vicia benghalensis 4356
Vicia cracca 591
Vicia dasycarpa 6051
Vicia dumetorum 2462
Vicia ervilia 609
Vicia faba 788
Vicia faba minor 5028
Vicia faba *var.* equina 2864
Vicia gracilis 5004
Vicia grandiflora 4885
Vicia hirsuta 5531
Vicia lutea 6111
Vicia macrocarpa 568
Vicia monanthos 3930

Vicia narbonensis 3794
Vicia pannonica 2893
Vicia pisiformis 4097
Vicia sativa 1554
Vicia sepium 2760
Vicia sylvatica 6046
Vicia tenuifolia 755
Vicia tetrasperma 2343
Vicia villosa 2693
Victoria 5830
Vigna 1650
Vigna catjang 1059
Vigna sesquipedalis 6066
Vigna sinensis 1393
Villaresia 5734
Vinca 4140
Vinca major 563
Vinca minor 1493
Vincetoxicum officinale 5916
Viola 5737
Viola alba 4008
Viola arenaria 4691
Viola arvensis 2238
Viola biflora 5661
Viola canina 1880
Violaceae 5740
Viola cornuta 2856
Viola elatior 5431
Viola hirta 2694
Viola lutea 2136
Viola mirabilis 6019
Viola odorata 5392
Viola palustris 3521
Viola riviniana 4507
Viola sylvestris 5412
Viola tricolor 5954
Virgilia 5747
Virgilia lutea 213
Virola surinamensis 398
Viscaria alpina 306
Viscaria viscosa 1250
Viscum 3631
Viscum album 2113
Vismia 5758
Vitaceae 2576
Vitex 1120
Vitex agnuscastus 3296
Vitis 2575
Vitis amurensis 227
Vitis berlandieri 6002
Vitis cordifolia 2386
Vitis davidi 777
Vitis labrusca 2347
Vitis palmata 1057
Vitis riparia 4503
Vitis rotundifolia 3756
Vitis vinifera 2104
Vitis vinifera sylvestris 6032
Vitis vulpina 2386
Voandzeia subterranea 1574
Vulpia bromoides 803
Vulpia myuros 4423

Wachendorfia 5761
Wahlenbergia 4511
Waldsteinia 458
Wallenia 5767
Wallichia 5770
Washingtonia 5790
Weigela 5861
Welwitschia 5865
Wendlandia 5866
Widdringtonia 5930
Widdringtonia schwarzi 4763
Widdringtonia whitei 3637
Wigandia 5935
Wistaria 6011
Wistaria floribunda 3078
Wistaria floribunda macrobotrys 3365
Wistaria frutescens 212
Wistaria multijuga 3365
Wistaria sinensis 1211
Wisteria 6011
Wolffia 6017
Woodsia 6043
Woodsia alpina 157
Woodsia glabella 5043
Woodsia ilvensis 4625
Woodwardia 1110
Wulfenia 6059

Xanthium 1306
Xanthium echinatum 486
Xanthium spinosum 5155
Xanthoceras 6095
Xanthoceras sorbifolium 4868
Xanthorrhiza 6105
Xanthorrhoea 2591
Xanthosoma 3441
Xanthoxylum 4295
Xeranthemum 2923
Xeranthemum annuum 1446
Xerocomus badius 485
Ximenia americana 5428
Xylia dolabriformis 865
Xylia xylocarpa 865
Xylocarpus 6060
Xylosma 6061
Xyris 6084

Yucca 6120
Yucca aloifolia 110
Yucca elephantipes 846
Yucca filamentosa 24
Yucca gloriosa 3692

Zaluzianskya 3852
Zamia 1581
Zannichellia 4266
Zannichellia palustris 1504
Zantedeschia 950
Zantedeschia aethiopica 1374
Zanthorrhiza 6105
Zanthoxylum 4295
Zanthoxylum americanum 1506
Zanthoxylum bungei 2282
Zanthoxylum fagara 3312
Zanthoxylum flavum 6094
Zanthoxylum fraxineum 1506
Zanthoxylum piperatum 3063
Zanthoxylum pterota 3312
Zanthoxylum simullans 2282
Zauschneria 2259
Zea mays 3434
Zea mays *convar.* amylacea 2307
Zea mays *convar.* amyleosaccharata 5215
Zea mays *convar.* ceratina 5849
Zea mays *convar.* vulgaris 2295
Zea mays *var.* dentiformis 1835
Zea mays *var.* everta 4268
Zea mays *var.* identata 1835
Zea mays *var.* saccharata 5301
Zea mays *var.* tunicata 4233
Zelkova 5800
Zelkova acuminata 3080
Zelkova carpinifolia 2021
Zelkova cuspidata 3080
Zelkova keaki 3080
Zelkova serrata 3080
Zelkova sinica 1216
Zephyranthes 6126
Zingiber 2489
Zingiberaceae 2490
Zingiber officinale 1419
Zinnia 6129
Zinnia elegans 1568
Zinnia multiflora 4466
Zizania 5956
Zizania aquatica 266
Ziziphora 6130
Zizyphus 3115
Zizyphus chloroxylon 1321
Zizyphus jujuba 1452
Zostera 1997
Zosteraceae 1998
Zostera marina 1407
Zoysia 3242
Zoysia japonica 3052
Zygadenus 1819
Zygomycetes 6131
Zygophyllaceae 499
Zygophyllum 498

ЧИСЛЕННЫЙ УКАЗАТЕЛЬ РУССКИХ НАЗВАНИЙ РАСТЕНИЙ

RUSSIAN NUMERICAL INDEX

1 зверобой чашечковидный; зверобой чашечный
2 абака; конопля манильская; манилла; пенька манильская
3 абелия
4 аброма
5 аброния
6 митрагина крупнолистная; митрагина прилистниковая
7 абутилон; грудника; канатник
8 банан абиссинский; банан декоративный; банан эфиопский
9 комифора абиссинская
10 акация
11 акантопанакс
12 акант(ус); аканф
13 чертополох акантовидный; чертополох акантолистный; чертополох колючий
14 акантовые
15 ахименес
16 лютик аконитолистный
17 аконит аптечный; аконит реповидный; аконит сборный; аконит ядовитый; борец аптечный; борец синий
18 акрокомия
19 акронихия
20 акростихум
21 актинидия
22 актинидиевые
23 актиномицеты; грибы лучистые; грибки лучистые
24 игла адамова; юкка виргинская; юкка волокнистая; юкка нитевидная; юкка нитчатая
25 микростилис; мякотница; стагачка
26 орляк обыкновенный
27 ужовник
28 ужовниковые
29 аденандра
30 адинандра
31 адлай; коикс; коикс слёзы Иовы; слёзки богородицины; слёзки Иова; слёзы Иовы; слёзник
32 адлумия
33 адонис; горицвет; желтоцвет
34 адзуки; фасоль адзуки; фасоль угловатая; фасоль лучистая
35 эхмея
36 кокорыш
37 барбарис этненский
38 терминалия пышная; терминалия хвалебная
39 афзелия африканская
40 агапантус африканский
41 турраэнтус
42 дальбергия сенегальская; дальбергия чернодревесная
43 можжевельник стройный
44 дерево бальзамическое африканское; комифора африканская
45 паркия африканская
46 пальма гвинейская масличная; пальма масличная гвинейская
47 птерокарпус габонский; птерокарпус Сойакси
48 пиптадения африканская
49 спармания африканская
50 гребенщик африканский
51 ольдфильдия африканская
52 салат алжирский; федия
53 афзелия; афцелия
54 агапант(ус)
55 гелидиум
56 грацилярия лишайниковатая; мох цейлонский
57 печерица; гриб пластинчатый
58 агариковые; грибы агариковые
59 агава
60 агератум; долгоцветка; целестина
61 аглая
62 аглаонема
63 агримония; приворот; репейничек; репешок; репяшок
64 кумарчик
65 айлант
66 симарубовые
67 кассава сладкая; маниок сладкий
68 бийльбергия
69 ямс клубненосный; ямс луковичный
70 аизоон
71 камнеломка вечнозелёная
72 ажгон; айован; айован душистый
73 акебия
74 блигия
75 блигия африканская; финган
76 берхемия лазящая
77 ива буреющая
78 роза Альберта
79 альбиц(ц)ия; альбизия
80 ель Алькокка; ель двуцветная
81 ольха
82 крушина ольхолистная
83 альдрованда
84 алектрион
85 дуб галловый
86 сосна алеппская; сосна иерусалимская
87 смирния
88 вздутоплодник; даная ветвистая; пузыресеменник; пузыресемянник
89 кассия остролистная; сенна остролистная; лист александрийский
90 ал(ь)фал(ь)фа; буркун красный; люцерна посевная
91 аистник цикутный; аистник цикутовый; аистник цикутолистный; буськи; грабельник цикутолистный; журавельник цикутовый; журавельник цикутолистный
92 водоросли
93 ясень двойственный
94 пихта алжирская; пихта нубийская; пихта нумидийская
95 частуховые
96 бескильница
97 алканет; ал(ь)кан(н)а
98 барбарис канадский
99 каштан карликовый; каштан низкорослый
100 можжевельник аллигаторовый; можжевельник толстокорый
101 радиола льновидная
102 перец гвоздичный; перец душистый; перец ямайский; пимента
103 акация устрашающая
104 миндаль обыкновенный; миндаль настоящий; дерево миндальное; миндаль
105 вишня железистая

106 ива миндальная; ива миндалелистная; ива трёхтычинковая; лоза; лозина миндальная
107 груша миндалевидная
108 алоказия
109 алое; алой; алоэ; бабушник
110 юкка алоелистная
111 сольданелля; сольданелла
112 ветреница альпийская
113 астра альпийская
114 азалия полярная; азалия стелющаяся; лаузелеурия лежачая; лаузелеурия стелющаяся;луазелеурия стелющаяся
115 бартсия альпийская
116 колокольчик альпийский
117 чистец альпийский
118 мятлик альпийский
119 сверция многолетняя
120 камнеломка ручейная
121 крушина альпийская
122 ясколка альпийская
123 пиретрум альпийский; хризантема альпийская
124 двулепестник альпийский; двулепестник горный; репей дикий; цирцея альпийская
125 вьюнец; клематис альпийский; княжик альпийский; ломонос альпийский
126 клевер альпийский
127 плаун альпийский
128 смородина альпийская
129 щавель альпийский
130 мелколепестник альпийский
131 горец альпийский; горец горный; кислец; таран
132 незабудка альпийская
133 лисохвост альпийский
134 луговик альпийский
135 ястребинка альпийская
136 василис(т)ник альпийский
137 астрагал альпийский
138 кисличник двухстолбчатый
139 ярутка горная
140 мак альпийский
141 толокнянка альпийская
142 горькуша альпийская
143 арабис альпийский; резуха альпийская
144 проломник альпийский
145 роза альпийская; шиповник альпийский
146 ситник альпийский
147
148 душевик альпийский; душёвка альпийская
149 смолёвка альпийская
150 вероника альпийская
151 таволга альпийская
152 нежник альпийский; солнцецвет альпийский
153 зубровка альпийская
154 тимофеевка альпийская
155 льнянка альпийская
156 кипрей(ник) альпийский
157 вудсия альпийская; вудсия горная
158 альпиния
159 ллойдия
160 язвенник горный
161 жимолость альпийская
162 подорожник альпийский
163 ива голая
164 альзик; клевер гибридный; клевер красно-белый; клевер розовый; клевер шведский
165 ал(ь)стония
166 альстремерия
167 боярышник алтайский
168 селезёночник обыкновенный; селезёночник очереднолистный
169 алтей; шток-роза; альтей
170 стрифнодендрон
171 геухера; хейхера; гейхера
172 алиссум; бурачок; каменник
173 мухоморовые
174 амарант; бархатник; аксамитник; щирица
175 амарантовые; щирицевые
176 амариллис; хипеаструм;гип(п)еаструм
177 амариллисовые
178 спондиас цитеры
179 сола
180 манго вонючее; манго пахучее; манго вонючий; манго пахучий
181 амбербоа
182 антоспермум
183 авокадо; авокадо-груша; аллигаторова груша; персея приятнейшая
184 бук крупнолистный; бук обыкновенный
185 бересклет вьющийся; краснопузырник древогубец
186 смородина американская
187 каштан американский; каштан высокий; каштан зубчатый
188 водосбор канадский
189 элеокарпус американский
190 бузина американская; бузина канадская
191 вяз американский; вяз белый; ильм американский; ильм белый
192 хвойник американский; эфедра американская
193 чемерица зелёная
194 лещина американская
195 женьшень американский; женьшень пятилистный; желтокорень китайский; стосил
196 падуб тусклый
197 хмелеграб американский; хмелеграб виргинский
198 граб американский; граб каролинский
199 липа американская; липа чёрная
200 дерево мангровое; древокорень мангле; шандал-дерево
201 бор развесистый; бор раскидистый
202 рябина американская
203 гварея; гварея трёхтысячевидная
204 саррацениевые
205 платан американский; платан западный; чинар американский; чинар западный; сикомор
206 слива американская
207 апиос клубневый
208 смородина печальная
209 бекмания обыкновенная
210 дерево амбровое; ликвидамбар смолоносный; ликвидамбар стираксовый
211 частуха обыкновенная; частуха подорожниковая

212 вистария кустарниковая
213 виргилия жёлтая; кладрастис жёлтый
214 рядовка фиолетовая; синявка
215 овсяница аметистовая
216 амми
217 аммобиум
218 амоора
219 гледичия аргентинская
220 аморфофалл(ус)
221 ампелопсис; виноградовник
222 жерушник земноводный
223 амсония
224 виноградовник короткоцветоножковый
225 барбарис амурский
226 бархат амурский; дерево амурское бархатное; дерево амурское пробковое; дерево бархатное; дерево пробковое амурское
227 виноград амурский
228 сирень амурская
229 липа амурская
230 гин(н)ала; клён гиннала; клён приречный; клён речной
231 рябина амурская
232 бирючина амурская
233 анабазис; ежовник
234 анакамптис
235 ананас
236 анхиета; фиалка бразильская
237 птерокарпус дальбергиевидный; птерокарпус палисандровидный
238 можжевельник распростёртый опушённый
239 андрахна
240 андреевые
241 андромеда; подбел
242 анемон(а); ветреница
243 прострел
244 ломонос горный
245 ангелика; дудник; дягель; дягиль
246 андира
247 дурман душистый
248 скрытосеменные; скрытосемянные
249 лук угловатый
250 тополь каролинский; тополь ребристый
251 ангустура; галипея лекарственная
252 овёс бесплодный; овёс декоративный; овёс дикий красный; овёс стерильный
253 анис (обыкновенный); бедренец-анис; ганус
254 бадьян; иллициум
255 анизохилус
256 аннато; бикса аннатовая; дерево орлеанское; орлеан
257 аннеслея
258 мятлик однолетний
259 мелколепестник однолетний; стенактис однолетний
260 горечавка горькая
261 аира волосовидная
262 дивало однолетнее
263 левкой однолетний
264 очиток однолетний
265 колосок душистый остистый; пахучеколосник остистый
266 рис болотный; рис водяной; рис дикий; рис индийский; рис канадский; рис озёрный; трава рисовая; тускарора; цицания водяная
267 донник индийский; донник мелкоцветковый
268 бук антарктический; нотофагус антарктический
269 венечник
270 антуриум
271 язвенник
272 питеколобиум
273 мшанка безлепестная
274 афананта
275 афеландра; афеляндра
276 афиллофоровые
277 арракач; ар(р)акача съедобная
278 апоногетоновые
279 апофиллум
280 апороселла
281 апозерис
282 яблоневые
283 мята круглолистная
284 никандра; хмель огородный
285 никандра пузыревидная; никандра физалиевидная; никандра физалисовидная
286 айва яблоковидная
287 роза яблочная; шиповник мохнатый; шиповник яблочный
288 яблоня
289 слива абрикосовая; слива абрикосовидная слива Симонова
290 аптандра
291 апулея
292 кофе арабский; дерево кофейное аравийское; дерево кофейное арабское
293 жасмин арабский; жасмин индийский; жасмин-самбак
294 кат(а); катх съедобный
295 аралия
296 дерево сандараковое; тетраклинис членистый; туя алжирская
297 араукария
298 араукариевые
299 алоэ древовидное
300 биота; туя
301 архегониальные
302 архимицеты; грибы хитридиевые
303 архитея
304 мятлик арктический
305 княженика; костяника арктическая; малина арктическая; мамура; поленика; поляника
306 лихнис альпийский;смолка альпийская
307 кассиопа гипновидная
308 крупка фладницийская
309 мелколепестник одноцветковый
310 мшанка Линнеева; мшанка моховидная
311 ива арктическая
312 арктотис
313 арека; пальма арековая
314 аретуза
315 аргания (крепкодревесная); дерево железное
316 геонома Шотта
317 кипарис аризонский
318 сосна аризонская; сосна жёлтая аризонская
319 ломонос Арманда
320 сосна Арманда
321 слива курдистанская
322 арнебия

323 арника; баранник
324 боярышник Арнольда
325 аракача
326 пельтандра
327 стрелолист
328 триостренник болотный
329 аррорут; маранта
330 марантовые
331 артишок (посевной)
332 артишок китайский; артишок японский; стахис клубненосный; хороги; чистец клубненосный
333 аронник; аройник; арум; борода ааронова
334 ароидные; аронниковые
335 арундинелла
336 ассафетида; вонючка; смолоносница вонючая; ферула вонючая; ферула ассафетида
337 аскомицеты; грибы асковые; грибы сумчатые
338 ясень
339 перламутровка
340 кодонопсис
341 лимонница гладкая
342 ирга азиатская
343 тоддалия азиатская
344 рёмерия
345 барбарис азиатский
346 женьшень обыкновенный; корень-человек; панакс женьшень
347 луносемянник даурский;плющ амурский
348 подорожник азиатский
349 пуходревник; хлопчатник древовидный
350 аспараг(ус); спаржа
351 аспергилл(ус); плесень леечная
352 асфодель; асфоделюс; асфодил
353 аспидистра
354 аспидиум; папоротник-аспидиум
355 бегония рекс; царь-бегония
356 чай ассамский; куст чайный ассамский
357 драцена астелия
358 астра
359 астровые
360 астильба; астильбе
361 астрокарпус
362 аталантия
363 дирка болотная
364 кедр атласский
365 пальма-атталея; атталея
366 обриеция
367 аукуба
368 аврикула; аурикула; первоцвет-аврикула
369 аврикуляриевые
370 евгения миртолистная
371 ливистона австралийская; ливистона южная
372 сифонодон южный
373 водоросль австралийская морская съедобная
374 роза вонючая; роза жёлтая; шиповник вонючий
375 ракитник австрийский
376 змееголовник австрийский
377 жерушник австрийский
378 лён австрийский
379 дороникум австрийский
380 сосна австрийская; сосна чёрная (австрийская); сосна столбовая
381 автобазидиомицеты
382 зимовник; безвременник; колхикум
383 пролеска осенняя
384 болотник осенний; красовласка осенняя
385 гравилат
386 авицения
387 шильник; шильница
388 аксирис
389 азалия; азалея
390 боярышник-азароль; боярышник испанский; боярышник плодовый; боярышник понтийский; дуляна
391 ацолла
392 незабудка азорская
393 спрекелия
394 бархатцы высокие; бархатцы крупноцветковые; бархатцы крупноцветные; бархатцы прямостоячие
395 табак деревенский; табак-махорка; махорка
396 бабассу
397 качим метельчатый; кучерявка; перекати-поле; перекатун
398 вирола суринамская
399 акация аравийская
400 ива вавилонская; ива плакучая
401 рдест маленький
402 хельсиния
403 энокарпус
404 баккаурея
405 бакхарис
406 вероника Бахофена
407 алария съедобная
408 баель; баэль; эгле мармеладное; дерево энгленовое
409 паспалум помеченный
410 реедия бразильская
411 платония
412 баланофоровые
413 костёр кистевидный
414 кипарис болотный; кипарис лысый
415 меум; корень медвежий
416 мятлик Бальфура
417 ель Бальфура
418 клён Гельдрейха
419 сосна Веймутова румелийская; сосна румелийская
420 кольник округлый;кольник шаровидный
421 лук круглоголовый
422 неслия метельчатая
423 ширококолокольчик
424 колокольчик крупноцветный; колокольчик китайский крупноцветный;колокольчик японский крупноцветный
425 белокудренник
426 флиндерсия австралийская
427 мелисса
428 тополь крупнолистный;тополь седой
429 дерево бальзамное; дерево бальзамовое; мироксилон
430 амирис бальзамический; амирис ямайский; дерево розовое
431 момордика (бальзамическая)
432 бальзамник; пихта бальзамическая; пихта канадская
433 бальзаминовые
434 момордика харантская; огурчик пёсий виноградолистный

435 ива грушанколистная
436 ситник балтийский
437 бамбуза; бамбук
438 кониограмма японская
439 банан
440 банановые
441 актеа; воронец; вороняжка
442 банксия
443 роза Банкса; роза банксиевая; роза Бэнкса; шиповник Бэнкса
444 баниан; баньян; смоковница бенгальская; фикус бенгальский
445 адансония; баобаб; дерево обезьянье хлебное
446 мальпигия гранатоволистная
447 кактус пейреския; крыжовник барбадосский; пересекия колючая; пересекия шиповатая
448 гиппеаструм
449 ятропа куркас; ятрофа куркас
450 лиций обыкновенный
451 стрифнодендрон-барбатимао
452 барбарис
453 барбарисовые
454 ячмень
455 ячмень культурный многорядный; ячмень многорядный; ячмень обыкновенный (многорядный); ячмень посевной; ячмень четырёхрядный; ячмень четырёхсторонний
456 ежовник-куриное просо; ежовник-петушье просо; просо мучнистое; просо куриное; просо настоящее; просо петушье; просо посевное; пайза; просяник
457 стеркулия утёсная
458 вальдштейния
459 бартония
460 бартсия; зубчатка; бартшия
461 лишайники базидиальные
462 базидиомицеты; грибы базидиальные
463 базилик; базилика
464 душёвка; щебрушка
465 цимбалярия
466 эшинантус
467 белотал; вязинник; лоза; ива болотная; ива коноплянная; ива корзиночная; ива прутовидная; ива прутьевидная
468 дерево масляное
469 кадило медоволистное; кадило мелиссолистное
470 беллиум
471 датиска коноплевая; пенька критская
472 люцерна гибридная
473 клевер лупиновидный; клевер люпиновидный; клевер лупиновый; клевер люпиновый
474 дерево розовое
475 вероника ложная; вероника ненастоящая
476 ленец; ленолистник
477 дерево даммаровое
478 птицемлечник пиренейский
479 батрахоспермум
480 акронихия австралийская; акронихия Бауэра
481 баугиния; баухиния
482 горечавка баварская
483 восковник; восковница; мирика
484 восковниковые
485 гриб польский
486 дурнишник американский; дурнишник шиповатый
487 аммофил(а); песколюб
488 слива морская
489 полынь звёздчатая
490 клинтония
491 нертера
492 майник
493 аденантера
494 купырь
495 диаррена
496 ринхоспора; очеретник
497 фасоль
498 парнолистник
499 парнолистниковые; парнолистные
500 толокнянка аптечная; толокнянка медвежья; виноград медвежий; ушко медвежье; ягода медвежья
501 кизильник Даммера
502 колокольчик бородатый
503 борода лешего; лишайник бородатый; косы ведьмины
504 многобородник
505 дуб карликовый
506 кортуза
507 морозник вонючий
508 калофиллум; дерево розовое
509 вероника поручейная; вероника поточная; вероника-поточник; вероника ручейная; ибунка; поточник
510 подмаренник
511 ясменник подмаренниковый; ясменник сизый
512 монарда
513 бук
514 буковые
515 гриб печёночный; печёночница; фистулина печёночная
516 казуарина
517 казуариновые
518 живокость высокая
519 пчелка; офрис пчелоносная
520 бурак; свёкла
521 череда
522 роза Беггера; шиповник Беггера
523 бегония
524 бегониевые
525 белладонна; красавка (обыкновенная); красавка белладонна; одурь сонная
526 амариллис белладонна
527 дерево габонское тюльпанное
528 кампанула; колокольчик
529 колокольчиковые
530 увулярия
531 прутняк веничный; кохия веничная
532 гамбир
533 дерево лаковое
534 полевица
535 мята бергамотная; мята лимонная
536 бергамот; цитрус бергамия; цитрус бергамот
537 бергения; бадан
538 бергия
539 берлиния
540 тополь берлинский
541 аррорут вестиндский; маранта вестиндская; маранта тростниковая
542 аджирык; злак пальчатый; зуб собачий;

свинорой пальчатый; трава бермудская; трава собачья
543 можжевельник барбадосский; можжевельник бермудский
544 волдырник ягодоплодный; волдырник ягодный
545 берула; беруля
546 вереск метлистый; эрика метловидная
547 вишня Бессея
548 арека катеху; пальма бетель(ная); орех арековый; орех нищих; пальма арековая; пальма катеху; пинанг; бетель (пальма)
549 бетель (перечный)
550 буквица; буковица; чистец
551 кипарис гималайский; кипарис надутый; кипарис Дукло
552 тюльпан Биберштейна
553 босвелия Картера; ладан
554 дуб пробковый западный
555 двойчатка
556 лжетсуга крупноплодная; лжетсуга крупношишечная
557 ракитник головчатый
558 чай почечный; ортосифон тычиночный
559 черноголовка крупноцветковая;черноголовка крупноцветная
560 шалфей крупноцветковый; шалфей крупноцветный
561 липа крупнолистная; липа плосколистная; липа широколистная
562 клён крупнолистный
563 барвинок большой
564 ива крупнолистная
565 гортензия крупнолистная
566 антидесма буниус; дерево саламандровое
567 бигнония
568 вика крупносемянная
569 трясунка большая
570 герань крупнокорневищная
571 чубушник крупноцветковый
572 рыжей яровой; рыжик посевной; рыжик яровой; рожь индийская
573 гледичия крупноколючковая
574 крапива большая; крапива двудомная; крапива обыкновенная
575 пеларгониум крупноцветный
576 авер(р)оа-билимби; билимби
577 таволга Биллиарда; таволга Бильярда
578 берёза
579 берёзовые
580 груша берёзолистная
581 таволга берёзолистная
582 абабок; берёзовик (обыкновенный); гриб подберёзовый; гриб чёрный; обабок; подберёзовик (обыкновенный); подобабик; подобабок
583 губка берёзовая
584 стрелиция
585 капуста полевая; репак; сурепица
586 первоцвет мучнистый; примула мучнистая
587 сераделла
588 лядвенец рогатый
589 гнездовковые
590 гнездовка
591 горох мышиный; горошек мышиный
592 кирказон ломоносовидный; кирказон обыкновенный; кирказон понтийский
593 кирказоновые
594 ерика средиземноморская
595 сосна мелкоигольчатая; сосна мягкоигольчатая
596 сныть обыкновенная
597 миндаль горький
598 гриб жёлчный; гриб белый ложный
599 пикрамния
600 сердечник
601 щавель туполистный
602 мелколепестник едкий; мелколепестник зонтичноцветный
603 латук дикий; латук ядовитый; молокан ядовитый; салат вонючий
604 глистник; паслён горько-сладкий; паслён сладко-горький
605 гикори горький; кария сердцевидная; орех горький
606 панус вяжущий; панус сморщивающийся
607 сочевичник весенний; чина весенняя
608 древогубец; краснопузырник; целаструс
609 горошек горький; горошек чёткообразный; чечевица французская
610 желтоцвет; сурепица обыкновенная; сурепка обыкновенная
611 трихилия
612 биксовые
613 ясень чёрный
614 белокудренник чёрный; кудренник чёрный; кудри пёстрые
615 бамбук чёрный
616 воронец колосовидный; воронец колосистый; воронец волосистый
617 полевица гигантская
618 кизильник черноплодный
619 ежевика
620 беламканда китайская
621 схенус чернеющий; схенус черноватый
622 тамус
623 канариум чёрный
624 ежевика американская; ежевика чёрная; малина американская; малина ежевикообразная; малина западная; малина североамериканская; малина чёрная
625 василёк чёрный
626 вишня чёрная дикая;черёмуха поздняя
627 багновка чёрная; вороника чёрная; шикша чёрная
628 подгруздок чернеющий; сыроежка чернеющая
629 рудбекия шершавая
630 чемерица чёрная
631 жимолость чёрная
632 калина грушелистная;калина сливолистная
633 белена чёрная
634 дуб мерилендский
635 родотипус красивый; розовик керриевидный; розовик японский
636 акация белая; акация американская; лжеакация; робиния лжеакация; робиния ложноакация; рябина лжеакация; рябина ложноакация
637 клён чёрный
638 буркунчик; люцерна хмелевая;люцерна хмелевидная; полуклевер; трилистник жёлтый
639 кольник чёрный

640 тут чёрный; шелковица чёрная
641 коровяк чёрный
642 горчица чёрная
643 паслён чёрный; поздника
644 дуб (американский) красильный; дуб бархатистый; дуб бархатный
645 сочевичник чёрный; чина чёрная
646 перец чёрный
647 осокорь; тополь чёрный
648 дерево палисандровое; палисандр
649 козелец испанский; козелец сладкий корень; корень сладкий; корень чёрный; скорцонер (испанский)
650 осока тёмная; осока темноодетая
651 костенец чёрный
652 глауциум рогатый; мак рогатый;мачёк рогатый
653 ель чёрная
654 ива чёрная
655 акация (чёрная) австралийская; акация чернодревесная; дерево чёрное (австралийское)
656 волдырник
657 пузырник; цистоптерис
658 араужия
659 клекачка; клокичка; стафилея
660 клекачковые; клокичковые
661 везикария; пузырник
662 пузырник
663 куколица широколистная; смолёвка-хлопушка; щелкунец
664 пузырчатка
665 пузырчатковые
666 фукус мелкопузырчатый; фукус пузырчатый
667 ламинария; ляминария
668 бикукулла; диелитра; диэлитра; дицентра; сердечки
669 остро-пёстро; расторопша остропёстрая; расторопша пятнистая
670 кникус
671 бенедикт аптечный; волчец кудрявый; волчец благословленный; кникус аптечный; кникус благословленный; кардобенедикт; чертополох благословленный
672 блетилла
673 лютик ядовитый
674 осока пузырчатая
675 жминда (головчатая);марь головчатая
676 ваточник кюрасао
677 ирезине
678 гемантус; хемантус
679 гриб сатанинский
680 герань кроваво-красная; герань кровянокрасная; журавельник кровянокрасный
681 сангвинария; сангуинария
682 сангвинария канадская; сонгвинария канадская
683 глог; дёрен красный; дёрен кроваво-красный; кизил кровяной; кизил свидина; свидина (кроваво-красная); спиж
684 щавель кровяной
685 роза мягкая; шиповник мягкий
686 сыроежка сине-жёлтая
687 ясень голубой; ясень четырёхгранный
688 кедр атласский голубой
689 кариоптерис
690 колокольчик круглолистный
691 мертензия
692 вакциниум; черника; ягодник
693 ива чернеющая; лоза волчья
694 заразиха пурпуровая
695 ель синяя
696 Иван-да-Марья; марьянник дубравный
697 страстоцвет светло-синий
698 сизюринхий; сисюринхий
699 кипарис голубой Говея
700 мятлик
701 водоросли сине-зелёные; водоросли циановые
702 тонконог сизый
703 люпин синий; люпин узколистный
704 сеслерия голубая
705 голубой очный цвет; очный цвет голубой
706 бородач; осенчук
707 пырей Смита
708 мак голубой тибетский; меконопсис
709 синеголовник альпийский
710 х(о)устония
711 гуньба; донник голубой; донник синий; пажитник голубой
712 рдест туполистный
713 мухомор серо-розовый
714 маточник болотный; остерикум болотный
715 костолом; нартеций
716 вахта; павун
717 голубика; гонобобель; пьяница
718 герань болотная
719 болотница болотная
720 кальмия многолистная; кальмия сизая
721 жерушник болотный
722 очный цвет нежный
723 андромеда-дубровник; андромеда дубровниколистная; подбел дубровник; подбел многолистный
724 схенус; очерёт
725 козелец низкий; козелец приземистый
726 гречишник бальджуанский
727 болетус
728 бол(ь)тония
729 бомарея
730 бомбакс
731 дорема аммиачная; дорема аммониакальная; дорема камеденосная; ошак
732 посконник
733 костянкоплодник
734 бора́го; бурачник; огуречник; трава огуречная
735 бурачниковые
736 дриобаланопс
737 пихта Борнмюллера
738 баниан;баньян; смоковница священная; фикус священный
739 каштан конский мелкоцветный
740 шероховатка
741 стеркулия
742 бугенвиллея
743 корень мыльный; мыльнянка аптечная; мальнянка лекарственная; сапонария аптечная; сапонария лекарственная
744 домбея точечная
745 роза бурбонская
746 ангрекум
747 бувардия

748 актинидия аргута; актинидия изогнутая; актинидия острая; актинидия острозубчатая; кишмиш крупный
749 букс(ус); дерево буксовое; самшит
750 клён американский; клён ясенелистный; негундо; неклён (виргинский)
751 самшитовые
752 азара мелколистная
753 бойкиния
754 птерис
755 горошек тонколистный
756 пшеница ветвистая
757 асфоделюс белый; асфодил белый
758 цезальпиния
759 котула
760 многорядник Брауна
761 евгения бразильская
762 мирокарпус густолиственный
763 уруть бразильская
764 дальбергия чёрная; дерево бразильское палисандровое
765 орех американский; орех бразильский
766 шинус терпентиннолистный
767 тритринакс бразильский
768 дерево красное (бразильское); дерево красное (фернамбуковое); фернамбук; дерево фернамбуковое
769 дерево хлебное
770 бросимум
771 гладиолус гентский; шпажник гентский
772 ель Бревера
773 абрикос альпийский
774 кизильник войлочный
775 спирея сливолистная
776 бриделия
777 виноград Давида
778 тристания свёрнутая; тристания скрученная
779 пихта калифорнийская;пихта прелестная
780 мышей; щетинник
781 жабрей; медов(н)ик; пикульник-жабрей; пикульник колючий; пикульник обыкновенный
782 чертогон; чертополох
783 горлюха румяниковидная; горлюха румянковидная
784 девясил британский
785 тимофеевка метельчатая
786 папоротник ломкий пузырчатый; пузырник ломкий
787 ива ломкая; ива хрупкая
788 бобы (кормовые); бобы русские
789 тургения широколистная
790 денежка широкопластинчатая; коллибия широкопластинчатая; монетка широколистная
791 камнеломка круглолистная
792 дремлик широколистный
793 бересклет широколистный
794 гроздовник многораздельный
795 лаванда широколистная
796 мятлик Ше; мятлик Шэ
797 таволга широколистная
798 ятрышник широколистный
799 купена широколистная
800 броколи; броколь; капуста спаржевая (цветная)
801 бродиея
802 костёр
803 вульпия белкохвостая
804 бромелия
805 бересклет американский
806 манник водяной; манник водный; поручейница водная; поручейница водяная
807 самолюс; северница
808 самолюс Валеранда; самолюс обыкновенный; северница обыкновенная
809 жарновец; ракитник; цитизус
810 сорго веничное; сорго метельчатое; сорго развесистое; сорго техническое
811 льнянка дроколистная
812 заразиха
813 заразиховые
814 броваллия
815 водоросли бурые
816 очеретник бурый
817 василёк горькуша; василёк луговой
818 просо ветвистое
819 тополь печальный
820 масленик; маслёнок (поздний); масляник (поздний)
821 брунера
822 дазилирион волосистоконечный
823 брюсселька; капуста брюссельская
824 бриония; переступень
825 мохообразные
826 диделотия африканская
827 баросма
828 каштан конский
829 подорожник ланцетный; подорожник ланцетовидный; подорожник ланцетолистный
830 жестёр; жостер; крушина; крушинник
831 крушинные; крушиновые
832 гречиха
833 гречишные
834 шефердия
835 паслён клювовидный; паслён клювообразный
836 бухлое; бухлоэ
837 клопогон; цимицифуга
838 дубница; дубровка; живучка
839 зюзник
840 анхуза; воловик; "язык воловий"
841 лютик клубневой; лютик клубненосный; лютик луковичный
842 лилия бульбоносная; лилия красная
843 бульбофиллум
844 мятлик луковичный
845 ситник луковичный
846 юкка слоновая
847 слива ненастоящая; слива чёрная мелкая; тернослив
848 мимусопс
849 анона сетчатая
850 бодяк ланцетолистный
851 бульнезия
852 камыш; изолепис
853 манго душистое; манго душистый
854 грибы головнёвые; устилягиновые
855 араукария Бидвилла
856 купырь обыкновенный
857 трава золотушная; череда трёхразде-

льная
858 огурец волосистый; сициос
859 лопух; лопушник
860 коз(е)лец; трагус
861 роза бургундская
862 эхинодорус
863 дерево индийское розовое; сандал индийский; птерокарпус индийский
864 птерокарпус крупноплодный
865 ксилия деревянистоплодная; ксилия рубанковидная
866 цедрела сингапурская
867 кровохлёбка
868 черноголовник
869 эрехтитес
870 дуб крупножелудевый; дуб крупноплодный; дуб репейник
871 ежеголовка; ежеголовник
872 ежеголовковые; ежеголовниковые
873 гертнериа
874 бурсария
875 бурзера; бурсера
876 бурзеровые; бурсеровые
877 грыжник; трава колосовая
878 кошельки цельнолистные
879 лапчатка курильский чай; лапчатка кустарниковая; чай курильский (кустарниковый)
880 диервил(л)а
881 фасоль обыкновенная кустовая
882 настурция карликовая; настурция малая; настурция мелкая; настурция низкорослая; капуцин малый
883 патиссон; тыква кустовая; тыква мозговая; тыква фигурная
884 перец ветвистый; перец испанский; перец красный; перец кустарниковый; перец многолетний; перец однолетний; перец овощной; перец стручковый (однолетний); перец ягодный
885 иглица колючая; иглица понтийская; иглица шиповатая; иглица шиповая; рускус колючий; тёрн мышиный
886 бутея
887 золотушник; льнянка обыкновенная
888 белокопытник; подбел
889 лютик; ранункул(юс)
890 лютиковые
891 корилопсис малоцветковый
892 буд(д)ле(й)я
893 схизантус; шизантус
894 ваточник клубневой; ваточник клубненосный
895 любка двулистная; фиалка ночная
896 хризалидокарпус
897 орех масляный; орех серый
898 дерево масляное
899 дерево сальное; пентадесма масличная
900 жирянка
901 денежка каштаново-коричневая
902 цефалантус; цефаляntус
903 люцерна дисковидная; люцерна круглая; люцерна округлая
904 гиббертия
905 бирсонима
906 капуста
907 дерево капустное
908 крестоцветные
909 пальма капустная американская; пальметта-сабаль; сабаль пальметто; сабаль пальмовидный
910 роза столепестная; роза столистная; центифолия
911 погостемон
912 дерево шоколадное; какао (настоящее); шоколадник
913 кахрис
914 кактус
915 кактусовые
916 кадия
917 урена
918 гриб цезарский; поганка цезарева
919 златолист каймито; каимито; яблоко звёздное
920 дерево каяпутовое; мелалеука бледнодревесная; мелалеука чайная
921 физостигма
922 горлянка; тыква бутылочная (обыкновенная); калебаса;тыква-горлянка
923 монодора мускатная; орех мускатный ямайский
924 дерево горлянковое; дерево калабас(с)овое; кресченция
925 каладиум
926 клиноног; пахучка
927 душевик аптечный; пахучка аптечная
928 калатея
929 календула; ноготки
930 орех калифорнийский
931 каштан конский калифорнийский; павия калифорнийская
932 крушина калифорнийская
933 люцерна маленькая; люцерна щетинистая; люцерна щетинистоволосистая
934 хвойник калифорнийский
935 лещина калифорнийская
936 кедр калифорнийский ладанный; кедр речной калифорнийский; кедр речной сбежистый
937 можжевельник калифорнийский
938 лавр калифорнийский; умбеллюлярия калифорнийская
939 дуб траволистный
940 дерево перечное калифорнийское
941 дарлингтония
942 тополь волосистоплодный
943 мак калифорнийский; полынёк калифорнийский; эшолция калифорнийская
944 бирючина овальнолистная
945 секвойя вечнозелёная
946 камнеломка дернистая
947 торрея калифорнийская
948 восковница калифорнийская
949 белокрыльник
950 зантедешия; рихардия; ричардия
951 каллиантемум
952 джузгун; жузгун; кандым
953 калотропис
954 якорцы
955 колумба; ятрориза дланевидная
956 калипсо
957 калипсо луковичное; калипсо луковичный
958 камассия; квамассия
959 бадьян камбоджийский
960 камелия
961 джантак; жантак; колючка верблюжья

(обыкновенная); трава верблюжья; чагерак верблюжий
962 стальник ползучий
963 антемис; пупавка
964 магнолия Кэмбелла
965 камфоросма; трава камфарная
966 дерево камфарное; коричник камфарный; лавр камфарный
967 горицвет; лихнис
968 бафия
969 голубика канадская; черника канадская
970 мятлик сплюснутый
971 золотарник канадский
972 боярышник канадский
973 гемлок восточный; гемлок канадский; тсуга канадская
974 уруть мутовчатая
975 бодяк полевой; осот розовый
976 зараза водяная канадская; элодея канадская
977 лапортея канадская
978 слива канадская; слива чёрная
979 тополь дельтовидный; тополь канадский
980 тис канадский
981 канегра; щавель североамериканский
982 дорикниум
983 финик канарский
984 двукисточник; канареечник
985 канарейник; канареечник канарский; канареечник птичий
986 настурция иноземная; настурция канарская
987 дуб Мирбека
988 сосна канарская
989 канариум
990 конофолис
991 кабралея бразильская
992 гумбанг мягкий; дерево свечное; лумбанг; орех чилийский; тунг молуккский
993 иберийка; иберис; перечник; разнолепестка; разнолепестник; разнолистник; стенник
994 арундинария; камышовка
995 канелла
996 канна
997 канновые
998 дыня десертная; дыня канталупа; дыня культурная; канталупа
999 колокольчик средний
1000 лисичка
1001 лисичковые
1002 дуб золоточешуйчатый; дуб калифорнийский вечнозелёный
1003 алоэ дикое; алоэ колючее; алоэ ферокс
1004 элеодендрон капский
1005 диморфотека; ноготки африканские
1006 стрептокарпус
1007 каперсник; каперсы; каперцы
1008 молочай масличный; молочай чиновидный
1009 каперсовые; каперцовые
1010 ситник головчатый
1011 черёмуха поздняя иволистная
1012 аверроа; карамбол(а)
1013 тмин (обыкновенный)
1014 кардамон
1015 артишок испанский; кардон
1016 кобрезия осоковидная
1017 колючелистник; колючник
1018 карлудовика; карлюдовика
1019 гвоздика голландская; гвоздика садовая
1020 осока сизая
1021 карнауба; пальма бразильская (восковая); пальма восконосная; пальма восковая бразильская коперниция восконосная
1022 лилия карниольская
1023 дерево рожковое; дерево стручковое; кароб; каруба дикая; рожечник; рожки цареградские; рожки царьградские; стручок цареградский; хлеб дивий; хлеб Иванов
1024 ясень каролинский
1025 кария североамериканская
1026 колокольчик карпатский
1027 смородина карпатская
1028 лапка кошачья карпатская
1029 карпезиум
1030 живучка ползучая; трава живучая
1031 лапчатка низкая; лапчатка стелющаяся
1032 моллуго; моллюго; мутовчатка
1033 аизоацневые; аизовые
1034 кактус ластовневый; стапелия
1035 морковь
1036 зонтичные
1037 гвоздика картузианская
1038 фукус пильчатый
1039 пихта миловидная
1040 крушина американская; крушина Пурша
1041 каскариль
1042 анакард(ия); дерево анакардовое
1043 анакардиевые; анакардовые; сумаховые; фисташковые
1044 гледичия каспийская
1045 лотос каспийский
1046 кассава; маниок(а); маниот
1047 дерево коричное китайское; корица китайская; коричник кассия; коричник китайский
1048 кассина
1049 кассиопа; кассиопея
1050 клещевина; рицин(ус)
1051 клещевина занзибарская; клещевина обыкновенная; рицина; рицинник
1052 жасмин крупноцветковый; жасмин крупноцветный
1053 катальпа
1054 куколица; смолёвка; силена; силене; хлопушка
1055 лепчица; липушник; подмаренник цепкий
1056 акация катеху
1057 виноград лапчатый
1058 тополь китайский
1059 вигна катьянг; вигна коноплевидная
1060 мята кошачья; котовник кошачий; котовник мятный
1061 гипохерис; пазник
1062 палочник; рогоз; чакан
1063 рогозовые
1064 дубровник горький; дубровник кошачий; майоран кошачий
1065 каттлея
1066 ольха восточная; ольха сердцелистная

1067 василёк восточный
1068 окопник кавказский
1069 борщевик пушистый
1070 волчеягодник кавказский; волчник кавказский
1071 лещина понтийская; орешник понтийский
1072 карказ кавказский
1073 граб кавказский
1074 касатик кавказский
1075 дороникум кавказский
1076 липа кавказская; липа пушистостолбчатая
1077 лапина кавказская; лапина ясенелистная; орех кавказский; птерокария рябинолистная
1078 жасмин садовый; чубушник кавказский
1079 капуста цветная
1080 цеанотус
1081 маниок Глацийова; маниока каучуконосная
1082 дерево кедровое; кедр
1083 кедр ливанский
1084 цедрела
1085 чистотел
1086 сельдерей
1087 филлокладус
1088 цельзия
1089 василёк
1090 золототысячник
1091 кентрантус; центрантус
1092 агава американская; агава древовидная; алоэ американское; столетник
1093 боровик; гриб белый; коровка
1094 головчатка; цефалярия
1095 цефалоцереус
1096 камнеломка австралийская
1097 рогач
1098 багрян(н)ик; церцис
1099 церкоспора
1100 кактус колонновидный; кактус-цереус; цереус
1101 монстера любимая; монстера пышная; филодендрон пробитый
1102 кактус свисающий; кактус церопегия; церопегия
1103 цетрария
1104 калаба; калофиллум обыкновенный
1105 дерево коричное (цейлонское); корица цейлонская; коричник цейлонский
1106 сансевиерия цейлонская
1107 дерево атласное (индийское); дерево красное; хлороксилон
1108 ахирантес; самоцвет; соломоцвет
1109 низмянка
1110 вудвардия
1111 лилия халцедонская
1112 хамедорея
1113 дубровник обыкновенный; дубровник пурпуровый; растигор
1114 гриб подорешник; подорешник; чешуйчатка травяная; опёнок летний
1115 лисичка настоящая; лисичка обыкновенная; сплоень
1116 мангольд
1117 скипетр царский
1118 горчица дикая; горчица полевая
1119 водоросли харовые; лучицы
1120 витекс; прутняк
1121 гиднокарпус; хиднокарпус
1122 мирлитон; огурец мексиканский; сехиум; чайот(а); шушу
1123 костёр кровельный
1124 дерево миробалановое
1125 гаультерия лежачая; гаультерия распростёртая; чай канадский
1126 рябчик шахматный
1127 сидальцея
1128 берека; глоговина; дерево красное; рябина-глоговина;рябина лечебная
1129 гвоздика сизая
1130 баррингтония
1131 вейник наземный
1132 анона черимола; анона черимойя
1133 вишня
1134 кизильник Цабеля
1135 лох многоцветковый
1136 овения вишнелистная
1137 бутень
1138 костёр ржаной
1139 каштан
1140 гриб каштановый
1141 дуб каштанолистный
1142 дуб горный
1143 слива узколистная; слива Чикаса
1144 горох бараний; горох турецкий; нагут; материнка; нут; нут культурный; нут обыкновенный; нут посевной; нут рогообразный
1145 ясколка
1146 звёздочка; звездчатка-мокрица;звездчатка средняя; мокрица
1147 цикорий; цикорник
1148 гуннера чилийская
1149 кедр речной чилийский
1150 земляника чилийская
1151 липучка масляная; мадия посевная
1152 лапагерия
1153 повилика душистая
1154 майтенус чилийский
1155 лоаза
1156 сосна Жерарда; сосна Жерардова
1157 колокольчик пирамидальный
1158 астра садовая; каллистефус
1159 мелия; ясенка
1160 мелия аzедарах; мелия ацедарах; мелия гималайская; мелия иранская
1161 мурра(й)я экзотическая
1162 кипарис азиатский болотный
1163 куннингамия
1164 антидесма
1165 гвоздика китайская;гвоздика Фишера
1166 сассапариль китайская
1167 аралия китайская
1168 ясень китайский
1169 астильбе китайская
1170 рододендрон мягкий
1171 берёза китайская
1172 костянкоплодник Шверина
1173 крушина китайская;крушина полезная
1174 вишня уссурийская; вишня японская
1175 катальпа яйцевидная; катальпа яйцевиднолистная;катальпа японская
1176 астрагал китайский
1177 вяз мелколистный
1178 ливистона китайская

1179 лещина китайская; орешник китайский
1180 яблоня замечательная
1181 айва китайская; хеномелес китайский
1182 купальница китайская
1183 каркас китайский
1184 тсуга китайская; цуга китайская
1185 гибискус китайская роза; роза китайская; мальва сирийская
1186 кария китайская; орешник китайский
1187 гледичия китайская
1188 каштан конский китайский
1189 куммеровия клиновидная; леспедеца заострённая; леспедеца клиновидная; леспедеца шелковистая; леспедеза клиновидная
1190 сирень китайская
1191 лимонник китайский
1192 магония Форчуна
1193 нейлия китайская
1194 стеркулия платанолистная; фирмиана
1195 сосна китайская; сосна погребальная
1196 цефалотаксус Фортуна
1197 первоцвет китайский; примула китайская
1198 бирючина китайская
1199 багрянник китайский; багрянник японский
1200 роза бенгальская; роза индийская; роза китайская; роза чайная китайская
1201 таволга китайская
1202 странвезия Давида
1203 сумах китайский; сумах яванский
1204 дерево восковое китайское; дерево китайское восковое; дерево китайское сальное; дерево сальное; сапиум салоносный
1205 цедрела китайская
1206 торрея большая
1207 лириодендрон китайский; дерево тюльпанное китайское
1208 орех китайский
1209 тыква восковая
1210 лапина узкокрылая
1211 вистария китайская; глициния китайская
1212 гамамелис китайский; гамамелис мягкий
1213 дереза китайская
1214 ямс посевной; ямс съедобный
1215 тис китайский
1216 дзельква китайская
1217 грудника Авиценны; грудника Теофраста; канатник Теофраста
1218 хюкразия
1219 лук-резанец; лук-скорода; резанец; скорода; шнит-лук
1220 зеленоцвет
1221 хлорантовые
1222 хлорелла
1223 стеркулиевые
1224 арония
1225 альхорнея
1226 морозник чёрный
1227 держи-дерево (обыкновенное); держи-дерево Христовы Тернии; драч; тернии христовы; палиурус держи-дерево
1228 хроококковые
1229 златоцвет; нивяник; пиретрум; ромашник; хризантема
1230 сыть съедобная; миндаль земляной; чуфа
1231 хускуеа
1232 языкоцветные
1233 эвкалипт Генна
1234 кедр остиндский; цедрела душистая
1235 пихта киликийская
1236 дерево хинное; цинхона
1237 пепельник; цинерария
1238 дерево коричное; коричник; лавр коричный; лавр цейлонский; циннамомум
1239 девясил блоший; девясил растопыренный
1240 роза коричная; роза майская; шиповник коричневый; шиповник коричный
1241 диоскорея батат; ямс китайский
1242 лапчатка; потентилла
1243 лимон дикий; цедрат; цитрон; цитрус медийский
1244 нард; цитронелла
1245 арбус кормовой; арбус цукатный
1246 цитрус
1247 цитрусовые
1248 кладантус
1249 кладония
1250 смолка клейкая
1251 додонея аптечная
1252 робиния клейкая
1253 кларкия; эухаридиум
1254 шалфей мускатный
1255 коровяк зопниковидный; коровяк лекарственный; коровяк мохнатый; коровяк шерстистый
1256 клоповник пронзённолистный; ласковец
1257 стрептопус стеблеобъемлющий
1258 спорынья
1259 пилея
1260 жгунец; клематис; княжик; лозинка; ломонос
1261 клетровые
1262 пеллея
1263 лапчатка скальная
1264 джемсия американская
1265 энтада ползучая
1266 лигодий; лигодиум
1267 фикус карликовый
1268 микания лазающая; микания ползучая; микания цепкая
1269 гортензия черешковая
1270 говорушка
1271 ложнопокровница
1272 тунбергия
1273 морошка (приземистая)
1274 говорушка серая
1275 корица гвоздичная
1276 смородина душистая
1277 льнянка двураздельная
1278 клевер; хмелёк
1279 заразиха малая
1280 повилика клеверная
1281 колюрия
1282 заразиха гвоздичная

1283 дерево гвоздичное
1284 булавоносец
1285 ликоподий; мох ползучий; мох стелющийся; плаун
1286 ликоподиевые; плауновые
1287 говорушка булавоногая
1288 ежовка; пшеница-ежовка; пшеница карликовая; пшеница круглозёрная; пшеница мягкая карликовая; пшеница мягкая яровая; пшеница плотная
1289 вероника длиннолистная
1290 клюзия
1291 горечавка Клюзия
1292 клевер клубковидный; клевер скрученный
1293 щавель клубковатый
1294 мальва красивая; просвирняк китайский; просвирняк красивый; просвирняк мутовчатый; просвирник мутовчатый
1295 диоклея
1296 сосна приморская
1297 щитница
1298 ива длинноножковая
1299 казуарина аргентинская
1300 акация Бейли
1301 кобея
1302 кобрезия
1303 кокаиновые
1304 эритроксилон
1305 кактус кошенильный; опунция кошенильная
1306 дурнишник; ксантиум
1307 целозия
1308 погремок петушиный гребешок
1309 ежа
1310 ежовник
1311 эритрина петушья
1312 крапивна
1313 боярышник петуший гребень; боярышник петушья шпора; боярышник шпорцевый
1314 кокос; орех кокосовый; пальма кокосовая
1315 слива кокосовая; хризобаланус
1316 кодиум
1317 целогине
1318 дерево кофейное; кофе
1319 бундук; гимнокладус; гимноклядус
1320 императа цилиндрическая
1321 зизифус ямайский
1322 стеблелист
1323 кола
1324 плющ колхидский
1325 колеус
1326 катран; крамбе; крамбекатран
1327 клён весёлый; клён светлый
1328 ветры волчьи; дождевик "заячья картошка"
1329 негниючник колёсиковидный
1330 коллеция
1331 коллинзия; коллинсия
1332 колломия
1333 коллибия; монетка
1334 арбуз горький; арбуз дикий; арбуз колоцинт; колоквинт; колоцинт; тыква горькая; яблоко заморское
1335 полевица волосовидная; полевица обыкновенная; полевица тонкая
1336 смородина колорадская
1337 сосна съедобная
1338 ель голубая; ель колючая
1339 мать-и-мачеха
1340 колубрина
1341 аквилегия; водосбор; акилей; колокольчики водосборные; орлики
1342 василис(т)ник водосборолистный
1343 кольза
1344 комбретум
1345 окопник
1346 ужовник обыкновенный; язычник
1347 сантполия; фиалка африканская
1348 парило; репейник аптечный; репейник обыкновенный; репейничек аптечный; репешок обыкновенный; репяшок обыкновенный
1349 леснина; яблоко-кислица; яблоня дикая; яблоня домашняя; яблоня культурная; яблоня карликовая; яблоня лесная; яблоня низкая; яблоня низкорослая;яблоня садовая
1350 абрикос; абрикос дикий; абрикос обыкновенный; абрикос культурный
1351 спаржа аптечная; спаржа лекарственная; спаржа обыкновенная;спаржа овощная
1352 аспидистра высокая; плектогине
1353 безвременник осенний; зимовник осенний; крокус осенний; осенник; песобой; шафран луговой
1354 гравилат городской; гравилат обыкновенный; любим-трава
1355 балата
1356 мелисса аптечная; мелисса лекарственная; мята лимонная (аптечная); трава пчелиная
1357 бамбуза обыкновенная
1358 банан браминов; бана десертный; банан яблочный
1359 дрожжи пивные (верхового брожения)
1360 свёкла обыкновенная
1361 буквица аптечная; буквица лекарственная; бетоника
1362 пузырник древовидный; пузырник обыкновенный
1363 дицентра великолепная; дицентра китайская
1364 сизюринхий узколистный; сисюринхий узколистный
1365 вахта трёхлистная; вахта трилистная; третьина; трилистник водяной; трифоль; трефоль
1366 бораго лекарственный; бурачник лекарственный; огуречник аптечный; огуречник лекарственный; трава огуречная лекарственная
1367 дерево камфарное; дерево камфорное (зондское); дриобаланопс душистый
1368 буксус обыкновенный; пальма кавказская; самшит вечнозелёный
1369 жестёр слабительный; жостер слабительный; крушина слабительная
1370 гречиха культурная; гречиха посевная
1371 анхуза лекарственная
1372 грыжник гладкий; грыжник голый
1373 дерево горлянковое обыкновенное
1374 арум кукушечный; рихардия африкан-

ская; калла
1375 квамаш съедобный
1376 камелия обыкновенная; камелия японская
1377 каперцы колючие; каперсник колючий
1378 акажу; анакард(ия) акажу; анакард западный; анакардия западная; кажу; орех кешью
1379 аррорут бразильский; кассава горькая; маниок горький; маниок полезнейший; маниок съедобный; маниот съедобный; тапиока
1380 рогоз широколистный; тырса
1381 цикорий дикий; цикорий корневой; цикорий обыкновенный; цикорий полевой
1382 астра китайская; астра садовая китайская; каллистефус китайский
1383 куннингамия китайская; куннингамия ланцетная
1384 вишня виргинская; черёмуха виргинская
1385 пепельник багровый
1386 колеус Блумего; колеус пёстрый
1387 капуста морская; катран перистый; катран приморский
1388 мать-и-мачеха обыкновенная; трава камчужная
1389 окопник аптечный; окопник лекарственный
1390 дерево коралловое
1391 куколь обыкновенный; куколь полевой; куколь посевной
1392 космос раздельнолистный
1393 вигна китайская; горох кормовой; горох коровий; долихос китайский
1394 сирень индийская
1395 крокус весенний; крокус голландский; шафран весенний
1396 жабник германский; жабник немецкий
1397 лженарцисс; нарцисс ложнонарциссовый; нарцисс ложный; нарцисс трубчатый; псевдонарцисс
1398 одуванчик аптечный; одуванчик лекарственный; одуванчик обыкновенный
1399 укроп душистый; укроп огородный; укроп пахучий
1400 повилика тимьянная; повилика тимьяновая; повилика чебрецовая
1401 дугласия тис(с)олистная; лжетсуга тис(с)олистная; пихта Дугласова
1402 драцена южная; кордилина кустарниковая
1403 ряска малая; ряска маленькая; ряска меньшая
1404 аристолохия крупнолистная; кирказон крупнолистный
1405 ложнодождевик обыкновенный
1406 эдельвейс альпийский
1407 взморник морской; зостера большая; зостера морская; камка; трава морская
1408 ослинник дву(х)летний; рапутник; свеча ночная; энотера дву(х)летняя
1409 резак обыкновенный
1410 гребень петуший; гребешок петуший; целозия гребенчатая
1411 укроп аптечный; фенхель аптечный; фенхель лекарственный; фенхель обыкновенный
1412 дерево фиговое; инжир; смоковница обыкновенная; фига; ягода винная
1413 зелье солодкое; зелье солодьковое
1414 квит; хеномелес лагенария
1415 мирабилис Ялапа; ночецвет(ник); красавица ночная
1416 наперстянка красная; наперстянка лекарственная; наперстянка пурпурная; наперстянка пурпуровая
1417 петрушка зеленная; петрушка обыкновенная; петрушка огородная; петрушка садовая
1418 ферула обыкновенная
1419 имбирь лекарственный
1420 шпажник большой; шпажник обыкновенный
1421 амарант шарообразный; гомфрена головчатая; гомфрена шаровидная
1422 авдотка; купава европейская; купальница европейская
1423 мордовник круглоголовый; мордовник шароголовый; мордовник широкоголовый
1424 глоксиния
1425 галега лекарственная; козлятник аптечный; козлятник лекарственный; трава козья
1426 дрок колючий европейский; улекс европейский
1427 лук гадючий гроздевидный
1428 воробейник аптечный; воробейник лекарственный
1429 котяник; котовник ползучий; будра плющевидная
1430 крестовник обыкновенный
1431 гойяба
1432 диптерокарпус бальзамический
1433 гипсофила изящная
1434 каркас западный; цельтис западный
1435 лён кукушкин обыкновенный; политрихум обыкновенный
1436 гелиотроп перувианский
1437 ветреница-перелеска
1438 горичник аптечный; горичник лекарственный
1439 гледичия колючая; гледичия обыкновенная; гледичия сладкая; гледичия трёхколючковая; гледичия трёхостренная; гледичия трёхшипная
1440 большеголовник альпийский; хмель обыкновенный; хмелица
1441 вязовик трёхлистный; кожанка (трёхлистная); птелея трёхлистная
1442 мята конская; шандра белая; шандра обыкновенная
1443 каштан конский обыкновенный
1444 корень чёрный; чернокорень аптечный; чернокорень лекарственный
1445 гиацинт восточный
1446 ксерантемум однолетний; сухоцвет однолетний
1447 пальма костяная; пальма слоновая; фителефас крупноплодный
1448 канавалия саблевидная

1449 асфоделина жёлтая; златок
1450 жасмин аптечный; жасмин белый;жасмин лекарственный; жасмин настоящий
1451 зопник кустарниковидный; зопник кустарниковый; шалфей иерусалимский
1452 зизифус обыкновенный; унаби обыкновенное; унаби юйюба; финик китайский; ююба унаби; ягода грудная европейская; ягода грудная французская
1453 можжевельник обыкновенный
1454 манжетка обыкновенная
1455 скрученник осенний
1456 лавр дикий; лавровишня аптечная; лавровишня лекарственная; лавровишня обыкновенная
1457 чечевица культурная; чечевица обыкновенная; чечевица пищевая; чечевица съедобная
1458 клевер японский; куммеровия полосатая; леспедеза полосатая; леспедеца полосатая
1459 лакричник голый; солодка гладкая; солодка голая
1460 гваяк; гваякум лекарственный; дерево гваяковое; дерево бакаутовое (железное)
1461 сирень обыкновенная
1462 медуница аптечная; медуница лекарственная; трава медвежья
1463 крап(п); марена красильная;марсин
1464 манго индийский
1465 маршанция многообразная; мох обыкновенный печёночный
1466 калужница болотная
1467 альгароба; дерево москитовое; мимозка серёжкоцветная; прозопис серёжкоцветный
1468 резеда душистая; резеда пахучая
1469 ваточник сирийский; ласточник
1470 истод обыкновенный
1471 губастик крапчатый;мимулюс жёлтый
1472 луносемянник канадский; мениспермум канадский; плющ канадский
1473 сморчок настоящий; сморчок обыкновенный
1474 вьюнок пурпурный; вьюнок садовый; ипомея пурпурная; фарбитис пурпурный
1475 пустырник обыкновенный; пустырник сердечный; трава сердечная
1476 шампиньон культивируемый; шампиньон обыкновенный
1477 дерево мирровое; комифора мира
1478 капуцин большой; красоля; кресс испанский; настурция большая; настурция высокая; настурция садовая
1479 бородавник обыкновенный
1480 дерево мускатное (душистое); мускат; орех мускатный; орех мушкатный
1481 овёс посевной
1482 олеандр обыкновенный
1483 дерево оливковое; маслина европейская; маслина культурная
1484 душица обыкновенная; майоран дикий; майоран зимний
1485 азимина трёхлопастная
1486 бруссонетия бумажная; бруссонеция бумажная; дерево бумажное;шелковица бумажная
1487 груша дикая; груша обыкновенная
1488 анафалис жемчужный; бисерница
1489 щитолистник обыкновенный
1490 пеон аптечный; пеон лекарственный; пион аптечный; пион лекарственный
1491 гайлярдия крупноцветная
1492 перилла базиликовая; перилла многолетняя
1493 барвинок малый; гроб-трава
1494 хурма виргинская
1495 петунья гибридная
1496 зимолюбка зонтичная
1497 пиптадения обыкновенная
1498 бузгунча; дерево фисташковое; фисташка настоящая
1499 саррацения пурпурная
1500 молочай красивейший; молочай прекрасный; поинсеттия; поинсеция
1501 сумах канадский; сумах укореняющийся; сумах ядовитый
1502 лаконос американский; фитолакка американская;фитолякка американская;фитолякка десятитычинковая
1503 корень зобный; многоножка обыкновенная; папоротник сладкий; папоротник солодковый; ранник;сладкокорень
1504 занникелия болотная
1505 портулак декоративный; портулак крупноцветковый; портулак крупноцветный
1506 зантоксилум американский; перец американский
1507 опунция обыкновенная; опунция одноиглистая
1508 бирючина обыкновенная
1509 портулак овощной; портулак огородный
1510 бессмертник белый; лапка кошачья двудомная
1511 тиллея водная; тиллея водяная
1512 айва обыкновенная; айва продолговатая
1513 амброзия полыннолистная
1514 тростник обыкновенный
1515 нежник обыкновенный; солнцецвет монетолистный; солнцецвет обыкновенный
1516 вейнрута; рута душистая; рута обыкновенная; рута пахучая; рута садовая
1517 ситник расходящийся
1518 зольник; курай обыкновенный; поташник; солянка калийная; солянка русская
1519 клевер турецкий; эспарцет виколистный; эспарцет посевной
1520 сассафрас лекарственный
1521 панданус полезный; вакуа
1522 ложечница аптечная; ложечница арктическая; ложечница лекарственная; трава ложечная лекарственная; трава цынготная
1523 облепиха крушиновая;облепиха крушиновидная;облепиха обыкновенная

1524 салат морской
1525 черноголовка обыкновенная
1526 птиценожка; сераделла посевная
1527 дерево жёлтое; дерево физетовое; желтинник; рай-дерево; скумпия кожевенная; сумах венецианский; сумах дубильный
1528 зев львиный большой; зев львиный садовый
1529 жимолость виргинская; снежноягодник белый; снежноягодник ветвистый; снежноягодник кистевидный; снежноягодник кистистый; ягода снежная кистистая
1530 подснежник белый; подснежник снеговой; подснежник снежный
1531 осот овощной; осот огородный
1532 вероника полевая
1533 ситняг болотный
1534 платицериум
1535 брандушки белые; птицемлечник зонтичный
1536 арум змеиный
1537 зверобой обыкновенный; зверобой пронзённолистный; зверобой пронзённый
1538 левкой обыкновенный; левкой садовый; левкой седой; маттиола садовая
1539 подсолнечник культурный масличный; подсолнечник обыкновенный; подсолнечник однолетний
1540 каликантус флоридский; каликантус цветущий
1541 нефролепис возвышенный
1542 пуговичник; пижма дикая рябина; пижма обыкновенная; поповник обыкновенный; рябина дикая; рябишник
1543 куст чайный (китайский)
1544 дерево джатовое; дерево тиковое; тик
1545 армерия приморская
1546 тимьян душистый; тимьян обыкновенный; чабрец песчаный
1547 тигридия павония
1548 табак виргинский; табак обыкновенный; табак турецкий
1549 помидор съедобный; томат культурный
1550 кампсис укореняющийся; текома укореняющаяся; трубоцвет
1551 имбирь жёлтый; куркума длинная
1552 болдырьян; валериана аптечная;валериана лекарственная;валерьяна аптечная; маун аптечный
1553 бухарник шерстистый; трава медовая
1554 вика кормовая; вика обыкновенная; вика плоско-семянная; вика посевная; вика сорнополевая; вика яровая; горошек посевной
1555 румянка обыкновенная; синяк обыкновенный
1556 желтофиоль обыкновенный; желтофиоль садовый; лакфиоль садовый; хейрантус садовый
1557 гиацинт водный; гиацинт водяной
1558 дерево восковое; плющ восковой
1559 квебрахо белое
1560 падуб мутовчатый
1561 просо волосовидное
1562 гамамелис виргинский
1563 дрок красильный
1564 абсент; абсинт; вермут; глистник; полынь-абсент; полынь горькая
1565 миллефоль; тысячелистник обыкновенный
1566 кислица прямая
1567 зелёнка пронзённолистная
1568 цинния изящная; цинния стройная
1569 губка разноцветная
1570 сложноцветные
1571 кондалия
1572 кондуранго
1573 рудбекия
1574 бобы земляные; воандзея подземная; горох земляной; орех земляной бамбарский
1575 кассия узколистная
1576 смолёвка коническая
1577 гигрофор конический
1578 сморчок конический
1579 хвойные
1580 конъюгаты; сцеплянки
1581 замия; цамия
1582 копайба; копайфера лекарственная
1583 копайфера
1584 коперниция
1585 акалифа
1586 навозниковые
1587 эритрина
1588 грибы рогатиковые; рогатиковые
1589 ежовик коралловидный
1590 пеон коралловый
1591 ладьян
1592 зубянка клубненосная; зубянка луковичная
1593 спартина
1594 кордия
1595 осока длиннокорневищная; осока струннокоренная; осока струнокорневая
1596 ленок; лептосине; кореопсис
1597 киндза; киндзя; кишнец; кориандр
1598 зелье вонючее; кишнец посевной; кориандр посевной
1599 кориария
1600 кориариевые
1601 пихта аризонская
1602 берест пробковый; вяз пробковый; ильм пробковый; карагач
1603 дуб пробковый
1604 бархат; дерево пробковое
1605 лейтнерия
1606 дюбуазия
1607 подмаренник трёхрогий
1608 лютик полевой
1609 златоцвет посевной; нивяник посевной; поповник посевной; ромашка посевная
1610 агростемма; куколь
1611 дёрен мужской; дёрен обыкновенный; дёрен съедобный; кизил мужской; кизил настоящий; кизил обыкновенный; кизил съедобный
1612 шпажник посевной
1613 василёк голубой; василёк обыкновенный; василёк полевой; василёк посевной; василёк синий

1614 воробейник полевой
1615 вяз граболистный роговидный
1616 мак самосейка
1617 свербига
1618 валерианелла; валериан(н)ица; валерьян(н)ица
1619 зев львиный полевой
1620 торица полевая; шпергель
1621 корнутия
1622 корригиоля
1623 мох караковый; мох корсиканский глистогонный
1624 мшанка шилолистная
1625 сосна чёрная калабрийская
1626 песчанка балеарская
1627 роза щитконосная; шиповник щитконосный
1628 хохлатка
1629 корифа
1630 космос
1631 псидиум лавролистный
1632 калуфер; колуфер; кануфер; конуфер; пижма душистая; пижма кануфер; пижма пахучая; пиретрум бальзамический; хризантема бальзамическая
1633 кизильник
1634 роза белая
1635 хлопчатник
1636 лопух войлочный;лопух паутинистый; лопух паутинский; лопушник паутинистый
1637 пушица
1638 онопордум; татарник
1639 котиледон
1640 пырей
1641 чубушник Культера
1642 сосна Культера
1643 колокольчик крапиволистный; трава горляная
1644 бобы бархатные
1645 брусника
1646 дерево коровье; дерево молочное
1647 кубышка
1648 бутень одуряющий
1649 борщев(н)ик
1650 вигна
1651 медуница узколистная
1652 ключики; первоцвет аптечный;первоцвет весенний; первоцвет жёлтый; первоцвет истинный; первоцвет лекарственный; примула аптечная; примула настоящая
1653 мыльнянка коровья;тысячеголов(ник)
1654 дерево молочное индийское
1655 марьянник
1656 шлюмбергерия
1657 карапа
1658 клюква
1659 абобра
1660 абобра тонколистная
1661 крассула; толстянка
1662 остролодка; остролодочник
1663 скабиоза бледно-жёлтая; скабиоза жёлтая; скабиоза желтоватая;скабиоза светло-жёлтая
1664 виноград девичий; виноград дикий
1665 полевица болотная; полевица ползучая; полевица приморская
1666 лютик ползучий
1667 лапчатка ползучая
1668 лисохвост вздутый
1669 воробейник пурпурно-синий
1670 качим ползучий
1671 можжевельник распростёртый
1672 омфалодес настоящий; пупочник весенний
1673 кислица рогатая; кислица рожковая
1674 пилюльница шариконосная
1675 гудайера ползучая
1676 овсяница песчаная
1677 зверобой распростёртый; зверобой стелющийся
1678 лап(к)а воронья простёртая
1679 ива ползучая; ива стелющаяся; ницелоз; тальник ползучий
1680 куст креозотовый
1681 гребневик гребенчатый; гребневик обыкновенный
1682 эльшольция гребенчатая
1683 баран
1684 житняк гребенчатый; житняк гребневидный; пырей гребенчатый
1685 папоротник гребенчатый; щитовник гребенчатый
1686 птерис критский
1687 ива голубая
1688 золотобородник цикадовый
1689 бедренец крымский
1690 оносма крымская
1691 плющ крымский
1692 липа крымская; липа ярко-зелёная
1693 перловник крымский
1694 клевер инкарнатный; клевер малиновый; клевер Молинери; клевер мясокрасный; клевер пунцовый
1695 схизостилис
1696 кринум
1697 лопастник курчавый
1698 мята курчавая
1699 крокус; шафран
1700 горечавка перекрестнолистная; перелёт соколий
1701 ерика крестолистная; эрика крестолистная
1702 бигнония усиковая
1703 крестовница
1704 подмаренник крестовидный
1705 кротон
1706 багновка; водян(и)ка; вороника; шикша
1707 ворониковые; ёрниковые; шикшевые
1708 подорожник перистый
1709 вербезина; вербесина
1710 хризантема увенчанная
1711 венец царский; рябчик садовый; рябчик турецкий; рябчик царский
1712 вязель
1713 вязель обыкновенный;вязель пёстрый
1714 крупина
1715 криптантус
1716 криптокорина
1717 споровые; тайнобрачные
1718 криптомерия
1719 багульник болотный; одурь болотная
1720 ктенанте
1721 пальма королевская; пальма ореодокса
1722 кубеба
1723 сердечник луговой

1724 огурец (грядовой);огурец посевной
1725 подсолнечник огурцеобразный; подсолнечник слабый
1726 дерево огуречное; магнолия длиннозаострённая; магнолия остроконечная
1727 гнафалиум; сушеница
1728 кмин
1729 кмин тминный; кмин тминовый; кумин; тмин волошский; тмин римский
1730 араукария Куннингама
1731 казуарина Куннингама
1732 купания
1733 ниремергия
1734 пецицовые
1735 шерст(н)як
1736 куфея; парсония
1737 катананхе
1738 куркулиго
1739 горец щавелелистный
1740 чертополох курчавый
1741 щавель курчавый
1742 схизеации; схизеевые; схицейные
1743 мальва курчавая; просвирник курчавый; просвирняк курчавый
1744 рдест курчавый
1745 смородина
1746 ожика изогнутая
1747 тыква мускатная;тыква мускусная
1748 ямс куш-куш; ямс трёхнадрезный
1749 качим постенный; качим стенной
1750 молочай многоцветный
1751 ан(н)она
1752 аноновые
1753 леерсия
1754 стефанандра надрезаннолистная
1755 рудбекия раздельнолистная; рудбекия рассечённая; рудбекия рассеянная; шар золотой
1756 яснотка гибридная
1757 герань рассечённая
1758 дубровник метельчатый
1759 шнитт-салат
1760 ворсянка раздельная; ворсянка раздельнолистная; ворсянка разрезная
1761 канареечник клубневой
1762 кювьера
1763 цианантус
1764 саговник; цикада
1765 саговниковые; цикадовые
1766 дряква; фиалка альпийская;цикламен
1767 циклантовые
1768 циклантера
1769 коллетия колючая
1770 телигоновые
1771 циноморий
1772 ципелла
1773 кипарис
1774 молочай кипарисный; молочай кипарисовидный; молочай кипарисовый
1775 кипарисовые
1776 сантолина волосистовойлочная; сантолина кипарисовидная; сантолина кипарисовниковидная
1777 каллитрис; сандарак
1778 кедр короткохвойный
1779 дуб ольхолистный
1780 циртантус
1781 подладанник
1782 георгина; георгиния; далия
1783 берёза даурская
1784 мирикария даурская
1785 роза даурская; шиповник даурский
1786 дуб зубчатый
1787 маргаритка
1788 олеария
1789 паспалум расширенный
1790 пиретрум цинерариелистный; поповник цинерариелистный; ромашка далматская
1791 льнянка далматская
1792 роза дамасская; роза казанлыкская роза полиантная; роза полиантовая; роза ремонтантная
1793 звёздоплодник
1794 вечерница ночная фиалка; красавица ночная
1795 агатис; даммара; каури
1796 одуванчик
1797 колокольчик сборный; колокольчик скученный; трава приточная
1798 дантония
1799 горечавка паннонская
1800 волчеягодник; волчник; лаврушка
1801 верба; ива волчниковая; синетал; шелюга (жёлтая); шелюжник
1802 схима Валлиха; шима Валлиша
1803 ятрышник обожжённый
1804 плевел одуряющий; плевел опьяняющий
1805 барбарис Дарвина
1806 колоказия съедобная; таро
1807 пальма финиковая; финик
1808 пальма финиковая культурная; финик пальчатый; финик финиколистный
1809 финик дикий; хурма обыкновенная; хурма кавказская
1810 датиска
1811 датура; дурман
1812 кетелеерия Давида
1813 фотиния Давидсона
1814 коммелина; лазорник
1815 крас(н)однев; лилейник; лилия жёлтая; хемерокаллис
1816 боб калабарский; физостигма ядовитая
1817 тапсия; трава злая
1818 зеленчук; яснотка
1819 зигаденус
1820 мухомор белый; мухомор зелёный; поганка бледная
1821 дербянка колосистая; папоротник ребристый
1822 пухонос дернистый
1823 боб бархатный
1824 богатень; лядвенец; тетрагонолобус; четырёхкрыльник
1825 лён обыкновенный с растрескивающимися коробочками; лён-прыгунец; лён-скакунец
1826 дейнболлия
1827 рыжей; рыжик (красный); рыжик сосновый
1828 адиантум клиновидный
1829 дендробиум
1830 денгамия
1831 мелкоплодник ольховолистный;мелкоплодник ольхолистный; рябина

ольхолистная
1832 зараза водяная зубчатая
1833 гренландия густая; рдест густолистный; рдест плотный
1834 донник зубчатый
1835 кукуруза зубовидная
1836 кедр гималайский
1837 гвоздика армериевидная
1838 череш; шарыш; ширяш; эремурус
1839 еремостахис; пустынноколосник; эремостахис
1840 губовник линейный
1841 водоросли десмидиевые
1842 мухомор вонючий
1843 дейция
1844 раувольфия
1845 ноголистник пальчатый; подофилл щитовидный
1846 пробосцидея
1847 эхинопанакс
1848 ямс слоновый Ривьера; аморфофаллус Ривьера
1849 аралия колючая; аралия шиповатая
1850 маслина американская
1851 диалиум
1852 диапензия; диапенсия
1853 водоросли диатомовые; водоросли кремнёвые; водоросли кремнистые; диатомеи
1854 дихапеталиум
1855 двудольные
1856 сосна белая калифорнийская; сосна Сабина
1857 укроп
1858 дил(л)ения
1859 рядовка серая
1860 осока двудомная
1861 диосма
1862 диптерокарповые
1863 дискостигма
1864 атрактилис
1865 осока расставленная
1866 бадьян; диктамнус; ясенец
1867 цезальпиния дубильная; цезальпиния душистая; цезальпиния кожевенная; цезальпиния красильная; цезальпиния центрально-американская
1868 трава заячья; щавель
1869 додартия; додарция
1870 кускута; повилика; трава войлочная
1871 повиликовые
1872 кендырь
1873 апоциновые; кутровые
1874 крапива жгучая
1875 роза собачья; шиповник обыкновенный; шиповник собачий
1876 капуста собачья; перелеска многолетняя; пролеска многолетняя; пролесник многолетний
1877 гребенник; гребневик; гребник
1878 зуб собачий обыкновенный; кандык собачий зуб; кандык южноевропейский; кендик; кендык
1879 свинорой
1880 фиалка собачья
1881 дёрен; кизил
1882 дёренные; дёреновые; килизовые
1883 долихос
1884 лунник однолетний
1885 домбея
1886 лансиум домашний; лансиум обыкновенный
1887 орех малдивский; пальма сейшельская
1888 дугласия; лжетсуга; псевдотсуга
1889 таволга Дугласа
1890 думпальма; пальма "дум"
1891 скабиоза голубиная
1892 герань мягкая; герань нежная
1893 лапчатка средняя
1894 физалис рубчиковидный
1895 пикульник пушистый
1896 осока шерстистая
1897 крупка
1898 драцена
1899 кордилина; лилия-кордилина
1900 дерево драконовое; драконник; драцена драконовая
1901 змееголовник
1902 дерево сандаловое антильское;птерокарпус драконовый
1903 дримис
1904 камнеломка поникшая
1905 смолёвка повислая
1906 спороболус; спороболюс
1907 лабазник шестилепестный; орешки земляные
1908 мухоловка
1909 золототысячник зонтичный
1910 очанка лекарственная; очанка Росткова; очанка Ростковиуса; очанка ростковская; трава очная
1911 дымянка аптечная; дымянка лекарственная
1912 дерево жёлтое; хлорофора красильная
1913 авран аптечный; авран лекарственный; зелье лихорадочное; лихорадочник; трава лихорадочная
1914 семя вшивое; сабадилла аптечная; сабадилла лекарственная
1915 стиракс аптечный; стиракс лекарственный; дерево стираксовое лекарственное
1916 купена аптечная; купена лекарственная; купена обыкновенная
1917 вероника аптечная; вероника лекарственная
1918 аир болотный; аир обыкновенный; аир тростниковый; корень ирный; корень чёрный
1919 флокс Друм(м)онда;флокс однолетний
1920 дриада; трава куропаточья
1921 вязовик; пестрец; трутовик пёстрый
1922 дубоизия
1923 многокоренник
1924 ряска
1925 рясковые
1926 горец вьюнковый; горец вьющийся; повитель
1927 дуриан; дурьян
1928 дуб-железняк; дуб зимний; дуб-зимняк; зимняк; дуб сидячецветный; дуб скальный
1929 сорго зерновое; сорго комовое; сорго скученное
1930 арнаутка; гарновка; кубанка; пшеница твёрдая
1931 герань красно-бурая;герань тёмная
1932 аристолохия; кирказон

1933 кумаруна
1934 лук стареющий(ся)
1935 берёза карликовая; ёра полярная; ёрник; стланец березовый; стланик березовый
1936 банан Кавендиша; банан карликовый; банан китайский
1937 дёрен шведский; кизил шведский
1938 сушеница лежачая
1939 бересклет карликовый; бересклет низкий
1940 панакс трёхлистниковый
1941 вьюнок трёхцветный; красавица дневная
1942 утёсник карликовый; утёсник малый
1943 касатик крымский; касатик маленький; касатик низкий; касатик равнодольный; касатик равнолопастный; петушки степные; петушок степной
1944 офиопогон японский
1945 мальва круглолистная;просвирняк низкий;просвирняк круглолистный
1946 астранция малая
1947 можжевелоягодник; разумовския
1948 сабаль Адансона; сабаль малый
1949 ирга колосистая; ирга метельчатая
1950 болотница маленькая
1951 молочай маленький; молочай малый
1952 кипрей(ник) тёмно-зелёный
1953 дикия
1954 алканна красильная
1955 серпуха красильная
1956 вайда красильная; синильник
1957 ясменник красильный
1958 дерево алойное; каламбак
1959 анхуза Баррелиера
1960 ладьян трёхнадрезный
1961 дейция крупноцветковая
1962 незабудка жестковолосистая; незабудка холмовая
1963 аира ранняя
1964 чешуйчатка раннеспелая
1965 кудряш
1966 сурепица весенняя; сурепка весенняя
1967 энтеролобиум
1968 трясинник
1969 звездовик; звёздочка земляная
1970 геоглоссум
1971 звездчатка ланцетовидная; звездчатка ланцетолистная
1972 лилия длинноцветковая; лилия длинноцветная
1973 негниючка; туя западная
1974 бакхарис галимолистный
1975 орех чёрный (восточный)
1976 боярышник круглолистный
1977 лиственница (восточная) американская
1978 тополь дельтовидный; тополь канадский; тополь треугольнолистный
1979 багрянник канадский
1980 дерево карандашное; можжевельник американский; можжевельник виргинский
1981 бересклет тёмно-багряный; бересклет тёмно-пурпуровый
1982 сосна Веймутова
1983 аррорут восточноиндийский; аррорут остиндский; куркума узколистная
1984 дальбергия широколистная; дерево розовое индийское
1985 бородач кровеостанавливающий
1986 хурмовые; эбеновые
1987 дерево чёрное; дерево эбеновое; хурма эбенова
1988 эхеверия
1989 эхинацея; эхинация
1990 эхинокактус
1991 эхиноцереус
1992 леонтоподиум; эдельвейс
1993 аррорут квинслендский; канна съедобная
1994 строчок обыкновенный; строчок съедобный
1995 рябина Мугеотта
1996 огурец змеевидный; огурец змеиный; трихозант(ес)
1997 взморник; зостера
1998 взморниковые; зостеровые
1999 тайник овальнолистный; тайник овальный
2000 берсим; клевер александрийский; клевер берсим;клевер египетский
2001 гифене; думпальма настоящая; пальма-гифаена
2002 ива египетская
2003 тюльпан Эйхлера
2004 однозернянка (культурная); пшеница культурная однозернянка;пшеница однозернянка; пшеница однозернянка культурная
2005 лофира африканская; лофира крупнолистная
2006 лох
2007 ива сизоватая
2008 элеокарпус
2009 бузина; самбук
2010 девясил большой; девясил высокий; девясил обыкновенный
2011 маслёнок лиственный; маслёнок лиственничный; масляник лиственный; масляник лиственничный
2012 ферония слоновая
2013 колоказия; колокасия; таро
2014 элефантопус
2015 елевзина; елевзине; элевзина
2016 диерама
2017 боярышник Эльвангера
2018 берест; вяз; ильм
2019 вязовые; ильмовые
2020 ежевика вязолистная; ежевика ильмолистная
2021 дзельква граболистная
2022 зараза водяная; чума водяная; элодея
2023 осока удлинённая
2024 эльсгольция; эльшольция
2025 эмбелия
2026 филлантус лекарственный
2027 эмботриум
2028 сыроежка едкая; сыроежка жгучеедкая
2029 двузернянка (культурная); полба волжская; полба двузёрная; полба ненастоящая; полба обыкновенная; полуполба; пшеница двузернянка; пшеница волжская; эммер

2030 овения
2031 двулепестник; колдун-трава; трава колдуновая; цирцея; черноцвет
2032 салат эндивий; цикорий зимний;цикорий салатный; цикорий-эндивий; салат цикорный;эндивий (зимний)
2033 сосна Энгельмана
2034 ель Энгельмана
2035 диптерокарпус шишковатый
2036 бук Энглера
2037 маргаритка многолетняя
2038 боярышник колючий; боярышник обыкновенный
2039 остролист; падуб остролистный
2040 блющ обыкновенный; плющ обыкновенный
2041 дуб обыкновенный; дуб летний; дуб черешчатый
2042 первоцвет бесстебельный; первоцвет обыкновенный; примула обыкновенная
2043 ложечница английская
2044 росянка английская
2045 тис европейский; тис ягодный
2046 энтада
2047 грибы энтомофторовые
2048 эпакрис
2049 бесцветник; горянка
2050 дремлик
2051 зелёнка; зеленушка; рядовка зелёная
2052 сосна калабрийская
2053 ржанец; рожки маточные; рожки чёрные
2054 эриогонум
2055 эриофиллум
2056 незабудочник
2057 берёза каменная; берёза Эрмана
2058 синеголовник
2059 эритеа
2060 эскал(л)ония
2061 спарта; альфа-трава
2062 альфа; ковыль тончайший; спарто; трава альфа; эспарто
2063 гвизоция абиссинская; нуг масличный; подсолнечник абиссинский; рамтила
2064 пшеница абиссинская
2065 эвкалипт
2066 эвхарис
2067 крепкоплодник
2068 евгения
2069 бересклет; клещина
2070 молоко волчье; молочай; эуфорбия
2071 купена многоцветковая; купена многоцветная
2072 ольха клейкая; ольха чёрная
2073 ясень высокий; ясень европейский; ясень обыкновенный
2074 осина; тополь дрожащий
2075 барбарис обыкновенный
2076 песколюб песчаный; песколюбка песчаная
2077 бук европейский; бук лесной
2078 черёмуха (кистевая); черёмуха обыкновенная
2079 горец змеиный; горлец; гречиха горлец; змеевик; корень змеиный; шейки раковые; шейки рачьи
2080 ежевика кустарниковая; куманика; росяника
2081 смородина чёрная
2082 боярышник чёрный
2083 клекачка перистая; клокичка перистая; стафилея перистая
2084 люпин волосистый; люпин жестководосистый; люпин жесткоопушённый; люпин лохматый
2085 зюзник европейский; крапчатка
2086 каштан благородный; каштан европейский; каштан настоящий;каштан обыкновенный; каштан посевной; каштан съедобный
2087 водосбор обыкновенный; голубки;колокольчики садовые; коломбина; орлики обыкновенные
2088 валерианелла колосковая; валерианелла малая; валерианица малая; валерьянница малая; валерьянница огородная; маш-салат; салат полевой
2089 кизильник обыкновенный; кизильник цельнокрайный
2090 кубышка жёлтая
2091 калина красная; калина обыкновенная
2092 дряква европейская
2093 ежевика сизая; куманика; ожина
2094 волоснец песчаный; колосняк песчаный; песчаник; рожь полярная; ячмень гигантский; ячмень песчаный
2095 бузина чёрная
2096 бересклет европейский; бересклет обыкновенный
2097 перо водяное; турча болотная
2098 лещина обыкновенная; орех лесной американский; орешник (лесной); орешник обыкновенный
2099 жимолость лесная; жимолость обыкновенная; жимолость пушистая; ягода волчья
2100 берёзка; вьюнок полевой
2101 золотарник "золотая розга"; золотарник обыкновенный; золотель; розга золотая; лоза золотая
2102 астра льнолистная; грудница обыкновенная; ленок (обыкновенный); солонечник ленок
2103 крыжовник европейский; крыжовник культурный
2104 виноград винный; виноград виноносный; виноград винородный; виноград евразиатский; виноград европейский; виноград культурный; виноград настоящий; виноград старосветский
2105 ольха горная
2106 дерево крапивное; каркас южный
2107 хмелеграб обыкновенный
2108 граб европейский;граб обыкновенный
2109 башмачок венерин настоящий
2110 лиственница европейская; лиственница опадающая
2111 липа европейская; липа голландская; липа обыкновенная
2112 лабазник вязолистный
2113 дубянка; омела белая
2114 бурка; рябина ликёрная; рябина обыкновенная

2115 ковыль обыкновенный; ковыль перистый
2116 пульсатилла
2117 мытник болотный; трава вшивая
2118 клоповник широколистный; кресс широколистный; хрен солнечный
2119 грушанка круглолистная
2120 бузина кистистая; бузина красная; бузина обыкновенная
2121 подлесник европейский
2122 корень пьяный; скополия карниолийская
2123 подорожник морской; подорожник приморский
2124 прибрежник одноцветковый; прибрежница одноцветковая
2125 седмичник европейский
2126 липучка ежёвая; липучка ежовая
2127 земляника лесная
2128 валонея; дуб австрийский; дуб турецкий; дуб бургундский
2129 вербена аптечная; вербена лекарственная
2130 бешеница водяная; вех ядовитый; цикута ядовитая
2131 берёза (белая) плакучая; берёза бородавчатая; берёза обыкновенная; берёза повислая
2132 кувшинка белая; лилия водяная
2133 копытень европейский;ладан конский
2134 ветреница дубравная
2135 люпин жёлтый
2136 фиалка жёлтая
2137 эврикома
2138 пальма капустная
2139 эвакс
2140 дрёма белая; дрёма беловатая;дрёма луговая; лихнис белый
2141 онагра; ослинник; энотера
2142 кипрейные; онагриковые; онагровые; ослинниковые
2143 роза месячная
2144 иберийка вечнозелёная; стенник вечнозелёный
2145 кастанопсис
2146 бересклет японский
2147 шиповник вечнозелёный
2148 очиток гибридный
2149 гелихризум; цмин
2150 эверния
2151 эводия
2152 очанка
2153 мак Аргемона; мак колючий
2154 гриб гвоздичный; кольцо ведьмино; опёнок луговой
2155 резак
2156 володушка серповидная
2157 чубушник Фальконера
2158 штернбергия жёлтая
2159 кульбаба осенняя
2160 икотник
2161 туевик
2162 тофильдия
2163 нотофагус
2164 коротконожка
2165 лисичка ложная
2166 прицепник
2167 кипарисник; кипарисовик;лжекипарис
2168 рыжей; рыжик
2169 лаперузия
2170 чемерица
2171 аморфа
2172 гил(л)ения
2173 мальвacтрум
2174 бемерия; бомерия; рами
2175 орех мускатный фальшивый
2176 трищетинник
2177 элеодендрон
2178 хедеома
2179 линдерния
2180 рябинник; сорбария
2181 мирикария
2182 адиантум нежный
2183 ливистона; пальма веерная Ливистона
2184 кабомба
2185 фарзетия; фарсетия
2186 лебеда поникшая; лебеда раскидистая
2187 фатсия
2188 веснянка
2189 гриб олений; плютей олений
2190 зуб собачий; кандык; эритрониум
2191 пихта Факсона
2192 целозия серебристая
2193 турча
2194 волкобой; волчеягодник волчье лыко; волчеягодник обыкновенный; волчник обыкновенный; волчник смертельный; корень волчий; лыко волчье; пережуй-лычко; сирень лесная; ягодки
2195 фейхоа; трава ананасная
2196 фелиция
2197 кругонос; папоротник язычный
2198 плеврогина
2199 фендлера
2200 укроп волошский; фенхель
2201 нигелла; чернушка
2202 рдест гребенчатый
2203 пажитник греческое сено; пажитник сенной; клевер греческий; сено греческое; шамбала
2204 аспарагус пушистоперистый; спаржа плюмажная; спаржа пушистая; спаржа пушистоперистая
2205 стангерия
2206 папоротники
2207 ферульник; ферулярго
2208 вульпия; овсяница
2209 бычок; валуй; сыроежка вонючая; сыроежка ядовитая
2210 златоцвет девичий; пиретрум девичий; поповник девичий; ромашка девичья; трава маточная
2211 пэдерия
2212 осока малоцветковая; осока мелкоцветковая
2213 болотница малоцветковая
2214 лён белосемянный; лён долгунец;лён культурный; лён-кудряш; лён обыкновенный; лён посевной
2215 лён новозеландский
2216 щавель красивый
2217 амзинцкия
2218 лиродревесник;цитарексилум
2219 костёр полевой
2220 пупавка полевая
2221 ясколка дернистая
2222 незабудка пашенная; незабудка по-

левая; незабудка средняя
2223 лук виноградный
2224 ястребинка луговая
2225 хвощ полевой
2226 манжетка полевая; невзрачница (полевая)
2227 жерардия; шерардия
2228 жерардия полевая; шерардия полевая
2229 мята полевая
2230 чистец полевой
2231 горох полевой; пелюшка
2232 вередник; денежник; ярутка полевая
2233 клоповник полевой; кресс полевой
2234 роза пашенная; роза полевая
2235 короставник полевой
2236 осот жёлтый; осот полевой
2237 вероника пашенная
2238 фиалка полевая
2239 ожика полевая; ожика равнинная
2240 инжир; смоква; смоковница
2241 лютик-чистяк; чистяк весенний; чистяк лютичный
2242 буквица водяная; норичник
2243 норичниковые
2244 аррорут индийский; нианца; цванга
2245 гифомицеты
2246 лещина; орех лесной; орешник
2247 тонковласник
2248 тонколистник
2249 гименофилловые; тонколистниковые
2250 клевер узколистный
2251 омежник водный; омежник водяной; раздулка; укроп водяной
2252 осока пальчатая
2253 вероника трёхлистная
2254 росичка
2255 гвоздика песчаная
2256 ель финская
2257 пихта
2258 плаун-баранец
2259 цаушнерия
2260 кассиопа четырёхрядная
2261 пираканта
2262 иван-чай; кипрей узколистный; конопля лесная; хаменерий узколистный; чай капорский
2263 стенокарпус
2264 анамирта
2265 кариота; пальма-кариота
2266 дробянки
2267 фитцро(й)я
2268 сумах пятилистниковый
2269 гребенщик пятитычинковый; тамарикс пятитычинковый
2270 флакоуртия; флакуртия
2271 пихта чешуйчатая
2272 адонис огненный; адонис пламенный; горицвет огненный
2273 анемона блестящая
2274 стеркулия кленолистная
2275 сумах копаловый
2276 коровяк медвежье ухо; коровяк обыкновенный; ухо медвежье
2277 пиреномицеты
2278 чина лесная
2279 чина красная
2280 сыть; циперус
2281 орех сердцевидный; орех японский
2282 перец Бунге; перец китайский
2283 рдест взморниколистный; рдест зостеролистный
2284 блисмус сжатый
2285 флаверия
2286 лён
2287 повилика льновая; повилика льняная
2288 лёновые; льновые
2289 радиола
2290 подорожник блошный; подорожник песчаный; колосеница; семя блошное
2291 блошница; пуликария
2292 мелколепестник; стенактис; тонколучник; эригерон; блошник
2293 осока блошная
2294 флиндерсия
2295 кукуруза кремнистая
2296 дескурения Софии
2297 болот(н)оцветник; лимнантемум; нимфейник
2298 рдест плавающий
2299 элизма плавающая
2300 фенхель итальянский; фенхель флорентинский
2301 касатик флорентийский
2302 тюльпан лесной
2303 торрея тис(с)олистная
2304 дурман древесный; дурман древовидный
2305 шпорник многолетний; дельфиниум гибридный; живокость гибридная
2306 пиретрум красный; пиретрум мясокрасный; пиретрум розовый; пиретрум ярко-красный; ромашка кавказская; ромашка (персидская) розовая
2307 кукуруза крахмалистая
2308 ясень белый; ясень маннитовый; ясень манноносный; ясень сахарный; ясень цветочный
2309 дерево корнелиевое; дёрен флоридский; дёрен цветочный; дёрен цветущий; кизил флоридский; циноксилон цветущий
2310 цветковые; явнобрачные
2311 миндаль трёхлопастный; слива трёхлопастная
2312 хеномелес
2313 сусак
2314 сусак зонтичный; хлебница
2315 сусаковые
2316 гибискус северный; гибискус троичный; гибискус тройчатый
2317 киксия
2318 жабник
2319 гриб царский; мухомор (красный)
2320 офрис мухоносная
2321 тиарелла
2322 свёкла кормовая
2323 фонтанезия
2324 пангиум съедобный
2325 зноиха; кокорыш обыкновенный; кокорыш собачья петрушка; петрушка собачья
2326 шампиньон лесной
2327 незабудка
2328 смолёвка вильчатая
2329 костенец северный

2330 василёк рогатый; васильки рогатые; грабельки; живокость полевая; живокость посевная; носики комаровы; рогульки; сокирки;топорики; черевички
2331 дикран
2332 можжевельник формозский
2333 берёза Форреста
2334 крас(н)однев рыжий; сарана
2335 форстерония
2336 ожика Форстера
2337 форзиция; форсиция
2338 кетелеерия Форчуна
2339 пальма веерная (высокая); трахикарпус высокий; трахикарпус Форчуна
2340 гумея красивая
2341 мирабилис
2342 никтагиновые; ночецветниковые; ночецветные
2343 горошек четырёхсемянный
2344 гребенщик четырёхтычинковый; тамарикс четырёхтычинковый
2345 мятлик болотный; мятлик трёхцветковый; мятлик шафранный
2346 наперстянка
2347 виноград Изабеллы; виноград лабруска; виноград североамериканский; катауба
2348 осока лисья
2349 батлачёк; батлачик; батлачок; лисехвостник; лисохвост(ник)
2350 ячмень гривастый; ячмень гривоподобный
2351 клевер красноватый;клевер красный
2352 гоми; могар; просо головчатое; просо итальянское; просо татарское; щетинник итальянский
2353 сосна Бальфура
2354 миндаль ломкий
2355 репейник пахучий; репейничек пахучий
2356 альбиц(ц)ия душистая
2357 драцена душистая
2358 сумах душистый
2359 малина душистая; малина пахучая
2360 плюмерия
2361 трава сайгачья; франкения
2362 франкениевые
2363 босвел(л)ия
2364 пихта Фразера
2365 фреезия
2366 тополь Фремонта
2367 лён французский
2368 копеечник тёмный
2369 бархатцы мелкоцветковые; бархатцы отклонённые; бархатцы распростёртые; тагетес распростёртый
2370 калликарпа
2371 роза французская; шиповник галльский; шиповник французский
2372 смолёвка английская
2373 щавель щитковидный; шавель щитковый
2374 гребенщик французский; тамарикс французский
2375 ностоковые
2376 рдест Фриса
2377 схизокодон
2378 болотноцветник кувшинковый; плавун нимфейный
2379 дерево снежное; хионантус
2380 рябчик; фритиллярия
2381 водокрас; лягушатник; лягушечник
2382 водокрас лягушечный; водокрас обыкновенный
2383 водокрасовые
2384 пололепестник зелёный
2385 баран-гриб
2386 виноград лисий
2387 крокантемум
2388 крестовник Фукса
2389 фуксия
2390 ворсянка посевная; ворсянка сукновалов; ворсянка сукновальная; шишка ворсовальная; чесалка
2391 клён японский
2392 дымянка
2393 дымянковые
2394 фунария
2395 говорушка ворончатая
2396 хлорофора
2397 дейция шероховатая; дейция шершавая
2398 симфония габонская
2399 лук гусиный
2400 гайлярдия
2401 галакс
2402 сыть длинная
2403 галипея
2404 трипсакум
2405 гарциния
2406 архангелика; ангелика аптечная; дудник лекарственный; дягиль аптечный; дягиль лекарственный
2407 бальзамин садовый; недотрога бальзаминовая; недотрога бальзаминовидная
2408 свёкла столовая
2409 кровохлёбка аптечная; кровохлёбка лекарственная; кровохлёбка обыкновенная
2410 морковь огородная;морковь посевная
2411 сельдерей лиственный; сельдерей листовой; сельдерей стеблевой; сельдерей черешковый
2412 клоповник посевной; кресс огородный; кресс посевной; кресс садовый; кресс-салат; перечник огородный; перечник посевной
2413 георгина изменчивая
2414 баклажан (обыкновенный);паслён баклажан; демьянка; бадиджан
2415 тмин чёрный; чернушка посевная
2416 гардения
2417 латук огородный; латук посевной; латук-салат; молокан посевной; салат огородный; салат посевной
2418 любисток аптечный; любисток лекарственный; зоря; корень любистоковый
2419 шампиньон двуспоровый
2420 лук культурный; лук обыкновенный; лук репчатый
2421 лебеда садовая; лебеда салатная
2422 пастернак посевной
2423 горох обыкновенный;горох посевной; горох садовый; горошек зелёный
2424 слива домашняя; слива европейская;

слива обыкновенная; чернослив
2425 редис; редька европейская; редька огородная; редька посевная
2426 рапонтика; ревень овощной; ревень огородный; ревень туполистный; ревень черноморский
2427 шалфей аптечный; шалфей лекарственный
2428 ирга круглолистная; ирга обыкновенная; ирга овалолистная; ирга овальная
2429 щавель кислый; щавель настоящий; щавель обыкновенный; щавель огородный
2430 рододендрон жестковолосистый
2431 таволга острая
2432 лук-чеснок; чеснок (посевной)
2433 чесночник аптечный; чесночник лекарственный
2434 гриб чесночный; чесночник
2435 бадьян-ясенец; ясенец белый
2436 гастерия
2437 гастеромицеты
2438 гаура
2439 лиатрис
2440 гацания
2441 гельземий; гельземиум
2442 живучка женевская
2443 генипап
2444 генипа
2445 генциана; горечавка
2446 горечавковые
2447 герань; ерань; журавельник
2448 гераневые; гераниевые; журавельниковые
2449 хвойник Жерарда; эфедра Жерарда
2450 резуха Жерара
2451 гербера
2452 ромашка аптечная; ромашка лекарственная; ромашка дикая; ромашка настоящая; ромашка обыкновенная
2453 дубровник; паклун
2454 вероника дубравка; вероника дубравная; вероника дубровка; дубравка; дубровка
2455 спирея вязолистная; спирея дубровколистная; таволга дубровколистная
2456 мирикария германская
2457 горечавка германская; горечавка немецкая
2458 ирис германский;касатик германский
2459 асперуга лежачая; острица лежачая; острица простёртая
2460 анациклус; ромашка немецкая
2461 бухарник мягкий; просяник
2462 горошек зарослевый; горошек кустарниковый
2463 дрок германский
2464 геснерия
2465 геснериевые
2466 аморфофалл(ус) гигантский
2467 дендрокаламус
2468 островския
2469 брунсвигия Жозефины
2470 драцена австралийская; кордилина южная
2471 смолоносница; ферула
2472 овсяница высокая; овсяница гигантская; овсяница исполинская
2473 лещина крупная; орех ломбардский; фундук (крупноплодный)
2474 лук приречный; лук причесночный; лук-рокамболь; рокамболь; чеснок змеиный
2475 гранадилла гигантская; пассифлора гигантская; пассифлора крупная; страстоцвет крупный; страстоцвет настоящий; страстоцвет четырёхугольный
2476 лофант
2477 макроцистис
2478 лилия гигантская;лилия исполинская
2479 дождевик-великан
2480 амброзия трёхнадрезная; амброзия трёхраздельная
2481 арундо
2482 арундо тростниковый; камыш гигантский
2483 веллингтония; дерево мамонтово; секвойя гигантская
2484 туя гигантская; туя складчатая
2485 губка горбатая
2486 дерево фиалковое (австралийское)
2487 гилия
2488 пластинниковые; пластинчатые; шампиньоновые
2489 имбирь; цингибер
2490 имбирные
2491 гедихиум; хедихиум
2492 гинкго
2493 гинкго дву(х)лопастный; гинкго китайское
2494 женьшень; панакс
2495 аралиевые
2496 сольданелла альпийская
2497 лютик ледниковый; оксиграфис обыкновенный
2498 крупка ледниковая
2499 гладиолус; шпажник
2500 солерос
2501 глейхения
2502 глейхениевые
2503 гомфрена
2504 иберийка зонтичная; перечник зонтичный; стенник зонтичный
2505 шаровницевые
2506 глобулярия; шаровница
2507 купальница
2508 мордовник; эхинопс
2509 вьюнок; калистегия; повой
2510 клеродендрон
2511 глориоза
2512 хионодокса
2513 крушина ломкая;крушина ольховидная; крушинник ломкий; крушинник ольховидный
2514 бирючина блестящая; бирючина лоснящаяся
2515 глоксиния
2516 мокруха еловая
2517 бодлак; коленница; эгилёпс;эгилопс
2518 стальник пашенный;стальник полевой
2519 арункус; волжанка; таволжник
2520 булавница; рогатик
2521 галега; козлятник
2522 курчавка
2523 бредина; ива-бредина; ива козья; ракита
2524 годеция

2525 лилия золотистая
2526 ясень золотистый
2527 бамбук золотистый; бамбук-листоколосник; листоколосник золотистый
2528 пупавка красильная; ромашка жёлтая; ромашка красильная
2529 бобовник "золотой дождь"; бобовник анагировидный; бобовник анагиролистный; дождь золотой обыкновенный; ракитник-золотой дождь
2530 лапчатка золотистая
2531 рогатик золотистый
2532 ломонос тангутский
2533 оронтиум
2534 хуннемания
2535 смородина белая; смородина декоративная; смородина золотая; смородина золотистая
2536 щавель морской; щавель приморский
2537 бузина золотистая
2538 лён жёлтый
2539 лжелиственница; ложнолиственница; лиственница ложная
2540 вербейник обыкновенный
2541 мимулюс крапчатый
2542 бузульник
2543 золотарник; солидаго
2544 бозеа; бозея
2545 селезёночник
2546 гидрастис; желтокорень; печать золотая
2547 гидрастис канадский;желтокорень канадский
2548 кассия дудка; кассия стручковая; кассия трубковидная; кассия южноазиатская; мёд дивий
2549 хризогонум
2550 ламаркия
2551 бурачок скалистый; бурачок скальный
2552 лютик золотистый
2553 очиток едкий
2554 полынёк; эшолтция; э(с)шольция; эшшольция
2555 дерево мыльное; кельрейтерия
2556 коптис
2557 гомфия
2558 аренга сахарная; пальма сахарная
2559 марь Доброго Генриха; марь цельнолистная
2560 крыжовник
2561 лебеда; марь
2562 лебедовые; маревые
2563 елевзина индийская
2564 гордония
2565 чубушник Гордона
2566 золотохворост; улекс; утёсник
2567 лагенария
2568 тыква
2569 тыквенные
2570 снедь-трава; сныть
2571 кипарис Говена; кипарис Говея; кипарис калифорнийский
2572 бутел(о)уа; трава бизонья
2573 пихта великая
2574 масляник зернистый; масляник летний; маслёнок зернистый; маслёнок летний
2575 виноград; лоза виноградная
2576 виноградные; виноградовые
2577 ботрихиум; гроздовник; папоротник-гроздовник
2578 грейпфрут; помпельмус гроздевидный
2579 лук гадючий
2580 лагерстремия; лягерстремия
2581 злаки; злаковые
2582 касатик злаковидный; касатик злаковый
2583 роза полевая; шиповник полевой
2584 песчанка злаколистная; песчанка траволистная
2585 крас(н)однев малый
2586 золотарник злаколистный
2587 чина посевная
2588 гвоздика перистая
2589 лимодорум
2590 лимодорум клубневой
2591 акароидин; дерево травяное; желтосмолка; ксанторрея
2592 чина злаколистная
2593 частуха Валенберга; частуха злаковая
2594 берёза жёлтая высокая; берёза тополелистная
2595 булавоносец седой
2596 ива сизая
2597 тополь седеющий; тополь сереющий; тополь серый
2598 ива пепельная; ива серая
2599 колокольчик широколистный
2600 камыш озёрный; схеноплектус озёрный
2601 лопух большой; лопух крупный; репей(ник); репейник исполинский; лопушник большой; лопушник съедобный; собашник
2602 сольданелла горная
2603 амми большой
2604 заразиха большая
2605 любка зеленоцветковая; любка зеленоцветная
2606 ластовица; трава ластовичная; чистотел большой
2607 повилика европейская
2608 бедренец большой
2609 альпиния Галанга; корень галанговый
2610 хвощ большой
2611 астранция большая
2612 капуста заячья большая; очиток большой
2613 ожика лесная
2614 наперстянка шерстистая
2615 лавр благородный
2616 обвойник греческий
2617 пихта греческая
2618 можжевельник высокий
2619 микромерия греческая
2620 синюха голубая; синюха лазоревая; синюха лазуревая; синюха лазурная; синюха обыкновенная
2621 водоросли зелёные
2622 ясень зелёный; ясень ланцетный; ясень ланцетолистный
2623 павой; сарсапарель; сассапариль; смилакс
2624 гоми; мышей зелёный;мышей сизый; чумиза; щетинник зелёный

2625 хвойник зелёный
2626 вошерия
2627 венгерка итальянская; ренклод; слива ренклод
2628 морозник зелёный
2629 сыроежка зеленоватая
2630 чернокорень немецкий
2631 плесень зелёная
2632 кистевик зелёный; пенициллиум сизый; плесень сизая
2633 грушанка зеленоватая; грушанка зеленоцветковая
2634 костенец зелёный
2635 земляника зелёная; земляника полуница; полуница
2636 телесперма
2637 акация низбегающая; акация сбегающая
2638 гревиллея
2639 гревия
2640 воробейник
2641 повилика Гронова; повилика Громовиуса
2642 касатик Гроссгейма
2643 многоветочник карликовый
2644 вишенник; вишня дикая; вишня кустарниковая; вишня лесостепная; вишня степная; вишня шпанка кустарниковая; слива кустарниковая
2645 вишня жидовская; вишня пёсья; вишня пузырная; вишня пузырчатая; можжуха; мозжуха; физалис
2646 ломонос прямой
2647 будра
2648 бляшконоска; пельтигера собачья
2649 сочевичник клубненосный; чина клубневая; чина клубненосная; чина клубеньковая
2650 живучка ёлочковидная
2651 крестовник
2652 ожика боровая; ожика дубровная
2653 грумичама
2654 эритеа съедобная
2655 гуарана; пауллиния гуарана
2656 гуава;гуайава; гуайява; гуаява; гуйава; гуйява; псидиум
2657 гвайюла; гваюла; гуайюла; гуаюла; гвайюла серебристая; гваюла серебристая; мариола; партениум серебристый
2658 нерине сарнийская
2659 шима
2660 дерево крабовое; карапа гвианская
2661 просо гвинейское; просо гигантское; просо крупное; трава гвинейская
2662 акация сенегальская
2663 гринделия
2664 лабданум
2665 кастилл(о)а
2666 ландольфия
2667 гуннера
2668 дву(х)крылоплодник; диптерокарпус
2669 гуммигутовые
2670 гуцмания
2671 голосеменные; голосемянные
2672 гипсолюбка; гипсофила; качим
2673 поводник; пололепестник
2674 дерево каменное; каркас
2675 хакетия
2676 лён кукушкин; политрихум
2677 аира; айра; луговик; щучка
2678 осока волосовидная
2679 репейник волосистый; репейничек волосистый
2680 эхинопсилон волосистый
2681 костёр ветвистый; костер переменчивый; костёр разветвлённый; костёр смешанный
2682 грыжник волосистый
2683 пальчатка; просо кровяное; росичка кровяная
2684 горечавка реснитчатая
2685 дрок волосистый
2686 сирень мохнатая; сирень опушённая
2687 овёс опушённый; овёс пушистый; овсец опушённый
2688 резуха шершавая
2689 ладанник волосистый
2690 осока волосистая; осока ворсистая; осока коротковолосистая;осока мохнатая
2691 очиток волосистый
2692 ворсянка волосистая
2693 вика озимая; вика мохнатая; вика песчаная; вика чёрная (мохнатая); горошек мохнатый
2694 фиалка опушённая
2695 кипрей(ник) волосистый; кипрей мохнатый
2696 ожика волосистая
2697 ива копьевидная
2698 реедия обыкновенная
2699 гамелия
2700 гамильтония
2701 жестковолосница; жестколосница
2702 жестковолосница твёрдая; жестколосница твёрдая
2703 спирея волосистая; таволга войлочная
2704 ситник искривлённый;ситник склонённый; ситник склоняющийся
2705 ложнодождевик
2706 ложнодождевиковые
2707 кринум длиннолистный
2708 плевел расставленный
2709 вероника лежачая; вероника простёртая
2710 конрингия
2711 давал(л)ия
2712 лапка заячья; осока заячья
2713 зайцехвост; лагурус
2714 адраспан; гармала обыкновенная; могильник; хутчик; юзерлык
2715 роза Гаррисона
2716 зерлик; листовик сколопендровый; сколопендриум обыкновенный;стоног; язык олений
2717 трюфель олений
2718 пентстемон гибридный
2719 тордилиум
2720 меченосница; секуригера
2721 картофель мадагаскарский
2722 земляника высокая; земляника мускатная; земляника мускусная; клубника высокая; клубника настоящая
2723 кульбаба
2724 шишечник; шишник
2725 баркаузия; скерда

2726 ястребинка
2727 горлюха ястребинковидная
2728 гавортия
2729 боярышник
2730 клён боярышниколистный
2731 прангос высокий; прангос кормовой; юган
2732 ольха морщинистая
2733 стеркулия вонючая
2734 капуста (настоящая) кочанная
2735 латук-салат кочанный; салат кочанный
2736 бадан сердцелистный
2737 шаровница сердцелистная
2738 граб сердцелистный
2739 чубушник сердцелистный
2740 телекия красивая
2741 кардиосперм(ум)
2742 ерика; эрика
2743 подмаренник герцинский; подмаренник скальный
2744 вереск
2745 вересковые
2746 зиглингия лежачая
2747 ситник растопыренный
2748 калистегия заборная; повой заборный
2749 эхинопсис
2750 гребневик шиповатый
2751 падуб остролистный колючий
2752 гиднум выемчатый; ежовик жёлтый
2753 прицепник морковновидный; прицепник морковный; каукалис
2754 авран
2755 горец кустарниковый
2756 клён полевой; паклён
2757 гулявник аптечный; гулявник лекарственный; сухоребрик
2758 чистец однолетний
2759 купырник; торилис; чермеш
2760 горошек заборный
2761 боярышник Гельдрейха
2762 сосна Гельдрейха
2763 геликония
2764 гелиопсис
2765 гелиотроп
2766 геллебор(ус); морозник; смертоед
2767 лопастник; строчок
2768 грибы сморчковые; сморчковые
2769 гемлок; тсуга; цуга
2770 гирчевник; гирчовник
2771 омежник шафранный; омежник шафрановый; омежник шафраноподобный
2772 конопля
2773 дерганец; замашка; конопля индийская; конопля культурная; конопля посевная; посконь
2774 заразиха ветвистая
2775 апоцинум коноплевидный; кендырь коноплевидный; кендырь конопляный; хлопок дикий
2776 посконник коноплевидный;посконник коноплёвый; посконник конопляный; седач; седаш (коноплевидный)
2777 коноплёвые
2778 петушник; пикульник
2779 живучка кровельная; молодило кровельное
2780 яснотка стеблеобъемлющая
2781 генекен; икстль; конопля юкатанская; хенекен
2782 лавсония; хенна; хна
2783 перелеска; печёночник; печёночница
2784 дорикниум травянистый
2785 зелье курье; перелеска однолетняя; пролеска однолетняя; пролесник однолетний
2786 глаз вороний обыкновенный; глаз вороний четырёхлистный; глаз вороний четырелистный; одноягодник четырёхлистный
2787 герань Роберта
2788 хатьма трёхмесячная
2789 аистник; грабельник; журавельник
2790 шафран Гейфеля
2791 гевеа; гевея; хевея
2792 пачули
2793 туевик поникающий; туевик японский
2794 гибиск(ус)
2795 гикори; кария
2796 блюберри; голубика американская; голубика зонтичнокистевая; черника щитковая; ягодник кистевой
2797 зензивер; мальва лесная; просвирник лесной; просвирняк лесной
2798 ежевика гигантская
2799 камелия масленосная
2800 пихта гималайская; пихта замечательная
2801 можжевельник гималайский; можжевельник отогнутый
2802 лиственница гималайская; лиственница Гриффитса
2803 сирень гималайская; сирень Эмоди
2804 подофилл гималайский; подофилл Эмода
2805 ревень Эмоди
2806 ель гималайская; ель индийская
2807 сосна Веймутова гималайская; сосна гималайская Веймутова
2808 мелицитус веткоцветный
2809 элеокарпус зубчатый; элеокарпус новозеландский
2810 дурман индийский; дурман метель
2811 лотос индийский; лотос орехоносный
2812 кипарисовик туполистный
2813 гиппократия
2814 дуб испанский; дуб пробковый ложный
2815 алтей жестковолосый; алтей шершавоволосый; алтей шершаволистный
2816 шандра
2817 желтушник сероватый; желтушник серо-зелёный; желтушник серый
2818 икотник серо-зелёный
2819 чубушник широколистный
2820 тефрозия
2821 солнцецвет седой
2822 гоферия
2823 белена
2824 горечник; горичник
2825 борщевик обыкновенный
2826 илекс; падуб
2827 падубовые
2828 многорядник

2829 алтей розовый; алтей-штокроза; мальва махровая; мальва садовая; рожа; шток-роза розовая
2830 мальва раздельнолистная; просвирняк штокрозовый
2831 майтенус падуболистный
2832 дуб вечнозелёный; дуб каменный; дуб падуболистный
2833 гомалиум
2834 лунария; лунник
2835 петрушечник; сизон
2836 скрытница канадская
2837 меликокка
2838 опёнок настоящий; опёнок осенний
2839 мелиант
2840 гледичия
2841 жимолость; лоницера
2842 жимолостные
2843 восковник; воскоцветник
2844 щетинник мутовчатый
2845 барбарис Гукера
2846 ива Гукера
2847 хмель
2848 клевер пашенный; клевер полевой; клевер посевной; клевер шуршащий; хмелёк полевой
2849 хмелеграб
2850 додонея
2851 вязовик; кожанка; птелея
2852 граб
2853 клён граболистный
2854 цератозамия
2855 дыня декоративная красноплодная; огурец африканский
2856 фиалка рогатая
2857 кипрей(ник) Горнемана
2858 вороночник рожковидный; лисичка серая
2859 глауциум; гляуциум; мачёк; мачок; рогомак
2860 кушир; роголистник
2861 роголистник погружённый; роголистник тёмно-зелёный
2862 роголистниковые
2863 коллинзония; коллинсония
2864 боб конский; бобы конские; бобы кормовые
2865 желудниковые; конскокаштановые
2866 фенхель конский
2867 трёхкосточник; триостеум
2868 алектория гривастая
2869 мята дикая; мята длиннолистная
2870 шампиньон полевой
2871 жеруха хреновая; хрен (деревенский); хрен дикий; хрен обыкновенный
2872 дерево хренное; моринга
2873 пеларгоний поперечнополосатый
2874 гиппокрепис; подковник
2875 хвощ
2876 казуарина хвощевидная; казуарина хвощелистная
2877 хвощёвые
2878 мелколепестник канадский
2879 циноглоссум; чернокорень; язык песий
2880 гриб домовый; губка домовая
2881 многорядник серповидный
2882 живучка; молодило; семпервивум
2883 дерево кокаиновое; кока; куст(арник) кокаиновый; эритроксилон кока
2884 гайлюссакия; гейлюссакия; гейлюссация
2885 гугония
2886 гумея
2887 нивяник круглолистный
2888 клевер венгерский; клевер жёлтый; клевер паннонский
2889 касатик пёстрый; петушки
2890 сирень венгерская;сирень угорская
2891 эспарцет песчаный
2892 вероника широколистная
2893 вика венгерская; вика паннонская; горошек венгерский
2894 дакридиум
2895 двусеменник; двусемянник
2896 гиацинт
2897 боб гиацинтовый; боб индийский; бобы гиацинтовые; долихос лябляб; долихос обыкновенный;лобия (египетская)
2898 канатник гибридный
2899 гиднофитум
2900 гиднум; ежовик; колчак
2901 грибы ежовиковые; ежевиковые; ежовиковые
2902 гидрангея; гортензия
2903 схизофрагма; шизофрагма
2904 гигрофор; мокрица
2905 гигрофоровые
2906 гипекоум
2907 ложноопёнок
2908 гипнум; мох гипновый
2909 гиссоп; иссоп
2910 иссоп аптечный; иссоп лекарственный
2911 верблюдка иссополистная
2912 дербенник иссополистный
2913 касатик иберийский
2914 икак(о); слива золотая
2915 фиппсия
2916 лишайник исландский; мох исландский; цетрария исландская
2917 ледянник; мезембриантемум кристаллический; мезембриантемум хрустальный; трава ледяная; трава хрустальная
2918 сансевиерия цилиндрическая
2919 хрящецветник; хрущецветник
2920 бассия широколистная; мадука индийская
2921 лютик иллирийский
2922 ежовик пёстрый; саркодон черепитчатый
2923 бессмертник; сухоцвет(ник)
2924 инкарвиллея
2925 кедр ладанный; кедр речной; либоцедрус
2926 можжевельник ладанный
2927 аталантия однолистная
2928 трема восточная
2929 баниан индийский; смоковница индийская
2930 полиальтия
2931 мышей волосистый; полевичка волосистая
2932 земляника индийская
2933 азалия индийская; рододендрон

индийский
2934 канна индийская
2935 снежноягодник красноплодный; снежноягодник круглолистный; снежноягодник обыкновенный;снежноягодник округлый
2936 опунция индийская (фига); смоква индийская; тупа; фига индийская
2937 босвел(л)ия (пильчатая)
2938 монция
2939 горчица сарептская
2940 вертляница; подъельник
2941 флакуртовые; флакуртиевые
2942 зелье рвотное; лобелия вздутая; лобелия надутая;лобелия одутлая
2943 калофиллум круглоплодный
2944 дерево каучуковое; фикус (каучуконосный)
2945 ороксилум
2946 бобы индийские
2947 индиго; индигоноска; индигофера
2948 аморфа кустарниковая; аморфа кустовидная; индиго бастардное; индиго ложное
2949 инга
2950 гриб чернильный; навозник серый
2951 навозник
2952 пырей промежуточный;пырей средний
2953 грушанка средняя
2954 плаун годичный; плаун годовалый
2955 девясил
2956 свинушка бурая; свинушка тонкая
2957 ипекакуана; корень рвотный
2958 ирис; касатик
2959 ирисовые; касатиковые
2960 кар(р)аген; хондрус
2961 караген кудрявый; мох ирландский; хондрус кудрявый
2962 ветреница алтайская
2963 хлорофора высокая
2964 метросидерос
2965 вернония; трава железная
2966 дерево железное; мезуа железная
2967 железница; сидеритис
2968 изидиум
2969 изолома
2970 равноплодник
2971 ольха сердцевидная
2972 аронник белокрылый; аронник итальянский; арум белокрылый
2973 астра дикая; астра звёздчатая; астра ромашковая; астра степная
2974 жостер вечнозелёный; крушина вечнозелёная
2975 анхуза итальянская
2976 ломонос фиолетовый
2977 валерианелла бугорчатая; валерьянница бугорчатая
2978 кипарис вечнозелёный; кипарис пирамидальный
2979 дуб густой; дуб Фрайнетто
2980 плевел итальянский; плевел многоцветковый; плевел многоцветный; райграс итальянский; райграс многоукосный
2981 смолёвка итальянская
2982 пролеска итальянская
2983 пиния; сосна итальянская; сосна пиния
2984 льнянка итальянская
2985 кайя иворская; кайя иворензис
2986 фителефас
2987 сифонодон
2988 плющ
2989 сциндапсус
2990 дряква персидская; цикламен персидский
2991 пеларгониум плющелистный
2992 вероника плющелистная
2993 иксия
2994 иксиолирион
2995 иксора
2996 дерево палисандровое; якаранда
2997 канавалия
2998 дерево хлебное восточноиндийское; дерево хлебное джек; дерево хлебное индийское; дерево хлебное разнолистное; дерево хлебное цельнолистное; джек(-дерево);джекфрут; кемпедек; чемпедек; як-дерево
2999 аризема
3000 клематис Жакмана
3001 сосна Банкса; сосна Банксова; сосна растопыренная
3002 джексония
3003 асфоделина
3004 костенец зонтичный
3005 ипомея пурга; ипомея слабительная; ипомея ялапа; ялапа (настоящая)
3006 платан Линдена
3007 калабюра; мунтингия (Калабура)
3008 гибискус индийский
3009 квассия ямайская; пикрена высокая
3010 джамболан; евгения ямболана; сизигиум яванский; слива яванская; слива явская; ямболан
3011 джемсия
3012 ольха японская
3013 анемон японский; анемона японская
3014 бадьян настоящий; дерево анисовое; иллициум анисовый; иллициум обрядный; сикими
3015 абрикос китайский; абрикос японский; абрикос муме; муме
3016 аралия высокая
3017 туя Стэндыша; туя японская
3018 ардизия японская
3019 аукуба японская; дерево золотое
3020 банан японский
3021 барбарис Тунберга; барбарис японский
3022 бук японский
3023 сосна Тунберга
3024 костёр японский
3025 крушина японская
3026 будле(й)я японская
3027 каштан городчатый; каштан японский
3028 виноград девичий триостренный; виноград дикий триостренный
3029 криптомерия японская
3030 смородина японская
3031 лжетсуга японская
3032 эводия сизая
3033 коротконожка перистая
3034 аралия японская; фатсия японская
3035 лещина Зибольда; орешник Зибольда
3036 горец осотоконечный; горец тонкозаострённый
3037 яблоня обильноцветковая; яблоня

обильноцветущая; яблоня многоцветковая
3038 айва низкая Маулея; айва цветочная; айва японская; хеномелес Маулея; хеномелес японский
3039 можжевельник лежачий
3040 тсуга разнолистная; цуга разнолистная
3041 гледичия японская
3042 жимолость японская
3043 хмель японский
3044 хмелеграб японский
3045 граб японский
3046 каштан конский японский
3047 касатик Кемпфера; касатик японский
3048 багрянник японский; круглолистник японский
3049 керия японская
3050 дерево лаковое; сумах лаконосный
3051 лиственница тонкочешуйчатая; лиственница японская
3052 цоизия японская
3053 лилия японская
3054 самшит японский
3055 магония японская
3056 клён веерный; клён дланевидный; клён пальмовидный
3057 ежовник хлебный; просо китайское; просо японское; пайза
3058 рябина японская
3059 софора японская
3060 функия ланцетолистная
3061 слива ивообразная; слива иволистная; слива китайская; слива японская
3062 тополь Максимовича
3063 перец японский
3064 первоцвет японский; примула японская
3065 бирючина японская
3066 говения сладкая; сладконожник
3067 сосна густоцветковая; сосна густоцветная
3068 роза многоцветковая
3069 скимия японская
3070 спирея мозолистая; спирея японская; таволга японская
3071 стаунтония шестилистная
3072 кедр карликовый; стланик кедровый
3073 аир злаковый
3074 торрея орехоносная
3075 сирень японская
3076 берёза японская
3077 лапина сумахолистная
3078 вистария обильноцветущая; глициния обильноцветущая
3079 гамамелис японский
3080 дзельква японская
3081 эвкалипт западноавстралийский
3082 букашник
3083 жасмин
3084 мурра(й)я; муррея
3085 канариум обыкновенный
3086 бишопия яванская; бишофия
3087 ливистона круглолистная
3088 тарриэтия яванская
3089 очиток отогнутый
3090 абрус; боб розовый
3091 роза иерихонская
3092 сушеница жёлто-белая; сушеница желтовато-белая
3093 ятрышник болотный
3094 бульва; груша земляная; подсолнечник клубненосный; репа волошская; топинамбур
3095 паслён лжеперечный; паслён ложноперечный; паслён перцеподобный
3096 кудрявец; марь душистая
3097 зопник
3098 цеструм
3099 родотипус; розовик
3100 деррис
3101 уши иудины; "ухо иудино"
3102 дурман вонючий; дурман обыкновенный; дурман татуля
3103 гумай; сорго алеппское; сорго многолетнее;трава джонсонова; чумай
3104 колча; хвойник; эфедра
3105 гнетум
3106 трава кузьмичёва; хвойник двухколосковый; эфедра дву(х)колосковая; трава эфедра
3107 гнетовые; хвойниковые; эфедровые
3108 манизурис
3109 амбач; эшиномена
3110 жонкилия; жонкилла; жонкиль; нарцисс-жёлтый; нарцисс-жонкиль
3111 альтернантера; очереднопыльник
3112 юануллоа
3113 пальма слоновая; юбея
3114 багрянник обыкновенный; багрянник стручковатый; багрянник стручковый; дерево иудейское; дерево иудино; церцис европейский
3115 зизифус; унаби; ю(й)юба
3116 ситниковидные
3117 дерево железное; сидероксилон
3118 арча; можжевельник
3119 кентрантус красный; центрантус красный
3120 нагловатка; наголоватка; юринея
3121 юстиция
3122 джут
3123 кадсура
3124 шотия
3125 энцефалартос
3126 кливия
3127 каки; слива финиковая; хурма восточная; хурма японская
3128 браунколь; капуста кормовая; капуста кудрявая; капуста листовая
3129 кальмия
3130 рябчик камчатский
3131 темеда
3132 алоэ пёстрый
3133 сорго китайское; сорго японское
3134 дерево капоковое; дерево сырное; дерево хлопчатое; дерево шерстяное; капок; сейба пятитычинковая; сумаум; цейба
3135 коринокарпус
3136 гребенщик опушённый
3137 кипарис кашмирский
3138 хемантус Екатерины
3139 агатис новозеландский; каури новозеландская
3140 ава-перец; кава(-перец);перец кава
3141 пихта Каваками
3142 кайея

3143 валерьянница килеватая
3144 лук килеватый
3145 ламинариевые
3146 гибискус кенаф; гибискус коноплёвый; джут африканский; джут яванский; кенаф; конопля декан; пенька бомбейская
3147 цимбалярия постенная; цимбалярия стенная
3148 кеннедия
3149 кентия; кенция
3150 мятлик луговой
3151 бундук двудомный;бундук канадский; гимнокладус двудомный; гимнокладус канадский; дерево кентуккское кофейное; шикот; шипот
3152 капуста кергуэленская
3153 дуб кермесовый; дуб кермесоносный; дуб хермесовый
3154 кернера
3155 чертополох Кернера
3156 кер(р)ия
3157 кетелеерия
3158 фасоль волокнистая; фасоль зерновая; фасоль настоящая; фасоль обыкновенная; фасоль почечная
3159 клевер заячий; трава зольная;язвенник Линнея; язвенник обыкновенный
3160 ястребинка превысокая; ястребинка стройная
3161 ястребинка флорентийская
3162 мандарин благородный;мандарин кинг
3163 киренгешома
3164 слива полусердцевидная
3165 клейнховия
3166 клюгия
3167 короставник
3168 дивала; дивало
3169 клевер полосатый
3170 купырник узловатый; торилис узловатый
3171 мшанка узловатая
3172 горец; гречишник; полигонум
3173 келерия; тонконог
3174 кёнигия
3175 голорепа; капуста кормовая мозговая; капуста листовая культурная; капуста репная; кольраби; репа венгерская
3176 актинидия коломикта; кишмиш обыкновенный; коломикта; крыжовник амурский
3177 туя корейская
3178 барбарис корейский
3179 пиретрум сибирский; поповник сибирский
3180 княжик корейский
3181 пихта корейская
3182 каркас корейский
3183 куммеровия прилистниковая; леспедеца корейская; леспедеца прилистниковая
3184 кедр корейский; сосна кедровая корейская; сосна кедровая маньчжурская; сосна корейская
3185 тополь корейский
3186 калина корейская
3187 ель корейская
3188 кудзу; пуерария; пуэрария
3189 кинкан; кумкват
3190 гагения абиссинская; куссо; хагения абиссинская
3191 лабульбениевые
3192 дождь золотой
3193 сосна Бунге
3194 дидискус
3195 млечник
3196 лопастник ямчатый
3197 скрученник
3198 аденофора; бубенчик(и)
3199 кочедыжник (женский); папоротник женский
3200 ятрышник пурпурный
3201 рапис
3202 манжетка
3203 башмачек; башмачок (венерин);орхидея-циприпедиум; сапожки
3204 лагозерис
3205 фунтумия каучуконосная
3206 лаготис; ляготис
3207 дерево хлебное Лакооха; хлебоплодник Лакуша
3208 лаллеманция; ляллеманция
3209 боярышник Ламберта
3210 кальмия узколистная
3211 валерианелла колосковая; валерианелла овощная; валерианница овощная; валерьянница овощная
3212 марь белая; марь обыкновенная
3213 арнозерис; баранец
3214 кипрей Лами; кипрей(ник) Ля́ми
3215 кроталярия ланцетовидная
3216 лонхокарпус
3217 лансиум
3218 лантана
3219 мытник лапландский
3220 рододендрон лапландский
3221 ива лапландская; ива лопарская; куропатник; куропаточник
3222 лиственница
3223 лардизабала
3224 клюква крупноплодная; черника крупноплодная
3225 клевер равнинный
3226 дельфиниум; живокость; разноцвет; шпорник
3227 гладыш; лазерпициум; лазурник
3228 лазия
3229 латания; пальма веерная
3230 крест Петров
3231 лавр
3232 лавровые
3233 дуб лавролистный
3234 тополь лавролистный
3235 ладанник лавролистный
3236 верболоз; ива пятитычинковая; чернолоз; чернотал
3237 калина вечнозелёная; калина лавролистная
3238 лавровишня
3239 лаванда
3240 глистогон; сантолина
3241 порфира
3242 цоизия
3243 кипарисовник белый; кипарисовник Лавзона; кипарисовник Лавсона; лжекипарис Лавзона
3244 порховка свинцово-серая
3245 левкена

3246 плюмбаговые; свинчатковые
3247 филлокактус; эпифиллум; эпифиллюм
3248 полимния
3249 филлантус
3250 мякотница болотная
3251 молочай лозный; молочай острый; молочай съедобный
3252 ежеголовник маленький
3253 ива круглолистная
3254 бадан толстолистный; бадан узколистный; чай монгольский
3255 ломонос виорна
3256 телефоровые
3257 кассандра; хамедафна; хамедафне
3258 роза кожистолистная; шиповник кожистолистный
3259 дирка
3260 крушина ливанская
3261 дуб ливанский
3262 леканора
3263 лекокия
3264 лилия Ледебура
3265 багульник
3266 лук-пор(р)ей; порей
3267 бобовые
3268 дерево лимонное; лимон
3269 крас(н)однев жёлтый
3270 лемонграс; сорго лимонное; трава лимонная
3271 вербена лимонная; липпия лимонная; липпия трёхлистная
3272 морозник восточный
3273 чечевица
3274 пилолистник
3275 дерево мастиковое; дерево мастичное; кустарник мастиковый; фисташка-лентискус; фисташка мастиковая; фисташка мастичная
3276 ароникум; дороникум
3277 леопольдия
3278 гриб-зонтик
3279 куммеровия; леспедеза; леспедеца; леспедеция
3280 воскоцветник малый
3281 ла(к)тук; молокан; мульгедиум; салат
3282 левкорхис; леукорхис
3283 гуза; хлопчатник африканский;хлопчатник афроазиатскмй
3284 дармина; полынь цитварная
3285 льюзия
3286 порезник
3287 дерево кофейное либерийское; кофе либерийский
3288 либертия
3289 ликания
3290 лишайник
3291 лакрица; лакричник; солодка
3292 ликуала
3293 бакаут; дерево бакаутовое; дерево гваяковое; гваякум; гуаяк
3294 лигустикум; бороздоплодник
3295 сирень
3296 дерево авраамово; прутняк обыкновенный; целомудренник
3297 ломонос раскидистый
3298 гвоздика пышная
3299 шалфей мутовчатый
3300 лилия
3301 лилейные
3302 бубенчики лилиелистные
3303 лук моли; моли
3304 магнолия лилейная; магнолия лилиецветная
3305 ландыш
3306 ландыш майский
3307 ландышник; офиопогон
3308 лириопе
3309 бобы лимские; лима; фасоль лима; фасоль лимская
3310 сосна гибкая; сосна кедровая калифорнийская
3311 лайм (настоящий); лиметта; лимон лиметта; лимон сладкий; цитрус лиметта; цитрус померанцеволистный; ляйм
3312 цантоксилум мартиникский
3313 линделофия
3314 липа
3315 липовые
3316 гибискус липолистный; дерево пробковое антильское
3317 гревия липолистная
3318 рябинник Линдлея
3319 дерево кружевное; лагетта
3320 леонотис
3321 физостегия
3322 леонтица; леонтице
3323 краекучник
3324 лип(п)ия
3325 лишайник красильный; лишайник лакмусовый; мох красильный; мох лакмусовый; рокцелля красильная; рочелла красильная
3326 галинсога мелкоцветковая; галинсога мелкоцветная
3327 липа мелколистная; липа сердцевидная
3328 чубушник мелколистный
3329 карагана мелколистная
3330 кольраушия побегоносная
3331 мышей малый; мышей мятликовидный; полевичка малая
3332 люцерна маленькая
3333 овёс короткий
3334 рыжей мелкоплодный; рыжик мелкоплодный
3335 трясунка малая
3336 канареечник малый
3337 сераделла (очень) маленькая
3338 звездчатка злаковая; звездчатка злаковидная; звездчатка злачная; звездчатка травяная; трава пьяная
3339 ива деревцевидная
3340 литтония
3341 капуста заячья; очиток "заячья капуста"; очиток телефиум;скрипун
3342 дуб виргинский; дуб каролинский
3343 перелеска благородная; печёночница благородная; печёночница обыкновенная
3344 ангалоний
3345 сауроматум
3346 ремнелепестник
3347 саурурус
3348 сауруровые
3349 многорядник лопастный
3350 лобелия

3351 лобелиевые
3352 сосна ладанная
3353 астрагал
3354 робиния
3355 сосна Муррея; сосна (скрученная) широкохвойная
3356 липарис Лёзеля; лосняк Лёзеля
3357 ягода логанова
3358 логаниевые; чилибуховые
3359 дерево кампешевое; дерево сандаловое; дерево сандальное; сандал синий
3360 тополь итальянский; тополь пирамидальный
3361 платан клёнолистный
3362 камнеломка тенистая
3363 гулявник ирио
3364 осока растянутая
3365 глициния японская
3366 сосна (длиннохвойная) болотная
3367 мак сомнительный
3368 колба; лук победный; черемша
3369 герань голубиная
3370 вербейник
3371 дербенниковые
3372 лофира
3373 фрима
3374 овёс колючий; овёс тощий; овёс щетинистый
3375 ериоботрия; эриоботрия
3376 ериоботрия японская; башмала японская; локва; мушмула (японская) обыкновенная
3377 ремнецветник
3378 арум пятнистый; аронник пятнистый; змей-трава; клещинец
3379 квебрахо Лоренца
3380 лоропеталум
3381 лотос
3382 мытник лесной
3383 вшивица; мытник
3384 любисток
3385 полевичка; тефф
3386 чернушка дамасская; чернушка "девица в зелени"
3387 амарант хвостатый; щирица метельчатая; щирица хвостатая
3388 лжелиственница Кемфера; лжелиственница китайская
3389 многоветочник низкий
3390 берёза малорослая
3391 сушеница болотная;сушеница топяная
3392 василисник малый
3393 лёгочница; медуница;трава лёгочная
3394 лобария лёгочная; лобария лёгочнообразная
3395 боб волчий; лупин; люпин
3396 дубовик; поддубник; поддубовик
3397 личжи; нефелиум
3398 личжи китайский;нефелиум китайский
3399 трюфель олений
3400 дербенник; литрум; плакун; подбережник
3401 роза Макартова;роза прицветниковая
3402 платимисциум
3403 дуб македонский
3404 льнянка македонская
3405 кипарис Макнаба
3406 макрозамия
3407 гвоздичник мадагаскарский
3408 рафия мадагаскарская
3409 дерево путешественников; равенала мадагаскарская
3410 марена
3411 мареновые
3412 персея индийская
3413 буссенгоя; буссингоа
3414 базелловые
3415 антолиза
3416 мадия
3417 лилия белая; лилия бледная
3418 дерево земляничное; земляничник
3419 острица;асперуга
3420 меруя
3421 барбарис самшитолистный
3422 фуксия
3423 магнолия
3424 магнолиевые
3425 лимонник; шизандра
3426 антипка; вишня антипка; вишня магалебская; вишня чернильная; кучина; магалебка; черёмуха-антипка; черёмуха-магалебка; черёмуха душистая; черешня антипка
3427 свиетения
3428 мелиациевые; мелиевые
3429 магония
3430 адиантум; волосатик; папоротник-адиантум
3431 костенец волосовидный
3432 гинкговые
3433 гвоздика травяная; гвоздика травянка; травянка
3434 кукуруза (обыкновенная); маис
3435 эремурус величественный
3436 лилия малабарская
3437 тыква черносеменная
3438 атхатоба; орешник индийский
3439 рандия
3440 дерево хлопковое; семул; цейба
3441 ксантозома; ксантосома; танье
3442 ливистона кохинхинская
3443 дерево гуттаперчевое; паляквиум
3444 малькольмия
3445 многокучник большой; папоротник лесной; папоротник мужской; щитник; щитовник мужской
3446 ятрышник мужской
3447 мальва; просвирник; просвирняк
3448 мальвовые; просвирниковые; просвирняковые
3449 дыравка; малопа
3450 мальпигия
3451 зорька ("барская смесь"); зорька татарское мыло; лихнис татарское мыло; лихнис халцедонский; мыло татарское; "смесь барская"
3452 маммея
3453 абрикос антильский; абрикос тропический; маммея американская
3454 кактус маммиллярия; кактус сосочковый; мам(м)иллярия
3455 гиппомана; манцинелла;маншинелла
3456 лещина маньчжурская;орешник маньчжурский
3457 абрикос маньчжурский; абрикос монгольский
3458 ясень маньчжурский
3459 смородина маньчжурская
3460 кирказон маньчжурский

3461 вяз лопастный
3462 пихта цельнолистная
3463 липа маньчжурская
3464 клён маньчжурский
3465 аконит пёстрый
3466 орех маньчжурский
3467 тладианта сомнительная
3468 апельсин навель; апельсин пупочный; мандарин (итальянский); танжерин
3469 зелье сонное; мандрагора
3470 манеттия
3471 свёкла кормовая
3472 дерево манговое; манго
3473 мангольд; свёкла листовая
3474 манглис; мангостан; мангустан
3475 древокорень; ризофора
3476 манник
3477 леканора съедобная; манна (лишайниковая)
3478 руеллия
3479 многоплодник
3480 толокнянка
3481 клён
3482 клёновые
3483 вишня-гриот; вишня мараскиновая
3484 негниючник
3485 мараттиевые
3486 циссус
3487 маршанция
3488 сосенка водяная; хвостник
3489 сосенка водяная; хвостник обыкновенный
3490 торичник окаймлённый
3491 роза Юндзилла; шиповник Юндзилла
3492 азадирахта индийская; мелия индийская
3493 бархатцы; тагетес
3494 калохортус; марипоза
3495 чина морская; чина приморская
3496 полынь приморская
3497 майоран
3498 горох мозговой
3499 марсдения
3500 подмаренник болотный
3501 чистец болотный
3502 пятилистник; сабельник болотный
3503 плаун болотный; плаун заливаемый
3504 папоротник болотный; щитовник болотный
3505 скерда болотная
3506 хвощ болотный
3507 алтей аптечный; алтей лекарственный; рожа дикая; корень алтейный; трава алтейная
3508 калужница
3509 горичник болотный; петрушка болотная
3510 чина болотная
3511 горец перечный; горчак; перец водяной; трава лягушечья; трава лягушечная
3512 людвигия болотная
3513 крестовник водяной
3514 солерос травянистый
3515 шейхцерия болотная
3516 осока островатая
3517 осот болотный
3518 вероника щитковая
3519 бодяк болотный
3520 валерьяна болотная; валерьяна двудомная
3521 фиалка болотная
3522 кипрей(ник) болотный
3523 марсилиевые
3524 лилия кудреватая;лилия мартагон; лилия саранка; лилия чалмовая; мартагон; сарана большая; саранка; кудри царские
3525 плаунок Мартенза; селягинелля Мартенза
3526 калимерис
3527 мартиния
3528 кемквот; кинкан круглый; кинкан маруми; кинкан японский; кумкват круглый; кумкват японский; кумкват маруми; фортунелла японская
3529 алонсоа
3530 сосна Массонова
3531 астранция; звездовка
3532 горичник настурциевый; горичник царский корень; корень царский; царь-зелье
3533 белоус (прямой); белоус торчащий; нард вытянутый; щеть
3534 матико; перец узколистный
3535 дереза лебедолистная; дереза обыкновенная; лиций обыкновенный
3536 лофоспермум; маурандия
3537 пальма винная; пальма маврикиева
3538 папеда ежеиглистая; лима
3539 ноголист(ник); подофилл(ум); подофиллюм
3540 майтенус
3541 матрикария; ромашка
3542 пупавка вонючая; пупавка обыкновенная; пупавка собачья; ромашка вонючая; ромашка собачья
3543 нанноропс
3544 вишня птичья; вишня сладкая; черешня (птичья)
3545 рексия
3546 костёр прямой; костёр прямостоячий; костёр стоячий
3547 овсяница луговая
3548 лимнантес
3549 батлачек луговой; батлачок луговой; лисохвост луговой
3550 горечавка полевая
3551 герань луговая;журавельник луговой
3552 хвощ луговой
3553 овёс луговой; овсец луговой
3554 прострел луговой
3555 чина луговая
3556 василис(т)ник
3557 брандушка
3558 шалфей луговой
3559 козлобородник луговой
3560 камнеломка зернистая
3561 сивец луговой; чертогон луговой
3562 лабазник
3563 ком(м)ифора бальзамная
3564 медеола
3565 люцерна
3566 ревень аптечный; ревень китайский; ревень лекарственный; ревень московский; ревень русский
3567 мединилла
3568 алоэ барбадосское; алоэ обыкновенное

3569 пиретрум щитковый; пиретрум щитконосный; поповник щитковидный; поповник щитковый; поповник щитконосный
3570 бузина карликовая; бузина травянистая; бузинник
3571 коровяк выемчатый; коровяк извилистый
3572 пальма веерная низкая; пальма веерная европейская; пальма карликовая; пальма хамеропс; хамеропс низкий; хамеропс приземистый; пальмито
3573 кермек обыкновенный
3574 козелец разрезной; козелец рассечённый
3575 анагирис; вонючка
3576 просвирняк мавританский
3577 мушмула
3578 мушмула германская
3579 дерево каепутовое; дерево кайюпутовое; дерево каяпутовое; мелалеука
3580 меланкониевидные; меланкониевые
3581 меластомовые
3582 перловник
3583 кадило
3584 мелокактус
3585 паслён арбузолистный
3586 мелотрия
3587 ментцелия
3588 хопея
3589 пролеска; пролесник
3590 анизоптера
3591 кактус лофофора; лофофора Вильямса
3592 мезембриантемум
3593 дерево мескитовое; мескито; мимозка; прозопис
3594 агератум мексиканский
3595 кальцеолария мексиканская; кошельки мексиканские
3596 клевер мексиканский; ричардия шершавая
3597 кипарис лузитанский
3598 можжевельник повислый
3599 можжевельник мексиканский
3600 чубушник мексиканский
3601 платан мексиканский
3602 милла
3603 теосинт мексиканский
3604 бравоа
3605 ваниль душистая; ваниль плосколистная
3606 сосна Веймутова мексиканская
3607 смородина Мейера
3608 волчниковые; тимелеевые; ягодковые
3609 мибора
3610 михелия
3611 микромерия
3612 берёза Миддендорфа
3613 красоднев Миддендорфа
3614 резеда
3615 резедовые
3616 микания
3617 хиококка
3618 синадениум
3619 астрагал сладколистный; астрагал солодколистный
3620 асклепиас; ваточник; трава эскулапова
3621 асклепиадовые; ластовенные; ластовневые; ласточниковые
3622 горечавка ластовневая
3623 истодовые
3624 бор
3625 мимоза
3626 мимозовые
3627 клайтония пронзённолистная; клейтония пронзённолистная; шпинат кубинский
3628 мята
3629 губоцветные
3630 мирабель
3631 омела
3632 лорантовые; омеловые; ремнецветниковые; ремнецветные
3633 романцовия
3634 мителла
3635 митрария
3636 кольник; фитеума
3637 виддрингтония Уайта
3638 мниум; мох звездчатый
3639 эхиноцистис
3640 гикори белая; гикори белый; гикори косматый; кария белая; кария опушённая; орех насмешливый
3641 чубушник
3642 липа каменная; филирея
3643 дюшенея
3644 мелинис; трава моласская; трава паточная
3645 змееголовник молдавский; маточник
3646 молиния
3647 молюцелла; трава молукская
3648 спондиас
3649 пихта сильная
3650 берёза Максимовича
3651 вербейник луговой чай; вербейник монетчатый; чай луговой
3652 хвойник хвощевидный; хвойник хвощевый
3653 дуб монгольский
3654 тополь душистый
3655 губастик; мимулус; мимулюс
3656 ятрышник обезьяний
3657 дерево горшечное; лецитис
3658 араукария чилийская
3659 катасетум
3660 аконит; борец
3661 однодольные; односемядольные
3662 монодора
3663 монстера
3664 раффлезия
3665 кипарис крупноплодный
3666 сосна замечательная
3667 кипарис болотный мексиканский
3668 сосна мексиканская; сосна Монтезумы
3669 клён монпелийский; клён трёхлопастный
3670 калоникцион
3671 луносемянник; мениспермум
3672 кукольвановые; луносемянниковые; мениспермовые
3673 гроздовник ключ-трава; гроздовник обыкновенный; гроздовник полулунный; ключ-трава
3674 молиния голубая; синявка
3675 сеслерия
3676 морея

3677 лютик альпийский
3678 сморчок
3679 вишня кислая
3680 каштаноспермум; каштаносеменник австралийский
3681 морина
3682 ипомея; фарбитис
3683 вьюнковые
3684 адоксовые
3685 роза моховая; роза моховидная
3686 смолёвка бесстебельная
3687 хризантема индийская
3688 тимьян полевой; тимьян ползучий; трава богородская; чабрец обыкновенный; щебрец
3689 пустырник
3690 коровяк молевый; коровяк тараканий; коровяк тараканный; коровяк тёмный; трава молевая
3691 мукор; плесень
3692 юкка величественная; юкка прекрасная; юкка славная
3693 бурачок горный
3694 арника горная; баранка; баранник горный
3695 мелкоплодник; рябина
3696 папоротник горный пузырчатый; пузырник горный
3697 василёк горный
3698 кунция
3699 лютик горный
3700 жирянка альпийская
3701 астрагал хлопунец; хлопунец
3702 клевер белоголовка; клевер горный
3703 можжевельник обыкновенный горный; можжевельник сибирский
3704 вязель горный
3705 дубровник горный
3706 филлодоце
3707 цуга калифорнийская
3708 многорядник копьевидный
3709 манжетка альпийская
3710 кальмия широколистная;лавр горный; лавр американский
3711 медуница горная; медуница красная
3712 церкокарпус
3713 порезник горный
3714 мята горная; пикнантемум
3715 горичник горный; петрушка горная; поречник; трава прорезная
3716 чина горная
3717 подорожник горный
3718 осока горная
3719 кисличник
3720 вероника горная
3721 зверобой горный
3722 валериана горная
3723 кипрей(ник) горный
3724 фисташка атлантическая
3725 мятлик горный; мятлик рыхлый
3726 дриада восьмилепестковая; дриада восьмилепестная; трава (куропаточья) восьмилепестная
3727 кипарис (китайский) плакучий
3728 касатик сузианский
3729 ячмень заячий; ячмень мышиный
3730 костенец
3731 чистец германский
3732 ясколка скученная; ясколка скученноцветковая
3733 резуховидка Таля; резушка Таля
3734 ястребинка волосистая
3735 лисохвост мышехвост(н)иковый; лисохвост полевой
3736 мышехвостник
3737 акрокомия южноамериканская
3738 горичник олений
3739 мукоровые
3740 мукуна
3741 колеантус
3742 колеантус тонкий
3743 гетерантера
3744 осока топяная
3745 лужайник; лужница
3746 сосна-жереп
3747 полынь обыкновенная; полынь-чернобыль; чернобыль(ник)
3748 мюленбергия
3749 дерево тутовое; тут(а); тутовник; шелковица
3750 тутовые
3751 акация безжилковая; мульга
3752 коровяк
3753 лук картофельный
3754 маш; фасоль золотистая; фасоль кустовая однолетняя
3755 мунго; урд; фасоль маш; фасоль мунго; фасоль азиатская
3756 виноград круглолиственный; виноград круглолистный; мюскадина
3757 альбиция противоглистная
3758 гриб
3759 чертополох поникший
3760 мальва мускусная; просвирняк мускусный
3761 амбрет; гибискус мускусный
3762 дыня (мускатная);огурец дыня
3763 бровник
3764 губастик мускусный; мимулюс мускусный
3765 адокса
3766 адокса мускусная; бесславник; мускатница; мускусница
3767 роза мускусная
3768 гварея
3769 пахлук; тысячелистник мускатный; тысячелистник мускусный
3770 дерево касторовое африканское;рицинодендрон африканский
3771 горчица
3772 желтушник; хейриния; эризимум
3773 полёвка
3774 мицена
3775 миопорум
3776 комбретовые
3777 алыча; вишнеслива; миробалан; слива вишнелистная
3778 мирокарпус
3779 дерево мирровое; ком(м)ифора
3780 мирзиновые
3781 мирт(а)
3782 кориария миртолистная
3783 миртовые
3784 черника (обыкновенная); черница; ягодник черника
3785 ива миртолистная
3786 наяда
3787 наядовые
3788 приноготковник; приноготов(н)ик
3789 овёс (мелкозёрный) голый; овёс

татарский
3790 нандина
3791 лилия терракотовая
3792 калина канадская
3793 пеннисетум красный; трава слоновая
3794 вика нарбонская; вика римская; вика фрацузская;горошек нарбонский
3795 нарцисс
3796 ветреница нарциссоцветковая
3797 папоротник дубравный; щитовник буковый
3798 ясень узколистный
3799 мятлик узколистный
3800 мочёнка; рогоз узколистный
3801 пираканта узколистная
3802 лён узколистный
3803 скерда кровельная
3804 ястребинка зонтичная
3805 пикульник узколистный
3806 василисник узколистный
3807 таволга белая
3808 липа каменная узколистная; филирея узколистная
3809 желтозелье
3810 горошек узколистный
3811 поручейник узколистный
3812 жеруха;жерушник;капуцин;настурция
3813 капуциновые; настурциевые
3814 палаквиум
3815 омфалодес; пупочник
3816 трава пупочная; умбиликус
3817 сурепица
3818 нектарин; персик-нектарин
3819 ковыль
3820 можжевельник твёрдый
3821 рапидофиллум; рапидофиллюм
3822 болотница игольчатая
3823 дрок английский
3824 нейлия
3825 немезия
3826 немофила
3827 коричник железконосный; лавр ложнокамфарный
3828 непентес
3829 непентесовые; непентовые
3830 котов(н)ик; непета
3831 нефрома
3832 нерине
3833 ива сетчатая
3834 дыня обыкновенная; дыня сетчатая
3835 крапива
3836 крапивные
3837 лебеда стенная; марь стенная
3838 ятропа; ятрофа
3839 астра Новая Англия
3840 цеанотус американский
3841 робиния новомексиканская
3842 астра виргинская; астра Новая Бельгия; астра новобельгийская
3843 конопля новозеландская; лён новозеландский прядильный; пенька новозеландская; формиум
3844 тетрагония; шпинат новозеландский
3845 боярышник Брайндера
3846 зябра; кадило бабье; пикульник красивый
3847 бондук; цезальпиния аптечная
3848 самандура индийская
3849 гвизоция; нуг
3850 дрёма ночная
3851 никтантес
3852 никтериния; ночница
3853 паслён
3854 паслёновые
3855 пихта равночешуйчатая
3856 клён Нико
3857 пузыреплодник; пузырчатник
3858 ломонос шерстистый
3859 пальма нипа
3860 бородавник
3861 нителла
3862 нитрария; селитрянка
3863 пузатик
3864 паркия
3865 пихта благородная
3866 череда поникшая
3867 сирень поникшая
3868 перловник поникающий
3869 смолёвка поникающая; смолёвка поникшая
3870 птицемлечник поникший
3871 нолана
3872 нолина
3873 нонеа; нон(н)ея
3874 кипарисовник нутканский
3875 роза нутканская
3876 пихта кавказская;пихта Нордман(н)а
3877 араукария высокая
3878 подмаренник северный
3879 катальпа западная; катальпа красивая; катальпа прекрасная
3880 ежевика лежачая; ежевика плетевидная
3881 тайник сердцевидный
3882 смородина колосковая; смородина красная; смородина обыкновенная; смородина пушистая; смородина скандинавская
3883 дуб (красный) северный
3884 щитовник горный
3885 клён обыкновенный; клён остролистный; клён платановидный
3886 ель высокая; ель европейская; ель обыкновенная; ель западная
3887 лапчатка норвежская
3888 крупка норвежская
3889 носток
3890 кермек выемчатый; лаванда морская
3891 ива туполистная
3892 золотарник поздний
3893 сыть круглая; саламалик
3894 мускатник
3895 миристицевые; мускатниковые
3896 кария мускатная
3897 пеларгония душистая
3898 корень рвотный; орех рвотный;орешник рвотный; стрихнос рвотный орех; целибуха; чилибуха
3899 дуб
3900 губка дубовая
3901 щитовник Линнея
3902 рябина гибридная
3903 дазистома
3904 овёс
3905 райграс французский
3906 ситник туполепестный
3907 охна
3908 окотеа
3909 гребенщик одесский; тамарикс одесский

3910 одонтоглоссум
3911 эдогоний
3912 евгения малаккская; яблоня малайская
3913 каштан конский гладкий; каштан конский голый
3914 оидиум
3915 лён кудряш
3916 пальма масличная; пальма масляная
3917 тельфайрия
3918 кислица клубневая; кислица клубненосная; ока
3919 слива изящная
3920 аукумения; аукумея
3921 абельмош; бамия; бамья; гибискус съедобный; гомбо; окра
3922 льнянка канадская
3923 дерево божье; полынь божье дерево; полынь лечебная
3924 стрелолист китайский; стрелолист обыкновенный; стрелолист стрелолистный
3925 нериум; олеандр
3926 лоховые
3927 маслина
3928 маслинные; маслиновые; масличные
3929 онцидиум
3930 горошек одноцветковый
3931 лук
3932 громовик; оносма
3933 мак опийный; мак снотворный
3934 опопанакс
3935 селезёночник супротивнолистный
3936 афеландра золотистая
3937 питоспорум волнистый; смолосемянник волнистый
3938 молочай; подмолочник; подорешник
3939 лилия клубненосная красная; лилия шафранная
3940 гриб подосиновик; осинов(н)ик; подосиновик (красный)
3941 буддле(й)я изменчивая
3942 ястребинка оранжево-красная
3943 ежа быкновенная; ежа сборная; ежа скученная
3944 орхидея
3945 орхид(ей)ные; ятрышниковые
3946 ятрышник
3947 магония падуболистная
3948 биота восточная; туя восточная
3949 бук восточный
3950 пузырник восточный; пузырник кровавый
3951 вишня мелкопильчатая; сакура
3952 ломонос восточный
3953 смородина восточная
3954 дубровник восточный
3955 каркас Турнефора
3956 граб восточный; грабинник; грабник
3957 клён критский
3958 эджевортия бумажная
3959 платан восточный; платан обыкновенный; чинар (восточный)
3960 мак восточный
3961 кунжут восточный; кунжут индийский; кунжут культурный; сезам
3962 шлемник восточный
3963 спирея средняя; таволга средняя
3964 ель восточная; ель кавказская
3965 ликвидамбар восточный; дерево стираксовое
3966 полынь венечная; полынь метельчатая
3967 душица
3968 симаруба горькая
3969 телефиум; хлябник
3970 толстянковые
3971 лжеапельсин; маклюра
3972 апельсин-оседж; дерево лжеапельсиновое; маклюра оранжевая
3973 османтус
3974 чистоуст
3975 осмундовые; чистоустовые
3976 страусник обыкновенный
3977 фил(л)антус кислый
3978 оттелия
3979 ouратея
3980 кинкан маргарита; кинкан нагами; кинкан овальный; кумкват овальный; нагами кинкан; фортунелла жемчужная
3981 ива яйцевиднолистная
3982 дуб лировидный; дуб лирообразный
3983 болотница яйцевидная
3984 пентаклетра крупнолистная
3985 кислица; оксалис
3986 телекия
3987 нивяник обыкновенный; поповник
3988 первоцвет высокий;примула высокая
3989 горлюха; горчак
3990 алоэ бородавчатое
3991 оксилобус
3992 вешенка обыкновенная; вешенка рожковидная; опёнок уховидный
3993 сколимус
3994 пахицентрия
3995 пахизандра
3996 тсуга западная; цуга западная
3997 пихта Лоуа
3998 тис(с) коротколистный
3999 падук; птерокарпус
4000 педерота
4001 гебелия; софора
4002 пагудия; пахудия
4003 кастил(л)ейя
4004 клён красивый
4005 капуста китайская; паккой;пак-хой
4006 палафоксия
4007 бурачок чашечный
4008 фиалка белая
4009 крестовник болотный
4010 осока бледноватая
4011 каликантус плодовитый; каликантус плодующий
4012 кипрей розовый
4013 лапша грибная; рогатик жёлтый
4014 альстония школьная
4015 палиурус
4016 пальма
4017 сабаль
4018 пальмовые
4019 борассус пальмира; делет; пальма борассус; пальма пальмира; пальма пальмировая; пальма пальмирская; пальма лантар
4020 кортадерия
4021 гинериум; трава пампаская; трава пампасовая
4022 кастилла каучуконосная
4023 панкраций; панкрациум

4024 пангиум
4025 флакоуртия кохинхинская
4026 просо
4027 кельрейтерия метельчатая
4028 гидрангея метельчатая; гортензия метельчатая
4029 аконит метельчатый
4030 осока метельчатая
4031 мильтония
4032 мухомор пантерный; мухомор серый
4033 панус
4034 дынниковые; кариковые; папаевые
4035 дерево дынное (обыкновенное); карика папайа; папайа; папайя
4036 клён бумажный; клён серый
4037 берёза бумажная
4038 эдгеворция; эджевортия;эджеворция
4039 бруссонетия; бруссонеция
4040 орхидея-пафиния
4041 мотыльковые
4042 хохлатник
4043 папирус
4044 дикориния (параензис)
4045 кресс бразильский; кресс масляный; шпилант (травянистый)
4046 парадизка; райка; яблоко райское; яблоня восточная; яблоня кавказская; яблоня парадизка; яблоня райская
4047 симаруба сизая
4048 йерба-матэ; дерево чайное парагвайское; мате; матэ; падуб парагвайский; чай парагвайский; чай-матэ
4049 араукария бразильская; араукария узколистная
4050 гевея бразильская; хевея бразильская
4051 гриб-зонтик пёстрый; скрипица пёстрая
4052 паринари(ум)
4053 одноягодник
4054 подмаренник парижский
4055 двулепестник парижский; цирцея парижская
4056 листовка; пармелия
4057 пармелиевые
4058 белозор
4059 клиантус
4060 мириофиллюм; уруть
4061 железняк; парротия
4062 паррия
4063 петрушка
4064 пастернак
4065 гва(й)юла; партениум
4066 митчела; мичелла
4067 паспалум
4068 звезда кавалерская; пассифлора; страстоцвет
4069 пассифлоровые; страстоцветные
4070 гранадилла; гренадилья; пассифлора плодовая; пассифлора съедобная; страстоцвет съедобный
4071 фитцройя патагонская
4072 шпинат английский; щавель английский; щавель шпинатный
4073 валерьяна каменная; патриния
4074 павлония; пауллиния
4075 павловния; пауловния
4076 паветта
4077 азимина; дерево азиминовое
4078 свинуха; свинушка
4079 пайена
4080 горох
4081 сесбания египетская
4082 пави; персик (обыкновенный)
4083 колокольчик персиколистный
4084 ива миндалевидная
4085 гуилельма замечательная; пальма персиковая
4086 мицена колпачковидная
4087 арахис
4088 арахис культурный; арахис подземный; миндаль иерусалимский; орех земляной; орех подземный
4089 груша; дерево грушевое
4090 экзохорда
4091 китайка; яблоня китайская; яблоня сливолистная
4092 бажра; дохн; пеннисетум сизый; просо африканское; просо негритянское
4093 мшанка
4094 анафалис
4095 карагана; караганник; чапыжник; чилига; чилижник
4096 мхи сфагновые; мхи торфяные; сфагновые
4097 горошек гороховидный
4098 латирус; чина
4099 сочевичник
4100 кария пекан; орех-пекан; пекан
4101 кунжутовые; педалиевые; сезамовые
4102 гармала
4103 пеларгоний; пеларгониум; пеларгония
4104 кирказон крупноцветный
4105 постенница
4106 дизоксилон
4107 осока висячая
4108 пенициллярия
4109 кистевик; пеницилл(иум)
4110 пеннизетум; пеннисетум
4111 сердечник шершавый
4112 ярутка
4113 мята болотная; мята блошница; мята пулегиевая; полей
4114 водолюб; гидрокотиле; гидрокотиль; щитолистник
4115 пентакме
4116 вереск сицилийский
4117 пент(а)стемон
4118 пеон; пион
4119 пеперомия
4120 дерево перечное; перец (настоящий)
4121 перечные
4122 мята анлгийская; мята перечная мята холодка; холодка
4123 груздь (перечный)
4124 морковник
4125 шинус
4126 клоповник; кресс
4127 клоповник крупковидный; кресс крупка
4128 марсилея; марсилия
4129 лён многолетний; лён сибирский
4130 лунник многолетний; лунник оживающий
4131 дивало многолетнее
4132 латук многолетний

4133 чина широколистная
4134 трясунка средняя
4135 плевел многолетний; райграс английский; райграс пастбищный; райграс многолетний пастбищный; райграс многолетний английский
4136 пе(й)реския
4137 трюфель настоящий; трюфель французский; трюфель чёрный
4138 перилла; судза
4139 перистрофа
4140 барвинок; могильница
4141 пернетия
4142 перовския
4143 бегония всегда цветущая; бегония месячная
4144 персея
4145 крушина красильная
4146 лютик азиатский
4147 клевер опрокинутый; клевер перевёрнутый; клевер персидский;клевер шабдар; шабдар; шафтал
4148 ирис персидский
4149 сирень персидская
4150 боккаут; дерево железное; парротия персидская
4151 орех волошский; орех грецкий
4152 диоспирос; персимон; хурма
4153 дерево бальзамовое перувианское
4154 фабиана
4155 бузина перувианская
4156 вишня ананасная; вишня перуанская; вишня перувианская; вишня песья съедобная; томат земляничный; физалис перуанский; физалис съедобный; физалис ягодный
4157 петрея
4158 капуста пекинская; петсай; пецай
4159 молочай бутерлаковидный; молочай бутерлаковый; молочай садовый
4160 петуния; петунья
4161 фацелия; эвтока
4162 весёлковые
4163 пыльцеголовник; цефалянтера
4164 адонис осенний; горец осенний;глазки павлиньи; уголёк в огне
4165 филодендрон
4166 пламенник; флокс
4167 синюховые
4168 можжевельник красноплодный
4169 фотиния
4170 листоколосник; филлостахис
4171 фитокренум
4172 фитофтора
4173 такковые
4174 понтедерия
4175 понтедериевые
4176 арктерика
4177 горох голубиный; кайанус; ка(й)янус; каянус индийский; каянус каян; катьянг
4178 клитория
4179 гикори гладкий; гикори свиной; кария голая
4180 гименомицеты; грибы шляпочные
4181 чистяк
4182 осока шариконосная
4183 пилюльница
4184 пилокарпус
4185 очный цвет
4186 бедренец
4187 пинанга
4188 черёмуха пенсильванская
4189 пихта западногималайская
4190 сосна
4191 ананас культурный; ананас настоящий; ананас посевной
4192 ананасные; ананасовые;бромелиевые
4193 ромашка безлепестная; ромашка душистая; ромашка пахучая; ромашка ромашковидная
4194 сосновые
4195 виктория ананасная; земляника ананасная; земляника крупноплодная; земляника садовая;клубника ананасная; клубника ананасовая
4196 ватерия индийская
4197 гвоздика
4198 гвоздичные
4199 горец головчатый
4200 дуб болотный
4201 шерстестебельник
4202 зимолюбка; химафила
4203 пиптадения
4204 пиптантус
4205 фисташка
4206 вишня суринамская (питанга); питанга
4207 саррацения
4208 сосна жёсткая
4209 конопля маврикийская; конопля маврициева
4210 питтоспорум; смолосемянник
4211 плагиохила
4212 кореопсис красильный; ленок красильный
4213 остролодочник полевой
4214 тополь Саржента
4215 платан
4216 платановые
4217 клён белый; клён ложноплатановый; клён явор; сикомора; явор
4218 подорожник
4219 банан кухонный; банан плодовой; банан подорожник; банан райский; банан фруктовый; смоковница райская; банан пизанг; пизанг
4220 подорожниковые
4221 госта; функия
4222 платистемон
4223 плюхея
4224 слива
4225 зубнина; корень свинцовый; свинцовка; свинчатник
4226 вишенник; ивишень; подвишенник; подвишень; садовик
4227 ломонос жгучий
4228 ериантус; шерстняк; эриантус
4229 кедр речной новозеландский; кедр речной перистый
4230 боккония; маклейя
4231 можжевельник крупноплодный
4232 тис головчатый; цефалотаксус
4233 кукуруза плёнчатая; кукуруза чешуйчатая
4234 триостренник
4235 ногоплодник; подокарпус
4236 ногоплодниковые
4237 нарцисс белый; нарцисс поэтический

4238 бородатка
4239 пуанзеция
4240 гастролобиум
4241 болиголов крапчатый; болиголов пятнистый; омег пятнистый
4242 сумах восковой; сумах лаковый
4243 лаконос
4244 лаконосные; фитол(л)якковые
4245 ива полярная
4246 синюха
4247 пшеница полоникум; пшеница польская; рожь гималайская
4248 нарцисс букетный; нарцисс константинопольский; нарцисс-тацетта
4249 хруплявник
4250 истод
4251 многоножка; скулатник; широколиственница
4252 многоножковые
4253 трутовиковые
4254 гриб древесный; трутовик
4255 гранат; гранатник (обыкновенный); дерево гранатовое
4256 роза помпонная
4257 анона обыкновенная
4258 щавель водяной
4259 дерево камедное; сосна жёлтая; сосна жёлтая западная
4260 нитчатка; спирогира
4261 болотник прудовый; красовласка прудовая
4262 рдест
4263 рдестовые
4264 азалия жёлтая; азалия понтийская; одурь кавказская; рододендрон жёлтый; рододендрон понтийский
4265 дуб понтийский
4266 занихелия; занникеллия
4267 майяка
4268 кукуруза лопающаяся; кукуруза мелкосемянная
4269 разноорешник
4270 тополь
4271 ладанник тополелистный
4272 мак
4273 анемона садовая
4274 маковые
4275 центролобиум
4276 ракитник белый
4277 лавровишня лузитанская; лавровишня португальская
4278 дуб лузитанский
4279 дуб звездчатый; дуб малый
4280 берёза Потанина
4281 картофель (культурный); паслён клубненосный
4282 апиос
4283 джут длинноплодный; джут огородный длинноплодный
4284 лук огородный
4285 календула лекарственная; ноготки аптечные; ноготки лекарственные; ноготки садовые
4286 пшеница английская; пшеница тургидум; пшеница тучная
4287 костёр бесплодный
4288 ситник тонкий
4289 мучнеросные; мучнисторосные
4290 клевер американский степной; петалостемум
4291 кендырь сибирский; пуховник
4292 келерия гребенчатая
4293 клоповник безлепестный; клоповник безлистный
4294 скрытница
4295 зантоксилум; зантоксилюм; ксантоксилум
4296 дерево красное фернамбуковое; дерево пернамбуковое; дерево фернамбуковое; цезальпиния ежовая; цезальпиния шиповатая
4297 комфрей; окопник жёсткий; окопник острый; окопник шершавый
4298 дерево карандашное; можжевельник испанский; можжевельник красноватый; можжевельник красный
4299 латук компасный; латук лесной;молокан компасный; салат дикий
4300 колючезонтичник; колюченосник
4301 кактус опунция; опунция
4302 аргемон(а); аргемоне
4303 осока колосистая
4304 шпинат огородный; шпинат овощной
4305 осот жёсткий; осот колючий; осот шероховатый
4306 акантолимон; ёжик
4307 первоцвет; примула
4308 первоцветные
4309 горец восточный; персикария восточная
4310 принсепия
4311 притчардия
4312 бирючина; лигуструм
4313 мшанка лежачая
4314 макротомия
4315 просо метельчатое; просо обыкновенное; просо посевное
4316 амарант жминдовидный; щирица жминдовидная
4317 буркун; горец птичий; гречиха птичья; гусятница; спорыш
4318 протея
4319 псевдобактерия
4320 арктоус
4321 папоротниковые; папоротникообразные
4322 птихосперма
4323 берёза пушистая
4324 дуб пушистый
4325 дождевик
4326 дождевиковые
4327 пулазан (плодовый)
4328 агава магуэй; агава сальма; агава тёмно-зелёная
4329 пампельмус; пом(м)ело; помпельмус; шеддок; пуммело
4330 тыква обыкновенная; тыква летняя
4331 цекропия
4332 якорцы земляные; якорцы стелящиеся
4333 губка лиственничная
4334 кротон слабительный
4335 лён слабительный
4336 абрикос волосистоплодный; абрикос фиолетовый; абрикос чёрный;абрикос шероховатоплодный
4337 кобея лазающая
4338 барбарис Бретшнейдера
4339 ракитник пурпурный; ракитник пурпуровый
4340 белокопытник гибридный; белокопы-

тник лекарственный; подбел гибридный; подбел лекарственный; подбел лечебный; подбел обыкновенный
4341 ель пурпурная
4342 герань пурпурная
4343 крапива глухая красная; яснотка пурпуровая
4344 рудбекия пурпурная
4345 бук лесной тёмно-пунцовый
4346 горечавка пурпурная
4347 клевер альпийский
4348 пельтогине
4349 дербенник иволистный; дербенник плакун; плакун иволистный; плакун-трава; подбережник
4350 астрагал датский
4351 коровяк фиолетовый
4352 желтолоз(ник); ива пурпурная; ива пурпуровая; тальник
4353 косогорник красный; латук красный
4354 очиток пурпурный;очиток пурпуровый
4355 амбербоа туранская
4356 вика бенгальская
4357 синяк подорожниковый
4358 портулак
4359 портулаковые
4360 вероника иноземная
4361 портулакария
4362 пушкиния
4363 лапка кошачья
4364 ситовник
4365 тиллея
4366 ива травянистая
4367 кипарис (вечнозелёный) пирамидальный
4368 коровяк пирамидальный
4369 можжевельник китайский
4370 астра пиренейская
4371 горечавка пиренейская
4372 герань пиренейская
4373 аконит Антора; аконит противоядный
4374 дуб пиренейский
4375 мак пиренейский
4376 златоцвет; золотоцвет; пиретрум
4377 грушанка
4378 грушанковые
4379 пырей ползучий; ржанец
4380 осина американская; тополь осинообразный
4381 трясунка
4382 квассия
4383 пикразма; пикрасма
4384 кактус крупноцветный; кактус "царица ночи"; царица ночи; цереус крупноцветный
4385 киндаль; макадамия; орех австралийский
4386 галинзога; галинсога
4387 изоэтес; полушилица; полушник; расходник
4388 изоэтовые; полушниковые
4389 айва; дерево квитовое
4390 квиноа; квиноя; киноа; киноя; лебеда кино; лебеда перуанская; лебеда рисовая; марь чилийская
4391 квисквалис
4392 алоэ раздвоенное
4393 касатик вылощенный; касатик гладкий
4394 клевер пашенный; котики; лапки заячьи
4395 редька
4396 раф(ф)ия
4397 раф(ф)лезиевые
4398 горицвет-кукушкин цвет; зорька-дрёма; кокушник; кукушник; коронария кукушкин-цвет; слёзки кукушкины; цвет кукушкин
4399 дагусса; елевзина коракан; коракан; раги; токусса; такусса
4400 амброзия; амврозия
4401 желтуха; крестовник-желтуха; крестовник луговой; крестовник Якова
4402 куперия
4403 любка
4404 дерево дождевое; саман; энтеролобиум саман
4405 брунфельзия; францисцея
4406 говения; дерево конфетное
4407 рамалина
4408 колокольчик раскидистый
4409 нефелиум репейниковый; рамбутан
4410 конопля китайская; крапива китайская; рами белое
4411 рамонда
4412 рамондия
4413 флакуртия Рамонтша
4414 колокольчик рапунцель;колокольчик репчатовидный; колокольчик репчатый; рапункул; рапунцель
4415 лук медвежий; лук-черемша;черемша
4416 раулия
4417 рапс
4418 редечник; репник
4419 расамала
4420 малина
4421 трахилобий
4422 цереус плетевидный
4423 вульпия мышехвостная; овсяница мышехвостная
4424 каламус
4425 кроталярия
4426 погремок
4427 гроздовник виргинский
4428 гудайера
4429 косогорник; пренантес
4430 погремок
4431 склерия
4432 ольха красная
4433 багрянки; водоросли багрянки; водоросли красные
4434 ясень пенсильванский; ясень пушистый
4435 зубчатка бартсиевидная; зубчатка красная; зубчатка осенняя
4436 лавр красный; персея бурбонская; персея красная
4437 кизильник кистецветковый; кизильник кистецветный
4438 свёкла столовая красная
4439 переступень двудомный
4440 можжевельник Пинчота
4441 каштан конский красный; каштан конский розоцветный; павия красная; каштан конский павия
4442 клён кавказский (высокогорный); клён высокогорный;клён Траутветтера; клён Траутфеттера

4443 блисмус рыжий; камыш ржавый
4444 капуста красная; капуста краснокочанная
4445 лапагерия розовая
4446 арония арбутусолистная
4447 дятлина (красная); кашка красная; клевер красный; клевер луговой; конюшина; трилистник (луговой)
4448 моховик пёстрый
4449 эскаллония красная
4450 овсяница красная
4451 пихта великолепная;пихта красивая
4452 боярышник кроваво-красный
4453 медунка; пикульник ладанниковый; пикульник ладанный
4454 роза сизая; шиповник краснолистный
4455 клён красный
4456 шелковица американская;тут красный
4457 овёс византийский; овёс византийский красный; овёс (посевной) средиземноморский
4458 дёрен американский; дёрен отпрысковый; кизил отпрысковый;кизил побегоносный; свидина отпрысковая; свидина порослевая
4459 капсикум
4460 сосна смолистая
4461 дерево квебраховое; квебрахо; к(в)ебрачо
4462 малина европейская; малина обыкновенная
4463 лахнантес
4464 амарант колосистый; подсвекольник; щирица запрокинутая
4465 ель красная
4466 цинния многоцветковая
4467 очиток краснеющий
4468 рогатик гроздевой
4469 полевица белая; полевица белая высокая
4470 клён рыжеватожилковатый
4471 ива красная
4472 тростник
4473 двукисточник тростниковидный;двукисточник тростниковый; камыш степной; канареечник тростниковый; трава канареечная; трава шёлковая
4474 овсяница тростниковая; овсяница тростниковидная
4475 вейник
4476 лилия королевская
4477 лапина Редера
4478 ремания
4479 кладония альпийская; лишайник олений; мох белый олений; мох олений; ягель
4480 рейнекия
4481 кокушник
4482 осока раздвинутая
4483 костёр униоловидный; роговик униоловидный
4484 волчуг; стальник; трава воловья
4485 кемпферия
4486 большеголовник; корень маралий; рапонтикум
4487 реедия
4488 кактус рипсалис; рипсалис
4489 ризоктония
4490 рододендрон
4491 ропалостилис
4492 ревень
4493 эвкалипт прутовидный
4494 риччия
4495 рис
4496 рис посевной
4497 леерсия рисовидная
4498 пимелея
4499 оризопсис; рисовидка
4500 тетрапанакс
4501 ива Ричардсона
4502 подорожник большой
4503 виноград американский; виноград прибрежный; рипария
4504 берёза чёрная
4505 тростянка
4506 подостемон
4507 фиалка Ривина; фиалка Ривин(и)уса
4508 боярышник Робсона
4509 лишайник рокцелла; рокцелла; рочелла
4510 петрокаллис
4511 валенбергия
4512 криптограмма
4513 крушина скалистая; крушина скальная
4514 иберийка скальная
4515 кизильник горизонтальный
4516 резуха
4517 башенница; вяжечка
4518 кардаминопсис песчаный
4519 крупка альпийская
4520 смородина каменная; смородина камневая
4521 вяз ветвистый; вяз кистевидный; вяз Томаса
4522 вечерница; гесперис;фиалка ночная
4523 иберийка горькая; перечник горький; разнолепестник горький; стенник горький
4524 живокость Аякса; живокость Аяксова; сокирки Аяксовы
4525 индау посевной; рокет-салат; эрука посевная
4526 рогачка
4527 проломник
4528 петрофитум
4529 каландриния
4530 рододендрон ржаволистный; рододендрон ржавый
4531 ладанник
4532 ладанниковые
4533 осока скальная
4534 вероника кустарничковая
4535 гирофора
4536 лишайник-умбиликария
4537 мох фиалковый; камень фиалковый
4538 фукус
4539 аскофиллум
4540 полынь каменная; полынь скалистая; полынь скальная
4541 дугласия сизая; лжетсуга сизая
4542 роэлла
4543 роджерия
4544 соймида
4545 латук-ромен; латук-ромэн; латук римский; ромэн; эндивий летний
4546 пупавка благородная; ромашка римская
4547 крапива шариконосная

4548 полынь понтийская
4549 ронделетия
4550 денежка длинноножковая; коллибия длинноножковая; монетка коротко-корневая
4551 вольфия бескорневая
4552 рестио
4553 абрус; чёточник
4554 роза; шиповник
4555 акация мохнатая; акация розовая; акация щетинистая;робиния щетинистоволосая; робиния щетинисто-волосистая; робиния щетиноволосая
4556 джамбоза; помароза; ямбоза
4557 рододендрон большой; рододендрон древовидный; рододендрон крупнейший
4558 зорька-аксамитки; горицвет кожистый; лихнис кожистый
4559 волчник-боровик; волчеягодник-боровик
4560 розанные; розовые; розоцветные
4561 сабатия
4562 гибискус розелла; гибискус сабдарифа; кислица ямайская; розелла; розель; салат-розель
4563 розмарин
4564 розмарин аптечный; розмарин лекарственный
4565 родомирт
4566 герань розовая; пеларгоний розовый
4567 корень розовый; родиола розовая; родянка
4568 пеларгониум головчатый
4569 валериана высокая
4570 дальбергия
4571 сильфиум; сильфия
4572 таволга Ростгорна
4573 ротала
4574 пальма ползучая; пальма ротанговая; ротанг; тростник индийский; тростник испанский
4575 ривина
4576 клевер жестковолосый; клевер шероховатый; клевер шершавый
4577 скерда дву(х)летняя; скрипуха
4578 чина мохнатая
4579 мятлик обыкновенный
4580 роза карликовая; роза миниатюрная
4581 ива ушастая
4582 ситник сплюснутый
4583 ясень круглолистный; ясень мелколистный
4584 герань круглолистная
4585 киксия ложная
4586 белена белая
4587 чина круглолистная
4588 росянка круглолистная
4589 володушка круглолистная; ласковцы
4590 ива округлая
4591 джут короткоплодный; джут круглоплодный
4592 роксбургия
4593 осмунда; чистоуст величавый; чистоуст королевский
4594 пальма капустная
4595 павловния войлочная; павловния пушистая; пауловния войлочная
4596 недотрога железистая; недотрога Ройля
4597 рута
4598 анемонелла василистниколистная
4599 рутовые
4600 камнеломка трёхпалая; камнеломка трёхпальчатая
4601 айфанес
4602 роза морщинистая; шиповник морщинистый
4603 репник морщинистый
4604 руизия
4605 флакуртия зондская
4606 плаун булавовидный; плаун обыкновенный; зелёнка; колтунник; "ножки заячьи"
4607 смолёвка Рупрехта
4608 ситник
4609 ситниковые
4610 хондрилла обыкновенная; хондрилла ситниковидная
4611 миндаль грузинский;миндаль карликовый; миндаль низкий; миндаль степной; орех заячий; персик степной; черёмуха карликовая
4612 лиций русский
4613 кок-сагыз; одуванчик кок-сагыз
4614 вяз гладкий; вяз обыкновенный; ильм гладкий
4615 мордовник русский
4616 лох узколистный; джида; пшат
4617 карагана кустарниковая; карагана степная; чилига кустарниковая; чилига степная; чилижник кустарниковый
4618 баялыч; курай; солянка
4619 гриб трюфельный; трюфель русский чёрный; трюфель летний
4620 ива русская
4621 сыроежка
4622 сыроежковые
4623 грибы ржавчинные; ржавчинные
4624 схенус ржавый
4625 вудсия волосистая; вудсия северная; вудсия эльбская
4626 брюква
4627 василёк русский
4628 руишия
4629 змееголовник аргунский; змееголовник Руйша
4630 рожь культурная; рожь посевная
4631 плевел; райграс
4632 осока плевельная
4633 сабадилла
4634 нандина домашняя
4635 пихта мексиканская; пихта священная
4636 сафлор
4637 сафлор красильный; сафлор культурный; трава картамова
4638 шафран посевной
4639 сальвия; шалфей
4640 полынь; артемизия
4641 полынь полевая; полынь равнинная; нехворощ
4642 гетероспата высокая
4643 дерево саговое;саговник поникающий
4644 коелококкус; целококкус
4645 цереус гигантский
4646 эспарцет
4647 асцирум (зверобойниковидный)

4648 венечник лилиаго; венечник лилейный; венечник лилиецветковый
4649 парадизея
4650 бобы игнатьевские; игнаций; игнация
4651 зверобой
4652 зверобоецветные; зверобойные
4653 сигезбекия
4654 баугиния шерстистая
4655 пихта сахалинская
4656 горец сахалинский; гречиха сахалинская
4657 ель Глена
4658 кервель листовой; кервель настоящий; кервель обыкновенный; кервель посевной; кервель салатный; кервель; купырь бутенелистный; купырь садовый; снедок
4659 барбарис агрегатный
4660 малина великолепная; малина замечательная; малина превосходная
4661 сал; сорея кистевая; шорея мощная
4662 козлобородник
4663 лебеда
4664 костелецкия
4665 торичник солончаковый
4666 ситник Герарда; ситник Жерарда
4667 галимодендрон; чемыш; чингиль
4668 батис
4669 сальвадора
4670 ладанник шалфеелистный
4671 сальвиния
4672 сальвиниевые
4673 самандура
4674 критмум; серпник
4675 критмум морской
4676 аденантера павлинья; дерево коралловое
4677 дерево сандаловое; дерево санталовое; сандал
4678 сандаловые; санталовые
4679 дерево сандаловое; дерево сандальное; птерокарпус сандаловый; птерокарпус санталовый; сандал красный
4680 ива длиннолистная
4681 хура трескающаяся
4682 ценхрус
4683 вишня карликовая; вишня песчаная
4684 слива Ватсона
4685 лилия ландышевая натальская; сандерсония
4686 цератиола
4687 груша китайская песочная
4688 осока песчаная
4689 торичник
4690 тимофеевка песчаная
4691 фиалка скальная
4692 песчанка
4693 лечуха; подлесник
4694 санзеверия; сансевиерия; сансевьера; сансевьерия
4695 сандорик
4696 сандорик индийский
4697 санвиталия
4698 энтандрофрагма
4699 сапиум
4700 ахрас; дерево сапотиловое; дерево сапотовое; саподилла
4701 гуттаперчевые; сапотовые
4702 дерево саппановое; цезальпиния ост-индская; цезальпиния саппан
4703 грибы сапролегниевые
4704 сарака
4705 саргассум
4706 барбарис Саржента
4707 можжевельник Саржента
4708 таволга Саржента
4709 ель Саржента
4710 саза
4711 камелия гвоздичная; камелия горная; камелия масличная; камелия сасанква; камелия эвгенольная
4712 сассафрас
4713 императа
4714 жасмин японский
4715 магнолия Суланжа
4716 дерево колбасное; кигелия
4717 сос(с)юрея
4718 эхитес
4719 можжевельник донской;можжевельник казацкий; можжевельник казачий
4720 душевик
4721 сатурея; чабёр
4722 капуста савойская; савой; сафой
4723 ястребинка савойская
4724 кипарисовик горохоносный; лжекипарис горохоносный
4725 кариокар орехоносный; орех савара
4726 дербенка; дербянка
4727 марискус; меч-трава
4728 серпуха
4729 саксаул
4730 камнеломка; разрыв-трава; саксифрага
4731 пельтифиллум
4732 камнеломковые
4733 бедренец камнеломка
4734 василёк скабиозный; василёк скабиозовидный; василёк скабиозовый; василёк шероховатый
4735 вдовушка; вдовушки; скабиоза
4736 букашник многолетний
4737 скребница
4738 пилолистник чешуйчатый; языкотруб садовый
4739 гриб шпальный
4740 чешуйчатка обыкновенная
4741 вьюнок стрелолистный; скамоний
4742 валлота пурпурная
4743 пираканта колючая; пираканта красная; пираканта ярко-красная
4744 кадсура японская
4745 кливия оранжевая; кливия суриковая; кливия суриково-красная
4746 дуб американский шарлаховый; дуб кошенильный; дуб пунцоволистный; дуб шарлаховый
4747 очный цвет пашенный; очный цвет полевой; полевой очный цвет
4748 бобы красные; бобы многоцветковые; бобы огненные; бобы турецкие; фасоль кроваво-красная; фасоль многоцветковая; фасоль огненная; фасоль огненно-красная
4749 сальвия блестящая; шалфей великолепный; шалфей голый
4750 золотобородник
4751 ромашка непахучая

4752 жасмин непахучий; чубушник непахучий
4753 роза гвоздичная; роза эллиптическолистная; шиповник гвоздичный
4754 шейхцерия
4755 шейхцериевые
4756 колокольчик одноцветковый
4757 шмиделия
4758 берёза железная; берёза Шмидта
4759 дейция Шнейдера
4760 чубушник Шренка
4761 ель тяньшаньская; ель Шренка
4762 лук Шуберта
4763 виддрингтония Шварца
4764 склеротиния
4765 скополия
4766 вязель скудноцветный; вязель эмеровый
4767 личинник; скорпионница
4768 тофильдия болотная
4769 ракитник метельчатый; ракитник метлистый; жарновец метельчатый
4770 онопордум колючий; онопордум обыкновенный; татарник колючий; татарник обыкновенный
4771 вяз горный; вяз шероховатый; вяз шершавый; ильм горный; ильм шершавый; ильм шотландский
4772 вереск обыкновенный; рыскун
4773 бобовник альпийский; дождь золотой альпийский; ракитник альпийский
4774 лигустикум северный; лигустикум шотландский
4775 сосна лесная; сосна обыкновенная
4776 роза колючейшая; роза шотландская; шиповник колючий; шиповник бедренцов(олистн)ый
4777 хвош зимующий
4778 урера
4779 пандан(ус)
4780 пандановые; панданусовые
4781 геликтерес
4782 псоралея
4783 ложечница; трава ложечная
4784 циботиум
4785 свёкла дикая; свёкла приморская
4786 сланоягодник; халорагис
4787 облепиха
4788 кокколоба
4789 синеголовник морской; синеголовник приморский
4790 хлопчатник барбадосский; хлопчатник египетский; хлопчатник южноамериканский
4791 родимения
4792 струна морская; хорда нитчатая
4793 гониолимон; кермек
4794 ульва
4795 циртостахис
4796 глаукс; млечник
4797 халидрис
4798 ургинея
4799 горчица морская
4800 ситник морской
4801 камыш приморский; клубнекамыш морской; нюнька
4802 сведа морская; сведа приморская; содник морской
4803 бескильница морская; пуцинелла приморская
4804 повой круглолистный
4805 молочай прибрежный
4806 ламинария пальчатая; ламинария пальчаторассеченная
4807 ламинария японская
4808 дримария
4809 секуринега
4810 осока
4811 осоковые
4812 людвигия
4813 сведа; содник; суеда; чорак
4814 плаунок; папоротник водяной; селагинелла; селягинелла; селягинелля
4815 селагинелловые; селягинеллевые
4816 лойник; черноголовка
4817 гирча; селин
4818 гинериум серебристый; злак пампасовый
4819 плевропогон; плеуропогон
4820 финик изогнутый
4821 кайя сенегальская
4822 дерево мыльное сенегальское
4823 кассия; сенна
4824 цезальпиниевые
4825 нептуния
4826 оноклея чувствительная
4827 мимоза стыдливая
4828 ховея
4829 секвойя
4830 ель сербская; оморика
4831 козелец
4832 амеланхир; ирга
4833 рябина домашняя; рябина садовая
4834 кунжут
4835 сесбания
4836 жабрица
4837 лапчатка семилисточковая;лапчатка тюрингенская; лапчатка тюрингская
4838 ирга канадская
4839 овсяница разнолистная
4840 геонома
4841 гикори крупнопочечный; кария овальная; кария косматая; пекан яйцевидный твёрдокорый
4842 ястребинка шерстистая
4843 навозник белый
4844 лук Аскалона; лук-шалот; сорокозубка; ша(р)лот
4845 дуб курчавый
4846 болотовик; козляк; решетник
4847 парохетус
4848 спаржа остролистная
4849 якаранда остролистная
4850 верба красная; верболоз; ива краснолистная; ива остролистная; краснотал; ива шелюга; шелюга красная
4851 киксия повойничек; киксия повойничковая
4852 ши
4853 поплавок влагалищный; поплавок серый; поплавок шафранный
4854 ацена
4855 кустарник овечий; пентция
4856 уроспермум
4857 букашник горный
4858 овсяница бороздчатая; овсяница

овечья; типчак; трава овечья
4859 щавелёк; щавель воробьиный
4860 кохлоспермум
4861 гребень Венерин; скандикс Венерин гребень; скандикс гребенчатый
4862 сумка пастушья обыкновенная; сумочник пастуший
4863 гриб-зонтик мелкощитовидный
4864 пел(ь)тария; щит
4865 дуб гонтовый; дуб черепитчатый; дуб черепичный
4866 рдест блестящий
4867 говорушка беловатая
4868 желторог рябинолистный; ксантоцерас рябинолистный; орех чекалкин
4869 бафия красивая
4870 галодуле
4871 дряква; дряквенник; додекатеон
4872 сореа; сорея; шорея
4873 гринделия высокая; гринделия исполинская
4874 можжевельник прибрежный; можжевельник приморский
4875 лебеда прибрежная
4876 сосна скрученная
4877 триостренник морской; триостренник приморский
4878 лук морской
4879 литторела; прибрежник;прибрежница
4880 брахиэлитрум
4881 шортия
4882 сосна жёлтая; сосна ежовая
4883 клевер изящный; клевер нарядный
4884 кроталярия
4885 горошек крупноцветковый
4886 гибискус сирийский; роза азиатская; роза сирийская; роза Шарона
4887 изотома
4888 володушка кустарниковая
4889 сиббальдия
4890 сиббальдия распростёртая
4891 абрикос сибирский
4892 астра сибирская
4893 барбарис сибирский
4894 колокольчик сибирский
4895 смолёвка сибирская
4896 клематис сибирский;княжик сибирский
4897 хохлатка сибирская
4898 борщевник сибирский
4899 крэб сибирский; ранетка пурпурная; яблоня (дикая) сибирская; яблоня корнесобственная; яблоня Палласова; яблоня ягодная
4900 смородина двуиглая; смородина скальная
4901 кизил сибирский; кизил татарский; свидина сибирская; свидина татарская
4902 крупка сибирская
4903 берест приземистый; вяз приземистый; ильм приземистый;ильмовник
4904 пихта сибирская
4905 герань сибирская
4906 купальница азиатская; огоньки
4907 касатик сибирский
4908 лиственница сибирская
4909 истод сибирский
4910 акация сибирская; акация жёлтая; карагана высокая; карагана древовидная; караганник древовидный; челыжник
4911 флокс сибирский
4912 зверобой изящный
4913 чингиль серебристый
4914 кермек Гмелина; кермек Гмелинов
4915 ель сибирская
4916 пролеска сибирская; сцилла сибирская
4917 кедр сибирский; сосна кедровая сибирская; сосна сибирская
4918 сибирка
4919 шалфей Сибторпа
4920 сумах кожевенный; сумах красильный; сумах обыкновенный; сумах сицилийский; дерево кожевенное
4921 буркун; джонушка; люцерна жёлтая; люцерна серповидная; юморка
4922 молочай серповидный
4923 чешуехвостник
4924 чешуехвостник согнутый
4925 грудинка; просвирнячок; сида
4926 рамишия однобокая; рамишия однобочная
4927 овёс венгерский; овёс восточный; овёс египетский; овёс косой; овёс турецкий
4928 сидеродендрон
4929 тополь Зибольда
4930 барбарис Зибольда
4931 бук Зи(е)больда
4932 дейция Зибольда
4933 тсуга Зибольда; тсуга японская; цуга Зибольда
4934 клён Зибольда
4935 калина Зибольда
4936 орех айлантолистный; орех Зибольда
4937 зиглингия
4938 можжевельник западный
4939 копайфера копалоносная
4940 боб лимский; бобы каролинские; бобы лимские мелкоплодные; бобы сива; лима мелкоплодная; фасоль полулунная
4941 сиверсия
4942 ель восточно-гималайская; ель шиповатая
4943 ива силезская
4944 баобабовые; бомбаксовые
4945 гревиллея крупная
4946 фунтумия
4947 гаррия
4948 акация шёлковая; альбиция ленкоранская; альбиццая шёлковая
4949 обвойник
4950 галезия; халезия
4951 лох серебристый
4952 ягода буйволова
4953 аргитамния
4954 лапчатка серебристая
4955 пихта белая; пихта гребенчатая; пихта европейская; пихта обыкновенная
4956 мискантус; трава серебряная
4957 акация беловатая; акация подбеленная; акация серебристая
4958 крестовник акантолистный; пепельник приморский; цинерария ковровая; цинерария летняя; цине-

рария морская; цинерария приморская
4959 аира гвоздичная
4960 боярышник восточный
4961 леукофиллум
4962 аргиролобий; аргиролобиум
4963 подсолнечник серебристолистный
4964 дерево серебряное; лейкодендрон серебристый
4965 липа венгерская; липа войлочная; липа серебристая
4966 клён сахаристый; клён серебристый
4967 липа длинночерешковая
4968 актинидия многобрачная; актинидия носатая; актинидия полигама; актинидия полигамная; актинидия трёхдомная
4969 лапка гусиная; лапчатка гусиная; трава гусиная
4970 осока сероватая
4971 симаруба
4972 ясень неправильный
4973 боярышник однопестичный; боярышник одностолбчатый
4974 можжевельник чешуйчатый
4975 сифония
4976 агава сизалевая; агава сизаловая; сизал(ь)
4977 дальбергия сиссоо; дерево палисандровое сиссоо
4978 гулявник
4979 ель ситхинская
4980 ива ситхинская
4981 ячмень культурный шестирядный; ячмень шестигранный; ячмень шестирядный
4982 повойничек шеститычинковый
4983 хондрил(л)а; хондрилля
4984 скимия
4985 водосбор Скиннера
4986 поручейник сахарный (корень); корень сахарный
4987 шлемник
4988 клопогон вонючий; ромашка клоповая; ромашка клоповная
4989 менцизия
4990 симплокарпус
4991 смородина железистая
4992 дуранта
4993 сосна караибская; сосна карибская; сосна Эллиота
4994 жабник маленький; жабник малый
4995 дейция изящная; дейция красивая
4996 коротконожка лесная
4997 рябчик горный; рябчик тонкий
4998 володушка тончайшая
4999 горец малый
5000 кроталярия промежуточная
5001 лён тонколистный
5002 наяда гибкая
5003 овёс бородатый
5004 горошек изящный
5005 щирица гибридная
5006 миксомицеты; грибы слизевые; грибы слизистые; слизевики
5007 двурядка тонколистная
5008 вейник незамечаемый; вейник незамеченный
5009 василисник простой
5010 педилантус
5011 кальцеолария; кальцеолярия; кошельки
5012 вяз ржавый; ильм сибирский
5013 слонея
5014 слива колючая; тёрн; терновик
5015 бекмания
5016 бейльшмидия
5017 подмаренник трёхдольный; подмаренник трёхраздельный; подмаренник трёхнадрезный
5018 кровохлёбка малая; черноголовник кровохлёбковый; пимпинель
5019 клюква болотная; клюква обыкновенная
5020 лопух малый; лопушник малый
5021 сердечник мелкоцветковый; сердечник мелкоцветный
5022 энотера короткоиглая
5023 недотрога мелкоцветковая
5024 кипрей(ник) мелкоцветковый; кипрей мелкоцветный
5025 роза мелкоцветковая; шиповник мелкоцветковый
5026 герань маленькая; герань мелкая
5027 крутай; мордовник русский;мордовник обыкновенный
5028 бобы мелкоплодные
5029 арнозерис малый; баранец малый
5030 редис
5031 смилаксовые
5032 рисовидка
5033 дерево париковое; скумпия
5034 костёр безостый
5035 колючелистник бесстебельный; колючник бесстебельный
5036 гипохерис голый; пазник гладкий; пазник голый
5037 росичка кровоостанавливающая; росичка линейная
5038 скерда волосовидная; скерда зелёная; скерда колосовидная
5039 гидрангея древовидная; гортензия древовидная
5040 берест листоватый; вяз граболистный; вяз листоватый; вяз полевой; ильм листоватый
5041 пальма саговая
5042 сумах гладкий; сумах голый
5043 вудсия гладкая; вудсия гладковатая
5044 коккулус; коккулюс; коломбо; кукольван
5045 трихозант(ес); трихозантус
5046 дыня змеевидная; дыня извилистая; дыня изогнутая; дыня огуречная; дыня повислая; тарра
5047 синеголовник полевой; синеголовник равнинный
5048 стальник жёлтый
5049 пиратинера гвианская
5050 зев львиный
5051 бальзамин; недотрога
5052 гелениум
5053 птероксилон
5054 ахиллеа-зоря; птармика; трава чихотная; тысячелистник птармика; тысячелистник чихотный; чихотник обыкновенный
5055 камнеломка снежная
5056 дерево стираксовое; стиракс
5057 снежник; снежноягодник; ягода

снежная
5058 рябина двуцветная; рябина китайская; рябина пёстрая
5059 лютик снеговой
5060 галантус; подснежник
5061 ветреница лесная
5062 амариллис альпийский; белоцветник
5063 горечавка снежная
5064 ясколка шерстистая
5065 грушанка малая
5066 груша снежная
5067 спирея городчатая; таволга городчатая
5068 невиузия
5069 лапчатка снежная
5070 квил(л)айя; килайя
5071 дерево мыльное; квил(л)айя мыльная; кора мыльная
5072 дерево мыльное; сапиндус
5073 каштановые; сапиндовые
5074 мыльница; мыльнянка; сапонария; трава мыльная
5075 бегония полуклубневая
5076 алоэ сокотрина; алоэ сокотринское
5077 акант мягколистный
5078 костёр мягкий
5079 певник болотный; ятрышник шлемовидный; ятрышник шлемоносный
5080 клематис цельнолистный; ломонос цельнолистный
5081 смилацина
5082 купена; печать Соломонова
5083 солорина
5084 сорбарония
5085 сорго сахарное
5086 сорго
5087 сорго обыкновенное; сорго африканское зерновое; хлеб каффрский
5088 кисличные
5089 ревень дланевидный; ревень пальчатый
5090 дазилирион
5091 вишня обыкновенная; вишня садовая; вишня чёрная поздняя
5092 апельсин горький; апельсин бигардия; апельсин кислый; бигардия; померанец (горький);цитрус бигардия
5093 анона игольчатая; анона колючая
5094 оксидендрон
5095 андромеда древовидная; оксидендрум древовидный
5096 люпин изменчивый; люпин пестроцветный
5097 каталпа бигнониевидная; каталпа сиренелистная
5098 лавр тюльпанный; магнолия крупноцветковая; магнолия крупноцветная
5099 адиантум Венерин волос; Венерин волос; Венерины волосы; папоротник "волосы Венеры"
5100 тополь бальзамический
5101 дуб красный; дуб серповидный; дуб серполистный
5102 дерево мыльное антильское; дерево мыльное обыкновенное; сапиндус мыльный; ягода мыльная
5103 сивец южный; синяк изогнутый; скабиоза южная
5104 вереск восковой; восковница восконосная; мирт восконосный
5105 осот
5106 бобы соевые; соя (культурная);соя щетинистая
5107 початковые
5108 буксус балеарский
5109 костёр мадридский
5110 пихта испанская
5111 касатик испанский
5112 блошак; златокорень (испанский); сколимус испанский; кардуль
5113 платан испанский
5114 очиток испанский
5115 спарманния
5116 тимелея; тимелия
5117 ардизия
5118 лебеда копьевидная; лебеда копьелистная
5119 дориантес
5120 мята зелёная; мята колосовая
5121 лютик жгучий; прыщинец
5122 ольха белая; ольха серая; елоха
5123 вероника
5124 польба (настоящая); пшеница-спельта; спел(ь)та; полба
5125 мох белый болотный; мох торфяной; сфагнум
5126 сфериальные
5127 дерево бензойное
5128 клеома; клеоме; паучник
5129 гинандропсис
5130 гименокаллис; лилия нильская; хименокаллис
5131 боэргавия
5132 брассия
5133 живучка паутинная
5134 традесканция
5135 ком(м)елиновые
5136 спигелия
5137 острокильница чернеющая; ракитник чернеющий
5138 уруть колосистая
5139 кольник колосистый; кольник колосовидный
5140 вероника колосистая; крест андреев
5141 ожика колосистая
5142 брукенталия
5143 пилокарпус перистолистный
5144 бойсдувалия
5145 болотница; ситняг
5146 трищетинник колосистый
5147 корилопсис колосковый
5148 шпинат
5149 гнетум гнемон
5150 акантофеникс
5151 акант колючий
5152 бурачок колючий
5153 щирица колючая
5154 бактрис; пальма-бактрис
5155 дурнишник колючий; ксантиум колючий
5156 наяда морская
5157 пальма саговая Румфа
5158 костус
5159 валлиснерия спиральная
5160 спирея; таволга
5161 асплений; костенец; папоротник-костенец
5162 спилантес; шпилант

5163 ятрышник пятнистый
5164 гипохерис короткокорневой; гипохерис укореняющийся; пазник укоренившийся; пазник укореняющийся
5165 василёк пятнистый
5166 крапива глухая пятнистая; яснотка крапчатая; яснотка пятнистая
5167 молочай пятнистый
5168 горец почечуйный; почечуй(ник); трава блошная; трава геморойная; трава почечуйная
5169 вербейник пятнистый; вербейник точечный
5170 люцерна арабская; люцерна аравийская; люцерна пятнистая
5171 лептохлоя
5172 ларрея
5173 кендырь проломниковый
5174 вяз раскидистый
5175 желтушник выгрызенный; желтушник растопыренный
5176 купырник полевой
5177 прострел луговой раскрытый; прострел поникающий; прострел раскрытый; прострел распростёртый; прострел сон-трава; сон-трава
5178 аспарагус Шпренгера; спаржа Шпренгера
5179 адонис весенний; горицвет весенний; горицвет черногорка;желтоцвет весенний; черногорка
5180 мухомор весенний; поганка весенняя
5181 клайтония; клейтония
5182 лапчатка весенняя
5183 веснянка весенняя; крупка весенняя
5184 норичник весенний
5185 горечавка весенняя
5186 эрика мясо-красная; эрика румяная
5187 душёвка обыкновенная; душёвка чабрецевидная; пахучка острая; пахучка остролистная; пахучка полевая
5188 белоцветник весенний
5189 ель
5190 шпороцветник
5191 молочайные
5192 волчеягодник лавролистный;волчник лавролистный
5193 галения
5194 кнеорум; оливник
5195 надбородник безлистный
5196 торица
5197 горох спаржевый; лядвенец четырёхлопастный; тетрагонолобус пурпурный; четырёхкрыльник пурпуровый
5198 зверобой четырёхгранный
5199 зверобой крылатый
5200 пролеска; ряст; сцилла
5201 марзанка; ясменник розовый
5202 огурец бешеный (обыкновенный)
5203 бересклетовые; краснопузырниковые
5204 мох дубовый
5205 папоротник олений рог
5206 дерево уксусное; сумах коротковолосый; сумах оленьерогий; сумах пушистый; сумах уксусный; уксусник
5207 трагус кистевидный
5208 лебеда стебельчатая; лебеда черешчатая
5209 берула прямая; беруля прямая
5210 житняк пустынный; житняк узколистный
5211 златолист; хризофиллум; хризофиллюм
5212 туррея
5213 астрофитум; кактус звездчатый
5214 лук гадючий кистевидный; лук гадючий кистистый; кистецветник обыкновенный; мускари
5215 кукуруза крахмалисто-сахарная;кукуруза крахмально-сахарная
5216 пентас
5217 ряска трёхдольная;ряска тройчатая
5218 седмичник; троечница
5219 звёздоплодник частуховидный
5220 квамоклит
5221 алетрис
5222 осока звездчатая
5223 пролеска прелестная
5224 трахелоспермум
5225 носток обыкновенный
5226 леукокринум
5227 магнолия звездчатая
5228 орнитогалум; птицемлечник
5229 хирония
5230 ясколка полевая
5231 камнеломка звездчатая
5232 василёк колючеголовый
5233 гигрофила
5234 астрониум
5235 ал(ь)зина; звездчатка; мокрица алзина; мокричник; стеллария
5236 ясколка трёхстолбиковая
5237 стонтония; стаунтония
5238 бодяк бесстебельный
5239 стенанциум
5240 стеногинум
5241 стефанандра
5242 штернбергия
5243 качим Стевена
5244 стевия
5245 стюартия
5246 липучка; турица
5247 крестовник клейкий
5248 качим вязкий
5249 шалфей клейкий
5250 стикта
5251 жесткомятлик
5252 пырей волосистый; пырей волосоносный
5253 пальма-ириартея
5254 дракункул
5255 полевичка крупная; полевичка крупноколосковая; полевичка опушённая
5256 весёлка вонючая; весёлка обыкновенная; весёлка рядовая;фаллус; яйцо чёртово
5257 двурядка стенная
5258 левкой; маттиола
5259 стоксия
5260 костяника (каменистая)
5261 крылотычинник; этионема
5262 розеточница
5263 горноколосник
5264 очиток; родиола; седум
5265 кунила
5266 тарриетия; тарриеция; тарриэтия
5267 лучица; хара

5268 касатик безлист(н)ый
5269 стираксовые
5270 странвезия
5271 корригиоля береговая; языки воробьиные
5272 земляника; клубника
5273 марь многолистная
5274 клевер земляниковидный; клевер земляничный; пустоягодник
5275 вишня алкекенги; вишня полевая; вишня Франшетти; вишня пёсья обыкновенная; вишня пёсья Франшетта; мохунка; мошнуха; физалис Франшетта; физалис обыкновенный
5276 дерево земляничное крупноплодное; земляничник крупноплодный
5277 камнеломка корнеотпрысковая; камнеломка отпрысковая; камнеломка ползучая
5278 каликантовые
5279 иммортель; бессмертник; цмин прицветниковый
5280 наперстянка жёлтая
5281 лук торчащий
5282 осока береговая
5283 кроталярия остроконечная; кроталярия полосатая
5284 клён пенсильванский
5285 стромбозия
5286 буррерия
5287 строфант
5288 строфария
5289 стилидиум
5290 люффа цилиндрическая; тыква мочальная; тыква цилиндрическая
5291 клевер подземный
5292 клевер сомнительный; клевер маленький
5293 кола блестящая; кола заострённая; кола культурная; кола остролистная; орех кола; орешник-кола
5294 сорго суданское; суданка; трава суданская
5295 мытник судетский
5296 ожика судетская
5297 анона чешуйчатая; канель
5298 свёкла сахарная; свекловица сахарная
5299 ламинария сахарная
5300 тростник сахарный (благородный); тростник сахарный культурный
5301 кукуруза сахарная; кукуруза сладкая столовая
5302 финик лесной
5303 поллиния
5304 клён сахарный
5305 аренга
5306 сосна Ламберта; сосна сахарная ; сосна сахарная гигантская
5307 лагунария
5308 протея медоносная
5309 сорго сахарное
5310 копеечник венцевидный; копеечник короноподобный; сулла; эспарцет испанский; хедизарум
5311 клевер бледно-жёлтый; клевер светло-жёлтый
5312 трутовик серножёлтый
5313 кожевник; сумах; шмальция
5314 каламус драконовый
5315 дерево бензоиновое; стиракс бензоин(овый)
5316 дорема
5317 адонис летний; горицвет летний
5318 кипарис летний; кохия
5319 гальтония
5320 флокс метельчатый
5321 чабёр садовый
5322 белоцветник летний
5323 клетра
5324 лжедурнишник; циклахена
5325 росница; росянка
5326 росянковые
5327 люпин многолетний
5328 молочай солнцегляд; солнцегляд
5329 подсолнечник
5330 ломонос джунгарский
5331 кроталярия индийская; кроталярия ситниковая; кроталярия ситниковидная; пенька индийская
5332 гелиамфора
5333 аккроклиниум; гелиптерум; стеблеклонник
5334 нежник; солнцегляд; солнцецвет
5335 серьяния
5336 берхемия
5337 филлоспадикс
5338 дерево квассиевое; кассия горькая; квассия горькая
5339 свейнсона
5340 ластовенник; ластовень; ластовник; цинанхум
5341 подмаренник топяной
5342 ежевика щетиноволос(н)ая
5343 дуб дубильный; дуб коровий; дуб Мишо
5344 хвощ топяной
5345 ваточник инкарнатный; ваточник розово-малиновый
5346 гелониас; хелониас
5347 дуб дву(х)цветный
5348 цикнохес
5349 астрагал боэтийский
5350 рябина промежуточная; рябина скандинавская
5351 акация Фарнеза; акация Фарнеси
5352 алиссум душистый; алиссум европейский; алиссум морской; бурачок приморский; конига морская; конига приморская
5353 ломонос метельчатый
5354 базилик душистый; базилик камфарный; базилик обыкновенный;душки
5355 лавр белый; лавр душистый; магнолия виргинская; магнолия сизая
5356 жимолость голубая; жимолость синяя
5357 берёза вишнёвая; берёза вишенная
5358 пальма-брахея мексиканская
5359 роза ржавая; роза эглантерия; шиповник ржавчинный; шиповник эглантерия
5360 маниок сладкий
5361 кервель испанский; кервель мускусный; миррис душистый; миррис пахучий
5362 донник
5363 мартиния душистая
5364 комптония
5365 аир; лепеха
5366 восковник обыкновенный;восковница

обыкновенная; мирт болотный
5367 зубровка; лядник
5368 зубровка душистая; зубровка пахучая; лядник душистый; лядник пахучий; чаполоть; чополоть
5369 ликвидамбар
5370 жимолость душистая; жимолость каприфоль; жимолость козья; каприфоль
5371 касатик бледный
5372 млечник сладковатый
5373 симплокос
5374 симплоковые
5375 душица; майоран летний; майоран садовый; трава колбасная
5376 жасмин грунтовой; жасмин дикий; жасмин ночной; жасмин садовый; чубушник венечный
5377 апельсин; апельсин настоящий; апельсин сладкий; дерево апельсиновое
5378 горошек душистый; чина душистая
5379 подорожник средний
5380 батат; ипомея батат; картофель сладкий; патат
5381 осмориза
5382 сансевиерия гвинейская; сансевиерия кистецветковая
5383 скабиоза тёмно-пурпурная
5384 подмаренник трёхцветковый
5385 говорушка душистая; говорушка пахучая
5386 каликантус; чашецветник
5387 итея
5388 белолист; белолиственник
5389 колос пахучий; колосок душистый обыкновенный; пахучеколосник
5390 гедизарум; копеечник
5391 калина душистая
5392 фиалка душистая
5393 гвоздика бородатая; гвоздика турецкая
5394 смолёвка армериевидная
5395 пахучка; ясменник душистый; ясменник пахучий
5396 полынь однолетняя
5397 сверция; трипутник
5398 свёкла листовая артишоковая
5399 незабудка крупноцветная
5400 сосна горная
5401 проломник швейцарский
5402 кедр европейский; сосна кедровая (европейская); сосна шведская
5403 просо прутовидное; просо прутьевидное
5404 ряска горбатая
5405 нефролепис
5406 канавалия мечевидная; канавалия мечелистная
5407 девясил мечелистный
5408 инжир ослиный; сикамор античный; сикомор; смоковница дикая; фикус-сикомор
5409 акация длиннолистная
5410 таволжник обыкновенный
5411 хвощ лесной
5412 фиалка лесная
5413 спайнолепестные; сростнолепестные
5414 симфония
5415 иксиолирион горный
5416 ясень сирийский
5417 головчатка сирийская; махобели
5418 можжевельник косточковый
5419 пальма медовая; юбея величавая; юбея замечательная
5420 камыш Табернемонтана; схеноплектус Табернемонтана
5421 такка
5422 артаботрис
5423 талаума
5424 корифа зонтичная; корифа теневая; пальма зонтичная; пальма тени; пальма тал(л)иповая
5425 курослеп; лютик едкий
5426 донник высокий
5427 овёс высокий; райграс высокий; райграс французский высокий
5428 ксимения американская
5429 шлемник высокий; шлемник высочайший
5430 стюартия-монодельфа
5431 фиалка высокая
5432 пырей русский; пырей удлинённый
5433 тамаринд
5434 тамаринд индийский;финик индийский
5435 бисерник; гребенчук; гребенщик; жидовник; тамарикс; тамариск
5436 бисерниковые; гребенщиковые; тамариксовые
5437 стефанандра Танаки
5438 чина танжерская
5439 камнеплодник; пазания
5440 пижма
5441 пиретрум крупнолистный
5442 дескурайния; дескурения
5443 фацелия пижмолистная
5444 флоуренсия
5445 полынь эстрагон; трава драгун; эстрагон; тургун
5446 гречиха татарская; кырлык
5447 эвкалипт голубой
5448 фитцройя Арчера; фитцройя тасманийская
5449 многоплодник альпийский
5450 лук гадючий хохлатый; леопольдия хохлатая; леопольдия хохолковая
5451 эмилия
5452 катран татарский
5453 головчатка татарская
5454 дёрен белый; кизил белый; свидина белая
5455 жимолость татарская
5456 клён татарский; черноклён
5457 лебеда татарская
5458 гониолимон татарский; кермек татарский
5459 смолёвка татарская
5460 иберийка крымская
5461 ясколка Биберштейна
5462 онопордум крымский
5463 таксодиевые
5464 чайные
5465 тектона
5466 ива двуцветная; ива филиколистная
5467 дерево чайное; чай
5468 роза чайная (настоящая)
5469 ворсянка
5470 клевер мелкоцветковый
5471 ворсянковые; мориновые
5472 лептосперм(ум)

5473 полевичка абиссинская; полевичка тефф; тефф абиссинский; трава абиссинская
5474 трава телеграфная
5475 теосинт; теозинт(а); теозинте
5476 дерево кевовое; дерево скипидарное; дерево терпентинное; дерево терпентиновое
5477 тетрацера
5478 фасоль остролистная; тепари
5479 орех скальный
5480 тринакс
5481 капуста собачья; телигонум
5482 телимитра
5483 термопсис
5484 виноград девичий садовый
5485 боярышник ярко-красный
5486 хопея душистая
5487 бодяк; колютик; хамепейце
5488 заразиха сетчатая
5489 пентаце
5490 ежевика канадская
5491 бамбук обыкновенный
5492 стальник колючий
5493 володушка
5494 ярутка пронзённая
5495 рдест пронзённолистный; рдест стеблеобъемлющий
5496 туиния
5497 клевер нитевидный; клевер тонкостебельный
5498 ситник нитевидный
5499 аристида; триостница
5500 шпажник болотный
5501 повойничек трёхтычинковый
5502 армерия; гвоздичник; статице
5503 трахелиум
5504 краснoднев Тунберга
5505 кудзу лопастная; пуэрария волосистая; пуэрария опушённая; пуэрария Тунберга
5506 спирея Тунберга; таволга Тунберга
5507 рожа собачья; хатьма тюрингенская
5508 ива изящная
5509 тимиан; тимиян; тимьян; чабрец
5510 змееголовник тимьянолистный; змееголовник тимьяноцветковый; змееголовник тимьяноцветный
5511 песчанка тимьянолистная
5512 вероника тимьянная; вероника тимьянолистная
5513 лещина тибетская
5514 нейлия тибетская
5515 апейба; тибурбу
5516 десмодиум; трилистник клещевой
5517 верблюдка
5518 акнида; конопля виргинская
5519 лайя; лэйа
5520 тигридия
5521 тизанолена
5522 лилия тигровая
5523 ель изящная; ель японская
5524 алектрион высокий
5525 тилландсия
5526 тиллециевые
5527 аржанец; тимофеевка
5528 аржанец луговой; палошник луговой; тимофеевка луговая; трава Тимофеева
5529 трутовик настоящий
5530 мышехвостник маленький
5531 горошек волосистый; горошек шершаволистный
5532 ишносифон
5533 линария; льнянка
5534 трициртис
5535 ситник лягушечий; ситник лягушечный
5536 табак
5537 питтоспоровые
5538 питтоспорум тобира; смолосемянник тобира
5539 тоддалия
5540 кариота жгучая; пальма винная; пальма жгучая; пальма "рыбий хвост"; пальма-тодди; тодди-пальма
5541 дерево бальзамное толу; дерево бальзамовое; дерево бальзамическое толюйское
5542 помидор; томат
5543 роза войлочная; шиповник войлочный
5544 курослеп болотный; лютик длиннолистный; лютик языколистный
5545 диптерикс
5546 папоротник игольчатый; папоротник щетинистый; папоротник щитник; щитовник шиповатый
5547 амми виснага; амми зубная
5548 зубянка
5549 крест петров чешуйчатый; чешуйник
5550 первоцвет комнатный; примула обконика
5551 книфофия; тритома
5552 амирис; бальзамник
5553 торения
5554 яблоня Зибольда
5555 дубровка; завязник; калган; лапчатка прямостоячая; лапчатка прямостоящая; лапчатка-узик; узик
5556 торре(й)я
5557 сосна Торрея
5558 тестудинария
5559 дорстения
5560 недотрога жёлтая; недотрога "не тронь меня"; недотрога обыкновенная; не-тронь-меня; прыгун
5561 тоулиция
5562 турнефорция
5563 вероника персидская; вероника Турнефора
5564 губка-люффа; губка растительная; люффа
5565 резуха башенная
5566 башенница голая; вяжечка голая; вяжечка гладкая; резуха гладкая
5567 тоц(ц)ия
5568 астрагаль камеденосный
5569 эпигея
5570 перловник трансильванский
5571 лозинка белая; лозинка обыкновенная; ломонос виноградолистный; ломонос вьющийся; ломонос обыкновенный
5572 равенала
5573 желтушник левкойный
5574 конрингия восточная
5575 трекулия
5576 лишайник тилланзия; мох бородатый; мох испанский; мох луизиан-

ский; мох флоридский
5577 циссус
5578 альфитония
5579 альфитония высокая
5580 папоротник древовидный; циатея
5581 вереск древовидный; вереск крупный средиземноморский; ерика древовидная; эрика древовидная
5582 лаватера; хатьма
5583 люцерна древовидная
5584 айлант высочайший; айлант железистый; айлант желёзковый; айлант китайский ясень; ясень китайский
5585 пеон древовидный;пион древовидный; пион полукустарниковый
5586 липа каменная широколистная; филирея широколистная
5587 рододендрон древовидный
5588 бузина древовидная
5589 цифомандра свекловичная
5590 камыш трёхгранный; схеноплектус трёхгранный
5591 лахеналия; лилия трёхцветная южноафриканская
5592 лимон трёхлисточковый; понцирус трёхлисточковый; трифолиата
5593 пажитник
5594 триллиум
5595 триния
5596 трёхзубка; триодия
5597 астра плавневая;астра солончаковая
5598 тритринакс
5599 монтбреция; тритония
5600 миндаль тропический; терминалия; терминалия катаппа
5601 алоэ настоящее
5602 незабудка болотная
5603 индигоноска красильная
5604 колос благовонный; лаванда аптечная; лаванда настоящая; лаванда обыкновенная
5605 мхи зелёные
5606 мирт обыкновенный
5607 анис китайский; бадьян звездчатый; бадьян китайский; анис звездчатый; дерево анисовое; иллициум настоящий
5608 трюфель
5609 трюфелевые
5610 текома
5611 кампсис
5612 бигнониевые
5613 жимолость вечнозелёная
5614 табебуйя
5615 сифонантус
5616 тладианта
5617 аньу; настурция клубненосная
5618 тубероза
5619 окопник клубненосный
5620 зопник клубненосный
5621 бегония клубневая
5622 картофель испанский гладкий;уллюко клубневый; уллюко клубненосный
5623 раяния
5624 сальпиглоссис; трубоязычник; языкотруб
5625 омежник дудчатый; омежник трубчатый
5626 астрокариум
5627 незабудка дернистая
5628 луговик дернистый; щучка дернистая
5629 кизляк кистецветковый; кизляк кистецветный; наумбургия кистецветная
5630 осока дернистая
5631 диффенбахия
5632 лилия африканская
5633 тюльпан
5634 каттлея цитрусовая
5635 дерево тюльпанное; лириодендрон
5636 дерево тюльпанное виргинское; лириодендрон тюльпанный
5637 амарант белый; щирица белая
5638 гулявник высокий; гулявник высочайший
5639 лебеда розовая
5640 дерево тунговое китайское; тунг китайский; тунг Форда
5641 туника
5642 нисса
5643 ниссовые
5644 ясень туркестанский
5645 экзохорда Альберта
5646 лещина древовидная; орех медвежий; орешник медвежий
5647 скипидарник; фисташка дикая; фисташка туполистная; фисташник
5648 куркума
5649 турнера
5650 репа (культурная); репа огородная; репа японская; сурепица яровая; турнепс
5651 бутень клубненосный; бутень луковичный; кервель клубневой; кервель корневой; кервель луковичный; кервель репный; репа кервельная
5652 петрушка корневая
5653 трава лакмусовая; хрозофора
5654 талласия
5655 хелоне
5656 осока высокая
5657 листера; офрис; тайник
5658 липарис; лосняк
5659 плагиантус
5660 линнея
5661 фиалка дву(х)цветковая
5662 джефферсония
5663 камнеломка супротивнолистная
5664 пролеска двулистная; сцилла двулистная
5665 диасция
5666 геликодисерос
5667 крупка седая
5668 вереск серый; ерика сизая; эрика сизая
5669 стрептопус
5670 паркия двухшаровидная; паркия суданская
5671 майник двулистный
5672 ячмень дву(х)рядный; ячмень крупный; ячмень культурный двурядный
5673 очиток ложный
5674 уапака
5675 аралия сердцевидная; спаржа японская; удо
5676 котовник украинский
5677 сыть разнолистная

5678 двулистник
5679 магнолия зонтичная; магнолия трёхлепестковая
5680 хедисцепе
5681 сциадопитис
5682 пихта японская зонтичная; сциадопитис мутовчатый
5683 мартиниевые
5684 антиарис африканский; антиарис ядовитый; антияр; анчар; упас
5685 хлопчатник американский; хлопчатник волосистый
5686 герань холмовая; журавельник холмовой
5687 чистец прямой
5688 купырник японский
5689 молочай прямой
5690 остролодочник уральский
5691 рябинник рябинолистный; спирея рябинолистная
5692 урзиния; урсиния
5693 уснея
5694 груша уссурийская
5695 валериана; валерьяна; маун; мяун
5696 валериановые; валерьяновые; мауновые
5697 дуб греческий; дуб крупночешуйчатый
5698 ванда
5699 ваниль
5700 моховик жёлто-бурый; пестрец
5701 хвощ пёстрый
5702 рдест злаковидный; рдест разнолистный
5703 паспалум
5704 ватерия
5705 кабачок
5706 козлобородник пореелистный; козлобородник порейолистный; козлобородник пор(р)ейнолистный
5707 пихта Бура; пихта Вича
5708 полевица собачья (опушённая)
5709 гриб зимний; опёнок зимний
5710 бухарник
5711 лимнохарис
5712 клён бархатистый; клён бархатный; клён величественный
5713 гинура
5714 геттарда
5715 диалиум гвинейский
5716 хатьма древовидная
5717 ворсянка лесная
5718 дионея; диония
5719 зеркало девичье; легузия зеркало Венеры; спекулярия
5720 шалфей вербеновидный; шалфей вербеновый; шалфей вербенолистный
5721 ветреница весенняя; прострел весенний
5722 барбарис весенний
5723 ломонос усатый
5724 желтостебельник; колосок душистый; колосок пахучий
5725 осока ранняя
5726 вероника весенняя
5727 гамамелис весенний
5728 вербена
5729 вербеновые
5730 ангиоптерис
5731 вика; горошек
5732 ветиверия
5733 калина
5734 вилларезия
5735 клён завитой
5736 базелла
5737 фиалка
5738 иохрома
5739 паутинник фиолетовый; приболотник фиолетовый
5740 фиалковые
5741 рёмерия гибридная
5742 петунья фиолетовая
5743 шалфей лесной
5744 румянка; синяк; эхиум
5745 рупрехтия
5746 помадерис
5747 виргилия
5748 виноград вьющийся; виноград девичий пятилистный; виноград дикий пятилистный; партеноциссус пятилисточковый
5749 клоповник виргинский
5750 сосна виргинская
5751 роза виргинская
5752 традесканция виргинская
5753 таволга виргинская
5754 земляника виргинская; клубника виргинская
5755 итея виргинская
5756 ломонос виргинский
5757 дрёма липкая
5758 висмия
5759 горец живородящий; корень сердечный
5760 кальцеолария гибридная; кошелки гибридные
5761 вахендорфия
5762 ожика Валенберга
5763 кривокучник
5764 эперуа
5765 огурец волосистый угловатый;огурец угловатый; сициос угловатый
5766 крупка стенная
5767 валления
5768 желтофиоль; лакфиоль; хейрантус
5769 ястребинка стенная; ястребинка стеночная
5770 валлихия
5771 можжевельник Валлиха
5772 латук стенной; мицелис стенной
5773 мокрединник; повой подстенный; постенница аптечная; постенница лекарственная; стенница
5774 резуха беловатая;резуха кавказская
5775 двурядка
5776 костенец постенный; костенец рута постенная; рута постенная
5777 орех
5778 ореховые
5779 цитроид
5780 традесканция бассейновая; традесканция белоцветковая
5781 спараксис
5782 дербенник лозный; дербенник прутовидный; плакун лозовый; плакун прутовидный
5783 лап(к)а воронья
5784 ветры волчьи; дождевик шиповатый
5785 веррукария
5786 барбарис бородавчатый

5787 бересклет бородавчатый; бородавочник
5788 чубушник бородавчатый
5789 люпин многолистный
5790 вашингтония
5791 бадан водяной; гравилат прибрежный; гравилат речной
5792 шильник водный; шильник водяной
5793 альдрованда пузырчатая
5794 каштан водяной; орех болотный;орех водяной (плавающий); рогульник (плавающий); чилим
5795 мягковолосник водный; мягковолосник водяной
5796 брункресс;жеруха аптечная; жеруха водная; жеруха лекарственная; хрен водяной; кресс водяной
5797 лютик водяной; шелковник
5798 щавель воднощавельный; щавель прибрежный
5799 омежник
5800 дзельква;дзелькова;зельква;планера
5801 планера водная
5802 болотник весенний
5803 цератоптерис
5804 норичник водяной
5805 лисохвост коленчатый
5806 дубровник чесночный; чеснок дикий; чеснок заячий; скордия
5807 апоногетон
5808 вех; цикута
5809 кария водяная
5810 эйх(х)орния
5811 бакопа
5812 горец земноводный; трава щучья; утевник
5813 водолюб; гидрофил(лум)
5814 водолистниковые; гидрофилляциевые
5815 пассифлора лавролистная; страстоцвет лавролистный
5816 пистия
5817 кувшинка; ненюфар; нимфея
5818 кувшинковые; нимфейные
5819 лобелия Дортман(н)а
5820 манник наплывающий; манник плавающий; манник ручьевой; трава манная; трава утиная
5821 арбуз (обыкновенный); арбуз столовый; арбуз съедобный; кавун
5822 мята водная; мята водяная
5823 мох водяной; фонтиналис
5824 плесени водяные; фикомицеты
5825 лужайник водяной; лужница водяная
5826 сеточка водяная
5827 дуб чёрный
5828 поручейник
5829 частуха
5830 виктория
5831 гидроклеис
5832 юссиея
5833 бутерлак
5834 осока водяная
5835 бразения
5836 телорез
5837 телорез алоэвидный; телорез обыкновенный
5838 вероника водная
5839 болотник; звёздочка водяная; красноволоска; красовласка
5840 болотниковые; красовласковые
5841 повойник; повойничек
5842 повойничковые
5843 ежовник; луговик извилистый
5844 ревень волнистый
5845 бенинказа
5846 пальма восковая андийская; пальма восковая перувианская; цероксилон андийский
5847 хойя
5848 дерево восковое; дерево лаковое японское; сумах сочный
5849 кукуруза восковая; кукуруза восковидная
5850 лаковица розовая
5851 хикама
5852 гордовина; калина гордовина; калина цельнолистная
5853 ясколка пятитычинковая
5854 вероника изящная
5855 зеленчук жёлтый
5856 дрок испанский; метельник
5857 бобров(н)ик; дрок испанский обыкновенный; метельник прутьевидный; прутовник ситниковидный; шильник ситниковидный
5858 кандолея
5859 бук лесной плакучий
5860 форзиция поникшая
5861 вейгела; вейгелия
5862 желтянка; резеда желтенькая; резеда желтоватая; резеда красильная; трава красильная; трескучник; церва; черва
5863 батун; лук-батун; лук дудчатый; лук зимний; лук-татарка; лук трубчатый; катависса; татарка
5864 меконопсис кумберлендский; меконопсис уэльский
5865 вельвичия
5866 вендландия
5867 папоротник орляк
5868 лиственница западная (американская)
5869 ангурия; огурец антильский; огурец вест-индский; корнишон пикульный
5870 бальза (заячья); бальса; дерево бальзовое; охрома
5871 акажу; дерево амарантовое; дерево красное; дерево махагониевое; махагони
5872 лядвенец болотный; лядвенец топяной
5873 пшеница
5874 пшеница летняя; пшеница мягкая; пшеница обыкновенная
5875 дазилирион Вилера
5876 троходендрон
5877 люхея
5878 ясень американский; ясень остроконечный
5879 сосна белоствольная
5880 очеретник белый
5881 боярка белая; рябина-ария
5882 подмаренник мягкий
5883 бриония белая; переступень белый
5884 белокопытник белый
5885 капуста белокочанная
5886 канариум белый
5887 кипарисовик туеобразный

5888 лапчатка белая
5889 дятлина белая; клевер белый (ползучий); клевер ползучий; кашка
5890 агатис белый
5891 крапива глухая; яснотка белая
5892 дур(р)а белая; джугара; сорго поникшее
5893 лотос египетский
5894 чемерица белая
5895 пихта одноцветная
5896 дерево снежное; хионантус виргинский
5897 магнолия обратноовальная; магнолия обратнояйцевидная; магнолия туполистная; магнолия японская
5898 офиопогон Ябуран
5899 люпин белый; люпин древнеегипетский; люпин тёплый
5900 резеда белая
5901 петунья ночецветковая
5902 тут белый; шелковница белая;шовкун
5903 коровяк метельчатый
5904 горчица белая (английская); горчица английская
5905 дуб белый
5906 тополь белый; тополь-белолистка
5907 аспидосперма
5908 сандал белый; чефрас
5909 казимироя съедобная
5910 ракитник многоцветковый
5911 чистец лесной
5912 аморфофаллус колокольчатый; ямс слоновый
5913 ель белая; ель канадская; ель серебристая
5914 рдест длиннейший
5915 очиток белый
5916 ластовенник обыкновенный; ластовень аптечный; ластовень лекарственный; ластовень обыкновенный; цинанхиум лекарственный
5917 донник белый; клевер бухарский
5918 двугнёздка; кардария
5919 дихромена
5920 трюфель белый; трюфель итальянский
5921 базелла-шпинат; шпинат индийский; шпинат малабарский
5922 белолоз; ветла; ива белая; ива серебристая
5923 груздь сухой; подгруздок белый; сухарь; сыроежка белая
5924 хрящецветник мутовчатый
5925 мытник мутовчатый
5926 подорожник индийский
5927 купена мутовчатая
5928 ива черниковидная; ива черничная
5929 роза Вихуры
5930 виддрингтония
5931 кермек широколистный
5932 белозор болотный
5933 руппия
5934 руппия морская
5935 вигандия
5936 пахучка обыкновенная
5937 строфостилес
5938 кривоцвет
5939 кривоцвет пашенный; кривоцвет полевой
5940 капуста огородная
5941 белокрыльник болотный; калла болотная; хлебница
5942 баркан; морковь двулетняя; морковь дикая; морковь культурная (каротиновая); морковь обыкновенная
5943 сельдерей душистый; сельдерей пахучий
5944 валлиснерия
5945 психотрия
5946 пастернак дикий; пастернак лесной
5947 копытень
5948 баптизия; баптисия
5949 исертия
5950 слива алыча растопыренная; слива растопыренная
5951 овёс дикий; овёс живой; овёс пустой; овсюг; полетай
5952 малахра
5953 маслина дикая
5954 братки; глазки анютины; трёхцветка; фиалка трёхцветная
5955 редька дикая; редька полевая
5956 цицания
5957 волоснец; вострец; колосняк;элимус
5958 кассия приморская
5959 яблоня венечная; яблоня душистая
5960 акнистус
5961 двузернянка дикая; зандури; пшеница двузернянковидная
5962 лютик лесной
5963 корилопсис Виллъмотта
5964 ива
5965 ивовые
5966 стенокарпус ивовый
5967 квебрахо красное
5968 астра иволистная
5969 эвкалипт иволистный; эвкалипт миндальный
5970 девясил иволистный
5971 груша иволистная
5972 спирея иволистная; таволга иволистная
5973 латук иволистный
5974 дуб ивовидный; дуб иволистный
5975 кипрей(ник); хаменерий; хамениериум
5976 барбарис Вильсона
5977 дейция Вильсона
5978 пихта Вильсона
5979 таволга Вильсона
5980 ель Вильсона
5981 метлица обыкновенная; метлица полевая
5982 хлорис
5983 пальма пеньковая; трахикарпус
5984 аристотелия
5985 бурути; пальма бурути
5986 дрожжи винные
5987 птерокактус
5988 каркас крылатый; птероцелтис
5989 вяз крылатый; вяз крыловаточерешковый; ильм крылатый
5990 бересклет крылатый
5991 аммобиум крылатый
5992 лофира крылатая; лофира пирамидальная
5993 ямс белый; ямс крылатый
5994 перистоголовник; птероцефалюс
5995 алария
5996 крылоорешина; крылоорешник; лапина; птерокария
5997 торица пятитычинковая

5998 весенник; любник; эрантис
5999 сурепка
6000 смородина кровяно-красная
6001 терескен
6002 виноград Берландье; виноград горный; виноград зимний
6003 гаультерия
6004 корилопсис
6005 брюква; бушма; рапс
6006 чабёр горный
6007 дримис Винтера; корица магелландская
6008 тыква большая; тыква гигантская; тыква крупноплодная
6009 химонант скороспелый
6010 мюленбекия
6011 вистария; глициния
6012 гамамелис; хамамелис; орех волшебный
6013 гамамелидовые; гамамелиевые
6014 вайда
6015 дрок
6016 дереза; лиций; повий
6017 вольфия
6018 аконит волкобойный; аконит волчий; борец волкобойный; борец лисий
6019 фиалка удивительная
6020 лимон персидский
6021 подмаренник лесной
6022 жимолость вьющаяся; жимолость немецкая; каприфоль немецкий
6023 мятлик боровой; мятлик дубравный; мятлик лесной
6024 герань лесная
6025 сушеница лесная
6026 щитовидник; щитовник
6027 норичник узловатый; норичник шишковатый
6028 дубровник чесноковый; дубровник шалфейный
6029 ангелика лесная; дудник лесной; дягиль лесной
6030 купырь лесной
6031 камыш лесной
6032 виноград лесной
6033 незабудка лесная
6034 крестовник лесной
6035 перловник одноцветковый
6036 лапортея
6037 одноцветка крупноцветковая; одноцветка крупноцветная; одноцветка одноцветковая
6038 вербейник дубравный
6039 цинна
6040 ясменник
6041 ожика
6042 осока лесная
6043 вудсия
6044 кислица обыкновенная
6045 звездчатка дубровная; звездчатка лесная
6046 горошек лесной
6047 живика пушистая; чистец шерстистый
6048 крупка войлочная
6049 волвянка; волжанка; волнушка; волнушка розовая
6050 жасмин розовый
6051 вика пушистоплодная
6052 терминалия шерстистая
6053 ива пушистопобеговая; ива шерстистопобеговая
6054 ива мохнатая; ива шерстистая
6055 дивина; коровяк высокий; коровяк скипетровидный; коровяк тапсовидный
6056 козыльник; марь амброзиевидная; марь благородная;чай иезуитский
6057 колпак (кольчатый)
6058 ладанник курчавый
6059 вульфения
6060 ксилокарпус
6061 ксилозма
6062 диоскорея; иньям; ям(с)
6063 пахиризус
6064 диоскорейные
6065 актинидия китайская
6066 бобы спаржевые; фасоль спаржевая
6067 деревей; тысячелистник
6068 куст чайный апалашский; падуб рвотный; чай рвотный
6069 грибок бродильный; грибок дрожжевой; дрожжи настоящие
6070 дрожжи; грибы дрожжевые; сахаромицеты
6071 ель аянская
6072 шлемник хмелевой
6073 незабудка разноцветная
6074 гравилат прямой
6075 подмаренник настоящий
6076 берёза жёлтая
6077 каштан конский восьмитычинковый; каштан жёлтый американский; павия жёлтая
6078 василёк подсолнечный
6079 головчатка альпийская
6080 хохлатка жёлтая
6081 моховик зелёный; подмошник
6082 ястребинка многоцветковая; ястребинка обильноцветущая
6083 бессмертник песчаный; гелихризум жёлтый; цмин жёлтый; цмин песчаный
6084 ксирис
6085 жерушник лесной
6086 касатик аировидный; касатик болотный; касатик водный; касатик водяной; касатик жёлтый; касатик ложно-аировый; касатик ложный
6087 рейнвардтия
6088 тыква яйцевидная
6089 наперстянка крупноцветковая; наперстянка сомнительная
6090 щетинник сизый
6091 горечавка жёлтая
6092 смолёвка зеленоцветковая
6093 боярышник жёлтый
6094 дерево шёлковое
6095 ксантоцерас
6096 глауциум жёлтый; мачёк жёлтый
6097 камнеломка козлиная; оч(к)и царские
6098 василисник жёлтый
6099 кувшинка жёлтая; кувшинка мексиканская
6100 резеда жёлтая
6101 момбин жёлтый; спондиас жёлтый; слива испанская тропическая
6102 камнеломка жёстколистная
6103 паслён волосистый; паслён жёлтый;

паслён мохнатый
6104 чина жёлтая
6105 ксантор(р)иза
6106 осока жёлтая
6107 ветла блестяще-жёлтая;ива жёлтая
6108 донник аптечный; донник жёлтый; донник лекарственный
6109 овёс желтеющий; овёс желтоватый; овёс золотистый; трищетинник желтоватый; трищетинник жёлтый
6110 ибицелла жёлтая
6111 горошек жёлтый
6112 бадьян мелколистный
6113 баптизия красильная; индиго дикое
6114 кладрастис
6115 ветреница лютиковая
6116 зелёнка; хлора
6117 дерево тисовое; негной-дерево; тис(с)
6118 тис(с)овые
6119 иланг-иланг; кананга (душистая)
6120 юкка
6121 мордовник банатский; мордовник венгерский
6122 магнолия голая; магнолия обнажё-нная; магнолия Юлана
6123 таволга Цабеля
6124 тельфайрия восточноафриканская; тельфайрия стоповидная
6125 корень цитварный; куркума цитва-рная
6126 зефирантес
6127 чубушник Цейера
6128 кашка лесная; клевер средний
6129 цин(н)ия
6130 зизифора
6131 зигомицеты

АЛФАВИТНЫЙ УКАЗАТЕЛЬ РУССКИХ НАЗВАНИЙ РАСТЕНИЙ

RUSSIAN ALPHABETICAL INDEX

А а

абабок 582
абака 2
абелия 3
абельмош 3921
абобра тонколистная 1660
абрикос 1350
абрикос альпийский 773
абрикос антильский 3453
абрикос волосистоплодный 4336
абрикос дикий 1350
абрикос китайский 3015
абрикос культурный 1350
абрикос муме 3015
абрикос маньчжурский 3457
абрикос монгольский 3457
абрикос обыкновенный 1350
абрикос сибирский 4891
абрикос тропический 3453
абрикос фиолетовый 4336
абрикос чёрный 4336
абрикос шероховатоплодный 4336
абрикос японский 3015
аброма 4
аброния 5
абрус 3090
абрус 4553
абсент 1564
абсинт 1564
абутилон 7
ава-перец 3140
авдотка 1422
авероа-билимби 576
аверроа 1012
аверроа-билимби 576
авицения 386
авокадо 183
авокадо-груша 183
авран 2754
авран аптечный 1913
авран лекарственный 1913
аврикула 368
аврикуляриевые 369
автобазидиомицеты 381
агава 59
агава американская 1092
агава древовидная 1092
агава магуэй 4328
агава сальма 4328
агава сизалевая 4976
агава сизаловая 4976
агава тёмно-зелёная 4328
агапант(ус) 54
агапантус африканский 40
агариковые 58
агатис 1795
агатис белый 5890
агатис новозеландский 3139
агератум 60
агератум мексиканский 3594
аглаонема 62
аглая 61
агримония 63
агростемма 1610
адансония 445
аденандра 29
аденантера 493
аденантера павлинья 4676
аденофора 3198
аджирык 542
адзуки 34
адиантум 3430
адиантум Венерин волос 5099
адиантум клиновидный 1828
адиантум нежный 2182
адинандра 30
адлай 31
адлумия 32
адокса 3765
адокса мускусная 3766
адоксовые 3684
адонис 33
адонис весенний 5179
адонис летний 5317
адонис огненный 2272
адонис осенний 4164
адонис пламенный 2272
адраспан 2714
ажгон 72
азадирахта индийская 3492
азалея 389
азалия 389
азалия жёлтая 4264
азалия индийская 2933
азалия полярная 114
азалия понтийская 4264
азалия стелющаяся 114
азара мелколистная 752
азимина 4077
азимина трёхлопастная 1485
аизоацневые 1033
аизовые 1033
аизоон 70
аир 5365
аир болотный 1918
аир злаковый 3073
аир обыкновенный 1918
аир тростниковый 1918
аира 2677
аира волосовидная 261
аира гвоздичная 4959
аира ранняя 1963
аистник 2789
аистник цикутный 91
аистник цикутовый 91
аистник цикутолистный 91
айва 4389
айва китайская 1181
айва низкая Маулея 3038
айва обыкновенная 1512
айва продолговатая 1512
айва цветочная 3038
айва яблоковидная 286
айва японская 3038
айлант 65
айлант высочайший 5584
айлант железистый 5584
айлант желёзковый 5584
айлант китайский ясень 5584
айован 72
айован душистый 72
айра 2677
айфанес 4601
акажу 1378
акажу 5871
акалифа 1585
акант 12
акант колючий 5151
акант мягколистный 5077

акантовые 14
акантолимон 4306
акантопанакс 11
акантофеникс 5150
акантус 12
аканф 12
акароидин 2591
акация 10
акация австралийская 655
акация американская 636
акация аравийская 399
акация безжилковая 3751
акация Бейли 1300
акация белая 636
акация беловатая 4957
акация длиннолистная 5409
акация жёлтая 4910
акация катеху 1056
акация мохнатая 4555
акация низбегающая 2637
акация подбеленная 4957
акация розовая 4555
акация сбегающая 2637
акация сенегальская 2662
акация серебристая 4957
акация сибирская 4910
акация устрашающая 103
акация Фарнеза 5351
акация Фарнеси 5351
акация чёрная австралийская 655
акация чернодревесная 655
акация шёлковая 4948
акация шетинистая 4555
аквилегия 1341
акебия 73
акилей 1341
акнида 5518
акнистус 5960
аконит 3660
аконит Антора 4373
аконит аптечный 17
аконит волкобойный 6018
аконит волчий 6018
аконит метельчатый 4029
аконит пёстрый 3465
аконит противоядный 4373
аконит реповидный 17
аконит сборный 17
аконит ядовитый 17
акроклиниум 5333
акрокомия 18
акрокомия южноамериканская 3737
акронихия 19
акронихия австралийская 480
акронихия Бауэра 480
акростихум 20
аксамитник 174
аксирис 388
актеа 441
актинидиевые 22
актинидия 21
актинидия аргута 748
актинидия изогнутая 748
актинидия китайская 6065
актинидия коломикта 3176
актинидия многобрачная 4968
актинидия носатая 4968
актинидия острая 748
актинидия острозубчатая 748
актинидия полигама 4968
актинидия полигамная 4968
актинидия трёхдомная 4968
актиномицеты 23
алария 5995
алария съедобная 407
алектория гривастая 2868
алектрион 84
алектрион высокий 5524
алетрис 5221
алзина 5235
алиссум 172
алиссум душистый 5352
алиссум европейский 5352
алиссум морской 5352
алкана 97
алканет 97
алканна 97
алканна красильная 1954
алое 109
алой 109
алоказия 108
алонсоа 3529
алоэ 109
алоэ американское 1092
алоэ барбадосское 3568
алоэ бородавчатое 3990
алоэ дикое 1003
алоэ древовидное 299
алоэ колючее 1003
алоэ настоящее 5601
алоэ обыкновенное 3568
алоэ пёстрый 3132
алоэ раздвоенное 4392
алоэ сокотрина 5076
алоэ сокотринское 5076
алоэ ферокс 1003
алстония 165
алтей 169
алтей аптечный 3507
алтей жестковолосый 2815
алтей лекарственный 3507
алтей розовый 2829
алтей шершавоволосый 2815
алтей шершаволистный 2815
алтей-штокроза 2829
алфалфа 90
алыча 3777
альбизия 79
альбиция 79
альбиция душистая 2356
альбиция ленкоранская 4948
альбиция противоглистная 3757
альбицция 79
альбицция душистая 2356
альбицция шёлковая 4948
альгароба 1467
альдрованда 83
альдрованда пузырчатая 5793
альзик 164
альзина 5235
алькaнна 97
альпиния 158
альпиния Галанга 2609
альстония 165
альстония школьная 4014
альстремерия 166
альтей 169
альтернантера 3111
альфа 2062
альфальфа 90

альфа-трава 2061
альфитония 5578
альфитония высокая 5579
альхорнея 1225
амарант 174
амарант белый 5637
амарант жминдовидный 4316
амарант колосистый 4464
амарант хвостатый 3387
амарант шарообразный 1421
амарантовые 175
амариллис 176
амариллис альпийский 5062
амариллис белладонна 526
амариллисовые 177
амбач 3109
амбербоа 181
амбербоа туранская 4355
амбрет 3761
амброзия 4400
амброзия полыннолистная 1513
амброзия трёхнадрезная 2480
амброзия трёхраздельная 2480
амврозия 4400
амеланхир 4832
амзинцкия 2217
амирис 5552
амирис бальзамический 430
амирис ямайский 430
амми 216
амми большой 2603
амми виснага 5547
амми зубная 5547
аммобиум 217
аммобиум крылатый 5991
аммофил 487, аммофила 487
амоора 218
аморфа 2171
аморфа кустарниковая 2948
аморфа кустовидная 2948
аморфофалл 220
аморфофалл гигантский 2466
аморфофаллус 220
аморфофаллус гигантский 2466
аморфофаллус колокольчатый 5912
аморфофаллус Ривьера 1848
ампелопсис 221
амсония 223
анабазис 233
анагирис 3575
анакамптис 234
анакард 1042
анакард акажу 1378
анакард западный 1378
анакардиевые 1043
анакардия 1042
анакардия акажу 1378
анакардия западная 1378
анакардовые 1043
анамирта 2264
ананас 235
ананас культурный 4191
ананас настоящий 4191
ананас посевной 4191
ананасные 4192, ананасовые 4192
анафалис 4094
анафалис жемчужный 1488
анациклус 2460
ангалоний 3344
ангелика 245
ангелика аптечная 2406
ангелика лесная 6029
ангиоптерис 5730
ангрекум 746
ангурия 5869
ангустура 251
андира 246
андрахна 239
андреевые 240
андромеда 241
андромеда древовидная 5095
андромеда-дубровник 723
андромеда дубровниколистная 723
анемон 242
анемон японский 3013
анемона 242
анемона блестящая 2273
анемона садовая 4273
анемона японская 3013
анемонелла василистниколистная 4598
анизоптера 3590
анизохилус 255
анис 253
анис звездчатый 5607
анис китайский 5607
анис обыкновенный 253
аннато 256
аннеслея 257
ан(н)она 1751
анона игольчатая 5093
анона колючая 5093
анона обыкновенная 4257
анона сетчатая 849
анона черимойя 1132
анона черимола 1132
анона чешуйчатая 5297
аноновые 1752
антемис 963
антиарис африканский 5684
антиарис ядовитый 5684
антидесма 1164
антидесма буниус 566
антипка 3426
антияр 5684
антолиза 3415
антоспермум 182
антуриум 270
анхиета 236
анхуза 840
анхуза Баррелиера 1959
анхуза итальянская 2975
анхуза лекарственная 1371
анчар 5684
аньу 5617
апейба 5515
апельсин 5377
апельсин бигардия 5092
апельсин горький 5092
апельсин кислый 5092
апельсин навель 3468
апельсин настоящий 5377
апельсин-оседж 3972
апельсин пупочный 3468
апельсин сладкий 5377
апиос 4282
апиос клубневый 207
апозерис 281
апоногетон 5807
апоногетоновые 278
апороселла 280

апофиллум 279
апоциновые 1873
апоцинум коноплевидный 2775
аптандра 290
апулея 291
арабис альпийский 143
аракача 325
аракача съедобная 277
аралиевые 2495
аралия 295
аралия высокая 3016
аралия китайская 1167
аралия колючая 1849
аралия сердцевидная 5675
аралия шиповатая 1849
аралия японская 3034
араужия 658
араукариевые 298
араукария 297
араукария Бидвилла 855
араукария бразильская 4049
араукария высокая 3877
араукария Куннингама 1730
араукария узколистная 4049
араукария чилийская 3658
арахис 4087
арахис культурный 4000
арахис подземный 4088
арбус 5821
арбус горький 1334
арбус дикий 1334
арбус колоцинт 1334
арбус кормовой 1245
арбус обыкновенный 5821
арбус столовый 5821
арбус съедобный 5821
арбус цукатный 1245
аргания 315
аргания крепкодревесная 315
аргемон(а) 4302
аргемоне 4302
аргиролобий 4962, аргиролобиум 4962
аргитамния 4953
ардизия 5117
ардизия японская 3018
арека 313
арека катеху 548
аренга 5305
аренга сахарная 2558
аретуза 314
аржанец 5527
аржанец луговой 5528
аризема 2999
аристида 5499
аристолохия 1932
аристолохия крупнолистная 1404
аристотелия 5984
арктерика 4176
арктотис 312
арктоус 4320
армерия 5502
армерия приморская 1545
арнаутка 1930
арнебия 322
арника 323
арника горная 3694
арнозерис 3213
арнозерис малый 5029
ароидные 334
аройник 333
ароникум 3276
арония 1224
арония арбутусолистная 4446
аронник 333
аронник белокрылый 2972
аронник итальянский 2972
аронник пятнистый 3378
аронниковые 334
арракач 277
арракача съедобная 277
аррорут 329
аррорут бразильский 1379
аррорут вестиндский 541
аррорут васточноиндийский 1983
аррорут индийский 2244
аррорут квинслендский 1993
аррорут остиндский 1983
артаботрис 5422
артемизия 4640
артишок 331
артишок испанский 1015
артишок китайский 332
артишок посевной 331
артишок японский 332
арум 333
арум белокрылый 2972
арум змеиный 1536
арум кукушечный 1374
арум пятнистый 3378
арундинария 994
арундинелла 335
арундо 2481
арундо тростниковый 2482
арункус 2519
архангелика 2406
архегониальные 301
архимицеты 302
архитея 303
арча 3118
асклепиадовые 3621
асклепиас 3620
аскомицеты 337
аскофиллум 4539
аспараг(ус) 350
аспарагус пушистоперистый 2204
аспарагус Шпренгера 5178
аспергилл(ус) 351
асперуга 3419
асперуга лежачая 2459
аспидистра 353
аспидистра высокая 1352
аспидиум 354
аспидосперма 5907
асплений 5161
ассафетида 336
астильба 360, астильбе 360
астильбе китайская 1169
астра 358
астра альпийская 113
астра виргинская 3842
астра дикая 2973
астра звёздчатая 2973
астра иволистная 5968
астра китайская 1382
астра льнолистная 2102
астра Новая Англия 3839
астра Новая Бельгия 3842
астра новобельгийская 3842
астра пиренейская 4370
астра плавненая 5597

астра ромашковая 2973
астра садовая 1158
астра садовая китайская 1382
астра сибирская 4892
астра солончаковая 5597
астра степная 2973
астргал 3353
астрагал альпийский 137
астрагал боэтийский 5349
астрагал датский 4350
астрагал камеденосный 5568
астрагал китайский 1176
астрагал сладколистный 3619
астрагал солодколистный 3619
астрагал хлопунец 3701
астранция 3531
астранция большая 2611
астранция малая 1946
астровые 359
астрокариум 5626
астрокарпус 361
астрониум 5234
астрофитум 5213
асфоделина 3003
асфоделина жёлтая 1449
асфодель 352
асфоделюс 352
асфоделюс белый 757
асфодил 352
асфодил белый 757
асцирум 4647
асцирум зверобойниковидный 4647
аталантия 362
аталантия однолистная 2927
атрактилис 1864
атталея 365
атхатоба 3438
аукуба 367
аукуба японская 3019
аукумения 3920
аурикула 368
афананта 274
афеландра 275
афеландра золотистая 3936
афеляндра 275
афзелия 53
афзелия африканская 39
афиллофоровые 276
афцелия 53
ахиллеа-зоря 5054
ахименес 15
ахирантес 1108
ахрас 4700
ацена 4854
ацолла 391

Б б

бабассу 396
бабушник 109
багновка 1706
багновка чёрная 627
багряник 1098
багрянки 4433
багрянник 1098
багрянник канадский 1979
багрянник китайский 1199
багрянник обыкновенный 3114
багрянник стручковатый 3114
багрянник стручковый 3114
багрянник японский 1199
багрянник японский 3048
багульник 3265
багульник болотный 1719
бадан 537
бадан водяной 5791
бадан сердцелистный 2736
бадан толстолистный 3254
бадан узколистный 3254
бадиджан 2414
бадьян 254
бадьян 1866
бадьян звездчатый 5607
бадьян камбоджийский 959
бадьян китайский 5607
бадьян мелколистный 6112
бадьян настоящий 3014
бадьян-ясенец 2435
баель 408
бажра 4092
базелла 5736
базелла-шпинат 5921
базелловые 3414
базидиомицеты 462
базилик 463
базилик душистый 5354
базилик камфарный 5354
базилик обыкновенный 5354
базилика 463
бакаут 3293
баккаурея 404
баклажан 2414
баклажан обыкновенный 2414
бакопа 5811
бактрис 5154
бакхарис 405
бакхарис галимолистный 1974
баланофоровые 412
балата 1355
бальза 5870
бальза заячья 5870
бальзамин 5051
бальзамин садовый 2407
бальзаминовые 433
бальзамник 432
бальзамник 5552
бальса 5870
бамбуза 437
бамбуза обыкновенная 1357
бамбук 437
бамбук золотистый 2527
бамбук-листоколосник 2527
бамбук обыкновенный 5491
бамбук чёрный 615
бамия 3921, бамья 3921
банан 439
банан абассинский 8
банан браминов 1358
банан декоративный 8
банан десертный 1358
банан Кавендиша 1936
банан карликовый 1936
банан китайский 1936
банан кухонный 4219
банан пизанг 4219
банан плодовой 4219
банан подорожник 4219
банан райский 4219
банан фруктовый 4219
банан эфиопский 8

банан яблочный 1358
банан японский 3020
банановые 440
баниан 444
баниан 738
баниан индийский 2929
банксия 442
баньян 444
баньян 738
баобаб 445
баобабовые 4944
баптизия 5948
баптизия красильная 6113
баптисия 5948
баран 1683
баран-гриб 2385
баранец 3213
баранец малый 5029
баранка 3694
баранник 323
баранник горный 3694
барбарис 452
барбарис агрегатный 4659
барбарис азиатский 345
барбарис амурский 225
барбарис бородавчатый 5786
барбарис Бретшнейдера 4330
барбарис весенний 5722
барбарис Вильсона 5976
барбарис Гукера 2845
барбарис Дарвина 1805
барбарис Зибольда 4930
барбарис канадский 98
барбарис корейский 3178
барбарис обыкновенный 2075
барбарис самшитолистный 3421
барбарис Саржента 4706
барбарис сибирский 4893
барбарис Тунберга 3021
барбарис этненский 37
барбарис японский 3021
барбарисовые 453
барвинок 4140
барвинок большой 563
барвинок малый 1493
баркан 5942
баркаузия 2725
баросма 827
баррингтония 1130
бартония 459
бартсия 460
бартсия альпийская 115
бартшия 460
бархат 1604
бархат амурский 226
бархатник 174
бархатцы 3493
бархатцы высокие 394
бархатцы крупноцветковые 394
бархатцы крупноцветные 394
бархатцы мелкоцветковые 2369
бархатцы отклонённые 2369
бархатцы прямостоячие 394
бархатцы распростёртые 2369
бассия широколистная 2920
батат 5380
батис 4668
батлачёк 2349
батлачек луговой 3549
батлачик 2349
батлачок 2349
батлачок луговой 3549
батрахоспермум 479
батун 5863
баугиния 481
баугиния шерстистая 4654
баухиния 481
бафия 968
бафия красивая 4869
башенница 4517
башенница голая 5566
башмала японская 3376
башмачек 3203
башмачок 3203
башмачок венерин 3203
башмачок венерин настоящий 2109
баэль 408
баялыч 4618
бегониевые 524
бегония 523
бегония всегда цветущая 4143
бегония клубневая 5621
бегония месячная 4143
бегония полуклубневая 5075
бегония рекс 355
бедренец 4186
бедренец-анис 253
бедренец большой 2608
бедренец камнеломка 4733
бедренец крымский 1689
безвременник 382
безвременник осенний 1353
бейльшмидия 5016
бекмания 5015
бекмания обыкновенная 209
беламканда китайская 620
белена 2823
белена белая 4586
белена чёрная 633
белладонна 525
беллиум 470
белозор 4058
белозор болотный 5932
белокопытник 888
белокопытник белый 5884
белокопытник гибридный 4340
белокопытник лекарственный 4340
белокрыльник 949
белокрыльник болотный 5941
белокудренник 425
белокудренник чёрный 614
белолист 5388
белолиственник 5388
белолоз 5922
белотал 467
белоус 3533
белоус прямой 3533
белоус торчащий 3533
белоцветник 5062
белоцветник весенний 5188
белоцветник летний 5322
бемерия 2174
бенедикт аптечный 671
бенинказа 5845
бергамот 536
бергения 537
бергия 538
берёза 578
берёза белая плакучая 2131
берёза бородавчатая 2131

берёза бумажная 4037
берёза вишенная 5357
берёза вишнёвая 5357
берёза даурская 1783
берёза железная 4758
берёза жёлтая 6076
берёза жёлтая высокая 2594
берёза каменная 2057
берёза карликовая 1935
берёза китайская 1171
берёза Максимовича 3650
берёза малорослая 3390
берёза Миддендорфа 3612
берёза обыкновенная 2131
берёза плакучая 2131
берёза повислая 2131
берёза Потанина 4280
берёза пушистая 4323
берёза тополелистная 2594
берёза Форреста 2333
берёза чёрная 4504
берёза Шмидта 4758
берёза Эрмана 2057
берёза японская 3076
берёзка 2100
берёзовик 582
берёзовик обыкновенный 582
берёзовые 579
берека 1128
бересклет 2069
бересклет американский 805
бересклет бородавчатый 5787
бересклет вьющийся 185
бересклет европейский 2096
бересклет карликовый 1939
бересклет крылатый 5990
бересклет низкий 1939
бересклет обыкновенный 2096
бересклет тёмно-багряный 1981
бересклет тёмно-пурпуровый 1981
бересклет широколистный 793
бересклет японский 2146
бересклетовые 5203
берест 2018
берест листоватый 5040
берест приземистый 4903
берест пробковый 1602
берлиния 539
берсим 2000
берула 545
берула прямая 5209
беруля 545
беруля прямая 5209
берхемия 5336
берхемия лазящая 76
бескильница 96
бескильница морская 4803
бесславник 3766
бессмертник 2923
бессмертник 5279
бессмертник белый 1510
бессмертник песчаный 6083
бесцветник 2049
бетель 548
бетель 549
бетель пальма 548
бетель перечный 549
бетоника 1361
бешеница водяная 2130
бигардия 5092
бигнониевые 5612
бигнония 567
бигнония усиковая 1702
бикса аннатовая 256
биксовые 612
бикукулла 668
билимби 576
билльбергия 68
биота 300
биота восточная 3948
бирсонима 905
бирючина 4312
бирючина амурская 232
бирючина блестящая 2514
бирючина китайская 1198
бирючина лоснящаяся 2514
бирючина обыкновенная 1508
бирючина овальнолистная 944
бирючина японская 3065
бисерник 5435
бисерниковые 5436
бисерница 1488
бишопия яванская 3086
бишофия 3086
блетилла 672
блигия 74
блигия африканская 75
блисмус рыжий 4443
блисмус сжатый 2284
блошак 5112
блошник 2292
блошница 2291
блюберри 2796
блющ обыкновенный 2040
бляшконоска 2648
боб бархатный 1823
боб волчий 3395
боб гиацинтовый 2897
боб индийский 2897
боб калабарский 1816
боб конский 2864
боб лимский 4940
боб розовый 3090
бобовник альпийский 4773
бобовник анагировидный 2529
бобовник анагиролистный 2529
бобовник "золотый дождь" 2529
бобовые 3267
бобров(н)ик 5857
бобы 788
бобы бархатные 1644
бобы гиацинтовые 2897
бобы земляные 1574
бобы игнатьевские 4650
бобы индийские 2946
бобы каролинские 4940
бобы конские 2864
бобы кормовые 788
бобы кормовые 2864
бобы красные 4748
бобы лимские 3309
бобы лимские мелкоплодные 4940
бобы мелкоплодные 5028
бобы многоцветковые 4748
бобы огненные 4748
бобы русские 788
бобы сива 4940
бобы соевые 5106
бобы спаржевые 6066
бобы турецкие 4748

богатень 1824
бодлак 2517
бодяк 5487
бодяк бесстебельный 5238
бодяк болотный 3519
бодяк ланцетолистный 850
бодяк полевой 975
бозеа 2544
бозея 2544
бойкиния 753
бойсдувалия 5144
боккаут 4150
боккония 4230
болдырьян 1552
болетус 727
болиголов крапчатый 4241
болиголов пятнистый 4241
болотник 5839
болотник весенний 5802
болотник осенний 384
болотник прудовый 4261
болотниковые 5840
болотница 5145
болотница болотная 719
болотница игольчатая 3822
болотница маленькая 1950
болотница малоцветковая 2213
болотница яйцевидная 3983
болотноцветник 2297
болотноцветник кувшинковый 2378
болотовик 4846
болотоцветник 2297
болтония 728
больтония 728
большеголовник 4486
большеголовник альпийский 1440
бомарея 729
бомбакс 730
бомбаксовые 4944
бомерия 2174
бондук 3847
бор 3624
бор развесистый 201
бор раскидистый 201
бораго 734
бораго лекарственный 1366
борассус пальмира 4019
борец 3660
борец аптечный 17
борец волкобойный 6018
борец лисий 6018
борец синий 17
боровик 1093
борода ааронова 333
борода лешего 503
бородавник 3860
бородавник обыкновенный 1479
бородавочник 5787
бородатка 4238
бородач 706
бородач кровеостанавливающий 1985
бороздоплодник 3294
борщевик 1649
борщевик обыкновенный 2825
борщевик пушистый 1069
борщевник 1649
борщевник сибирский 4898
босвелия 2363
босвелия 2937
босвелия Картера 553
босвеллия 2363
босвеллия 2937
босвеллия пильчатая 2937
ботрихиум 2577
боэргавия 5131
боярка белая 5881
боярышник 2729
боярышник-азароль 390
боярышник алтайский 167
боярышник Арнольда 324
боярышник Брайндера 3845
боярышник восточный 4960
боярышник Гельдрейха 2761
боярышник жёлтый 6093
боярышник испанский 390
боярышник канадский 972
боярышник колючий 2038
боярышник кроваво-красный 4452
боярышник круглолистный 1976
боярышник Ламберта 3209
боярышник обыкновенный 2038
боярышник однопестичный 4973
боярышник одностолбчатый 4973
боярышник петуший гребень 1313
боярышник петушья шпора 1313
боярышник плодовый 390
боярышник понтийский 390
боярышник Робсона 4508
боярышник чёрный 2082
боярышник шпорцевый 1313
боярышник Эльвангера 2017
боярышник ярко-красный 5485
бравоа 3604
бразения 5835
брандушка 3557
брандушки белые 1535
брассия 5132
братки 5954
браунколь 3128
брахиэлитрум 4880
бредина 2523
бриделия 776
бриония 824
бриония белая 5883
броваллия 814
бровник 3763
бродиея 801
броколи 800
броколь 800
бромелиевые 4192
бромелия 804
бросимум 770
брукенталия 5142
брунера 821
брункресс 5796
брунсвигия Жозефины 2469
брунфельзия 4405
брусника 1645
бруссонетия 4039
бруссонетия бумажная 1486
бруссонеция 4039
бруссонеция бумажная 1486
брюква 4626
брюква 6005
брюсселька 823
бубенчик 3198
бубенчики 3198
бубенчики лилиелистные 3302
бувардия 747
бугенвиллея 742

буддлейя 892
буддлейя изменчивая 3941
буддлейя японская 3026
буддлея 892
буддлея изменчивая 3941
буддлея японская 3026
будлея 892
будра 2647
будра плющевидная 1429
бузгунча 1498
бузина 2009
бузина американская 190
бузина древовидная 5588
бузина золотистая 2537
бузина канадская 190
бузина карликовая 3570
бузина кистистая 2120
бузина красная 2120
бузина обыкновенная 2120
бузина перувианская 4155
бузина травянистая 3570
бузина чёрная 2095
бузинник 3570
бузульник 2542
бук 513
бук антарктическии 268
бук восточный 3949
бук европейский 2077
бук Зи(е)больда 4931
бук крупнолистный 184
бук лесной 2077
бук лесной плакучий 5859
бук лесной тёмно-пунцовый 4345
бук обыкновенный 184
бук Энглера 2036
бук японский 3022
букашник 3082
букашник горный 4857
букашник многолетний 4736
буквица 550
буквица аптечная 1361
буквица водяная 2242
буквица лекарственная 1361
буковица 550
буковые 514
букс 749
буксус 749
буксус балеарский 5108
буксус обыкновенный 1368
булавница 2520
булавоносец 1284
булавоносец седой 2595
бульбофиллум 843
бульва 3094
бульнезия 851
бундук 1319
бундук двудомный 3151
бундук канадский 3151
бурак 520
бурачник 734
бурачник лекарственный 1366
бурачниковые 735
бурачок 172
бурачок горный 3693
бурачок колючий 5152
бурачок приморский 5352
бурачок скалистый 2551
бурачок скальный 2551
бурачок чашечный 4007
бурзера 875
бурзеровые 876
бурка 2114
буркун 4317
буркун 4921
буркун красный 90
буркунчик 638
буррерия 5286
бурсария 874
бурсера 875
бурсеровые 876
бурути 5985
буссенгоя 3413
буссингоа 3413
буськи 91
бутелоуа 2572
бутелуа 2572
бутень 1137
бутень клубненосный 5651
бутень луковичный 5651
бутень одуряющий 1648
бутерлак 5833
бутея 886
бухарник 5710
бухарник мягкий 2461
бухарник шерстистый 1553
бухлое 836
бухлоэ 836
бушма 6005
бычок 2209

В в

вайда 6014
вайда красильная 1956
вакуа 1521
вакциниум 692
валенбергия 4511
валериана 5695
валериана аптечная 1552
валериана высокая 4569
валериана горная 3722
валериана лекарственная 1552
валерианелла 1618
валерианелла бугорчатая 2977
валерианелла колосковая 2088
валерианелла колосковая 3211
валерианелла малая 2088
валерианелла овощная 3211
валерианица 1618
валерианица малая 2088
валерианница 1618
валерианница овощная 3211
валериановые 5696
валерьяна 5695
валерьяна аптечная 1552
валерьяна болотная 3520
валерьяна двудомная 3520
валерьяна каменная 4073
валерьяница 1618
валерьянница 1618
валерьянница бугорчатая 2977
валерьянница килеватая 3143
валерьянница малая 2088
валерьянница овощная 3211
валерьянница огородная 2088
валерьяновые 5696
валления 5767
валлиснерия 5944
валлиснерия спиральная 5159
валлихия 5770

валлота пурпурная 4742
валонея 2128
валуй 2209
вальдштейния 458
ванда 5698
ваниль 5699
ваниль душистая 3605
ваниль плосколистная 3605
василёк 1089
василёк восточный 1067
василёк голубой 1613
василёк горный 3697
василёк горькуша 817
василёк колючеголовый 5232
василёк луговой 817
василёк обыкновенный 1613
василёк подсолнечный 6078
василёк полевой 1613
василёк посевной 1613
василёк пятнистый 5165
василёк рогатый 2330
василёк русский 4627
василёк синий 1613
василёк скабиозный 4734
василёк скабиозовидный 4734
василёк скабиозовый 4734
василёк чёрный 625
василёк шероховатый 4734
василисник 3556
василисник альпийский 136
василисник водосборолистный 1342
василисник жёлтый 6098
василисник малый 3392
василисник простой 5009
василисник узколистный 3806
василистник 3556
василистник альпийский 136
василистник водосборолистный 1342
васильки рогатые 2330
ватерия 5704
ватерия индийская 4196
ваточник 3620
ваточник инкарнатный 5345
ваточник клубневой 894
ваточник клубненосный 894
ваточник кюрасао 676
ваточник розово-малиновый 5345
ваточник сирийский 1469
вахендорфия 5761
вахта 716
вахта трёхлистная 1365
вахта трилистная 1365
вашингтония 5790
вдовушка 4735
вдовушки 4735
везикария 661
вейгела 5861
вейгелия 5861
вейник 4475
вейник наземный 1131
вейник незамечаемый 5008
вейник незамеченный 5008
вейнрута 1516
веллингтония 2483
вельвичия 5865
венгерка итальянская 2627
вендландия 5866
Венерин волос 5099
Венерины волосы 5099
венец царский 1711
венечник 269
венечник лилейный 4648
венечник лилиаго 4648
венечник лилиецветковый 4648
верба 1801
верба красная 4850
вербезина 1709
вербейник 3370
вербейник дубравный 6038
вербейник луговой чай 3651
вербейник монетчатый 3651
вербейник обыкновенный 2540
вербейник пятнистый 5169
вербейник точечный 5169
вербена 5728
вербена аптечная 2129
вербена лекарственная 2129
вербена лимонная 3271
вербеновые 5729
вербесина 1709
верблюдка 5517
верблюдка иссополистная 2911
верболоз 3236
верболоз 4850
вередник 2232
вереск 2744
вереск восковой 5104
вереск древовидный 5581
вереск крупный средиземноморский 5581
вереск метлистый 546
вереск обыкновенный 4772
вереск серый 5668
вереск сицилийский 4116
вересковые 2745
вермут 1564
вернония 2965
вероника 5123
вероника альпийская 150
вероника аптечная 1917
вероника Бахофена 406
вероника весенняя 5726
вероника водная 5838
вероника горная 3720
вероника длиннолистная 1289
вероника дубравка 2454
вероника дубравная 2454
вероника дубровка 2454
вероника изящная 5854
вероника иноземная 4360
вероника колосистая 5140
вероника кустарничковая 4534
вероника лежачая 2709
вероника лекарственная 1917
вероника ложная 475
вероника ненастоящая 475
вероника пашенная 2237
вероника персидская 5563
вероника плющелистная 2992
вероника полевая 1532
вероника поручейная 509
вероника поточная 509
вероника-поточник 509
вероника простёртая 2709
вероника ручейная 509
вероника тимьянная 5512
вероника тимьянолистная 5512
вероника трёхлистная 2253
вероника Турнефора 5563
вероника широколистная 2892
вероника щитковая 3518

веррукария 5785
вертляница 2940
весёлка вонючая 5256
весёлка обыкновенная 5256
весёлка рядовая 5256
весёлковые 4162
весенник 5998
веснянка 2188
веснянка весенняя 5183
ветиверия 5732
ветла 5922
ветла блестяще-жёлтая 6107
ветреница 242
ветреница алтайская 2962
ветреница альпийская 112
ветреница весенняя 5721
ветреница дубравная 2134
ветреница лесная 5061
ветреница лютиковая 6115
ветреница нарциссоцветковая 3796
ветреница-перелеска 1437
ветры волчьи 1328
ветры волчьи 5784
вех 5808
вех ядовитый 2130
вечерница 4522
вечерница ночная фиалка 1794
вешенка обыкновенная 3992
вешенка рожковидная 3992
вздутоплодник 88
взморник 1997
взморник морской 1407
взморниковые 1998
вигандия 5935
вигна 1650
вигна катьянг 1059
вигна китайская 1393
вигна коноплевидная 1059
виддрингтония 5930
виддрингтония Уайта 3637
виддрингтония Шварца 4763
вика 5731
вика бенгальская 4356
вика венгерская 2893
вика кормовая 1554
вика крупносемянная 568
вика мохнатая 2693
вика нарбонская 3794
вика обыкновенная 1554
вика озимая 2693
вика паннонская 2893
вика песчаная 2693
вика плоско-семянная 1554
вика посевная 1554
вика пушистоплодная 6051
вика римская 3794
вика сорнополевая 1554
вика французская 3794
вика чёрная 2693
вика чёрная мохнатая 2693
вика яровая 1554
виктория 5830
виктория ананасная 4195
вилларезия 5734
виноград 2575
виноград американский 4503
виноград амурский 227
виноград Берландье 6002
виноград винный 2104
виноград виноносный 2104
виноград винородный 2104
виноград вьющийся 5748
виноград горный 6002
виноград Давида 777
виноград девичий 1664
виноград девичий пятилистный 5748
виноград девичий садовый 5484
виноград девичий триостренный 3028
виноград дикий 1664
виноград дикий пятилистный 5748
виноград дикий триостренный 3028
виноград евразиатский 2104
виноград европейский 2104
виноград зимний 6002
виноград Изабеллы 2347
виноград круглолиственный 3756
виноград круглолистный 3756
виноград культурный 2104
виноград лабруска 2347
виноград лапчатый 1057
виноград лесной 6032
виноград лисий 2386
виноград медвежий 500
виноград настоящий 2104
виноград прибрежный 4503
виноград североамериканский 2347
виноград старосветский 2104
виноградные 2576
виноградовник 221
виноградовник короткоцветоножковый 224
виноградовые 2576
виргилия 5747
виргилия жёлтая 213
вирола суринамская 398
висмия 5758
вистария 6011
вистария китайская 1211
вистария кустарниковая 212
вистария обильноцветущая 3078
витекс 1120
вишенник 2644
вишенник 4226
вишнеслива 3777
вишня 1133
вишня алкекенги 5275
вишня ананасная 4156
вишня антипка 3426
вишня Бессея 547
вишня виргинская 1384
вишня-гриот 3483
вишня дикая 2644
вишня железистая 105
вишня жидовская 2645
вишня карликовая 4683
вишня кислая 3679
вишня кустарниковая 2644
вишня лесостепная 2644
вишня магалебская 3426
вишня мараскиновая 3483
вишня мелкопильчатая 3951
вишня обыкновенная 5091
вишня перуанская 4156
вишня перувианская 4156
вишня песчаная 4683
вишня пёсья 2645
вишня пёсья обыкновенная 5275
вишня пёсья съедобная 4156
вишня пёсья Франшетта 5275
вишня полевая 5275
вишня птичья 3544

вишня пузырная 2645
вишня пузырчатая 2645
вишня садовая 5091
вишня сладкая 3544
вишня степная 2644
вишня суринамская 4206
вишня суринамская питанга 4206
вишня уссурийская 1174
вишня Франшетти 5275
вишня чёрная дикая 626
вишня чёрная поздняя 5091
вишня чернильная 3426
вишня шпанка кустарниковая 2644
вишня японская 1174
воандзея подземная 1574
водокрас 2381
водокрас лягушечный 2382
водокрас обыкновенный 2382
водокрасовые 2383
водолистниковые 5814
водолюб 4114
водолюб 5813
водоросли 92
водоросли багрянки 4433
водоросли бурые 815
водоросли десмидиевые 1841
водоросли диатомовые 1853
водоросли зелёные 2621
водоросли красные 4433
водоросли кремнёвые 1853
водоросли кремнистые 1853
водоросли сине-зелёные 701
водоросли харовые 1119
водоросли циановые 701
водоросль австралийская морская съедобная 373
водосбор 1341
водосбор канадский 188
водосбор обыкновенный 2087
водосбор Скиннера 4985
водяника 1706
водянка 1706
волвянка 6049
волдырник 656
волдырник ягодный 544
волдырник ягодоплодный 544
волжанка 2519
волжанка 6049
волкобой 2194
волнушка 6049
волнушка розовая 6049
воловик 840
володушка 5493
володушка круглолистная 4589
володушка кустарниковая 4888
володушка серповидная 2156
володушка тончайшая 4998
волосатик 3430
волоснец 5957
волоснец песчаный 2094
волчец благословленный 671
волчец кудрявый 671
волчеягодник 1800
волчеягодник-боровик 4559
волчеягодник волчье лыко 2194
волчеягодник кавказский 1070
волчеягодник лавролистный 5192
волчеягодник обыкновенный 2194
волчник 1800
волчник-боровик 4559
волчник кавказский 1070
волчник лавролистный 5192
волчник обыкновенный 2194
волчник смертельный 2194
волчниковые 3608
волчуг 4484
вольфия 6017
вольфия бескорневая 4551
вонючка 336
вонючка 3575
воробейник 2640
воробейник аптечный 1428
воробейник лекарственный 1428
воробейник полевой 1614
воробейник пурпурно-синий 1669
воронец 441
воронец волосистый 616
воронец колосистый 616
воронец колосовидный 616
вороника 1706
вороника чёрная 627
вороничковые 1707
вороночник рожковидный 2858
вороняжка 441
ворсянка 5469
ворсянка волосистая 2692
ворсянка лесная 5717
ворсянка посевная 2390
ворсянка раздельная 1760
ворсянка раздельнолистная 1760
ворсянка разрезная 1760
ворсянка сукновалов 2390
ворсянка сукновальная 2390
ворсянковые 5471
восковник 483
восковник 2843
восковник обыкновенный 5366
восковниковые 484
восковница 483
восковница восконосная 5104
восковница калифорнийская 948
восковница обыкновенная 5366
воскоцветник 2843
воскоцветник малый 3280
вострец 5957
вошерия 2626
вудвардия 1110
вудсия 6043
вудсия альпийская 157
вудсия волосистая 4625
вудсия гладкая 5043
вудсия гладковатая 5043
вудсия горная 157
вудсия северная 4625
вудсия эльбская 4625
вульпия 2208
вульпия белкохвостая 803
вульпия мышехвостная 4423
вульфения 6059
вшивица 3383
вьюнец 125
вьюнковые 3683
вьюнок 2509
вьюнок полевой 2100
вьюнок пурпурный 1474
вьюнок садовый 1474
вьюнок стрелолистный 4741
вьюнок трёхцветный 1941
вяжечка 4517
вяжечка гладкая 5566

вяжечка голая 5566
вяз 2018
вяз американский 191
вяз белый 191
вяз ветвистый 4521
вяз гладкий 4614
вяз горный 4771
вяз граболистный 5040
вяз граболистный роговидный 1615
вяз кистевидный 4521
вяз крылатый 5989
вяз крыловаточерешковый 5989
вяз листоватый 5040
вяз лопастный 3461
вяз мелколистный 1177
вяз обыкновенный 4614
вяз полевой 5040
вяз приземистый 4903
вяз пробковый 1602
вяз ржавый 5012
вяз раскидистый 5174
вяз Томаса 4521
вяз шероховатый 4771
вяз шершавый 4771
вязель 1712
вязель горный 3704
вязель обыкновенный 1713
вязель пёстрый 1713
вязель скудноцветный 4766
вязель эмеровый 4766
вязинник 467
вязовик 1921
вязовик 2851
вязовик трёхлистный 1441
вязовые 2019

Г г

гавортия 2728
гагения абиссинская 3190
гайлюссакия 2884
гайлярдия 2400
гайлярдия крупноцветная 1491
галакс 2401
галантус 5060
галега 2521
галега лекарственная 1425
галезия 4950
галения 5193
галимодендрон 4667
галинзога 4386
галинсога 4386
галинсога мелкоцветковая 3326
галинсога мелкоцветная 3326
галипея 2403
галипея лекарственная 251
галодуле 4870
гальтония 5319
гамамелидовые 6013
гамамелиевые 6013
гамамелис 6012
гамамелис весенний 5727
гамамелис виргинский 1562
гамамелис китайский 1212
гамамелис мягкий 1212
гамамелис японский 3079
гамбир 532
гамелия 2699
гамильтония 2700
ганус 253
гардения 2416
гармала 4102
гармала обыкновенная 2714
гарновка 1930
гаррия 4947
гарциния 2405
гастерия 2436
гастеромицеты 2437
гастролобиум 4240
гаультерия 6003
гаультерия лежачая 1125
гаультерия распростёртая 1125
гаура 2438
гацания 2440
гвайюла 2657
гвайюла 4065
гвайюла серебристая 2657
гварея 203
гварея 3768
гварея трёхтысячевидная 203
гваюла 2657
гваюла 4065
гваюла серебристая 2657
гваяк 1460
гваякум 3293
гваякум лекарственный 1460
гвизоция 3849
гвизоция абиссинская 2063
гвоздика 4197
гвоздика армериевидная 1837
гвоздика бородатая 5393
гвоздика голландская 1019
гвоздика картузианская 1037
гвоздика китайская 1165
гвоздика перистая 2588
гвоздика песчаная 2255
гвоздика пышная 3298
гвоздика садовая 1019
гвоздика сизая 1129
гвоздика травяная 3433
гвоздика травянка 3433
гвоздика турецкая 5393
гвоздика Фишера 1165
гвоздичник 5502
гвоздичник мадагаскарский 3407
гвоздичные 4198
гебелия 4001
гевеа 2791
гевея 2791
гевея бразильская 4050
гедизарум 5390
гедихиум 2491
гейлюссакия 2884
гейлюссация 2884
гейхера 171
гелениум 5052
гелиамфора 5332
гелидиум 55
геликодисерос 5666
геликония 2763
геликтерес 4781
гелиопсис 2764
гелиотроп 2765
гелиотроп перувианский 1436
гелиптерум 5333
гелихризум 2149
гелихризум жёлтый 6083
геллебор 2766
геллеборус 2766
гелониас 5346

гельземий 2441
гельземиум 2441
гемантус 678
гемлок 2769
гемлок восточный 973
гемлок канадский 973
генекен 2781
генипа 2444
генипап 2443
генциана 2445
геоглоссум 1970
геонома 4840
геонома Шотта 316
георгина 1782
георгина изменчивая 2413
георгиния 1782
гераневые 2448
гераниевые 2448
герань 2447
герань болотная 718
герань голубиная 3369
герань красно-бурая 1931
герань кроваво-красная 680
герань кровянокрасная 680
герань круглолистная 4584
герань крупнокорневищная 570
герань лесная 6024
герань луговая 3551
герань маленькая 5026
герань мелкая 5026
герань мягкая 1892
герань нежная 1892
герань пиренейская 4372
герань пурпурная 4342
герань рассечённая 1757
герань Роберта 2787
герань розовая 4566
герань сибирская 4905
герань тёмная 1931
герань холмовая 5686
гербера 2451
гертнериа 873
геснериевые 2465
геснерия 2464
гесперис 4522
гетерантера 3743
гетероспата высокая 4642
геттарда 5714
геухера 171
гиацинт 2896
гиацинт водный 1557
гиацинт водяной 1557
гиацинт восточный 1445
гиббертия 904
гибиск 2794
гибискус 2794
гибискус индийский 3008
гибискус кенаф 3146
гибискус китайская роза 1185
гибискус коноплёвый 3146
гибискус липолистный 3316
гибискус мускусный 3761
гибискус розелла 4562
гибискус сабдарифа 4562
гибискус северный 2316
гибискус сирийский 4886
гибискус съедобный 3921
гибискус троичный 2316
гибискус тройчатый 2316
гигрофила 5233
гигрофор 2904
гигрофор конический 1577
гигрофоровые 2905
гиднокарпус 1121
гиднофитум 2899
гиднум 2900
гиднум выемчатый 2752
гидрангея 2902
гидрангея древовидная 5039
гидрангея метельчатая 4028
гидрастис 2546
гидрастис канадский 2547
гидроклеис 5831
гидрокотиле 4114
гидрокотиль 4114
гидрофил 5813
гидрофиллум 5813
гидрофилляциевые 5814
гикори 2795
гикори белая 3640
гикори белый 3640
гикори гладкий 4179
гикори горький 605
гикори косматый 3640
гикори крупнопочечный 4841
гикори свиной 4179
гиления 2172
гилия 2487
гилления 2172
гименокаллис 5130
гименомицеты 4180
гименофилловые 2249
гимнокладус 1319
гимнокладус двудомный 3151
гимнокладус канадский 3151
гимноклядус 1319
гинала 230
гинандропсис 5129
гинериум 4021
гинериум серебристый 4818
гинкго 2492
гинкго дву(х)лопастный 2493
гинкго китайское 2493
гинкговые 3432
гиннала 230
гинура 5713
гипеаструм 176
гипекоум 2906
гипнум 2908
гипохерис 1061
гипохерис голый 5036
гипохерис короткокорневой 5164
гипохерис укореняющийся 5164
гиппеаструм 176
гиппеаструм 448
гиппократия 2813
гиппокрепис 2874
гиппомана 3455
гипсолюбка 2672
гипсофила 2672
гипсофила изящная 1433
гирофора 4535
гирча 4817
гирчевник 2770
гирчовник 2770
гиссон 2909
гифене 2001
гифомицеты 2245
гладиолус 2499
гладиолус гентский 771

гладыш 3227
глаз вороний обыкновенный 2786
глаз вороний четырелистный 2786
глаз вороний четырёхлистный 2786
глазки анютины 5954
глазки павлиньи 4164
глаукс 4796
глауциум 2859
глауциум жёлтый 6096
глауциум рогатый 652
гледичия 2840
гледичия аргентинская 219
гледичия каспийская 1044
гледичия китайская 1187
гледичия колючая 1439
гледичия крупноколючковая 573
гледичия обыкновенная 1439
гледичия сладкая 1439
гледичия трёхколючковая 1439
гледичия трёхостренная 1439
гледичия трёхшипная 1439
гледичия японская 3041
глейхениевые 2502
глейхения 2501
глистник 604
глистник 1564
глистогон 3240
глициния 6011
глициния китайская 1211
глициния обильноцветущая 3078
глициния японская 3365
глобулярия 2506
глог 683
глоговина 1128
глоксиния 1424
глоксиния 2515
глориоза 2511
гляуциум 2859
гнафалиум 1727
гнездовка 590
гнездовковые 589
гнетовые 3107
гнетум 3105
гнетум гнемон 5149
говения 4406
говения сладкая 3066
говорушка 1270
говорушка беловатая 4867
говорушка булавоногая 1287
говорушка ворончатая 2395
говорушка душистая 5385
говорушка пахучая 5385
говорушка серея 1274
годеция 2524
гойяба 1431
головчатка 1094
головчатка альпийская 6079
головчатка сирийская 5417
головчатка татарская 5453
голорепа 3175
голосеменные 2671
голосемянные 2671
голубика 717
голубика американская 2796
голубика зонтичнокистевая 2796
голубика канадская 969
голубки 2087
голубой очный цвет 705
гомалиум 2833
гомбо 3921
гоми 2352
гоми 2624
гомфия 2557
гомфрена 2503
гомфрена головчатая 1421
гомфрена шаровидная 1421
гониолимон 4793
гониолимон татарский 5458
гонобобель 717
гордовина 5852
гордония 2564
горец 3172
горец альпийский 131
горец восточный 4309
горец вьюнковый 1926
горец вьющийся 1926
горец головчатый 4199
горец горный 131
горец живородящий 5759
горец земноводный 5812
горец змеиный 2079
горец кустарниковый 2755
горец малый 4999
горец осенний 4164
горец осотоконечный 3036
горец перечный 3511
горец почечуйный 5168
горец птичий 4317
горец сахалинский 4656
горец тонкозаострённый 3036
горец щавелелистный 1739
горечавка 2445
горечавка баварская 482
горечавка весенняя 5185
горечавка германская 2457
горечавка горькая 260
горечавка жёлтая 6091
горечавка Клюзия 1291
горечавка ластовневая 3622
горечавка немецкая 2457
горечавка паннонская 1799
горечавка перекрестнолистная 1700
горечавка пиренейская 4371
горечавка полевая 3550
горечавка пурпурная 4346
горечавка реснитчатая 2684
горечавка снежная 5063
горечавковые 2446
горечник 2824
горицвет 33
горицвет 967
горицвет весенний 5179
горицвет кожистый 4558
горицвет-кукушкин цвет 4398
горицвет летний 5317
горицвет огненный 2272
горицвет черногорка 5179
горичник 2824
горичник аптечный 1438
горичник болотный 3509
горичник горный 3715
горичник лекарственный 1438
горичник настурциевый 3532
горичник олений 3738
горичник царский корень 3532
горлец 2079
горлюха 3989
горлюха румяниковидная 783
горлюха румянковидная 783
горлюха ястребинковидная 2727

горлянка 922
горноколосник 5263
горох 4080
горох бараний 1144
горох голубиный 4177
горох земляной 1574
горох кормовой 1393
горох коровий 1393
горох мозговой 3498
горох мышиный 591
горох обыкновенный 2423
горох полевой 2231
горох посевной 2423
горох садовый 2423
горох спаржевый 5197
горох турецкий 1144
горошек 5731
горошек венгерский 2893
горошек волосистый 5531
горошек гороховидный 4097
горошек горький 609
горошек душистый 5378
горошек жёлтый 6111
горошек заборный 2760
горошек зарослевый 2462
горошек зелёный 2423
горошек изящный 5004
горошек крупноцветковый 4885
горошек кустарниковый 2462
горошек лесной 6046
горошек мохнатый 2693
горошек мышиный 591
горошек нарбонский 3794
горошек одноцветковый 3930
горошек посевной 1554
горошек тонколистный 755
горошек узколистный 3810
горошек чёткообразный 609
горошек четырёхсемянный 2343
горошек шершаволистный 5531
гортензия 2902
гортензия древовидная 5039
гортензия крупнолистная 565
гортензия метельчатая 4028
гортензия черешковая 1269
горчак 3511
горчак 3989
горчица 3771
горчица английская 5904
горчица белая 5904
горчица белая английская 5904
горчица дикая 1118
горчица морская 4799
горчица полевая 1118
горчица сарептская 2939
горчица чёрная 642
горькуша альпийская 147
горянка 2049
госта 4221
гоферия 2822
граб 2852
граб американский 198
граб восточный 3956
граб европейский 2108
граб кавказский 1073
граб каролинский 198
граб обыкновенный 2108
граб сердцелистный 2738
граб японский 3045
грабельки 2330
грабельник 2789
грабельник цикутолистный 91
грабинник 3956
грабник 3956
гравилат 385
гравилат городской 1354
гравилат обыкновенный 1354
гравилат прибрежный 5791
гравилат прямой 6074
гравилат речной 5791
гранадилла 4070
гранадилла гигантская 2475
гранат 4255
гранатник 4255
гранатник обыкновенный 4255
грацилярия лишайниковатая 56
гребенник 1877
гребенчук 5435
гребенщик 5435
гребенщик африканский 50
гребенщик одесский 3909
гребенщик опушённый 3136
гребенщик пятитычинковый 2269
гребенщик французский 2374
гребенщик четырёхтычинковый 2344
гребенщиковые 5436
гребень Венерин 4861
гребень петуший 1410
гребешок петуший 1410
гребневик 1877
гребневик гребенчатый 1681
гребневик обыкновенный 1681
гребневик шиповатый 2750
гребник 1877
гревиллея 2638
гревиллея крупная 4945
гревия 2639
гревия липолистная 3317
грейпфрут 2578
гренадилья 4070
гренландия густая 1833
гречиха 832
гречиха горлец 2079
гречиха культурная 1370
гречиха посевная 1370
гречиха птичья 4317
гречиха сахалинская 4656
гречиха татарская 5446
гречишник 3172
гречишник бальджуанский 726
гречишные 833
гриб 3758
гриб белый 1093
гриб белый ложный 598
гриб гвоздичный 2154
гриб домовый 2880
гриб древесный 4254
гриб жёлчный 598
гриб зимний 5709
гриб-зонтик 3278
гриб-зонтик мелкощитовидный 4863
гриб-зонтик пёстрый 4051
гриб каштановый 1140
гриб олений 2189
гриб печёночный 515
гриб пластинчатый 57
гриб подберёзовый 582
гриб подорешник 1114
гриб подосиновик 3940
гриб польский 485

гриб сатанинский 679
гриб трюфельный 4619
гриб царский 2319
гриб цезарский 918
гриб чернильный 2950
гриб чёрный 582
гриб чесночный 2434
гриб шпальный 4739
грибки лучистые 23
грибок бродильный 6069
грибок дрожжевой 6069
грибы агариковые 58
грибы асковые 337
грибы базидиальные 462
грибы головнёвые 854
грибы дрожжевые 6070
грибы ежовиковые 2901
грибы лучистые 23
грибы ржавчинные 4623
грибы рогатиковые 1588
грибы сапролегниевые 4703
грибы слизевые 5006
грибы слизистые 5006
грибы сморчковые 2768
грибы сумчатые 337
грибы хитридиевые 302
грибы шляпочные 4180
грибы энтомофторовые 2047
гринделия 2663
гринделия высокая 4873
гринделия исполинская 4873
гроб-трава 1493
гроздовник 2577
гроздовник виргинский 4427
гроздовник ключ-трава 3673
гроздовник многораздельный 794
гроздовник обыкновенный 3673
гроздовник полулунный 3673
громовик 3932
грудинка 4925
грудника 7
грудника Авиценны 1217
грудника Теофраста 1217
грудница обыкновенная 2102
груздь 4123
груздь перечный 4123
груздь сухой 5923
грумичама 2653
груша 4089
груша аллигаторова 183
груша берёзолистная 580
груша дикая 1487
груша земляная 3094
груша иволистная 5971
груша китайская песочная 4687
груша миндалевидная 107
груша обыкновенная 1487
груша снежная 5066
груша уссурийская 5694
грушанка 4377
грушанка зеленоватая 2633
грушанка зеленоцветковая 2633
грушанка круглолистная 2119
грушанка малая 5065
грушанка средняя 2953
грушанковые 4378
грыжник 877
грыжник волосистый 2682
грыжник гладкий 1372
грыжник голый 1372
гуава 2656
гуайава 2656
гуайюла 2657
гуайява 2656
гуарана 2655
гуаюла 2657
гуаява 2656
гуаяк 3293
губастик 3655
губастик крапчатый 1471
губастик мускусный 3764
губка берёзовая 583
губка горбатая 2485
губка домовая 2880
губка дубовая 3900
губка лиственничная 4333
губка-люффа 5564
губка разноцветная 1569
губка растительная 5564
губовник линейный 1840
губоцветные 3629
гугония 2885
гудайера 4428
гудайера ползучая 1675
гуза 3283
гуилельма замечательная 485
гуйава 2656
гуйява 2656
гулявник 4978
гулявник аптечный 2757
гулявник высокий 5638
гулявник высочайший 5638
гулявник ирио 3363
гулявник лекарственный 2757
гумай 3103
гумбанг мягкий 992
гумея 2886
гумея красивая 2340
гуммигутовые 2669
гуннера 2667
гуннера чилийская 1148
гуньба 711
гусятница 4317
гуттаперчевые 4701
гуцмания 2670

Д д

давалия 2711
даваллия 2711
дагусса 4399
дазилирион 5090
дазилирион Вилера 5875
дазилирион волосистоконечный 822
дазистома 3903
дакридиум 2894
далия 1782
дальбергия 4570
дальбергия сенегальская 42
дальбергия сиссоо 4977
дальбергия чёрная 764
дальбергия чернодревесная 42
дальбергия широколистная 1984
даммара 1795
даная ветвистая 88
дантония 1798
дарлингтония 941
дармина 3284
датиска 1810
датиска коноплёвая 471

датура 1811
двойчатка 555
двугнёздка 5918
двудольные 1855
двузернянка 2029
двузернянка дикая 5961
двузернянка культурная 2029
двукисточник 984
двукисточник тростниковидный 4473
двукисточник тростниковый 4473
двукрылоплодник 2668
двулепестник 2031
двулепестник альпийский 124
двулепестник горный 124
двулепестник парижский 4055
двулистник 5678
двурядка 5775
двурядка стенная 5257
двурядка тонколистная 5007
двусеменник 2895
двусемянник 2895
двухкрылоплодник 2668
девясил 2955
девясил блоший 1239
девясил большой 2010
девясил британский 784
девясил высокий 2010
девясил иволистный 5970
девясил мечелистный 5407
девясил обыкновенный 2010
девясил растопыренный 1239
дейнболлия 1026
дейция 1843
дейция Вильсона 5977
дейция Зибольда 4932
дейция изящная 4995
дейция красивая 4995
дейция крупноцветковая 1961
дейция шероховатая 2397
дейция шершавая 2397
дейция Шнейдера 4759
делет 4019
дельфиниум 3226
дельфиниум гибридный 2305
демьянка 2414
денгамия 1830
дендробиум 1829
дендрокаламус 2467
денежка длинноножковая 4550
денежка каштаново-коричневая 901
денежка широкопластинчатая 790
денежник 2232
дербенка 4726
дербенник 3400
дербенник иволистный 4349
дербенник иссополистный 2912
дербенник лозный 5782
дербенник плакун 4349
дербенник прутовидный 5782
дербенниковые 3371
дербянка 4726
дербянка колосистая 1821
дерганец 2773
деревей 6067
дерево авраамово 3296
дерево азиминовое 4077
дерево алойное 1958
дерево амарантовое 5871
дерево амбровое 210
дерево амурское бархатное 226
дерево амурское пробковое 226
дерево анакардовое 1042
дерево анисовое 3014
дерево анисовое 5607
дерево апельсиновое 5377
дерево атласное 1107
дерево атласное индийское 1107
дерево бакаутовое 1460
дерево бакаутовое 3293
дерево бакаутовое железное 1460
дерево бальзамическое африканское 44
дерево бальзамическое толюйское 5541
дерево бальзамное 429
дерево бальзамное толу 5541
дерево бальзамовое 429
дерево бальзамовое 5541
дерево бальзамовое перувианское 4153
дерево бальзовое 5870
дерево бархатное 226
дерево бензоиновое 5315
дерево бензойное 5127
дерево божье 3923
дерево бразильское палисандровое 764
дерево буксовое 749
дерево бумажное 1486
дерево восковое 1558
дерево восковое 5848
дерево восковое китайское 1204
дерево габонское тюльпанное 527
дерево гваяковое 1460
дерево гваяковое 3293
дерево гвоздичное 1283
дерево горлянковое 924
дерево горлянковое обыкновенное 1373
дерево горшечное 3657
дерево гранатовое 4255
дерево грушевое 4089
дерево гуттаперчевое 3443
дерево даммаровое 477
дерево джатовое 1544
дерево дождевое 4404
дерево драконовое 1900
дерево дынное 4035
дерево дынное обыкновенное 4035
дерево железное 315
дерево железное 2966
дерево железное 3117
дерево железное 4150
дерево жёлтое 1527
дерево жёлтое 1912
дерево земляничное 3418
дерево земляничное крупноплодное 5276
дерево золотое 3019
дерево индийское розовое 863
дерево иудейское 3114
дерево иудино 3114
дерево каепутовое 3579
дерево кайюпутовое 3579
дерево калабас(с)овое 924
дерево камедное 4259
дерево каменное 2674
дерево кампешевое 3359
дерево камфарное 966
дерево камфарное 1367
дерево камфарное зондское 1367
дерево капоковое 3134
дерево капустное 907
дерево карандашное 1980
дерево карандашное 4298
дерево касторовое африканское 3770

дерево каучуковое 2944
дерево каяпутовое 920
дерево каяпутовое 3579
дерево квассиевое 5338
дерево квебраховое 4461
дерево квитовое 4389
дерево кевовое 5476
дерево кедровое 1082
дерево кентуккское кофейное 3151
дерево китайское восковое 1204
дерево китайское сальное 1204
дерево кожевенное 4920
дерево кокаиновое 2883
дерево колбасное 4716
дерево конфетное 4406
дерево коралловое 1390
дерево коралловое 4676
дерево коричное 1105
дерево коричное 1238
дерево коричное китайское 1047
дерево коричное цейлонское 1105
дерево корнелиевое 2309
дерево коровье 1646
дерево кофейное 1318
дерево кофейное арабское 292
дерево кофейное аравийское 292
дерево кофейное либерийское 3287
дерево крабовое 2660
дерево крапивное 2106
дерево красное 768
дерево красное 1107
дерево красное 1128
дерево красное 5871
дерево красное бразилийское 768
дерево красное фернамбуковое 768
дерево красное фернамбуковое 4296
дерево кружевное 3319
дерево лаковое 533
дерево лаковое 3050
дерево лаковое японское 5848
дерево лжеапельсиновое 3972
дерево лимонное 3268
дерево мамонтово 2483
дерево манговое 3472
дерево мангровое 200
дерево масляное 468
дерево масляное 898
дерево мастиковое 3275
дерево мастичное 3275
дерево махагониевое 5871
дерево мескитовое 3593
дерево миндальное 104
дерево миробалановое 1124
дерево мирровое 1477
дерево мирровое 3779
дерево молочное 1646
дерево молочное индийское 1654
дерево москитовое 1467
дерево мускатное 1480
дерево мускатное душистое 1480
дерево мыльное 2555
дерево мыльное 5071
дерево мыльное 5072
дерево мыльное антильское 5102
дерево мыльное обыкновенное 5102
дерево мыльное сенегальское 4822
дерево обезьянье хлебное 445
дерево огуречное 1726
дерево оливковое 1483
дерево орлеанское 256
дерево палисандровое 648
дерево палисандровое 2996
дерево палисандровое сиссоо 4977
дерево париковое 5033
дерево перечное 4120
дерево перечное калифорнийское 940
дерево пернамбуковое 4296
дерево пробковое 1604
дерево пробковое амурское 226
дерево пробковое антильское 3316
дерево путешественников 3409
дерево рожковое 1023
дерево розовое 430
дерево розовое 474
дерево розовое 508
дерево розовое индийское 1984
дерево саговое 4643
дерево саламандровое 566
дерево сальное 899
дерево сальное 1204
дерево сандаловое 3359
дерево сандаловое 4677
дерево сандаловое 4679
дерево сандаловое антильское 1902
дерево сандальное 3359
дерево сандальное 4679
дерево сандараковое 296
дерево санталовое 4677
дерево сапотиловое 4700
дерево сапотовое 4700
дерево саппановое 4702
дерево свечное 992
дерево серебряное 4964
дерево скипидарное 5476
дерево снежное 2379
дерево снежное 5896
дерево стираксовое 3965
дерево стираксовое 5056
дерево стираксовое лекарственное 1915
дерево стручковое 1023
дерево сырное 3134
дерево терпентинное 5476
дерево терпентиновое 5476
дерево тиковое 1544
дерево тисовое 6117
дерево травяное 2591
дерево тунговое китайское 5640
дерево тутовое 3749
дерево тюльпанное 5635
дерево тюльпанное виргинское 5636
дерево тюльпанное китайское 1207
дерево уксусное 5206
дерево фернамбуковое 768
дерево фернамбуковое 4296
дерево фиалковое 2486
дерево фиалковое австралийское 2486
дерево фиговое 1412
дерево физетовое 1527
дерево фисташковое 1498
дрeво хинное 1236
дерево хлебное 769
дерево хлебное восточноиндийское 2998
дерево хлебное джек 2998
дерево хлебное индийское 2998
дерево хлебное Лакооха 3207
дерево хлебное разнолистное 2998
дерево хлебное цельнолистное 2998
дерево хлопковое 3440
дерево хлопчатое 3134
дерево хренное 2872

дерево чайное 5467
дерево чайное парагвайское 4048
дерево чёрное 655
дерево чёрное 1987
дерево чёрное австралийское 655
дерево шёлковое 6094
дерево шерстяное 3134
дерево шоколадное 912
дерево эбеновое 1987
дерево энгленовое 408
дереза 6016
дереза китайская 1213
дереза лебедолистная 3535
дереза обыкновенная 3535
дёрен 1881
дёрен американский 4458
дёрен белый 5454
дёрен красный 683
дёрен кроваво-красный 683
дёрен мужской 1611
дёрен обыкновенный 1611
дёрен отпрысковый 4458
дёрен съедобный 1611
дёрен флоридский 2309
дёрен цветочный 2309
дёрен цветущий 2309
дёрен шведский 1937
дёренные 1882
дёреновые 1882
держи-дерево 1227
держи-дерево обыкновенное 1227
держи-дерево Христовы Тернии 1227
деррис 3100
дескурайния 5442
дескурения 5442
дескурения Софии 2296
десмодиум 5516
джамбоза 4556
джамболан 3010
джантак 961
джек 2998
джек-дерево 2998
джексония 3002
джекфрут 2998
джемсия 3011
джемсия американская 1264
джефферсония 5662
джида 4616
джонушка 4921
джугара 5892
джузгун 952
джут 3122
джут африканский 3146
джут длинноплодный 4283
джут короткоплодный 4591
джут круглоплодный 4591
джут огородный длинноплодный 4283
джут яванский 3146
дзелква 5800
дзелква граболистная 2021
дзелква китайская 1216
дзелква японская 3080
дзелкова 5800
диалиум 1851
диалиум гвинейский 5715
диапензия 1852
диапенсия 1852
диаррена 495
диасция 5665
диатомеи 1853
дивала 3168
дивало 3168
дивало многолетнее 4131
дивало однолетнее 262
дивина 6055
диделотия африканская 826
дидискус 3194
диелитра 668
диерама 2016
диервила 880
диервилла 880
дизоксилон 4106
дикия 1953
дикориния 4044
дикориния параензис 4044
дикран 2331
диктамнус 1866
диления 1858
дилления 1858
диморфотека 1005
диоклея 1295
дионея 5718
диония 5718
диоскорейные 6064
диоскорея 6062
диоскорея батат 1241
диосма 1861
диоспирос 4152
диптерикс 5545
диптерокарповые 1862
диптерокарпус 2668
диптерокарпус бальзамический 1432
диптерокарпус шишковатый 2035
дирка 3259
дирка болотная 363
дискостигма 1863
диффенбахия 5631
дихапеталиум 1854
дихромена 5919
дицентра 668
дицентра великолепная 1363
дицентра китайская 1363
диэлитра 668
додартия 1869
додарция 1869
додекатеон 4871
додонея 2850
додонея аптечная 1251
дождевик 4325
дождевик-великан 2479
дождевик "заячья картошка" 1328
дождевик шиповатый 5784
дождевиковые 4326
дождь золотой 3192
дождь золотой альпийский 4773
дождь золотой обыкновенный 2529
долгоцветка 60
долихос 1883
долихос китайский 1393
долихос лябляб 2897
долихос обыкновенный 2897
домбея 1885
домбея точечная 744
донник 5362
донник аптечный 6108
донник белый 5917
донник высокий 5426
донник голубой 711
донник жёлтый 6108
донник зубчатый 1834

донник индийский 267
донник лекарственный 6108
донник мелкоцветковый 267
донник синий 711
дорема 5316
дорема аммиачная 731
дорема аммониакальная 731
дорема камеденосная 731
дориантес 5119
дорикниум 982
дорикниум травянистый 2784
дороникум 3276
дороникум австрийский 379
дороникум кавказский 1075
дорстения 5559
дохн 4092
драконник 1900
дракункул 5254
драцена 1898
драцена австралийская 2470
драцена астелия 357
драцена драконовая 1900
драцена душистая 2357
драцена южная 1402
драч 1227
древогубец 608
древокорень 3475
древокорень мангле 200
дрёма белая 2140
дрёма беловатая 2140
дрёма липкая 5757
дрёма луговая 2140
дрёма ночная 3850
дремлик 2050
дремлик широколистный 792
дриада 1920
дриада восьмилепестковая 3726
дриада восьмилепестная 3726
дримария 4808
дримис 1903
дримис Винтера 6007
дриобаланопс 736
дриобаланопс душистый 1367
дробянки 2266
дрожжи 6070
дрожжи винные 5986
дрожжи настоящие 6069
дрожжи пивные 1359
дрожжи пивные верхового брожения 1359
дрок 6015
дрок английский 3823
дрок волосистый 2685
дрок германский 2463
дрок испанский 5856
дрок испанский обыкновенный 5857
дрок колючий европейский 1426
дрок красильный 1563
дряква 1766
дряква 4871
дряква европейская 2092
дряква персидская 2990
дряквенник 4871
дуб 3899
дуб австрийский 2128
дуб американский красильный 644
дуб американский шарлаховый 4746
дуб бархатистый 644
дуб бархатный 644
дуб белый 5905
дуб болотный 4200
дуб бургундский 2128
дуб вечнозелёный 2832
дуб виргинский 3342
дуб галловый 85
дуб гонтовый 4865
дуб горный 1142
дуб греческий 5697
дуб густой 2979
дуб дву(х)цветный 5347
дуб дубильный 5343
дуб-железняк 1928
дуб звездчатый 4279
дуб зимний 1928
дуб-зимняк 1928
дуб золоточешуйчатый 1002
дуб зубчатый 1786
дуб ивовидный 5974
дуб иволистный 5974
дуб испанский 2814
дуб калифорнийский вечнозелёный 1002
дуб каменный 2832
дуб карликовый 505
дуб каролинский 3342
дуб каштанолистный 1141
дуб кермесовый 3153
дуб кермесоносный 3153
дуб коровий 5343
дуб кошенильный 4746
дуб красильный 644
дуб красный 5101
дуб красный северный 3883
дуб крупножелудевый 870
дуб крупноплодный 870
дуб крупночешуйчатый 5697
дуб курчавый 4845
дуб лавролистный 3233
дуб летний 2041
дуб ливанский 3261
дуб лировидный 3982
дуб лирообразный 3982
дуб лузитанский 4278
дуб македонский 3403
дуб малый 4279
дуб мерилендский 634
дуб Мирбека 987
дуб Мишо 5343
дуб монгольский 3653
дуб обыкновенный 2041
дуб ольхолистный 1779
дуб падуболистный 2832
дуб пиренейский 4374
дуб понтийский 4265
дуб пробковый 1603
дуб пробковый западный 554
дуб пробковый ложный 2814
дуб пунцоволистный 4746
дуб пушистый 4324
дуб репейник 870
дуб северный 3883
дуб серповидный 5101
дуб серполистный 5101
дуб сидячецветный 1928
дуб скальный 1928
дуб траволистный 939
дуб турецкий 2128
дуб Фрайнетто 2979
дуб хермесовый 3153
дуб черепитчатый 4865
дуб черепичный 4865
дуб черешчатый 2041

дуб чёрный 5827
дуб шарлаховый 4746
дубница 838
дубовик 3396
дубоизия 1922
дубравка 2454
дубровка 838
дубровка 2454
дубровка 5555
дубровник 2453
дубровник восточный 3954
дубровник горный 3705
дубровник горький 1064
дубровник кошачий 1064
дубровник метельчатый 1758
дубровник обыкновенный 1113
дубровник пурпуровый 1113
дубровник чесноковый 6028
дубровник чесночный 5806
дубровник шалфейный 6028
дубянка 2113
дугласия 1888
дугласия сизая 4541
дугласия тис(с)олистная 1401
дудник 245
дудник лекарственный 2406
дудник лесной 6029
дуляна 390
думпальма 1890
думпальма настоящая 2001
дура белая 5892
дуранта 4992
дуриан 1927
дурман 1811
дурман вонючий 3102
дурман древесный 2304
дурман древовидный 2304
дурман душистый 247
дурман индийский 2810
дурман метель 2810
дурман обыкновенный 3102
дурман татуля 3102
дурнишник 1306
дурнишник американский 486
дурнишник колючий 5155
дурнишник шиповатый 486
дурра белая 5892
дурьян 1927
душевик 4720
душевик альпийский 148
душевик аптечный 927
душёвка 464
душёвка альпийская 148
душёвка обыкновенная 5187
душёвка чабрецевидная 5187
душица 3967
душица 5375
душица обыкновенная 1484
душки 5354
дымянка 2392
дымянка аптечная 1911
дымянка лекарственная 1911
дымянковые 2393
дынниковые 4034
дыня 3762
дыня декоративная красноплодная 2855
дыня десертная 998
дыня змеевидная 5046
дыня извилистая 5046
дыня изогнутая 5046
дыня канталупа 998
дыня культурная 998
дыня мускатная 3762
дыня обыкновенная 3834
дыня огуречная 5046
дыня повислая 5046
дыня сетчатая 3834
дыравка 3449
дюбуазия 1606
дюшенея 3643
дягель 245
дягиль 245
дягиль аптечный 2406
дягиль лекарственный 2406
дягиль лесной 6029
дятлина 4447
дятлина белая 5889
дятлина красная 4447

Е е

евгения 2068
евгения бразильская 761
евгения малаккская 3912
евгения миртолистная 370
евгения ямболана 3010
ежа 1309
ежа обыкновенная 3943
ежа сборная 3943
ежа скученная 3943
ежевика 619
ежевика американская 624
ежевика вязолистная 2020
ежевика гигантская 2798
ежевика ильмолистная 2020
ежевика канадская 5490
ежевика кустарниковая 2080
ежевика лежачая 3880
ежевика плетевидная 3880
ежевика сизая 2093
ежевика чёрная 624
ежевика щетиноволос(н)ая 5342
ежевиковые 2901
ежеголовка 871
ежеголовковые 872
ежеголовник 871
ежеголовник маленький 3252
ежеголовниковые 872
ёжик 4306
ежовик 2900
ежовик жёлтый 2752
ежовик коралловидный 1589
ежовик пёстрый 2922
ежовиковые 2901
ежовка 1288
ежовник 233
ежовник 1310
ежовник 5843
ежовник куриное просо 456
ежовник петушье просо 456
ежовник хлебный 3057
елевзина 2015
елевзина индийская 2563
елевзина коракан 4399
елевзине 2015
елоха 5122
ель 5189
ель Алькокка 80
ель аянская 6071
ель Бальфура 417

ель белая 5913
ель Бревера 772
ель Вильсона 5980
ель восточная 3964
ель восточно-гималайская 4942
ель высокая 3886
ель гималайская 2806
ель Глена 4657
ель голубая 1338
ель двуцветная 80
ель европейская 3886
ель западная 3886
ель изящная 5523
ель индийская 2806
ель кавказская 3964
ель канадская 5913
ель колючая 1338
ель корейская 3187
ель красная 4465
ель обыкновенная 3886
ель пурпурная 4341
ель Саржента 4709
ель сербская 4830
ель серебристая 5913
ель сибирская 4915
ель синяя 695
ель ситхинская 4979
ель тяньшаньская 4761
ель финская 2256
ель чёрная 653
ель шиповатая 4942
ель Шренка 4761
ель Энгельмана 2034
ель японская 5523
ёра полярная 1935
ерань 2447
еремостахис 1839
ериантус 4228
ерика 2742
ерика древовидная 5581
ерика крестолистная 1701
ерика сизая 5668
ерика средиземноморская 594
ериоботрия 3375
ериоботрия японская 3376
ёрник 1935
ёрниковые 1707

Ж ж

жабник 2318
жабник германский 1396
жабник маленький 4994
жабник малый 4994
жабник немецкий 1396
жабрей 781
жабрица 4836
жантак 961
жарновец 809
жарновец метельчатый 4769
жасмин 3083
жасмин аптечный 1450
жасмин арабский 293
жасмин белый 1450
жасмин грунтовой 5376
жасмин дикий 5376
жасмин индийский 293
жасмин крупноцветковый 1052
жасмин крупноцветный 1052
жасмин лекарственный 1450
жасмин настоящий 1450
жасмин непахучий 4752
жасмин ночной 5376
жасмин розовый 6050
жасмин садовый 1078
жасмин садовый 5376
жасмин-самбак 293
жасмин японский 4714
жгунец 1260
железница 2967
железняк 4061
желтинник 1527
желтозелье 3809
желтокорень 2546
желтокорень канадский 2547
желтокорень китайский 195
желтолоз 4352
желтолозник 4352
желторог рябинолистный 4868
желтосмолка 2591
желтостебельник 5724
желтофиоль 5768
желтофиоль обыкновенный 1556
желтофиоль садовый 1556
желтоцвет 33
желтоцвет 610
желтоцвет весенний 5179
желтуха 4401
желтушник 3772
желтушник выгрызенный 5175
желтушник левкойный 5573
желтушник растопыренный 5175
желтушник сероватый 2817
желтушник серо-зелёный 2817
желтушник серый 2817
желтянка 5862
желудниковые 2865
женьшень 2494
женьшень американский 195
женьшень обыкновенный 346
женьшень пятилистный 195
жерардия 2227
жерардия полевая 2228
жеруха 3812
жеруха аптечная 5796
жеруха водная 5796
жеруха лекарственная 5796
жеруха хреновая 2871
жерушник 3812
жерушник австрийский 377
жерушник болотный 721
жерушник земноводный 222
жерушник лесной 6085
жестёр 830
жестёр слабительный 1369
жестковолосница 2701
жестковолосница твёрдая 2702
жесткоколосница 2701
жесткоколосница твёрдая 2702
жесткомятлик 5251
живика пушистая 6047
живокость 3226
живокость Аякса 4524
живокость Аяксова 4524
живокость высокая 518
живокость гибридная 2305
живокость полевая 2330
живокость посевная 2330
живучка 838
живучка 2882

живучка ёлочковидная 2650
живучка женевская 2442
живучка кровельная 2779
живучка паутинная 5133
живучка ползучая 1030
жидовник 5435
жимолостные 2842
жимолость 2841
жимолость альпийская 161
жимолость вечнозелёная 5613
жимолость виргинская 1529
жимолость вьющаяся 6022
жимолость голубая 5356
жимолость душистая 5370
жимолость каприфоль 5370
жимолость козья 5370
жимолость лесная 2099
жимолость немецкая 6022
жимолость обыкновенная 2099
жимолость пушистая 2099
жимолость синяя 5356
жимолость татарская 5455
жимолость чёрная 631
жимолость японская 3042
жирянка 900
жирянка альпийская 3700
житняк гребенчатый 1684
житняк гребневидный 1684
житняк пустынный 5210
житняк узколистный 5210
жминда 675
жминда головчатая 675
жонкилия 3110
жонкилла 3110
жонкиль 3110
жостер 830
жостер вечнозелёный 2974
жостер слабительный 1369
жузгун 952
журавельник 2447
журавельник 2789
журавельник кровяно-красный 680
журавельник луговой 3551
журавельник холмовой 5686
журавельник цикутовый 91
журавельник цикутолистный 91
журавельниковые 2448

З з

завязник 5555
зайцехвост 2713
замашка 2773
замия 1581
зандури 5961
занихелия 4266
занникеллия 4266
занникеллия болотная 1504
зантедешия 950
зантоксилум 4295
зантоксилум американский 1506
зантоксилюм 4295
зараза водяная 2022
зараза водяная зубчатая 1832
зараза водяная канадская 976
заразиха 812
заразиха большая 2604
заразиха ветвистая 2774
заразиха гвоздичная 1282
заразиха малая 1279
заразиха пурпуровая 694
заразиха сетчатая 5488
заразиховые 813
звезда кавалерская 4068
звездовик 1969
звездовка 3531
звёздоплодник 1793
звёздоплодник частуховидный 5219
звёздочка 1146
звёздочка водяная 5839
звёздочка земляная 1969
звездчатка 5235
звездчатка дубровная 6045
звездчатка злаковая 3338
звездчатка злаковидная 3338
звездчатка злачная 3338
звездчатка ланцетовидная 1971
звездчатка ланцетолитная 1971
звездчатка лесная 6045
звездчатка-мокрица 1146
звездчатка средняя 1146
звездчатка травяная 3338
зверобоецветные 4652
зверобой 4651
зверобой горный 3721
зверобой изящный 4912
зверобой крылатый 5199
зверобой обыкновенный 1537
зверобой пронзённолистный 1537
зверобой пронзённый 1537
зверобой распростёртый 1677
зверобой стелющийся 1677
зверобой чашечковидный 1
зверобой чашечный 1
зверобой четырёхгранный 5198
зверобойные 4652
зев львиный 5050
зев львиный большой 1528
зев львиный полевой 1619
зев львиный садовый 1528
зелёнка 2051
зелёнка 4606
зелёнка 6116
зелёнка пронзённолистная 1567
зеленоцвет 1220
зеленушка 2051
зеленчук 1818
зеленчук жёлтый 5855
зелье вонючее 1598
зелье курье 2785
зелье лихорадочное 1913
зелье рвотное 2942
зелье солодкое 1413
зелье солодьковое 1413
зелье сонное 3469
зелька 5800
земляника 5272
земляника ананасная 4195
земляника виргинская 5754
земляника высокая 2722
земляника зелёная 2635
земляника индийская 2932
земляника крупноплодная 4195
земляника лесная 2127
земляника мускатная 2722
земляника мускусная 2722
земляника полуница 2635
земляника садовая 4195
земляника чилийская 1150
земляничник 3418

земляничник крупноплодный 5276
зензивер 2797
зеркало девичье 5719
зерлик 2716
зефирантес 6126
зигаденус 1819
зиглингия 4937
зиглингия лежачая 2746
зигомицеты 6131
зизифора 6130
зизифус 3115
зизифус обыкновенный 1452
зизифус ямайский 1321
зимняк 1928
зимовник 382
зимовник осенний 1353
зимолюбка 4202
зимолюбка зонтичная 1496
злак пальчатый 542
злак пампасовый 4818
злаки 2581
злаковые 2581
златок 1449
златокорень 5112
златокорень испанский 5112
златолист 5211
златолист каймито 919
златоцвет 1229
златоцвет 4376
златоцвет девичий 2210
златоцвет посевной 1609
змеевик 2079
змееголовник 1901
змееголовник австрийский 376
змееголовник аргунский 4629
змееголовник молдавский 3645
змееголовник Руйша 4629
змееголовник тимьянолистный 5510
змееголовник тимьяноцветковый 5510
змееголовник тимьяноцветный 5510
змей-трава 3378
зноиха 2325
золотарник 2543
золотарник злаколистный 2586
золотарник "золотая розга" 2101
золотарник канадский 971
золотарник обыкновенный 2101
золотарник поздний 3892
золотель 2101
золотобородник 4750
золотобородник цикадовый 1688
золототысячник 1090
золототысячник зонтичный 1909
золотохворост 2566
золотоцвет 4376
золотушник 887
зольник 1518
зонтичные 1036
зопник 3097
зопник клубненосный 5620
зопник кустарниковидный 1451
зопник кустарниковый 1451
зорька 3451
зорька-аксамитки 4558
зорька "барская смесь" 3451
зорька-дрёма 4398
зорька татарское мыло 3451
зоря 2418
зостера 1997
зостера большая 1407
зостера морская 1407
зостеровые 1998
зуб собачий 2190
зуб собачий обыкновенный 1878
зубница 4225
зубровка 5367
зубровка альпийская 153
зубровка душистая 5368
зубровка пахучая 5368
зубчатка 460
зубчатка бартсиевидная 4435
зубчатка красная 4435
зубчатка осенняя 4435
зубянка 5548
зубянка клубненосная 1592
зубянка луковичная 1592
зюзник 839
зюзник европейский 2085
зябра 3846

И и

иберийка 993
иберийка вечнозелёная 2144
иберийка горькая 4523
иберийка зонтичная 2504
иберийка крымская 5460
иберийка скальная 4514
иберис 993
ибицелла жёлтая 6110
ибунка 509
ива 5964
ива арктическая 311
ива белая 5922
ива болотная 467
ива-бредина 2523
ива буреющая 77
ива вавилонская 400
ива волчниковая 1801
ива голая 163
ива голубая 1687
ива грушанколистная 435
ива Гукера 2846
ива двуцветная 5466
ива деревцевидная 3339
ива длиннолистная 4680
ива длинноножковая 1298
ива египетская 2002
ива жёлтая 6107
ива изящная 5508
ива козья 2523
ива конопляная 467
ива копьевидная 2697
ива корзиночная 467
ива красная 4471
ива краснолистная 4850
ива круглолистная 3253
ива крупнолистная 564
ива лапландская 3221
ива ломкая 787
ива лопарская 3221
ива миндалевидная 4084
ива миндалелистная 106
ива миндальная 106
ива миртолистная 3785
ива мохнатая 6054
ива округлая 4590
ива остролистная 4850
ива пепельная 2598
ива плакучая 400

ива ползучая 1679
ива полярная 4245
ива прутовидная 467
ива прутьевидная 467
ива пурпурная 4352
ива пурпуровая 4352
ива пушистопобеговая 6053
ива пятитычинковая 3236
ива Ричардсона 4501
ива русская 4620
ива серая 2598
ива серебристая 5922
ива сетчатая 3833
ива сизая 2596
ива сизоватая 2007
ива силезская 4943
ива ситхинская 4980
ива стелющаяся 1679
ива травянистая 4366
ива трёхтычинковая 106
ива туполистная 3891
ива ушастая 4581
ива филиколистная 5466
ива хрупкая 787
ива чёрная 654
ива чернеющая 693
ива черниковидная 5928
ива черничная 5928
ива шелюга 4850
ива шерстистая 6054
ива шерстистопобеговая 6053
ива яйцевиднолистная 3901
Иван-да-Марья 696
иван-чай 2262
ивишень 4226
ивовые 5965
игла адамова 24
иглица колючая 885
иглица понтийская 885
иглица шиповатая 885
иглица шиповая 885
игнаций 4650
игнация 4650
изидиум 2968
изолепис 852
изолома 2969
изотома 4887
изоэтес 4387
изоэтовые 4388
икак 2914
икако 2914
икотник 2160
икотник серозелёный 2818
иксиолирион 2994
иксиолирион горный 5415
иксия 2993
иксора 2995
икстль 2781
иланг-иланг 6119
илекс 2826
иллициум 254
иллициум анисовый 3014
иллициум настоящий 5607
иллициум обрядный 3014
ильм 2018
ильм американский 191
ильм белый 191
ильм гладкий 4614
ильм горный 4771
ильм крылатый 5989
ильм листоватый 5040
ильм приземистый 4903
ильм пробковый 1602
ильм сибирский 5012
ильм шершавый 4771
ильм шотландский 4771
ильмовник 4903
ильмовые 2019
имбирные 2490
имбирь 2489
имбирь жёлтый 1551
имбирь лекарственный 1419
иммортель 5279
императа 4713
императа цилиндрическая 1320
инга 2949
индау посевной 4525
индиго 2947
индиго бастардное 2948
индиго дикое 6113
индиго ложное 2948
индигоноска 2947
индигоноска красильная 5603
индигофера 2947
инжир 1412
инжир 2240
инжир ослиный 5408
инкарвиллея 2924
иньям 6062
иохрома 5738
ипекакуана 2957
ипомея 3682
ипомея батат 5380
ипомея пурга 3005
ипомея пурпурная 1474
ипомея слабительная 3005
ипомея ялапа 3005
ирга 4832
ирга азиатская 342
ирга канадская 4838
ирга колосистая 1949
ирга круглолистная 2428
ирга метельчатая 1949
ирга обыкновенная 2428
ирга овалолистная 2428
ирга овальная 2428
ирезине 677
ирис 2958
ирис германский 2458
ирис персидский 4148
ирисовые 2959
исертия 5949
иссоп 2909
иссоп аптечный 2910
иссоп лекарственный 2910
истод 4250
истод обыкновенный 1470
истод сибирский 4909
истодовые 3623
итея 5387
итея виргинская 5755
ишносифон 5532

Й й

йерба-матэ 4048

К к

кабачок 5705

кабомба 2184
кабралея бразильская 991
кава 3140
кава-перец 3140
кавун 5821
кадило 3583
кадило бабье 3846
кадило медоволистное 469
кадило мелиссолистное 469
кадия 916
кадсура 3123
кадсура японская 4744
кажу 1378
казимироя съедобная 5909
казуарина 516
казуарина аргентинская 1299
казуарина Куннингама 1731
казуарина хвощевидная 2876
казуарина хвощелистная 2876
казуариновые 517
каимито 919
кайанус 4177
кайея 3142
кайя иворензис 2985
кайя иворская 2985
кайя сенегальская 4821
кайянус 4177
какао 912
какао настоящее 912
каки 3127
кактус 914
кактус звездчатый 5213
кактус колонновидный 1100
кактус кошенильный 1305
кактус крупноцветный 4384
кактус ластовневый 1034
кактус лофофора 3591
кактус маммиллярия 3454
кактус опунция 4301
кактус пейреския 447
кактус рипсалис 4488
кактус свисающий 1102
кактус сосочковый 3454
кактус "царица ночи" 4384
кактус-цереус 1100
кактус церопегия 1102
кактусовые 915
калаба 1104
калабюра 3007
каладиум 925
каламбак 1958
каламус 4424
каламус драконовый 5314
каландриния 4529
калатея 928
калган 5555
калебаса 922
календула 929
календула лекарственная 4285
каликантовые 5278
каликантус 5386
каликантус плодовитый 4011
каликантус плодующий 4011
каликантус флоридский 1540
каликантус цветущий 1540
калимерис 3526
калина 5733
калина вечнозелёная 3237
калина гордовина 5852
калина грушелистная 632
калина душистая 5391
калина Зибольдова 4935
калина канадская 3792
калина корейская 3186
калина красная 2091
калина лавролистная 3237
калина обыкновенная 2091
калина сливолистная 632
калина цельнолистная 5852
калипсо 956
калипсо луковичное 957
калипсо луковичный 957
калистегия 2509
калистегия заборная 2748
калла 1374
калла болотная 5941
каллиантемум 951
калликарпа 2370
каллистефус 1158
каллистефус китайский 1382
каллитрис 1777
калоникцион 3670
калотропис 953
калофиллум 508
калофиллум круглоплодный 2943
калофиллум обыкновенный 1104
калохортус 3494
калужница 3508
калужница болотная 1466
калуфер 1632
кальмия 3129
кальмия многолистная 720
кальмия сизая 720
кальмия узколистная 3210
кальмия широколистная 3710
кальцеолария 5011
кальцеолария гибридная 5760
кальцеолария мексиканская 3595
кальцеолярия 5011
камассия 958
камелия 960
камелия гвоздичная 4711
камелия горная 4711
камелия масличная 4711
камелия маслоносная 2799
камелия обыкновенная 1376
камелия сасанква 4711
камелия эвгенольная 4711
камелия японская 1376
каменник 172
камень фиалковый 4537
камка 1407
камнеломка 4730
камнеломка австралийская 1096
камнеломка вечнозелёная 71
камнеломка дернистая 946
камнеломка жёстколистная 6102
камнеломка звездчатая 5231
камнеломка зернистая 3560
камнеломка козлиная 6097
камнеломка корнеотпрысковая 5277
камнеломка круглолистная 791
камнеломка отпрысковая 5277
камнеломка ползучая 5277
камнеломка поникшая 1904
камнеломка ручейная 120
камнеломка снежная 5055
камнеломка супротивнолистная 5663
камнеломка тенистая 3362
камнеломка трёхпалая 4600

камнеломка трёхпальчатая 4600
камнеломковые 4732
камнеплодник 5439
кампанула 528
кампсис 5611
кампсис укореняющийся 1550
камфоросма 965
камыш 852
камыш гигантский 2482
камыш лесной 6031
камыш озёрный 2600
камыш приморский 4801
камыш ржавый 4443
камыш степной 4473
камыш Табернемонтана 5420
камыш трёхгранный 5590
камышовка 994
канавалия 2997
канавалия мечевидная 5406
канавалия мечелистная 5406
канавалия саблевидная 1448
кананга 6119
кананга душистая 6119
канареечник 984
канареечник канарский 985
канареечник клубневой 1761
канареечник малый 3336
канареечник птичий 985
канареечник тростниковый 4473
канарейник 985
канариум 989
канариум белый 5886
канариум обыкновенный 3085
канариум чёрный 623
канатник 7
канатник гибридный 2898
канатник Теофраста 1217
кандолея 5858
кандык 2190
кандык собачий зуб 1878
кандык южноевропейский 1878
кандым 952
канегра 981
канелла 995
канель 5297
канна 996
канна индийская 2934
канна съедобная 1993
канновые 997
канталупа 998
кануфер 1632
каперсник 1007
каперсник колючий 1377
каперсовые 1009
каперсы 1007
каперцовые 1009
каперцы 1007
каперцы колючие 1377
капок 3134
каприфоль 5370
каприфоль немецкий 6022
капсикум 4459
капуста 906
капуста белокочанная 5885
капуста брюссельская 823
капуста заячья 3341
капуста заячья большая 2612
капуста кергуэленская 3152
капуста китайская 4005
капуста кормовая 3128
капуста кормовая мозговая 3175
капуста кочанная 2734
капуста красная 4444
капуста краснокочанная 4444
капуста кудрявая 3128
капуста листовая 3128
капуста листовая культурная 3175
капуста морская 1387
капуста настоящая кочанная 2734
капуста огородная 5940
капуста пекинская 4158
капуста полевая 585
капуста репная 3175
капуста савойская 4722
капуста собачья 1876
капуста собачья 5481
капуста спаржевая 800
капуста спаржевая цветная 800
капуста цветная 1079
капуцин 3812
капуцин большой 1478
капуцин малый 882
капуциновые 3813
карагана 4095
карагана высокая 4910
карагана древовидная 4910
карагана кустарниковая 4617
карагана мелколистная 3329
карагана степная 4617
караганник 4095
караганник древовидный 4910
карагач 1602
караген 2960
караген кудрявый 2961
карамбол 1012
карамбола 1012
карапа 1657
карапа гвианская 2660
кардаминопсис песчаный 4518
кардамон 1014
кардария 5918
кардиосперм 2741
кардиоспермум 2741
кардобенедикт 671
кардон 1015
кардуль 5112
карика папайа 4035
кариковые 4034
кариокар орехоносный 4725
кариоптерис 689
кариота 2265
кариота жгучая 5540
кария 2795
кария белая 3640
кария водяная 5809
кария голая 4179
кария китайская 1186
кария косматая 4841
кария мускатная 3896
кария овальная 4841
кария опушённая 3640
кария пекан 4100
кария северокаролинская 1025
кария сердцевидная 605
каркас 2674
каркас западный 1434
каркас кавказский 1072
каркас китайский 1183
каркас корейский 3182
каркас крылатый 5988

каркас Турнефора 3955
каркас южный 2106
карлудовика 1018
карлюдовика 1018
карнауба 1021
кароб 1023
карпезиум 1029
карраген 2960
картофель 4281
картофель испанский гладкий 5622
картофель культурный 4281
картофель мадагаскарский 2721
картофель сладкий 5380
каруба дикая 1023
касатик 2958
касатик аировидный 6086
касатик безлист(н)ый 5268
касатик бледный 5371
касатик болотный 6086
касатик водный 6086
касатик водяной 6086
касатик вылощенный 4393
касатик германский 2458
касатик гладкий 4393
касатик Гроссгейма 2642
касатик жёлтый 6086
касатик злаковидный 2582
касатик злаковый 2582
касатик иберийский 2913
касатик испанский 5111
касатик кавказский 1074
касатик Кемпфера 3047
касатик крымский 1943
касатик ложно-аировой 6086
касатик ложный 6086
касатик маленький 1943
касатик низкий 1943
касатик пёстрый 2889
касатик равнодольный 1943
касатик равнолопастный 1943
касатик сибирский 4907
касатик сузианский 3728
касатик флорентийский 2301
касатик японский 3047
касатиковые 2959
каскариль 1041
кассава 1046
кассава горькая 1379
кассава сладкая 67
кассандра 3257
кассина 1048
кассиопа 1049
кассиопа гипновидная 307
кассиопа четырёхрядная 2260
кассиопея 1049
кассия 4823
кассия горькая 5338
кассия дудка 2548
кассия остролистная 89
кассия приморская 5958
кассия стручковая 2548
кассия трубковидная 2548
кассия узколистная 1575
кассия южноазиатская 2548
кастанопсис 2145
кастилейя 4003
кастилла 2665
кастилла каучуконосная 4022
кастиллейя 4003
кастиллоа 2665
кат 294
ката 294
катависса 5863
катальпа 1053
катальпа бигнониевидная 5097
катальпа западная 3879
катальпа красивая 3879
катальпа прекрасная 3879
катальпа сиренелистная 5097
катальпа яйцевидная 1175
катальпа яйцевиднолистная 1175
катальпа японская 1175
катананхе 1737
катасетум 3659
катауба 2347
катран 1326
катран перистый 1387
катран приморский 1387
катран татарский 5452
каттлея 1065
каттлея цитрусовая 5634
катх съедобный 294
катьянг 4177
каукалис 2753
каури 1795
каури новозеландская 3139
кахрис 913
качим 2672
качим вязкий 5248
качим метельчатый 397
качим ползучий 1670
качим постенный 1749
качим Стевена 5243
качим стенной 1749
кашка 5889
кашка красная 4447
кашка лесная 6128
каштан 1139
каштан американский 187
каштан благородный 2086
каштан водяной 5794
каштан высокий 187
каштан городчатый 3027
каштан европейский 2086
каштан жёлтый американский 6077
каштан зубчатый 187
каштан карликовый 99
каштан конский 828
каштан конский восьмитычинковый 6077
каштан конский гладкий 3913
каштан конский голый 3913
каштан конский калифорнийский 931
каштан конский китайский 1188
каштан конский красный 4441
каштан конский мелкоцветный 739
каштан конский обыкновенный 1443
каштан конский павия 4441
каштан конский розовоцветный 4441
каштан конский японский 3046
каштан настоящий 2086
каштан низкорослый 99
каштан обыкновенный 2086
каштан посевной 2086
каштан съедобный 2086
каштан японский 3027
каштановые 5073
каштаносеменник австралийский 3680
каштаноспермум 3680
каянус 4177
каянус индийский 4177

каянус каян 4177
квамассия 958
квамаш съедобный 1375
квамоклит 5220
квассия 4382
квассия горькая 5338
квассия ямайская 3009
квебрахо 4461
квебрахо белое 1559
квебрахо красное 5967
квебрахо Лоренца 3379
квебрачо 4461
квил(л)айя 5070
квил(л)айя мыльная 5071
квиноа 4390
квиноя 4390
квисквалис 4391
квит 1414
кебрачо 4461
кедр 1082
кедр атласский 364
кедр атласский голубой 688
кедр гималайский 1836
кедр европейский 5402
кедр калифорнийский ладанный 936
кедр карликовый 3072
кедр корейский 3184
кедр коротхвойный 1778
кедр ладанный 2925
кедр ливанский 1083
кедр остиндский 1234
кедр речной 2925
кедр речной калифорнийский 936
кедр речной новозеландский 4229
кедр речной перистый 4229
кедр речной сбежистый 936
кедр речной чилийский 1149
кедр сибирский 4917
келерия 3173
келерия гребенчатая 4292
кельрейтерия 2555
кельрейтерия метельчатая 4027
кемквот 3528
кемпедек 2998
кемпферия 4485
кенаф 3146
кендик 1878
кендык 1878
кендырь 1872
кендырь коноплевидный 2775
кендырь конопляный 2775
кендырь проломниковый 5173
кендырь сибирский 4291
кёнигия 3174
кеннедия 3148
кентия 3149
кентрантус 1091
кентрантус красный 3119
кенция 3149
кервель 4658
кервель испанский 5361
кервель клубневой 5651
кервель корневой 5651
кервель листовой 4658
кервель луковичный 5651
кервель мускусный 5361
кервель настоящий 4658
кервель обыкновенный 4658
кервель посевной 4658
кервель репный 5651
кервель салатный 4658
керия 3156
керия японская 3049
кермек 4793
кермек выемчатый 3890
кермек Гмелина 4914
кермек Гмелинов 4914
кермек обыкновенный 3573
кермек татарский 5458
кермек широколистный 5931
кернера 3154
керрия 3156
кетелеерия 3157
кетелеерия Давида 1812
кетелеерия Форчуна 2338
кигелия 4716
кизил 1881
кизил белый 5454
кизил кровяной 683
кизил мужской 1611
кизил настоящий 1611
кизил обыкновенный 1611
кизил отпрысковый 4458
кизил побегоносный 4458
кизил свидина 683
кизил сибирский 4901
кизил съедобный 1611
кизил татарский 4901
кизил флоридский 2309
кизил шведский 1937
кизиловые 1882
кизильник 1633
кизильник войлочный 774
кизильник горизонтальный 4515
кизильник Даммера 501
кизильник кистецветковый 4437
кизильник кистецветный 4437
кизильник обыкновенный 2089
кизильник Цабеля 1134
кизильник цельнокрайный 2089
кизильник черноплодный 618
кизляк кистецветковый 5629
кизляк кистецветный 5629
киксия 2317
киксия ложная 4585
киксия повойничек 4851
киксия повойничковая 4851
килайя 5070
киндаль 4385
киндза 1597
киндзя 1597
кинкан 3189
кинкан круглый 3528
кинкан маргарита 3980
кинкан маруми 3528
кинкан нагами 3980
кинкан овальный 3980
кинкан японский 3528
киноа 4390
киноя 4390
кипарис 1773
кипарис азиатский болотный 1162
кипарис аризонский 317
кипарис болотный 414
кипарис болотный мексиканский 3667
кипарис вечнозелёный 2978
кипарис вечнозелёный пирамидальный 4367
кипарис гималайский 551

кипарис Говена 2571
кипарис Говея 2571
кипарис голубой Говея 699
кипарис Дукло 551
кипарис калифорнийский 2571
кипарис кашмирский 3137
кипарис китайский плакучий 3727
кипарис крупноплодный 3665
кипарис летний 5318
кипарис лузитанский 3597
кипарис лысый 414
кипарис Макнаба 3405
кипарис надутый 551
кипарис пирамидальный 2978
кипарис пирамидальный 4367
кипарис плакучий 3727
кипарисник 2167
кипарисовик 2167
кипарисовик горохоносный 4724
кипарисовик туеобразный 5887
кипарисовик туполистный 2812
кипарисовник белый 3243
кипарисовник Лавзона 3243
кипарисовник Лавсона 3243
кипарисовник нутканский 3874
кипарисовые 1775
кипрей 5975
кипрей альпийский 156
кипрей болотный 3522
кипрей волосистый 2695
кипрей Горнемана 2857
кипрей горный 3723
кипрей Лами 3214
кипрей Лями 3214
кипрей мелкоцветковый 5024
кипрей мелкоцветный 5024
кипрей мохнатый 2695
кипрей розовый 4012
кипрей тёмно-зелёный 1952
кипрей узколистный 2262
кипрейник 5975
кипрейник альпийский 156
кипрейник болотный 3522
кипрейник волосистый 2695
кипрейник Горнемана 2857
кипрейник горный 3723
кипрейник Лями 3214
кипрейник мелкоцветковый 5024
кипрейник тёмно-зелёный 1952
кипрейные 2142
киренгешома 3163
кирказон 1932
кирказон крупнолистный 1404
кирказон крупноцветный 4104
кирказон ломоносовидный 592
кирказон маньчжурский 3460
кирказон обыкновенный 592
кирказон понтийский 592
кирказоновые 593
кислец 131
кислица 3985
кислица клубневая 3918
кислица клубненосная 3918
кислица обыкновенная 6044
кислица прямая 1566
кислица рогатая 1673
кислица рожковая 1673
кислица ямайская 4562
кисличник 3719
кисличник двухстолбчатый 138
кисличные 5088
кистевик 4109
кистевик зелёный 2632
кистецветник обыкновенный 5214
китайка 4091
кишмиш крупный 748
кишмиш обыкновенный 3176
кишнец 1597
кишнец посевной 1598
кладантус 1248
кладония 1249
кладония альпийская 4479
кладрастис 6114
кладрастис жёлтый 213
клайтония 5181
клайтония пронзённолистная 3627
кларкия 1253
клевер 1278
клевер александрийский 2000
клевер альпийский 126
клевер альпийский 4347
клевер американский степной 4290
клевер белоголовка 3702
клевер белый 5889
клевер белый ползучий 5889
клевер берсим 2000
клевер бледно-жёлтый 5311
клевер бухарский 5917
клевер венгерский 2888
клевер гибридный 164
клевер горный 3702
клевер греческий 2203
клевер египетский 2000
клевер жёлтый 2888
клевер жестковолосый 4576
клевер заячий 3159
клевер земляниковидный 5274
клевер земляничный 5274
клевер изящный 4883
клевер инкарнатный 1694
клевер клубковидный 1292
клевер красно-белый 164
клевер красноватый 2351
клевер красный 2351
клевер красный 4447
клевер луговой 4447
клевер лупиновидный 473
клевер лупиновый 473
клевер люпиновидный 473
клевер люпиновый 473
клевер маленький 5292
клевер малиновый 1694
клевер мексиканский 3596
клевер мелкоцветковый 5470
клевер Молинери 1694
клевер мясокрасный 1694
клевер нарядный 4883
клевер нитевидный 5497
клевер опрокинутый 4147
клевер паннонский 2888
клевер пашенный 2848
клевер пашенный 4394
клевер перевёрнутый 4147
клевер персидский 4147
клевер подземный 5291
клевер полевой 2848
клевер ползучий 5889
клевер полосатый 3169
клевер посевной 2848
клевер пунцовый 1694

клевер равнинный 3225
клевер розовый 164
клевер светло-жёлтый 5311
клевер скрученный 1292
клевер сомнительный 5292
клевер средний 6128
клевер тонкостебельный 5497
клевер турецкий 1519
клевер узколистный 2250
клевер шабдар 4147
клевер шведский 164
клевер шероховатый 4576
клевер шершавый 4576
клевер шуршащий 2848
клевер японский 1458
клейнховия 3165
клейтония 5181
клейтония пронзённолистная 3627
клекачка 659
клекачка перистая 2083
клекачковые 660
клематис 1260
клематис альпийский 125
клематис Жакмана 3000
клематис сибирский 4896
клематис цельнолистный 5080
клён 3481
клён американский 750
клён бархатистый 5712
клён бархатный 5712
клён белый 4217
клён боярышниколистный 2730
клён бумажный 4036
клён веерный 3056
клён величественный 5712
клён весёлый 1327
клён высокогорный 4442
клён Гельдрейха 418
клён гиннала 230
клён граболистный 2853
клён дланевидный 3056
клён завитой 5735
клён Зибольда 4934
клён кавказский 4442
клён кавказский высокогорный 4442
клён красивый 4004
клён красный 4455
клён критский 3957
клён крупнолистный 562
клён ложноплатановый 4217
клён маньчжурский 3464
клён монпелийский 3669
клён Нико 3856
клён обыкновенный 3885
клён остролистный 3885
клён пальмовидный 3056
клён пенсильванский 5284
клён платановидный 3885
клён полевой 2756
клён приречный 230
клён речной 230
клён рыжеватожилковатый 4470
клён сахаристый 4966
клён сахарный 5304
клён светлый 1327
клён серебристый 4966
клён серый 4036
клён татарский 5456
клён Траутветтера 4442
клён Траутфеттера 4442
клён трёхлопастный 3669
клён чёрный 637
клён явор 4217
клён японский 2391
клён ясенелистный 750
клёновые 3482
клеома 5128
клеоме 5128
клеродендрон 2510
клетра 5323
клетровые 1261
клещевина 1050
клещевина занзибарская 1051
клещевина обыкновенная 1051
клещина 2069
клещинец 3378
клиантус 4059
кливия 3126
кливия оранжевая 4745
кливия суриковая 4745
кливия суриково-красная 4745
клиноног 926
клинтония 490
клитория 4178
клокичка 659
клокичка перистая 2083
клокичковые 660
клоповник 4126
клоповник безлепестный 4293
клоповник безлистный 4293
клоповник виргинский 5749
клоповник крупковидный 4127
клоповник полевой 2233
клоповник посевной 2412
клоповник пронзённолистный 1256
клоповник широколистный 2118
клопогон 837
клопогон вонючий 4988
клубнекамыш морской 4801
клубника 5272
клубника ананасная 4195
клубника ананасовая 4195
клубника виргинская 5754
клубника высокая 2722
клубника настоящая 2722
клюгия 3166
клюзия 1290
клюква 1658
клюква болотная 5019
клюква крупноплодная 3224
клюква обыкновенная 5019
ключики 1652
ключ-трава 3673
кмин 1728
кмин тминный 1729
кмин тминовый 1729
кнеорум 5194
кникус 670
кникус аптечный 671
кникус благословленный 671
книфофия 5551
княженика 305
княжик 1260
княжик альпийский 125
княжик корейский 3180
княжик сибирский 4896
кобея 1301
кобея лазающая 4337
кобрезия 1302
кобрезия осоковидная 1016

ковыль 3819
ковыль обыкновенный 2115
ковыль перистый 2115
ковыль тончайший 2062
кодиум 1316
кодонопсис 340
коелококкус 4644
кожанка 1441
кожанка 2851
кожанка трёхлистная 1441
кожевник 5313
козелец 860
козелец 4831
козелец испанский 649
козелец низкий 725
козелец приземистый 725
козелец разрезной 3574
козелец рассечённый 3574
козелец сладкий корень 649
козлец 860
козлобородник 4662
козлобородник луговой 3559
козлобородник пореелистный 5706
козлобородник порейнолистный 5706
козлобородник порейолистный 5706
козлобородник поррейнолистный 5706
козляк 4846
козлятник 2521
козлятник аптечный 1425
козлятник лекарственный 1425
козыльник 6056
коикс 31
коикс слёзы Иовы 31
кока 2883
кокаиновые 1303
кокколоба 4788
коккулус 5044
коккулюс 5044
кокорыш 36
кокорыш обыкновенный 2325
кокорыш собачья петрушка 2325
кокос 1314
кок-сагыз 4613
кокушник 4398
кокушник 4481
кола 1323
кола блестящая 5293
кола заострённая 5293
кола культурная 5293
кола остролистная 5293
колба 3368
колдун-трава 2031
колеантус 3741
колеантус тонкий 3742
коленница 2517
колеус 1325
колеус Блумего 1386
колеус пёстрый 1386
коллетия колючая 1769
коллеция 1330
коллибия 1333
коллибия длинноножковая 4550
коллибия широкопластинчатая 790
коллинзия 1331
коллинзония 2863
коллинсия 1331
коллинсония 2863
колломия 1332
колоказия 2013
колоказия съедобная 1806
колокасия 2013
колоквинт 1334
колокольчик 528
колокольчик альпийский 116
колокольчик бородатый 502
колокольчик карпатский 1026
колокольчик китайский крупноцветный 424
колокольчик крапиволистный 1643
колокольчик круглолистный 690
колокольчик крупноцветный 424
колокольчик одноцветковый 4756
колокольчик персиколистный 4083
колокольчик пирамидальный 1157
колокольчик рапунцель 4414
колокольчик раскидистый 4408
колокольчик репчатовидный 4414
колокольчик репчатый 4414
колокольчик сборный 1797
колокольчик сибирский 4894
колокольчик скученный 1797
колокольчик средний 999
колокольчик широколистный 2599
колокольчик японский крупноцветный 424
колокольчики водосборные 1341
колокольчики садовые 2087
колокольчиковые 529
коломбина 2087
коломбо 5044
коломикта 3176
колос благовонный 5604
колос пахучий 5389
колосеница 2290
колосняк 5957
колосняк песчаный 2094
колосок душистый 5724
колосок душистый обыкновенный 5389
колосок душистый остистый 265
колосок пахучий 5724
колоцинт 1334
колпак 6057
колпак кольчатый 6057
колтунник 4606
колубрина 1340
колумба 955
колуфер 1632
колхикум 382
колча 3104
колчак 2900
кольза 1343
кольник 3636
кольник колосистый 5139
кольник колосовидный 5139
кольник округлый 420
кольник чёрный 639
кольник шаровидный 420
кольраби 3175
кольраушия побегоносная 3330
кольцо ведьмино 2154
колюрия 1281
колютик 5487
колючезонтичник 4300
колючелистник 1017
колючелистник бесстебельный 5035
колюченосник 4300
колючка верблюжья 961
колючка верблюжья обыкновенная 961
колючник 1017
колючник бесстебельный 5035
комбретовые 3776
комбретум 1344

комелиновые 5135
комифора 3779
комифора абиссинская 9
комифора африканская 44
комифора бальзамная 3563
комифора мира 1477
коммелина 1814
коммелиновые 5135
коммифора 3779
коммифора бальзамная 3563
комптония 5364
комфрей 4297
кондалия 1571
кондуранго 1572
конига морская 5352
конига приморская 5352
кониограмма японская 438
коноплёвые 2777
конопля 2772
конопля виргинская 5518
конопля декан 3146
конопля индийская 2773
конопля китайская 4410
конопля культурная 2773
конопля лесная 2262
конопля маврикийская 4209
конопля маврициева 4209
конопля манильская 2
конопля новозеландская 3843
конопля посевная 2773
конопля юкатанская 2781
конофолис 990
конрингия 2710
конрингия восточная 5574
конскокаштановые 2865
конуфер 1632
конъюгаты 1580
конюшина 4447
копайба 1582
копайфера 1583
копайфера копалоносная 4939
копайфера лекарственная 1582
копеечник 5390
копеечник венцевидный 5310
копеечник короноподобный 5310
копеечник тёмный 2368
копперниция 1584
коперниция восконосная 1021
коптис 2556
копытень 5947
копытень европейский 2133
кора мыльная 5071
коракан 4399
кордилина 1899
кордилина кустарниковая 1402
кордилина южная 2470
кордия 1594
корень алтейный 3507
корень волчий 2194
корень галанговый 2609
корень змеиный 2079
корень зобный 1503
корень ирный 1918
корень любистоковый 2418
корень маралий 4486
корень медвежий 415
корень мыльный 743
корень пьяный 2122
корень рвотный 2957
корень рвотный 3898
корень розовый 4567
корень сахарный 4986
корень свинцовый 4225
корень сердечный 5759
корень сладкий 649
корень царский 3532
корень цитварный 6125
корень-человек 346
корень чёрный 649
корень чёрный 1444
корень чёрный 1918
кореопсис 1596
кореопсис красильный 4212
кориандр 1597
кориандр посевной 1598
кориариевые 1600
кориария 1599
кориария миртолистная 3782
корилопсис 6004
корилопсис Вилльмотта 5963
корилопсис колосковый 5147
корилопсис малоцветковый 891
коринокарпус 3135
корифа 1629
корифа зонтичная 5424
корифа теневая 5424
корица гвоздиная 1275
корица китайская 1047
корица магелландская 6007
корица цейлонская 1105
коричник 1238
коричник железконосный 3827
коричник камфарный 966
коричник кассия 1047
коричник китайский 1047
коричник цейлонский 1105
корнишон пикульный 5869
корнутия 1621
коровка 1093
коровяк 3752
коровяк выемчатый 3571
коровяк высокий 6055
коровяк зопниковидный 1255
коровяк извилистый 3571
коровяк лекарственный 1255
коровяк медвежье ухо 2276
коровяк метельчатый 5903
коровяк молевый 3690
коровяк мохнатый 1255
коровяк обыкновенный 2276
коровяк пирамидальный 4368
коровяк скипетровидный 6055
коровяк тапсовидный 6055
коровяк тараканий 3690
коровяк тараканный 3690
коровяк тёмный 3690
коровяк фиолетовый 4351
коровяк чёрный 641
коровяк шерстистый 1255
коронария кукушкин-цвет 4398
короставник 3167
короставник полевой 2235
коротконожка 2164
коротконожка лесная 4996
коротконожка перистая 3033
корригиоля 1622
корригиоля береговая 5271
кортадерия 4020
кортуза 506
космос 1630

космос раздельнолистный 1392
косогорник 4429
косогорник красный 4353
костелецкия 4664
костенец 3730
костенец 5161
костенец волосовидный 3431
костенец зелёный 2634
костенец зонтичный 3004
костенец постенный 5776
костенец рута постенная 5776
костенец северный 2329
костенец чёрный 651
костёр 802
костёр безостый 5034
костёр бесплодный 4287
костёр ветвистый 2681
костёр кистевидный 413
костёр кровельный 1123
костёр мадридский 5109
костёр мягкий 5078
костёр переменчивый 2681
костёр полевой 2219
костёр прямой 3546
костёр прямостоячий 3546
костёр разветвлённый 2681
костёр ржаной 1138
костёр смешанный 2681
костёр стоячий 3546
костёр униоловидный 4483
костёр японский 3024
костолом 715
костус 5158
костяника 5260
костяника арктическая 305
костяника каменистая 5260
костянкоплодник 733
костянкоплодник Шверина 1172
косы ведьмины 503
котики 4394
котиледон 1639
котовик 3830
котовник 3830
котовник кошачий 1060
котовник мятный 1060
котовник ползучий 1429
котовник украинский 5676
котула 759
котяник 1429
кофе 1318
кофе арабский 292
кофе либерийский 3287
кохия 5318
кохия веничная 531
кохлоспермум 4860
кочедыжник 3199
кочедыжник женский 3199
кошельки 5011
кошельки гибридные 5760
кошельки мексиканские 3595
кошельки цельнолистные 878
краекучник 3323
крамбе 1326
крамбекатран 1326
крап 1463
крапива 3835
крапива большая 574
крапива глухая 5891
крапива глухая красная 4343
крапива глухая пятнистая 5166
крапива двудомная 574
крапива жгучая 1874
крапива китайская 4410
крапива обыкновенная 574
крапива шариконосная 4547
крапивна 1312
крапивные 3836
крапп 1463
крапчатка 2085
красавица дневная 1941
красавица ночная 1415
красавица ночная 1794
красавка 525
красавка белладонна 525
красавка обыкновенная 525
красноволоска 5839
красноднев 1815
красноднев жёлтый 3269
красноднев малый 2585
красноднев рыжий 2334
красноднев Тунберга 5504
краснопузырник 608
краснопузырник древогубец 185
краснопузырниковые 5203
краснотал 4850
красовласка 5839
красовласка осенняя 384
красовласка прудовая 4261
красовласковые 5840
красоднев 1815
красоднев жёлтый 3269
красоднев малый 2585
красоднев Миддендорфа 3613
красоднев рыжий 2334
красоля 1478
крассула 1661
крепкоплодник 2067
кресс 4126
кресс бразильский 4045
кресс водяной 5796
кресс испанский 1478
кресс крупка 4127
кресс масляный 4045
кресс огородный 2412
кресс полевой 2233
кресс посевной 2412
кресс садовый 2412
кресс-салат 2412
кресс широколистный 2118
крест андреев 5140
крест Петров 3230
крест петров чешуйчатый 5549
крестовник 2651
крестовник акантолистный 4958
крестовник болотный 4009
крестовник водяной 3513
крестовник-желтуха 4401
крестовник клейкий 5247
крестовник лесной 6034
крестовник луговой 4401
крестовник обыкновенный 1430
крестовник Фукса 2388
крестовник Якова 4401
крестовница 1703
крестоцветные 908
кресченция 924
кривокучник 5763
кривоцвет 5938
кривоцвет пашенный 5939
кривоцвет полевой 5939

кринум 1696
кринум длиннолистный 2707
криптантус 1715
криптограмма 4512
криптокорина 1716
криптомерия 1718
криптомерия японская 3029
критмум 4674
критмум морской 4675
кровохлёбка 867
кровохлёбка аптечная 2409
кровохлёбка лекарственная 2409
кровохлёбка малая 5018
кровохлёбка обыкновенная 2409
крокантемум 2387
крокус 1699
крокус весенний 1395
крокус голландский 1395
крокус осенний 1353
кроталярия 4425
кроталярия 4884
кроталярия индийская 5331
кроталярия ланцетовидная 3215
кроталярия остроконечная 5283
кроталярия полосатая 5283
кроталярия промежуточная 5000
кроталярия ситниковая 5331
кроталярия ситниковидная 5331
кротон 1705
кротон слабительный 4334
круглолистник японский 3048
кругонос 2197
крупина 1714
крупка 1897
крупка альпийская 4519
крупка весенняя 5183
крупка войлочная 6048
крупка ледниковая 2498
крупка норвежская 3888
крупка седая 5667
крупка сибирская 4902
крупка стенная 5766
крупка фладницийская 308
крутай 5027
крушина 830
крушина альпийская 121
крушина американская 1040
крушина вечнозелёная 2974
крушина калифорнийская 932
крушина китайская 1173
крушина красильная 4145
крушина ливанская 3260
крушина ломкая 2513
крушина ольховидная 2513
крушина ольхолистная 82
крушина полезная 1173
крушина Пурша 1040
крушина скалистая 4513
крушина скальная 4513
крушина слабительная 1369
крушина японская 3025
крушинник 830
крушинник ломкий 2513
крушинник ольховидный 2513
крушинные 831
крушиновые 831
крыжовник 2560
крыжовник амурский 3176
крыжовник барбадосский 447
крыжовник европейский 2103
крыжовник культурный 2103
крылоорешина 5996
крылоорешник 5996
крылотычинник 5261
крэб сибирский 4899
ксантиум 1306
ксантиум колючий 5155
ксантозома 3441
ксантоксилум 4295
ксанториза 6105
ксанторрея 2591
ксанторриза 6105
ксантосома 3441
ксантоцерас 6095
ксантоцерас рябинолистный 4868
ксерантемум однолетний 1446
ксилия деревянистоплодная 865
ксилия рубанковидная 865
ксилозма 6061
ксилокарпус 6060
ксимения американская 5428
ксирис 6084
ктенанте 1720
кубанка 1930
кубеба 1722
кубышка 1647
кубышка жёлтая 2090
кувшинка 5817
кувшинка белая 2132
кувшинка жёлтая 6099
кувшинка мексиканская 6099
кувшинковые 5818
кудзу 3188
кудзу лопастная 5505
кудренник чёрный 614
кудри пёстрые 614
кудри царские 3524
кудрявец 3096
кудряш 1965
куколица 1054
куколица широколистная 663
куколь 1610
куколь обыкновенный 1391
куколь полевой 1391
куколь посевной 1391
кукольван 5044
кукольвановые 3672
кукуруза 3434
кукуруза восковая 5849
кукуруза восковидная 5849
кукуруза зубовидная 1835
кукуруза крахмалистая 2307
кукуруза крахмалисто-сахарная 5215
кукуруза крахмально-сахарная 5215
кукуруза кремнистая 2295
кукуруза лопающаяся 4268
кукуруза мелкосемянная 4268
кукуруза обыкновенная 3434
кукуруза плёнчатая 4233
кукуруза сахарная 5301
кукуруза сладкая столовая 5301
кукуруза чешуйчатая 4233
кукушник 4398
кульбаба 2723
кульбаба осенняя 2159
куманика 2080
куманика 2093
кумаруна 1933
кумарчик 64
кумин 1729

кумкват 3189
кумкват круглый 3528
кумкват маруми 3528
кумкват овальный 3980
кумкват японский 3528
куммеровия 3279
куммеровия клиновидная 1189
куммеровия полосатая 1458
куммеровия прилистниковая 3183
кунжут 4834
кунжут восточный 3961
кунжут индийский 3961
кунжут культурный 3961
кунжутовые 4101
кунила 5265
куннингамия 1163
куннингамия китайская 1383
куннингамия ланцетная 1383
кунция 3698
купава европейская 1422
купальница 2507
купальница азиатская 4906
купальница европейская 1422
купальница китайская 1182
купания 1732
купена 5082
купена аптечная 1916
купена лекарственная 1916
купена многоцветковая 2071
купена многоцветная 2071
купена мутовчатая 5927
купена обыкновенная 1916
купена широколистная 799
куперия 4402
купырник 2759
купырник полевой 5176
купырник узловатый 3170
купырник японский 5688
купырь 494
купырь бутенелистный 4658
купырь лесной 6030
купырь обыкновенный 856
купырь садовый 4658
курай 4618
курай обыкновенный 1518
куркулиго 1738
куркума 5648
куркума длинная 1551
куркума узколистная 1983
куркума цитварная 6125
куропатник 3221
куропаточник 3221
курослеп 5425
курослеп болотный 5544
курчавка 2522
кускута 1870
куссо 3190
куст кокаиновый 2883
куст креозотовый 1680
куст чайный 1543
куст чайный апалашский 6068
куст чайный китайский 1543
кустарник кокаиновый 2883
кустарник мастиковый 3275
кустарник овечий 4855
кутровые 1873
куфея 1736
кучерявка 397
кучина 3426
кушир 2860
кырлык 5446
кювьера 1762

Л л

лабазник 3562
лабазник вязолистный 2112
лабазник шестилепестный 1907
лабданум 2664
лабульбениевые 3191
лаванда 3239
лаванда аптечная 5604
лаванда морская 3890
лаванда настоящая 5604
лаванда обыкновенная 5604
лаванда широколистная 795
лаватера 5582
лавр 3231
лавр американский 3710
лавр белый 5355
лавр благородный 2615
лавр горный 3710
лавр дикий 1456
лавр душистый 5355
лавр калифорнийский 938
лавр камфарный 966
лавр коричный 1238
лавр красный 4436
лавр ложнокамфарный 3827
лавр тюльпанный 5098
лавр цейлонский 1238
лавровишня 3238
лавровишня аптечная 1456
лавровишня лекарственная 1456
лавровишня лузитанская 4277
лавровишня обыкновенная 1456
лавровишня португальская 4277
лавровые 3232
лаврушка 1800
лавсония 2782
лагенария 2567
лагерстремия 2580
лагетта 3319
лагозерис 3204
лаготис 3206
лагунария 5307
лагурус 2713
ладан 553
ладан конский 2133
ладанник 4531
ладанник волосистый 2689
ладанник курчавый 6058
ладанник лавролистный 3235
ладанник тополелистный 4271
ладанник шалфеелистный 4670
ладанниковые 4532
ладьян 1591
ладьян трёхнадрезный 1960
лазерпициум 3227
лазия 3228
лазорник 1814
лазурник 3227
лайм 3311
лайм настоящий 3311
лайя 5519
лаковица розовая 5850
лаконос 4243
лаконос американский 1502
лаконосные 4244
лакрица 3291

лакричник 3291
лакричник голый 1459
лактук 3281
лакфиоль 5768
лакфиоль садовый 1556
лаллеманция 3208
ламаркия 2550
ламинариевые 3145
ламинария 667
ламинария пальчатая 4806
ламинария пальчаторассеченная 4806
ламинария сахарная 5299
ламинария японская 4807
ландольфия 2666
ландыш 3305
ландыш майский 3306
ландышник 3307
лансиум 3217
лансиум домашний 1886
лансиум обыкновенный 1886
лантана 3218
лапа воронья 5783
лапа воронья простёртая 1678
лапагерия 1152
лапагерия розовая 4445
лаперузия 2169
лапина 5996
лапина кавказская 1077
лапина Редера 4477
лапина сумахолистная 3077
лапина узкокрылая 1210
лапина ясенелистная 1077
лапка воронья 5783
лапка воронья простёртая 1678
лапка гусиная 4969
лапка заячья 2712
лапка кошачья 4363
лапка кошачья альпийская 142
лапка кошачья двудомная 1510
лапка кошачья карпатская 1028
лапки заячьи 4394
лапортея 6036
лапортея канадская 977
лапчатка 1242
лапчатка белая 5888
лапчатка весенняя 5182
лапчатка гусиная 4969
лапчатка золотистая 2530
лапчатка курильский чай 879
лапчатка кустарниковая 879
лапчатка низкая 1031
лапчатка норвежская 3887
лапчатка ползучая 1667
лапчатка прямостоячая 5555
лапчатка прямостоящая 5555
лапчатка семилисточковая 4837
лапчатка серебристая 4954
лапчатка скальная 1263
лапчатка снежная 5069
лапчатка средняя 1893
лапчатка стелющаяся 1031
лапчатка тюрингенская 4837
лапчатка тюрингская 4837
лапчатка-узик 5555
лапша грибная 4013
лардизабала 3223
ларрея 5172
ласковец 1256
ласковцы 4589
ластовенник 5340
ластовенник обыкновенный 5916
ластовенные 3621
ластовень 5340
ластовень аптечный 5916
ластовень лекарственный 5916
ластовень обыкновенный 5916
ластовица 2606
ластовневые 3621
ластовник 5340
ласточник 1469
ласточниковые 3621
латания 3229
латирус 4098
латук 3281
латук дикий 603
латук иволистный 5973
латук компасный 4299
латук красный 4353
латук лесной 4299
латук многолетний 4132
латук огородный 2417
латук посевной 2417
латук римский 4545
латук-ромен 4545
латук-ромэн 4545
латук-салат 2417
латук-салат кочанный 2735
латук стенной 5772
латук ядовитый 603
лаузелеурия лежачая 114
лаузелеурия стелющаяся 114
лахеналия 5591
лахнантес 4463
лебеда 2561
лебеда 4663
лебеда кино 4390
лебеда копьевидная 5118
лебеда копьелистная 5118
лебеда перуанская 4390
лебеда поникшая 2186
лебеда прибрежная 4875
лебеда раскидистая 2186
лебеда рисовая 4390
лебеда розовая 5639
лебеда садовая 2421
лебеда салатная 2421
лебеда стебельчатая 5208
лебеда стенная 3837
лебеда татарская 5457
лебеда черешчатая 5208
лебедовые 2562
левкена 3245
левкой 5258
левкой обыкновенный 1538
левкой однолетний 263
левкой садовый 1538
левкой седой 1538
левкорхис 3282
лёгочница 3393
легузия зеркало Венеры 5719
ледянник 2917
леерсия 1753
леерсия рисовидная 4497
лейкодендрон серебристый 4964
лейтнерия 1605
леканора 3262
леканора съедобная 3477
лекокия 3263
лемонграс 3270
лён 2286

лён австрийский 378
лён белосемянный 2214
лён долгунец 2214
лён жёлтый 2538
лён-кудряш 2214
лён кудряш 3915
лён кукушкин 2676
лён кукушкин обыкновенный 1435
лён культурный 2214
лён многолетний 4129
лён новозеландский 2215
лён новозеландский прядильный 3843
лён обыкновенный 2214
лён обыкновенный с растрескивающимися коробочками 1825
лён посевной 2214
лён-прыгунец 1825
лён сибирский 4129
лён-скакунец 1825
лён слабительный 4335
лён тонколистный 5001
лён узколистный 3802
лён французский 2367
ленец 476
лёновые 2288
ленок 1596
ленок 2102
ленок красильный 4212
ленок обыкновенный 2102
ленолистник 476
леонотис 3320
леонтица 3322
леонтице 3322
леонтоподиум 1992
леопольдия 3277
леопольдия хохлатая 5450
леопольдия хохолковая 5450
лепеха 5365
лептосине 1596
лептосперм 5472
лептоспермум 5472
лептохлоя 5171
лепчица 1055
леснина 1349
леспедеза 3279
леспедеза клиновидная 1189
леспедеза полосатая 1458
леспедеца 3279
леспедеца заострённая 1189
леспедеца клиновидная 1189
леспедеца корейская 3183
леспедеца полосатая 1458
леспедеца прилистниковая 3183
леспедеца шелковистая 1189
леспедеция 3279
леукокринум 5226
леукорхис 3282
леукофиллум 4961
лецитис 3657
лечуха 4693
лещина 2246
лещина американская 194
лещина древовидная 5646
лещина Зибольдa 3035
лещина калифорнийская 935
лещина китайская 1179
лещина крупная 2473
лещина маньчжурская 3456
лещина обыкновенная 2098
лещина понтийская 1071
лещина тибетская 5513
лжеакация 636
лжеапельсин 3971
лжедурнишник 5324
лжекипарис 2167
лжекипарис горохоносный 4724
лжекипарис Лавзона 3243
лжелиственница 2539
лжелиственница Кемпфера 3388
лжелиственница китайская 3388
лженарцисс 1397
лжетсуга 1888
лжетсуга крупноплодная 556
лжетсуга крупношишечная 556
лжетсуга сизая 4541
лжетсуга тис(с)олистная 1401
лжетсуга японская 3031
лиатрис 2439
либертия 3288
либоцедрус 2925
ливистона 2183
ливистона австралийская 371
ливистона китайская 1178
ливистона кохинхинская 3442
ливистона круглолистная 3087
ливистона южная 371
лигодий 1266
лигодиум 1266
лигустикум 3294
лигустикум северный 4774
лигустикум шотландский 4774
лигуструм 4312
ликания 3289
ликвидамбар 5369
ликвидамбар восточный 3965
ликвидамбар смолоносный 210
ликвидамбар стираксовый 210
ликоподиевые 1286
ликоподий 1285
ликуала 3292
лилейник 1815
лилейные 3301
лилия 3300
лилия африканская 5632
лилия белая 3417
лилия бледная 3417
лилия бульбоносная 842
лилия водяная 2132
лилия гигантская 2478
лилия длинноцветковая 1972
лилия длинноцветная 1972
лилия жёлтая 1815
лилия золотистая 2525
лилия исполинская 2478
лилия карниольская 1022
лилия клубненосная красная 3939
лилия-кордилина 1899
лилия королевская 4476
лилия красная 842
лилия кудреватая 3524
лилия ландышевая натальская 4685
лилия Ледебура 3264
лилия малабарская 3436
лилия мартагон 3524
лилия нильская 5130
лилия саранка 3524
лилия терракотовая 3791
лилия тигровая 5522
лилия трёхцветная южноафриканская 5591
лилия халцедонская 1111

лилия чалмовая 3524
лилия шафранная 3939
лилия японская 3053
лима 3309
лима 3538
лима мелкоплодная 4940
лиметта 3311
лимнантемум 2297
лимнантес 3548
лимнохарис 5711
лимодорум 2589
лимодорум клубневой 2590
лимон 3268
лимон дикий 1243
лимон лиметта 3311
лимон персидский 6020
лимон сладкий 3311
лимон трёхлисточковый 5592
лимонник 3425
лимонник китайский 1191
лимонница гладкая 341
линария 5533
линделофия 3313
линдерния 2179
линнея 5660
липа 3314
липа американская 199
липа амурская 229
липа венгерская 4965
липа войлочная 4965
липа голландская 2111
липа длинночерешковая 4967
липа европейская 2111
липа кавказская 1076
липа каменная 3642
липа каменная узколистная 3808
липа каменная широколистная 5586
липа крупнолистная 561
липа крымская 1692
липа маньчжурская 3463
липа мелколистная 3327
липа обыкновенная 2111
липа плосколистная 561
липа пушистостолбчатая 1076
липа сердцевидная 3327
липа серебристая 4965
липа чёрная 199
липа широколистная 561
липа ярко-зелёная 1692
липарис 5658
липарис Лёзеля 3356
липия 3324
липовые 3315
липпия 3324
липпия лимонная 3271
липпия трёхлистная 3271
липучка 5246
липучка ежёвая 2126
липучка ежовая 2126
липучка масляная 1151
липушник 1055
лириодендрон 5635
лириодендрон китайский 1207
лириодендрон тюльпанный 5636
лириопе 3308
лиродревесник 2218
лисехвостник 2349
лисичка 1000
лисичка ложная 2165
лисичка настоящая 1115
лисичка обыкновенная 1115
лисичка серая 2858
лисичковые 1001
лисохвост 2349
лисохвост альпийский 133
лисохвост вздутый 1668
лисохвост коленчатый 5805
лисохвост луговой 3549
лисохвост мышехвостиковый 3735
лисохвост мышехвостниковидный 3735
лисохвост полевой 3735
лисохвостник 2349
лист александрийский 89
лиственница 3222
лиственница американская 1977
лиственница восточная американская 1977
лиственница гималайская 2802
лиственница Гриффитса 2802
лиственница европейская 2110
лиственница западная 5868
лиственница западная американская 5868
лиственница ложная 2539
лиственница опадающая 2110
лиственница сибирская 4908
лиственница тонкочешуйчатая 3051
лиственница японская 3051
листера 5657
листовик сколопендровый 2716
листовка 4056
листоколосник 4170
листоколосник золотистый 2527
литрум 3400
литтония 3340
литторела 4879
лихнис 967
лихнис альпийский 306
лихнис белый 2140
лихнис кожистый 4558
лихнис татарское мыло 3451
лихнис халцедонский 3451
лихорадочник 1913
лиций 6016
лиций обыкновенный 450
лиций обыкновенный 3535
лиций русский 4612
личжи 3397
личжи китайский 3398
личинник 4767
лишайник 3290
лишайник бородатый 503
лишайник исландский 2916
лишайник красильный 3325
лишайник лакмусовый 3325
лишайник олений 4479
лишайник рокцелла 4509
лишайник тилланзия 5576
лишайник-умбиликария 4536
лишайники базидиальные 461
ллоидия 159
лоаза 1155
лобария лёгочная 3394
лобария лёгочнообразная 3394
лобелиевые 3351
лобелия 3350
лобелия вздутая 2942
лобелия Дортман(н)а 5819
лобелия надутая 2942
лобелия одутлая 2942
лобия 2897
лобия египетская 2897

логаниевые 3358
ложечница 4783
ложечница английская 2043
ложечница аптечная 1522
ложечница арктическая 1522
ложечница лекарственная 1522
ложнодождевик 2705
ложнодождевик обыкновенный 1405
ложнодождевиковые 2706
ложнолиственница 2539
ложноопёнок 2907
ложнопокровница 1271
лоза 106
лоза 467
лоза виноградная 2575
лоза волчья 693
лоза золотая 2101
лозина миндальная 106
лозинка 1260
лозинка белая 5571
лозинка обыкновенная 5571
лойник 4816
локва 3376
ломонос 1260
ломонос альпийский 125
ломонос Арманда 319
ломонос винорадолистный 5571
ломонос виорна 3255
ломонос виргинский 5756
ломонос восточный 3952
ломонос вьющийся 5571
ломонос горный 244
ломонос джунгарский 5330
ломонос жгучий 4227
ломонос метельчатый 5353
ломонос обыкновенный 5571
ломонос прямой 2646
ломонос раскидистый 3297
ломонос тангутский 2532
ломонос усатый 5723
ломонос фиолетовый 2976
ломонос цельнолистный 5080
ломонос шерстистый 3858
лоницера 2841
лонхокарпус 3216
лопастник 2767
лопастник курчавый 1697
лопастник ямчатый 3196
лопух 859
лопух большой 2601
лопух войлочный 1636
лопух крупный 2601
лопух малый 5020
лопух паутинистый 1636
лопух паутинский 1636
лопушник 859
лопушник большой 2601
лопушник малый 5020
лопушник паутинистый 1636
лопушник съедобный 2601
лорантовые 3632
лоропеталум 3380
лосняк 5658
лосняк Лёзеля 3356
лотос 3381
лотос египетский 5893
лотос индийский 2811
лотос каспийский 1045
лотос орехоносный 2811
лофант 2476
лофира 3372
лофира африканская 2005
лофира крупнолистная 2005
лофира крылатая 5992
лофира пирамидальная 5992
лофоспермум 3536
лофофора Вильямса 3591
лох 2006
лох многоцветковый 1135
лох серебристый 4951
лох узколистный 4616
лоховые 3926
луазелеурия стелющаяся 114
луговик 2677
луговик альпийский 134
луговик дернистый 5628
луговик извилистый 5843
лужайник 3745
лужайник водяной 5825
лужница 3745
лужница водяная 5825
лук 3931
лук Аскалона 4844
лук-батун 5863
лук виноградный 2223
лук гадючий 2579
лук гадючий гроздевидный 1427
лук гадючий кистевидный 5214
лук гадючий кистистый 5214
лук гадючий хохлатый 5450
лук гусиный 2399
лук дудчатый 5863
лук зимний 5863
лук картофельный 3753
лук килеватый 3144
лук круглоголовый 421
лук культурный 2420
лук медвежий 4415
лук моли 3303
лук морской 4878
лук обыкновенный 2420
лук огородный 4284
лук победный 3368
лук пор(р)ей 3266
лук приречный 2474
лук причесночный 2474
лук-резанец 1219
лук репчатый 2420
лук-рокамболь 2474
лук-скорода 1219
лук стареющий(ся) 1934
лук-татарка 5863
лук торчащий 5281
лук трубчатый 5863
лук угловатый 249
лук-черемша 4415
лук-чеснок 2432
лук-шалот 4844
лук Шуберта 4762
лумбанг 992
лунария 2834
лунник 2834
лунник многолетний 4130
лунник однолетний 1884
лунник оживающий 4130
луносемянник 3671
луносемянник даурский 347
луносемянник канадский 1472
луносемянниковые 3672
лупин 3395

лучица 5267
лучицы 1119
лыко волчье 2194
льновые 2288
льнянка 5533
льнянка альпийская 155
льнянка далматская 1791
льнянка двураздельная 1277
льнянка дроколистная 811
льнянка итальянская 2984
льнянка канадская 3922
льнянка македонская 3404
льнянка обыкновенная 887
льюзия 3285
лэйа 5519
любим-трава 1354
любисток 3384
любисток аптечный 2418
любисток лекарственный 2418
любка 4403
любка двулистная 895
любка зеленоцветковая 2605
любка зеленоцветная 2605
любник 5998
людвигия 4812
людвигия болотная 3512
люпин 3395
люпин белый 5899
люпин волосистый 2084
люпин древнеегипетский 5899
люпин жёлтый 2135
люпин жестковолосистый 2084
люпин жесткоопушённый 2084
люпин изменчивый 5096
люпин лохматый 2084
люпин многолетний 5327
люпин многолистный 5789
люпин пестроцветный 5096
люпин синий 703
люпин тёплый 5899
люпин узколистный 703
лютик 889
лютик азиатский 4146
лютик аконитолистный 16
лютик альпийский 3677
лютик водяной 5797
лютик горный 3699
лютик длиннолистный 5544
лютик едкий 5425
лютик жгучий 5121
лютик золотистый 2552
лютик иллирийский 2921
лютик клубневой 841
лютик клубненосный 841
лютик ледниковый 2497
лютик лесной 5962
лютик луковичный 841
лютик полевой 1608
лютик ползучий 1666
лютик снеговой 5059
лютик-чистяк 2241
лютик ядовитый 673
лютик языколистный 5544
лютиковые 890
люффа 5564
люффа цилиндрическая 5290
люхея 5877
люцерна 3565
люцерна арабская 5170
люцерна аравийская 5170
люцерна гибридная 472
люцерна дисковидная 903
люцерна древовидная 5583
люцерна жёлтая 4921
люцерна круглая 903
люцерна маленькая 933
люцерна маленькая 3332
люцерна округлая 903
люцерна посевная 90
люцерна пятнистая 5170
люцерна серповидная 4921
люцерна хмелевая 638
люцерна хмелевидная 638
люцерна щетинистая 933
люцерна щетинистоволосистая 933
лягерстремия 2580
ляготис 3206
лягушатник 2381
лягушечник 2381
лядвенец 1824
лядвенец болотный 5872
лядвенец рогатый 588
лядвенец топяной 5872
лядвенец четырёхлопастный 5197
лядник 5367
лядник душистый 5368
лядник пахучий 5368
ляйм 3311
ляллеманция 3208
ляминария 667

М м

магалебка 3426
магнолиевые 3424
магнолия 3423
магнолия виргинская 5355
магнолия голая 6122
магнолия длиннозаострённая 1726
магнолия звездчатая 5227
магнолия зонтичная 5679
магнолия крупноцветковая 5098
магнолия крупноцветная 5098
магнолия Кэмбелла 964
магнолия лилейная 3304
магнолия лилиецветная 3304
магнолия обнажённая 6122
магнолия обратноовальная 5897
магнолия обратнояйцевидная 5897
магнолия остроконечная 1726
магнолия сизая 5355
магнолия Суланжа 4715
магнолия трёхлепестковая 5679
магнолия туполистная 5897
магнолия Юлана 6122
магнолия японская 5897
магония 3429
магония падуболистная 3947
магония Форчуна 1192
магония японская 3055
мадия 3416
мадия посевная 1151
мадука индийская 2920
маис 3434
майник 492
майник двулистный 5671
майоран 3497
майоран дикий 1484
майоран зимний 1484
майоран кошачий 1064

майоран летний 5375
майоран садовый 5375
майтенус 3540
майтенус падуболистный 2831
майтенус чилийский 1154
майяка 4267
мак 4272
мак альпийский 140
мак Аргемона 2153
мак восточный 3960
мак голубой тибетский 708
мак калифорнийский 943
мак колючий 2153
мак опийный 3933
мак пиренейский 4375
мак рогатый 652
мак самосейка 1616
мак снотворный 3933
мак сомнительный 3367
макадамия 4385
маклейя 4230
маклюра 3971
маклюра оранжевая 3972
маковые 4274
макрозамия 3406
макротомия 4314
макроцистис 2477
малахра 5952
малина 4420
малина американская 624
малина арктическая 305
малина великолепная 4660
малина душистая 2359
малина европейская 4462
малина ежевикообразная 624
малина замечательная 4660
малина западная 624
малина обыкновенная 4462
малина пахучая 2359
малина превосходная 4660
малина североамериканская 624
малина чёрная 624
малопа 3449
мальва 3447
мальва красивая 1294
мальва круглолистная 1945
мальва курчавая 1743
мальва лесная 2797
мальва махровая 2829
мальва мускусная 3760
мальва раздельнолистная 2830
мальва садовая 2829
мальва сирийская 1185
мальваструм 2173
мальвовые 3448
малькольмия 3444
мальпигия 3450
мальпигия гранатоволистная 446
мамиллярия 3454
маммея 3452
маммея американская 3453
маммиллярия 3454
мамура 305
манглис 3474
манго 3472
манго вонючее 180
манго вонючий 180
манго душистое 853
манго душистый 853
манго индийский 1464
манго пахучее 180
манго пахучий 180
мангольд 1116
мангольд 3473
мангостан 3474
мангустан 3474
мандарин 3468
мандарин благородный 3162
мандарин итальянский 3468
мандарин кинг 3162
мандрагора 3469
манеттия 3470
манжетка 3202
манжетка альпийская 3709
манжетка обыкновенная 1454
манжетка полевая 2226
манизурис 3108
манилла 2
маниок 1046
маниок Глацийова 1081
маниок горький 1379
маниок полезнейший 1379
маниок сладкий 67
маниок сладкий 5360
маниок съедобный 1379
маниока 1046
маниока каучуконосная 1081
маниот 1046
маниот съедобный 1379
манна 3477
манна лишайниковая 3477
манник 3476
манник водный 806
манник водяной 806
манник наплывающий 5820
манник плавающий 5820
манник ручьевой 5820
манцинелла 3455
маншинелла 3455
маранта 329
маранта вестиндская 541
маранта тростниковая 541
марантовые 330
мараттиевые 3485
маргаритка 1787
маргаритка многолетняя 2037
маревые 2562
марена 3410
марена красильная 1463
мареновые 3411
марзанка 5201
мариола 2657
марипоза 3494
марискус 4727
марсдения 3499
марсилея 4128
марсилиевые 3523
марсилия 4128
марсин 1463
мартагон 3524
мартиниевые 5683
мартиния 3527
мартиния душистая 5363
маршанция 3487
маршанция многообразная 1465
марь 2561
марь амброзиевидная 6056
марь белая 3212
марь благовонная 6056
марь головчатая 675

марь Доброго Генриха 2559
марь душистая 3096
марь многолистная 5273
марь обыкновенная 3212
марь стенная 3837
марь цельнолистная 2559
марь чилийская 4390
марьянник 1655
марьянник дубравный 696
масленик 820
маслёнок 820
маслёнок зернистый 2574
маслёнок летний 2574
маслёнок лиственничный 2011
маслёнок лиственный 2011
маслёнок поздний 820
маслина 3927
маслина американская 1850
маслина дикая 5953
маслина европейская 1483
маслина культурная 1483
маслинные 3928
маслиновые 3928
масличные 3928
масляник 820
масляник зернистый 2574
масляник летний 2574
масляник лиственничный 2011
масляник лиственный 2011
масляник поздний 820
мате 4048
материнка 1144
матико 3534
маточник 3645
маточник болотный 714
матрикария 3541
маттиола 5258
маттиола садовая 1538
мать-и-мачеха 1339
мать-и-мачеха обыкновенная 1388
матэ 4048
маун 5695
маун аптечный 1552
мауновые 5696
маурандия 3536
махагони 5871
махобели 5417
махорка 395
мачёк 2859
мачёк жёлтый 6096
мачёк рогатый 652
мачок 2859
маш 3754
маш-салат 2088
мёд дивий 2548
медеола 3564
мединилла 3567
медовик 781
медовник 781
медуница 3393
медуница аптечная 1462
медуница горная 3711
медуница красная 3711
медуница лекарственная 1462
медуница узколистная 1651
медунка 4453
мезембриантемум 3592
мезембриантемум кристаллический 2917
мезембриантемум хрустальный 2917
мезуа железная 2966
меконопсис 708
меконопсис кумберлендский 5864
меконопсис уэльский 5864
мелалеука 3579
мелалеука бледнодревесная 920
мелалеука чайная 920
меланкониевидные 3580
меланкониевые 3580
меластомовые 3581
мелиант 2839
мелиациевые 3428
мелиевые 3428
меликокка 2837
мелинис 3644
мелисса 427
мелисса аптечная 1356
мелисса лекарственная 1356
мелицитус веткоцветный 2808
мелия 1159
мелия азедарах 1160
мелия ацедарах 1160
мелия гималайская 1160
мелия индийская 3492
мелия иранская 1160
мелколепестник 2292
мелколепестник альпийский 130
мелколепестник едкий 602
мелколепестник зонтичноцветный 602
мелколепестник канадский 2878
мелколепестник однолетний 259
мелколепестник одноцветковый 309
мелкоплодник 3695
мелкоплодник ольховолистный 1831
мелкоплодник ольхолистный 1831
мелокактус 3584
мелотрия 3586
мениспермовые 3672
мениспермум 3671
мениспермум канадский 1472
ментцелия 3587
менцизия 4989
мертензия 691
меруя 3420
мескито 3593
метельник 5856
метельник прутьевидный 5857
метлица обыкновенная 5981
метлица полевая 5981
метросидерос 2964
меум 415
меченосница 2720
меч-трава 4727
мибора 3609
микания 3616
микания лазающая 1268
микания ползучая 1268
микания цепкая 1268
микромерия 3611
микромерия греческая 2619
микростилис 25
миксомицеты 5006
милла 3602
миллефоль 1565
мильтония 4031
мимоза 3625
мимоза стыдливая 4827
мимозка 3593
мимозка серёжкоцветная 1467
мимозовые 3626
мимулус 3655

мимулюс 3655
мимулюс жёлтый 1471
мимулюс крапчатый 2541
мимулюс мускусный 3764
мимусопс 848
миндаль 104
миндаль горький 597
миндаль грузинский 4611
миндаль земляной 1230
миндаль иерусалимский 4088
миндаль карликовый 4611
миндаль ломкий 2354
миндаль настоящий 104
миндаль низкий 4611
миндаль обыкновенный 104
миндаль степной 4611
миндаль трёхлопастный 2311
миндаль тропический 5600
миопорум 3775
мирабель 3630
мирабилис 2341
мирабилис Ялапа 1415
мирзиновые 3780
мирика 483
мирикария 2181
мирикария германская 2456
мирикария даурская 1784
мириофиллюм 4060
миристицевые 3895
мирлитон 1122
миробалан 3777
мирокарпус 3778
мирокарпус густолиственный 762
мироксилон 429
миррис душистый 5361
миррис пахучий 5361
мирт 3781
мирт болотный 5366
мирт восконосный 5104
мирт обыкновенный 5606
мирта 3781
миртовые 3783
мискантус 4956
мителла 3634
митрагина крупнолистная 6
митрагина прилистниковая 6
митрария 3635
митчела 4066
михелия 3610
мицелис стенной 5772
мицена 3774
мицена колпачковидная 4086
мичелла 4066
млечник 3195
млечник 4796
млечник сладковатый 5372
мниум 3638
многобородник 504
многоветочник карликовый 2643
многоветочник низкий 3389
многокоренник 1923
многокучник большой 3445
многоножка 4251
многоножка обыкновенная 1503
многоножковые 4252
многоплодник 3479
многорядник 2828
многорядник Брауна 760
многорядник копьевидный 3708
многорядник лопастный 3349
многорядник серповидный 2881
могар 2352
могильник 2714
могильница 4140
можжевелоягодник 1947
можжевельник 3118
можжевельник аллигаторовый 100
можжевельник американский 1980
можжевельник барбадосский 543
можжевельник бермудский 543
можжевельник Валлиха 5771
можжевельник виргинский 1980
можжевельник высокий 2618
можжевельник гималайский 2801
можжевельник донской 4719
можжевельник западный 4938
можжевельник испанский 4298
можжевельник казацкий 4719
можжевельник казачий 4719
можжевельник калифорнийский 937
можжевельник китайский 4369
можжевельник косточковый 5418
можжевельник красноватый 4298
можжевельник красноплодный 4168
можжевельник красный 4298
можжевельник крупноплодный 4231
можжевельник ладанный 2926
можжевельник лежачий 3039
можжевельник мексиканский 3599
можжевельник обыкновенный 1453
можжевельник обыкновенный горный 3703
можжевельник отогнутый 2801
можжевельник Пинчота 4440
можжевельник повислый 3598
можжевельник прибрежный 4874
можжевельник приморский 4874
можжевельник распростёртый 1671
можжевельник распростёртый опушённый 238
можжевельник Саржента 4707
можжевельник сибирский 3703
можжевельник стройный 43
можжевельник твёрдый 3820
можжевельник толстокорый 100
можжевельник формозский 2332
можжевельник чешуйчатый 4974
можжуха 2645
мозжуха 2645
мокрединник 5773
мокрица 1146
мокрица 2904
мокрица алзина 5235
мокричник 5235
мокруха еловая 2516
моли 3303
молиния 3646
молиния голубая 3674
моллуго 1032
моллюго 1032
молодило 2882
молодило кровельное 2779
молоко волчье 2070
молокан 3281
молокан компасный 4299
молокан посевной 2417
молокан ядовитый 603
молочай 2070
молочай 3938
молочай бутерлаковидный 4159
молочай бутерлаковый 4159

молочай кипарисный 1774
молочай кипарисовидный 1774
молочай кипарисовый 1774
молочай красивейший 1500
молочай лозный 3251
молочай маленький 1951
молочай малый 1951
молочай масличный 1008
молочай многоцветный 1750
молочай острый 3251
молочай прекрасный 1500
молочай прибрежный 4805
молочай прямой 5689
молочай пятнистый 5167
молочай садовый 4159
молочай серповидный 4922
молочай солнцегляд 5328
молочай съедобный 3251
молочай чиновидный 1008
молочайные 5191
молюцелла 3647
момбин жёлтый 6101
момордика 431
момордика бальзамическая 431
момордика харантская 434
монарда 512
монетка 1333
монетка короткокорневая 4550
монетка широколистная 790
монодора 3662
монодора мускатная 923
монстера 3663
монстера любимая 1101
монстера пышная 1101
монтбреция 5599
монция 2938
мордовник 2508
мордовник банатский 6121
мордовник венгерский 6121
мордовник круглоголовый 1423
мордовник обыкновенный 5027
мордовник русский 4615
мордовник русский 5027
мордовник шароголовый 1423
мордовник широкоголовый 1423
морея 3676
морина 3681
моринга 2872
мориновые 5471
морковник 4124
морковь 1035
морковь двулетняя 5942
морковь дикая 5942
морковь культурная 5942
морковь культурная каротиновая 5942
морковь обыкновенная 5942
морковь огородная 2410
морковь посевная 2410
морозник 2766
морозник вонючий 507
морозник восточный 3272
морозник зелёный 2628
морозник чёрный 1226
морошка 1273
морошка приземистая 1273
мотыльковые 4041
мох белый болотный 5125
мох белый олений 4479
мох бородатый 5576
мох водяной 5823
мох гипновый 2908
мох дубовый 5204
мох звездчатый 3638
мох ирландский 2961
мох исландский 2916
мох испанский 5576
мох караковый 1623
мох корсиканский глистогонный 1623
мох красильный 3325
мох лакмусовый 3325
мох луизианский 5576
мох обыкновенный печёночный 1465
мох олений 4479
мох ползучий 1285
мох стелющийся 1285
мох торфяной 5125
мох фиалковый 4537
мох флоридский 5576
мох цейлонский 56
моховик жёлто-бурый 5700
моховик зелёный 6081
моховик пёстрый 4448
мохообразные 825
мохунка 5275
мочёнка 3800
мошнуха 5275
мукор 3691
мукоровые 3739
мукуна 3740
мульга 3751
мульгедиум 3281
муме 3015
мунго 3755
мунтингия 3007
мунтингия Калабура 3007
муррайя 3084
муррайя экзотическая 1161
муррая 3084
муррая экзотическая 1161
муррея 3084
мускари 5214
мускат 1480
мускатник 3894
мускатниковые 3895
мускатница 3766
мускусница 3766
мутовчатка 1032
мухоловка 1908
мухомор 2319
мухомор белый 1820
мухомор весенний 5180
мухомор вонючий 1842
мухомор зелёный 1820
мухомор красный 2319
мухомор пантерный 4032
мухомор серо-розовый 713
мухомор серый 4032
мухоморовые 173
мучнеросные 4289
мучнисторосные 4289
мушмула 3577
мушмула германская 3578
мушмула японская 3376
мушмула японская обыкновенная 3376
мхи зелёные 5605
мхи сфагновые 4096
мхи торфяные 4096
мшанка 4093
мшанка безлепестная 273
мшанка лежачая 4313

мшанка Линнеева 310
мшанка моховидная 310
мшанка узловатая 3171
мшанка шилолистная 1624
мыло татарское 3451
мыльница 5074
мыльнянка 5074
мыльнянка аптечная 743
мыльнянка коровья 1653
мыльнянка лекарственная 743
мытник 3383
мытник болотный 2117
мытник лапландский 3219
мытник лесной 3382
мытник мутовчатый 5925
мытник судетский 5295
мышей 780
мышей волосистый 2931
мышей зелёный 2624
мышей малый 3331
мышей мятликовидный 3331
мышей сизый 2624
мышехвостник 3736
мышехвостник маленький 5530
мюленбекия 6010
мюленбергия 3748
мюскадина 3756
мягковолосник водный 5795
мягковолосник водяной 5795
мякотница 25
мякотница болотная 3250
мята 3628
мята английская 4122
мята бергамотная 535
мята блошница 4113
мята болотная 4113
мята водная 5822
мята водяная 5822
мята горная 3714
мята дикая 2869
мята длиннолистная 2869
мята зелёная 5120
мята колосовая 5120
мята конская 1442
мята кошачья 1060
мята круглолистная 283
мята курчавая 1698
мята лимонная 535
мята лимонная 1356
мята лимонная аптечная 2356
мята перечная 4122
мята полевая 2229
мята пулегиевая 4113
мята холодка 4122
мятлик 700
мятлик альпийский 118
мятлик арктический 304
мятлик Бальфура 416
мятлик болотный 2345
мятлик боровой 6023
мятлик горный 3725
мятлик дубравный 6023
мятлик лесной 6023
мятлик луговой 3150
мятлик луковичный 844
мятлик обыкновенный 4579
мятлик однолетний 258
мятлик рыхлый 3725
мятлик сплюснутый 970
мятлик трёхцветковый 2345
мятлик узколистный 3799
мятлик шафранный 2345
мятлик Ше 796
мятлик Шэ 796
мяун 5695

Н н

навозник 2951
навозник белый 4843
навозник серый 2950
навозниковые 1586
нагами кинкан 3980
нагловатка 3120
наголоватка 3120
нагут 1144
надбородник безлистный 5195
нандина 3790
нандина домашняя 4634
наннеропс 3543
наперстянка 2346
наперстянка жёлтая 5280
наперстянка красная 1416
наперстянка крупноцветковая 6089
наперстянка лекарственная 1416
наперстянка пурпурная 1416
наперстянка пурпуровая 1416
наперстянка сомнительная 6089
наперстянка шерстистая 2614
нард 1244
нард вытянутый 3533
нартеций 715
нарцисс 3795
нарцисс белый 4237
нарцисс букетный 4248
нарцисс жёлтый 3110
нарцисс-жонкиль 3110
нарцисс константинопольский 4248
нарцисс ложнонарциссовый 1397
нарцисс ложный 1397
нарцисс поэтический 4237
нарцисс-тацетта 4248
нарцисс трубчатый 1397
настурциевые 3813
настурция 3812
настурция большая 1478
настурция высокая 1478
настурция иноземная 986
настурция канарская 986
настурция карликовая 882
настурция клубненосная 5617
настурция малая 882
настурция мелкая 882
настурция низкорослая 882
настурция садовая 1478
наумбургия кистецветная 5629
наяда 3786
наяда гибкая 5002
наяда морская 5156
наядовые 3787
невзрачница 2226
невзрачница полевая 2226
невиузия 5068
негниючка 1973
негниючник 3484
негниючник колёсиковидный 1329
негной-дерево 6117
негундо 750
недотрога 5051
недотрога бальзаминовая 2407

недотрога бальзаминовидная 2407
недотрога железистая 4596
недотрога жёлтая 5560
недотрога мелкоцветковая 5023
недотрога "не тронь меня" 5560
недотрога обыкновенная 5560
недотрога Ройля 4596
нежник 5334
нежник альпийский 152
нежник обыкновенный 1515
незабудка 2327
незабудка азорская 392
незабудка альпийская 132
незабудка болотная 5602
незабудка дернистая 5627
незабудка жестковолосистая 1962
незабудка крупноцветная 5399
незабудка лесная 6033
незабудка пашенная 2222
незабудка полевая 2222
незабудка разноцветная 6073
незабудка средняя 2222
незабудка холмовая 1962
незабудочник 2056
нейлия 3824
нейлия китайская 1193
нейлия тибетская 5511
неклён 750
неклён виргинский 750
нектарин 3818
немезия 3825
немофила 3826
ненюфар 5817
непентес 3828
непентесовые 3829
непентовые 3829
непета 3830
нептуния 4825
нерине 3832
нерине сарнийская 2658
нериум 3925
нертера 491
неслия метельчатая 422
не-тронь-меня 5560
нефелиум 3397
нефелиум китайский 3398
нефелиум репейниковый 4409
нефролепис 5405
нефролепис возвышенный 1541
нефрома 3831
нехворощ 4641
нианца 2244
нивяник 1229
нивяник круглолистный 2887
нивяник обыкновенный 3987
нивяник посевной 1609
нигелла 2201
низмянка 1109
никандра 284
никандра пузыревидная 285
никандра физалиевидная 285
никандра физалисовидная 285
никтагиновые 2342
никтантес 3851
никтериния 3852
нимфейник 2297
нимфейные 5818
нимфея 5817
нирембергия 1733
нисса 5642
ниссовые 5643
нителла 3861
нитрария 3862
нитчатка 4260
ницелоз 1679
ноголист 3539
ноголистник 3539
ноголистник пальчатый 1845
ногоплодник 4235
ногоплодник альпийский 5449
ногоплодниковые 4236
ноготки 929
ноготки аптечные 4285
ноготки африканские 1005
ноготки лекарственные 4285
ноготки садовые 4285
"ножки заячьи" 4606
нолана 3871
нолина 3872
нонеа 3873
нонея 3873
ноннея 3873
норичник 2242
норичник весенний 5184
норичник водяной 5804
норичник узловатый 6027
норичник шишковатый 6027
норичниковые 2243
носики комаровы 2330
носток 3889
носток обыкновенный 5225
ностоковые 2375
нотофагус 2163
нотофагус антарктический 268
ночецвет 1415
ночецветник 1415
ночецветниковые 2342
ночецветные 2342
ночница 3852
нуг 3849
нуг масличный 2063
нут 1144
нут культурный 1144
нут обыкновенный 1144
нут посевной 1144
нут рогообразный 1144
нюнька 4801

О о

обабок 582
обвойник 4949
обвойник греческий 2616
облепиха 4787
облепиха крушиновая 1523
облепиха крушиновидная 1523
облепиха обыкновенная 1523
обриеция 366
овения 2030
овения вишнелистная 1136
овёс 3904
овёс бесплодный 252
овёс бородатый 5003
овёс венгерский 4927
овёс византийский 4457
овёс византийский красный 4457
овёс восточный 4927
овёс высокий 5427
овёс голый 3789
овёс декоративный 252

овёс дикий 5951
овёс дикий красный 252
овёс египетский 4927
овёс желтеющий 6109
овёс желтоватый 6109
овёс живой 5951
овёс золотистый 6109
овёс колючий 3374
овёс короткий 3333
овёс косой 4927
овёс луговой 3553
овёс мелкозёрный голый 3789
овёс опушённый 2687
овёс посевной 1481
овёс посевной средиземноморский 4457
овёс пустой 5951
овёс пушистый 2687
овёс средиземноморский 4457
овёс стерильный 252
овёс татарский 3789
овёс тощий 3374
овёс турецкий 4927
овёс щетинистый 3374
овсец луговой 3553
овсец опушённый 2687
овсюг 5951
овсяница 2208
овсяница аметистовая 215
овсяница бороздчатая 4858
овсяница высокая 2472
овсяница гигантская 2472
овсяница исполинская 2472
овсяница красная 4450
овсяница луговая 3547
овсяница мышехвостная 4423
овсяница овечья 4858
овсяница песчаная 1676
овсяница разнолистная 4839
овсяница тростниковая 4474
овсяница тростниковидная 4474
огоньки 4906
огурец 1724
огурец антильский 5869
огурец африканский 2855
огурец бешеный 5202
огурец бешеный обыкновенный 5202
огурец вест-индский 5869
огурец волосистый 858
огурец волосистый угловатый 5765
огурец грядовой 1724
огурец дыня 3762
огурец змеевидный 1996
огурец змеиный 1996
огурец мексиканский 1122
огурец посевной 1724
огурец угловатый 5765
огуречник 734
огуречник аптечный 1366
огуречник лекарственный 1366
огурчик пёсий виноградолистный 434
однодольные 3661
однозернянка 2004
однозернянка культурная 2004
односемядольные 3661
одноцветка крупноцветковая 6037
одноцветка крупноцветная 6037
одноцветка одноцветковая 6037
одноягодник 4053
одноягодник четырёхлистный 2786
одонтоглоссум 3910
одуванчик 1796
одуванчик аптечный 1398
одуванчик кок-сагыз 4613
одуванчик лекарственный 1398
одуванчик обыкновенный 1398
одурь болотная 1719
одурь кавказская 4264
одурь сонная 525
ожика 6041
ожика боровая 2652
ожика Валенберга 5762
ожика волосистая 2696
ожика дубровная 2652
ожика изогнутая 1746
ожика колосистая 5141
ожика лесная 2613
ожика полевая 2239
ожика равнинная 2239
ожика судетская 5296
ожика Форстера 2336
ожина 2093
оидиум 3914
ока 3918
окопник 1345
окопник аптечный 1389
окопник жёсткий 4297
окопник кавказский 1068
окопник клубненосный 5619
окопник лекарственный 1389
окопник острый 4297
окопник шершавый 4297
окотеа 3908
окра 3921
оксалис 3985
оксиграфис обыкновенный 2497
оксидендрон 5094
оксидендрум древовидный 5095
оксилобус 3991
олеандр 3925
олеандр обыкновенный 1482
олеария 1788
оливник 5194
ольдфильдия африканская 51
ольха 81
ольха белая 5122
ольха восточная 1066
ольха горная 2105
ольха клейкая 2072
ольха красная 4432
ольха морщинистая 2732
ольха серая 5122
ольха сердцевидная 2971
ольха сердцелистная 1066
ольха чёрная 2072
ольха японская 3012
омег пятнистый 4241
омежник 5799
омежник водный 2251
омежник водяной 2251
омежник дудчатый 5625
омежник трубчатый 5625
омежник шафранный 2771
омежник шафрановый 2771
омежник шафраноподобный 2771
омела 3631
омела белая 2113
омеловые 3632
оморика 4830
омфалодес 3815
омфалодес настоящий 1672

онагра 2141
онагриковые 2142
онагровые 2142
оноклея чувствительная 4826
онопордум 1638
онопордум колючий 4770
онопордум крымский 5462
онопордум обыкновенный 4770
оносма 3932
оносма крымская 1690
онцидиум 3929
опёнок зимний 5709
опёнок летний 1114
опёнок луговой 2154
опёнок настоящий 2838
опёнок осенний 2838
опёнок уховидный 3992
опопанакс 3934
опунция 4301
опунция индийская 2936
опунция индийская фига 2936
опунция кошенильная 1305
опунция обыкновенная 1507
опунция одноиглистая 1507
орех 5777
орех австралийский 4385
орех айлантолистный 4936
орех американский 765
орех арековый 548
орех болотный 5794
орех бразильский 765
орех водяной 5794
орех водяной плавающий 5794
орех волошский 4151
орех волшебный 6012
орех горький 605
орех грецкий 4151
орех заячий 4611
орех земляной 4088
орех земляной бамбарский 1574
орех Зибольда 4936
орех кавказский 1077
орех калифорнийский 930
орех кешью 1378
орех китайский 1208
орех кокосовый 1314
орех кола 5293
орех лесной 2246
орех лесной американский 2098
орех ломбардский 2473
орех малдивский 1887
орех маньчжурский 3466
орех масляный 897
орех медвежий 5646
орех мускатный 1480
орех мускатный фальшивый 2175
орех мускатный ямайский 923
орех мушкатный 1480
орех насмешливый 3640
орех нищих 548
орех-пекан 4100
орех подземный 4088
орех рвотный 3898
орех савара 4725
орех сердцевидный 2281
орех серый 897
орех скальный 5479
орех чекалкин 4868
орех чёрный 1975
орех чёрный восточный 1975
орех чилийский 992
орех японский 2281
ореховые 5778
орешки земляные 1907
орешник 2098
орешник 2246
орешник Зибольда 3035
орешник индийский 3438
орешник китайский 1179
орешник китайский 1186
орешник-кола 5293
орешник лесной 2098
орешник маньчжурский 3456
орешник медвежий 5646
орешник обыкновенный 2098
орешник понтийский 1071
орешник рвотный 3898
оризопсис 4499
орлеан 256
орлики 1341
орлики обыкновенные 2087
орляк обыкновенный 26
орнитогалум 5228
ороксилум 2945
оронтиум 2533
ортосифон тычиночный 558
орхидейные 3945
орхидея 3944
орхидея-пафиния 4040
орхидея-циприпедиум 3203
орхидные 3945
осенник 1353
осенчук 706
осина 2074
осина американская 4380
осиновик 3940
осиновник 3940
ослинник 2141
ослинник дву(х)летний 1408
ослинниковые 2142
османтус 3973
осмориза 5381
осмунда 4593
осмундовые 3975
осока 4810
осока береговая 5282
осока бледноватая 4010
осока блошиная 2293
осока висячая 4107
осока водяная 5834
осока волосистая 2690
осока волосовидная 2678
осока ворсистая 2690
осока высокая 5656
осока горная 3718
осока двудомная 1860
осока дернистая 5630
осока длиннокорневищная 1595
осока жёлтая 6106
осока заячья 2712
осока звездчатая 5222
осока колосистая 4303
осока коротковолосистая 2690
осока лесная 6042
осока лисья 2348
осока малоцветковая 2212
осока мелкоцветковая 2212
осока метельчатая 4030
осока мохнатая 2690
осока островатая 3516

осока пальчатая 2252
осока песчаная 4688
осока плевельная 4632
осока пузырчатая 674
осока раздвинутая 4482
осока ранняя 5725
осока расставленная 1865
осока растянутая 3364
осока сероватая 4970
осока сизая 1020
осока скальная 4533
осока струннокоренная 1595
осока струнокорневая 1595
осока тёмная 650
осока темноодетая 650
осока топяная 3744
осока удлинённая 2023
осока шариконосная 4182
осока шерстистая 1896
осоковые 4811
осокорь 647
осот 5105
осот болотный 3517
осот жёлтый 2236
осот жёсткий 4305
осот колючий 4305
осот овощной 1531
осот огородный 1531
осот полевой 2236
осот розовый 975
осот шероховатый 4305
остерикум болотный 714
острица 3419
острица лежачая 2459
острица простёртая 2459
островския 2468
острокильница чернеющая 5137
остролист 2039
остролодка 1662
остролодочник 1662
остролодочник полевой 4213
остролодочник уральский 5690
остро-пёстро 669
оттелия 3978
оуратея 3979
офиопогон 3307
офиопогон Ябуран 5898
офиопогон японский 1944
офрис 5657
офрис мухоносная 2320
офрис пчелоносная 519
охна 3907
охрома 5870
очанка 2152
очанка лекарственная 1910
очанка Росткова 1910
очанка Ростковиуса 1910
очанка ростковская 1910
очереднопыльник 3111
очерёт 724
очеретник 496
очеретник белый 5880
очеретник бурый 816
очи царские 6097
очиток 5264
очиток белый 5915
очиток большой 2612
очиток волосистый 2691
очиток гибридный 2148
очиток едкий 2553
очиток "заячья капуста" 3341
очиток испанский 5114
очиток краснеющий 4467
очиток ложный 5673
очиток однолетний 264
очиток отогнутый 3089
очиток пурпурный 4354
очиток пурпуровый 4354
очиток телефиум 3341
очки царские 6097
очный цвет 4185
очный цвет голубой 705
очный цвет нежный 722
очный цвет пашенный 4747
очный цвет полевой 4747
ошак 731

П п

паветта 4076
пави 4082
павия жёлтая 6077
павия калифорнийская 931
павия красная 4441
павловния 4075
павловния войлочная 4595
павловния пушистая 4595
павлония 4074
павой 2623
павун 716
пагудия 4002
падуб 2826
падуб мутовчатый 1560
падуб остролистный 2039
падуб остролистный колючий 2751
падуб парагвайский 4048
падуб рвотный 6068
падуб тусклый 196
падубовые 2827
падук 3999
пажитник 5593
пажитник голубой 711
пажитник греческое сено 2203
пажитник сенной 2203
пазания 5439
пазник 1061
пазник гладкий 5036
пазник голый 5036
пазник укоренившийся 5164
пазник укореняющийся 5164
пайена 4079
пайза 456
пайза 3057
паккой 4005
паклён 2756
паклун 2453
пак-хой 4005
палаквиум 3814
палафоксия 4006
палисандр 648
палиурус 4015
палиурус держи-дерево 1227
палочник 1062
палошник луговой 5528
пальма 4016
пальма арековая 313
пальма арековая 548
пальма-атталея 365
пальма-бактрис 5154
пальма бетель 548

пальма бетельная 548
пальма борассус 4019
пальма бразильская 1021
пальма бразильская восковая 1021
пальма-брахея мексиканская 5358
пальма бурути 5985
пальма веерная 2339
пальма веерная 3229
пальма веерная высокая 2339
пальма веерная европейская 3572
пальма веерная Ливистона 2183
пальма веерная низкая 3572
пальма винная 3537
пальма винная 5540
пальма восковая андийская 5846
пальма восковая бразильская 1021
пальма восковая перувианская 5846
пальма восконосная 1021
пальма гвинейская масличная 46
пальма-гифаена 2001
пальма "дум" 1890
пальма жгучая 5540
пальма зонтичная 5424
пальма-ириартея 5253
пальма кавказская 1368
пальма капустная 2138
пальма капустная 4594
пальма капустная американская 909
пальма-кариота 2265
пальма карликовая 3572
пальма катеху 548
пальма кокосовая 1314
пальма королевская 1721
пальма костяная 1447
пальма лантар 4019
пальма маврикиева 3537
пальма масличная 3916
пальма масличная гвинейская 46
пальма масляная 3916
пальма медовая 5419
пальма нипа 3859
пальма ореодокса 1721
пальма пальмира 4019
пальма пальмировая 4019
пальма пальмирская 4019
пальма пеньковая 5983
пальма персиковая 4085
пальма ползучая 4574
пальма ротанговая 4574
пальма "рыбий хвост" 5540
пальма саговая 5041
пальма саговая Румфа 5157
пальма сахарная 2558
пальма сейшельская 1887
пальма слоновая 1447
пальма слоновая 3113
пальма тал(л)ипотовая 5424
пальма тени 5424
пальма-тодди 5540
пальма финиковая 1807
пальма финиковая культурная 1808
пальма хамеропс 3572
пальметта-сабаль 909
пальмито 3572
пальмовые 4018
пальчатка 2683
паляквиум 3443
пампельмус 4329
панакс 2494
панакс женьшень 346
панакс трёхлистниковый 1940
пангиум 4024
пангиум съедобный 2324
пандан 4779
пандановые 4780
панданус 4779
панданус полезный 1521
панданусовые 4780
панкраций 4023
панкрациум 4023
панус 4033
панус вяжущий 606
панус сморщивающийся 606
папаевые 4034
папайа 4035
папайя 4035
папеда ежеиглистая 3538
папирус 4043
папоротник-адиантум 3430
папоротник-аспидиум 354
папоротник болотный 3504
папоротник водяной 4814
папоротник "волосы Венеры" 5099
папоротник горный пузырчатый 3696
папоротник гребенчатый 1685
папоротник-гроздовник 2577
папоротник древовидный 5580
папоротник дубравный 3797
папоротник женский 3199
папоротник игольчатый 5546
папоротник-костенец 5161
папоротник лесной 3445
папоротник ломкий пузырчатый 786
папоротник мужской 3445
папоротник олений рог 5205
папоротник орляк 5867
папоротник ребристый 1821
папоротник сладкий 1503
папоротник солодковый 1503
папоротник щетинистый 5546
папоротник щитник 5546
папоротник язычный 2197
папоротники 2206
папоротниковые 4321
папоротникообразные 4321
парадизея 4649
парадизка 4046
парило 1348
паринари 4052
паринариум 4052
паркия 3864
паркия африканская 45
паркия двухшаровидная 5670
паркия суданская 5670
пармелиевые 4057
пармелия 4056
парнолистник 498
парнолистниковые 499
парнолистные 499
парохетус 4847
паррия 4062
парротия 4061
парротия персидская 4150
парсония 1736
партениум 4065
партениум серебристый 2657
партеноциссус пятилисточковый 5748
паслён 3853
паслён арбузолистный 3585
паслён баклажан 2414

паслён волосистый 6103
паслён горько-сладкий 604
паслён жёлтый 6103
паслён клубненосный 4281
паслён клювовидный 835
паслён клювообразный 835
паслён лжеперечный 3095
паслён ложноперечный 3095
паслён мохнатый 6103
паслён перцеподобный 3095
паслён сладко-горький 604
паслён чёрный 643
паслёновые 3854
паспалум 4067
паспалум 5703
паспалум помеченный 409
паспалум расширенный 1789
пассифлора 4068
пассифлора гигантская 2475
пассифлора крупная 2475
пассифлора лавролистная 5815
пассифлора плодовая 4070
пассифлора съедобная 4070
пассифлоровые 4069
пастернак 4064
пастернак дикий 5946
пастернак лесной 5946
пастернак посевной 2422
патат 5380
патиссон 883
патриния 4073
пауллиния 4074
пауллиния гуарана 2655
пауловния 4075
пауловния войлочная 4595
паутинник фиолетовый 5739
паучник 5128
пахизандра 3995
пахиризус 6063
пахицентрия 3994
пахлук 3769
пахудия 4002
пахучеколосник 5389
пахучеколосник остистый 265
пахучка 926
пахучка 5395
пахучка аптечная 927
пахучка обыкновенная 5936
пахучка острая 5187
пахучка остролистная 5187
пахучка полевая 5187
пачули 2792
певник болотный 5079
педалиевые 4101
педерота 4000
педилантус 5010
пейреския 4136
пекан 4100
пекан яйцевидный твёрдокорый 4841
пеларгоний 4103
пеларгоний поперечнополосатый 2873
пеларгоний розовый 4566
пеларгониум 4103
пеларгониум головчатый 4568
пеларгониум крупноцветный 575
пеларгониум плющелистный 2991
пеларгония 4103
пеларгония душистая 3897
пеллея 1262
пелтария 4864
пельтандра 326
пельтария 4864
пельтария собачья 2648
пельтифиллум 4731
пельтогине 4348
пелюшка 2231
пеницилл 4109
пенициллиум 4109
пенициллиум сизый 2632
пенициллярия 4108
пеннизетум 4110
пеннисетум 4110
пеннисетум красный 3793
пеннисетум сизый 4092
пентадесма масличная 899
пентаклетра крупнолистная 3984
пентакме 4115
пентас 5216
пентастемон 4117
пентаце 5489
пентстемон 4117
пентстемон гибридный 2718
пентция 4855
пенька бомбейская 3146
пенька индийская 5331
пенька критская 471
пенька манильская 2
пенька новозеландская 3843
пеон 4118
пеон аптечный 1490
пеон древовидный 5585
пеон коралловый 1590
пеон лекарственный 1490
пепельник 1237
пепельник багровый 1385
пепельник приморский 4958
пеперомия 4119
первоцвет 4307
первоцвет-аврикула 368
первоцвет аптечный 1652
первоцвет бесстебельный 2042
первоцвет весенний 1652
первоцвет высокий 3988
первоцвет жёлтый 1652
первоцвет истинный 1652
первоцвет китайский 1197
первоцвет комнатный 5550
первоцвет лекарственный 1652
первоцвет мучнистый 586
первоцвет обыкновенный 2042
первоцвет японский 3064
первоцветные 4308
пережуй-лычко 2194
перекати-поле 397
перекатун 397
перелеска 2783
перелеска благородная 3343
перелеска многолетняя 1876
перелеска однолетняя 2785
перелёт соколий 1700
переския 4136
переския колючая 447
переския шиповатая 447
переступень 824
переступень белый 5883
переступень двудомный 4439
перец 4120
перец американский 1506
перец Бунге 2282
перец ветвистый 884

перец водяной 3511
перец гвоздичный 102
перец душистый 102
перец испанский 884
перец кава 3140
перец китайский 2282
перец красный 884
перец кустарниковый 884
перец многолетний 884
перец настоящий 4120
перец овощной 884
перец однолетний 884
перец стручковый 884
перец стручковый однолетний 884
перец узколистный 3534
перец чёрный 646
перец ягодный 884
перец ямайский 102
перец японский 3063
перечник 993
перечник горький 4523
перечник зонтичный 2504
перечник огородный 2412
перечник посевной 2412
перечные 4121
перилла 4138
перилла базиликовая 1492
перилла многолетняя 1492
перистоголовник 5994
перистрофа 4139
перламутровка 339
перловник 3582
перловник крымский 1693
перловник одноцветковый 6035
перловник поникающий 3868
перловник трансильванский 5570
пернетия 4141
перо водяное 2097
перовския 4142
персея 4144
персея бурбонская 4436
персея индийская 3412
персея красная 4436
персея приятнейшая 183
персик 4082
персик-нектарин 3818
персик обыкновенный 4082
персик степной 4611
персикария восточная 4309
персимон 4152
песколюб 487
песколюб песчаный 2076
песколюбка песчаная 2076
пособой 1353
пестрец 1921
пестрец 5700
песчаник 2094
песчанка 4692
песчанка балеарская 1626
песчанка злаколистная 2584
песчанка тимьянолистная 5511
песчанка траволистная 2584
петалостемум 4290
петрея 4157
петрокаллис 4510
петрофитум 4528
петрушечник 2835
петрушка 4063
петрушка болотная 3509
петрушка горная 3715
петрушка зеленная 1417
петрушка корневая 5652
петрушка обыкновенная 1417
петрушка огородная 1417
петрушка садовая 1417
петрушка собачья 2325
петсай 4158
петуния 4160
петунья 4160
петунья гибридная 1495
петунья ночецветковая 5901
петунья фиолетовая 5742
петушки 2889
петушки степные 1943
петушник 2778
петушок степной 1943
пецай 4158
пецицовые 1734
печать золотая 2546
печать Соломонова 5082
печёночник 2783
печёночница 515
печёночница 2783
печёночница благородная 3343
печёночница обыкновенная 3343
печерица 57
пижма 5410
пижма дикая рябина 1542
пижма душистая 1632
пижма кануфер 1632
пижма обыкновенная 1542
пижма пахучая 1632
пизанг 4219
пикнантемум 3714
пикразма 4383
пикрамния 599
пикрасма 4383
пикрена высокая 3009
пикульник 2778
пикульник-жабрей 781
пикульник колючий 781
пикульник красивый 3846
пикульник ладанниковый 4453
пикульник ладанный 4453
пикульник обыкновенный 781
пикульник пушистый 1895
пикульник узколистный 3805
пилея 1259
пилокарпус 4184
пилокарпус перистолистный 5143
пилолистник 3274
пилолистник чешуйчатый 4738
пилюльница 4183
пилюльница шариконосная 1674
пимелея 4498
пимента 102
пимпинель 5018
пинанг 548
пинанга 4187
пиния 2983
пион 4118
пион аптечный 1490
пион древовидный 5585
пион лекарственный 1490
пион полукустарниковый 5585
пиптадения 4203
пиптадения африканская 48
пиптадения обыкновенная 1497
пиптантус 4204
пираканта 2261

пираканта колючая 4743
пираканиа красная 4743
пираканта узколистная 3801
пираканта ярко-красная 4743
пиратинера гвианская 5049
пиреномицеты 2277
пиретрум 1229
пиретрум 4376
пиретрум альпийский 123
пиретрум бальзамический 1632
пиретрум девичий 2210
пиретрум красный 2306
пиретрум крупнолистный 5441
пиретрум мясокрасный 2306
пиретрум розовый 2306
пиретрум сибирский 3179
пиретрум цинерариелистный 1790
пиретрум щитковый 3569
пиретрум щитконосный 3569
пиретрум ярко-красный 2306
пистия 5816
питанга 4206
питеколобиум 272
питоспорум волнистый 3937
питтоспоровые 5537
питтоспорум 4210
питтоспорум тобира 5538
пихта 2257
пихта алжирская 94
пихта аризонская 1601
пихта бальзамическая 432
пихта белая 4955
пихта благородная 3865
пихта Борнмюллера 737
пихта Бура 5707
пихта великая 2573
пихта великолепная 4451
пихта Вильсона 5978
пихта Вича 5707
пихта гималайская 2800
пихта гребенчатая 4955
пихта греческая 2617
пихта Дугласова 1401
пихта европейская 4955
пихта замечательная 2800
пихта западногималайская 4189
пихта испанская 5110
пихта Каваками 3141
пихта кавказская 3876
пихта калифорнийская 779
пихта канадская 432
пихта киликийская 1235
пихта корейская 3181
пихта красивая 4451
пихта Лоуа 3997
пихта мексиканская 4635
пихта миловидная 1039
пихта Нордман(н)а 3876
пихта нубийская 94
пихта нумидийская 94
пихта обыкновенная 4955
пихта одноцветная 5895
пихта прелестная 779
пихта равночешуйчатая 3855
пихта сахалинская 4655
пихта священная 4635
пихта сибирская 4904
пихта сильная 3649
пихта Факсона 2191
пихта Фразера 2364
пихта цельнолистная 3462
пихта чешуйчатая 2271
пихта японская зонтичная 5682
плавун нимфейный 2378
плагиантус 5659
плагиохила 4211
плакун 3400
плакун иволистный 4349
плакун лозовый 5782
плакун прутовидный 5782
плакун-трава 4349
пламенник 4166
планера 5800
планера водная 5801
пластинниковые 2488
пластинчатые 2488
платан 4215
платан американский 205
платан восточный 3959
платан западный 205
платан испанский 5113
платан клёнолистный 3361
платан Линдена 3006
платан мексиканский 3601
платан обыкновенный 3959
платановые 4216
платимисциум 3402
платистемон 4222
платицериум 1534
платония 411
плаун 1285
плаун альпийский 127
плаун-баранец 2258
плаун болотный 3503
плаун булавовидный 4606
плаун годичный 2954
плаун годовалый 2954
плаун заливаемый 3503
плаун обыкновенный 4606
плауновые 1286
плаунок 4814
плаунок Мартенза 3525
плевел 4631
плевел итальянский 2980
плевел многолетний 4135
плевел многоцветковый 2980
плевел многоцветный 2980
плевел одуряющий 1804
плевел опьяняющий 1804
плевел расставленный 2708
плеврогина 2198
плевропогон 4819
плектогине 1352
плесени водяные 5824
плесень 3691
плесень зелёная 2631
плесень леечная 351
плесень сизая 2632
плеуропогон 4819
плюмбаговые 3246
плюмерия 2360
плютей олений 2189
плюхея 4223
плющ 2988
плющ амурский 347
плющ восковой 1558
плющ канадский 1472
плющ колхидский 1324
плющ крымский 1691
плющ обыкновенный 2040

повий 6016
повилика 1870
повилика Гронова 2641
повилика Гроновиуса 2641
повилика душистая 1153
повилика европейская 2607
повилика клеверная 1280
повилика льновая 2287
повилика льняная 2287
повилика тимьянная 1400
повилика тимьяновая 1400
повилика чебрецовая 1400
повиликовые 1871
повитель 1926
поводник 2673
повой 2509
повой заборный 2748
повой круглолистный 4804
повой подстенный 5773
повойник 5841
повойничек 5841
повойничек трёхтычинковый 5501
повойничек шеститычинковый 4982
повойничковые 5842
поганка бледная 1820
поганка весенняя 5180
поганка цезарева 918
погостемон 911
погремок 4426
погремок 4430
погремок петушиный гребешок 1308
подбел 888
подбел 241
подбел гибридный 4340
подбел дубровник 723
подбел лекарственный 4340
подбел лечебный 4340
подбел многолетний 723
подбел обыкновенный 4340
подбережник 3400
подбережник 4349
подберёзовик 582
подберёзовик обыкновенный 582
подвишенник 4226
подвишень 4226
подгруздок белый 5923
подгруздок чернеющий 628
поддубник 3396
поддубовик 3396
подковник 2874
подладанник 1781
подлесник 4693
подлесник европейский 2121
подмаренник 510
подмаренник болотный 3500
подмаренник герцинский 2743
подмаренник крестовидный 1704
подмаренник лесной 6021
подмаренник мягкий 5882
подмаренник настоящий 6075
подмаренник парижский 4054
подмаренник северный 3878
подмаренник скальный 2743
подмаренник топяной 5341
подмаренник трёхдольный 5017
подмаренник трёхнадрезный 5017
подмаренник трёхраздельный 5017
подмаренник трёхрогий 1607
подмаренник трёхцветковый 5384
подмаренник цепкий 1055
подмолочник 3938
подмошник 6081
подобабик 582
подобабок 582
подокарпус 4235
подорешник 1114
подорешник 3938
подорожник 4218
подорожник азиатский 348
подорожник альпийский 162
подорожник блошный 2290
подорожник большой 4502
подорожник горный 3717
подорожник индийский 5926
подорожник ланцетный 829
подорожник ланцетовидный 829
подорожник ланцетолистный 829
подорожник морской 2123
подорожник перистый 1708
подорожник песчаный 2290
подорожник приморский 2123
подорожник средний 5379
подорожниковые 4220
подосиновик 3940
подосиновик красный 3940
подостемон 4506
подофилл 3539
подофилл гималайский 2804
подофилл щитовидный 1845
подофилл Эмода 2804
подофиллум 3539
подофиллюм 3539
подсвекольник 4464
подснежник 5060
подснежник белый 1530
подснежник снеговой 1530
подснежник снежный 1530
подсолнечник 5329
подсолнечник абиссинский 2063
подсолнечник клубненосный 3094
подсолнечник культурный масличный 1539
подсолнечник обыкновенный 1539
подсолнечник огурцеобразный 1725
подсолнечник однолетний 1539
подсолнечник серебристолистный 4963
подсолнечник слабый 1725
подъельник 2940
поздника 643
поинсеттия 1500
поинсеция 1500
полба 5124
полба волжская 2029
полба двузёрная 2029
полба настоящая 5124
полба ненастоящая 2029
полба обыкновенная 2029
полевица 534
полевица белая 4469
полевица белая высокая 4469
полевица болотная 1665
полевица волосовидная 1335
полевица гигантская 617
полевица обыкновенная 1335
полевица ползучая 1665
полевица приморская 1665
полевица собачья 5708
полевица собачья опушённая 5708
полевица тонкая 1335
полевичка 3385
полевичка абиссинская 5473

полевичка волосистая 2931
полевичка крупная 5255
полевичка крупноколосковая 5255
полевичка малая 3331
полевичка опушённая 5255
полевичка тефф 5473
полёвка 3773
полевой очный цвет 4747
полей 4113
поленика 305
полиальтия 2930
полигонум 3172
полимния 3248
политрихум 2676
политрихум обыкновенный 1435
поллиния 5303
пололепестник 2673
пололепестник зелёный 2384
полетай 5951
полуклевер 638
полуница 2635
полуполба 2029
полушилица 4387
полушник 4387
полушниковые 4388
полынёк 2554
полынёк калифорнийский 943
полынь 4640
полынь-абсент 1564
полынь божье дерево 3923
полынь веничная 3966
полынь горькая 1564
полынь звёздчатая 489
полынь каменная 4540
полынь лечебная 3923
полынь метельчатая 3966
полынь обыкновенная 3747
полынь однолетняя 5396
полынь полевая 4641
полынь понтийская 4548
полынь приморская 3496
полынь равнинная 4641
полынь скалистая 4540
полынь скальная 4540
полынь цитварная 3284
полынь-чернобыль 3747
полынь эстрагон 5445
польба 5124
польба настоящая 5124
поляника 305
помадерис 5746
помароза 4556
помело 4329
померанец 5092
померанец горький 5092
помидор 5542
помидор съедобный 1549
поммело 4329
помпельмус 4329
помпельмус гроздевидный 2578
понтедериевые 4175
понтедерия 4174
понцирус трёхлисточковый 5592
поплавок влагалищный 4853
поплавок серый 4853
поплавок шафранный 4853
поповник 3987
поповник девичий 2210
поповник обыкновенный 1542
поповник посевной 1609
поповник сибирский 3179
поповник цинерариелистный 1790
поповник щитковидный 3569
поповник щитковый 3569
поповник щитконосный 3569
порезник 3286
порезник горный 3713
порей 3266
поречник 3715
портулак 4358
портулак декоративный 1505
портулак крупноцветковый 1505
портулак крупноцветный 1505
портулак овощной 1509
портулак огородный 1509
портулакария 4361
портулаковые 4359
поручейник 5828
поручейник сахарный 4986
поручейник сахарный корень 4986
поручейник узколистный 3811
поручейница водная 806
поручейница водяная 806
порфира 3241
порховка свинцово-серая 3244
посконник 732
посконник коноплевидный 2776
посконник коноплёвый 2776
посконник конопляный 2776
посконь 2773
постенница 4105
постенница аптечная 5773
постенница лекарственная 5773
поташник 1518
потентилла 1242
поточник 509
початковые 5107
почечуй 5168
почечуйник 5168
прангос высокий 2731
прангос кормовой 2731
пренантес 4429
приболотник фиолетовый 5739
прибрежник 4879
прибрежник одноцветковый 2124
прибрежница 4879
прибрежница одноцветковая 2124
приворот 63
примула 4307
примула аптечная 1652
примула высокая 3988
примула китайская 1197
примула мучнистая 586
примула настоящая 1652
примула обконика 5550
примула обыкновенная 2042
примула японская 3064
приноготковник 3788
приноготовик 3788
приноготовник 3788
принсепия 4310
притчардия 4311
прицепник 2166
прицепник морковновидный 2753
прицепник морковный 2753
пробосцидея 1846
прозопис 3593
прозопис серёжкоцветный 1467
пролеска 3589
пролеска 5200

пролеска двулистная 5664
пролеска итальянская 2982
пролеска многолетняя 1876
пролеска однолетняя 2785
пролеска осенняя 383
пролеска прелестная 5223
пролеска сибирская 4916
пролесник 3589
пролесник многолетний 1876
пролесник однолетний 2785
проломник 4527
проломник альпийский 144
проломник швейцарский 5401
просвирник 3447
просвирник курчавый 1743
просвирник лесной 2797
просвирник мутовчатый 1294
просвирниковые 3448
просвирняк 3447
просвирняк китайский 1294
просвирняк красивый 1294
просвирняк круглолистный 1945
просвирняк курчавый 1743
просвирняк лесной 2797
просвирняк мавританский 3576
просвирняк мускусный 3760
просвирняк мутовчатый 1294
просвирняк низкий 1945
просвирняк штокрозовый 2830
просвирняковые 3448
просвирнячок 4925
просо 4026
просо африканское 4092
просо ветвистое 818
просо волосовидное 1561
просо гвинейское 2661
просо гигантское 2661
просо головчатое 2352
просо итальянское 2352
просо китайское 3057
просо кровяное 2683
просо крупное 2661
просо куриное 456
просо метельчатое 4315
просо мучнистое 456
просо настоящее 456
просо негритянское 4092
просо обыкновенное 4315
просо петушье 456
просо посевное 456
просо посевное 4315
просо прутовидное 5403
просо прутьевидное 5403
просо татарское 2352
просо японское 3057
прострел 243
прострел весенний 5721
прострел луговой 3554
прострел луговой раскрытый 5177
прострел поникающий 5177
прострел раскрытый 5177
прострел распростёртый 5177
прострел сон-трава 5177
просяник 456
просяник 2461
протея 4318
протея медоносная 5308
прутняк 1120
прутняк веничный 531
прутняк обыкновенный 3296
прутовник ситниковидный 5857
прыгун 5560
прыщинец 5121
псевдобактерия 4319
псевдонарцисс 1397
псевдотсуга 1888
псидиум 2656
псидиум лавролистный 1631
психотрия 5945
псоралея 4782
птармика 5054
птелея 2851
птелея трёхлистная 1441
птерис 754
птерис критский 1686
птерокактус 5987
птерокария 5996
птерокария рябинолистная 1077
птерокарпус 3999
птерокарпус габонский 47
птерокарпус дальбергиевидный 237
птерокарпус драконовый 1902
птерокарпус индийский 863
птерокарпус крупноплодный 864
птерокарпус палисандровидный 237
птерокарпус сандаловый 4679
птерокарпус санталовый 4679
птерокарпус Сойакси 47
птероксилон 5053
птероцелтис 5988
птероцефалюс 5994
птихосперма 4322
птицемлечник 5228
птицемлечник зонтичный 1535
птицемлечник пиренейский 478
птицемлечник поникший 3870
птиценожка 1526
пуанзеция 4239
пуговичник 1542
пуерария 3188
пузатник 3863
пузыреплодник 3857
пузыресеменник 88
пузыресемянник 88
пузырник 657
пузырник 661
пузырник 662
пузырник восточный 3950
пузырник горный 3696
пузырник древовидный 1362
пузырник кровавый 3950
пузырник ломкий 786
пузырник обыкновенный 1362
пузырчатка 664
пузырчатковые 665
пузырчатник 3857
пулазан 4327
пулазан плодовый 4327
пуликария 2291
пульсатилла 2116
пуммело 4329
пупавка 963
пупавка благородная 4546
пупавка вонючая 3542
пупавка красильная 2528
пупавка обыкновенная 3542
пупавка полевая 2220
пупавка собачья 3542
пупочник 3815
пупочник весенний 1672

пустоягодник 5274
пустынноколосник 1839
пустырник 3689
пустырник обыкновенный 1475
пустырник сердечный 1475
пуховник 4291
пуходревник 349
пухонос дернистый 1822
пуцинелла приморская 4803
пушица 1637
пушкиния 4362
пуэрария 3188
пуэрария волосистая 5505
пуэрария опушённая 5505
пуэрария Тунберга 5505
пчелка 519
пшат 4616
пшеница 5873
пшеница абиссинская 2064
пшеница английская 4286
пшеница ветвистая 756
пшеница волжская 2029
пшеница двузернянка 2029
пшеница двузернянковидная 5961
пшеница-ежовка 1288
пшеница карликовая 1288
пшеница круглозёрная 1288
пшеница культурная однозернянка 2004
пшеница летняя 5874
пшеница мягкая 5874
пшеница мягкая карликовая 1288
пшеница мягкая яровая 1288
пшеница обыкновенная 5874
пшеница однозернянка 2004
пшеница однозернянка культурная 2004
пшеница плотная 1288
пшеница полоникум 4247
пшеница польская 4247
пшеница-спельта 5124
пшеница твёрдая 1930
пшеница тургидум 4286
пшеница тучная 4286
пыльцеголовник 4163
пырей 1640
пырей волосистый 5252
пырей волосоносный 5252
пырей гребенчатый 1684
пырей ползучий 4379
пырей промежуточный 2952
пырей русский 5432
пырей Смита 707
пырей средний 2952
пырей удлинённый 5432
пьяница 717
пэдерия 2211
пятилистник 3502

Р р

равенала 5572
равенала мадагаскарская 3409
равноплодник 2970
раги 4399
радиола 2289
радиола льновидная 101
раздулка 2251
разнолепестка 993
разнолепестник 993
разнолепестник горький 4523
разнолистник 993
разноорешник 4269
разноцвет 3226
разрыв-трава 4730
разумовския 1947
райграс 4631
райграс английский 4135
райграс высокий 5427
райграс итальянский 2980
райграс многолетний английский 4135
райграс многолетний пастбищный 4135
райграс многоукосный 2980
райграс пастбищный 4135
райграс французский 3905
райграс французский высокий 5427
рай-дерево 1527
райка 4046
ракита 2523
ракитник 809
ракитник австрийский 375
ракитник альпийский 4773
ракитник белый 4276
ракитник головчатый 557
ракитник-золотой дождь 2529
ракитник метельчатый 4769
ракитник метлистый 4769
ракитник многоцветковый 5910
ракитник пурпурный 4339
ракитник пурпуровый 4339
ракитник чернеющий 5137
рамалина 4407
рамбутан 4409
рами 2174
рами белое 4410
рамишия однобокая 4926
рамишия однобочная 4926
рамонда 4411
рамондия 4412
рамтила 2063
рандия 3439
ранетка пурпурная 4899
ранник 1503
ранункул 889
ранункулюс 889
рапидофиллум 3821
рапидофиллюм 3821
рапис 3201
рапонтика 2426
рапонтикум 4486
рапс 6005
рапс 4417
рапункул 4414
рапунцель 4414
рапутник 1408
расамала 4419
растигор 1113
расторопша остропёстрая 669
расторопша пятнистая 669
расходник 4387
раувольфия 1844
раулия 4416
рафия 4396
рафия мадагаскарская 3408
рафлезиевые 4397
раффия 4396
раффлезиевые 4397
раффлезия 3664
раяния 5623
рдест 4262
рдест блестящий 4866
рдест взморниколистный 2283

рдест гребенчатый 2202
рдест густолистный 1833
рдест длиннейший 5914
рдест злаковидный 5702
рдест зостеролистный 2283
рдест курчавый 1744
рдест маленький 401
рдест плавающий 2298
рдест плотный 1833
рдест пронзённолистный 5495
рдест разнолистный 5702
рдест стеблеобъемлющий 5495
рдест туполистный 712
рдест Фриса 2376
рдестовые 4263
ревень 4492
ревень аптечный 3566
ревень волнистый 5844
ревень дланевидный 5089
ревень китайский 3566
ревень лекарственный 3566
ревень московский 3566
ревень овощной 2426
ревень огородный 2426
ревень пальчатый 5089
ревень русский 3566
ревень туполистный 2426
ревень черноморский 2426
ревень Эмоди 2805
редечник 4418
редис 2425
редис 5030
редька 4395
редька дикая 5955
редька европейская 2425
редька огородная 2425
редька полевая 5955
редька посевная 2425
реедия 4487
реедия бразильская 410
реедия обыкновенная 2698
резак 2155
резак обыкновенный 1409
резанец 1219
резеда 3614
резеда белая 5900
резеда душистая 1468
резеда жёлтая 6100
резеда желтенькая 5862
резеда желтоватая 5862
резеда красильная 5862
резеда пахучая 1468
резедовые 3615
резуха 4516
резуха альпийская 143
резуха башенная 5565
резуха беловатая 5774
резуха гладкая 5566
резуха Жерара 2450
резуха кавказская 5774
резуха шершавая 2688
резуховидка Таля 3733
резушка Таля 3733
рейнвардтия 6087
рейнекия 4480
рексия 3545
ремания 4478
рёмерия 344
рёмерия гибридная 5741
ремнелепестник 3346
ремнецветник 3377
ремнецветниковые 3632
ремнецветные 3632
ренклод 2627
репа 5650
репа венгерская 3175
репа волошская 3094
репа кервельная 5651
репа культурная 5650
репа огородная 5650
репа японская 5650
репак 585
репей 2601
репей дикий 124
репейник 2601
репейник аптечный 1348
репейник волосистый 2679
репейник исполинский 2601
репейник обыкновенный 1348
репейник пахучий 2355
репейничек 63
репейничек аптечный 1348
репейничек волосистый 2679
репейничек пахучий 2355
репешок 63
репешок обыкновенный 1348
репник 4418
репник морщинистый 4603
репяшок 63
репяшок обыкновенный 1348
рестио 4552
решетник 4846
ржавчинные 4623
ржанец 2053
ржанец 4379
ривина 4575
ризоктония 4489
ризофора 3475
ринхоспора 496
рипария 4503
рипсалис 4488
рис 4495
рис болотный 266
рис водяной 266
рис дикий 266
рис индийский 266
рис канадский 266
рис озёрный 266
рис посевной 4496
рисовидка 4499
рисовидка 5032
рихардия 950
рихардия африканская 1374
рицин 1050
рицина 1051
рицинник 1051
рицинодендрон африканский 3770
рицинус 1050
ричардия 950
ричардия шершавая 3596
риччия 4494
робиния 3354
робиния клейкая 1252
робиния лжеакация 636
робиния ложноакация 636
робиния новомексиканская 3841
робиния щетинистоволосая 4555
робиния щетинистоволосистая 4555
робиния щетиноволосая 4555
рогатик 2520

рогатик гроздевой 4468
рогатик жёлтый 4013
рогатик золотистый 2531
рогатиковые 1588
рогач 1097
рогачка 4526
роговик униоловидный 4483
рогоз 1062
рогоз узколистный 3800
рогоз широколистный 1380
рогозовые 1063
роголистник 2860
роголистник погружённый 2861
роголистник тёмно-зелёный 2861
роголистниковые 2862
рогомак 2859
рогульки 2330
рогульник 5794
рогульник плавающий 5794
роджерия 4543
родимения 4791
родиола 5264
родиола розовая 4567
рододендрон 4490
рододендрон большой 4557
рододендрон древовидный 4557
рододендрон древовидный 5587
рододендрон жёлтый 4264
рододендрон жестковолосистый 2430
рододендрон индийский 2933
рододендрон крупнейший 4557
рододендрон лапландский 3220
рододендрон мягкий 1170
рододендрон понтийский 4264
рододендрон ржаволистный 4530
рододендрон ржавый 4530
родомирт 4565
родотипус 3099
родотипус красивый 635
родянка 4567
рожа 2829
рожа дикая 3507
рожа собачья 5507
рожечник 1023
рожки маточные 2053
рожки цареградские 1023
рожки царьградские 1023
рожки чёрные 2053
рожь гималайская 4247
рожь индийская 572
рожь культурная 4630
рожь полярная 2094
рожь посевная 4630
роза 4554
роза азиатская 4886
роза Альберта 78
роза альпийская 145
роза Банкса 443
роза банксиевая 443
роза Беггера 522
роза белая 1634
роза бенгальская 1200
роза бурбонская 745
роза бургундская 861
роза Бэнкса 443
роза виргинская 5751
роза Вихуры 5929
роза войлочная 5543
роза вонючая 374
роза Гаррисона 2715
роза гвоздичная 4753
роза дамасская 1792
роза даурская 1785
роза жёлтая 374
роза иерихонская 3091
роза индийская 1200
роза казанлыкская 1792
роза карликовая 4580
роза китайская 1185
роза китайская 1200
роза кожистолистная 3258
роза колючейшая 4776
роза коричная 1240
роза майская 1240
роза Макартова 3401
роза мелкоцветковая 5025
роза месячная 2143
роза миниатюрная 4580
роза многоцветковая 3068
роза морщинистая 4602
роза моховая 3685
роза моховидная 3685
роза мускусная 3767
роза мягкая 685
роза нутканская 3875
роза пашенная 2234
роза полевая 2234
роза полевая 2583
роза полиантная 1792
роза полиантовая 1792
роза помпонная 4256
роза прицветниковая 3401
роза ремонтантная 1792
роза ржавая 5359
роза сизая 4454
роза сирийская 4886
роза собачья 1875
роза столепестная 910
роза столистная 910
роза французская 2371
роза чайная 5468
роза чайная китайская 1200
роза чайная настоящая 5468
роза Шарона 4886
роза шотландская 4776
роза щитконосная 1627
роза эглантерия 5359
роза эллиптическолистная 4753
роза Юндзилла 3491
роза яблочная 287
розанные 4560
розга золотая 2101
розелла 4562
розель 4562
розеточница 5262
розмарин 4563
розмарин аптечный 4564
розмарин лекарственный 4564
розовик 3099
розовик керриевидный 635
розовик японский 635
розовые 4560
розоцветные 4560
рокамболь 2474
рокет-салат 4525
роксбургия 4592
рокцелла 4509
рокцелля красильная 3325
романцовия 3633
ромашка 3541

ромашка аптечная 2452
ромашка безлепестная 4193
ромашка вонючая 3542
ромашка далматская 1790
ромашка девичья 2210
ромашка дикая 2452
ромашка душистая 4193
ромашка жёлтая 2528
ромашка кавказская 2306
ромашка клоповая 4988
ромашка клоповная 4988
ромашка красильная 2528
ромашка лекарственная 2452
ромашка настоящая 2452
ромашка немецкая 2460
ромашка непахучая 4751
ромашка обыкновенная 2452
ромашка пахучая 4193
ромашка персидская розовая 2306
ромашка посевная 1609
ромашка римская 4546
ромашка розовая 2306
ромашка ромашковидная 4193
ромашка собачья 3542
ромашник 1229
ромэн 4545
ронделетия 4549
ропалостилис 4491
росичка 2254
росичка кровоостанавливающая 5037
росичка кровяная 2683
росичка линейная 5037
росница 5325
росяника 2080
росянка 5325
росянка английская 2044
росянка круглолистная 4588
росянковые 5326
ротала 4573
ротанг 4574
рочелла 4509
рочелла красильная 3325
роэлла 4542
рудбекия 1573
рудбекия пурпурная 4344
рудбекия раздельнолистная 1755
рудбекия рассечённая 1755
рудбекия рассеянная 1755
рудбекия шершавая 629
руеллия 3478
руизия 4604
руишия 4628
румянка 5744
румянка обыкновенная 1555
руппия 5933
руппия морская 5934
рупрехтия 5745
рускус колючий 885
рута 4597
рута душистая 1516
рута обыкновенная 1516
рута пахучая 1516
рута постенная 5776
рута садовая 1516
рутовые 4599
рыжей 1827
рыжей 2168
рыжей мелкоплодный 3334
рыжей яровой 572
рыжик 1827
рыжик 2168
рыжик красный 1827
рыжик мелкоплодный 3334
рыжик посевной 572
рыжик сосновый 1827
рыжик яровой 572
рыскун 4772
рябина 3695
рябина американская 202
рябина амурская 231
рябина-ария 5881
рябина гибридная 3902
рябина-глоговина 1128
рябина двуцветная 5058
рябина дикая 1542
рябина домашняя 4833
рябина китайская 5058
рябина лечебная 1128
рябина ликёрная 2114
рябина лжеакация 636
рябина ложноакация 636
рябина Мугеотта 1995
рябина обыкновенная 2114
рябина ольхолистная 1831
рябина пёстрая 5058
рябина промежуточная 5350
рябина садовая 4833
рябина скандинавская 5350
рябина японская 3058
рябинник 2180
рябинник Линдлея 3318
рябинник рябинолистный 5691
рябишник 1542
рябчик 2380
рябчик горный 4997
рябчик камчатский 3130
рябчик садовый 1711
рябчик тонкий 4997
рябчик турецкий 1711
рябчик царский 1711
рябчик шахматный 1126
рядовка зелёная 2051
рядовка серая 1859
рядовка фиолетовая 214
ряска 1924
ряска горбатая 5404
ряска малая 1403
ряска маленькая 1403
ряска меньшая 1403
ряска трёхдольная 5217
ряска тройчатая 5217
рясковые 1925
ряст 5200

С с

сабадилла 4633
сабадилла аптечная 1914
сабадилла лекарственная 1914
сабаль 4017
сабаль Адансона 1948
сабаль малый 1948
сабаль пальметто 909
сабаль пальмовидный 909
сабатия 4561
сабельник болотный 3502
савой 4722
саговник 1764
саговник поникающий 4643
саговниковые 1765

садовик 4226
саза 4710
саксаул 4729
саксифрага 4730
сакура 3951
сал 4661
саламалик 3893
салат 3281
салат алжирский 52
салат вонючий 603
салат дикий 4299
салат кочанный 2735
салат морской 1524
салат огородный 2417
салат полевой 2088
салат посевной 2417
салат-розель 4562
салат цикорный 2032
салат эндивий 2032
сальвадора 4669
сальвиниевые 4672
сальвиния 4671
сальвия 4639
сальвия блестящая 4749
сальпиглоссис 5624
саман 4404
самандура 4673
самандура индийская 3848
самбук 2009
самолюс 807
самолюс Валеранда 808
самолюс обыкновенный 808
самоцвет 1108
самшит 749
самшит вечнозелёный 1368
самшит японский 3054
самшитовые 751
санвиталия 4697
сангвинария 681
сангвинария канадская 682
сангуинария 681
сандал 4677
сандал белый 5908
сандал индийский 863
сандал красный 4679
сандал синий 3359
сандаловые 4678
сандарак 1777
сандерсония 4685
сандорик 4695
сандорик индийский 4696
санзеверия 4694
сансевиерия 4694
сансевиерия гвинейская 5382
сансевиерия кистецветковая 5382
сансевиерия цейлонская 1106
сансевиерия цилиндрическая 2918
сансевьера 4694
сансевьерия 4694
санталовые 4678
сантолина 3240
сантолина волосистовойлочная 1776
сантолина кипарисовидная 1776
сантолина кипарисовниковидная 1776
сантполия 1347
сапиндовые 5073
сапиндус 5072
сапиндус мыльный 5102
сапиум 4699
сапиум салоносный 1204
саподилла 4700
сапожки 3203
сапонария 5074
сапонария аптечная 743
сапонария лекарственная 743
сапотовые 4701
сарака 4704
сарана 2334
сарана большая 3524
саранка 3524
саргассум 4705
саркодон черепитчатый 2922
саррацениевые 204
саррацения 4207
саррацения пурпурная 1499
сарсапарель 2623
сассапариль 2623
сассапариль китайская 1166
сассафрас 4712
сассафрас лекарственный 1520
сатурея 4721
сауроматум 3345
сауруровые 3348
саурурус 3347
сафлор 4636
сафлор красильный 4637
сафлор культурный 4637
сафой 4722
сахаромицеты 6070
сведа 4813
сведа морская 4802
сведа приморская 4802
свейнсона 5339
свёкла 520
свёкла дикая 4785
свёкла кормовая 2322
свёкла кормовая 3471
свёкла листовая 3473
свёкла листовая артишоковая 5398
свёкла обыкновенная 1360
свёкла приморская 4785
свёкла сахарная 5298
свёкла столовая 2408
свёкла столовая красная 4438
свекловица сахарная 5298
свербига 1617
сверция 5397
сверция многолетняя 119
свеча ночная 1408
свидина 683
свидина белая 5454
свидина кроваво-красная 683
свидина отпрысковая 4458
свидина порослевая 4458
свидина сибирская 4901
свидина татарская 4901
свиетения 3427
свинорой 1879
свинорой пальчатый 542
свинуха 4078
свинушка 4078
свинушка бурая 2956
свинушка тонкая 2956
свинцовка 4225
свинчатковые 3246
свинчатник 4225
северница 807
северница обыкновенная 808
седач 2776
седаш 2776

седаш коноплевидный 2776
седмичник 5218
седмичник европейский 2125
седум 5264
сезам 3961
сезамовые 4101
сейба пятитычинковая 3134
секвойя 4829
секвойя вечнозелёная 945
секвойя гигантская 2483
секуригера 2720
секуринега 4809
селагинелла 4814
селагинелловые 4815
селезёночник 2545
селезёночник обыкновенный 168
селезёночник очереднолистный 168
селезёночник супротивнолистный 3935
селин 4817
селитрянка 3862
сельдерей 1086
сельдерей душистый 5943
сельдерей лиственный 2411
сельдерей листовой 2411
сельдерей пахучий 5943
сельдерей стеблевой 2411
сельдерей черешковый 2411
селягинелла 4814
селягинеллевые 4815
селягинелля 4814
селягинелля Мартенза 3525
семпервивум 2882
семул 3440
семя блошное 2290
семя вшивое 1914
сенна 4823
сенна остролистная 89
сено греческое 2203
сераделла 587
сераделла маленькая 3337
сераделла очень маленькая 3337
сераделла посевная 1526
сердечки 668
сердечник 600
сердечник луговой 1723
сердечник мелкоцветковый 5021
сердечник мелкоцветный 5021
сердечник шершавый 4111
серпник 4674
серпуха 4728
серпуха красильная 1955
серьяния 5335
сесбания 4835
сесбания египетская 4081
сеслерия 3675
сеслерия голубая 704
сеточка водяная 5826
сехиум 1122
сиббальдия 4889
сиббальдия распростёртая 4890
сибирка 4918
сиверсия 4941
сивец луговой 3561
сивец южный 5103
сигезбекия 4653
сида 4925
сидальцея 1127
сидеритис 2967
сидеродендрон 4928
сидероксилон 3117
сизал 4976
сизаль 4976
сизигиум яванский 3010
сизон 2835
сизюринхий 698
сизюринхий узколистный 1364
сикамор античный 5408
сикими 3014
сикомор 205
сикомор 5408
сикомора 4217
силена 1054
силене 1054
сильфиум 4571
сильфия 4571
симаруба 4971
симаруба горькая 3968
симаруба сизая 4047
симарубовые 66
симплокарпус 4990
симплоковые 5374
симплокос 5373
симфония 5414
симфония габонская 2398
синадениум 3618
синеголовник 2058
синеголовник альпийский 709
синеголовник морской 4789
синеголовник полевой 5047
синеголовник приморский 4789
синеголовник равнинный 5047
синетал 1801
синильник 1956
синюха 4246
синюха голубая 2620
синюха лазоревая 2620
синюха лазуревая 2620
синюха лазурная 2620
синюха обыкновенная 2620
синюховые 4167
синявка 214
синявка 3674
синяк 5744
синяк изогнутый 5103
синяк обыкновенный 1555
синяк подорожниковый 4357
сирень 3295
сирень амурская 228
сирень венгерская 2890
сирень гималайская 2803
сирень индийская 1394
сирень китайская 1190
сирень лесная 2194
сирень мохнатая 2686
сирень обыкновенная 1461
сирень опушённая 2686
сирень персидская 4149
сирень поникшая 3867
сирень угорская 2890
сирень Эмоди 2803
сирень японская 3075
сисюринхий 698
сисюринхий узколистный 1364
ситник 4608
ситник альпийский 146
ситник бальтийский 436
ситник Герарда 4666
ситник головчатый 1010
ситник Жерарда 4666
ситник искривлённый 2704

ситник луковичный 845
ситник лягушечий 5535
ситник лягушечный 5535
ситник морской 4800
ситник нитевидный 5498
ситник растопыренный 2747
ситник расходящийся 1517
ситник склонённый 2704
ситник склоняющийся 2704
ситник сплюснутый 4582
ситник тонкий 4288
ситник туполепестный 3906
ситниковидные 3116
ситниковые 4609
ситняг 5145
ситняг болотный 1533
ситовник 4364
сифонантус 5615
сифония 4975
сифонодон 2987
сифонодон южный 372
сициос 858
сициос угловатый 5765
скабиоза 4735
скабиоза бледно-жёлтая 1663
скабиоза голубиная 1891
скабиоза жёлтая 1663
скабиоза желтоватая 1663
скабиоза светло-жёлтая 1663
скабиоза тёмно-пурпурная 5383
скабиоза южная 5103
скамоний 4741
скандикс Венерин гребень 4861
скандикс гребенчатый 4861
скерда 2725
скерда болотная 3505
скерда волосовидная 5038
скерда дву(х)летняя 4577
скерда зелёная 5038
скерда колосовидная 5038
скерда кровельная 3803
скимия 4984
скимия японская 3069
скипетр царский 1117
скипидарник 5647
склерия 4431
склеротиния 4764
сколимус 3993
сколимус испанский 5112
сколопендриум обыкновенный 2716
скополия 4765
скополия карниолийская 2122
скордия 5806
скорода 1219
скорпионница 4767
скорцонер 649
скорцонер испанский 649
скребница 4737
скрипица пёстрая 4051
скрипун 3341
скрипуха 4577
скрученник 3197
скрученник осенний 1455
скрытница 4294
скрытница канадская 2836
скрытосеменные 248
скрытосемянные 248
скулатник 4251
скумпия 5033
скумпия кожевенная 1527
сладкокорень 1503
сладконожник 3066
сланоягодник 4786
слёзки богородицины 31
слёзки Иова 31
слёзки кукушкины 4398
слёзник 31
слёзы Иовы 31
слива 4224
слива абрикосовая 289
слива абрикосовидная 289
слива алыча растопыренная 5950
слива американская 206
слива Ватсона 4684
слива вишнелистная 3777
слива домашняя 2424
слива европейская 2424
слива золотая 2914
слива иволистная 3061
слива ивообразная 3061
слива изящная 3919
слива испанская тропическая 6101
слива канадская 978
слива китайская 3061
слива кокосовая 1315
слива колючая 5014
слива курдистанская 321
слива кустарниковая 2644
слива морская 488
слива ненастоящая 847
слива обыкновенная 2424
слива полусердцевидная 3164
слива растопыренная 5950
слива ренклод 2627
слива Симонова 289
слива трёхлопастная 2311
слива узколистная 1143
слива финиковая 3127
слива чёрная 978
слива чёрная мелкая 847
слива Чикаса 1143
слива яванская 3010
слива явская 3010
слива японская 3061
слизевики 5006
сложноцветные 1570
слонея 5013
смертоед 2766
"смесь барская" 3451
смилакс 2623
смилаксовые 5031
смилацина 5081
смирния 87
смоква 2240
смоква индийская 2936
смоковница 2240
смоковница бенгальская 444
смоковница дикая 5408
смоковница индийская 2929
смоковница обыкновенная 1412
смоковница райская 4219
смоковница священная 738
смолёвка 1054
смолёвка альпийская 149
смолёвка английская 2372
смолёвка армериевидная 5394
смолёвка бесстебельная 3686
смолёвка вильчатая 2328
смолёвка зеленоцветковая 6092
смолёвка итальянская 2981

смолёвка коническая 1576
смолёвка повислая 1905
смолёвка поникающая 3869
смолёвка поникшая 3869
смолёвка Рупрехта 4607
смолёвка сибирская 4895
смолёвка татарская 5459
смолёвка-хлопушка 663
смолка альпийская 306
смолка клейкая 1250
смолоносница 2471
смолоносница вонючая 336
смолосемянник 4210
смолосемянник волнистый 3937
смолосемянник тобира 5538
смородина 1745
смородина альпийская 128
смородина американская 186
смородина белая 2535
смородина восточная 3953
смородина двуиглая 4900
смородина декоративная 2535
смородина душистая 1276
смородина железистая 4991
смородина золотая 2535
смородина золотистая 2535
смородина каменная 4520
смородина камневая 4520
смородина карпатская 1027
смородина колорадская 1336
смородина колосковая 3882
смородина красная 3882
смородина кровяно-красная 6000
смородина маньчжурская 3459
смородина Мейера 3607
смородина обыкновенная 3882
смородина печальная 208
смородина пушистая 3882
смородина скальная 4900
смородина скандинавская 3882
смородина чёрная 2081
смородина японская 3030
сморчковые 2768
сморчок 3678
сморчок конический 1578
сморчок настоящий 1473
сморчок обыкновенный 1473
снедок 4658
снедь-трава 2570
снежник 5057
снежноягодник 5057
снежноягодник белый 1529
снежноягодник ветвистый 1529
снежноягодник кистевидный 1529
снежноягодник кистистый 1529
снежноягодник красноплодный 2935
снежноягодник круглолистный 2935
снежноягодник обыкновенный 2935
снежноягодник округлый 2935
сныть 2570
сныть обыкновенная 596
собашник 2601
содник 4813
содник морской 4802
соймида 4544
сокирки 2330
сокирки Аяксовы 4524
сола 179
солерос 2500
солерос травянистый 3514
солидаго 2543
солнцегляд 5328
солнцегляд 5334
солнцецвет 5334
солнцецвет альпийский 152
солнцецвет монетолистный 1515
солнцецвет обыкновенный 1515
солнцецвет седой 2821
солодка 3291
солодка гладкая 1459
солодка голая 1459
соломоцвет 1108
солонечник ленок 2102
солорина 5083
сольданелла 111
сольданелла альпийская 2496
сольданелла горная 2602
сольданелля 111
солянка 4618
солянка калийная 1518
солянка русская 1518
сонгвинария канадская 682
сон-трава 5177
сорбария 2180
сорбарония 5084
сорго 5086
сорго алеппское 3103
сорго африканское зерновое 5087
сорго веничное 810
сорго зерновое 1929
сорго китайское 3133
сорго комовое 1929
сорго лимонное 3270
сорго метельчатое 810
сорго многолетнее 3103
сорго обыкновенное 5087
сорго поникшее 5892
сорго развесистое 810
сорго сахарное 5085
сорго сахарное 5309
сорго скученное 1929
сорго суданское 5294
сорго техническое 810
сорго японское 3133
сореа 4872
сорея 4872
сорея кистевая 4661
сорокозубка 4844
сосенка водяная 3488
сосенка водяная 3489
сосна 4190
сосна австрийская 380
сосна алеппская 86
сосна аризонская 318
сосна Арманда 320
сосна Бальфура 2353
сосна Банкса 3001
сосна Банксова 3001
сосна белая калифорнийская 1856
сосна белоствольная 5879
сосна болотная 3366
сосна Бунге 3193
сосна Веймутова 1982
сосна Веймутова гималайская 2807
сосна Веймутова мексиканская 3606
сосна Веймутова румелийская 419
сосна виргинская 5750
сосна Гельдрейха 2762
сосна гибкая 3310
сосна гималайская Веймутова 2807

сосна горная 5400
сосна густоцветковая 3067
сосна густоцветная 3067
сосна длиннохвойная болотная 3366
сосна ежовая 4882
сосна жёлтая 4259
сосна жёлтая 4882
сосна жёлтая аризонская 318
сосна жёлтая западная 4259
сосна Жерарда 1156
сосна Жерардова 1156
сосна-жереп 3746
сосна жёсткая 4208
сосна замечательная 3666
сосна иерусалимская 86
сосна итальянская 2983
сосна калабрийская 2052
сосна канарская 988
сосна караибская 4993
сосна карибская 4993
сосна кедровая 5402
сосна кедровая европейская 5402
сосна кедровая калифорнийская 3310
сосна кедровая корейская 3184
сосна кедровая маньчжурская 3184
сосна кедровая сибирская 4917
сосна китайская 1195
сосна корейская 3184
сосна Культера 1642
сосна ладанная 3352
сосна Ламберта 5306
сосна лесная 4775
сосна Массонова 3530
сосна мексиканская 3668
сосна мелкоигольчатая 595
сосна Монтезумы 3668
сосна Муррея 3355
сосна мягкоигольчатая 595
сосна обыкновенная 4775
сосна пиния 2983
сосна погребальная 1195
сосна приморская 1296
сосна растопыренная 3001
сосна румелийская 419
сосна Сабина 1856
сосна сахарная 5306
сосна сахарная гигантская 5306
сосна сибирская 4917
сосна скрученная 4876
сосна скрученная широкохвойная 3355
сосна смолистая 4460
сосна столбовая 380
сосна съедобная 1337
сосна Торрея 5557
сосна Тунберга 3023
сосна чёрная 380
сосна чёрная австрийская 380
сосна чёрная калабрийская 1625
сосна шведская 5402
сосна широкохвойная 3355
сосна Эллиота 4993
сосна Энгельмана 2033
сосновые 4194
соссюрея 4717
сосюрея 4717
софора 4001
софора японская 3059
сочевичник 4099
сочевичник весенний 607
сочевичник клубненосный 2649
сочевичник чёрный 645
соя 5106
соя культурная 5106
соя щетинистая 5106
спайнолепестные 5413
спараксис 5781
спаржа 350
спаржа аптечная 1351
спаржа лекарственная 1351
спаржа обыкновенная 1351
спаржа овощная 1351
спаржа остролистная 4848
спаржа плюмажная 2204
спаржа пушистая 2204
спаржа пушистоперистая 2204
спаржа Шпренгера 5178
спаржа японская 5675
спарманния 5115
спарманния африканская 49
спарта 2061
спартина 1593
спарто 2062
спекулярия 5719
спелта 5124
спельта 5124
спигелия 5136
спиж 683
спилантес 5162
спирея 5160
спирея волосистая 2703
спирея вязолистная 2455
спирея городчатая 5067
спирея дубровколистная 2455
спирея иволистная 5972
спирея мозолистая 3070
спирея рябинолистная 5691
спирея сливолистная 775
спирея средняя 3963
спирея Тунберга 5506
спирея японская 3070
спирогира 4260
сплоень 1115
спондиас 3648
спондиас жёлтый 6101
спондиас цитеры 178
спороболус 1906
спороболюс 1906
споровые 1717
спорынья 1258
спорыш 4317
спрекелия 393
сростнолепестные 5413
стагачка 25
стальник 4484
стальник жёлтый 5048
стальник колючий 5492
стальник пашенный 2518
стальник полевой 2518
стальник ползучий 962
стангерия 2205
стапелия 1034
статице 5502
стаунтония 5237
стаунтония шестилистная 3071
стафилея 569
стафилея перистая 2083
стахис клубненосный 332
стеблеклонник 5333
стеблелист 1322
стевия 5244

стеллария 5235
стенактис 2292
стенактис однолетний 259
стенанциум 5239
стенник 993
стенник вечнозелёный 2144
стенник горький 4523
стенник зонтичный 2504
стенница 5773
стеногинум 5240
стенокарпус 2263
стенокарпус ивовый 5966
стеркулиевые 1223
стеркулия 741
стеркулия вонючая 2733
стеркулия клёнолистная 2274
стеркулия платанолистная 1194
стеркулия утёсная 457
стефанандра 5241
стефанандра надрезаннолистная 1754
стефанандра Танаки 5437
стикта 5250
стилидиум 5289
стиракс 5056
стиракс аптечный 1915
стиракс бензоин 5315
стиракс бензоиновый 5315
стиракс лекарственный 1915
стираксовые 5269
стланец березовый 1935
стланик березовый 1935
стланик кедровый 3072
стоксия 5259
столетник 1092
стоног 2716
стонтония 5237
стосил 195
странвезия 5270
странвезия Давида 1202
страстоцвет 4068
страстоцвет крупный 2475
страстоцвет лавролистный 5815
страстоцвет настоящий 2475
страстоцвет светло-синий 697
страстоцвет съедобный 4070
страстоцвет четырёхугольный 2475
страстоцветные 4069
страусник обыкновенный 3976
стрелиция 584
стрелолист 327
стрелолист китайский 3924
стрелолист обыкновенный 3924
стрелолист стрелолистный 3924
стрептокарпус 1006
стрептопус 5669
стрептопус стеблеобъемлющий 1257
стрифнодендрон 170
стрифнодендрон-барбатимао 451
стрихнос рвотный орех 3898
стромбозия 5285
строфант 5287
строфария 5288
строфостилес 5937
строчок 2767
строчок обыкновенный 1994
строчок съедобный 1994
струна морская 4792
стручок цареградский 1023
стюартия 5245
стюартия-монодельфа 5430
суданка 5294
судза 4138
суеда 4813
сулла 5310
сумаум 3134
сумах 5313
сумах венецианский 1527
сумах восковой 4242
сумах гладкий 5042
сумах голый 5042
сумах дубильный 1527
сумах душистый 2358
сумах канадский 1501
сумах китайский 1203
сумах кожевенный 4920
сумах копаловый 2275
сумах коротковолосый 5206
сумах красильный 4920
сумах лаковый 4242
сумах лаконосный 3050
сумах обыкновенный 4920
сумах оленьерогий 5206
сумах пушистый 5206
сумах пятилистниковый 2268
сумах сицилийский 4920
сумах сочный 5848
сумах укореняющийся 1501
сумах уксусный 5206
сумах яванский 1203
сумах ядовитый 1501
сумаховые 1043
сумка пастушья обыкновенная 4862
сумочник пастуший 4862
сурепица 585
сурепица 3817
сурепица весенняя 1966
сурепица обыкновенная 610
сурепица яровая 5650
сурепка 5999
сурепка весенняя 1966
сурепка обыкновенная 610
сусак 2313
сусак зонтичный 2314
сусаковые 2315
сухарь 5923
сухоребрик 2757
сухоцвет 2923
сухоцвет однолетний 1446
сухоцветник 2923
сушеница 1727
сушеница болотная 3391
сушеница жёлто-белая 3092
сушеница желтовато-белая 3092
сушеница лежачая 1938
сушеница лесная 6025
сушеница топяная 3391
сфагновые 4096
сфагнум 5125
сфериальные 5126
схеноплектус озёрный 2600
схеноплектус Табернемонтана 5420
схеноплектус трёхгранный 5590
схенус 724
схенус ржавый 4624
схенус чернеющий 621
схенус черноватый 621
схизантус 893
схизеации 1742
схизеевые 1742
схизокодон 2377

схизостилис 1695
схизофрагма 2903
схима Валлиха 1802
схицейные 1742
сцеплянки 1580
сциадопитис 5681
сциадопитис мутовчатый 5682
сцилла 5200
сцилла двулистная 5664
сцилла сибирская 4916
сциндапсус 2989
сыроежка 4621
сыроежка белая 5923
сыроежка вонючая 2209
сыроежка едкая 2028
сыроежка жгучеедкая 2028
сыроежка зеленоватая 2629
сыроежка сине-жёлтая 686
сыроежка чернеющая 628
сыроежка ядовитая 2209
сыроежковые 4622
сыть 2280
сыть длинная 2402
сыть круглая 3893
сыть разнолистная 5677
сыть съедобная 1230

Т т

табак 5536
табак виргинский 1548
табак деревенский 395
табак-махорка 395
табак обыкновенный 1548
табак турецкий 1548
табебуйя 5614
таволга 5160
таволга альпийская 151
таволга белая 3807
таволга берёзолистная 581
таволга Биллиарда 577
таволга Бильярда 577
таволга Вильсона 5979
таволга виргинская 5753
таволга войлочная 2703
таволга городчатая 5067
таволга дубровколистная 2455
таволга Дугласа 1889
таволга иволистная 5972
таволга китайская 1201
таволга острая 2431
таволга Ростгорна 4572
таволга Саржента 4708
таволга средняя 3963
таволга Тунберга 5506
таволга Цабеля 6123
таволга широколистная 797
таволга японская 3070
таволжник 2519
таволжник обыкновенный 5410
тагетес 3493
тагетес распростёртый 2369
тайник 5657
тайник овальнолистный 1999
тайник овальный 1999
тайник сердцевидный 3881
тайнобрачные 1717
такка 5421
такковые 4173
таксодиевые 5463
такусса 4399
талассия 5654
талаума 5423
тальник 4352
тальник ползучий 1679
тамарикс 5435
тамарикс одесский 3909
тамарикс пятитычиночный 2269
тамарикс французский 2374
тамарикс четырёхтычинковый 2344
тамариксовые 5436
тамаринд 5433
тамаринд индийский 5434
тамариск 5435
тамус 622
танжерин 3468
танье 3441
тапиока 1379
тапсия 1817
таран 131
таро 1806
таро 2013
тарра 5046
тарриетия 5266
тарриеция 5266
тарриэтия 5266
тарриэтия яванская 3088
татарка 5863
татарник 1638
татарник колючий 4770
татарник обыкновенный 4770
текома 5610
текома укореняющаяся 1550
тектона 5465
телекия 3986
телекия красивая 2740
телесперма 2636
телефиум 3969
телефоровые 3256
телигоновые 1770
телигонум 5481
телимитра 5482
телорез 5836
телорез алоэвидный 5837
телорез обыкновенный 5837
тельфайрия 3917
тельфайрия восточноафриканская 6124
тельфайрия стоповидная 6124
темеда 3131
теозинт 5475
теозинта 5475
теозинте 5475
теосинт 5475
теосинт мексиканский 3603
тепари 5478
терескен 6001
терминалия 5600
терминалия катаппа 5600
терминалия пышная 38
терминалия хвалебная 38
теяминалия шерстистая 6052
термопсис 5483
тёрн 5014
тёрн мышиный 885
тернии христовы 1227
терновник 5014
тернослив 847
тестудинария 5558
тетрагония 3844
тетрагонолобус 1824

тетрагонолобус пурпурный 5197
тетраклинис членистый 296
тетрапанакс 4500
тетрацера 5477
тефрозия 2820
тефф 3385
тефф абиссинский 5473
тиарелла 2321
тибурбу 5515
тигридия 5520
тигридия павония 1547
тизанолена 5521
тик 1544
тилландсия 5525
тиллециевые 5526
тиллея 4365
тиллея водная 1511
тиллея водяная 1511
тимелеевые 3608
тимелея 5116
тимелия 5116
тимиан 5509
тимиян 5509
тимофеевка 5527
тимофеевка альпийская 154
тимофеевка луговая 5528
тимофеевка метельчатая 785
тимофеевка песчаная 4690
тимьян 5509
тимьян душистый 1546
тимьян обыкновенный 1546
тимьян полевой 3688
тимьян ползучий 3688
типчак 4858
тис 6117
тис головчатый 4232
тис европейский 2045
тис канадский 980
тис китайский 1215
тис коротколистный 3998
тис ягодный 2045
тисовые 6118
тисс 6117
тисс коротколистный 3998
тиссовые 6118
тладианта 5616
тладианта сомнительная 3467
тмин 1013
тмин волошский 1729
тмин обыкновенный 1013
тмин римский 1729
тмин чёрный 2415
тоддалия 5539
тоддалия азиатская 343
тодди-пальма 5540
токусса 4399
толокнянка 3480
толокнянка альпийская 141
толокнянка аптечная 500
толокнянка медвежья 500
толстянка 1661
толстянковые 3970
томат 5542
томат земляничный 4156
томат культурный 1549
тонковласник 2247
тонколистник 2248
тонколистниковые 2249
тонколучник 2292
тонконог 3173
тонконог сизый 702
топинамбур 3094
тополь 4270
тополь бальзамический 5100
тополь-белолистка 5906
тополь белый 5906
тополь берлинский 540
тополь волосистоплодный 942
тополь дельтовидный 979
тополь дельтовидный 1978
тополь дрожащий 2074
тополь душистый 3654
тополь Зибольда 4929
тополь итальянский 3360
тополь канадский 979
тополь канадский 1978
тополь каролинский 250
тополь китайский 1058
тополь корейский 3185
тополь крупнолистный 428
тополь лавролистный 3234
тополь Максимовича 3062
тополь осинообразный 4380
тополь печальный 819
тополь пирамидальный 3360
тополь ребристый 250
тополь Саржента 4214
тополь седеющий 2597
тополь седой 428
тополь сереющий 2597
тополь серый 2597
тополь треугольнолистный 1978
тополь Фремонта 2366
тополь чёрный 647
топорики 2330
тордилиум 2719
торения 5553
торилис 2759
торилис узловатый 3170
торица 5196
торица полевая 1620
торица пятитычинковая 5997
торичник 4689
торичник окаймлённый 3490
торичник солончаковый 4665
торрейя 5556
торрея 5556
торрея большая 1206
торрея калифорнийская 947
торрея орехоносная 3074
торрея тисолистная 2303
торрея тиссолистная 2303
тоулиция 5561
тофильдия 2162
тофильдия болотная 4768
тоция 5567
тоцция 5567
трава абиссинская 5473
трава алтейная 3507
трава альфа 2062
трава ананасная 2195
трава бермудская 542
трава бизонья 2572
трава блошная 5168
трава богородская 3688
трава верблюжья 961
трава войлочная 1870
трава воловья 4484
трава восьмилепестная 3726
трава вшивая 2117

трава гвинейская 2661
трава геморойная 5168
трава горляная 1643
трава гусиная 4969
трава джонсонова 3103
трава драгун 5445
трава железная 2965
трава живучая 1030
трава заячья 1868
трава злая 1817
трава золотушная 857
трава зольная 3159
трава камфарная 965
трава камчужная 1388
трава канареечная 4473
трава картамова 4637
трава козья 1425
трава колбасная 5375
трава колдуновая 2031
трава колосовая 877
трава красильная 5862
трава кузьмичёва 3106
трава куропаточья 1920
трава куропаточья восьмилепестная 3726
трава лакмусовая 5653
трава ластовичная 2606
трава лёгочная 3393
трава ледяная 2917
трава лимонная 3270
трава лихорадочная 1913
трава ложечная 4783
трава ложечная лекарственная 1522
трава лягушечная 3511
трава лягушечья 3511
трава манная 5820
трава маточная 2210
трава медвежья 1462
трава медовая 1553
трава моласская 3644
трава молевая 3690
трава молукская 3647
трава морская 1407
трава мыльная 5074
трава овечья 4858
трава огуречная 734
трава огуречная лекарственная 1366
трава очная 1910
трава пампаская 4021
трава пампасовая 4021
трава паточная 3644
трава почечуйная 5168
трава приточная 1797
трава прорезная 3715
трава пупочная 3816
трава пчелиная 1356
трава пьяная 3338
трава рисовая 266
трава сайгачья 2361
трава сердечная 1475
трава серебряная 4956
трава слоновая 3793
трава собачья 542
трава суданская 5294
трава телеграфная 5474
трава Тимофеева 5528
трава утиная 5820
трава хрустальная 2917
трава цынготная 1522
трава чихотная 5054
трава шёлковая 4473
трава щучья 5812
трава эскулапова 3620
трава эфедра 3106
травянка 3433
трагус 860
трагус кистевидный 5207
традесканция 5134
традесканция бассейновая 5780
традесканция белоцветковая 5780
традесканция виргинская 5752
трахелиум 5503
трахелоспермум 5224
трахикарпус 5983
трахикарпус высокий 2339
трахикарпус Форчуна 2339
трахилобий 4421
трекулия 5575
трема восточная 2928
трескучник 5862
третьина 1365
трефоль 1365
трёхзубка 5596
трёхкосточник 2867
трёхцветка 5954
трилистник 4447
трилистник водяной 1365
трилистник жёлтый 638
трилистник клещевой 5516
трилистник луговой 4447
триллиум 5594
тринакс 5480
триния 5595
триодия 5596
триостеум 2867
триостница 5499
триостренник 4234
триостренник болотный 328
триостренник морской 4877
триостренник приморский 4877
трипсакум 2404
трипутник 5397
тристания свёрнутая 778
тристания скрученная 778
тритома 5551
тритония 5599
тритринакс 5598
тритринакс бразильский 767
трифолиата 5592
трифоль 1365
трихилия 611
трихозант 1996
трихозант 5045
трихозантес 1996
трихозантес 5045
трихозантус 5045
трициртис 5534
трищетинник 2176
трищетинник желтоватый 6109
трищетинник жёлтый 6109
трищетинник колосистый 5146
троечница 5218
тростник 4472
тростник индийский 4574
тростник испанский 4574
тростник обыкновенный 1514
тростник сахарный 5300
тростник сахарный благородный 5300
тростник сахарный культурный 5300
тростянка 4505
троходендрон 5876

трубоцвет 1550
трубоязычник 5624
трутовик 4254
трутовик настоящий 5529
трутовик пёстрый 1921
трутовик серножёлтый 5312
трутовиковые 4253
трюфелевые 5609
трюфель 5608
трюфель белый 5920
трюфель итальянский 5920
трюфель летний 4619
трюфель настоящий 4137
трюфель олений 2717
трюфель олений 3399
трюфель русский чёрный 4619
трюфель французский 4137
трюфель чёрный 4137
трясинник 1968
трясунка 4381
трясунка большая 569
трясунка малая 3335
трясунка средняя 4134
тсуга 2769
тсуга западная 3996
тсуга Зибольда 4933
тсуга канадская 973
тсуга китайская 1184
тсуга разнолистная 3040
тсуга японская 4933
тубероза 5618
туевик 2161
туевик поникающий 2793
туевик японский 2793
туиния 5496
тунбергия 1272
тунг китайский 5640
тунг молуккский 992
тунг Форда 5640
туника 5641
тупа 2936
тургения широколистная 789
тургун 5445
турица 5246
турнепс 5650
турнера 5649
турнефорция 5562
турраэнтус 41
туррея 5212
турча 2193
турча болотная 2097
тускарора 266
тут 3749
тут белый 5902
тут красный 4456
тут чёрный 640
тута 3749
тутовник 3749
тутовые 3750
туя 300
туя алжирская 296
туя восточная 3948
туя гигантская 2484
туя западная 1973
туя корейская 3177
туя складчатая 2484
туя Стэндыша 3017
туя японская 3017
тыква 2568
тыква большая 6008
тыква бутылочная 922
тыква бутылочная обыкновенная 922
тыква восковая 1209
тыква гигантская 6008
тыква-горлянка 922
тыква горькая 1334
тыква крупноплодная 6008
тыква кустовая 883
тыква летняя 4330
тыква мозговая 883
тыква мочальная 5290
тыква мускатная 1747
тыква мускусная 1747
тыква обыкновенная 4330
тыква фигурная 883
тыква цилиндрическая 5290
тыква черносеменная 3437
тыква яйцевидная 6088
тыквенные 2569
тырса 1380
тысячеголов 1653
тысячеголовник 1653
тысячелистник 6067
тысячелистник мускатный 3769
тысячелистник мускусный 3769
тысячелистник обыкновенный 1565
тысячелистник птармика 5054
тысячелистник чихотный 5054
тюльпан 5633
тюльпан Биберштейна 552
тюльпан лесной 2302
тюльпан Эйхлера 2003

У у

уапака 5674
увулярия 530
уголёк в огне 4164
удо 5675
ужовник 27
ужовник обыкновенный 1346
ужовниковые 28
узик 5555
укроп 1857
укроп аптечный 1411
укроп водяной 2251
укроп волошский 2200
укроп душистый 1399
укроп огородный 1399
укроп пахучий 1399
уксусник 5206
улекс 2566
улекс европейский 1426
уллюко клубневый 5622
уллюко клубненосный 5622
ульва 4794
умбеллюлярия калифорнийская 938
умбиликус 3816
унаби 3115
унаби обыкновенное 1452
унаби юйюба 1452
упас 5684
ургинея 4798
урд 3755
урена 917
урера 4778
урзиния 5692
уроспермум 4856
урсиния 5692
уруть 4060

уруть бразильская 763
уруть колосистая 5138
уруть мутовчатая 974
уснея 5693
устилягиновые 854
утевник 5812
утёсник 2566
утёсник карликовый 1942
утёсник малый 1942
"ухо иудино" 3101
"ухо медвежье" 2276
уши иудины 3101
ушко медвежье 500

Ф ф

фабиана 4154
фаллус 5256
фарбитис 3682
фарбитис пурпурный 1474
фарзетия 2185
фарсетия 2185
фасоль 497
фасоль адзуки 34
фасоль азиатская 3755
фасоль волокнистая 3158
фасоль зерновая 3158
фасоль золотистая 3754
фасоль кроваво-красная 4748
фасоль кустовая однолетняя 3754
фасоль лима 3309
фасоль лимская 3309
фасоль лучистая 34
фасоль маш 3755
фасоль многоцветковая 4748
фасоль мунго 3755
фасоль настоящая 3158
фасоль обыкновенная 3158
фасоль обыкновенная кустовая 881
фасоль огненная 4748
фасоль огненно-красная 4748
фасоль остролистная 5478
фасоль полулунная 4940
фасоль почечная 3158
фасоль спаржевая 6066
фасоль угловатая 34
фатсия 2187
фатсия японская 3034
фацелия 4161
фацелия пижмолистная 5443
федия 52
фейхоа 2195
фелиция 2196
фендлера 2199
фенхель 2200
фенхель аптечный 1411
фенхель итальянский 2300
фенхель конский 2866
фенхель лекарственный 1411
фенхель обыкновенный 1411
фенхель флорентинский 2300
фернамбук 768
ферония слоновая 2012
ферула 2471
ферула ассафетида 336
ферула вонючая 336
ферула обыкновенная 1418
ферульник 2207
феруляго 2207
фиалка 5737
фиалка альпийская 1766
фиалка африканская 1347
фиалка белая 4008
фиалка болотная 3521
фиалка бразильская 236
фиалка высокая 5431
фиалка дву(х)цветковая 5661
фиалка душистая 5392
фиалка жёлтая 2136
фиалка лесная 5412
фиалка ночная 895
фиалка ночная 4522
фиалка опушённая 2694
фиалка полевая 2238
фиалка Ривина 4507
фиалка Ривиниуса 4507
фиалка Ривинуса 4507
фиалка рогатая 2856
фиалка собачья 1880
фиалка трёхцветная 5954
фиалка скальная 4691
фиалка удивительная 6019
фиалковые 5740
фига 1412
фига индийская 2936
физалис 2645
физалис обыкновенный 5275
физалис перуанский 4156
физалис рубчиковидный 1894
физалис съедобный 4156
физалис Франшетта 5275
физалис ягодный 4156
физостегия 3321
физостигма 921
физостигма ядовитая 1816
фикомицеты 5824
фикус 2944
фикус бенгальский 444
фикус карликовый 1267
фикус каучуконосный 2944
фикус священный 738
фикус-сикомор 5408
филантус кислый 3977
филирея 3642
филирея узколистная 3808
филирея широколистная 5586
филлантус 3249
филлантус кислый 3977
филлантус лекарственный 2026
филлодоце 3706
филлокактус 3247
филлокладус 1087
филлоспадикс 5337
филлостахис 4170
филодендрон 4165
филодендрон пробитый 1101
финган 75
финик 1807
финик дикий 1809
финик изогнутый 4820
финик индийский 5434
финик канарский 983
финик китайский 1452
финик лесной 5302
финик пальчатый 1808
финик финиколистный 1808
фиппсия 2915
фирмиана 1194
фисташка 4205
фисташка атлантическая 3724

фисташка дикая 5647
фисташка-лентискус 3275
фисташка мастиковая 3275
фисташка мастичная 3275
фисташка настоящая 1498
фисташка туполистная 5647
фисташковые 1043
фисташник 5647
фистулина печёночная 515
фителефас 2986
фителефас крупноплодный 1447
фитеума 3636
фитокренум 4171
фитолакка американская 1502
фитоллякковые 4244
фитолякка американская 1502
фитолякка десятитычинковая 1502
фитолякковые 4244
фитофтора 4172
фитцройя 2267
фитцройя Арчера 5448
фитцройя патагонская 4071
фитцройя тасманийская 5448
фитцроя 2267
флаверия 2285
флакоуртия 2270
флакоуртия кохинхинская 4025
флакуртиевые 2941
флакуртия 2270
флакуртия зондская 4605
флакуртия Рамонтша 4413
флакуртовые 2941
флиндерсия 2294
флиндерсия австралийская 426
флокс 4166
флокс Друммонда 1919
флокс Друмонда 1919
флокс метельчатый 5320
флокс однолетний 1919
флокс сибирский 4911
флоуренсия 5444
фонтанезия 2323
фонтиналис 5823
форзиция 2337
форзиция поникшая 5860
формиум 3843
форсиция 2337
форстерония 2335
фортунелла жемчужная 3980
фортунелла японская 3528
фотиния 4169
фотиния Давидсона 1813
франкениевые 2362
франкения 2361
францисцея 4405
фреезия 2365
фрима 3373
фритиллярия 2380
фуксия 2389
фуксия 3422
фукус 4538
фукус мелкопузырчатый 666
фукус пильчатый 1038
фукус пузырчатый 666
фунария 2394
фундук 2473
фундук крупноплодный 2473
функия 4221
функия ланцетолистная 3060
фунтумия 4946
фунтумия каучуконосная 3205

Х х

хагения абиссинская 3190
хакетия 2675
халезия 4950
халидрис 4797
халорагис 4786
хамамелис 6012
хамедафна 3257
хамедафне 3257
хамедорея 1112
хаменерий 5975
хаменерий узколистный 2262
хаменериум 5975
хамепейце 5487
хамеропс низкий 3572
хамеропс приземистый 3572
хара 5267
хатьма 5582
хатьма древовидная 5716
хатьма трёхмесячная 2788
хатьма тюрингенская 5507
хвойник 3104
хвойник американский 192
хвойник двухколосковый 3106
хвойник Жерарда 2449
хвойник зелёный 2625
хвойник калифорнийский 934
хвойник хвощевидный 3652
хвойник хвощевый 3652
хвойниковые 3107
хвойные 1579
хвостник 3488
хвостник обыкновенный 3489
хвощ 2875
хвощ болотный 3506
хвощ большой 2610
хвощ зимующий 4777
хвощ лесной 5411
хвощ луговой 3552
хвощ пёстрый 5701
хвощ полевой 2225
хвощ топяной 5344
хвощёвые 2877
хевея 2791
хевея бразильская 4050
хедеома 2178
хедизарум 5310
хедисцепе 5680
хедихиум 2491
хейрантус 5768
хейрантус садовый 1556
хейриния 3772
хейхера 171
хелоне 5655
хелониас 5346
хельсиния 402
хемантус 678
хемантус Екатерины 3138
хемерокаллис 1815
хенекен 2781
хенна 2782
хеномелес 2312
хеномелес китайский 1181
хеномелес лагенария 1414
хеномелес Маулея 3038
хеномелес японский 3038
хиднокарпус 1121

хикама 5851
химафила 4202
хименокаллис 5130
химонант скороспелый 6009
хиококка 3617
хионантус 2379
хионантус виргинский 5896
хионодокса 2512
хипеаструм 176
хирония 5229
хлеб дивий 1023
хлеб Иванов 1023
хлеб каффрский 5087
хлебница 2314
хлебница 5941
хлебоплодник Лакуша 3207
хлопок дикий 2775
хлопунец 3701
хлопушка 1054
хлопчатник 1635
хлопчатник американский 5685
хлопчатник африканский 3283
хлопчатник афроазиатский 3283
хлопчатник барбадосский 4790
хлопчатник волосистый 5685
хлопчатник древовидный 349
хлопчатник египетский 4790
хлопчатник южноамериканский 4790
хлора 6116
хлорантовые 1221
хлорелла 1222
хлорис 5982
хлороксилон 1107
хлорофора 2396
хлорофора высокая 2963
хлорофора красильная 1912
хлябник 3969
хмелеграб 2849
хмелеграб американский 197
хмелеграб виргинский 197
хмелеграб обыкновенный 2107
хмелеграб японский 3044
хмелёк 1278
хмелёк полевой 2848
хмелица 1440
хмель 2847
хмель обыкновенный 1440
хмель огородный 284
хмель японский 3043
хна 2782
ховея 4828
хойя 5847
холодка 4122
хондрила 4983
хондрилла 4983
хондрилла обыкновенная 4610
хондрилла ситниковидная 4610
хондрилля 4983
хондрус 2960
хондрус кудрявый 2961
хопея 3588
хопея душистая 5486
хорда нитчатая 4792
хороги 332
хоустония 710
хохлатка 1628
хохлатка жёлтая 6080
хохлатка сибирская 4897
хохлатник 4042
хрен 2871
хрен водяной 222
хрен водяной 5796
хрен деревенский 2871
хрен дикий 2871
хрен обыкновенный 2871
хрен солнечный 2118
хризалидокарпус 896
хризантема 1229
хризантема альпийская 123
хризантема бальзамная 1632
хризантема индийская 3687
хризантема увенчанная 1710
хризобаланус 1315
хризогонум 2549
хризофиллум 5211
хризофиллюм 5211
хрозофора 5653
хроококковые 1228
хруплявник 4249
хрущецветник 2919
хрящецветник 2919
хрящецветник мутовчатый 5924
хуннемания 2534
хура трескающаяся 4681
хурма 4152
хурма виргинская 1494
хурма восточная 3127
хурма кавказская 1809
хурма обыкновенная 1809
хурма эбенова 1987
хурма японская 3127
хурмовые 1986
хускуеа 1231
хустония 710
хутчик 2714
хюкразия 1218

Ц ц

цамия 1581
цантоксилум мартиникский 3312
царица ночи 4384
царь-бегония 355
царь-зелье 3532
цаушнерия 2259
цванга 2244
цвет кукушкин 4398
цветковые 2310
цеанотус 1080
цеанотус американский 3840
цедрат 1243
цедрела 1084
цедрела душистая 1234
цедрела китайская 1205
цедрела сингапурская 866
цезальпиниевые 4824
цезальпиния 758
цезальпиния аптечная 3847
цезальпиния дубильная 1867
цезальпиния душистая 1867
цезальпиния ежовая 4296
цезальпиния кожевенная 1867
цезальпиния красильная 1867
цезальпиния ост-индская 4702
цезальпиния саппан 4702
цезальпиния центрально-американская 1867
цезальпиния шиповатая 4296
цейба 3134
цейба 3440

цекропия 4331
целаструс 608
целестина 60
целибуха 3898
целогине 1317
целозия 1307
целозия гребенчатая 1410
целозия серебристая 2192
целококкус 4644
целомудренник 3296
цельзия 1088
цельтис западный 1434
центифолия 910
центрантус 1091
центрантус красный 3119
центролобиум 4275
ценхрус 4682
цератиола 4686
цератозамия 2854
цератоптерис 5803
церва 5862
цереус 1100
цереус гигантский 4645
цереус крупноцветный 4384
цереус плетевидный 4422
церкокарпус 3712
церкоспора 1099
цероксилон андийский 5846
церопегия 1102
церцис 1098
церцис европейский 3114
цеструм 3098
цетрария 1103
цетрария исландская 2916
цефалантус 902
цефалотаксус 4232
цефалотаксус Фортуна 1196
цефалоцереус 1095
цефалянтера 4163
цефалянтус 902
цефалярия 1094
цианантус 1763
циатея 5580
циботиум 4784
цикада 1764
цикадовые 1765
цикламен 1766
цикламен персидский 2990
циклантера 1768
циклантовые 1767
циклахена 5324
цикнохес 5348
цикорий 1147
цикорий дикий 1381
цикорий зимний 2032
цикорий корневой 1381
цикорий обыкновенный 1381
цикорий полевой 1381
цикорий салатный 2032
цикорий-эндивий 2032
цикорник 1147
цикута 5808
цикута ядовитая 2130
цимбалярия 465
цимбалярия постенная 3147
цимбалярия стенная 3147
цимицифуга 837
цинанхиум 5340
цинанхиум лекарственный 5916
цингибер 2489
цинерария 1237
цинерария ковровая 4958
цинерария летняя 4958
цинерария морская 4958
цинерария приморская 4958
циния 6129
цинна 6039
циннамомум 1238
цинния 6129
цинния изящная 1568
цинния многоцветковая 4466
цинния стройная 1568
циноглоссум 2879
циноксилон цветущий 2309
циноморий 1771
цинхона 1236
ципелла 1772
циперус 2280
циртантус 1780
циртостахис 4795
цирцея 2031
цирцея альпийская 124
цирцея парижская 4055
циссус 3486
циссус 5577
цистоптерис 657
цитарексилум 2218
цитизус 809
цитроид 5779
цитрон 1243
цитронелла 1244
цитрус 1246
цитрус бергамия 536
цитрус бергамот 536
цитрус бигардия 5092
цитрус лиметта 3311
цитрус медийский 1243
цитрус померанцеволистный 3311
цитрусовые 1247
цифомандра свекловичная 5589
цицания 5956
цицания водяная 266
цмин 2149
цмин жёлтый 6083
цмин песчаный 6083
цмин прицветниковый 5279
цоизия 3242
цоизия японская 3052
цуга 2769
цуга западная 3996
цуга Зибольда 4933
цуга калифорнийская 3707
цуга китайская 1184
цуга разнолистная 3040

Ч ч

чабёр 4721
чабёр горный 6006
чабёр садовый 5321
чабрец 5509
чабрец обыкновенный 3688
чабрец песчаный 1546
чагерак верблюжий 961
чай 5467
чай ассамский 356
чай иезуитский 6056
чай канадский 1125
чай капорский 2262

чай курильский 879
чай курильский кустарниковый 879
чай луговой 3651
чай-матэ 4048
чай монгольский 3254
чай парагвайский 4048
чай почечный 558
чай рвотный 6068
чайные 5464
чайот 1122
чайота 1122
чакан 1062
чаполоть 5368
чапыжник 4095
частуха 5829
частуха Валенберга 2593
частуха злаковая 2593
частуха обыкновенная 211
частуха подорожниковая 211
частуховые 95
чашецветник 5386
челыжник 4910
чемерица 2170
чемерица белая 5894
чемерица зелёная 193
чемерица чёрная 630
чемпедек 2998
чемыш 4667
черва 5862
черевички 2330
череда 521
череда поникшая 3866
череда трёхраздельная 857
черёмуха 2078
черёмуха-антипка 3426
черёмуха виргинская 1384
черёмуха душистая 3426
черёмуха кистевая 2078
черёмуха карликовая 4611
черёмуха-магалебка 3426
черёмуха обыкновенная 2078
черёмуха пенсильванская 4188
черёмуха поздняя 626
черёмуха поздняя иволистная 1011
черемша 3368
черемша 4415
череш 1838
черешня 3544
черешня антипка 3426
черешня птичья 3544
чермеш 2759
черника 692
черника 3784
черника канадская 969
черника крупноплодная 3224
черника обыкновенная 3784
черника щитковая 2796
черница 3784
чернобыль 3747
чернобыльник 3747
черноголовка 4816
черноголовка крупноцветковая 559
черноголовка крупноцветная 559
черноголовка обыкновенная 1525
черноголовник 868
черноголовник кровохлёбковый 5018
черногорка 5179
черноклён 5456
чернокорень 2879
чернокорень аптечный 1444
чернокорень лекарственный 1444
чернокорень немецкий 2630
чернолоз 3236
чернослив 2424
чернотал 3236
черноцвет 2031
чернушка 2201
чернушка дамасская 3386
чернушка "девица в зелени" 3386
чернушка посевная 2415
чертогон 782
чертогон луговой 3561
чертополох 782
чертополох акантовидный 13
чертополох акантолистный 13
чертополох благословленный 671
чертополох Кернера 3155
чертополох колючий 13
чертополох курчавый 1740
чертополох поникший 3759
чесалка 2390
чеснок 2432
чеснок дикий 5806
чеснок заячий 5806
чеснок змеиный 2474
чеснок посевной 2432
чесночник 2434
чесночник аптечный 2433
чесночник лекарственный 2433
чёточник 4553
четырёхкрыльник 1824
четырёхкрыльник пурпуровый 5197
чефрас 5908
чечевица 3273
чечевица культурная 1457
чечевица обыкновенная 1457
чечевица пищевая 1457
чечевица съедобная 1457
чечевица французская 609
чешуехвостник 4923
чешуехвостник согнутый 4924
чешуйник 5549
чешуйчатка обыкновенная 4740
чешуйчатка раннеспелая 1964
чешуйчатка травяная 1114
чилибуха 3898
чилибуховые 3358
чилига 4095
чилига кустарниковая 4617
чилига степная 4617
чилижник 4095
чилижник кустарниковый 4617
чилим 5794
чина 4098
чина болотная 3510
чина весенняя 607
чина горная 3716
чина душистая 5378
чина жёлтая 6104
чина злаколистная 2592
чина клубеньковая 2649
чина клубневая 2649
чина клубненосная 2649
чина красная 2279
чина круглолистная 4587
чина лесная 2278
чина луговая 3555
чина морская 3495
чина мохнатая 4578
чина посевная 2587

чина приморская 3495
чина танжерская 5438
чина чёрная 645
чина широколистная 4133
чинар 3959
чинар американский 205
чинар восточный 3959
чинар западный 205
чингиль 4667
чингиль серебристый 4913
чистец 550
чистец альпийский 117
чистец болотный 3501
чистец германский 3731
чистец клубненосный 332
чистец лесной 5911
чистец однолетний 2758
чистец полевой 2230
чистец прямой 5687
чистец шерстистый 6047
чистотел 1085
чистотел большой 2606
чистоуст 3974
чистоуст величавый 4593
чистоуст королевский 4593
чистоустовые 3975
чистяк 4181
чистяк весенний 2241
чистяк лютичный 2241
чихотник обыкновенный 5054
чополоть 5368
чорак 4813
чубушник 3641
чубушник бородавчатый 5788
чубушник венечный 5376
чубушник Гордона 2565
чубушник кавказский 1078
чубушник крупноцветковый 571
чубушник Культера 1641
чубушник мексиканский 3600
чубушник мелколистный 3328
чубушник непахучий 4752
чубушник сердцелистный 2739
чубушник Фальконера 2157
чубушник Цейера 6127
чубушник широколистный 2819
чубушник Шренка 4760
чума водяная 2022
чумай 3103
чумиза 2624
чуфа 1230

Ш ш

шабдар 4147
шалот 4844
шалфей 4639
шалфей аптечный 2427
шалфей великолепный 4749
шалфей вербеновидный 5720
шалфей вербеновый 5720
шалфей вербенолистный 5720
шалфей голый 4749
шалфей иерусалимский 1451
шалфей клейкий 5249
шалфей крупноцветковый 560
шалфей крупноцветный 560
шалфей лекарственный 2427
шалфей лесной 5743
шалфей луговой 3558
шалфей мускатный 1254
шалфей мутовчатый 3299
шалфей Сибторпа 4919
шамбала 2203
шампиньон 1476
шампиньон двуспоровый 2419
шампиньон культивируемый 1476
шампиньон лесной 2326
шампиньон обыкновенный 1476
шампиньон полевой 2870
шампиньоновые 2488
шандал-дерево 200
шандра 2816
шандра белая 1442
шандра обыкновенная 1442
шар золотой 1755
шарлот 4844
шаровница 2506
шаровница сердцелистная 2737
шаровницевые 2505
шарыш 1838
шафран 1699
шафран весенний 1395
шафран Гейфеля 2790
шафран луговой 1353
шафран посевной 4638
шафтал 4147
шеддок 4329
шейки раковые 2079
шейки рачьи 2079
шейхцериевые 4755
шейхцерия 4754
шейхцерия болотная 3515
шелковица 3749
шелковица американская 4456
шелковица белая 5902
шелковица бумажная 1486
шелковица чёрная 640
шелковник 5797
шелюга 1801
шелюга жёлтая 1801
шелюга красная 4850
шелюжник 1801
шерардия 2227
шерардия полевая 2228
шероховатка 740
шерстестебельник 4201
шерстняк 1735
шерстняк 4228
шерстяк 1735
шефердия 834
ши 4852
шизандра 3425
шизантус 893
шизофрагма 2903
шикот 3151
шикша 1706
шикша чёрная 627
шикшевые 1707
шильник 387
шильник водный 5792
шильник водяной 5792
шильник ситниковидный 5857
шильница 387
шима 2659
шима Валлиша 1802
шинус 4125
шинус терпентиннолистный 766
шиповник 4554
шиповник альпийский 145

шиповник Беггера 522
шиповник бедренцоволистный 4776
шиповник бедренцовый 4776
шиповник Бэнкса 443
шиповник вечнозелёный 2147
шиповник войлочный 5543
шиповник вонючий 374
шиповник галльский 2371
шиповник гвоздичный 4753
шиповник даурский 1785
шиповник кожистолистный 3258
шиповник колючий 4776
шиповник коричневый 1240
шиповник коричный 1240
шиповник краснолистный 4454
шиповник мелкоцветковый 5025
шиповник морщинистый 4602
шиповник мохнатый 287
шиповник мягкий 685
шиповник обыкновенный 1875
шиповник полевой 2583
шиповник ржавчинный 5359
шиповник собачий 1875
шиповник французский 2371
шиповник щитконосный 1627
шиповник эглантерия 5359
шиповник Юндзилла 3491
шиповник яблочный 287
шипот 3151
ширококолокольчик 423
широколиственница 4251
ширяш 1838
шишечник 2724
шишка ворсовальная 2390
шишник 2724
шлемник 4987
шлемник восточный 3962
шлемник высокий 5429
шлемник высочайший 5429
шлемник хмелевой 6072
шлюмбергерия 1656
шмальция 5313
шмиделия 4757
шнит-лук 1219
шнитт-салат 1759
шовкун 5902
шоколадник 912
шорея 4872
шорея мощная 4661
шортия 4881
шотия 3124
шпажник 2499
шпажник болотный 5500
шпажник большой 1420
шпажник гентский 771
шпажник обыкновенный 1420
шпажник посевной 1612
шпергель 1620
шпилант 4045
шпилант 5162
шпилант травянистый 4045
шпинат 5148
шпинат английский 4072
шпинат индийский 5921
шпинат кубинский 3627
шпинат малабарский 5921
шпинат новозеландский 3844
шпинат овощной 4304
шпинат огородный 4304
шпорник 3226
шпорник многолетний 2305
шпороцветник 5190
штернбергия 5242
штернбергия жёлтая 2158
шток-роза 169
шток-роза розовая 2829
шушу 1122

Щ щ

щавелёк 4859
щавель 1868
щавель альпийский 129
щавель английский 4072
щавель водрощавелевый 5798
щавель водяной 4258
щавель воробьиный 4859
щавель кислый 2429
щавель клубковатый 1293
щавель красивый 2216
щавель кровяной 684
щавель курчавый 1741
щавель морской 2536
щавель настоящий 2429
щавель обыкновенный 2429
щавель огородный 2429
щавель прибрежный 5798
щавель приморский 2536
щавель североамериканский 981
щавель туполистный 601
щавель шпинатный 4072
щавель щитковидный 2373
щавель щитковый 2373
щебрец 3688
щебрушка 464
щелкунец 663
щетинник 780
щетинник зелёный 2624
щетинник итальянский 2352
щетинник мутовчатый 2844
щетинник сизый 6090
щеть 3533
щирица 174
щирица белая 5637
щирица гибридная 5005
щирица жминдовидная 4316
щирица запрокинутая 4464
щирица колючая 5153
щирица метельчатая 3387
щирица хвостатая 3387
щирицевые 175
щит 4864
щитник 3445
щитница 1297
щитовидник 6026
щитовник 6026
щитовник болотный 3504
щитовник буковый 3797
щитовник горный 3884
щитовник гребенчатый 1685
щитовник Линнея 3901
щитовник мужской 3445
щитовник шиповатый 5546
щитолистник 4114
щитолистник обыкновенный 1489
щучка 2677
щучка дернистая 5628

Э э

эбеновые 1986
эвакс 2139
эверния 2150
эвкалипт 2065
эвкалипт Генна 1233
эвкалипт голубой 5447
эвкалипт западноавстралийский 3081
эвкалипт иволистный 5969
эвкалипт миндальный 5969
эвкалипт прутовидный 4493
эводия 2151
эводия сизая 3032
эврикома 2137
эвтока 4161
эвхарис 2066
эгилёпс 2517
эгилопс 2517
эгле мармеладное 408
эдгеворция 4038
эдельвейс 1992
эдельвейс альпийский 1406
эджевортия 4038
эджевортия бумажная 3958
эджеворция 4038
эдогоний 3911
эйхорния 5810
эйххорния 5810
экзохорда 4090
экзохорда Альберта 5645
элевзина 2015
элеодендрон 2177
элеодендрон капский 1004
элеокарпус 2008
элеокарпус американский 189
элеокарпус зубчатый 2809
элеокарпус новозеландский 2809
элефантопус 2014
элизма плавающая 2299
элимус 5957
элодея 2022
элодея канадская 976
эльсгольция 2024
эльшольция 2024
эльшольция гребенчатая 1682
эмбелия 2025
эмботриум 2027
эмилия 5451
эммер 2029
эндивий 2032
эндивий зимний 2032
эндивий летний 4545
энокарпус 403
энотера 2141
энотера дву(х)летняя 1408
энотера короткоиглая 5022
энтада 2046
энтада ползучая 1265
энтандрофрагма 4698
энтеролобиум 1967
энтеролобиум саман 4404
энцефалартос 3125
эпакрис 2048
эперуа 5764
эпигея 5569
эпифиллум 3247
эпифиллюм 3247
эрантис 5998
эремостахис 1839
эремурус 1838
эремурус величественный 3435
эрехтитес 869
эриантус 4228
эригерон 2292
эризимум 3772
эрика 2742
эрика древовидная 5581
эрика крестолистная 1701
эрика метловидная 546
эрика мясо-красная 5186
эрика румяная 5186
эрика сизая 5668
эриоботрия 3375
эриогонум 2054
эриофиллум 2055
эритеа 2059
эритеа съедобная 2654
эритрина 1587
эритрина петушья 1311
эритроксилон 1304
эритроксилон кока 2883
эритрониум 2190
эрука посевная 4525
эскаллония 2060
эскаллония красная 4449
эскалония 2060
эспарто 2062
эспарцет 4646
эспарцет виколистный 1519
эспарцет испанский 5310
эспарцет песчаный 2891
эспарцет посевной 1519
эстрагон 5445
эсшольция 2554
этионема 5261
эуфорбия 2070
эухаридиум 1253
эфедра 3104
эфедра американская 192
эфедра дву(х)колосковая 3106
эфедра Жерарда 2449
эфедровые 3107
эхеверия 1988
эхинацея 1989
эхинация 1989
эхинодорус 862
эхинокактус 1990
эхинопанакс 1847
эхинопс 2508
эхинопсилон волосистый 2680
эхинопсис 2749
эхиноцереус 1991
эхиноцистис 3639
эхитес 4718
эхиум 5744
эхмея 35
эшинантус 466
эшиномена 3109
эшолтция 2554
эшолтция калифорнийская 943
эшольция 2554
эшшольция 2554

Ю ю

юануллоа 3112
юбея 3113
юбея величавая 5419

юбея замечательная 5419
юган 2731
юзерлык 2714
юйюба 3115
юкка 6120
юкка алоелистная 110
юкка величественная 3692
юкка виргинская 24
юкка волокнистая 24
юкка нитевидная 24
юкка нитчатая 24
юкка прекрасная 3692
юкка славная 3692
юкка слоновая 846
юморка 4921
юринея 3120
юссиея 5832
юстиция 3121
ююба 3115
ююба унаби 1452

Я я

яблоко заморское 1334
яблоко звёздное 919
яблоко-кислица 1349
яблоко райское 4046
яблоневые 282
яблоня 288
яблоня венечная 5959
яблоня восточная 4046
яблоня дикая 1349
яблоня дикая сибирская 4899
яблоня домашняя 1349
яблоня душистая 5959
яблоня замечательная 1180
яблоня Зибольда 5554
яблоня кавказская 4046
яблоня карликовая 1349
яблоня китайская 4091
яблоня корнесобственная 4899
яблоня культурная 1349
яблоня лесная 1349
яблоня малайская 3912
яблоня многоцветковая 3037
яблоня низкая 1349
яблоня низкорослая 1349
яблоня обильноцветковая 3037
яблоня обильноцветущая 3037
яблоня Палласова 4899
яблоня парадизка 4046
яблоня райская 4046
яблоня садовая 1349
яблоня сибирская 4899
яблоня сливолистная 4091
яблоня ягодная 4899
явнобрачные 2310
явор 4217
ягель 4479
ягода буйволова 4952
ягода винная 1412
ягода волчья 2099
ягода грудная европейская 1452
ягода грудная французская 1452
ягода логанова 3357
ягода медвежья 500
ягода мыльная 5102
ягода снежная 5057
ягода снежная кистистая 1529
ягодки 2194
ягодковые 3608
ягодник 692
ягодник кистевой 2796
ягодник черника 3784
язвенник 271
язвенник горный 160
язвенник Линнея 3159
язвенник обыкновенный 3159
"язык воловий" 840
язык олений 2716
язык песий 2879
языки воробьиные 5271
языкотруб 5624
языкотруб садовый 4738
языкоцветные 1232
язычник 1346
яйцо чёртово 5256
якаранда 2996
якаранда остролистная 4849
як-дерево 2998
якорцы 954
якорцы земляные 4332
якорцы стелящиеся 4332
ялапа 3005
ялапа настоящая 3005
ям 6062
ямбоза 4556
ямболан 3010
ямс 6062
ямс белый 5993
ямс китайский 1241
ямс клубненосный 69
ямс крылатый 5993
ямс куш-куш 1748
ямс луковичный 69
ямс посевной 1214
ямс слоновый 5912
ямс слоновый Ривьера 1848
ямс съедобный 1214
ямс трёхнадрезный 1748
ярутка 4112
ярутка горная 139
ярутка полевая 2232
ярутка пронзённая 5494
ясенец 1866
ясенец белый 2435
ясенка 1159
ясень 338
ясень американский 5878
ясень белый 2308
ясень высокий 2073
ясень голубой 687
ясень двойственный 93
ясень европейский 2073
ясень зелёный 2622
ясень золотистый 2526
ясень каролинский 1024
ясень китайский 1168
ясень китайский 5584
ясень круглолистный 4583
ясень ланцетный 2622
ясень ланцетолистный 2622
ясень маннитовый 2308
ясень манноносный 2308
ясень маньчжурский 3458
ясень мелколистный 4583
ясень неправильный 4972
ясень обыкновенный 2073
ясень остроконечный 5878
ясень пенсильванский 4434

ясень пушистый 4434
ясень сахарный 2308
ясень сирийский 5416
ясень туркестанский 5644
ясень узколистный 3798
ясень цветочный 2308
ясень чёрный 613
ясень четырёхгранный 687
ясколка 1145
ясколка альпийская 122
ясколка Биберштейна 5461
ясколка дернистая 2221
ясколка полевая 5230
ясколка пятитычинковая 5853
ясколка скученная 3732
ясколка скученноцветковая 3732
ясколка трёхстолбиковая 5236
ясколка шерстистая 5064
ясменник 6040
ясменник душистый 5395
ясменник красильный 1957
ясменник пахучий 5395
ясменник подмаренниковый 511
ясменник розовый 5201
ясменник сизый 511
яснотка 1818
яснотка белая 5891
яснотка гибридая 1756
яснотка крапчатая 5166
яснотка пурпуровая 4343
яснотка пятнистая 5166
яснотка стеблеобъемлющая 2780
ястребинка 2726
ястребинка альпийская 135
ястребинка волосистая 3734
ястребинка зонтичная 3804
ястребинка луговая 2224
ястребинка многоцветковая 6082
ястребинка обильноцветущая 6082
ястребинка оранжево-красная 3942
ястребинка превысокая 3160
ястребинка савойская 4723
ястребинка стенная 5769
ястребинка стеночная 5769
ястребинка стройная 3160
ястребинка флорентийская 3161
ястребинка шерстистая 4842
ятропа 3838
ятропа куркас 449
ятроориза дланевидная 955
ярофа 3838
ятрофа куркас 449
ятрышник 3946
ятрышник болотный 3093
ятрышник мужской 3446
ятрышник обезьяний 3656
ятрышник обожжённый 1803
ятрышник пурпурный 3200
ятрышник пятнистый 5163
ятрышник широколистный 798
ятрышник шлемовидный 5079
ятрышник шлемоносный 5079
ятрышниковые 3945
ячмень 454
ячмень гигантский 2094
ячмень гривастый 2350
ячмень гривоподобный 2350
ячмень дву(х)рядный 5672
ячмень заячий 3729
ячмень крупный 5672
ячмень культурный двурядный 5672
ячмень культурный многорядный 455
ячмень культурный шестирядный 4981
ячмень многорядный 455
ячмень мышиный 3729
ячмень обыкновенный 455
ячмень обыкновенный многорядный 455
ячмень песчаный 2094
ячмень посевной 455
ячмень четырёхрядный 455
ячмень четырёхсторонний 455
ячмень шестигранный 4981
ячмень шестирядный 4981